REMOTE SENSING
IN GEOLOGY

REMOTE SENSING IN GEOLOGY

Edited by

Barry S. Siegal
Earth Science Group
Ebasco Services, Inc.
Greensboro, North Carolina

Alan R. Gillespie
Jet Propulsion Laboratory
Pasadena, California

John Wiley & Sons   New York   Chichester   Brisbane   Toronto

COVER PHOTO

This picture, taken in March, 1969, from the Apollo 9 spacecraft at an altitude of 120 miles, covers northern Mexico and southern California (view is to the north). The Gulf of California is at lower right and the Pacific coast at left, with the Salton Sea just right of center. The Peninsular Ranges, trending north from the lower left, are an uplifted block of batholithic rocks bounded by normal faults on the east. The Mojave Desert is the light-colored triangular area at top center, and is bounded by the San Andreas fault on its southwest. The continuation of the San Andreas fault trends southeast along the east side of the Salton Sea. The southern part of the Basin and Range province is visible at upper right.

Apollo 9 photo AS 9-21-3263

P. D. Lowman

Copyright © 1980, by John Wiley & Sons, Inc.

All rights reserved. Published simultaneously in Canada.

Reproduction or translation of any part of this work beyond that permitted by Sections 107 and 108 of the 1976 United States Copyright Act without the permission of the copyright owner is unlawful. Requests for permission or further information should be addressed to the Permissions Department, John Wiley & Sons.

Library of Congress Cataloging in Publication Data:

Main entry under title:

Siegal, Barry S.
Remote sensing in geology

Includes index
   1. Geology—Remote sensing. I. Siegal, Barry S.,
1947-   II. Gillespie, Alan R.

QE33.2.R4R44   550'.28   79-17967
ISBN 0-471-79052-4

Printed in the United States of America

10 9 8 7 6 5 4 3 2 1

FOREWORD

Remote sensing, the ability to perceive and study an object from a distant location, is an area of human endeavor that is extending our capability to gather important resource and environmental information. In recent years, developments in technology have permitted us to sense data beyond the normal range of human perception, to obtain these data from the vantage points of space, which was not previously accessible to us, and to conduct data acquisition on a global basis. The combination of these factors adds up to a fairly revolutionary development in technology. Remote sensing techniques are now routinely used in fields as diverse as agriculture, oceanography, and planetary exploration. *Remote Sensing in Geology* is a timely book that presents an authoritative discussion of remote sensing techniques and their application to the geological sciences. The impetus for the development of many of the imaging techniques emphasized in this book has been the NASA space program, and I am pleased to note the application of these new techniques to difficult problems facing society today, such as mineral and petroleum exploration.

The contributing authors, drawn from industry, government, and universities, are recognized leaders in their individual disciplines, which together comprise a cross section of those geological sciences in which remote sensing is currently used. Remote sensing will play an increasingly important role in the geological sciences, and I am pleased to introduce *Remote Sensing in Geology* as both a summary of the state-of-the-art and as a textbook for future researchers.

James C. Fletcher
Former NASA Administrator

PREFACE

Remote Sensing in Geology provides students in geology who have some knowledge of physics and calculus with a complete introduction to most of the aspects of remote sensing that they will encounter later as professionals. It is not a compendium of separate, unrelated articles but is, instead, a joint effort by different authors. This format was chosen to acquaint students with the current research in remote sensing as it affects geology, without sacrificing the virtues of a limited-author textbook. The authors represent a broad spectrum of university, industry, and government affiliations and experience.

The authors present complete coverage of these fields rather than an intensive coverage of one aspect of a field. However, references to other works that explore certain aspects of remote sensing in depth have been provided for the interested student.

Remote Sensing in Geology is organized into four sections: the physics of the interaction of light with surfaces and the acquisition of data; optical and digital processing of these data to prepare them for analysis; interpretive techniques; and the application of remotely sensed data to different disciplines in geology.

The first section consists of three chapters. In Chapter 2, Graham Hunt provides an overview of the electromagnetic spectrum and the physical laws governing electromagnetic radiation and fields. The absorption, reflection, and emission of electromagnetic radiation by matter is the fundamental basis of remote sensing, and Hunt's discussion of the interaction of light with surfaces is a necessary framework for students, regardless of their particular interest in the field of remote sensing. In Chapter 3, Don Lowe summarizes the methods and equipment actually used to create images. Again, this chapter is of general interest to all students. In the last chapter of the section, Paul Lowman surveys the extension of conventional aerial photography to very high altitude photography from orbiting spacecraft. This chapter provides an historical

perspective and demonstrates how the concepts and instruments outlined in the beginning of this section have actually been put to use.

The second section deals with the preparation of images for interpretation once they have been taken. In Chapter 5, Jim Skaley discusses both conventional methods of photographic development and more exotic photo-optical techniques of image enhancement—ways to optimally display information for the photo interpreter. In Chapter 6, Alan Gillespie provides a survey of current methods used to enhance the digital images acquired by modern electronic imaging devices.

The third section consists of five chapters devoted to discussions of the use of different types of imagery. In Chapter 7, Harmer Weeden and Nanna Bolling discuss interpretation of aerial photographs, which is the workhorse of remote sensing. In Chapter 8, Anne Kahle reviews the physics of thermal infrared radiation and describes ways of computing geologically significant parameters, such as thermal inertia from measurements made in this part of the spectrum. Floyd Sabins then discusses conventional uses of images made from thermal radiation. The last two chapters explore the uses of the microwave region of the spectrum. Harold MacDonald describes applications of imaging radar systems, in which reflected microwave energy beamed at the scene from the imaging platform (usually aircraft) is used to construct pictures, and Thomas Schmugge discusses uses of naturally emitted microwave radiation.

The fourth section consists of 10 chapters that describe the applications of remote sensing techniques to solving problems in different disciplines. In the first chapter of the section, Gary Raines and Frank Canney discuss the special role of remote sensing over the major portion of the Earth which is covered by plants. Just as vegetation obscures much surface detail, so does it reveal much about the chemical and water content of the soil. Chapters 13 through 15 explore the role of remote sensing in geologic mapping. Michael Abrams and Barry Siegal emphasize methods of discriminating among different rock types using multispectral imagery, that is, registered images of a scene viewed in different spectral regions. David Gold describes the application of remote sensing to the mapping of geologic structure, and Paul Lowman and Larry Lattman emphasize the role of very high altitude or small-scale images in mapping morphologic characteristics of the Earth's surface.

Chapters 16 through 20 are concerned with special applications of remote sensing to geologic studies. John Elson discusses the use of remote sensing in the study of glaciers and glacial landforms—specialized aspects of geology of particular interest because of their dynamic character and their importance in understanding post-Pleistocene climatic trends.

In Chapter 17, Lawrence Rowan and Ernest Lathram present techniques of mineral exploration that are based on remote sensing. Exploration geology is one of the potentially most important applications of remote sensing, and was also one of the earliest—in the 1920s petroleum geologists used aerial photographs to help identify drilling sites.

The study of water resources has long been another important role of the geologist to society. Remote sensing has been particularly useful in monitoring rapidly changing phenomena such as floods and melting snowpacks. In Chapter 18, Vincent Salomonson and Albert Rango review all aspects of the application of remote sensing to water resources, from runoff prediction to water exploration.

The next chapter, by Wayne Pettyjohn, discusses the role of remote sensing in a new branch of geology—environmental geoscience. Remote sensing is particularly well suited for monitoring water and air pollution, as well as for long-term changes in vegetative cover.

Stephen Saunders and Thomas Mutch describe the role of remote sensing in extraterrestrial geology: a field often overlooked by traditional geologists, but nevertheless one that offers much to terrestrial geology, especially insights into the early history of the Earth. It is also a field that more than any of these has been responsible for the rapid increase in the sophistication of remote sensing.

The final chapter, by Alexander Goetz, explores remote sensing in the 1980s. It discusses the kinds of instrumentation and results we can realistically expect, and the programs that are already planned by the United States.

Of course, remote sensing would be of no practical importance if remotely sensed data were not readily available to the student and researcher. Consequently, the emphasis in this book is on photographs taken from aircraft and on images taken by the highly successful Landsat satellites. These data are available to all organizations and individuals from the EROS Data Center in Sioux Falls, South Dakota.

The techniques discussed in *Remote Sensing in Geology* will provide the basis for continued exploration and application of these data to problems in geology.

Barry S. Siegal   Greensboro, N. C.
Alan R. Gillespie   Pasadena, California
                                   September 1979

ACKNOWLEDGMENTS

This book is the collective work of many authors who have joined together to present a complete introduction and overview of the basic principles and research in geologic remote sensing. The information clearly represents the research of individuals and institutions over the past years, who unfortunately cannot be acknowledged here because of space limitations.

Many individuals made important contributions, critical reviews, or provided illustrative material. Dr. Louis F. Dellwig reviewed the entire manuscript, made constructive criticism, and helped to improve organizational structure and technical clarity.

Chapters 6, 7, 13, 20, and 21 are significantly based on the results of research carried out at the Jet Propulsion Laboratory, California Institute of Technology, under contract NAS7-100, sponsored by the National Aeronautics and Space Administration.

Dr. Ernest E. Hardy made important contributions to the organization of Chapter 5, and Professor Elmer S. Phillips offered valuable comments and source information on photographic methods. Dr. William Travers described geologic features for two illustrations, and Ruth Lyon, Brian McKelvey, and Douglas Payne provided technical assistance. The Resource Information Laboratory, Cornell University, provided source material and financial assistance.

All of the photographs (except as noted) and most of the line drawings used as illustrations for Chapter 6 were provided by the Image Processing Laboratory of the Jet Propulsion Laboratory, California Institute of Technology. Bruce Jakosky reviewed the manuscript.

Dr. Dennis O'Leary, United States Geological Survey, critically reviewed and improved Chapter 14. Illustrations and tables used in Chapter 14 were generated by the following persons and organizations: A. H. Strahler; S. W. Carey; R. H. Hardy, Jr.; C. R. Twidale; M. H. Podwysocki; W. S. Kowalik; A. F. Buddington; R. Black and M. Girard; M. H. Carr; E. S. O'Driscoll; the Dominion Observatory, Canada; T. W. Offield; M. R. Canich; R. R. Parizek; E. Werner; P. W. Guild; K. C. Burke and J. T. Wilson; P. D. Lawman, Jr.; P. Molnar and P. Tapponier, and the annotated compilation of Landsat images by N. M. Short, P. D. Lowman, Jr.; S. C. Freden; and W. A. Finch, Jr. N. Manula, Jr., and H. A. Pohn of the United States Geological Survey assisted with the Ronchi grid photographs.

The National Geographic Society graciously permitted the reproduction of the black-and-white version of their color mosaic of Landsat images of the United States in Chapter 15.

Mark Meier and an anonymous reader made constructive suggestions to Chapter 16. J. T. Parry offered advice on radar, and R. Yates provided line drawings. Photographs and images were provided by the National Air Photo Library, Ottawa, the Canada Center for Remote Sensing, and the United States Geological Survey. Research reported in Chapter 16 was supported by the National Research Council of Canada.

Chapter 20 is JPL Planetology Publication Number 79-326-54. The writing of Chapter 21 was made easier by the support of all the members of the Jet Propulsion Laboratory's Earth Resources Applications Group during the many periods when administrative duties precluded the author's active participation in ongoing research.

B. S. S.
A. R. G.

CONTENTS WITH CONTRIBUTING AUTHORS

1 Introduction   1

SECTION I BASIC CONCEPTS IN REMOTE SENSING   3

2 Electromagnetic Radiation: The Communication Link in Remote Sensing *Graham R. Hunt*   5

3 Acquisition of Remotely Sensed Data *Donald S. Lowe*   47

4 The Evolution of Geological Space Photography *Paul D. Lowman, Jr.*   91

SECTION II IMAGE PROCESSING AND ENHANCEMENT   117

5 Photooptical Techniques of Image Enhancement *James E. Skaley*   119

6 Digital Techniques of Image Enhancement *Alan R. Gillespie*   139

SECTION III INTERPRETATIVE TECHNIQUES   227

7 Fundamentals of Aerial Photography Interpretation *Harmer A. Weeden and Nanna B. Bolling*   229

8 Surface Thermal Properties *Anne B. Kahle*   257

9 Interpretation of Infrared Images *Floyd F. Sabins, Jr.*   275

10 Techniques and Applications of Imaging Radars *Harold C. MacDonald*   297

11 Techniques and Applications of Microwave Radiometry *Thomas Schmugge*   337

SECTION IV APPLICATIONS OF REMOTE SENSING TECHNIQUES TO GEOLOGIC PROBLEMS   363

12 Vegetation and Geology *Gary L. Raines and Frank C. Canney*   365

13 Lithologic Mapping *Michael Abrams and Barry S. Siegal*   381

14 Structural Geology *David P. Gold*   419

15 Geomorphology *Paul Lowman and Laurence Lattman*   485

16 Glacial Geology *John A. Elson*   505

17 Mineral Exploration *Lawrence C. Rowan and Ernest H. Lathram*   553

18 Water Resources *Vincent V. Salomonson and Albert Rango*   607

19 Environmental Geoscience *Wayne A. Pettyjohn*   635

20 Extraterrestrial Geology *R. Stephen Saunders and Thomas A. Mutch*   659

21 Geological Remote Sensing in the 1980s *Alexander F. H. Goetz*   679

Glossary   687

Index   695

ABOUT THE AUTHORS

MICHAEL ABRAMS

Michael Abrams received his B.S. degree in biology in 1970 and his M.S. degree in geology in 1972 from the California Institute of Technology. He has since been employed as a Scientist in the Geology and Geophysics Group at NASA's Jet Propulsion Laboratory, California Institute of Technology. His current interests center about the application of remote sensing information to geologic problems. These include lithologic mapping, identification of alteration associated with mineral and petroleum deposits, and determination of wavelength regions more appropriate to solving geological problems. Currently, he is investigator on a mineral exploration project, and co-investigator for a uranium alteration detection project and NASA's Heat Capacity Mapping Mission project.

NANNA B. BOLLING

Nanna Bolling received her B.S. in earth and space science in 1964 and her M.S. in geology in 1970, both from The Pennsylvania State University. She has worked with the Department of Civil Engineering and is presently an image analyst and a technical writer and editor for the Office for Remote Sensing of Earth Resources (ORSER) at The Pennsylvania State University. In this capacity, she has developed a system for annotation and cataloging extensive files of aerial photography and satellite imagery and has assisted researchers in a variety of fields with photograph and image interpretation as ground truth for computer classification of satellite data. She has edited and collaborated in the writing of well over 100 technical reports.

FRANK C. CANNEY

Frank Canney received a B.S. degree in chemistry in 1942 from the Massachusetts Institute of Technology. From 1942 to 1947 he served in the U.S. Navy as a gunnery officer in the Pacific theater of operations. From 1947 to 1951 he took graduate training in geology at M.I.T. and received his Ph.D.

in 1952. His entire professional career since then has been with the United States Geological Survey in Denver, Colorado where his research is devoted to the development of geochemical methods of exploration for concealed mineral deposits. He served as Chief of the Branch of Exploration Research from 1970 to 1973. His current assignment is the development of methods for the remote detection of geochemical anomalies in heavily vegetated regions.

JOHN A. ELSON

John A. Elson received a B.S. in geology from the University of Western Ontario. He lectured in geography for a year at McMaster University and obtained an M.S. degree with a thesis on hydrology. From 1946 to 1956 he was with the Geological Survey of Canada, except during educational leave that enabled him to obtain a Ph.D. in glacial geology from Yale. He joined the Department of Geological Sciences of McGill University in 1956. His research interests include glacial Lake Agassiz, glacial landforms and deposition, Pleistocene lakes and seas, and freeze-thaw processes in temperate climates. He developed one of the first photo-geology courses in Canada for students interested in mineral exploration and engineering geology. At McGill he has served terms as departmental director of graduate studies, and as chairman. He is active on national and international committees on Quaternary research.

ALAN R. GILLESPIE

Alan Gillespie studied geophysics at Stanford where he received his B.S. in 1969. He received his M.S. in geology at the California Institute of Technology in 1977. From 1969 until the present, he has been involved in remote sensing and computer image enhancement at the Jet Propulsion Laboratory, California Institute of Technology. He has worked on the 1969 and 1971 Mariner Mars Projects, and has subsequently been involved in remote sensing of terrestrial geology. He was instrumental in developing cartographic projection software and multispectral display and color image analysis techniques at JPL. His current interests in remote sensing lie in improving techniques of rock identification and geologic mapping.

ALEXANDER F. H. GOETZ

Alexander Goetz received his B.S. in physics, M.S. in geology, and Ph.D. in planetary science (1967) from the California Institute of Technology. He was a member of the technical staff at Bell Telephone Laboratories, Washington, D.C. from 1967 to 1970 where he worked on Apollo landing site geological experiments. He was a Principal Investigator on Apollo 8 and 12 orbital multispectral photography experiments. In 1970, Dr. Goetz joined the Jet Propulsion Laboratory where he continued his work in lunar studies and applied remote sensing techniques—particularly image processing developed for planetary work—to earthly problems. He was Principal Investigator in both Landsat and Skylab programs that developed methods for remote geological mapping and detection of mineralized zones using orbital multispectral images. Dr. Goetz was manager of the Planetology and Oceanography Section for two years and now is concentrating on the development of new instruments for spectroscopy and imaging of the Earth. The first of these will be flown on Orbital Flight Test-2 of Shuttle in 1981.

DAVID P. GOLD

After graduating with a B.Sc. Hons degree in geology and chemistry from the University of Natal in South Africa, Dr. Gold spent two years with the Union Corporation, Ltd., in mining geology, geophysical surveying, and mineral exploration. He completed a M.Sc. degree at the same time, and then continued graduate studies at McGill University in Montreal, Canada. During the summers he worked as party chief on regional mapping, mineral exploration, and other special projects (for example, impact craters) in a variety of terrains in eastern Quebec and the Canadian Arctic for various organizations. Following two years of teaching at Loyola College in Montreal, he took a postdoctoral position at The Pennsylvania State University. In 1968 he joined the faculty of the Geology Department at the university and currently is the chairman of the Geology Graduate Program in the Department of Geosciences. He is an active member of ORSER, an interdisciplinary institute of remote sensing at The Pennsylvania State University. His research and teaching interests are in structural geology, economic geology, remote sensing for structure and resources, emplacement of kimberlites and carbonatites, and cratering phenomena, and in the use of "plate tectonics" elements and setting for mineral exploration.

GRAHAM R. HUNT

Graham Hunt was born in Australia and received his B.Sc. (1951), B.Sc. Hons (1952), M.Sc. (1954), and Ph.D. (1958) all from Sydney University, Australia. After working at the Division of Coal Research, CSIRO (1959), he came to the United States to the Chemistry Department of Tufts University (1960-1961) and the Spectroscopy Department of M.I.T. (1962-1963). He joined the Air Force Cambridge Research Laboratory (1963-1975) and in 1975 the Petrophysics and Remote Sensing Branch of the United States Geological Survey, his current position. He has been a permanent consultant to Polaroid Corporation (1960-1970), Baird Atomic (1960-1961), Atlantic Gelatin Corporation (1961-1962), and Jarrell Ash (1963), and a visiting professor to Florida State University and the University of Vermont. His research interests have covered the areas of electronics, vibrational and rotational spectroscopy, vibration theory, molecular structure, lunar and planetary remote sensing from ground-based, balloon and rocketborne telescope-spectrometer systems, and he is currently interested in applying radiometric and spectroscopic techniques to terrestrial remote sensing. He holds four United States invention patents and has 88 papers published in scientific journals.

ANNE B. KAHLE

Anne Kahle received her B.S. degree in physics from the University of Alaska in 1955, and her M.S. degree in geophysics in 1962. She joined the staff of the Rand Corporation as a Physical Scientist in 1961 and worked on problems in geomagnetism and climate dynamics until 1974. She received her Ph.D. degree in meteorology at the University of California at Los Angeles in 1975. Dr. Kahle joined the Planetology and Oceanography Section of the Jet Propulsion Laboratory, California Institute of Technology, in August, 1974, working on modeling of the Earth's surface heating, and developing techniques to utilize remotely sensed thermal IR data. Since July 1975 she has been the supervisor of the Earth Applications and Climatology group. Current professional interests include geologic applications of remote sensing, using a combination of all available types of data.

ERNEST H. LATHRAM
Ernest Lathram received his B.A. degree from Miami University, Oxford, Ohio in 1940 and his M.S. degree from the University of Minnesota in 1942, both in geology. He served in the Armed Forces from 1942 to 1946 in photographic, photo interpretation, and intelligence capacities. He was a geologist of the United States Geological Survey from 1946 to retirement at the end of 1975. Except for four years during which he studied the marine geology of the United States continental shelves, his work has been focused on Alaska. This work has involved stratigraphic and structural mapping in northern Alaska, reconnaissance mapping and mineral resource analysis in southeastern Alaska, photogeologic mapping of selected areas, compilation of regional geologic maps, preparation of regional tectonic analyses and resource appraisals, and assignments as Assistant Chief, Alaskan Geology Branch; Assistant Chief; Office of Marine Geology; and EROS Representative for Pacific States and Alaska. His publications include detailed and general maps, and papers on remote sensing, structural geology, tectonics, and Arctic science. He is now a remote sensing and geologic consultant.

LAURENCE LATTMAN
Laurence Lattman holds a B.S. degree in chemical engineering and M.S. and Ph.D. degrees in geology. He worked for the Gulf Oil Corporation as a photogeologist from 1953 until 1957. From 1957 to 1975 he was on the faculty of The Pennsylvania State University and the University of Cincinnati. Since 1975 he has been Dean of the College of Mines and Mineral Industries at the University of Utah. He is the author of one book and approximately 50 articles on geomorphology and interpretation of imagery. He has been a consultant to oil and mining companies and to foreign governments on the use of imagery in oil and mineral exploration.

DONALD S. LOWE
Donald S. Lowe received his M.S. in physics from Duke University in 1948. Afterward, he joined the Naval Ordnance Laboratory where he was responsible for infrared and optical instrumentation development and spectroradiometric measurements of targets and backgrounds. In 1958, he joined the Bendix Aerospace Systems Division where he was responsible for developing airborne spectroradiometric instrumentation systems and remote sensing in Earth resources. He is presently the director of the Applications Division and Vice President of the Environmental Institute of Michigan. In his eight years at ERIM, he has pioneered the development of multispectral scanning systems and their application to Earth resource problems. Mr. Lowe has participated in various NAS/NRC panels, chairing the panel on Electromechanical Scanners of NASA's working group on Advance Scanners and Imagers of Earth Observations (1972). Since 1964, he has authored or coauthored 19 papers about Earth resource systems and technology as well as chapters of three textbooks on remote sensing.

PAUL D. LOWMAN, JR.
Paul Lowman is a geologist employed by the NASA Goddard Space Flight Center in the Geophysics Branch. He joined NASA in 1959 after graduate study at the University of Colorado, and worked on lunar geology, the origin of tekites, and lunar mission planning for several years. In 1962, he proposed an experiment using hand-held cameras in which the Mercury astronauts would take pictures of geologically important features. The Synoptic Terrain Photography Experiment was successfully carried out on Mercury, Gemini, and Apollo missions. Dr. Lowman was also Principal Investigator for the SO65 Multispectral Terrain Photography Experiment on Apollo 9, and took part in astronaut photographic training and picture screening for Skylab missions in 1973 to 1974. Since 1974, he has published papers on crustal evolution and comparative planetology, and was a coauthor of *Mission to Earth; Landsat View the World*. He is presently engaged in synthesis and application of satellite geophysical data. Dr. Lowman is author of *Lunar Panorama, Space Panorama*, and *The Third Planet*, as well as technical papers in the fields mentioned.

HAROLD C. MACDONALD
Harold C. MacDonald received a B.S. degree in geology from the State University of New York at Binghamton, and his M.S. and Ph.D. degrees in geology from the University of Kansas, Lawrence. Prior to his college education he served six years with the U. S. Air Force as Navigator, and SHORAN

Test Project officer. From 1962 to 1965 he was employed by the Sinclair Oil and Gas Company, Denver, Colorado, as a petroleum geologist conducting local and regional reservoir studies. While completing work toward the Ph.D. degree (1965-1969), he was a research assistant at the Remote Sensing Laboratory, Center for Research in Engineering Science, University of Kansas. From 1969 to 1970 he held the position of research associate at the University of Kansas where his research included the evaluation of the geoscience potential of side-looking radar systems. Dr. MacDonald is now Professor of Geology at the University of Arkansas, Fayetteville, Arkansas, and is also Associate Director of the Arkansas Water Resources Research Center. Dr. MacDonald has been an invited member of several NASA workshops, including Seasat and Space Shuttle.

THOMAS A. MUTCH

Thomas Mutch received his A.B. in history from Princeton University (1952), M.S. in geology from Rutgers (1957), and Ph.D. in geology from Princeton University (1960). He joined Brown University as an assistant professor in 1960, and presently holds the rank of professor in the Department of Geological Sciences. Dr. Mutch has served as a NASA consultant and Leader, Viking Lander Imaging Science Team. His research interests include planetary geology, general field geology, and stratigraphy. He is the author of numerous scientific articles, including two well-known books: *The Geology of the Moon* and *The Geology of Mars*.

WAYNE A. PETTYJOHN

Wayne A. Pettyjohn received his B.A. and M.A. degrees in geology from the University of South Dakota and the Ph.D. degree, also in geology, from Boston University. While serving as a hydrologist with the Water Resources Division of the United States Geological Survey in North Dakota, he studied law and was admitted to the bar in 1968. Since 1967 Dr. Pettyjohn has been a professor in the Department of Geology and Mineralogy at The Ohio State University. One of the original Landsat-1 investigators, he evaluated strip-mine reclamation and monitoring, which was developed into an operational procedure. He has published more than 100 reports, articles, manuals, and books.

GARY L. RAINES

Gary Raines received a B.A. degree in geophysics in 1968 from the University of California at Los Angeles and a Ph.D. in geology from the Colorado School of Mines in 1974. From 1974 through 1975 he worked as a National Research Fellow with the United States Geological Survey on the spectral properties of sedimentary rocks, and digital image processing. Since 1976 he has been working for the United States Geological Survey on the spectral properties of rocks and vegetation and the application of remote sensing techniques to mineral exploration and regional tectonics.

ALBERT RANGO

Albert Rango graduated with B.S. (1965) and M.S. (1966) degrees in meteorology from The Pennsylvania State University and received a Ph.D. (1969) in watershed management from Colorado State University. He has been a research hydrologist at NASA's Goddard Space Flight Center, an assistant professor of meteorology at The Pennsylvania State University, and a private water-resources consultant. His research interests are in watershed management, snow hydrology, geomorphology, floodplain management, and remote sensing of hydrologic parameters. In addition to these areas of interest, he has experience in the preparation of educational TV features, evaluation of the effects of weather modification, and coordination of and participation in field expeditions in support of remote sensing experiments. He has also served on committees of several professional organizations and has published numerous professional papers.

LAWRENCE C. ROWAN

After receiving his Ph.D. from the University of Cincinnati in 1964, Lawrence C. Rowan joined the Branch of Astrogeology of the United States Geological Survey in Flagstaff, Arizona. His principal research was in lunar geologic mapping using small-scale satellite images, and he also worked on the Geological Survey's Lunar Orbiter Program, coordinating phases of geologic analysis of the possible Apollo landing sites. In 1969, he joined a small group of geologists and geophysicists to form the Remote Sensing Geophysics Section, which now constitutes the Geological Survey's main research effort in this field. As Staff Geologist for

Remote Sensing from 1972 to 1974, Dr. Rowan coordinated the geologic remote sensing program during the Landsat-1 and Skylab experiments. His current research focuses mainly on spectral and spatial analysis of Landsat, Skylab, and aerial photographs and images, especially as applied to mineral exploration.

FLOYD F. SABINS, JR.

Floyd Sabins received his B.S. degree in geology from the University of Texas at Austin (1952) and his Ph.D. degree in geology from Yale University (1955). He is now a senior research associate at Chevron Oil Field Research Company, California, where he is applying remote sensing techniques for petroleum exploration. Dr. Sabins is an instructor for remote sensing application at the AAPG Petroleum Exploration School, where he recently received a Distinguished Lecturer award. Dr. Sabins also is an adjunct professor of Geology at the University of California at Los Angeles and the University of Southern California. He is a member of various professional societies and is the author of numerous publications, including *Remote Sensing—Principles and Interpretation*.

VINCENT C. SALOMONSON

Vincent Salomonson received a B.S. degree in agricultural engineering from Colorado State University, a B.S. degree in meteorology from the University of Utah, an M.S. degree in agricultural engineering from Cornell University, and a Ph.D. in atmospheric science from Colorado State University. He presently heads the Hydrospheric Sciences Branch in the Laboratory for Atmospheric Sciences at Goddard Space Flight Center and is the Project Scientist for the fourth in the series of Landsat satellites, Landsat-D. During the past eight years his studies have addressed the applications of space technology to meteorology and water resources management. In 1976, Dr. Salomonson was awarded the NASA Exceptional Scientific Achievement Medal for outstanding contributions in the practical applications of remote sensing data to the water resources field, for transferring these applications to the user community, and for developing and guiding NASA's water resources research program. Prior to coming to Goddard, he spent three years as a weather officer in the United States Air Force (1959-1962). He is the author of approximately 70 publications in scientific journals, conference proceedings, and NASA reports.

THOMAS SCHMUGGE

Thomas Schmugge received a B.S. degree in physics from the Illinois Institute of Technology and a Ph.D. in physics from the University of California at Berkeley in 1965. Dr. Schmugge's initial research was in the area of the microwave properties of earth ions in solids. He continued this research while teaching in the Physics Department at Trinity University from 1964 to 1970. His work on the use of microwave techniques for remote sensing began in 1970 as a National Academy of Sciences Senior Research Associate at Goddard Space Flight Center. Dr. Schmugge remains at Goddard working primarily on the application of passive microwave techniques for the remote sensing of soil moisture, sea ice, and snow. He was also involved in the early research on the use of thermal inertia techniques for soil moisture sensing, which aided in the definition of NASA's Heat Capacity Mapping Mission satellite.

BARRY S. SIEGAL

Barry Siegal received a B.S. degree in secondary education—earth science from the State University of New York, College at Oswego and M.S. and Ph.D. degrees in geology from The Pennsylvania State University. After completing his degree, Dr. Siegal held a National Research Council Resident Research Associateship at the Jet Propulsion Laboratory, California Institute of Technology where his research centered on the use of remote sensing data for lithologic mapping. He is now a remote sensing geologist with Ebasco Services, Inc., where he is working on power plant siting studies and uranium exploration. His recent studies include siting of the first nuclear power plant in the Republic of the Philippines, and the Laguna Verde Nuclear Power Plant, Mexico. Dr. Siegal is the author of numerous publications in remote sensing and is coauthor of *Earth Perspectives*. He has held teaching appointments at The Pennsylvania State University; State University of New York, College at Oswego; Bloomsburg State College; and California State University, Fullerton.

R. STEPHEN SAUNDERS

R. Stephen Saunders received his B.S. degree in geology from the University of Wisconsin and a Ph.D. in geology from Brown University. He is currently Manager of the Planetary Surfaces and Interiors Section at the Jet Propulsion Laboratory, California Institute of Technology. His current research centers on geologic interpretation of Mars' surface features from Mariner and Viking pictures, including quadrangle mapping (MC 19), geologic interpretation of Mars' gravity data, and structural mapping. Dr. Saunders has been a Planetology Program Principal Investigator since 1971 and a member of the Viking Lander Imaging Team, and the Viking Flight Operations Team. He is Development Project Scientist for the planned Venus Orbiting Imaging Radar mission. He has held previous appointments at the United States Geological Survey, Flagstaff, (1967-1970) and Brown University (1973).

JAMES E. SKALEY

James E. Skaley has B.S. and M.S. degrees in biology from the University of North Dakota. He served four and one-half years in the Air Force as a Project Officer in the Reconnaissance Branch at Rome Air Development Center. While in the Air Force he participated in the evaluation and interpretation of both photographic and electronic systems for military and civilian applications. After leaving the Air Force, he was employed by the Resource Information Laboratory at Cornell University as Project Coordinator on NASA contracts evaluating Landsat and Skylab imagery for land use in New York State. The emphasis of his research was on developing low cost, low technology photographic enhancement techniques to improve the usability of the imagery without making large capital investments. Mr. Skaley is currently performing research on the use of large-scale, multispectral photography to classify vegetation and wildlife habitat.

HARMER A. WEEDEN

Harmer A. Weeden, Emeritus Professor, at The Pennsylvania State University, is in charge of surveying and photogrammetry. His interest in photo interpretation dates back to 1948, when he started graduate work in this field at Cornell University. He received his Ph.D. from Cornell in 1965. His experience has been principally related to work with the Pennsylvania Department of Transportation, for which he has directed several studies. More recently he has participated in interdisciplinary research with the Office for Remote Sensing of Earth Resources at The Pennsylvania State University. Dr. Weeden has published numerous papers and reports on photo interpretation for highway engineers, remote sensing data analysis, and related topics.

Remote sensing is the science of gathering information describing distant objects or scenes—targets that could not be studied without instruments. Instruments permit the study of not only very distant objects, like the Moon and the planets, but also nearby objects, which are for some reason inaccessible or in hazardous locations. Thus, remote sensing is useful if we wish to study the geology of central Asia, which is politically and logistically difficult to study on the ground, or the depths of the ocean, which certainly is in a hazardous environment.

Instruments also extend the range of human perception from visible light to include most of the spectrum of electromagnetic energy as well as magnetic, gravitational, and particle radiation fields. Not only does this permit geologists to infer the composition of surface materials (from absorption bands in infrared radiation reflected from the surface), but enables them to determine some bulk or body properties, for example, thermal conductivity. The use of instruments such as radar allows us to study difficult terrains under adverse weather conditions, whereas the use of gravitational and magnetic fields enables us to infer subsurface structure.

Instruments also extend our perspective by providing observation at a variety of scales. When viewed in a regional or global context, tectonic and lithologic relationships often become clearer, thus permitting interpretation of features in relation to their regional setting. This perspective also allows discovery of regional relationships that may not have been apparent in the field, and consequently may not have been recorded on geologic maps.

Remotely sensed data are often acquired or displayed as images. The photograph is a familiar example. Because aerial photographs and other remotely sensed images are so important in geology and other fields, the term "remote sensing" is often used to mean "remote imaging." The emphasis in *Remote Sensing in Geology* is really on remote imaging.

Remote imaging had its beginnings in the middle of the last century with the invention of the photograph. One of the first applications of photography was to topographic mapping, and by the time of the Civil War, balloons may have been used as aerial platforms for cameras. However, two inventions were required before remote sensing became widely available: the gelatin photosensitive emulsion in 1871, which allowed delayed development of the latent image, and the airplane in 1903, which provided a fast and controllable camera platform for systematic observations.

Several important innovations in remote sensing were made in the first half of the twentieth century. These included the invention of multilayer color film and infrared-sensitive emulsions (which were also incorporated in false-color infrared films), nonimaging radar, and thermal sensors, and the use of sounding rockets to obtain very high altitude photographs (1).

However, it was not until the 1950s that remote sensing began to assume its modern character. Imaging radars and thermal infrared scanners were built and tested, and the first orbiting satellites were launched. The 1960s saw the first photographs of the Earth from manned orbiting satellites and the first hand-held photographs from the surface of the Moon. Systematic monitoring of the Earth's weather was performed by special satellites of the TIROS and NIMBUS series. Deep space probes returned high quality television pictures of Mars, while lunar orbiters mapped the back of the Moon in detail. By 1977, Mars orbiters had mapped the entire planet and two landers provided detailed pictures of its surface. Jupiter, Venus, and Mercury had all been mapped in some detail by space probes. However, the most important developments to the geologist were the systematic imaging of most of the Earth's surface by the three Landsat satellites, continued improvement of electronic cameras to permit high quality imaging in narrow bands collectively covering more of the electromagnetic spectrum than ever before, and widespread use of digital computers to assist in the analysis and interpretation of such images.

In 1978, two new satellites carrying remote imaging instruments were launched: Seasat, which utilizes an imaging radar system designed for oceanographic studies, and the misnamed "HCMM" or "Heat Capacity Mapping Mission" satellite, which provides repetitive thermal images of the Earth at moderate resolution. In the 1980's, they will be joined by Landsat-D, which will provide a "thematic mapper" acquring images in seven spectral channels, one of which is in the thermal infrared, and by "Stereosat," designed to gather medium resolution topographic data for the whole globe. Furthermore, there may be satellites flown that specialize in rock mapping by imaging in near-infrared light, and aircraft-borne electronic cameras will be more widely available than today. Finally, the space shuttle will provide the opportunity for a variety of remote sensing programs.

During the past few years, the role of remote imaging in geological exploration and research has been examined by many investigators. It is important to recognize that very few of these investigators regard remote sensing as a panacea. Few subjects can be studied better remotely than by competent field observers. However, relatively little of the land area of the Earth has been geologically mapped and even in the continental United States, there are regions where reconnaissance mapping would be welcome. Here, remote sensing has an important role to play because of cameras mounted on satellites or aircraft provide a quick, economical view of even the most difficult terrain.

Remote sensing is also widely used as an aid to mineral, groundwater, or petroleum exploration. Whereas in the past aerial images have been used chiefly for structural information or as base maps on which to plot information gathered in the field, with the advent of high quality images taken in infrared light, mineral anomalies are being detected directly.

Remote imaging has also played a vital role in increasing our understanding of the geology of other terrestrial planets. The issue here is especially clear-cut because man cannot now visit the surfaces of these planets to gather information first hand. Also, we should remember that the lunar samples returned by the Apollo astronauts have increased our knowledge of the Moon beyond that which we would have from images alone.

In the final analysis, the question is not whether remote sensing will ever replace field geology, nor whether field geology is superior to remote geology, but rather how to combine all avenues of investigation to more effectively solve geological problems and answer geological questions.

REFERENCES
1. Fisher, W. A., Badgley, P., Orr, P., and Zissis, G. J., 1975, History of remote sensing, *in* Reeves, R., Anson, A., and Landen, D., v. 1, chap. 2, Manual of remote sensing; Falls Church, Virginia, American Society of Photogrammetry, p. 27-50.

BASIC CONCEPTS IN REMOTE SENSING

A solid grasp of the role and potential significance of remote sensing as a tool for solving geologic problems requires a thorough understanding of the principles governing acquisition of the data. In the first section of *Remote Sensing in Geology,* readers are introduced or reacquainted with the physics of electromagnetic radiation. They are then exposed to a summary of the different kinds of instruments actually used in remote sensing to measure and record, usually in pictorial format, such data. An historical survey of the development and maturation of remote sensing as a science in the Space Age provides readers with an overview of the use of such instruments as well as a general understanding of their potential as research tools.

CHAPTER 2

ELECTROMAGNETIC RADIATION: THE COMMUNICATION LINK IN REMOTE SENSING

GRAHAM R. HUNT

2.1 INTRODUCTION

Remote sensing in geology is a formidable pursuit. Minerals, the fundamental materials of geology, are among the most complex of inorganic solids. They are combined in a variety of ways to form rocks, which are then altered and disintegrated to form soils in which complicated organic substances may thrive when a suitable environment of gases and liquids, as well as solids is present.

The task of the person using remote sensing is to acquire, usually from a distance, information about the nature of a given region containing some or all of the above constituents, and usually without altering the region in any way. Such information may be acquired by measuring (a) electromagnetic energy emitted, scattered, polarized, or reflected by elements of the scene; (b) force fields, such as gravity or magnetic, that have been created or modified by the scene; and (c) mechanical vibrations or waves, such as acoustical or seismic, emanating from, transmitted through, or reflected from, the scene. In this chapter, the discussion will be limited to considering the most versatile and commonly used of these sources of information, electromagnetic radiation. To effectively use such radiation, the observer must understand its nature and behavior so that he can manipulate it in the sense of generating, directing, collecting, and detecting it; and in order to extract and decipher the information about the region contained in the radiation, he must understand the ways in which electromagnetic radiation interacts with matter.

The purpose of this chapter is to present some fairly basic material dealing with the nature of electromagnetic radiation, and the way in which it interacts with matter. Macroscopic interactions, which can be generally described as geometrical optical effects, are largely ignored, while interactions on the atomic and molecular level are considered in some detail. Particular emphasis is placed upon the concept of discrete energy levels and the transitions between them.

The treatment of atoms is included as a logical starting point because of their particular relevance to the treatment of ions in crystal fields and because nomenclature commonly encountered can be introduced. Rotational energy levels are very briefly considered because of the relevance to atmospheric gases that are encountered in terrestrial remote sensing.

The wavelength range discussed in any detail here is limited to the regions included by the visible and midinfrared, because that is the principal range where electromagnetic radiation truly interacts with matter and is absorbed and emitted. Certain spectroscopic disciplines are not discussed here, even though they can provide extremely useful data concerning minerals and rocks. Included in these are: gamma and x ray, submillimeter microwave, Raman, Mossbauer, and nuclear magnetic resonance spectroscopy. The former two lie outside the region of particular interest here, and the applications of the latter three are extremely restricted in direct remote sensing.

Many of the areas treated in this chapter traditionally fall more into the realm of the chemist and physicist rather than in the area of interest to the geologist, but the ever-increasing importance of remote sensing of the environment makes it imperative that the geologist acquire a basic understanding of these disciplines.

2.2 REMOTE SENSING

The term "object" is used quite generally throughout this chapter to indicate any sample, target, area or scene; the term "observer" is used to include the collecting, wavelength analyzing, detecting, and recording system.

Remote sensing is a very commonly used term, but it is rarely defined because of its largely self-explanatory nature. In the present context, the definition of remote sensing is restricted to mean the process of acquiring information about any object without physically contacting it in any way, regardless of whether the observer is immediately adjacent to the object or millions of miles away. It is further required that such sensing may be achieved in the absence of any matter in the intervening space between object and observer. Consequently, the information concerning the object must be available in a form that can be impressed on a carrier that can transport the information even through a complete vacuum. The information carrier, or communication link, in this case is electromagnetic energy.

The fact that very frequently there is some physical matter between the object and observer does not alter this definition, and the effects of its presence are to alter the speed and direction of propagation or to attenuate the information content of the radiation, all by interaction of the radiation with the intervening matter.

Using the above definition, of the five senses of man, only sight qualifies as a remote sensing process, because both the tactile sense and taste require physical contact with the object, smell requires transport of some small part of the object to the observer, and hearing requires the transmission of sound waves through a physically conducting medium, which may be a gas, liquid, or solid.

Remote sensing data basically consists of wavelength-intensity information, acquired by collecting the electromagnetic radiation leaving the object at specific wavelengths and measuring its intensity.

Geological remote sensing techniques are referred to as *active* or *passive*. In the present context, active techniques are defined as those in which man supplies the radiation from a specific artificial source, such as a laser or a radar antenna. All other techniques which utilize a naturally occurring source are defined as passive. This includes not only techniques that measure the radiation emitted from the object itself as a consequence of its internal kinetic energy (temperature) such as in the thermal infrared and microwave regions, but also techniques which require that the sun act as a source of reflected energy, such as in the visible and near infrared regions.

The desired information is contained in the radiation emitted, reflected, scattered, or transmitted by the object, because the object imposes its characteristic imprint on the wavelength-intensity relationship. In order to recognize the object, it is necessary to know how it affects the intensity of the radiation, and to do so requires an understanding of the nature of electromagnetic radiation and the way in which it interacts with matter. The rest of this chapter is devoted to providing such an understanding.

2.3 ELECTROMAGNETIC RADIATION

2.3.1 Historical Development

A basic difficulty encountered in discussing electromagnetic energy and its interaction with matter derives from the lack of a familiar object or macroscopic model that can be used as a reasonable analogue. The lack arises principally because of the dual nature of the behavior of electromagnetic energy. In some circumstances electromagnetic energy behaves as waves, while in other situations it displays the properties of particles.

The problem of the dual nature of electromagnetic energy has been apparent since early attempts to explain the nature of visible light, even though originally it was not realized that light was indeed just one form of electromagnetic energy. It will perhaps be informative to briefly outline the development of current theories before discussing the particular properties that are of most importance in remote sensing applications.

Newton (1642-1727), in agreement with many of his predecessors, regarded light as a stream of very small bodies emitted by shining substances, and this corpuscular view derived from the observation that in a uniform medium, light appears to travel in straight lines. Although he was aware of certain manifestations of the wave nature of light, Newton considered that if light were indeed a wave, it would propagate around corners, like sound.

At approximately the same time Huyghens (1629-1695) described light as a wave motion spreading out from a source in all directions, and contended that each point on a wave front may be considered to be a new source of spherical waves, and the envelope of those waves constitutes the wave front at a later time. This wave picture was supported by Young, who in 1802 first demonstrated the interference of light waves and enunciated the principle of superposition. He was able to quantitatively explain a large number of diffraction and polarization phenomena.

It was, however, Maxwell who, in formulating his electromagnetic theory in 1862, succeeded in characterizing the electric and magnetic fields and their relation to charges and current, and he expressed these relationships in a set of partial differential equations now known quite generally as Maxwell's equations. He showed that it was possible to have wave-like configurations of electric and magnetic fields which are entirely disconnected from their source, and which propagate throughout space at a constant speed. Subsequently, based on this theory, it was proved that visible light is composed of electromagnetic radiation of this type. Maxwell's equations explain a great variety of phenomena relating to propagation, dispersion, reflection, refraction, and interference of electromagnetic waves; but they do not explain the interaction of electromagnetic energy with matter on an atomic and molecular level.

The first contemporary suggestion of the discrete or quantum nature of electromagnetic energy is due to Planck. He found in 1900 that in order to calculate the correct distribution of energy emitted by a black body he could not assume that the constituent oscillators gain and lose energy continuously. Rather, he was forced to assume that a particular oscillator of frequency ν is able to exist only in discrete states whose energies are separated by the interval $h\nu$, where h is now known as the Planck constant. Planck's ideas were applied and extended shortly afterwards, as for example in Einstein's theories of the photoelectric effect and the heat capacity of solids.

The next major step in the extension of the idea of quantum restrictions was made by Bohr in 1913. Using the newly acquired knowledge of the nuclear atom enunciated by Rutherford and Thompson in 1911, he proposed a theory for the hydrogen atom based on the assumption that the electron moved in circular orbits around the nucleus such that its angular momentum was an integral multiple of $h/2\pi$. Thus, electrons in an atom occupy only discrete quantized states or orbits having different energies, and the lowest energy state is the ground or normal state. When an electron undergoes a transition from one state to another, it does so with the emission or absorption of packages of radiation, and these quanta of radiation were called photons.

The reality of the photon was shown not only by the photoelectric effect, but even more convincingly by the Compton effect, discovered in 1923. It was the Compton experiment in which x rays scattered by atoms display a new frequency ν_j, in addition to the incident frequency, ν_o, an observation that convinced physicists of the necessity of reconciling the wave and corpuscular theories.

The decisive step was provided in 1925 by de Broglie who proposed that waves accompanied particles, or particles had a wave nature such that $\lambda = h/mv$, where λ is the wavelength and mv (mass × velocity) is the momentum. De Broglie's work was responsible for showing that Bohr's theory of the atom did not sufficiently recognize the wave nature of the electron in the atom, and it also predicted that particles, such as electrons, should display diffraction effects. This prediction was soon verified in 1927 by Davisson and Germer who observed electron diffraction effects of precisely the type predicted by de Broglie.

However, it was Schrodinger in 1926 who formulated wave mechanics in terms of a wave equation. In considering de Broglie's hypothesis, Schrodinger was led quite unambiguously to the wave equation (which bears his name) and when he solved it he found that the results of Bohr for the hydrogen atom, the theory of linear oscillators, and many other problems of quantum mechanics were explained in a simple and elegant form.

The Schrodinger equation for atomic-molecular scale problems is not really derivable, and should be regarded as the counterpart of Newton's laws of motion for macroscopic bodies. It is used and accepted, not because of its derivation showing validity, but because when properly applied it yields correct results in agreement with observation and experiment. The particular importance of it in the present context is that it yields directly the allowed energy levels of an atomic or molecular system.

2.3.2 Nature of Radiation

As a result of development of understanding the nature of electromagnetic energy, which was outlined historically above, it is presently possible to furnish a consistent and unambiguous theoretical explanation for all optical phenomena using a combination of Maxwell's electromagnetic wave theory and modern quantum theory. Maxwell's theory deals primarily with the propagation and macroscopic optical effects of electromagnetic energy, while quantum theory is concerned with the atomic-molecular absorption and emission aspects of radiation.

The four differential equations that form the basis of electromagnetic theory are quite generally referred to as "Maxwell's equations," and they express in mathematical terms all the facts determined prior to 1860 by such workers as Coulomb, Oersted, Ampere, Biot, Savart, Henry, Faraday, and Gauss. They predict that electric and magnetic fields may exist in regions where no electric charges are present, and that when the fields at one point in space vary with time, then some variation of the fields must occur at every other point in space at some other time, and consequently, changes in the fields propagate throughout space. The propagation of such a disturbance is called an electromagnetic wave.

Because of the paramount importance of Maxwell's equations, they will be briefly introduced at this point.

2.3.3 Wave Nature: Maxwell's Equations

The electromagnetic state at a point in a vacuum can be specified by two vectors: **E**, the electric field (in volts per meter) and **H**, the magnetic field (in ampere turns per meter). These vector quantities are completely independent of each other in the static case, and are determined by the distribution of all charges and currents in space. In the dynamic case, however, the fields are not independent, but rather their space and time derivatives are interrelated as expressed by the curl (∇) equations

$$\nabla \times \mathbf{E} = -\mu_0 \frac{\partial \mathbf{H}}{\partial t} \tag{2.1}$$

$$\nabla \times \mathbf{H} = \epsilon_0 \frac{\partial \mathbf{E}}{\partial t} \tag{2.2}$$

where $\mu_0 \equiv$ permeability of the vacuum $= 4\pi \times 10^{-7}$ h/m and $\epsilon_0 \equiv$ permittivity of the vacuum $= 8.85 \times 10^{-12}$ farads/m.

The divergence conditions:

$$\nabla \cdot \mathbf{E} = 0 \tag{2.3}$$

$$\nabla \cdot \mathbf{H} = 0 \tag{2.4}$$

indicate that there is no charge at the point in question, and this is true in both the static and dynamic case. The four equations above are "Maxwell's equations" for a vacuum.

It can be seen that both fields satisfy the same formal partial differential equation

$$\nabla^2(X) = \frac{1}{c^2} \frac{\partial^2 (X)}{\partial t^2} \tag{2.5}$$

where $X = \mathbf{E}$ or $\mathbf{H}$, and $c = \frac{1}{\mu_0 \epsilon_0}$, and this is called the wave equation, which occurs in connection with many different kinds of physical phenomena. The major implication of the equation is that changes in the fields **E** or **H** propagate through space with a speed equal to the constant value c, which is known as the speed of light, and has a measured value of 2.9979×10^8 m/s.

The Maxwell curl equations are precisely the same for isotropic nonconducting media as they are for vacuum, except that the vacuum constants μ_0 and ϵ_0 are replaced by corresponding constants for the medium, denoted μ and ϵ.

It can be shown, for the case where the spatial variation occurs in the z direction, that the function $E_z t = E_0 \cos(k_z - \omega t)$ is a solution to the wave equation, provided $\nu = \frac{\omega}{k_z}$

This is the fundamental solution to the wave equation, and represents a plane harmonic wave, and the solution is of the same form for the magnetic field. It can be shown that the magnetic and electric components are perpendicular to each other and that these plane waves are both perpendicular to the direction of propagation (see Figure 2.1).

Poynting's theorem states that the time rate of flow of electromagnetic energy per unit area is given by a vector, called the Poynting vector, which is defined as the cross product of the electric and magnetic field vectors, and this vector specifies both the direction and energy flux of the wave.

In summary, it can be seen that all electromagnetic radiation is energy in transit and can be regarded as a wave motion, and it consists of inseparable oscillating electric and magnetic fields that are always mutually perpendicular to each other and to the direction of propagation, and this rate of propagation is constant in a vacuum.

2.3.4 Particulate Nature: Quantum Viewpoint

As far as electromagnetic radiation is concerned, the basic idea of quantum theory is that radiant energy is transmitted in indivisible packets whose

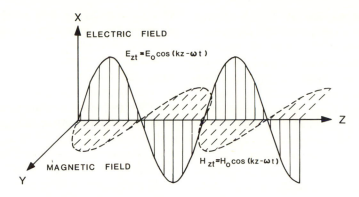

FIGURE 2.1. Electromagnetic wave; the electric (vertical) and
magnetic (horizontal) components are indicated.

energy is given in integral parts, of size $h\nu$, (where h
is Planck's constant = 6.6252×10^{-34} j-s, and ν is the
frequency of the radiation), and these are called
quanta or photons. In this basic way quantum theory
differs from Maxwell's theory, which implies that
energy is supplied continuously in a wave.

The dilemma of the simultaneous wave and particulate views of electromagnetic energy may be
conceptually resolved by considering that energy is
not supplied continuously throughout a wave, but
rather that it is carried by photons, and that the
classical wave theory does not give the intensity of
energy at a point in space, but gives the probability
of finding a photon at that point; thus the classical
concept of a wave yields to the idea that a wave
simply describes the probability path for the motion
of the individual photons.

The particular importance of the quantum approach for remote sensing is that it provides the
concept of *discrete* energy levels in materials, and
the values and arrangement of these levels is different for each different material. Information about a
given material is thus available in electromagnetic
radiation as a consequence of transitions between
these energy levels, a transition to a higher energy
level being caused by the absorption of energy, or
from a higher to a lower energy level being caused
by the emission of energy. The amounts of energy
either absorbed or emitted correspond precisely to
the energy difference between the two levels involved in the transition. Because the energy levels
are different for each material, the amounts of
energy a particular substance can absorb or emit
are different from those of any other materials. Consequently, the positions and intensities of the bands
in the spectrum of a given material are characteristic of that material.

2.4 ELECTROMAGNETIC SPECTRUM

The electromagnetic spectrum may be defined
as the ordering of the radiation according to wavelength, frequency, or energy.

The wavelength, denoted by λ, is the distance
between adjacent intensity maxima (for example) of
the electromagnetic wave, and consequently, it may
be expressed in any unit of length. Most commonly
wavelength is expressed in meters (m) or centimeters (cm); microns or micrometers (μ or μm =
10^{-4} cm); nanometers (nm = 10^{-7} cm); or Angstrom
units (Å = 10^{-8} cm).

The frequency, denoted by ν, is the number of
maxima of the electromagnetic wave which pass a
fixed point in a given time. Its relationship to
wavelength is simply

$$\nu = \frac{c}{\lambda}$$

where c is the speed of light. Frequency is commonly expressed in reciprocal centimeters, also
called wave numbers (cm^{-1}), or cycles per second
(cps), which are also called Hertz (Hz).

The energy of radiation, ϵ , is related to its
frequency (and consequently its wavelength) via
Planck's constant, (h = 6.6256×10^{-27} erg-s) such
that $\epsilon = h\nu$. The energy is commonly expressed in
ergs per molecule, calories per mole, or electron
volts (e.v.). The conversion factors to wave numbers
from these above mentioned energy units are 5036
$\times 10^{-15}$, 0.3499, and 8067, respectively.

The wavelength may assume any value, although for most practical purposes the spectrum is
usually presented between 10^{-16} and 10^{7} m, or from
the cosmic ray to the audio range. However, wavelengths as long as 10^{11} m have been detected by
sensitive magnetometers.

No matter what the wavelength of the electromagnetic radiation, it is all generated by electrically charged matter. However, there is no universal radiation generator that provides a useful intensity of radiation at all wavelengths for practical purposes, and there is no universal wavelength resolving instrument or universal detector. Consequently, the spectrum has been divided into regions that bear names related to the sources that produce it—such as the "ray" regions; or as extensions from the visible range—such as the *ultra*violet and the *infra*red regions; or according to the way in which wavelengths in a range are used—such as radio and television.

The extent of the wavelength ranges corresponding to these names were made somewhat arbitrarily, and the decision as to where the divisions should be was made mostly on the basis of limits imposed by the range of the human eye (visible), the properties of optical materials, and the response limits of various sources and detectors.

Figure 2.2 illustrates the wavelength ranges covered by each of these radiation types, together with the various effects that produce spectral features in each of these regions.

At the short wavelength end of the spectrum (gamma and x-ray region) the most common active sources of radiation are radioactivity and cathode discharge tubes; the detectors commonly used are ionization and phosphor detectors.

At the long wavelength end (microwave, radar and radio wave region), the typical sources of radiation are cavity resonators, oscillating dipoles, and electronic circuits; the radiation is detected using antennae and electronic circuits.

The principal concern here is with the visible and infrared regions, where a large variety of sources and detectors are available. These will be discussed separately.

2.4.1 Sources

Energy sources in the visible and infrared may be classified generally as either discontinuous or continuous.

Discontinuous sources emit either a single or a series of individual spectral lines or bands as a consequence of transitions between the discrete energy levels of the source materials. Examples of such sources are high voltage sparks, the glow discharge of vacuum tubes at low pressure, and certain low pressure arcs, such as the mercury arc.

Atmospheric gases emit in this manner, and the positions and intensities of their spectral emission lines are completely characteristic of the gas.

Perhaps the most valuable discontinuous source is the laser, which possesses the unique properties of being monochromatic, extremely intense, and providing coherent radiation. Recent advances now allow the wavelength to be tuned, which makes it particularly useful for many applications.

However, for general illuminating purposes, discontinuous sources are of limited usefulness, especially in remote sensing applications.

As the name suggests, continuous spectral sources radiate continuously over a wide range of wavelengths. The continuous sources are usually thermal devices in which the radiation is the result of high temperatures, and consequently, they are frequently hot solids.

The emitted radiation results from the acceleration of electrical charges within the material. Because the interaction of these charges is so complex, it is essentially impossible to apply electromagnetic theory, and so thermodynamical arguments have been applied to explain and predict the behavior of such bodies.

Certain emitting surfaces exist whose radiation characteristics are completely specified if their temperature is known. These are the ideal thermal radiators known as "black bodies." The spectral distribution of the radiation emitted by a black body is a smooth curve with a single maximum, and this spectral distribution is specified by Planck's law (see Section 2.3.1), given by

$$W_\lambda = \frac{c_1}{\lambda^5} \cdot \frac{1}{e^{\frac{c_2}{\lambda T}} - 1}$$

where

$W_\lambda \equiv$ spectral radiant emittance, $w/(cm^2\ \mu m)$
$\lambda \equiv$ wavelength in micrometers, μm
$h \equiv$ Planck's constant = $(6.256 \times 10^{-3}\ w/s^2)$
$T \equiv$ absolute temperature, $^\circ K$
$c \equiv$ velocity of light = $(2.9979 \times 10^{10}\ cm/s)$
$c_1 \equiv 2\pi\ hc^2$ = first radiation constant
$c_2 \equiv \dfrac{ch}{k}$ = second radiation constant
$k \equiv$ Boltzmann constant = $[(1.3805 \times 10^{-23}\ w/(s\ ^\circ k)]$

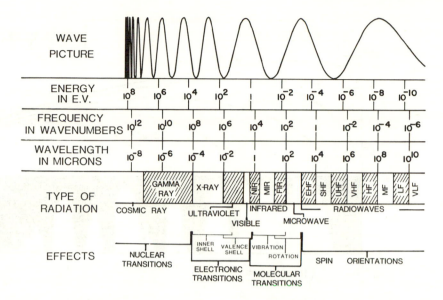

FIGURE 2.2. Electromagnetic spectrum, showing a picture of the wave, the wavelength ranges in various units, the types of radiation, and the atomic and molecular effects that produce spectral features in each range.

Black body or Planck curves for a series of emitters at temperatures between 200°K and 6000°K appear at the top of Fig. 2.3.

Integration over the entire wavelength range yields the expression for the radiant emittance, the energy flux radiated into a hemisphere above a 1 cm² of black body surface. The radiant emittance W, in watts per square centimeter is given by

$$W = \frac{2\pi^5 k^4 T^4}{15c^2 h^3} \ (= \sigma T^4)$$

where $\sigma \equiv$ Stefan-Boltzmann constant

$$5.669 \times 10^{-12} \ w/(cm^2 \cdot °K^4)$$

This expresses the Stefan-Boltzmann law, which indicates that the radiant emittance of a black body increases as the fourth power of the absolute temperature.

Differentiating Planck's law, and solving for the maximum yields the Wien's displacement law, expressed as

$$\lambda_{max} T = \alpha$$

where $\lambda_{max} \equiv$ wavelength of maximum emittance, and $\alpha = 2897.8°K$, which indicates that the wavelength of the maximum spectral emission occurs inversely as the absolute temperature.

Emissivity is a function of the type of material and the condition of the emitting surface; for a black body, the emissivity is 1. When the emissivity is less than 1, but constant at all wavelengths, the material is called a gray body. When the emissivity varies with wavelength, the material is called a selective radiator.

As stated above, there are numerous visible and infrared energy sources. Among the most practical artificial sources are the following: Nernst glower, globar, Welsbach mantle, carbon arc, tungsten filament, mercury arc, and cavity sources.

The most important natural source is, of course, the sun, which provides energy spectrally equivalent to a black body operating near 6000°K over the range of particular interest here. It is the chief source of energy for remote sensing purposes, although many natural surfaces also provide gray body emission. However, for terrestrial remote sensing purposes, the sun's energy is severely attenuated by the earth's atmosphere. The atmospheric transmission is indicated in the middle of Fig. 2.3.

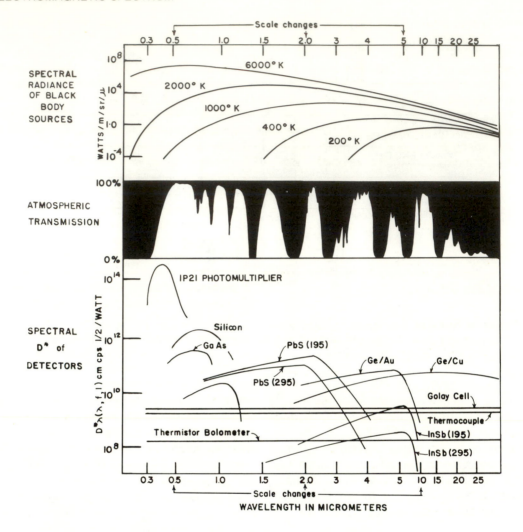

FIGURE 2.3. Spectral radiance curves for black bodies at various temperatures (top); atmospheric transmission spectrum, indicating atmospheric windows (middle); spectral D^* (dee star) curves for some detectors used in the visible and infrared ranges.

2.4.2 Detectors

There are numerous detectors available for use in the visible and infrared. The most extensively used detector for remote sensing purposes in the visible and very near infrared—out to about $0.9\,\mu$ m—is undoubtedly photographic film. However, because of the great variety of types and properties of available films, a discussion of them and the photographic process is beyond the scope of this chapter. The reader is therefore referred to standard texts on the subject, for example, Neblette (1), Manual of Photographic Interpretation (2), and Manual of Color Aerial Photography (3).

The nonphotographic film detectors may be classified into two main groups, namely thermal detectors and photodetectors.

THERMAL DETECTORS

The operation of these detectors depends upon the energy falling on them causing a temperature change, which results in a change in some physical characteristic. Among the thermal detectors are: thermistors, thermocouples, thermopiles, and Golay cells.

PHOTODETECTORS

Here, the individual photons falling on the detector cause a change in a physical property. They are of several types: photovoltaic, photoconductive, photoelectromagnetic, and photoemissive, of which the principal detector is the photomultiplier.

The performance of the above detectors is a function of many parameters, such as temperature, chopping frequency, field of view, detector area, and so forth, and the performance is specified in numerous ways. One common performance criterion is the spectral D^* (dee star). The spectral response, as indicated by the spectral D^*, for several of the above mentioned detectors is indicated at the bottom of Fig. 2.3. The difference in performance caused by changing the temperature from 195 to 295°K for lead sulphide and indium antinomide is indicated.

More detailed discussions of sources and detectors are available in standard texts (4, 5) and in chapter 3 of this volume.

2.4.3 Types of Spectra

The energy distribution in any material is fundamentally a question of the relative positions of all particles at a given time. The total energy of the system can be expressed as the sum of four different types of energy: translational, rotational, vibrational, and electronic. Because different amounts of energy are required to cause transitions within each of these different energy types, evidence for a transition will occur in a specific part of the spectrum.

Translational energy need not be considered here because it is generally very small and is not quantized, and the space available for this type of motion is unrestricted and usually undefined. The other three types of energy are limited in space to the size of the molecule, and are truly quantized.

Rotational energy is the kinetic energy of rotation of a molecule as a whole in space, and transitions between the rotational energy levels which produce rotational spectra require only small amounts of energy. Consequently, pure rotational transitions may be observed in the mid- and far-infrared and microwave regions, where the energy is very low and insufficient to cause or result from vibrations or electronic transitions.

Vibrational energy is involved with the movement of atoms relative to each other about a fixed position. Transitions here require greater amounts of energy than for rotations, and consequently evidence for such vibrational transitions occur in the mid and near infrared regions. However, because the energy necessary to cause vibrations is greater than that necessary to cause rotations, vibrational transitions are always accompanied by rotations in the situation where the material is free to rotate (as in gases).

Electronic energy is the energy required to cause the electrons about individual atoms, or located in bonds, to adopt a different configuration, and this requires even larger amounts of energy. Therefore, evidence for this type of transition is observed predominantly in the visible, ultraviolet, and x-ray regions. Again, the energy required is more than sufficient to cause vibrational and rotational transitions as well, and so evidence for these transitions appear together with those of electronic transitions. The energy levels, energy level diagrams and the transitions between these levels will be discussed in somewhat more detail below (see Section 2.6).

The spectral range between the ultraviolet and microwave, which corresponds to the energy transitions between electronic, vibrational, and rotational levels, can be considered to be the range where electromagnetic energy truly interacts with matter, and consequently they are the regions where most information about the nature of a material can be obtained from a study of the electromagnetic energy which has interacted with it. As a result, this is the range that is of prime interest for remote sensing purposes in the present discussion.

In regions of higher energy in the spectrum, such as the "ray" region (gamma, x-ray, and cosmic), the energy is sufficient to disrupt matter, causing electrons to be completely lost to the system or for bonds to be broken and molecules to fragment. In regions of lower energy, the longwave region—radar, radio, and audio—the interaction with matter is frequently quite feeble or transient and is quite often a consequence of macroscopic effects. Although these latter two regions yield extremely powerful data in remote sensing applications, (for example, from γ-ray spectroscopy and x-ray fluorescence of the lunar surface) discussion of these techniques lies outside the scope of the present treatment.

2.5 QUANTUM DESCRIPTIONS

It is necessary to adopt a quantum view for the interaction of electromagnetic energy with matter because, whereas the large objects of our ordinary experience move according to Newton's classical laws of motion, the small objects that make up atoms and atomic systems (the electrons and nuclei) move according to the principles of quantum mechanics.

2.5.1 Description of Matter

The quantum mechanical description of an atom or atomic system is made in terms of a wave function, or state function, which is usually designated by Ψ.

A characteristic state (also called an eigenstate, or a stationary state) is one that corresponds to a perfectly defined energy for the system, and a system may have many such states, which are usually different in energy. When two or more states have the same energy, those states are said to be degenerate.

The basic task of quantum mechanics is to determine methods of finding the wave functions for a given atom or atomic system, and the performance of this task involves obtaining solutions for the Schrodinger equation. As stated earlier, the Schrodinger equation is not strictly derivable, and should be considered to be the counterpart in quantum mechanics of Newton's $f = ma$ in classical mechanics.

The Schrodinger wave equation has the general form

$$\nabla^2 \psi + \frac{8\pi^2 m}{h}(E - U)\psi = 0$$

for a particle of mass m, where ψ is the wave function, h is Planck's constant, E is the eigenvalue, and U is a potential function.

The details of solving the equation are, except for the simplest systems, usually quite complicated; but the results are extremely important and easily understood, and most remote sensing concepts are based on such results.

It turns out that acceptable solutions to the Schrodinger equation are only possible if the energy, E, has certain definite discrete values. These values are the eigenvalues and are, in fact, the characteristic energy levels of that particular system. Because every system has a different arrangement of the energy values or energy levels, the arrangement of energy levels for any system—or

material—completely characterizes that system.

Information about any system in a remote sensing situation is only available as a consequence of the system changing from one of these characteristic energy levels to another, and such a change is called a transition. Such transitions most often take place as a result of absorption or emission of electromagnetic radiation. The various types of energy levels will be discussed in the next section.

2.5.2 Description of Interaction of Energy with Matter

Of particular interest in remote sensing of geology is the study of the interaction of energy with solids. They, in particular, may be characterized in terms of their optical constants, n, and k, the refractive index and extinction coefficients, respectively, which are functions of wavelength for a given material. The magnitude of the refractive index indicates the speed with which electromagnetic energy passes through the material, while the effect of the extinction coefficient is to introduce an exponential damping to the electromagnetic wave as it traverses the material.

The usefulness of the quantum mechanical approach is that it allows the stationary states of a system to be determined. These time independent stationary states are, in fact, the only allowed energy levels which the system can adopt. Such energy levels may be plotted to yield an energy level diagram that can be most useful in visualizing the transitions that produce features in a spectrum.

Once a system occupies a particular stationary state or energy level, it will remain in that state indefinitely until something causes it to change, thus causing a transition to another allowed energy level.

At present, we are primarily interested in those interactions that alter the energy content of the material, and the frequency-intensity relationship of the radiation, and so basically we are concerned with absorption and emission of electromagnetic radiation.

If we consider a system that is in the process of changing from one stationary state (energy level) to another, the state function during the transition is given by a linear combination of the two stationary states involved. This intermediate state function is called a coherent state, and the essential difference between a coherent and stationary state is that the energy of the stationary state is well defined while that of the coherent state is not.

The quantum mechanical description of, for example, a radiating system may be given as follows: During a transition from one energy level (E_1) to another (E_2), the system enters a coherent state and oscillates sinusoidally with a frequency (ν) determined by the energy difference between the two energy levels, that is

$$E_1 - E_2 = h\nu$$

Such a sinusoidal oscillation is accompanied by an oscillating electromagnetic field that constitutes the radiation.

Before electromagnetic radiation can interact with matter, there must be some way in which the matter can interact with either the electric or magnetic field of the radiation. In most transitions of interest in remote sensing situations, the interaction takes place between the electric field and matter. However, matter does react with the magnetic field and a great deal of very valuable information concerning a material is available as a result. For example, radiation in the microwave and radio frequency regions is used to study very low energy transitions that result from the reorientation of nuclear and electron spins when subjected to an applied magnetic field, as in the case of nuclear and electron spin resonance studies.

In order for the electric field of electromagnetic energy to interact with matter, the system must have a charge distribution that changes when the system makes the transition from the initial to the final state. Two stationary states may be induced to become coherent, or an induced transition may take place, due to the fact that when electromagnetic energy falls on a system, the oscillating field of the electromagnetic radiation may disturb the potential energy of the system in just the correct way to allow the transition. The electric field of the radiation, in fact, oscillates at the point occupied by the system with a frequency ν. Then the electric field in the x direction, E_x, can act on the dipole component μ_x to produce a change in the energy $E_x \mu_x$. This term adds to the potential energy of the system and consequently is responsible for causing the transition to take place.

2.6 ENERGY LEVELS

In most geological remote sensing situations the energy detected passes through the terrestrial atmosphere, and so it is necessary to have some

understanding of the spectral behavior of gases as well as solids. Gases differ from solids because they possess the ability to rotate. The discussion here will be limited to the energy levels and the transitions between them in molecules, except for a brief treatment of crystal field effects on ions in solids.

The total energy of a system is made up of the sum of three distinct types of energy—rotational, vibrational, and electronic—and each of these forms will be discussed separately below.

2.6.1 Rotational Energy Levels

These are only introduced because of their relevance to the atmospheric gases, and to illustrate the concept of discrete energy levels between which transitions occur to produce features in a spectrum.

The rotation of a gas molecule can be described in terms of its angular momentum, I, defined as

$$I = \Sigma_i m_i r_i^2$$

where m_i is the mass of the ith particle and r_i is its perpendicular distance from an axis of rotation. There are three principal axes of rotation (designated a, b, and c) whose origin is the center of gravity of the system, and the moments of inertia about these axes are designated I_a, I_b, and I_c.

Any gas molecule can be classified into one of four general types, depending upon the relationship between the values of its three principal moments. The molecule is called linear when $I_a = I_b$, and $I_c = 0$; a spherical top when $I_a = I_b = I_c$; a symmetric top when $I_a = I_b \neq I_c$; and an asymmetric top when $I_a \neq I_b \neq I_c$.

The energy levels for a linear or symmetric top are very simply calculated because only one quantum number, designated J, is required to specify the rotational state, and the energy of the levels are given by

$$E_{\text{ROT}} = \frac{h}{8\pi^2 cI} J(J + 1)$$

where J can take any integral value; that is, $J = 0, 1, 2, 3,$ and so forth.

Fig. 2.4 shows an energy level diagram calculated using this approach where the allowed transitions between the levels are indicated by vertical

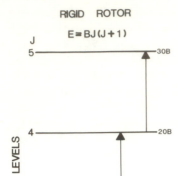

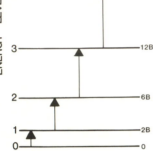

FIGURE 2.4. Rotational energy level diagram. The energy levels were calculated using a rigid-rotor model, where $B = \dfrac{h}{8\pi cI}$ is the rotational constant. The appearance of spectra produced as a result of transitions between these levels is included at the bottom.

arrows, and the form of the resultant spectrum is illustrated immediately below. The treatment for symmetric and asymmetric top molecules is more complex, and their treatment should be pursued through standard works on the subject, for example, King (6, 7).

2.6.2 Vibrational Energy Levels

The forms of the vibrations and the values of the permitted energy levels of a material are determined by the number and type of the constituent atoms, their spatial geometry, and the magnitude of the binding forces between them. The vibrations of a molecule consist of very small displacements of the atoms from their equilibrium positions.

In a molecule composed of N atoms there are $3N$ possible modes of motion because each atom has three degrees of freedom. Of these modes of motion, three constitute translations, and three constitute rotations of the molecule as a whole (except for linear molecules where there are only two rotations) and so there are $3N$-6 (or $3N$-5 for

linear molecules) possible independent types of vibrations.

The general motions of a system of atoms can be analyzed in terms of a set of internal coordinates. There is one particular such set called "normal coordinates," which is especially convenient for describing molecular vibrations, and indeed this set is necessary for carrying out quantum mechanical treatments. One particular value of the normal coordinate treatment is that the symmetry of the molecule allows great simplifications to be made in the computational process.

The energy levels of a linear harmonic oscillator are given by

$$E_v = (v + \tfrac{1}{2})\, h\nu \;,$$

where ν is the classical frequency of the system, and v is the vibrational quantum number which may take any integral value, that is, $v = 0, 1, 2 \dots$, and so forth. For a molecule with many classical frequencies, the energy is given by $E = (v_1 + \tfrac{1}{2})\,h\nu_1 + (v_2 + \tfrac{1}{2})h\nu_2 + \cdots + (v_{3N-6} + \tfrac{1}{2})h\nu_{3N-6}$.

The bands that occur in a spectrum as a result of transitions between vibrational energy levels are referred to as fundamentals, overtones, or combination tones. Fundamentals occur as a result of a transition from the ground state, where the value of all quantum numbers v_i is zero, to a state where all are zero except one, and its value is unity, that is, $v_i = 1$. This is, then, the fundamental of the ith vibrational mode, and features due to fundamental vibrations typically occur in the mid and far infrared regions of the spectrum, all at wavelengths longer than 3 μm. An overtone occurs when the transition is from the ground state to one in which $v_i = 2$, while all other quantum numbers remain zero. Combination tones occur when a transition takes place from the ground level (all $v_i = 0$) to a level whose energy is determined by the sum of two or more fundamental or overtone vibrations, that is, to a level where

$$E_v = (v_1 + \tfrac{1}{2})h\nu_1 + (v_2 + \tfrac{1}{2})\, h\nu_2$$

where v_1 and $v_2 \neq 0$. Features due to overtone and combination zones typically appear between 1.1 and 5 μm. The energy levels, normal coordinates, and spectrum resulting from the indicated transitions are shown in Fig. 2.5. More complete discussions of the vibrational process are available in

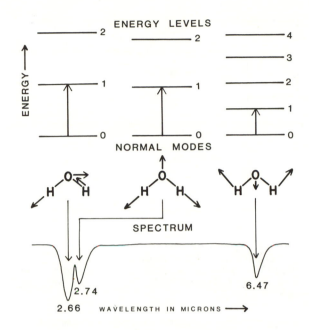

ENERGY LEVELS

NORMAL MODES

SPECTRUM

2.74

2.66 WAVELENGTH IN MICRONS ⟶

6.47

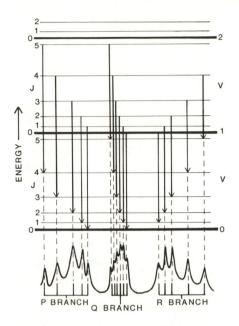

P BRANCH Q BRANCH R BRANCH

FIGURE 2.5. Vibrational energy level diagram for water, in which the levels of the three fundamental (v_3, v_1, and v_2) vibrations are separated horizontally. A representation of the form of these normal modes is illustrated diagrammatically near the center, and the appearance of the infrared absorption spectrum corresponding to the transitions from the ground to the first excited states is included at the bottom.

FIGURE 2.6. Vibration-rotation energy level diagram showing the allowed transitions (indicated with arrows) between the various energy levels. The resultant spectrum is shown below, displaying the typical P, Q, and R branches.

standard text books, that is, Wilson (8), Herzberg (9), and Barrow (10).

2.6.3 Vibration-Rotation Energy Levels

Normally, both the vibrational and rotational energy will change simultaneously because the amount of energy required to cause a vibrational transition is much more than sufficient to cause rotational transitions. Consequently, gas spectra usually consist of a large number of lines extending to both higher and lower frequencies from the frequency of the vibrational transition. These are due to the familiar P, Q, and R branches in vibrational gas spectra. The P branches appear at lower frequencies, R branches at higher frequencies and Q branches occur near the center. This is illustrated in Fig. 2.6.

Gas spectra of different molecules have quite different appearances, depending upon which selection rules apply in the particular case. Detailed studies of the associated rotational bands in the PQR branches can yield information on the moments of inertia of a molecule in both the ground and excited states.

2.6.4 Electronic Energy Levels

The subject of electronic energy levels is both extensive and complex because the treatment varies widely depending upon the nature of the material being considered.

In performing remote sensing, one encounters diatomic (O_2, N_2, CO, and so forth) and polyatomic (H_2O, CO_2, N_2O, and so forth) atmospheric gases, complex molecular organic materials (principally vegetation), and solid inorganic substances (rocks and soils). From a geological standpoint, much excellent data is available as a result of transitions between the electronic energy levels of ions embedded in inorganic solids. These transitions provide extremely valuable, though indirect, information concerning the bulk material. Consequently, remote sensing involves a consideration of the energy levels in many different types of matter.

An exact solution to the Schrodinger equation can only be obtained for an atom with a single electron. Solutions for polyelectronic atoms are achieved by making various approximations, such as ignoring the effects of the inner shell electrons.

For polyatomic molecules, consideration of the electronic states involves excursions into valence-bond and molecular-orbital theory, and, because the symmetry of the molecular skeleton becomes important, inclusion of group theoretical concepts is required. For ions located in solids, ligand and crystal field theoretical considerations become necessary.

Obviously, any treatment of the calculations involved in producing electronic energy level diagrams is well beyond the scope of the present work. The most that can be attempted here is to provide the barest background to the subject, give some idea of the nomenclature involved in specifying the energy levels, and include some examples of energy level diagrams that are of interest and importance in remote sensing applications. Atoms are discussed because they provide the logical starting point as well as a basis for discussion, and because of the particular relevance of ions in solids for remote sensing.

Quantum mechanics shows that the electrons in an atom can only occupy specific quantized orbits, and these can be specified in terms of three quantum numbers, designated n, the principal quantum number; L, the angular momentum quantum number; and m, the magnetic quantum number. For polyelectronic atoms a fourth quantum number, m_s, the spin quantum number is required. For the hydrogen atom, the electronic levels can be specified by just the principal quantum number (n), and the values of the levels are given by

$$E_{ni} = \frac{R}{n_i^{\,2}}$$

where R is a constant determined by the mass and the charge of the electron. The energy levels of hydrogen are shown on the left side of Fig. 2.7.

The spectroscopic designation of an electronic energy level for most polyatomic atoms can be given in terms of a quantum number L (corresponding to the individual electron angular momentum quantum number l) and a quantity called the multiplicity. States for which $L = 0, 1, 2, 3$, and so forth, are designated S, P, D, F, and so forth, while the multiplicity is defined as $2S + 1$, where S is the resultant spin. Since S can take ½, or whole integral values, $2S + 1$ takes values, $1, 2, 3$, and so forth. The electronic state is then designated ^{2S+1}L. When the value of $2S + 1$ is 1, 2, 3, 4, and 5, the states are called singlet, doublet, triplet, quadruplet, or quintet states.

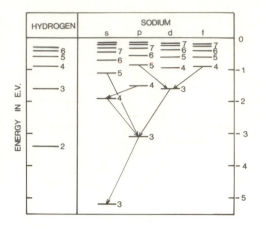

FIGURE 2.7. A partial electronic energy level diagram for the single electron atom, hydrogen appears on the left. On the right the levels for a polyelectronic atom, sodium, and some of the allowed transitions between these levels are indicated with arrows.

A partial energy level diagram for sodium is shown on the right-hand side of Fig. 2.7.

As distinct from atoms, many molecules and all diatomics possess the ability to both vibrate and rotate, and consequently vibrational and rotational energy levels are superimposed on each electronic energy level. Because the energy involved in an electronic transition is much greater than that required for vibrational and rotational transitions, the electronic transitions are always accompanied by vibrational, and where possible, by rotational transitions. Consequently, diatomic electronic spectroscopy is a particularly fruitful area for obtaining detailed information on moments of inertia and potential energy functions of the states involved. A partial energy level diagram for the diatomic molecule N_2 appears in Fig. 2.8.

For some polyatomic molecules, an electronic transition may be essentially located in a specific localized bond or group, such as in the $>C = O$ group, and in such a situation the analysis can be similar to that for diatomic molecules. For others, in particular the aromated hydrocarbons, the various electronic states are designated in terms of the states of the π electrons.

However, in general, the electronic states of polyatomic molecules are best described in terms of the state of the molecule as a whole for a given assignment of electrons in individual orbitals. The total electronic wave function must have a symmetry that is compatible with the symmetry of the molecular skeleton, and so it is the symmetry behavior that

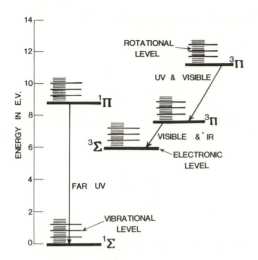

FIGURE 2.8. Partial electronic-vibration-rotation energy level diagram for the diatomic molecule, nitrogen. Only the allowed transitions between the electronic energy levels are indicated with arrows.

best describes the electronic states of a polyatomic molecule.

In order to have some feeling for the meaning of the symbols assigned to these states, it is necessary to briefly consider some aspects of symmetry groups. Such concepts are at least equally important in considering the vibrational motions of polyatomic molecules. The advantage of the symmetry group approach is that it allows hundreds of thousands of molecules to be classified into a small number of groups according to the number and nature of their symmetry elements. All molecular symmetries can be treated in terms of just five symmetry elements, and these elements can be explained in terms of the symmetry operation involved. The symmetry elements, the symbols designating them, and the corresponding symmetry operations are listed in Table 2.1.

It is found that only a relatively few combinations of these symmetry elements occur, and each combination of the elements is called a group. They are called either symmetry point groups or symmetry space groups. The point groups are so named because under all operations of the group, some point in the molecule remains fixed. Symmetry space groups are usually appropriate for crystals, and they are so called because translations are allowed (that is, no point remains fixed) where the entire unit cell is translated by an operation to another completely equivalent position in the crystal.

The behavior of the wave function with respect to each of the symmetry operations of a group leads to a classification of the energy levels according to its symmetry species. The possible symmetry species for each group appear in a table called a character table, and the reader is referred to a standard text for further explanation (11, 12).

The symmetry species are designated A or B, depending upon the symmetric or antisymmetric behavior, respectively, with respect to the principal element of symmetry, or E for double degenerate and F or T for triply degenerate states. Subscripts u and g refer to symmetric or antisymmetric behavior with respect to inversion when a center of inversion is present. The group theoretical treatment not only allows the states to be simply designated, but also provides selection rules that tell whether a transition between particular levels is allowed or forbidden.

As mentioned above, for remote sensing of geological materials, one of the more important considerations is that of the electronic energy levels of an ion in a solid. When an atom is embedded in a crystal lattice, either as a constituent or an impurity, one or more of its electrons may be shared by the solid as a whole, and is not associated with any particular atom. The energy levels of such electrons become smeared into regions called valence or conduction bands of the solid as a whole, and the atom becomes an ion. The bound electrons of the ion have quantized energy states associated with them.

In the case of rare earth atoms, the unfilled shells involve the deep-lying $4f$ electrons, which are well-shielded from outside influences; thus their energy levels remain essentially unchanged when embedded in a solid.

In the transition elements (for example, Ni, Cr, Fe, and so forth) however, it is the $3d$ shell electrons that primarily determine the energy levels, and these electrons are not shielded, so their energy levels are greatly perturbed by the external field of the crystal. In such a case it is the symmetry of the surrounding electric field that designates the energy levels.

In the spherical potential of the free ion, the $5d$ orbitals have identical energy levels, that is, they are fivefold degenerate, and the effect of applying a surrounding field is to resolve this degeneracy by changing the energies of some of the orbitals relative to others. However, the new energy states

TABLE 2.1.   SYMMETRY ELEMENTS AND SYMMETRY OPERATIONS

Symbol	Symmetry Element Description	Symmetry Operation
E	Identity	No change
σ	Plane of symmetry	Reflection through plane
i	Center of symmetry	Inversion in the center
C_n	Axis of symmetry	Rotation about the axis by $\left(\dfrac{360}{n}\right)^{\circ}$
S^n	Rotation-Reflection axis	Rotation about the axis by $\left(\dfrac{360}{n}\right)^{\circ}$ followed by reflection

must be compatible with the symmetry of the imposed field and, consequently, the energy levels are specified in terms of the symmetry species of the applied crystal field.

The way in which the degenerate energy levels of the isolated ion are resolved by application of a crystal field (called crystal field splitting) is indicated in Fig. 2.9 for the simplest case of an ion with only a single d electron, such as is the case for the Ti^{3+} ion. The five-fold ^2D level of the free ion, shown on the left, is split into a doubly degenerate ^2E_g and a triply degenerate ^2T_g level under the influence of an octahedral field, and

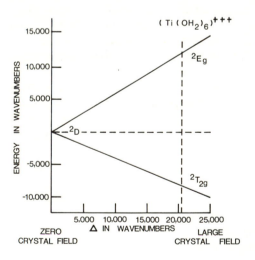

FIGURE 2.9. The electronic energy level diagram for a metal ion with a single d electron. The effect of increasing the strength of an octahedral crystal field on the ^2D state is illustrated. The five-fold degenerate ^2D state is split into a doubly degenerate ^2E_g and a triply degenerate $^2T_{2g}$ state, and their separation increases with increasing field strength. The position of the levels for the Ti^{3+} ion in the complex $(Ti(H_2O)_6)^{3+}$ is indicated by the vertical dotted line. A transition ^2E_g ^2Ti_g will produce a spectral feature at about 20,400 cm^{-1}

it can be seen that the separation of these two energy levels increases with increasing field strength.

A more complex, but only partial energy diagram appears in Fig. 2.10 for a d^5 ion, such as Mn^{2+}. The energy levels of the free ion are shown (where the states are designated using nomenclature appropriate for a polyelectronic atom) on the left-hand side, and the energy levels adopted as a function of field strength for an octahedral field on the right. It will be noted that the totally symmetric ^6S state is not split.

The energy levels for the same ion in different crystal fields are greatly different. Consequently, in observing transition between the energy levels of the ions in crystals, one obtains particularly useful information about the type of material in which the ion is contained, and even though this information is somewhat indirect, it is among the most powerful techniques available for remote sensing of solids.

2.7 REQUIREMENTS FOR REMOTE SENSING

The basic concept for remote sensing is that information about the object or target is contained in the electromagnetic radiation passing from the object to the observer, and it is this radiation that forms the communication link. The information is present in the form of a wavelength-intensity relationship, and when this is presented for more than one wavelength, and the wavelengths are ordered, it constitutes a spectrum.

The essential elements of a remote sensing system are the generation, interaction with the object, transmission, collection and wavelength separation, and the detection and recording of electro-

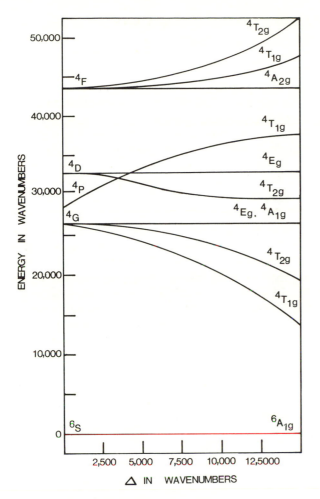

FIGURE 2.10. A partial energy level diagram for a d^5 ion, such as Mn^{2+}. Only the ground state (^6S and $^6T_{1g}$), and the quartet states are included. The way in which the quartet states split into A, E, and T states as a function of increasing octahedral field strength is indicated (61).

magnetic radiation. Each of these elements will be briefly discussed below.

2.7.1 Generation: The Source

As pointed out in Section 2.2, it is conventional to refer to a source of radiation as "active" when it is supplied by man, and as "passive" when the thermal energy of the object is the source of radiation. Active sources are of many types, depending upon the wavelength range being considered, but usually the main requirement is that they provide a large spectral radiance over the wavelength range of interest. A detailed discussion of artificial energy sources for visible and infrared ranges was presented in Section 2.4.1.

For remote sensing of geological materials, the prime natural source is, of course, the sun. The sun also provides the energy for cases in which the energy is absorbed, often completely redistributed internally, and finally emitted from the object.

2.7.2 Interaction

In the active situation, where radiation impinges on the object, information about the object is imposed on the radiation as a consequence of both macroscopic and atomic-molecular effects. The latter effects, specifically absorption and emission, where transitions between the various discrete energy levels of the material occur to either subtract or add intensity at specific wavelengths from or to the original radiation, have formed the bulk of the present discussion.

The magnitude and nature of the macroscopic effects, such as reflection, scattering, refraction, and so forth, which take place at boundaries, are largely a function not only of the chemicophysical makeup of the object, but also of its size and shape and the general environment in which it exists. All these parameters serve to modify the wavelength-intensity relationship of the radiation in predictable ways that are particularly useful. This is also true in the passive case where the energy emitted is modified in passing through the object's boundaries.

2.7.3 Transmission

Radiation will be unaffected in traveling between object and observer unless there is matter in the intervening space.

In terrestrial geological sensing, the transmission properties of the earth's atmosphere are extremely important, because they impose severe restrictions on what can be achieved.

The atmosphere is composed of a relatively constant mixture of homonuclear diatomic gases (O_2, N_2) with less constant but much smaller amounts of other molecular gases (H_2O, CO_2, CO, N_2O, O_3, CH_4, NH_3, and so forth) and highly variable quantities of liquid droplets (water in particular), and finely particulate solids, of which the most important are ice particles.

Because they possess no permanent dipole moment, the main constituents, the homonuclear diatomic gases, do not absorb in the visible and infrared, but their electronic absorption forms the cutoff of transmission in the ultraviolet.

Although the other molecular gases only constitute a very small percentage of the total atmospheric content, their individual vibrational and rotational absorption coefficients are extremely large, and features due to their absorption occur throughout the visible and infrared. Even the minimum amounts that are present completely absorb all the energy available in certain band passes, and so their presence completely precludes the acquisition of any information about an object in these regions. Between these regions are ranges more or less free of absorption, but these are referred to as the atmospheric windows, and their locations are indicated in the transmission spectrum presented in the central portion of Fig. 2.3. In Chapter 3 of this text, a more detailed treatment of atmospheric transmission is given, and the atmospheric transmission spectra, in which the windows are indicated, as well as the spectra of individual constituent gases are presented in Fig. 3.1, 3.2, and 3.3.

Even within the windows, the transmission is extremely sensitive and even small variations in local concentrations can result in large changes in transparency, and this is particularly apparent near the frames of the windows where the tails of features extend in.

As well as being governed by genuine molecular absorption, the transmission is also drastically affected by scattering, which is largely a function of particle size. For particles that are very small compared to the wavelength, and this includes individual molecules, the scattering intensity follows the Rayleigh law, that is, scattering is proportional to λ^{-4}. It is in fact this type of scattering by molecular species that causes the sky to appear blue.

The scattering behavior of larger particles, comparable in size to the wavelength of the radiation, is more complicated to describe, and requires the use of Mie theory (see Section 2.8.2). This type of scattering behavior is exhibited by the aerosol layer.

Because of the variability of the atmospheric transmission as a consequence of even small changes in the concentrations of both absorbers and scatterers, it is essential in most remote sensing activities to monitor the fluctuations in atmospheric transmission concurrently with making the measurements.

2.7.4 Wavelength Separation

Certain information may be obtained by collecting the total energy over a wide range of wavelengths, and this activity is referred to generally as radiometry. However, it is usually more fruitful to require that the total radiation be separated into its component wavelengths so that spectral information can be acquired.

Separation of radiation into spectral components may be achieved in many ways, and this is normally performed by making use of the macroscopic interaction of electromagnetic radiation with certain materials. The most common way is by *dispersion*, that is, by making use of the dispersive properties of a particular material by passing the radiation through a prism of the material, or by making use of diffraction properties by transmitting the radiation through or reflecting it from a grating. In addition, the transmission or scattering properties of materials may be used to design and construct filters to give specific limited wavelength ranges, or the interference properties of different thicknesses of various materials may be used to the same end. One particularly useful product is the circular variable filter, which is comprised of a wedge of material laid down on a substrate. The wavelength transmitted can be continuously varied as the wedge thickness is altered by rotating the wedge through the radiation beam.

Of special value in remote sensing applications is the use of interferometry in a form referred to as *Fourier transform spectroscopy*. Here the technique is based on the fact that the electric field as a function of distance is the Fourier transform of the electric field as a function of frequency (or wavelength). Thus, any interferogram formed from the radiation collected from an object can be mathematically converted to a conventional spectrum with some particularly attractive experimental advantages over the process of scanning with a monochromator. In particular, one advantage is that all frequencies are observed simultaneously (the Felgett advantage) and another is that no size limitation (other than the size of the optics) is imposed on the beam dimension (Jacquinot advantage).

2.7.5 Detection

As with the sources, different types of detectors are most effective in different wavelength ranges. The function of most detectors is to convert the electromagnetic radiation reaching them into the form of an electrical voltage or current proportional to the intensity of the radiation, which can then, usually after suitable amplification, be used to drive some recording device that presents the intensity data as a function of wavelength in a form suitable for interpretation.

The specific types of detectors suitable for use in the visible and infrared regions have been discussed in Section 2.4.2.

2.8 FORM OF REMOTELY SENSED DATA

This section is concerned with the way in which various parameters affect the form of the spectral data collected from a given material.

The overall appearance (the wavelength positions and relative intensities of the maxima and minima) of a spectrum is determined by numerous factors, and the most important of these are the experimental arrangement used in collecting the data, the phase and condition of the object, and various environmental parameters. All these factors must be carefully considered to avoid misinterpretation of remotely sensed data.

There are essentially three basic processes that allow an object to imprint information onto electromagnetic radiation, and quite generally, these are referred to as absorption, emission, and reflection (or scattering). In fact, provided the object possesses a well-defined boundary, as is the case with liquids and solids, contributions from all three of these processes are contained in the energy leaving the object. Thus, the radiation or energy detected, E_{total}, at any particular wavelength, may be written as

$$E_{total} = E_{source} - E_A + E_E - E_{RS}$$

where E_{source} is the energy from the source; E_A is the energy absorbed; E_E, the energy emitted; and E_{RS}, that reflected or scattered.

Experimental arrangements are chosen to maximize one of the last three components in the above expression, while at the same time minimizing the other two.

2.8.1 Experimental Arrangement

Although in most remote sensing situations the experimental arrangement is dictated by the existing conditions, there are three arrangements of source, object, and detector, which are referred to by the name of the process that they usually tend to maximize. These are discussed below.

ABSORPTION

The absorption arrangement is frequently called the transmission mode, and for it the object is located directly between the source and detector. This mode is particularly appropriate for gases and liquids where $E_{RS} \simeq 0$, and E_E is usually small for geometrical reasons. It can rarely be used in the study of solids in their natural environment, even though after multiple reflection and scattering, the data present is a result of the absorption process.

EMISSION

In the emission arrangement, the object acts as the source, and $E_{source} = 0$ in the above expression. The effect of E_{RS} depends on the condition of the sample, and for bulk solids, E_{RS} can, in fact, be the basis for the desired information. In particulate solids, as will be discussed below, E_A may also play an important role in providing required information.

REFLECTION

In the reflection arrangement, the angle formed by source-object-detector is less than 180°, and usually much less. Reflection is only possible when the object possesses well-defined boundaries, and the particular usefulness of the technique applies to the study of solids. Again, the condition of the object is the major factor determining the contribution of E_A, and the temperature of the object determines E_A. In some experiments it is possible to eliminate the effect of E_E, even when the object is emitting strongly, by modulating the source energy before it reaches the object, and this is common practice in laboratory experiments.

2.8.2 Phase and Condition of Object

Most substances may exist in the gas, liquid, or solid phase, but water is the only substance of interest in terrestrial geological remote sensing which is regularly encountered in all three phases.

GAS

The gases of interest are the atmospheric constituents, and the energy levels of gases have been discussed above. Their spectra display pure rotational fine structure at long wavelengths (in the infrared microwave regions), and with the exception of homonuclear diatomics that have no permanent dipole moment, gases also display vibration-rotation spectra in the near- and midinfrared regions. All gases display electronic spectra with vibrational and rotational structure superimposed, and this occurs in the ultraviolet and visible regions. It is the presence of all such absorption that severely attenuates the radiation in many regions where the spectra of materials of geological interest could provide very useful data.

The form of the gas absorption spectra is essentially unaffected by environmental parameters, such as pressure and temperature, over the ranges appropriate to the terrestrial atmosphere, and scattering in the form of Rayleigh scattering is essentially a minimal effect.

LIQUIDS

The only liquid of interest is water. The water molecule exhibits hydrogen bonding and readily forms dimers in the liquid form, and so it loses its ability to rotate freely. Consequently, the fine structure of the gas phase becomes smeared out into broad features in the spectrum. The liquid water spectrum is, however, extremely important in remote sensing because of its ubiquitous nature. Water is present terrestrially as seas, lakes, and rivers and is readily both physi- and chemiadsorbed on solids, as well as occurring in fluid inclusions in many rocks and soils. Together with the iron ion and the OH group, it provides the most common source of spectral features throughout the near-infrared and near-midinfrared regions of the spectrum.

SOLIDS

By far the most important phase for geological remote sensing is the solid phase, and the spectral behavior of solids as a function of object condition is of paramount importance.

Solids may exist in many different forms ranging from large blocks with smooth polished surfaces down to submicron-sized particles that vary in their shapes and modes of packing. Consequently, the appearance of the spectrum recorded from a solid is often governed by a balance between macroscopic and molecular type interaction with the radiation.

As stated above, the fundamental properties that determine the spectroscopic behavior of solids are the refractive index, n, and the extinction coefficient, k, which are wavelength dependent because of the optically active transitions between the

various energy levels of the material. Large and rapid variations in n occur together with large variations in k, and although this always happens, the phenomenon is referred to as "anomalous" dispersion.

The reflection, R, for small angles of incidence to flat surfaces is directly related to n and k by the Fresnel formula, namely

$$R = \frac{(n - 1)^2 + k^2}{(n + 1)^2 + k^2}$$

However, this is a quite special case, and when the material is reduced in dimension such that the size of the particle approaches that of the wavelength of the radiation, the spectral properties become dominated by the scattering characteristics of the material.

An adequate description of scattering requires a rigorous solution of Maxwell's equations, subject to the appropriate boundary conditions. For very small particles that have radii much smaller than the wavelength of the radiation, the particle is treated as if it is completely inside a homogeneous electric field, and Rayleigh's law adequately describes the scattering process, which essentially states that the scattered intensity is proportional to λ^{-4}. For very large particles, where the particle radius is much larger than the wavelength, geometric optics and Fraunhoffer diffraction theory may be used to describe the interaction effects.

It was specifically to explain the behavior of intermediate-sized particles, where the particle radius is approximately the same size as the radiation wavelength and consequently does not fit the requirements of the above two mentioned theories, that Mie diffraction theory was developed.

Mie introduced a quantity called the scattering cross section, Q_s, which is simply defined so that the energy intercepted by an area Q_s, normal to the beam represents the energy scattered by the particle. Thus, if there are N particles in a beam of length δ_x, the fractional energy lost will be $NQ_s\delta_x$, provided that multiple scattering is ignored. This cross section may be related to an effective absorption coefficient, α_s by

$$\alpha_s \equiv NQ_s$$

Q_s may be accurately evaluated for perfectly spherical particles. The mathematical expression for Q_s is complicated, except when the sphere is very small compared with the wavelength, in which case the expression reduces to Rayleigh's formula. Detailed discussion of the above-mentioned theories are available in standard texts (13-16), and will not be discussed further here.

Because there is no completely adequate physical theory that describes light scattering by real surfaces, some empirical interpolation function is commonly used to quantitatively express the scattering and reciprocity principle in terms of the three angles (i, the angle of incidence; ϵ, the emergence angle; and, α, the phase angle) that uniquely specify the scattering geometry. One such useful expression is given by the Minnaert "law" (17, 18), which states that

$$B \cos \epsilon = B_0 (\cos i \cos \epsilon)^\kappa$$

where B is the apparent brightness of the surface, B_0, the brightness of an ideal reflector and κ is a parameter that describes darkening at zero phase angle, where $i = \epsilon$. For a Lambert surface (which gives full cosine darkening) $\kappa = 1$, while for a specular reflector $\kappa \rightarrow \infty$.

From a geological remote sensing point of view, one of the particularly important effects of scattering is the production of transmission and emission maxima at wavelengths that are considerably removed from the positions of the maxima in $\kappa -$, that is, these maxima are not located at the resonance frequencies. These transmission and emission features are referred to as Christiansen frequencies or Christiansen peaks, and they are located at, or near the wavelength at which the refractive index of the material matches that of the medium in which it exists. Consequently, the location of the Christiansen peaks are as much a function of the medium as of the object's composition. Fortunately, all the media of remote sensing interest have very nearly the same refractive index with a value of approximately 1. The position and appearance of Christiansen peaks, especially in the midinfrared region, can provide particularly valuable diagnostic information concerning particulate minerals and rocks (19, 20).

The extreme variations that occur in the appearance of a spectral feature corresponding to a single specific optical transition in a solid material

is illustrated in Fig. 2.11 as a function of experimental arrangement (vertical displacements) and a limited range of object conditions (horizontal displacements). In the figure is illustrated the typical appearance of a single feature in a hypothetical solid material.

Each of the nine spectra shown in the figure will be discussed briefly.

Thin Polished Plate

The polished plate is optically thick only at the center of the spectral feature, and its surfaces are such that all reflections are specular, that is, all reflected energy leaves the object in a direction such that the angle of reflection equals the angle of the incident radiation.

(a) In the absorption (or transmission) arrangement, the spectrum reveals a deep minimum due to genuine absorption and the loss due to specular (reststrahlen) reflection from the surface back in the direction of the incident radiation, that is,

$$E_{total} = E_{source} - E_A - E_R$$

(b) In emission, the spectrum again reveals a minimum, which in this particular case can be expressed in terms of Kirchoff's law in its simplest form, that is,

$$E_E = E_{BB} (1 - R)$$

where E_{BB} is the energy emitted by a black body at the same temperature as the object, and R is the object reflectivity.

(c) The reflection spectrum displays a maximum at approximately the same frequency as the minima in absorption and emission. This is the typical reststrahlen effect.

Particulate Samples

The hypothetical particulate sample is chosen to be composed of particles whose radius is comparable to the wavelength of radiation, and consequently, considerable scattering takes place. The particulate material is considered in two conditions; (B1) that of a thin or single layer of particles where radiative interaction between particles is minimum, and where the thickness of the layer is equivalent to the thickness of the thin polished

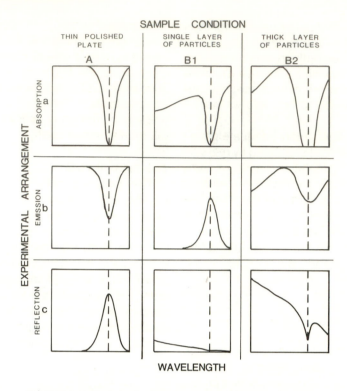

FIGURE 2.11. Variation in the appearance of the same feature in a given material as a function of experimental arrangement and sample condition. The vertical dotted lines indicate the position of the true resonance frequency of the transition.

plate discussed above; and (B2) that of a very thick layer of particles where radiative interaction between particles is maximized. It is this latter type of sample that is of prime importance in terrestrial remote sensing, because the majority of objects of special interest are particulate solids.

Thin Particulate Layer (a) The absorption or transmission spectrum again reveals a sharp minimum feature in the same position as for the thin plate, but, in addition, there has been a lowering of the energy transmitted throughout the spectrum due to generalized scattering out of the beam.

(b) In emission, the spectrum reveals a sharp maximum feature, which is opposite to the behavior of a thin plate. For particles this size, a black body distribution of energy cannot be established within the individual particles, and so it emits as it absorbs, behaving essentially like an individual molecule. Consequently, the spectrum is the sum of the emission from the individual particles. Further, because there is no radiative interaction between particles,

black body distribution of energy cannot be established in the bulk sample either. For a more complete discussion of single particle emission see Hunt and Logan (21).

(c) In the reflection arrangement, the spectrum records only that energy scattered back in the direction of the detector. Essentially none of this has penetrated the sample to produce an absorption feature, but that which was absorbed has been replaced fairly effectively by reststrahlen reflection. Consequently, a general low energy scattering curve is all that is recorded.

Thick Particulate Sample (a) In the absorption arrangement, as with other experimental arrangements for this sample, multiple radiative interactions become important. In transmission, the energy is multiply scattered in all directions, but the scattering reaches a minimum at the Christiansen frequency, and so the transmission at this wavelength reaches a maximum. This explains the displacement of the maximum, or Christiansen peak, to shorter wavelengths than the genuine absorption minimum in the spectrum.

(b) The explanation of the emission spectrum is quite complicated. The only energy available is emitted by the individual single particles of which the sample is composed, and these emit on the profile shown immediately to the left in Fig. 2.11 for emission from a thin layer. However, after each particle emits in this form, multiple radiative interactions take place in which this energy is reflected or reabsorbed, and as a result of such processes it is possible to establish a black body distribution of energy within the bulk of the object, but usually not near the surface. The appearance of the spectrum is then governed by the extent to which thermal gradients are established in from the object surface, which in turn is governed by environmental parameters. However, while the contrast in a spectrum may vary considerably, the form of the spectrum is usually such that it displays a Christiansen maximum at slightly shorter wavelengths than a broad absorption minimum that occurs near the resonance frequency. The appearance of the emission spectra of such samples, and the way in which they vary with environmental parameters has been discussed in some detail by Logan et al. (22) and Aronson and Emslie (23, 24), for example.

(c) The spectral features in the reflection from a thick particulate object are due to absorption. The energy is multiply scattered and, during this process, some of the energy that is eventually directed back towards the observer penetrates some of the particles and is absorbed. Most of the remotely sensed data in the visible and near infrared occur in this form.

2.8.3 Environmental Factors

Several environmental factors, such as packing, background temperature, pressure, and illumination angle have their effects on the appearance of some spectra, but the effects are only particularly significant in the case of particulate materials in the lunar environment.

PACKING

The effect of altering the packing density is to alter the appearance of the spectrum because it alters (i) the scattering properties, (ii) the thermal conductivity and radiation transfer mechanisms, and (iii) the surface area of the particles/unit volume, and these are all parameters that affect the energy emitted by the individual particles in the object, and also affect the type of steady energy distribution established within the object as a whole.

BACKGROUND TEMPERATURE

This affects the appearance of the spectrum because altering it alters the thermal gradients within the sample and alters the contribution from reflected energy from the surface.

PRESSURE

The effect of altering the pressure is caused by the fact that the thermal conductivity is altered between the particles when the pressure changes, and so the thermal gradients change. Large changes only occur at very low pressures, but it is the pressure effect that is responsible for allowing much better data with better spectral contrast to be collected from the lunar surface than from the earth's surface.

ILLUMINATION ANGLE

The illumination angle (in the case of the sun, the insolation angle) has its effect because it alters the depth scale on which energy is available to a given sample.

2.9 SPECTRAL SIGNATURES OF MINERALS AND ROCKS

By *spectral signature* is meant a single feature or a pattern of features whose shapes, locations, and relative intensities are completely characteristic of a particular material. For organic substances, the infrared spectrum is typically so characteristic it is quite generally referred to as a "fingerprint."

In remote sensing situations, a multitude of objects are frequently present in a scene. Among these, the most important are vegetation; water; natural surfaces of inorganic solids, such as minerals, rocks, and soils; and artificial or human-made objects such as roads and buildings. Only the spectral signatures of inorganic solids will be considered here, so that the principal concern is with the characterizing information contained in the visible and infrared spectra of minerals, rocks, and soils.

The intrinsic information available as a consequence of the chemistry and structure of a specific material is modified by its physical condition and environment, as discussed in the preceding section (2.8). In terrestrial remote sensing, atmospheric absorption and scattering properties restrict the wavelength ranges that may be investigated to the atmospheric "windows" (see Section 2.7.3), and other object parameters alter the form and dilute the quality and quantity of the intrinsic information. In addition, specific remote sensing parameters, such as lack of spatial resolution generally make both compositional- and physical-condition heterogeneity of the sample area unavoidable.

Rocks are assemblages of minerals, and so the spectrum of a rock is a composite of the individual spectra of its constituent minerals. The spectra of soils similarly are a composite of mineral spectra, but soils typically also contain considerable amounts of organic material, and often water, both of which contribute their spectra to the overall spectrum, often completely dominating it.

Because rotational processes are precluded in solids, the features comprising the spectral signatures of minerals, rocks, and soils are produced as a consequence of either electronic or vibrational processes. To occur, these two processes require different amounts of energy, and thus evidence for their occurrence appears in different regions of the spectrum.

Electronic processes require more energy than vibrational processes, and evidence in the form of bands, slope changes, and so forth, for purely electronic processes appears primarily in the visible range, whereas evidence for fundamental vibrational processes is evident in the mid and far infrared.

The range between 0.5 and 2.5 μm is referred to as the "near" infrared and is frequently called the "overtone region." In this range there is an overlap of features due to electronic transitions, and excitation of overtone and combination tone vibrations, but it is a range capable of providing very useful remote sensing data.

Because the information available derives from different processes, and because, experimentally, the regions are usually separated, the visible and near-infrared regions will be discussed together, and separately from the midinfrared.

2.9.1 Visible and Near-Infrared Spectra

Naturally occuring surfaces are usually rough or particulate, and so the spectra presented in this section are all for particulate materials. The spectra

shown are called bidirectional reflection spectra, which refers to the technique used to generate them rather than to the process that produces the features.

The bidirectional reflection technique involves a source illuminating the sample at the same angle to the surface plane as that from which the radiation leaving the sample is collected (25). However, in the visible and near infrared, the energy arriving is scattered, often multiply, in all directions by the sample, and some of this radiation penetrates the sample material, being preferentially absorbed at certain wavelengths before being scattered in the direction of the detection system. It is, in fact, this loss due to absorption that produces the features in the spectra, and thus the absorption process dominates in producing the features even though it is a reflection technique used to record the spectrum.

The particle size of the sample determines the degree of scattering, and consequently the amount of absorption that takes place. Quite generally for a largely transparent material, the smaller the particle size, the greater the overall "reflection," and also the less the contrast in the absorption features. For opaque materials, the smaller the particle size, the less the reflection. The effect of particle size on the spectral behavior of trans-opaque materials has been discussed by Salisbury and Hunt (26).

In the visible or optical range, the color imparted to a material may be due to the appearance of a specific absorption in this region, or it may be due to intense absorption in either or both the near infrared or ultraviolet, the wings of such absorptions extending backward or forward into the visible, providing essentially a restricted window. This will impart color to a material without it having a specific chromophoric group absorbing in the visible. An example is olivine, which displays a green color in the visible while its major absorptions occur at 1.1 μm and to shorter wavelengths than 0.4 μm.

The visible and near-infrared spectra of both minerals and rocks have received wide-spread attention in the literature (27-31) and much of this work has been carried out to resolve certain spectroscopic problems and provide specific answers to questions not directly related to remote sensing. For this reason, and because there are so many publications, they will not be individually referenced here, but the references in those which are will provide a good starting point for further investigation.

The largest collection of published visible and near-infrared spectra of minerals and rocks is available in a series of papers by Hunt and Salisbury (32-34) and Hunt et al. (35-42), which contain the spectra of well over 200 minerals and 150 rock samples; for each sample, the spectra of four particle size range (less than 5 μm, less than 74 μm, 74 to 250 μm, and 250 to 1200 μm) are presented. The discussion here is based primarily upon those spectra.

As stated above, in the region between 0.3 and 2.5 μm, both electronic and vibrational processes yield spectral features. Figure 2.12 displays the spectra of four particle size range samples of the mineral beryl, and it is included to illustrate the appearance of spectra containing well-defined features due to both electronic and vibrational transitions, as well as indicate the effect of particle size. Typically, the vibrational features are much sharper than the electronic features, which are usually quite broad.

The electronic and vibrational processes, and the locations and appearances of the spectral features they produce, are largely very different, and so they will be discussed separately.

2.9.2 Electronic Processes

The concept of discrete electronic energy levels for atoms, and for ions located in crystals was introduced in Section 2.6.4, and it was pointed out that transitions between these levels are respon-sible for the features in their spectra. Many of the bands in the visible and near-infrared spectra are due to crystal field effects, but in addition, there are other electronic processes that produce spectral features. Such features occur more rarely in the spectra of naturally occurring geological materials, and the processes that produce them will be discussed in order of decreasing frequency of occurrence, and consequently in order of decreasing usefulness in geological remote sensing situations.

It is important to realize that the most common and necessary ingredients of minerals and rocks, namely silicon, aluminum, and oxygen do not possess energy levels located in such a way that transitions between them can yield spectral features in the visible and near-infrared range. Consequently, no information concerning the bulk composition of geological materials is *directly* available in this range, but considerable *indirect* information is available as a consequence of the particular crystal structure imposing its effect upon the energy levels, and consequently of the spectra of specific ions present in the structure.

In naturally occurring geological materials, the vast majority of characterizing information is due to the presence of iron, which occurs ubiquitously. The reason that iron is so frequently present is that, apart from being the principal ingredient in many materials, ferrous and ferric ions substitute with some facility into the octahedral Al^{3+} and Mg^{2+} sites, and sometimes into the tetrahedral

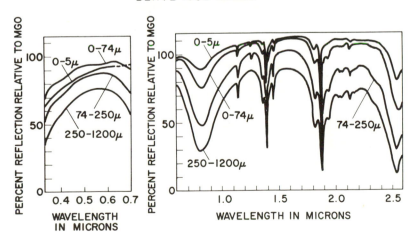

FIGURE 2.12. The bidirectional reflection spectra of four particle size range samples of the mineral beryl.

silicon sites as well, and the ready solubility of the iron ions in water accounts for their widespread distribution in terrestrial materials.

CRYSTAL FIELD EFFECTS

The electronic energy levels of an isolated ion are usually split and displaced when located in a solid. In the case of the transition metal ions, namely copper, nickel, cobalt, manganese, chromium, vanadium, titanium, scandium, and particularly iron, the unfilled d orbitals, which have essentially identical energies in the free ion, are split by interaction with surrounding ions and assume new (displaced) energy values.

These new energy values, and hence the transitions between them and consequently their spectra, are primarily determined by the valence state of the ion (for example, Fe^{2+} or Fe^{3+}), and by its coordination number and site symmetry. The spectrum is secondarily determined by the type of ligand formed (for example, metal-oxygen), the extent of distortion of the metal ion site from perfect, and the value of the metal-ligand interatomic distances. The effect of an octahedral crystal field on the divalent manganese ion produces the richest and most clearly resolved spectrum in the visible range. This is illustrated in Fig. 2.13 for the manganese carbonate mineral, rhodochrosite. An energy level diagram for the divalent manganese ion in an octahedral field is shown in Fig. 2.10. Features in the near-infrared spectrum of rhodochrosite are due to a transition in the ferrous ion (at ~ 1 μm)

and vibrations of the carbonate group at wavelengths larger than 1.5 μm.

The way in which the spectrum, due to a particular ion, is changed by locating it in different crystal fields is best illustrated for the case of the ferrous ion in the near-infrared range. All bands indicated in the spectra (Fig. 2.14) are due to spin-allowed transitions of the ferrous ion.

The spectrum of beryl, shown at the top of Fig. 2.14, displays an intense feature centered near 0.8 μm due to the allowed transition in a ferrous ion located in an aluminum octahedral site which is in sixfold coordination.

In bronzite, the ferrous ions are located in distorted octahedral sites, again sixfold coordinated, but in this case the spectrum displays two distinct features, one near 0.9 μm and the other near 1.8 μm. These features result because the distortion of the octahedral site causes the quintet states to split, and the new energy levels now permit two transitions of different energy to take place.

The spectrum of pigeonite provides the same pattern of two features, but here the energy levels have somewhat different values from those in bronzite, and so the bands are shifted.

In the olivine spectrum, three features are apparent in the very broad, overlapping set of bands centered near 1.0 μm. In olivine, there are two different distorted octahedral, sixfold coordinated sites that the ferrous ions can occupy, and ions in one site contribute to one of these features while ions in the other site contribute to the other two.

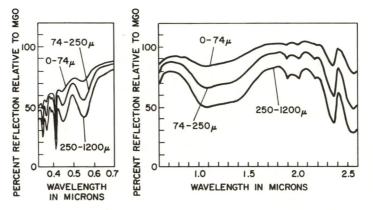

RHODOCHROSITE 67
CATAMARCA PROVENEE, ARGENTINA

FIGURE 2.13. The bidirectional reflection spectra of three particle size range samples of the mineral rhodochrosite.

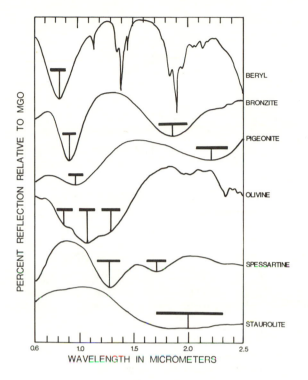

FIGURE 2.14. Location and appearance of features due to spin allowed transitions in the ferrous ion located in different crystal field environments. In the reflection spectra of particulate samples of these minerals, the ferrous ion is located in an aluminum octahedral six-fold coordinated site in beryl; in distorted octahedral six-fold coordinated sites in bronzite and pigeonite; in two distorted octahedral (centro, and noncentrosymmetric) six-fold coordinated sites in olivine; in an octahedral eight-fold coordinated site in spessartine; and in a tetrahedral site in staurolite. The ferrous ion bands are indicated by vertical lines.

In spessartine, two features, near 1.25 μm and 1.7 μm, are the results of transition in the ferrous ion in an octahedral site, but in this case the ion is eightfold coordinated.

The bottom spectrum, that of staurolite, displays a broad feature centered near 2.0 μm. This band is due to a transition in the ferrous ion in a tetrahedral site.

Although all the electronic features in the spectra of the above materials derived from the ferrous ion, the indirect information provided is about the type of crystal field in which it is located, and it is the type of indirect evidence that provides information concerning the bulk structure of minerals and rocks from crystal field effects.

Other transition metal ions provide similar information, although it is more rarely encountered. For example, the presence of copper is indicated by an intense feature near 0.8 μm, chromium by a feature near 0.55 μm with two sharper features near 0.35 μm and 0.45 μm, and nickel by features near 1.25, 0.74, and 0.4 μm.

CONDUCTION BAND AND CHARGE—TRANSFER TRANSITIONS

In some periodic lattices, the discrete energy levels of the shell electrons are broadened into energy bands by the proximity of other atoms. There are two such bands in which electrons may exist, and they are referred to as the "conduction" band and the "valence" band. Electrons possessing sufficient energy to exist in the higher energy conduction band have so much energy that they are not attached to any specific atom, but are free to wander throughout the crystal structure. They are referred to as "conduction" or "free" electrons. In the valence band, electrons, or the absence of electrons (holes), are located at specific atoms or bonds. Between these two energy bands is a zone of energies that electrons may not adopt, and this is called the "forbidden band" or "forbidden gap."

In dielectrics, the valence electrons are so tightly bound it requires very large amounts of energy to set them free and so the conduction band commences in vacuum ultraviolet, and may have very wide forbidden gaps. Metals, on the other hand, display very high conductivity, indicating that the electrons are very mobile, and so the forbidden gap is very narrow, or nonexistent, in which case the valence and conduction bands touch.

In semiconductors, the width of the forbidden band is intermediate between that for metals and dielectrics, and the edge of the conduction band is marked by the appearance of an intense absorption edge in the visible or near infrared. Examples of the appearance of such an absorption edge are shown in Fig. 2.15 for some sulphides. The sharpness of the absorption edge is a function of the purity and crystallinity of the material. For particulate materials, effects due to presence of boundaries, defects, and lack of periodicity in the crystal, as well as to the inclusion of impurities, will generally produce a much more sloped absorption edge than woud be observed in a pure single crystal.

Charge-transfer refers to the specific case where electrons migrate between neighboring ions, so that although they transfer from one atom to another, they do not enter a conduction band, but remain partially localized. Charge-transfer bands are gen-

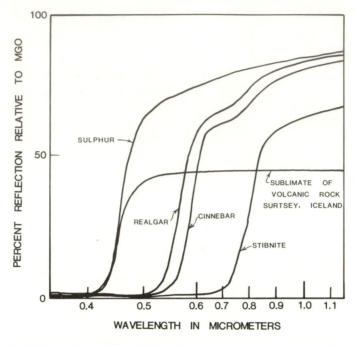

FIGURE 2.15. Visible and near-infrared bidirectional reflection spectra of particulate samples of stibnite, cinnabar, realgar, sulphur, and a sample of a sublimate volcanic rock, all of which display sharp absorption edge effects.

erally extremely intense, and the Fe-O transfer band is one of the most commonly observed features in the spectra of terrestrial minerals, being responsible for the steep falloff in intensity towards the blue in the visible spectra, particularly in the spectra of many weathered materials. The spectra of samples of hematite, goethite, and limonite are plotted in Fig. 2.16.

COLOR CENTERS

There are a limited number of materials that display features in the visible region, and consequently are colored, where the color cannot be explained in terms of their chemistry or by the presence of impurities. Such features are occasionally exhibited in the spectra of halides, and in particular by fluorites. In these cases, the features may be explained in terms of an electronic phenomenon called the "color center," which is caused by crystal defects. There are many types of defects, but the most common is the F-center. A more detailed discussion of the various types and the physics of color centers is available in the literature, (43, 44). Examples of the spectra produced by color centers in samples of fluorite are illustrated in Fig. 2.17.

2.9.3 Vibrational Processes

The theory of the vibrational process was discussed in Section 2.6.2. Spectral evidence for the fundamental vibrational transitions in all geologically important materials occurs in the mid-and far infrared regions, and the spectral signatures that appear there will be discussed in Section 2.9.4. However, spectral signatures due to vibrational processes do occur in the near-infrared range and these features are caused by overtones and combination tones of the fundamental modes.

An overtone occurs when a fundamental is excited with two or more quanta, so that features may appear at twice (or some integral multiple) of the fundamental frequency. Combination tone features may occur when two or more different fundamentals or overtones combine, and they are located at (or near) the sum of the frequencies of all the fundamental vibrations involved.

For strictly harmonic motions, spectral features due to all overtones and combinations are forbidden from appearing by the selection rules, but for most materials, they do appear to a greater or lesser degree because of anharmonicity, which is the departure from strictly harmonic motion.

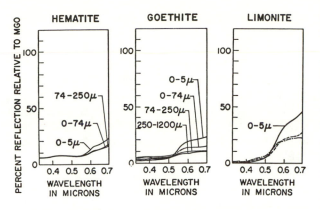

FIGURE 2.16. Visible reflection spectra of the iron oxide minerals.

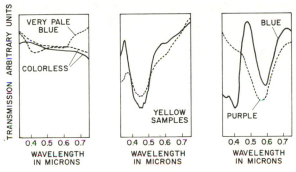

FIGURE 2.17. Visible transmission spectra of various samples of fluorite illustrating the variations in spectral response due to the presence of color centers in the crystal.

All the fundamental vibrational modes of silicon and aluminum with oxygen occur near 10 μm or at shorter wavelengths. Because the first overtone (which should be the most intense and occur near 5 μm or at longer wavelengths) is not observed, one does not expect to detect any direct evidence from them in the form of much higher order overtones in the near infrared.

What is observed in the near infrared are features due to the overtones and combinations involving materials that have very high fundamental frequencies. There are, however, relatively few groups that provide such features, and by far the most common of these are the near-infrared bands that involve the OH stretching mode.

Evidence for OH occurs so frequently in the spectra of terrestrial materials, not only because OH exists as part of the structure of a large number of

materials, but also because hydroxyl bands appear whenever water is present.

There are several ways in which molecular water may be associated with inorganic solids. For example, it may occur as free molecules trapped in interstices or pockets, as in milky quartz; it may occur singly or in clusters essential to the crystal lattice, as in hydrates such as gypsum; it may be present as water molecules located at specific sites in the crystal without being essential to its structure, as in zeolites; or it may be present as molecules physically absorbed on the surfaces of the solids. The many ways in which water can be associated with inorganic solids, and the variety of sites and environments which it can adopt, lead to some variation in the values of the fundamental frequencies, and consequently to variations in the frequencies of the overtone and combination bands. Thus, the exact location and appearance of the features that occur in the near-infrared spectra provide quite specific information about the nature of the bulk material in some cases.

The liquid water molecule has three fundamental vibrations; ν_1, the symmetric stretch, which appears at 3.106 μm; ν_2, the H-O-H bend occurring at 6.08 μm; and ν_3, the assymetric stretch at 2.903 μm. The overtones and combinations of these which appear prominently in the near infrared are $(\nu_2 + \nu_3)$ at ~ 1.875 μm, $(2\nu_2 + \nu_3)$ at ~ 1.454 μm; $(\nu_1 + \nu_3)$ at 1.38 μm; $(\nu_1 + \nu_2 + \nu_3)$ at 1.135 μm; and $(2\nu_1 + \nu_3)$ at 0.942 μm.

In the spectra of minerals and rocks, whenever water is present, two bands appear, one at 1.4 μm due to $2\nu_3$, and one at 1.9 μm due to $\nu_2 + \nu_3$. These may be either relatively sharp, indicating that the water molecules are located in well-defined, ordered sites, or they may be broad, indicating that they occupy unordered or several unequivalent sites. However, the presence of the 1.4 μm and 1.9 μm features together is completely diagnostic of the presence of molecular water.

The hydroxyl ion, OH', very frequently occurs in inorganic solids. There is only one oxygen-hydrogen stretching fundamental, and it is infrared active, and occurs near 2.77 μm, but its exact location depends upon the site on which it is located. The fact that OH groups may occupy different sites (and, hence, experience different potential fields) in a given material gives rise to several different features for the same motion. As we have said, the first overtone ($2\nu_{OH}$) of the OH stretch produces a feature near 1.4 μm, which is the most common feature present in the near-infrared spectra of terrestrial materials.

The fundamental OH stretching mode may form combination tones with other fundamentals, including lattice and librational modes. In particular, such combinations with fundamental X-OH bending modes, where X is Al or Mg, produces features near 2.2 or 2.3 μm, respectively.

Figure 2.18 displays typical spectra of minerals that contain OH groups shown at the top half of the figure, and of those that contain molecular water at the bottom half of the figure.

In addition to the water and OH features, several other specific groups contained in minerals display features characteristic of those groups in the near infrared.

The carbonate minerals display somewhat similar features between 1.6 and 2.5 μm, which are all due to combinations and overtones of the four fundamental internal vibrations of the planar CO_3^{2-} ion. These fundamental vibrations may be identified as ν_1, the symmetric C-O stretch; ν_2, the out-of-plane bend; ν_3, the doubly degenerate asymmetric stretch; and ν_4, the doubly degenerate in-plane bend. These fundamentals are responsible for producing five features in the near infrared, namely ($\nu_1 + 3\nu_3$) near 1.9 μm; ($2\nu_1 + 2\nu_3$) near 2.0 μm; ($\nu_1 + 2\nu_3 + \nu_4$) or ($3\nu_1 + 2\nu_4$) near 2.16 μm; $3\nu_3$ near 2.35 μm; and ($\nu_1 + 2\nu_3$) near 2.55 μm. The appearance of these bands are illustrated in Fig. 2.13 in the spectrum of rhodochrosite.

2.9.4 Midinfrared Spectra

Spectral signatures of inorganic solids in the midinfrared region are entirely the result of vibrational processes, and the most intense features result from excitation of the fundamental modes.

The midinfrared spectra of minerals have received widespread attention, and extensive reference to the early work is made in the publications of Lyons (45) and Lazarev (46), and several large collections of mineral spectra are available (47-50). Much less work has been published dealing with rock spectra, although several collections of such spectra are available (48, 51-55), while very little work has been performed on soils. This is not surprising, because of these, minerals are by far the most simple systems, and consequently are more tractable both experimentally and theoretically. Rocks, being composed of several minerals

render interpretation much more difficult, while soils, being composed of pulverized rock mixtures and their alteration products, also often contain abundant water and organic material and they yield spectra that are correspondingly even more complicated.

Even though some minerals are chemically relatively simple (as quartz SiO^2), analysis and interpretation of their midinfrared spectra is a formidable task. The difficulty is that theoretical techniques developed for isolated molecules cannot be simply transferred to the treatment of solids, and major problems arise when attempts are made to correctly account for the interaction of the internal vibrations of the anions with the lattice vibrations of the total solid. Consequently, only in rare instances, and for the simplest minerals, have analyses approaching complete satisfaction been achieved.

The vibrational motions of inorganic solids are typically so complex that only rarely can individual features in a spectrum be assigned to specific motions. However, an empirical approach, similar to the "group vibration" concept as applied to large organic molecules, has been most productive.

The spectral range that has received most attention is the 8 to 14 μm region, because it is the region in which evidence for the most intense "silicon-oxygen stretching" modes occur, and because it also fortuitously coincides with a region of minimum absorption in the earth's atmosphere. Also, the midinfrared is generally a region of particular interest because whatever information is acquired here relates directly to the bulk composition—the chemistry and structure of the material.

2.9.5 Silicate Signatures

Because the majority of terrestrial solids are silicates, the discussion here will be largely limited to the features that appear in their spectra with very limited reference to other substances, such as carbonates, sulphates, and oxides.

Figure 2.19 shows three typical spectra, recorded in different ways, of trachyte. The top two spectra are transmission spectra, the dashed curve being that of a finely particulate sample, and the solid curve that of the same sample embedded in a KBr pellet. The bottom curve is the reflection spectrum recorded from a polished surface of the rock. As can be readily seen, there is a wealth of spectral information provided, but the technique employed to generate these spectra tend to maximize

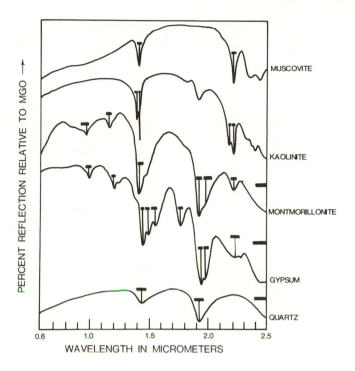

FIGURE 2.18. Near-infrared bidirectional reflection spectra of particulate samples which display typical vibrational overtone and combination tone features. Muscovite and kaolinite display sharp features (indicated by vertical lines) due to combinations and overtones involving the hydroxyl group, while in montmorillonite, gypsum, and quartz the features are due to the presence of water in various forms.

the detail for such a rock, and in the actual remote sensing situation such data is usually greatly degraded.

To facilitate the discussion of silicate spectral signatures, the wavelength range will be broken down into five segments where features of different kinds typically occur.

7 TO 9 μm

As was discussed more fully in Section 2.8.2, in the transmission and emission spectra of finely particulate silicates, a well-defined maximum occurs, and its location is referred to as the principal Christiansen frequency. This Christiansen peak is very obvious between 7 and 9 μm in silicates, because it occurs just prior to the onset of the intense absorption due to the silicon-oxygen stretching vibrations in a region otherwise devoid of any features.

The location of this peak migrates fairly systematically to longer wavelengths as the nature of the investigated material progresses from felsic to

intermediate to mafic to ultramafic. Consequently, because of this progressive migration, its location has been used as means of discriminating between different rock types, with particular emphasis on the remote sensing of the moon, which provides a unique environment suitable for optimizing the effect (22).

The appearance of the various features discussed in each of these five wavelength segments will be illustrated using mineral, rather than rock, spectra for clarity, and these are shown in Fig. 2.20. There the Christiansen frequency is indicated by a vertically ascending arrow for each mineral in the 7 to 9 μm range, and the location of this peak can be seen to be progressing to longer wavelengths as the mineral becomes increasingly more mafic.

8.5 TO 12 μm

The most intense absorption feature in the spectra of all silicates occurs near 10μm, and this region of the spectrum is quite generally referred

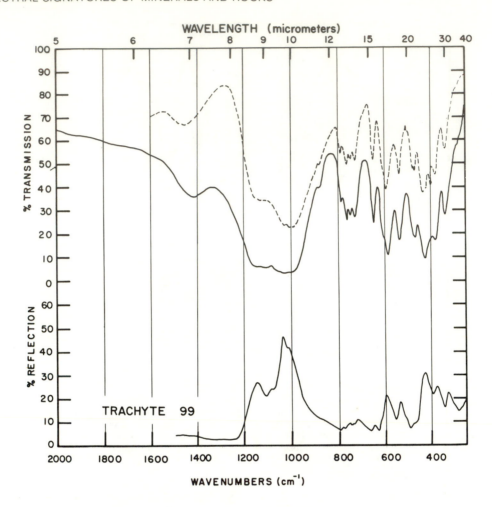

FIGURE 2.19. Midinfrared spectra of an igneous intermediate rock. The dotted curve is the transmission of a finely particulate sample in air. The top solid curve is the transmission of a finely particulate sample embedded in a potassium bromide pellet. The bottom solid curve is the reflection spectrum from a 1 × ½ in. polished surface of the rock.

to as the "Si-O stretching region."

Regardless of how adjacent SiO_4 tetrahedra share common oxygen atoms to form the various silicate structures, vibrations of the SiO_4 produce multiple bands in the 10 μm region due to the highly allowed asymmetric O-Si-O, Si-O-Si, O^--Si-O^-, and sometimes the symmetric O^--Si-O^- valence stretching motions, and in all these vibrations, it is the oxygen atoms that are most displaced from their equilibrium positions. The lack of resolution of these multiple features is partly due to the experimental technique, in which grinding the samples to a fine size range reduces the periodicity of the lattice, partly due to disorder in the lattice itself, and partly due to random substitution of Al for Si.

In a parallel manner to the way in which the Christiansen frequency maximum generally migrates to longer wavelengths as the material becomes more mafic, so also does the position of the minimum of this absorption band. Its location is quite sensitive to structure and, as can be seen in Fig. 2.20 (where it is indicated by a descending arrow), it occurs at shortest wavelengths for framework silicates, such as quartz and feldspars, and at longest wavelengths for materials in which the tetrahedra are isolated, as in the orthosilicates.

The only other feature to appear in this range is due to an H-O-Al bending mode, and it occurs distinctively near 11 μm, and is particularly evident in the spectra of aluminum bearing clay minerals.

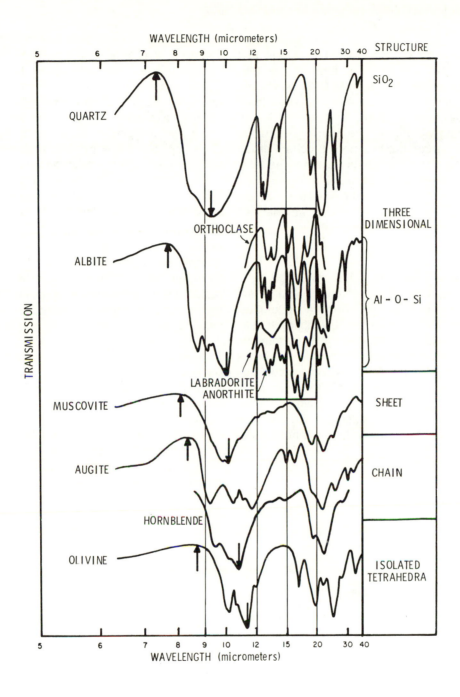

FIGURE 2.20. Segments of midinfrared transmission spectra of a series of minerals of different composition and/or structure. The wavelength range is divided into five segments and the diagnostic nature of the features that occur in each are discussed in the text.

12 TO 15 μm

The occurrence of features here indicate that (Si, Al)-O-(Si, Al) bridges have formed. The features are due to Si-O-Si and Al-O-Si symmetric stretches in which the oxygen is displaced least from its equilibrium position or they are due to Al-O′ stretching vibrations. The quartz spectrum is unique in this range, but the most important group of minerals to display features here are the feldspars. Even though all feldspars display bands, their spectra differ in this range from showing a single feature (labradorite) to a set of four well-defined bands (albite) as shown in Fig. 2.20.

15 TO 20 μm

This region is devoid of bands in a large number of silicates, and this constitutes a gap before features due to the bending modes that appear just past 20 μm. However, extremely characteristic bands due to the (Si, Al)-O-(Al, Si) symmetric stretching modes in feldspars occur here. Whereas the feldspar bands in the 12- to 15- μm range differed with different feldspars, those due to the feldspars in the 12- to 20- μm range are extremely similar to each other for different feldspars, and are quite characteristic. Other features do occur here, notably due to a symmetric Si-O-Si stretch in mafic minerals (see Fig. 2.20).

20 TO 40 μm

This region contains bands due to O-(Al, Si)-O, (Si, Al)-O-(Si, Al), and O⁻-(Al, Si)-O⁻ deformation and bending modes, and also due to (Al, Si-O⁻ metal valence stretching vibrations. There is very little justification for attempting to assign these bands to specific motions. Those calculations that do exist indicate that bands near 20 μm are due to bending of the (Al, Si)-O-(Si, Al) and O-(Si, Al)-O groups, or wagging and rocking of the O⁻-(Al, Si)-O⁻ group, but these motions (except for rocking) also produce features at longer wavelengths. Twisting motions of the O⁻-(Al, Si)-O⁻ groups produce bands near 30 μm, while ring puckering motions provide very long wavelength features.

Features due to metal-O valence stretches (when the metal ion is Mg, Ca, Fe, and so forth) typically occur between 15 and 30 μm. Despite the large number of different motions that occur in this region, it is typically referred to as the "Si-O bending region." The information available from the midinfrared spectra of silicates is summarized in Fig. 2.21.

In addition to indicating the positions of the absorption features and their typical causes, and the location of the principal Christiansen frequency maxima, the occurrence of other Christiansen frequencies are indicated at longer wavelengths. It is interesting to note that, whereas the principal Christiansen frequencies shift in a manner that can be correlated with rock type, the other observed Christiansen maxima do not, but this is largely a function of whether the region toward shorter wavelengths than the resonance frequency on which the feature is based is clear of bands or not. If it is not, the

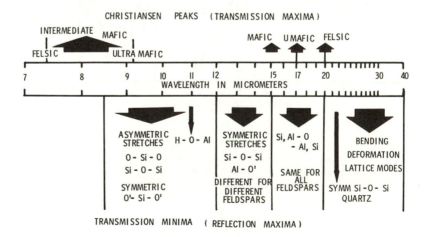

FIGURE 2.21. Diagrammatic illustration indicating the location of features and the types of vibrations that produce the spectral signatures of silicates in the midinfrared region.

Christiansen effect will not be apparent.

The spectral signatures of a suite of eight granites are shown, displaced for clarity, in Fig. 2.22. These spectra are obviously very similar, a fact made more apparent when they are superimposed. The similarity of the spectra for a class of rocks, such as the granites, allows a composite signature to be generated, which may be used as representative of all granites. Such a composite signature is shown as the dashed curved at the bottom of Fig. 2.22.

Such composites may be generated for each class of rock, and while the individual spectrum of each member usually closely resembles the composite, the composites for different classes are usually quite different. Figure 2.23 shows the composite signature of four different rock types, each generated in a similar manner to that shown as the dashed curve in Fig. 2.22. It is apparent from these curves that discrimination between these rock types could be readily made on the basis of their midinfrared spectral signatures. It should be remembered, however, that the spectral data presented here are transmission spectra of particulate samples, and this technique tends to optimize the intrinsic information. In the remote sensing situation, spectra of this quality are not available, and the information content is considerably degraded.

2.9.6 Nonsilicate Signatures

The discussion thus far has centered around silicate signatures, where the basic unit responsible for producing spectral bands was the SiO_4 tetrahedron. Other molecular units are present in naturally occurring materials, and the more frequently encountered of these are the carbonates, sulphates, phosphates, oxides, and hydroxides. Minerals containing these particular anions occur most frequently in sedimentary and metamorphic rocks,

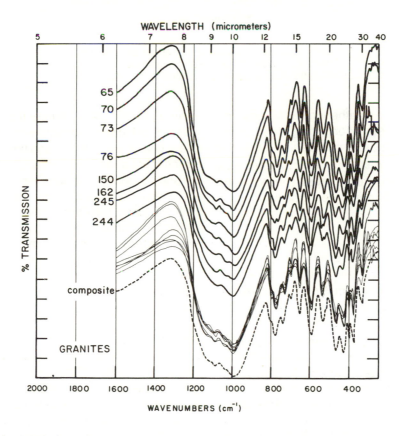

FIGURE 2.22. Midinfrared transmission spectra of a suite of eight granites. The individual spectra are presented, displaced vertically for clarity, in the top portion of the figure. Beneath this they are presented superimposed to indicate variability, and the bottom dashed curve represents a single composite "average" spectrum generated from all eight individual spectra.

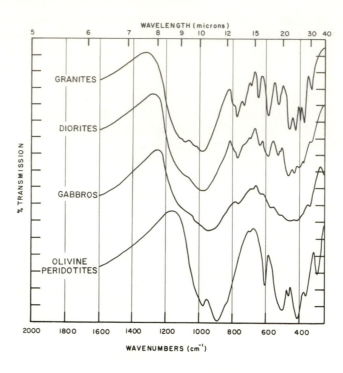

FIGURE 2.23. The midinfrared composite spectra for four different rock types are shown to illustrate the difference in intrinsic spectral signature as a function of rock type.

and in situations where weathering has taken place.

The spectra of minerals containing these groups are collected in Fig 2.24, and it can be seen that their spectral signatures differ greatly from each other, and from those of the silicates. These spectra will be discussed individually very briefly.

CARBONATES

The spectrum at the top of Fig. 2.24 is of calcite that contains the features that typify all carbonate spectra. The spectrum is dominated by features due to the internal vibrations of the CO_3^{2-} group.

The nature of the carbonate mineral spectra has been discussed extensively in the literature (for example, 56). As discussed earlier (Section 2.9.3), four features due to the fundamental modes typically appear in the midinfrared spectrum, and these are, v_3, the assymetric stretches near 7 μm, which are doubly degenerate; and the planar bends, v_4, which occur between 13 and 15 μm, also doubly degenerate. The other planar bend, v_2, is not degenerate, and appears between 11 and 12 μm,

while v_1, the totally symmetric C-O stretch, is infrared forbidden, but when "made allowed" appears near 9 μm. The reason that ranges are given for the location of some of these bands is that they are shifted, depending upon the nature of the cation, and in less symmetric structures than calcite, such as in aragonite, the degeneracies are removed, and doubled features appear for v_3 of v_4 and in addition v_1 appears as a weak feature.

Most of the carbonate bands occur in regions unoccupied by silicate bands, thus their spectral signature is very noticeable even when the carbonate is present in only minor quantities.

PHOSPHATES

The infrared spectra of phosphates have been considered in the literature (57, 58). The isolated group possesses tetrahedral symmetry and has nine normal modes. These are a symmetric stretch, v_1 which is infrared inactive but usually occurs near 10.3 μm; a doubly degenerate symmetric bend, v_2, near 28.5 μm; two asymmetric stretches, v_3, near 9.25 μm; and the symmetric bends, v_4, near 18 μm. Distortion from pure tetrahedral symmetry

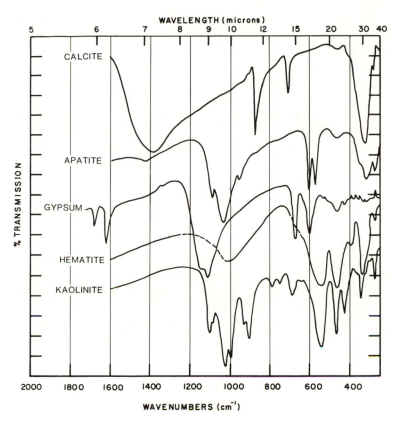

FIGURE 2.24. Midinfrared transmission spectra of a representative member of the nonsilicate mineral groups, which are commonly encountered in terrestrial remote sensing.

relieves the degeneracies so that more than just three features typically occur in phosphate signatures. The phosphate group is usually encountered in the form of apatite in sedimentary rocks.

SULPHATES

The symmetry of the sulphate ion is the same as that of the phosphates and the fundamental frequencies of the isolated ion are known from Raman studies (59) to be ν_1, near $10.2\,\mu m$; ν_2, near $22.2\,\mu m$; ν_3, near $9\,\mu m$; and ν_4, near $16\,\mu m$. The description of these modes is the same as for the phosphates, and in gypsum, the mineral used to illustrate sulphates in Fig. 2.24, the degeneracies are resolved. In the gypsum spectrum, the sharp features in the 6-μm region are caused by the fundamental bending mode of its constitutional water.

There are other minerals less frequently encountered whose spectra are determined by the same symmetry considerations as those of sul-

phates, and they are the arsenates and vanadates. Borates are more complex because they can assume different structures. However, the spectra of some borates with the calcite and aragonite structure have been analyzed (60).

OXIDES

Features due to the fundamental stretching modes of oxides typically occur at longer wavelengths than $5\mu m$, and so usually lie in a region occupied by bands due to silicon-oxygen bending modes. Of importance in remote sensing is the detection of the presence of iron oxides and the spectrum of hematite is shown in Fig. 2.24. There intense features due to the Fe-O fundamental stretching modes are apparent around $20\mu m$.

HYDROXIDES

The fundamental stretching modes of the OH group all occur around $3\,\mu m$ and the number and appearance of such features form a very powerful signature. However, because the region is also occupied with the OH stretching features of water, the OH signature in that range is of limited usefulness in the remote sensing situation. In the midinfrared, however, the fundamental bending modes of OH attached to various metal ions occur, and one particularly useful feature is that due to the fundamental bending mode when the OH is attached to aluminum. This occurs near 11 μm and it is very apparent in the spectrum of kaolinite, shown in Fig. 2.24. This spectrum also serves to indicate generally where the silicate and aluminate signatures occur relative to the features due to those of other molecular groups.

Of the above nonsilicate groups, carbonate is by far the most frequently encountered. Fig. 2.25 shows the spectra of a series of limestones. The top spectrum is essentially indistinguishable from pure calcite, but in increasing amounts, clays and quartz are present in the materials whose spectra are shown in progressing to the bottom of the figure,

and the presence of this clay and quartz in increasing quantities is indicated by the development of the features near 9μm, in the 13-μm region, and at wavelengths longer than 20μm.

2.10 CONCLUDING REMARKS

To a large extent, remote sensing in geology has been pursued on an empirical level, and this is not surprising when one considers the large number of variables operative in natural situations. To adopt a more analytical approach to geological remote sensing, an understanding of the basic nature of electromagnetic radiation and the specific ways in which it interacts with matter is required.

Commencing with a brief review of the way in which currently accepted theories were developed, this chapter provides an introduction to, and an outline of the nature of electromagnetic energy, from both a wave and a particle viewpoint. The component parts of the spectrum, the sources of the

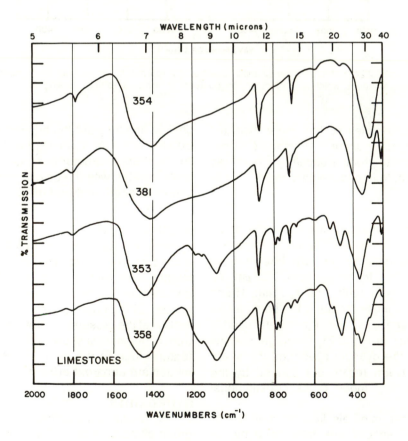

FIGURE 2.25. Midinfrared transmission spectra of a series of limestones.

radiations, and the techniques used for manipulating and detecting them are discussed.

Information about a remote object is obtained by examining the wavelength intensity relationship of the electromagnetic radiation arriving from it, and interpretation requires an understanding of the manner in which the radiation interacts with object. Two types of interaction are distinguished here: (a) those categorized as physical optical effects, where the radiation interacts at a boundary, resulting in a redirection of the energy by such processes as refraction, reflection, or polarization, and the magnitude of the effect depends on the refractive index, size, shape and texture of the material, and (b) those that fall into the realm of spectroscopy, where specific wavelengths determined by the chemistry, molecular structure, and internal forces of the material are either absorbed or emitted.

In this chapter, more emphasis is placed on the consideration of interactions that fall into the latter category. To discuss such interactions, the general concept that materials can only exist in discrete energy states, or specific energy ranges is introduced, together with the fact that absorption or emission of energy can only occur as a result of shifts between these energy states. Such shifts are referred to as transitions. Evidence for transitions appear in a spectrum as intensity maxima or minima that are referred to as bands of spectral features. Because the arrangement of the energy states is different for different materials, the spectrum produced as a result of transitions between the energy levels of a particular material is characteristic of it.

The nature of energy levels and the nomenclature employed to identify them is discussed. Although rotational processes are not operative in geological (solid) substances, these processes are considered because of their relevance to terrestrial atmospheric gases. Of direct importance are both the vibrational and electronic processes in solids, and because evidence for their occurrence appears mostly in the visible and infrared regions, these regions receive considerable attention. One particularly important source of geological information arises as a result of the energy levels of an ion being perturbed when it is embedded in a solid, due to crystal field effects.

The spectral signatures of geological materials, the intrinsic information available, and the physical causes for them in each spectral range are considered. Even though a complete interpretation for complex materials is rarely possible, and a somewhat empirical approach is resorted to, nevertheless, these signatures allow surprisingly good characterization of some substances and discrimination among others.

Although the results of laboratory investigations provide the optimum intrinsic information available concerning a particular material, and clearly delineate the spectral regions that offer the best remote sensing potential, it is necessary to know the ways in which physical conditions and environmental parameters alter the form of the data and degrade its quality in a real situation. The effects that these parameters have on the intrinsic information are discussed.

REFERENCES

1. Neblette, C. B., 1962, Photography, its materials and processes: New York, Van Nostrand, 500 p.
2. Colwell, R. N., ed., 1960, Manual of photographic interpretation: Falls Church, Virginia, American Society of Photogrammetry, 868 p.
3. Smith, J. T., Jr., ed., 1968, Manual of color aerial photography, Falls Church, Virginia, American Society of Photogrammetry, 550 p.
4. Kruse, P. W., McGlauchlin, L. D., and McQuistan, R. B., 1962, Elements of infrared technology: New York, Wiley, 448 p.
5. Smith, R. A., Jones, F. E., Chasmir, R. P., 1968, The detection and measurement of infrared radiation: London, Oxford University Press, 503 p.
6. King, G. W., Haimer, R. M., and Cross, P. C., 1943, The asymmetric rotor: I. calculation and symmetric classification of energy levels: Jour. Chem. Phys., v. 11, p. 27-42.
7. King, G. W., Haimer, R. M., and Cross, P. C., 1949, The asymmetric rotor: VII. extension of calculation of energy levels: Jour. Chem. Phys., v. 17, p. 826-836.
8. Wilson, E. B., Decius, J. C., and Cross, P. C., 1955, Molecular vibrations: New York, McGraw-Hill, 388 p.
9. Herzberg, G., 1945, Infrared and Raman spectra of polyatomic molecules: New York, Van Nostrand, 632 p.
10. Barrow, G. M., 1962, Introduction to molecular spectroscopy: New York, McGraw-Hill, 318 p.

11. Eyring, H., Walter, J., and Kimble, A. H., 1948, Quantum chemistry: New York, Wiley, 394 p.

12. Wigner, E. P., 1959, Group theory: New York, Academic, 372 p.

13. Born, M., and Wolf, E., 1964, principles of optics: New York, MacMillan, 808 p.

14. Stratton, J. A., 1941, Electromagnetic theory: New York, McGraw-Hill Co., 615 p.

15. Van de Hulst, 1957, Light scattering by small particles: New York, Wiley, 470 p.

16. Samuelson, R. E., 1967, The transfer of thermal infrared radiations in cloudy atmospheres: application to Venus: unpub. Ph.D. Dissertation, Georgetown Univ., Washington, D. C., 309 p.

17. Minnaert, M., 1941, The reciprocity principle in lunar photometry: Astrophys. Jour., v. 93, p. 403-410.

18. Young, A. T., and Collins, S. A., 1971, Photometric properties of the Mariner cameras and of selected regions on Mars: Jour. Geophysical Res., v. 76, p. 432-437.

19. Conel, J. E., 1969, Infrared emissivities in silicates: experimental results and a cloudy atmospheric model of spectral emission from condensed particulate mediums: Jour. Geophysical Res., v. 74, p. 1614-1634.

20. Logan, L. M., and Hunt, G. R., 1970, Emission spectra of particulate silicates under simulated lunar conditions: Jour. Geophysical Res., v. 75, p. 6539-6548.

21. Hunt, G. R., and Logan, L. M., 1972, Variation of single particle mid-infrared emission spectrum with particle size: Applied Optics, v. 11, p. 142-147.

22. Logan, L. M., Hunt, G. R., Salisbury, J. W., and Balsamo, S. R., 1973, Compositional implications of Christiansen frequency maxima for infrared remote sensing applications: Jour. Geophysical Res., v. 78, p. 4983-5003.

23. Aronson, J. R., and Emslie, A. G., 1973, Spectral reflectance and emittance of particulate materials, 2. applications and results: Applied Optics, v. 12, p. 2573-2584.

24. Aronson, J. R., and Emslie, A. G., 1975, Applications of infrared spectroscopy and radiative transfer to earth sciences, in Karr, C., ed., Infrared and Raman spectroscopy of lunar and terrestrial minerals: New York, Academic, p. 143-164.

25. Hunt, G. R., and Ross, H. P., 1967, A bidirectional reflectance accessory for spectroscopic measurements: Applied Optics, v. 6, p. 1687-1691.

26. Salisbury, J. W., and Hunt, G. R., 1968, Martian surface materials; effect of particle size on spectral behavior: Science, v. 161, p. 365-366.

27. Adams, J. B., 1975, Interpretation of visible and near infrared diffuse reflectance spectra of pyroxenes and other rock-forming minerals, in Karr, C., ed., Infrared and Raman spectroscopy of lunar and terrestrial minerals: New York, Academic, p. 91-116.

28. Ross, H. P., Adler, J. E. M., and Hunt, G. R., 1969, A statistical analysis of the reflectance of igneous rocks from 0.2-2.65 microns: Icarus, v. 11, p. 46-54.

29. Adams, J. B., and McCord, T. B., 1972, Electronic spectra of pyroxenes and interpretation of telescopic spectral curves of the Moon: Geochim. et Cosmochim. Acta, v. 3, p. 3021-3034.

30. Burns, R. G., 1970, Mineralogical applications of crystal field theory: Cambridge, Md., Cambridge U. P., 224 p.

31. Adams, J. B., and Filice, A. L., 1967, Spectral reflectance of 0.4 to 2.0 microns of silicate rock powders: Jour. Geophysical Res., v. 72, p. 5705-5715.

32. Hunt, G. R., and Salisbury, J. W., 1970, Visible and near-infrared spectra of minerals and rocks: I. silicate minerals: Modern Geology, v. 1, p. 283-300.

33. Hunt, G. R., and Salisbury, J. W., 1971, Visible and near-infrared spectra of minerals and rocks: II. carbonates: Modern Geology, v. 2, p. 23-30.

34. Hunt, G. R., and Salisbury, J. W., 1971, Visible and near-infrared spectra of minerals and rocks: XI. sedimentary rocks: Modern Geology, v. 5, p. 211-217.

35. Hunt, G. R., and Salisbury, J. W., 1971, Visible and near-infrared spectra of minerals and rocks XII. metamorphic rocks: Modern Geology, v. 5, p. 221-228.

36. Hunt, G. R., Salisbury, J. W., and Lenhoff, C. J., 1971, Visible and near-infrared spectra of minerals and rocks: III. oxides and hydroxides: Modern Geology, v. 2, p. 195-205.

37. Hunt, G. R., Salisbury, J. W., and Lenhoff, C. J., 1971, Visible and near-infrared spectra of minerals and rocks: IV. Sulphides and sulphates: Modern Geology, v. 3, p. 1-4.

38. Hunt, G. R., Salisbury, J. W., and Lenhoff, C. J.,

1972, Visible and near-infrared spectra of minerals and rocks: V. halides, phosphates, arsenates, vanadates, and borates: Modern Geology, v. 3, p. 121-132.

39. Hunt, G. R., Salisbury, J. W., and Lenhoff, C. J., 1973, Visible and near-infrared spectra of minerals and rocks: VI. additional silicates: Modern Geology, v. 4, p. 85-106.

40. Hunt, G. R., Salisbury, J. W., and Lenhoff, C. J., 1973, Visible and near-infrared spectra of minerals and rocks: VII. acidic igneous rocks: Modern Geology, v. 4, p. 217-224.

41. Hunt, G. R., Salisbury, J. W., and Lenhoff, C. J., 1974, Visible and near-infrared spectra of minerals and rocks: VIII. intermediate igneous rocks: Modern Geology, v. 4, p. 237-244.

42. Hunt, G. R., Salisbury, J. W., and Lenhoff, C. J., 1974, Visible and near-infrared spectra of minerals and rocks: IX. basic and ultrabasic rocks: Modern Geology, v. 5, p. 15-22.

43. Prizbaum, K., 1956, Irradiation colors and luminescence: London, Pergamon, 332 p.

44. Fowler, W. B., 1968, Physics of color centers: New York, Academic, 156 p.

45. Lyon, R. J. P., 1962, Minerals in the infrared—a critical bibliography: Palo Alto, Cal., Pub. of Stanford Res. Inst., 76 p.

46. Lazarev, A. N., 1972, Vibrational spectra and structure of silicate: New York, Pub. of Consultants Bureau, 302 p.

47. Lyon, R. J. P., 1962, Evaluation of infrared spectroscopy for compositional analysis of lunar and planetary soils: SRI Final Rpt., Contract NASR 49(04), 139 p.

48. Lyon, R. J. P., 1964, Evaluation of infrared spectrophotometry for compositional analysis of lunar and planetary soils: rough and powdered surfaces: SRI Final Rpt., Part II, Contract NASR 49(04), 175 p.

49. Moenke, H., 1966, Mineralspektren II: Berlin, Academic Vertug.

50. Sadtler, 1973, Minerals: infrared grating spectra: Philadelphia, Sadtler Res. Labs.

51. Vincent, R. K., 1973, A thermal infrared ratio imaging method for mapping compositional variations: unpub. Ph.D. Thesis, Univ. of Mich., Ann Arbor, Mich., 102 p.

52. Hunt, G. R., and Salisbury, J. W., 1974, Mid-infrared spectral behavior of igneous rocks: Environ. Res. Paper 496-AFCRL-TR-74-0625, 142 p.

53. Hunt, G. R., and Salisbury, J. W., 1975, Mid-infrared spectral behavior of sedimentary rocks: Environ. Res. Paper 520-AFCRL-TR-75-0256, 49 p.

54. Hunt, G. R., and Salisbury, J. W., 1976, Mid-infrared spectral behavior of metamorphic rocks: Environ. Res. Paper 543-AFCRL-TR-76-0003, 67 p.

55. Vincent R. K., Rowan, L. C. Gillespie, R. E., and Knapp, C., 1975, Thermal infrared spectra and chemical analyses of twenty-six igneous rock samples: Remote Sensing of Environ., 4. 4, p. 199-210.

56. Adler, H. H., and Kerr, P. F., 1963, Infrared absorption frequency trends for anomalous normal carbonates: Amer. Mineralogist, v. 48, p. 124-137.

57. Adler, H. H., 1968, Infrared spectra of phosphate minerals: splittings and frequency shifts associated with substitution of PO_4^{3-} for AsO_4^{3-} in mimetite: Amer. Mineralogist, v. 53, p. 1740-1744.

58. Kravitz, L. C., Kingsley, J. D., and Elkin, E. G., 1968, Raman and infrared studies of coupled PO_4^{3-} vibrations: Jour. Chem. Physics, v. 49, p. 4600-4616.

59. Kohlrausch, K. W. F., 1943, Ramansperkren: Hand und Jahrbuch der Chemischen Physik, v. 96, 399 p.

60. Steele, W. C., and Decius, J. C., 1956, Infrared absorption of lanthanum, scandium, and indium borate and the force constants of the borate ion: Jour. Chem. Physics, v. 25, p. p. 1184-1188.

61. Orgel, L. E., 1960, An introduction to transition-metal chemistry: London, Methuen, 186 p.

CHAPTER 3

ACQUISITION OF REMOTELY SENSED DATA
DONALD LOWE

3.1 INTRODUCTION

In July, 1972, NASA launched the *Earth Resources Technology Satellite* (ERTS), which produced moderately high-resolution imagery of the cloud-free Earth every 18 days. The ERTS program has produced imagery data in magnitudes hitherto unknown, and a large group of users has formed seeking to apply the imagery to problems in their disciplines. Geologists are one such set of users. Although the ERTS program (now called Landsat) uses sensors that generate multiband imagery in the visible and near-visible region, it has sparked interest in imaging the Earth in all regions of the electromagnetic spectrum.

This chapter describes *remote sensing instrumentation*: their features, principles of operation, limitations, and data procurement. It discusses the instrumental, environmental, and operational parameters which limit or affect the quality of data, considering these parameters only to the level necessary for the user to plan data-collection missions or order existing data with some assurance that resulting products will meet requirements and expectations. Consideration is also given to ancillary equipment needed for various sensor systems, factors that cause image distortions and means of eliminating them, and advanced sensor systems used or being developed.

Emphasis is primarily on imaging sensors operating from aircraft or space. In addition, more promising nonimaging sensors are described, because today's sensors that are capable of making observations only along a ground transect may be imaging sensors of tomorrow. In accord with the scope of this text, only sensors of electromagnetic radiation are considered. The reader is referred to standard texts for discussion on important geologic survey sensors of force fields, for example, magnetic and gravity.

3.2 CONVENTIONAL IMAGING SENSORS

As discussed in Chapter 2, the electromagnetic spectrum covers a wide range of frequencies or wavelengths. The response or sensitivity of detection devices depends on frequency (or wavelength), so that any single sensor is useful in only a limited portion of the spectrum. For example, the eye responds only to visible wavelengths. Therefore, it is not surprising that the design of imaging sensors varies radically with operational wavelength.

Imaging sensors can be classified according to the detection process, (for example, photographic or television), wavelength region of operation, or operational mode (that is, active or passive). An active sensor provides its own source of illumination, whereas, a passive sensor observes solar reflected energy or radiant emittance.

Sensors are classified here as optical, microwave and "other," a classification somewhat dependent on wavelength, as optical sensors operate in the ultraviolet, visible, or infrared region, and all employ reflective and/or refractive elements for imaging. Microwave sensors operate at millimeter wavelengths or larger and may or may not use reflective or refractive elements. The category of "other" is reserved for systems operating at extremely short wavelengths; these currently operate in a nonimaging mode.

3.2.1 Energy Available for Remote Sensing

Photographic and television sensors operate in the visible and near-visible region of the electromagnetic spectrum. Coincidentally or through evolution, the visible region of the electromagnetic spectrum lies in a relatively transparent region of the atmosphere where radiant emission from the sun is peaked. Figure 3.1 shows the spectral irradiance of the sun at the top of the atmosphere and after

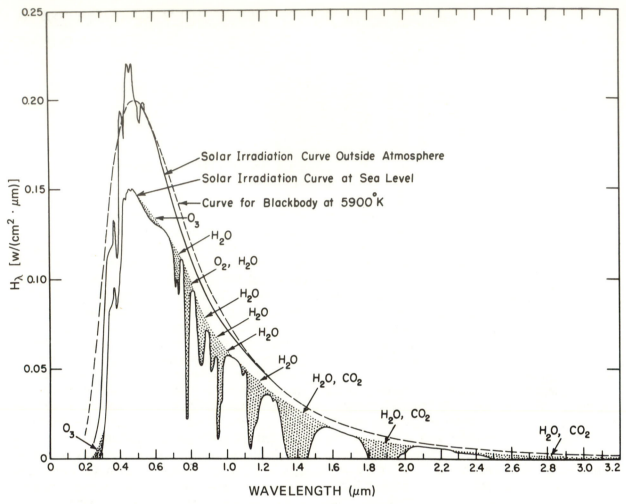

Figure. 3.1. Solar spectral irradiance (1).

transmission through the atmosphere; as illustrated, the solar curve is peaked at 0.5 μm (0.5×10^{-6} m or 5000 Å). The figure also illustrates how the intensity of solar irradiation is reduced by scattering and absorption as it passes through the atmosphere. There are no serious limitations to solar energy or atmospheric absorption in the region between 0.32 and 1.0 μm, where photographic and television systems are responsive.

In operating at longer wavelengths, in the infrared and microwave, it is necessary to use regions called *windows*, where the atmosphere is relatively transparent. Figure 3.2 shows the expected transmission in the infrared region through 1.6 km of atmosphere at sea level. The precise transmission along any atmospheric path is a complex function of atmospheric pressure and composition. Figure 3.3 shows

absorption bands of various atmospheric gases in the near-infrared region where thermal imagers operate. Atmospheric transmission models exist which permit calculation of the absorption for various horizontal, vertical, and slant paths of the atmosphere (2).

During daylight, the radiant energy from a scene is composed of two components: at wavelengths shorter than 3 μm, the energy is predominantly reflected sunlight; at longer wavelengths, beyond 4 μm, self-emission predominates. All objects radiate energy, the amount of radiation depending on the object's temperature. It is this radiant emission that thermal line scanners and passive microwave imagers detect. Figure 3.4 shows the spectral radiance of concrete measured during the day and at night. The solid curve shows daytime radiance measure-

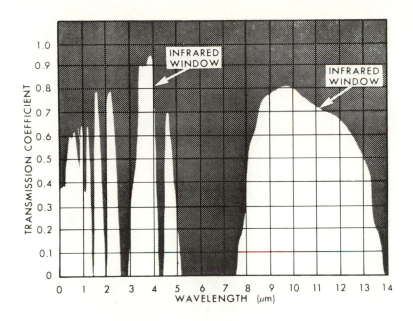

FIGURE 3.2. Typical transmission through 1.6-km atmospheric thickness.

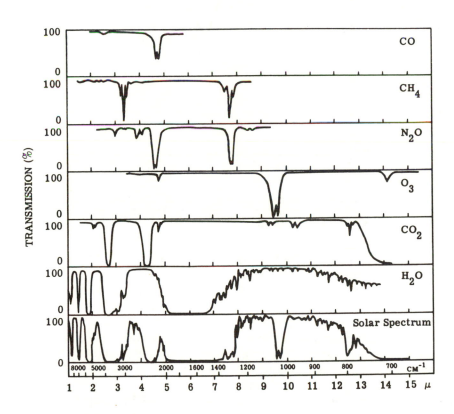

FIGURE 3.3. Comparison of the near-infrared solar spectrum with laboratory spectra of various atmospheric gases (1).

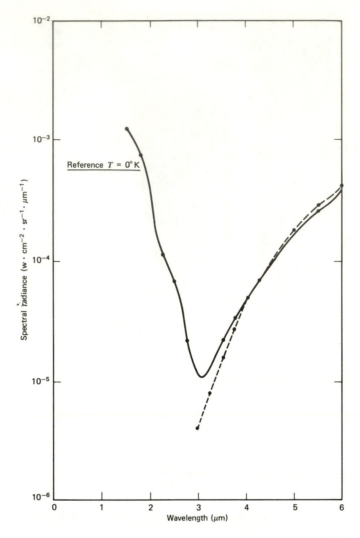

FIGURE 3.4. Spectral radiance of concrete observed during overcast (3). Explanation: ——— daytime radiance measurement; •- - - - -• nighttime radiance measurement.

ments, representing both reflected sunlight and thermal emission. The nighttime curve (dotted line) shows that emission below 3 μm is absent. Figure 3.5 shows the relative energy available for remote sensing in daylight in the 0.3 to 20 μm region in the absence of atmospheric absorption. The two short-wavelength curves were computed from the solar constant for materials having a diffuse reflection of 100% (ρ = 1) and 10% (ρ = 0.1). The long-wavelength curve is the black body emission from a surface at 300°K (81°F). These curves can be used to estimate the relative magnitude of reflected solar energy and thermal emission. The energy available for remote sensing in the visible region is about 10 times greater than that available in the thermal region.

The atmosphere is essentially opaque between 15 μm and 1 mm, but transmission beyond 3 cm is generally high. Figure 3.6 illustrates the atmospheric transmission for a 1-km path length in the transitional region of 0.8 to 30 mm. Notice the high transmission at 0.8 cm and longer.

3.2.2 Optical Imaging Systems

The photographic camera produces black-and-white or color images by a simple and relatively inexpensive process, which, like the eye, is limited in wavelength response. In particular, it cannot re-

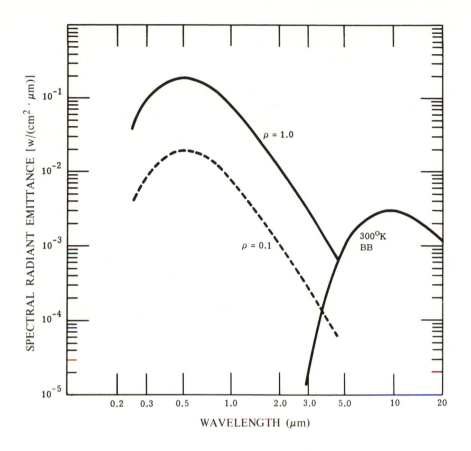

FIGURE 3.5. Energy available for remote sensing.

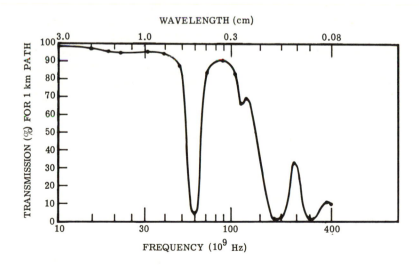

FIGURE 3.6. Spectral transmission in the microwave region at sea level.

spond to wavelengths beyond 1   m. Except for photography, all operational remote sensors are electro-optical, that is, they use detectors that convert observed radiation to electrical signals, which in turn are used to generate an image. While electro-optical sensors are generally more complex than photographic cameras, they may be preferred to photographic systems because of these factors: (1) the electrical signal can be transmitted over radio links, stored on magnetic tapes, and conveniently analyzed and processed by computer; (2) the detection process is reversible, permitting the detector to be reused, and (3) the detectors have a wider dynamic range than photographic film. Most optical imaging systems used in geologic surveys are framing cameras or line scanners.

FRAMING CAMERAS

Framing cameras register the relative position of elements in a scene with geometric precision, for all elements of the scene are observed at the same time, that is, during exposure. If the camera platform moves or rotates during exposure, the image may be smeared, but the elements remain fixed relative to one another. Variations in platform motion can generate serious geometric errors in line scanners.

In a framing camera, a scene is imaged on a film detector, as show in Fig. 3.7. If the scene is flat (no relief) and parallel to the camera film, the recorded image has near perfect geometric relationship to the scene. Distortion is introduced only by the lens or by the dimensional stability of the film. The angular width of the image, θ, is determined by the width of the film, d, and camera lens focal length, f:

$$\tan \theta/2 = \frac{d}{2f}$$

The magnification (or demagnification) of the image is given by the ratio of the camera lens focal length, f, to the perpendicular distance to the scene, h. The scale is independent of the width of the film.

In vertical imaging from moving aerial platforms, provision must be made for automatic exposure and advancement of film. Each scene point must be recorded in two or more images in succession for stereoviewing or topographic map generation.

Framing cameras may use photographic, photoemissive or photoconductive film. On photographic film, a latent image is made visible through chemical processing, whereas on photoconductive and photoemissive films, the latent image is scanned by an electron beam and the resulting electrical signal is used to generate a visible image on photographic film or on a television-type display.

Airborne framing cameras are almost exclusively photographic, but most space cameras use television systems. Television imaging systems can transmit the electrical output signal to earth for processing to visual images. Unlike photographic film, the photocathode of a television tube is reusable after exposure. Since imaging tubes have small photocathodes (with a maximum available diameter of 5 cm), the total number of resolution elements per frame is quite small compared to a camera system, using, for example, 9 in. (23 cm) film. Images produced on television systems often lack geometrical fidelity. Therefore, most airborne camera systems employ photographic film.

PHOTOGRAPHIC CAMERAS

Most black-and-white or color aerial film is sensitive from 0.36 to 0.72 μm, with aerial infrared film extending this range to 0.9 μm. Because of molecular scattering of the atmosphere at shorter wavelengths, most aerial photographs are taken through minus blue filters, that is, wavelengths shorter than 0.5 μm are absorbed.

Black-and-white photographic film is exposed to light within the bandpass defined by the spectral response of the film and any filter between the film and the scene. The exposure process does not recognize which color produces it; hence, all color information is lost. As a result, two adjacent objects of different color can appear as the same shade of gray in a black-and-white photograph. When this occurs, there is lack of contrast between objects. When a photographic mission is planned for a specific target, the investigator should select a film-filter combination to operate in the wavelength region of maximum contrast between the target and background of interest. For example, consider a mission designed to map soil, vegetation, and turbid water, which have the spectral brightness (a function of the reflectance and spectral radiance of sunlight) shown in Fig. 3.8. Operation in a spectral interval of 0.54-0.68 μm (band 1) is not the region to work in, because the integrated energy from each of these targets is approximately equal and would produce the same exposure on film (similar gray tones). Operation in a narrow band centered about 0.7 μm (band 2) would offer maximum tonal contrast. Vegetation would have the lightest tone on

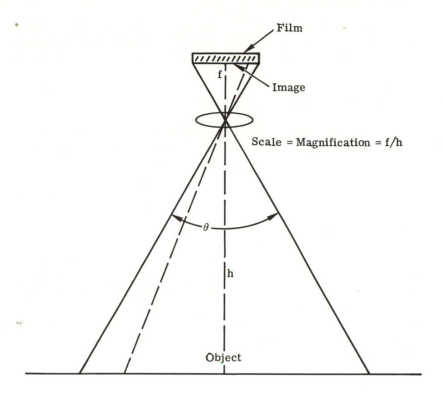

FIGURE 3.7. Geometry of a framing camera.

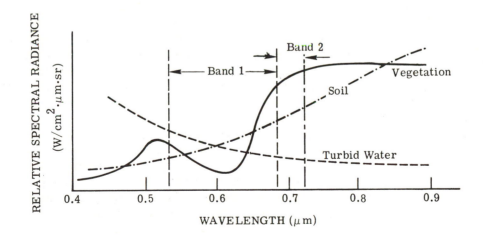

FIGURE 3.8. Relative spectral radiance of commonly found features with bands for minimum (band 1) and maximum (band 2) contrast among them.

a positive print, water would be the darkest.

The more target objects or backgrounds, the more difficult it is to produce the best contrast among them with a single film-filter combination. Therefore, it is convenient to operate several cameras as a unit, each camera being equipped with a different film-filter combination; these are called *multiband* or *multispectral cameras*. Most commonly, multiband cameras have four or nine bands (4, 5). The more bands, the more multiple images the user or interpreter must examine to find the one exhibiting the best contrast for an object or area. This time consuming procedure can be shortened by using a multispectral viewer, which permits registration and simultaneous projection of up to four images. Each image is projected in a different color light, with controls varying the intensity of each lamp independently. The resulting "false color" projected image often contrasts objects to such a degree that they may be identified by color alone. Subtle tonal variations and correlations are enhanced by color projection.

Multiband photography was one of the first techniques used to produce color images. Three-layered color films made this technique obsolete except for unusual situations requiring special film-filter combinations or dynamic ranges of exposure that cannot be accommodated with color film.

Aerial color film can be obtained as normal color (for example, Eastman Kodak Type 8442) or infrared color (for example, Kodak Type 8443) film. Reference (4) describes in detail the principles of operation of these films. In normal color film, a dye transfer technique is used, which makes the blue-exposed layer look blue; the green-exposed layer, green; and red-exposed layer, red. To the viewer, image colors appear to be reasonably faithful reproduction of the scene colors. In infrared color film, however, the colors are false, as a color must be assigned to the layer exposed by the invisible infrared radiation. The film is exposed by green, red, and infrared radiation, and the exposed layers are assigned the colors of blue, green, and red, respectively. Thus, vegetation, which reflects highly in the infrared, appears red on infrared color film.

TELEVISION CAMERAS

Television cameras operate in a black-and-white mode. As a result, color images are produced by the multiband camera technique, that is, signals from three filtered black-and-white camera tubes are mixed at the display to produce color. The spectral response of these tubes is limited to the visible and near-infrared regions, as illustrated in Fig. 3.9, which plots the spectral response of several common tubes.

Television image tubes are classified by the manner the latent image is produced and read out by an electron beam. Figure 3.10 is a schematic representation of a vidicon, one type of image tube, in which an image is formed on a photoconductive target. The target is coated on front with a transparent electrode that serves as the signal electrode. Conductivity of the photocathode target increases with illumination. An electron beam scans the photocathode in a known, methodical pattern. At any instant, the resulting output signal is a function of photocathode conductivity, which in turn depends on the image brightness where the electron beam falls. The image is reproduced in a cathode-ray tube (CRT) by scanning a beam of electrons over a phosphor in a pattern similar to that scanned by the electron beam of the image tube. The intensity of the electron beam is made proportional to the output signal from the vidicon. The light output from the phosphor is, in turn, a function of the electron beam intensity.

As mentioned earlier, television imaging systems have two major limitations compared with photographic systems. Photocathode targets are small; hence, the total number of resolution elements in an image is small. The geometric fidelity of the image depends on the accuracy to which the electron beam position can be controlled and is known, both in the scanning beam of the image tube or the scanning beam of the image reproduction system. The reader is referred to Refs. (6) and (7) for additional details on the operational principle of image tubes and their usefulness in earth observation systems.

LINE SCANNERS

Originally, line scanners were developed to generate imagery outside the spectral region of photographic film. Line scanners were primarily used for thermal infrared imaging, but now they are being used in multispectral sensing programs for the generation of multiband images from space platforms and for automatic classification of features, using spectral pattern recognition techniques.

The principles of airborne or spaceborne optical/mechanical scanners have been treated extensively in the open literature (8, 9) and will only be considered briefly in this chapter. As shown in Fig.

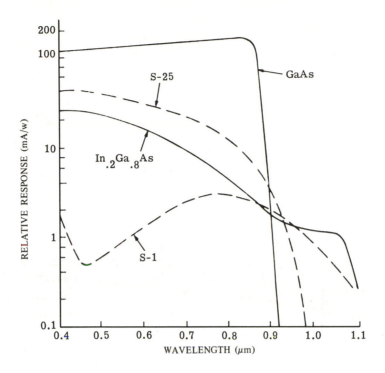

FIGURE 3.9. Spectral response of some photocathodes used in imaging devices (6).

3.11, the field of view of an infrared telescope (radiometer) is reflected from a rotating mirror so that it scans the ground in lines perpendicular to the flight path. The scan rate is adjusted so that succeeding scan lines are adjacent or overlapping as the sensor moves forward. Thus, the rotating mirror provides a scanning motion normal to the vehicle path while the forward motion advances the scan pattern. The output signal from the detector is amplified and used to modulate the intensity of a light source. The result is a pictorial image of the scene radiance.

The spectral region of operation of a line scanner is determined by the spectral transmission of its optics and filters and the spectral response of its detector. Scanners have been designed to operate in the ultraviolet, visible, near-infrared, and far-infrared regions. Figures 3.12 and 3.13 show the spectral response of detectors commonly used in line scanners and that such detectors operate in any of the atmospheric windows from 0.3 to 15 μm. No one detector, however, gives optimum performance throughout the entire region. Accordingly, scanners are usually designed so that the detector unit can be easily changed and the best detector used for a given spectral region of operation. It should be noted that detectors operating beyond 3μm are generally cooled. The longer the wavelength of operation, the lower the temperature of operation.

Since a line scanner generates an image by scanning the scene a line at a time, it is important that the velocity vector and attitude (roll, pitch, and yaw) of the sensor be known throughout the period of scanning if the image is to be reconstructed with geometric fidelity. Indeed, these parameters should be known and constant for perfect imaging. Figure 3.14 shows the rotational axes for roll, pitch and yaw and Fig. 3.15 shows the variations in the scan pattern as these parameters vary. If the writing beam scan pattern is kept constant during the

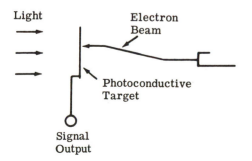

FIGURE 3.10. Schematic representation of vidicon (6).

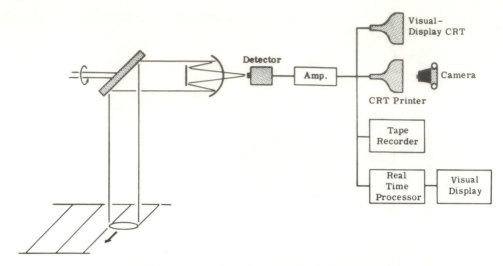

FIGURE 3.11. Schematic of an airborne line scanner.

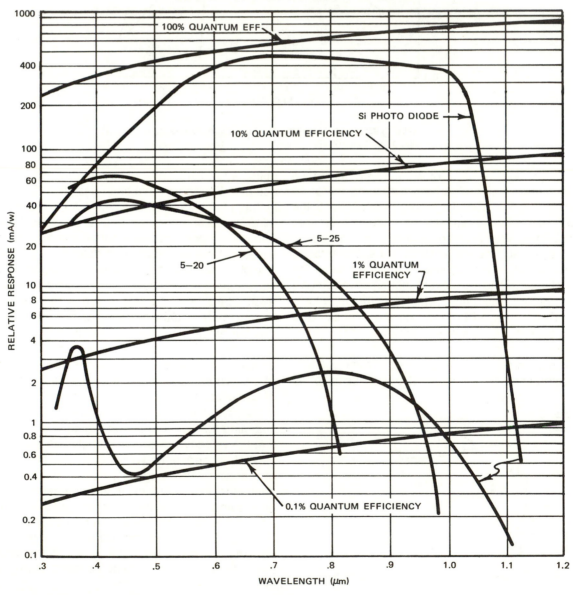

FIGURE 3.12. Spectral response of visible and near-visible detectors (after data from RCA).

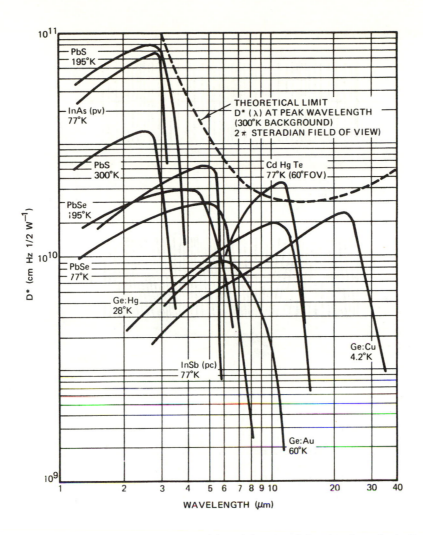

FIGURE 3.13. Spectral response of some infrared detectors (after data from Santa Barbara Research Center).

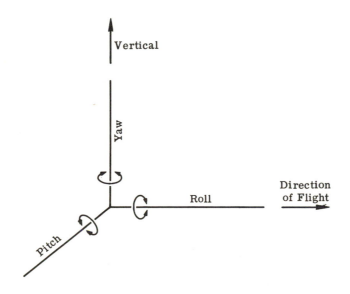

FIGURE 3.14. Rotational axes of roll, pitch, and yaw.

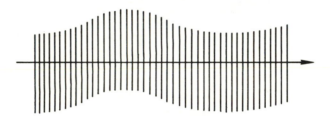

FIGURE 3.15. Gaps in scan coverage of the scene arising from variations in platform motion and attitude.

reconstruction and does not vary analogous to the scene scan pattern, distortions will occur. As seen in Fig. 3.15, if the forward velocity is not perfectly simulated, the image will be stretched or shortened in the flight direction. Pitch also causes stretching or shortening in the direction of flight. Yaw causes smearing or gaps in the data, which increase with distance from the ground track. Roll displaces objects with respect to the ground track; because of it, objects with straight edges parallel to the direction of flight are wavy. All these distortions are evident in scanner imagery made from aircraft; their magnitude depends on the extent of efforts to minimize them. Space platforms can be made remarkably stable, and scanner imagery from space, for example, Landsat, is surprisingly free from geometric distortions.

Images from scanners, like framing cameras, are two-dimensional representations of a three-dimensional scene. Hence parallax distortions, discussed in Section 3.4.3., occur if the scene is not perfectly flat.

MULTISPECTRAL SCANNERS

If a line scanner uses reflective optics, its wavelength region of operation is determined by the detector's spectral response and optical filters inserted in the scanner's optical path. Thus, a scanner can operate in any wavelength region, and multispectral operation is relatively easy to achieve. For example, two-band scanning requires only the addition of a dichroic beam splitter, a second detector, and a second data-recording channel. The bulk of the scanner system (optics, scan mirror

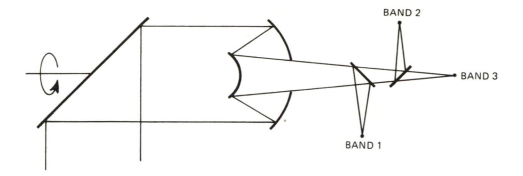

FIGURE 3.16. Multispectral scanner using dichroic beam splitters.

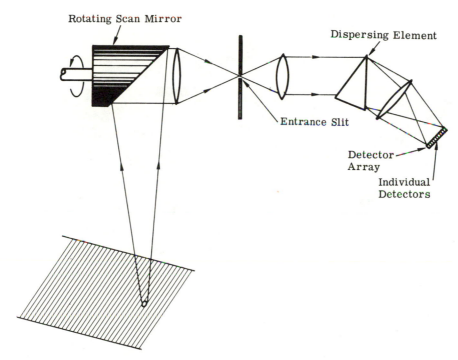

FIGURE 3.17. Schematic of multispectral dispersive scanner.

and motor control, operating console, and so forth) remains unchanged.

Figure 3.16 diagrams a multispectral scanner using dichroic beam splitters. It should be noted that the data output from the scanner is registered in both time and space, that is, radiation from each element of a scene is observed in all bands at the same time. As will be shown, this registration of the spectral information is an extremely important feature of multispectral scanners.

It is difficult to obtain a large number of spectral bands, using dichroic beam splitters. A preferred method of spectral separation uses a dispersing spectrometer. In a single band scanner, as shown in

Fig. 3.11, the detector image is scanned over the ground. For multispectral operation, the detector is replaced with the entrance slit of a spectrometer, as shown in Fig. 3.17. A detector array is placed in the image plane where electromagnetic radiation is dispersed in a spectrum, that is, separated according to wavelength. Each detector "sees" the slit, but each in a different wavelength region. Thus, as the slit image is scanned over the ground, each detector observes the radiation in its proper spectral region.

Figure 3.18 is a block diagram of a research multispectral scanner system. The data from each detector are recorded in parallel on a wideband tape

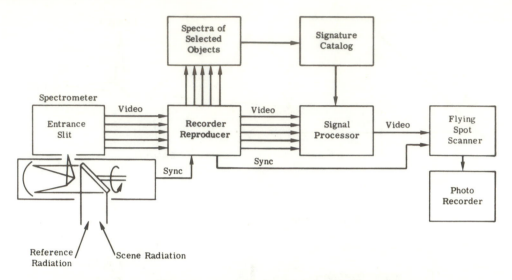

FIGURE 3.18. Schematic of multispectral scanner and data processor (8).

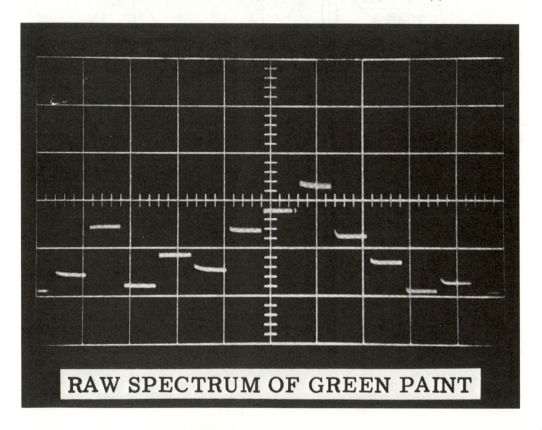

FIGURE 3.19. Raw spectrum of green paint (8).

recorder. If these signals were sampled at any instant, they would represent a raw spectrum of a resolution element of the scene. Figure 3.19 shows a sampled raw spectrum of a green-colored panel, as obtained from a 12-channel scanner; the vertical height represents the signal level from the 12 detectors. In that the signal levels from each detector undergo different amplifications, the raw signals are meaningless without calibration. By observing a source of known radiation character-

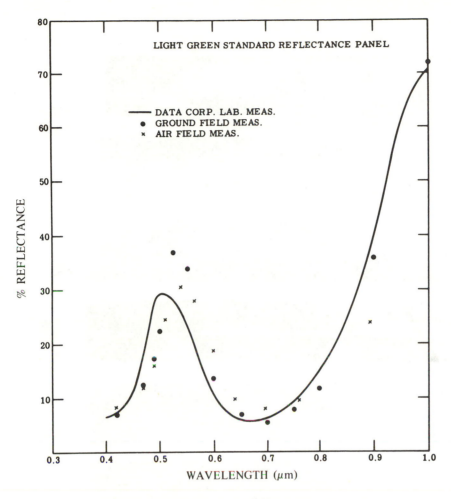

FIGURE 3.20. Reduced spectrum of green paint (8).

istics, the relationship between the output signal level and the radiant power input to each detector can be established to permit calibration. Figure 3.20 shows the reduced reflectance spectrum of the green panel, with reflectance measurements made by laboratory and field (*in situ*) spectrometers. The output signal from each detector element is a video signal corresponding to the scene brightness in the particular wavelength region of operation. These video signals can be used to generate multiband imagery in the manner described for the single band scanner. Figures 3.21 (*a*) and (*b*) show 18 images generated from tape-recorded video signals of a multispectral scanner.

As will be discussed in Chapter 6, tape-recorded multichannel video signals can be displayed and analyzed in a number of ways. The spectral data can be analyzed by high-speed digital computers to obtain summary statistics of the spectral character-

istics of objects and backgrounds and determine how the spectral signature varies with instrumental, operational, and environmental parameters. For example, Fig. 3.22 shows the effect of view angle on relative apparent radiance, in a given wavelength band, of a row of crops. The spectral signatures of objects and their backgrounds permit identification or selective enhancement or suppression of object brightness in a scene. In this manner, the multichannel video data can be fed to a signal processor designed to generate a single video signal; the intensity may be proportional to the probability that the spectrum is that of a given material or terrain.

3.2.3 Microwave Imagers

Like optical imagers, microwave imagers can be classed as active or passive. The active systems are radars; the passive systems are microwave scanning radiometers. These imagers are treated in

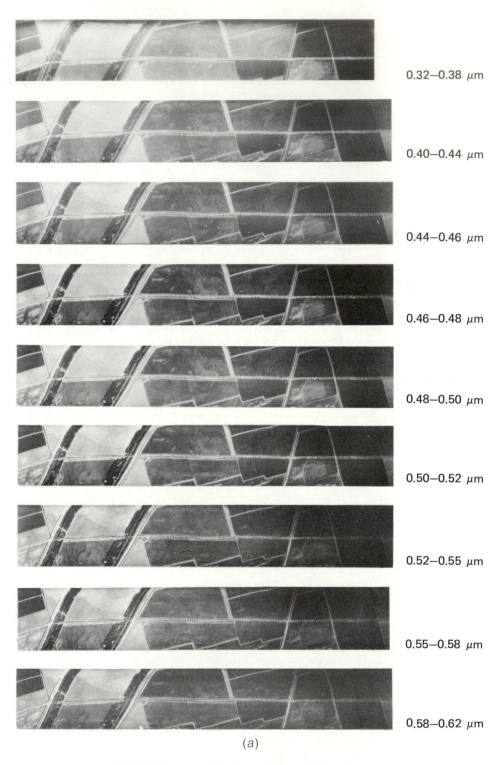

0.32—0.38 μm

0.40—0.44 μm

0.44—0.46 μm

0.46—0.48 μm

0.48—0.50 μm

0.50—0.52 μm

0.52—0.55 μm

0.55—0.58 μm

0.58—0.62 μm

(a)

FIGURE 3.21. (a) and (b) Channels of multispectral imagery.

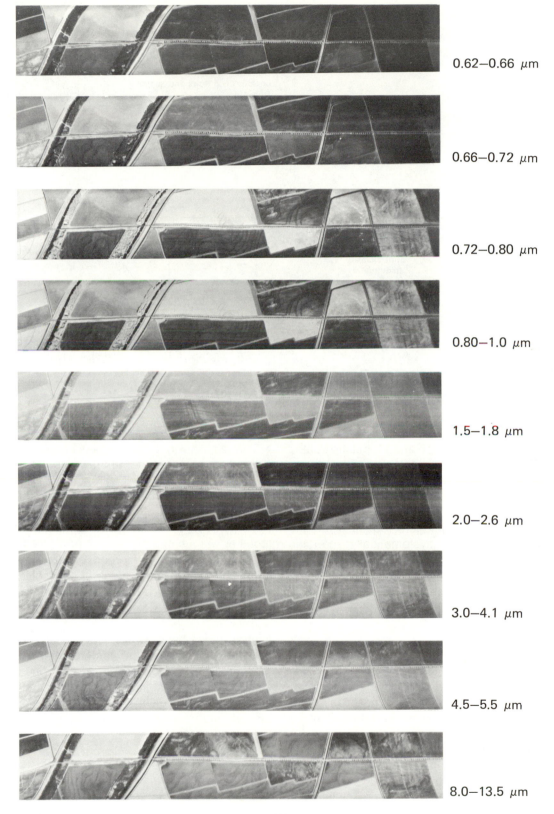

0.62—0.66 μm

0.66—0.72 μm

0.72—0.80 μm

0.80—1.0 μm

1.5—1.8 μm

2.0—2.6 μm

3.0—4.1 μm

4.5—5.5 μm

8.0—13.5 μm

(b)

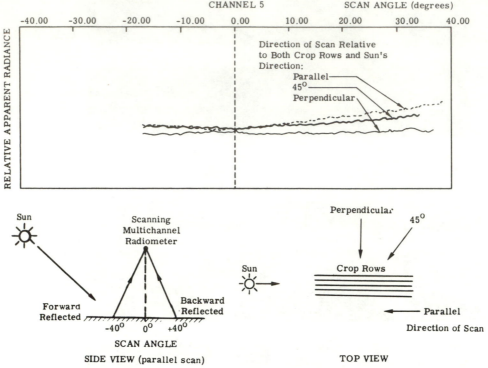

FIGURE 3.22. Effects of scan angle and scan direction.

greater detail in Chapters 10 and 11, but are treated here for completeness and comparison with other imagers.

PASSIVE MICROWAVE IMAGERS

Passive microwave imagers resemble the line scanner because they systematically scan the scene mechanically or electronically, and the output signal is used to construct a visible image. There are some notable differences in these systems, however, in components, performance, and terminology:

- The collector is an antenna rather than an optical system.
- The detector is a receiver.
- The spectral bandpass is determined by electronic filtering in the amplifier rather than by optical filtering of the radiation.
- The radiant energy being detected is low level compared with that detected by optical sensors.
- The longer wavelengths of microwaves require large diameter antennas for high resolution.

- The relatively long microwaves have a far superior ability to penetrate atmospheric clouds.

The impact of these major differences has hindered the operational use of passive microwave imaging systems. The potential of all weather operation together with unique microwave properties of matter has encouraged numerous research investigations, particularly in the areas of soil moisture, ice mapping, and surface conditions of water such as oil pollution and sea state (10).

Two major differences between an infrared imager and a passive microwave imager are angular resolution and the factors that determine measured surface temperature. The fundamental limit in angular resolution of an electromagnetic imaging system is given by the diffraction limit. For a circular entrance aperture, this limit is given by the Rayleigh criterion,

$$\theta = \frac{1.2\lambda}{D} \qquad (3.1)$$

where θ is the angular resolution limit in radians, λ is the wavelength of the radiation, and D is the collector diameter.

Infrared systems usually operate in bands centered at 11 μm or shorter wavelengths, while passive microwave systems operate at 3 mm or longer. Thus, the microwave wavelengths are at least 270 times longer than infrared wavelengths used for thermal imaging. An imager with a 20-cm aperture has a limiting resolution of about 66 μ rad (13 arc sec) for the 11-μm region and 18 mrad (1 deg of arc) for the 3-mm region.

The radiation from any object is determined by its temperature and its emissivity. For an opaque object, the emissivity is one minus the reflectance. The dependence of the spectral radiant emittance on temperature and wavelength is given by Planck's radiation law (8). To a first order approximation, the radiant emittance, W_r in the infrared (10-μm region) and microwave regions is given by the following equations:

$$W_r \text{ (infrared)} = K_1 \; \epsilon \; T^5 \text{ w/cm}^2$$
$$W_r \text{ (microwave)} = K_2 \; \epsilon \; T \text{ w/cm}^2$$

Since emission in the microwave region is linearly dependent upon both temperature, T, and emissivity, ϵ, fractional changes in these two parameters have equal effects on the radiant emission. A 10% change in emissivity between objects at room temperature (300°K) has the same effect as two identical objects with a 10% temperature difference (30°K). On the other hand, in the thermal infrared wavelength region, where the signal varies with T^5, a 10% change in emissivity corresponds to a 2% change in temperature, a 6°K temperature difference. Thus, microwave radiation measurements of surface temperature are more strongly influenced by changes in surface reflection (emissivity) than infrared measurements. Since reflection from smooth surfaces is highly polarized at non-normal angles (10), the microwave signal is highly dependent on polarization since many surfaces are smooth at these wavelengths.

Figure 3.23 shows images taken through heavy clouds with a passive scanning microwave radiometer (MICRAD) operating at 33.6 GHz (11). A visual photograph of the scene is included for comparison. The airborne scanner, with resolution of 1°, uses three rotating parabolic antennas and scans a 120° swathwidth. Notice the ability to distinguish terrain features such as roads, buildings, parks, and rivers.

ACTIVE MICROWAVE IMAGERS

Active microwave imaging systems can discern properties of targets and backgrounds which cannot be observed using passive techniques, and they generally have higher spatial resolution than their passive counterparts. These imagers are called *radars* (an acronym of *ra*dio *d*etection *a*nd *r*anging). Radars most commonly used in earth resource applications observe the terrain in a direction normal to their motion. Two-dimensional imaging of these *side looking* airborne systems (SLAR) requires forward motion of the sensor platform, as shown in Fig. 3.24. Two configurations are used in earth resources: real aperture SLAR systems and SAR (*Synthetic Aperture Radar*). The resolution achievable by SAR far exceeds that of the SLAR, particularly at long ranges.

The principle of operation of real aperture SLAR can best be understood with the aid of the schematic shown in Fig. 3.24 and the block diagram of Fig. 3.25. As seen in Fig. 3.24, a transmitter sends out a pulse through an antenna with a beam pattern of small azimuth, θ, but large elevation, ϕ. The azimuth beamwidth determines resolution in the direction of flight; resolution normal to the flight path (range resolution) is determined by return time of the pulse. As in all electronic imaging devices, the return signal is made visible by a display or recording system with a scan raster pattern that is an analog of the scene scan pattern (see Fig. 3.25). As in optical systems, the ground resolution in the direction of flight, Gr, is a function of range, R, that is, $Gr = R\theta$. Resolution perpendicular to the flight path is determined by the system's ability to resolve the return signal as a function of time. As shown in Fig. 3.24, a time resolution corresponding to a transmitted wavefront resolution of ΔR represents a ground resolution, d, where $d = \Delta R / \sin \phi$. Thus, as the elevation angle ϕ approaches zero, $\sin \phi$ approaches zero, and the ground resolution, d, approaches infinity. Therefore, range resolution improves as elevation angle is increased. Hence, the sensor looks to the side of the flight path rather than vertically as do optical imagers.

While range resolution improves as the view angle, ϕ, from the ground track (nadir) increases, the return signal decreases since the energy received from an extended source is proportional to the inverse square of the range. Furthermore, the total energy transmitted and received from a scene

MICRAD

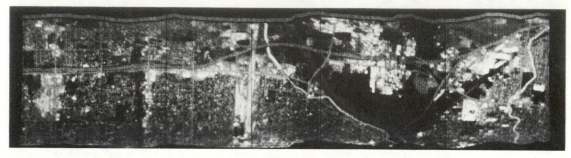

MICRAD

VISUAL PHOTOGRAPH

FIGURE 3.23. Micrad images of Bakersfield, California through 460 m of clouds (Altitude, 1070 m; slant range, 1220 to 2440 m).

element is directly proportional to the pulse strength and repetition rate. The pulse repetition rate varies inversely with range of operation, since all return signals from an outgoing pulse must be received before sending out another in order to avoid simultaneous returns from multiple scene elements. These factors require the designer to make trade-offs among performance parameters such as range of operation, depression angle, swathwidth, and ground resolution. An important feature of SLAR to the geologist is that relief is accented by shadowing normal to flight direction, an effect discussed in Section 3.4.3 and Chapter 10.

Microwave sensors can operate at different wavelengths. A surface that may appear smooth at a given wavelength may appear rough at shorter wavelengths. Radars with wavelengths ranging from 1 to 133 cm have been developed. Often general operational wavelength regions are indicated by *radar bands* identified by capital letters, as illustrated in Table 3.1.

SAR (*Synthetic Aperture Radar*) overcomes the angular resolution limitation of conventional radars (sometimes referred to as brute force radars), which is set by the diffraction limit. SAR takes advantage of the coherency of human-made microwaves. By recording the intensity, amplitude, and phase of the return pulse, sophisticated optical and/or electronic processing can reconstruct the signal from a small moving antenna to simulate that which would

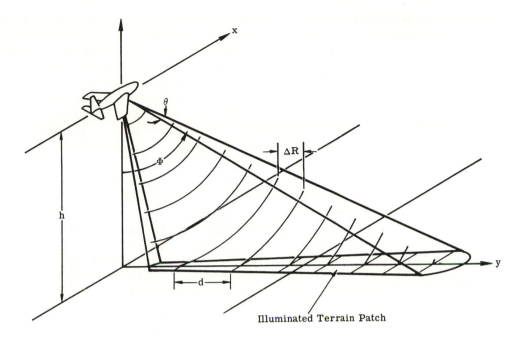

FIGURE 3.24. Viewing geometry of a side looking radar.

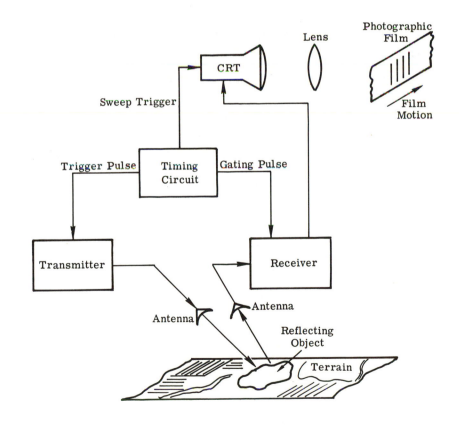

FIGURE 3.25. SLAR block diagram.

TABLE 3.1 RADAR BANDS AND THEIR ASSOCIATED FREQUENCIES

Band	Nominal Frequencies (MHz)	Nominal Wavelength (cm)
P	200-390	133-77
L	390-1,550	77-19
S	1,500-5,220	19-5.8
C	3,900-6,200	7.7-4.8
X	5,200-10,900	5.8-2.7
K	10,900-36,000	2.7-0.83

be received from a large antenna, provided the precise position of the moving antenna is known. It has been shown (12) that the limiting ground resolution, *Gr*, of SAR is given by

$$Gr = D/2, \tag{3.2}$$

where D is the antenna diameter.

Thus, the theoretical limit of ground resolution is independent of range and improves as the antenna is made smaller. Since amplitude of the return signal varies inversely with the square of the range and directly with the square of the antenna diameter, there are limits to the conditions under which the theoretical limit of resolution can be met. These limits are set by engineering capability, so that in principle (and practice) they can be met even from space.

3.2.4 γ-Ray Spectrometer

A spectral region of interest to geologists outside the optical and microwave region is that of shorter wavelengths associated with radioactivity. Current state-of-the-art of technology does not permit airborne imaging in the conventional sense, only measurement along a ground transect. A low resolution image can be reconstructed, however, from data collected by an airborne radiometer flying systematic transects over an area.

The principle of γ-ray spectroscopy is based on the fact that light is created when γ rays are absorbed by crystals such as thallium-activated sodium iodide. The amplitude (intensity) of the light is a measure of the energy (wavelength) of the γ rays, whereas pulse frequency is a measure of the amount of radioactivity. Figure 3.26 is a schematic of a γ-ray radiometer. A ring of photomultipliers

observe the light pulses induced in the crystal by the absorbed γ rays, and the amplitude and frequency of the pulses are measured by a wave-height analyzer. In practice, multiple crystals may be required to improve accuracy by increasing the signal-to-noise ratio; a reference crystal shielded by lead to block out ground radiation is used to record ambient γ rays emitted from radioactive gases.

3.3 ANCILLARY INSTRUMENTATION USED WITH IMAGING SENSORS

A number of subsystems are required for sensor operation and the most efficient generation of images. These instruments and subsystems are quite numerous and have varied functions, but for image interpretation and utilization, it is not essential that the user be familiar with all these functions. For example, a photointerpreter may not need to know the technical details of exposure for aerial photographs, but he does want the vendor to supply him with properly exposed images. He should be familiar with the various sensor instrumentation used for making the delivered product. As with most services, increased or tighter performance specifications require increased data collection and/or reproduction sophistication. This sophistication is generally, but not necessarily, reflected as increased cost. This section will briefly treat the various functions and subsystems found with imaging sensors. Because the same function may be performed differently in each of the sensors, a detailed description of them and their many associated subsystems is beyond the scope of this chapter. Therefore, only the following functions and techniques will be described in a rather general manner:

- Exposure control
- Stabilization and navigation
- Telemetry and magnetic tape recording
- Generation of imagery from electrical video data

3.3.1 Exposure Control

In framing cameras using film or image tubes, the exposure time determines (1) how much energy reaches the detector, (2) where the center of the image will be on the frame, and (3) how much smear will occur during the exposure if the system is moving or rotating and does not have image motion

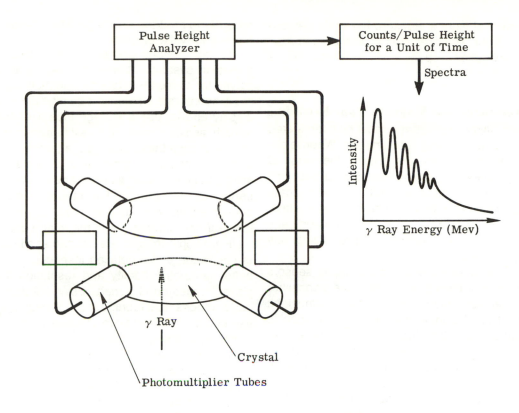

FIGURE 3.26. Schematic of a γ-ray spectrometer.

compensation. Thus, airborne and spaceborne cameras require manual or automatic exposure metering and intervalometers to set the time automatically between exposures and among cameras (if more than one is being used) to assure correct overlap between successive frames. Reference (13) gives a complete list of photographic cameras and camera accessories, with a description of their physical, electrical, and performance characteristics.

For electronic imagers, care must be taken to assure that the signal amplitude is greater than the system noise and that the signal level remains within the dynamic range of the signal handling electronics, including the recording systems. Detectors of line scanners and microwave receivers have large dynamic range, whereas most image television tubes readily saturate or bloom (spread in resolution) when their limited dynamic range is exceeded. For systems having large dynamic range, digital recording on magnetic tape is preferred to provide the required dynamic range in signal handling.

3.3.2 Stabilization and Navigation

All imaging sensors are designed for one axis to be vertical and another to be aligned in the direction of motion. This condition can be maintained if the sensor is installed on a gyrostabilized platform and its forward axis rotated into alignment with the forward motion of the system as determined by a drift meter. With some sensors, economic consideration prevents gyrostabilizing the entire sensor. In this case, the attitude variations (roll, pitch, and yaw), as measured by a reference gyro, can be used in the electronic image processing to correct for these variations. In the "after-the-fact" correction method, some areas may be underscanned and data may be missing, as illustrated in Fig. 3.15.

Navigation aids are particularly important when flight lines must be laid down with precision. This is always the case when a large area is to be covered by a mosaic. Optical scanners and microwave imagers need precise knowledge of the vehicle's ground velocity and altitude if the resulting image is to have the proper aspect ratio, that is, the scale in the direction of flight is equal to the scale perpendicular to flight. The precision of attitude stabilization and navigation in space systems is excellent, as exemplified by Landsat.

3.3.3 Telemetry and Magnetic Tape Recording

Imagery taken from space must be telemetered to Earth for processing into images. When electronic imaging systems are outside the range of receiving stations or are mounted in aircraft, the usual mode of operation is to tape record the data for transmission and/or image processing at a later date. Depending on data requirements for accuracy and dynamic range, telemetry and recording may be analog or digital.

Telemetry of video data from imaging sensors requires wide bandwidths. Hence, the transmitter operates at very high frequencies and must be within the line-of-sight of the receiver. The specific bandwidths attainable depend on the transmitter power, range, and diameter of the receiving antenna. The Landsat system is typical of the state-of-the-art of technology. The data telemetry links of Landsat use 20-w, S-band transmitters with either an analog FM bandwidth of 3.2 MHz (megacycles per second) or a 15 Mbps (megabits per second) digital data stream (14). A 10 m diameter antenna can lock on and receive the data as the satellite rises above the horizon at a range of 2900 km.

Data recording on magnetic tape is usually on wide-band instrumentation tape recorders. When recording in analog mode, FM recording is preferred to preserve low frequency information in the imagery. These recorders are usually 7-track with ½-in. tape and 14-track with 1-in. tape. The tape speed is 120 inches per second (ips) or reduced multiples of two, that is, 60 ips, 30 ips, 15 ips, and so forth. At the maximum speed of 120 ips, the FM bandwidth is 500 kHz per track and digital data can be recorded at 10,000 bits per inch (bpi) or greater per track. These digital tapes are known as high density digital tapes (HDDT) in contrast to computer compatible tapes (CCTs). At a tape speed of 120 ips, the 14 tracks can record better than 17 Mbps (megabits per second).

3.3.4 Generation of Imagery from Electrical Video Data

Imaging devices whose output signal is in electrical form must use a visual display or facsimile recorder to produce an image. The images may be black-and-white or color.

Display devices are similar to those used in television systems. When an image is to be displayed for a time for detailed inspection, the display must incorporate a data storage medium for refresh display and/or use a storage tube. Generally, the refresh display is preferred, as it has greater resolution and brightness than storage tubes or long persistent phosphor tubes. Image display devices are particularly useful when image analysis is to be performed by computer. The interactive display is a convenient and desirable means of communication between the computer and the image analyst.

"Hard copy" images are needed as a final record of the image or for detailed inspection; they have better geometric fidelity and resolution than displays. Images can be made using electronic, optical, or mechanical techniques. All imaging systems use one or a combination of these techniques.

Perhaps the most common image producing technique is the CRT (cathode-ray tube) strip printer, as illustrated in the radar system diagram shown in Fig. 3.25. In this system, an electron beam is scanned over a phosphor screen in a manner similar to that of the sensor scanning the Earth. The intensity of the electron beam is modulated by the signal from the sensor so that the output light from the phosphor is proportional to the sensor signal. In the infrared thermal mapper, the brightness of the image is proportional to the thermal emission, which is often directly related to surface temperature; in SLAR, it is proportional to the radar return. The line swept by the CRT is imaged on film moving at a rate to simulate the sensor moving over the Earth. The resulting image is an endless raster or strip map. However, one can artificially frame the data to make the format compatible with framing devices, and this is done with the Landsat scanner.

As an alternate to the above technique, the CRT can be deflected in two dimensions as in a display device. The horizontal sweep simulates the lateral scan of the sensor, while the vertical sweep simulates its forward motion.

Since phosphors bloom at high intensities and resolution is degraded, better performance can be obtained by writing directly on film with an electron beam. However, to obtain this improvement, the film must be placed in vacuum, as electron beams cannot be maintained in air because of collision and scattering by air molecules. Landsat images made

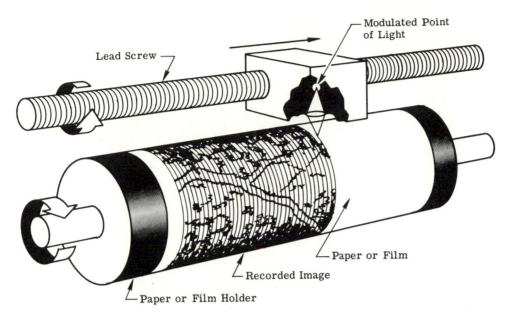

FIGURE 3.27. Schematic of a drum recorder.

by NASA are recorded on an EBR (*Electron Beam Recorder*).

Other recording techniques are variations of the drum recorder. Figure 3.27 is a schematic of a drum recorder. Film or paper is placed on a drum and a modulated light beam is focused on it. As the drum rotates one revolution, either the film or light source is advanced one scan line, usually by a screw mechanism. The light source can be a glow modulator, a light emitting diode, or a laser. For color, multiple light sources may be used.

Not all drum recorders require photographic film. One highly practical color recorder uses three colored ink jets directed at the paper on the drum; the ink flow is modulated by electrostatic deflection. With this process, which is not light sensitive and can operate in a fully illuminated room, a color image of an entire Landsat frame can be made in a matter of minutes.

3.4 FACTORS THAT AFFECT DATA UTILITY

An image is a picture of the radiance distribution of a scene. There are a number of system parameters, design factors, environmental parameters, and operational parameters which may affect or distort the geometric and radiometric quality of the image. This section discusses these factors so that

the user can specify his requirements to assure that the resulting data will meet his needs.

3.4.1 Performance Parameters That Affect Data Utility

Good imaging systems abound. A perfect imager will never exist, because there are a number of components in the imaging system which limit the accuracy of the process that transforms the "real world" scene to an image. For example, all images have degraded resolution. In a photographic camera, this resolution may be set by aberrations in the lens, but more likely it will be set by the structure (graininess) of the film. Resolution in television-type sensors is limited by a composite of degrading elements beginning with lens aberrations, followed by graininess in the photocathode, size of the electron beam readout, electronic bandwidth in the data handling chain, and element resolution of the image recording process.

Analysis of other sensors can quickly uncover similar factors. In the infrared line scanner, the major factors determining the resolution limit are the angular size of the detector, and the spot size of the film recorder. In microwave systems, the angular resolution limit is set generally by the diffraction of the collector aperture. In real aperture SLAR, it is the diffraction limit of the antenna in the flight direction and the range resolution normal to the flight path.

Because resolutions of the various sensors have different origins or depend on different parameters (for example, the resolution of television vidicon tubes depends on the illumination level at the photocathode, and operator adjustments), various ways of expressing resolution capability have evolved. Photographic systems use line pairs per millimeter; television systems are often described by the number of TV lines scanned and displayed. Line scanners use the instantaneous field-of-view, IFOV, which is the angular subtense of the detector. With Landsat data, many users refer to the ground resolution in terms of the picture element, or pixel, an area that may be defined as the product of the distance between samples and the width of the IFOV projection on the ground.

The different means for describing resolution among the sensors should not be a problem for the user if he understands what effect the described resolution has on the quality and usefulness of his imagery. It is decidedly a problem when one attempts to compare the performance capability of two different sensor types, for example, photographic versus television. Instrument designers and manufacturers are aware of this problem and have proposed a more uniform means of describing resolution. It now appears that the measure most likely to be adopted is the modulation transfer function, MTF. Roughly, the MTF describes the fidelity of representation of features of different sizes, or different spatial frequencies. Its use is elaborated in Chapters 5 and 6.

Another parameter affecting data utility is often called "sensitivity." In photographic systems, sensitivity may be made up of terms involving film speed, exposure, and gamma (a measure of film transfer characteristics). Sensitivity is related to the ability to detect two *large* objects with small differences in reflectance. Note the term "large objects." For small objects, this parameter quickly becomes dependent on resolution.

In electronic systems, the sensitivity is expressed in terms of signal-to-noise ratio, S/N, for a given set of radiance levels. Although this description may be useful to engineers, it has little value to the user. Therefore, a set of closely related terms has been devised to describe the performance capability of a sensor:

- *Noise Equivalent Temperature Difference, NETD or NEΔT:* This term is used for thermal and micro-wave scanners. It is defined as that temperature difference between two adjacent, large area black bodies at about 300°K, which will produce a signal difference just equal to the system noise. Figure 3.28 illustrates this concept.

- *Noise Equivalent Reflectance Difference, NE$\Delta\rho$:* This term is used for imagers using reflected radiation. For this term to be meaningful, the level of illumination must be specified (for example, sun angle of 45°, atmospheric transmission of 80%) and computed for reflectance differences of objects having an average reflectance of 20%. NE$\Delta\rho$ is that reflectance difference between two objects which produces a signal difference equal to the noise.

3.4.2 Radiometric Fidelity and Calibration

The photointerpreter often attempts to extract information from an image, using subtle tonal or color differences. The ability of a satellite to collect imagery over a large area with near identical viewing and illumination conditions improves the ability to observe and interpret these fine differences. With the advent of multispectral cameras and scanners, the ability to classify terrain features based on color and tone is now being realized, both manually and with computers.

To classify a scene based on reflected or emitted radiation, the sensor response to radiation must be precisely known, as well as the effects introduced by the atmosphere (path absorption, scattering, and radiance) and geometry of illumination and viewing. Because of variability in illuminating and viewing conditions, as well as stability of the sensor response, the sensor must be calibrated frequently during operation.

To calibrate a sensor radiometrically, the sensor must view a minimum of two known sources of radiation through its complete set of optics. These sources must be viewed periodically with a frequency sufficient to detect both short- and long-term changes in responsivity. In the case of Landsat the detectors observe a calibration light source and dark level once per scan line and the sun once per orbit. The stability of solar irradiance is used to detect and measure the stability of the calibration light source.

Once the calibration signals are recorded with the scene signals, their relative amplitudes are estab-

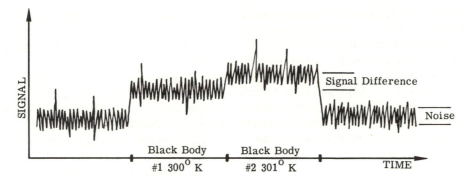

FIGURE 3.28. An A-Scope presentation of the signal from a scanner having a NETD of 1°K as it scans two black bodies in a uniform background.

lished, As long as the data handling processes (such as amplification) are linear or known, the relative amplitudes of signals will be maintained and the resulting tones in the image can be related to the radiance or irradiance at the entrance aperture of the sensor, as determined by reference to the calibration source radiation.

3.4.3 Photogrammetric Fidelity

In many applications, the investigator would like to make maps or overlay the resulting data on maps. These applications require that geometric distances between points in an image be proportional to the actual distances in the scene. To obtain this geometric accuracy in the images, a number of instrumental, operational and environmental conditions must be met. Rather than discuss the particular geometric errors in each type of imaging system, the cause and sources of such errors will be discussed in general terms.

Sensor Attitude

All sensors are designed with an axis that should be maintained vertically. If it is not maintained vertically or is not carefully compensated for in image reconstitution, serious geometric distortions can arise. For example, the scale of an image made with a vertical camera is given by the focal-length divided by the altitude, as illustrated in Fig. 3.7. From geometrical considerations, the ratio of the distance of a point in the image plane from the lens to the distance of the corresponding point in the scene is constant. When the camera is not vertical, as in an oblique image, the scale (or magnification) is variable throughout the image and depends on the position within the image as shown in Fig. 3.29.

Velocity-to-Height Ratio

For scanners and radars, the accuracy of the scale in the direction of the flight path is no better than the accuracy to which the velocity-to-height ratio of the sensor is known.

Lens Distortion

Lenses can introduce geometric aberrations known as pincushion and barrel distortions.

Scan Motion Stability

The scan pattern used in scanning a scene must be precisely known and simulated in the image recording process. Any variations in the scan pattern which are not accounted for will introduce geometric displacement or distortion. This is true for systems using mechanical or electron beam scanning in the imaging or recording processes. For example, the image processing facility for Landsat multispectral scanner accounts for variations in scan mirror velocity from scan line to scan line (14).

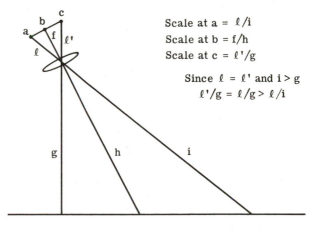

FIGURE 3.29. Illustration of variable scale in an oblique image.

The Landsat Vidicon system has reseau marks (crosses) etched on the face of the vidicon tube to identify geometric errors in the electron readout (14). In the vidicon, the scene is imaged over the fixed pattern of reseau marks; the two are merged on the photocathode. Any distortion introduced by the scan beam will distort both the image and the reseau pattern. Measures discussed in Chapter 4 to restore the reseau pattern to its original configuration will also correct the geometric distortion in the image.

Time

In electronic sensors, the signal with respect to scan position is handled as voltage varying with time. In SLAR, distances normal to the flight path are measured as time of return of the transmitted pulse. Errors in time introduced by the sensor or recording system correspond to errors in distance in the image.

Parallax Distortion

Parallax distortion occurs whenever the scene has vertical relief. As its name implies, the magnitude of this distortion or displacement varies with the look angle. All imaging sensors exhibit parallax distortion, but the magnitude and direction vary. The distortion can be attributed to the projection of an image of a three-dimensional scene on a two-dimensional (flat) film. An object above or below the average object plane will be displaced in the image plane. Figure 3.30 illustrates this displacement for a framing camera; displacement in the image plane is proportional to the height of the object and the tangent of the view angle as measured from the vertical. In a framing camera, buildings appear to lean radially outward from the center of the picture.

In scanners, parallax displacement also causes elevated objects to lean outward. Because the ''look'' (scan) angle always lies in a plane normal to the flight direction, objects lean outward (perpendicular) from the flight line, which would be a line down the center of the image.

In SLARs, parallax displacement causes elevated objects to lean toward the flight line of the radar. As seen in Fig. 3.31, the spherical wavefront strikes the top of the building at the same time it strikes the ground at position A. The two return pulses arrive at the receiver at the same time and are recorded as the same spot in the image. Sim-

ilarly, a point halfway up the building is imaged as the point B on the ground. The base is imaged as the correct position. Notice that the building shadows the ground behind it, and the radar receives no signal until point C on the ground is observed. Because the flight line of the SLAR is off the image, all elevated objects are displaced in one direction, toward the flight line.

It is interesting to note the difference between the sensors with regard to shadows. In sensors that observe sunlight, the shadows depend on sun angle. To avoid too much confusion in the scene, most photography is done with high sun angles. Thermal (infrared and microwave) scanners do not have shadows. SLAR images, as stated above, always have shadows normal to the flight path.

The parallax distortion of cameras, scanners, and radars are illustrated in Fig. 3.32. The original scene consists of an array of 13 flagpoles spaced as shown in Fig. 3.32(a). As discussed in the preceding paragraphs, the poles in a framing camera lean radially away from the center and the displacement of the tops increases with angle [Fig. 3.32(b)]. With scanners, the poles lean outward from a line down the center of the stripmap, and the displacement of the tops increases with distance from the flight path [Fig. 3.32(c)]. In SLARs, all the poles lean toward the flight path [Fig. 3.32(d)].

Parallax displacement provides a means for obtaining three-dimensional data from two-dimensional images. The conditions required for measuring height of stationary objects are that the objects be observable (contrasted) in two images taken from a fixed, known height with a known separation distance between exposures. As shown in Fig. 3.33, the height of an object above a datum plane (in Fig. 3.33, this plane passes through the base of the tree) is given by

$$h = \frac{H\,dP}{P + dP} \tag{3.3}$$

where

h = height of object
H = height of camera above the base of the object
P = separation distance between exposures
dP = total parallax displacement in the scene

The variables can be measured in the images; the actual height can be computed by dividing h by the scale (magnification) of the image. In di-

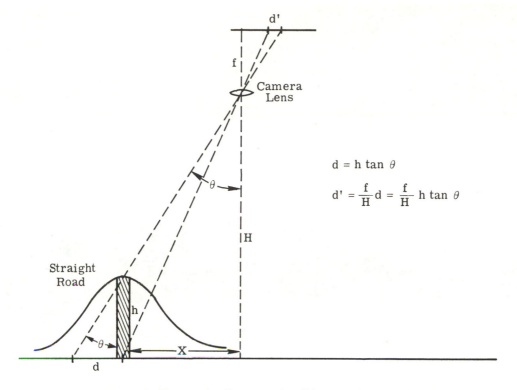

$$d = h \tan \theta$$

$$d' = \frac{f}{H} d = \frac{f}{H} h \tan \theta$$

(a) Elevation Distortion in Photography. Top mountain will image as if it were "leaning" away distance d from its base as shown in (b) below.

(b) Off-Axis perspective view. "Courtesy of W. Fisher, USGS"

(c) Near-Axis Perspective view. "Courtesy of W. Fisher, USGS"

FIGURE 3.30. Image distortions resulting from nonlevel terrain.

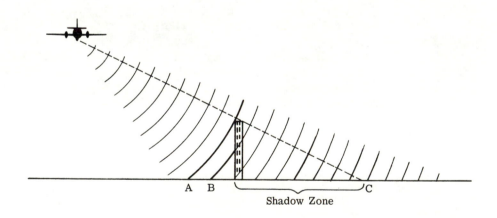

FIGURE 3.31. Parallax displacement in SLAR.

viding, remember that the scale is a ratio and hence a fraction of one. For a scale of 1:24,000, division by the scale is the same as multiplication by 24,000.

Two overlapping images are known as stereo pairs. These are obtained in aerial photography by setting the time between exposures to provide the desired overlap (usually 60%) in the direction of flight. Only the component of parallax displacement parallel to the direction of camera motion contributes to the determination of height (15), that is,

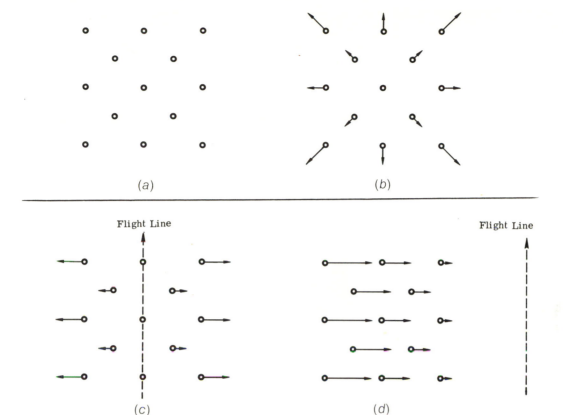

FIGURE 3.32. Relative parallax displacement of an array of flagpoles as observed with cameras, scanners, and SLARS.

dP_1 and dP_2 in Fig. 3.33 are the vector components of parallax displacement in the flight direction.

Three-dimensional viewing of stereo pairs is possible by viewing the images in a manner analogous to their generation, that is, the eyes represent the two camera positions and the images the scene. Techniques and aids for stereoviewing and plotting are described in the *Manual of Photographic Interpretation* (15) at a suitable level of understanding for the general user or photointerpreter. The *Manual of Photogrammetry* (16) should be read by those interested in precision mapping or measurements.

In principle, stereo pairs can be obtained with scanners and radars. The image pairs are obtained by overlap between adjacent, but parallel, flight lines. The separation distance between flight lines is more difficult to obtain and control than distance along a flight line. For this reason, quantitative stereoplotting is not commonplace with scanners and radars. The geometric relationship for obtaining height from stereo pairs of photographs is not valid for scanner and radar imagery.

3.4.4 Relative Merits of Space Versus Airborne Platforms

Except for synthetic aperture radar (SAR) systems, the ground resolution of all the sensors described in this chapter is inversely proportional to the distance of the sensor from the scene. Accordingly, spaceborne imagers operating at distances of hundreds of miles cannot begin to match resolutions attainable with airborne sensors without increasing the size of the sensor and/or introducing complex engineering modification such as the use of multielement detectors in a line scanner. The 80-m ground IFOV of the Landsat multispectral scanner requires an 8-in. (21-cm) diameter aperture and six detectors per channel. In contrast, a 5-in. (13-cm) aperture airborne scanner flying at an altitude of 5 km would have a ground resolution of 10 m using only one detector per channel.

Despite the fact that spaceborne imagers have poorer resolution than airborne sensors, they have numerous advantages. Some of these are the following:

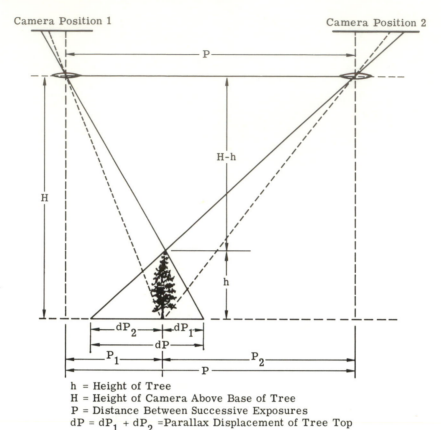

h = Height of Tree
H = Height of Camera Above Base of Tree
P = Distance Between Successive Exposures
$dP = dP_1 + dP_2$ = Parallax Displacement of Tree Top

Triangles defined by solid lines are similar, hence

$$\frac{h}{dP} = \frac{H - h}{P} ; \quad \text{or} \quad h = \frac{HdP}{P + dP}$$

FIGURE 3.33. Geometry and derivation of height to parallax relationship.

- Large area coverage is achieved with small look angles. Landsat has a 185-km swathwidth with a ± 6° scan angle. Parallax displacement with a view angle this small is negligible. Only two or three geometric control points may be required to convert 185 X 185 km image into a photomap of scale 1:250,000.
- The large area coverage with fixed illumination angle and near-normal viewing permits meaningful analysis of color and tone in the image.
- The fast moving platform provides frequent global coverage.
- Once in orbit, the cost for data collection is relatively small.

3.5 OPERATIONAL SENSORS AND DATA PROCUREMENT

Except for aerial photography, remote sensing is an emerging technology, and would-be users often find it difficult to obtain imagery or information about the characteristics of sensors and imagery. In this section, the performance characteristics of air and spaceborne sensor systems having data available to earth resource investigators are given as well as where and how one can obtain remote sensing data. Particular emphasis will be placed on Landsat multispectral scanner (MSS) data, because repetitive coverage of large portions of the Earth makes this system truly unique and invaluable.

3.5.1 Aircraft Sensors

Aerial photographic services can be procured on a routine basis. Conventional aerial camera systems have been tabulated by the military (13) and in manuals of the American Society of Photogrammetry. The *Manual of Photogrammetry* (16) discusses how to plan and estimate the cost of photographic missions. It also discusses procurement methods and provides a sample procurement form detailing performance specifications and acceptance criteria. The *Manual for Color Aerial Photography* (4) gives detailed descriptions and specifications for the more commonly used aerial cameras and color films. Appendix 1 of this manual treats multiband photography and color additive viewers for viewing and interpreting multiband images. Description of more recently developed multiband cameras and low cost "do it yourself" systems (17, 18) can be found in journal articles. The *Manual of Remote Sensing* (5) gives a far more comprehensive treatment of remote sensors than any other single publication. It treats cameras, scanners, and radars, both from aircraft and spacecraft installations.

Tables 3.2 and 3.3 tabulate some performance characteristics of a number of commercially available airborne infrared and multispectral line scanners. Infrared surveys can be procured from a number of aerial survey firms who either own or lease scanners for this purpose. However, procurement of multispectral scanner data and processing is not as readily available; means for obtaining such data are discussed in Section 3.5.3.

Airborne microwave scanners, real aperture SLARs, and SARs are not off-the-shelf commercially available items. Microwave scanners, real aperture SLARs, and SARs can be purchased, but so far sales outside the manufacturer's corporation have been limited to government procure-

TABLE 3.2.   COMMERCIALLY AVAILABLE INFRARED SCANNERS

Manufacturer Model	Resolution, mRad	Swath Width, °	V/H Max, Rad/sec	Spectral, Band μm	NE$\triangle$T, °K
TRW Hawker Siddeley	1.5	120	0.75	7.5-14	0.5
Daedalus DEI-100	2.5	120	0.3	1.0-5.5 8.0-14	0.3 0.2
Daedalus DS-1200	2.5	77	0.2	1.0-5.5 8.0-14	0.3 0.2
HRB Singer Reconofax IV	2 3	120 or 140	0.37 to 0.56	8.0-14	0.3 0.5
HRB Singer Reconofax VI	2 3	120 or 140	0.75 to 1.1	8.0-14	0.5 0.7
HRB Singer Reconofax XI	1.5	120	0.20	8-14	0.32 0.2
HRB Singer Reconofax XIIIA Model 13-21	1 and 2 (in same dewar)	120	1.6	8-14	0.3 at 1 mrad 0.2 at 2 mrad
Bendix T/M LN-3	2.5	120	0.25	1.0-5.5 8.0-14.0	0.3 0.2
Texas Inst. RS-310	1.5	90	0.3	4.5-5.5 8.0-13.0	0.48 0.20
Texas Inst. RS-14	1.0 3.0	80	0.2	8.0-13	0.5 0.1
Texas Inst. RS-18	2.5	100	0.2	1.0-5.5 8.0-14	0.2 0.1
Texas Inst. RS-25	1.5	100	0.3	4.5-5.5 8.0-14	0.15 0.15

TABLE 3.3.   AIRBORNE MULTISPECTRAL SCANNERS

Manufacturer Model / Parameters	ERIM M-7	Bendix Aerospace Systems Division Modular Multispectral Scanner M²S	Bendix Aerospace Systems Division Multispectral Data System	Daedalus Enterprises DS-1050	Actron Ind., Inc. HMS-564X	Texas Instruments RS-14	Texas Instruments RS-18 MS
Swathwidth (degrees)	90	100	80	77	51.3°	80	100
Resolution IFOV (mrad)	2 max	2.5	2.0	2.5	2	1 or 3	2.5
V/H max	0.12 or 0.20	0.25	0.18	0.2	0.1	0.2	0.3
Roll compensation	Signal ± 10°	Signal ± 10°	Signal ± 8°	Signal ± 10°	No. stabilized mount option	Signal ± 8°	
Calibration in flight	Black bodies, Q-I lamps. Sky	Black bodies, Q-I lamps. Sunlight	Black bodies, Q-H lamps. Sky ref.	Q-H lamps.	Lamp	Black bodies, Q-I lamps.	Black bodies, Q-I lamps.
Method of recording	Magnetic tape	Magnetic tape	14-track Digital Magnetic tape	7-track magnetic tape tape	Magnetic tape	CRT with film	Magnetic tape
Method of spectral separation	Beam divider, dichroic beam splitter, Filters, prism	Diffraction grating/Dichroic	Diffraction grating	Prism	Prism/beam splitter	Dichroic	Diffraction grating/dichroic
Spectral bands	0.32- 0.38 μm 0.40- 0.44 0.44- 0.46 0.46- 0.48 0.48- 0.50 0.50- 0.52 0.52- 0.55 0.55- 0.58 0.58- 0.62 0.62- 0.66 0.66- 0.72 0.72- 0.82 0.82- 0.96 1.0 - 1.4 1.5 - 1.8 2.0 - 2.6 8.0 -13.5	0.38- 0.44 μm 0.44- 0.49 0.49- 0.54 0.54- 0.58 0.58- 0.62 0.62- 0.66 0.66- 0.70 0.70- 0.74 0.76- 0.86 0.97- 1.05 8.0 -12.0	0.34- 0.40 μm 0.40- 0.44 0.46- 0.50 0.53- 0.57 0.57- 0.63 0.64- 0.68 0.71- 0.75 0.77- 0.81 0.82- 0.87 0.97- 1.06 1.06- 1.095 1.13- 1.17 1.18- 1.3 1.52- 1.73 2.1 - 2.4 3.54- 4.0 4.5 - 4.75 6.0 - 7.0 8.3 - 8.8 8.8 - 9.3 9.3 - 9.8 10.1 -11.0 11.0 -12.0 12.0 -13.0	0.38-0.42 μm 0.42-0.45 0.45-0.50 0.50-0.55 0.55-0.60 0.60-0.65 0.65-0.70 0.70-0.80 0.80-0.90 0.90-1.10	0.5- 0.6 μm 0.6- 0.7 0.7- 0.8 0.8- 1.1 10.4-12.6	0.3- 0.55 μm 0.7- 0.9 1.0- 1.5 1.5- 1.8 2.0- 2.5 3.0- 5.5 8.0-14 Only two of the bands can be used at one time	0.5- 0.6 μm 0.6- 0.7 0.7- 0.8 0.8- 1.1 8.0-14
	Any 12 bands can be recorded at one time.						

80

ment because of capital and operating costs. Table 3.4 lists some of the radar systems currently in use.

3.5.2 Space Sensors

Infrared and optical imaging sensors have played a major role in shaping the United States space program. These sensors are used in planetary and astronomical programs as well as in terrestrial work. In Earth observation programs, the first imagers were obtained for meterological purposes, but the initial meteorological satellites had poor resolution, were not earth stabilized, and gave infrequent coverage. Today, meterological satellites in geo-stationary orbit produce relatively high resolution (1 km) imagery every 20 minutes.

In the last few years, considerably higher resolution sensors have been orbited expressly for earth resource applications. Most important of these are Landsats 1, 2, and 3 and Seasat 1. Landsat carries optical imaging sensors, and Seasat carries a radar imaging sensor.

Most of the early optical imagers were television sensors, but optical mechanical scanners have been gaining in favor over the last decade. More than 20 line scanners have been orbited, and the performance characteristics of some of these of greater interest to earth resource users are given

TABLE 3.4. EXAMPLES OF RADAR SYSTEMS (Courtesy R. Larson ERIM)

Frequency	Name	Origin	Platform	Depression Angle	Swath (km)	Resolution (Az-Range)
Airborne Synthetic Aperture Systems						
150 MHz	UHF Imaging Radar	NASA/JPL	CV-990	$0 < \theta_d \leqslant 90°$	14	30 × 30 m
1.2 GHz	*L*-band Imaging Radar	NASA/JPL	CV-990	0°-90°	14	30 × 10 m
1.3 GHz	ERIM 4-channel SAR	ERIM	CV-580	45° ± 45°	5 (20)	6 × 6 m
X-band	ERIM 4-channel SAR	ERIM	CV-580	45° ± 45°	5 (20)	3 × 3 m
X-band	JSC *X*-band SAR	NASA/JSC	WB-57	Mode I 13°-50° Mode II 45°-60°	16	20 × 20 m
X-band	JPL *X*-band SAR	NASA/JPL	CV-990	$0 \leqslant \theta_d \leqslant 90°$	14	30 × 10 m
X-band	GEMS	Goodyear	Caravelle	Mode A 9°-45° Mode B 18°-54°	36	10 × 12 m
S-band	APS-94E (Mod.)	AES,[a] Canada	Electro	5°-48°	50	≈50 × 30 m
Airborne Real Aperture Systems						
X-band	APS-94	NASA/LRC	Coast Guard C-130	5°-48°	50	Range dependent (0.45°)
X-band	Motorola SLAR	Motorola	Gulfstream	5°-48°	50	Range dependent (0.45°)
X-band	K.U. SLAR	Univ. of Kansas	N/A	60°-85°	12	Range dependent (0.46°)
Spaceborne Synthetic Aperture Systems						
1.2 GHz	SEASAT-A	NASA/NOAA	Spacecraft	67°-73°	100 km	25 × 25 m

[a] Atmospheric and Environmental Services

in Table 3.5. While most users are seeking higher spatial resolution, the lower resolution meteorological scanners have been found useful because of their frequent, repetitive coverage and their ability to image in the thermal infrared and passive microwaves. These sensors are particularly useful for hydrologic studies (soil moisture, snow cover and conditions, rainfall, and sea ice mapping), heat budget studies, and heat capacity studies.

By far the most important satellite for earth resource observations is the Landsat series. Landsat 1 was launched in July 1972 and is now inoperable. The third in the series, Landsat 3, was launched in March 1978. The two imaging systems orbited in Landsat are a three-band return-beam vidicon (RBV) system and a four-band multispectral scanner (MSS) system (14). The RBV failed early in orbit in Landsat 1 and is on standby as a backup unit to the MSS in Landsat 2. The RBV system in Landsats 1 and 2 consists of three, boresighted television cameras as shown in Fig. 3.34. The cameras image an area of 185 X 185 km with 4125 scan lines (14) and are filtered for multiband operation. Their spectral responses are shown in Fig. 3.35. As discussed in Section 3.4.3, it was originally thought that the vidicon with reseau marks would yield better geometric accuracy for cartographic purposes than the MSS. However, the geometric accuracy of the scanner was much better than was anticipated. When the format is computer adjusted with ground control points, the images meet 1:250,000 map accuracy standards. Because the radiometric quality of the MSS data is much superior to the RBV, the RBV system of Landsat is not turned on but serves as a backup sensor in the event the MSS fails. The RBV system of Landsat 3 has been modified to operate in a higher resolution, black-and-white mode. The resulting images cover 99 X 99 km and have a ground resolution of about 35 m.

The specifications of the MSS of Landsats 1, 2, 3 are given in Table 3-5. It should be mentioned that NASA designates succeeding generations of spacecraft with letters until they are successfully orbited. Once in operation, the letter is changed to a corresponding number. Thus Landsat D, currently under development, will become Landsat 4. The MSSs on Landsats 1 and 2 are identical and Landsat 3 differs only in that it has a fifth channel, which was to operate in the thermal infrared. This fifth channel is a major engineering undertaking, because the thermal detectors must be cooled to about 100°K. The cooling was to be achieved by radiation coupling to space. Unfortunately, this channel has not successfully operated since launch.

Figure 3.36 is a schematic of the MSS of Landsat. Four linear arrays of six detectors are used. Each array is optically filtered to limit their wavelength region of operation. The filters, together with the detector spectral response, produce the following four spectral bands:

Band 4	0.5-0.6 μm
Band 5	0.6-0.7 μm
Band 6	0.7-0.8 μm
Band 7	0.8-1.1 μm

It should be noted that the Landsat system uses the designation of bands 1, 2, and 3 for the RBV sensor and bands 4, 5, 6, and 7 for the MSS.

Figure 3.36 shows how an oscillating mirror sweeps the six-element detector array in a direction normal to the orbital plane. Thus, six lines are scanned per sweep. The four filtered detector arrays are swept in succession and small time delays between the detectors bring the four bands of video data into registration. The spacecraft provides repetitive coverage of the earth every 18 days, making about 20 such cyclic passes per year. The repetitive coverage permits analysis of dynamic events such as seasonal variations in snow cover, surface water, and vegetation. The long life of the satellite assures a high probability of obtaining a cloud-free image of extremely cloudy areas.

Three imaging systems were orbited under Skylab as part of the Earth Resource Experiment Package: two photographic systems known as S190-A and S190-B and a multispectral scanner known as S192. Reference (19) gives a detailed description of these sensors and their performance specifications. The sixband multispectral photographic camera, S190-A, and the single-band earth terrain camera, S190-B, are also described in Ref. (19). The performance specifications of the multispectral scanner are given in Table 3.4. Unlike Landsat, which is capable of producing imagery daily, the Skylab program was of very limited duration, and imagery was obtained only over a limited number of test sites. The photography was excellent, but the scanner data had limited utility because of engineering problems.

In addition to optical sensors, the NIMBUS and Seasat satellites have orbited microwave imagers.

TABLE 3.5. CHARACTERISTICS OF SOME SPACECRAFT SCANNERS

Manufacturer Scanner model	ITT Aerospace Optical Div. High resolution infrared radiometer (HRIR)	Santa Barbara Research Center Medium resolution IR radiometer	Santa Barbara Research Center Spin scan cloud camera	Santa Barbara Research Center Temperature-humidity IR radiometer	Santa Barbara Research Center ITOS high resolution scanning radiometer
Satellite	Nimbus I, II, III	Nimbus II, III	ATS 1 ATS III	Nimbus IV	NOAA 1, ITOS B-E
Method of Scanning	Single 45° mirror	Single 45° mirror	Spinning spacecraft and tilting mirror	Single 45° mirror	Single 45° mirror
Scan Lines per second	0.8	0.133	1.66 1.66	0.8	0.8
Swathwidth (degrees)	180	180	15 17.3	180	180
Orbit	Sun synchronous (600 nmi)	Sun synchronous (600 nmi)	Earth synchronous	Sun synchronous (600 nmi)	Sun synchronous (790 nmi)
Resolution (mrad)	7.9	50	0.1 0.1	7(10.5-12.5 μm) 21(6.7 μm)	2.7 (0.55-0.73 μm) 5.6 (10.5-12.5 μm)
IFOV (km)	8.7	55	3.6 3.6	7.7 23.	4.0 8.2
Calibration in flight	Black body and outerspace	Black body, space, sun		Black body, space	Black body, space, lamps
Method of spectral separation	Filters	Filters	Filters	Dichroic	Dichroic
Spectral bands (s) (μm)	0.7-1.1 0.3-4.2	0.2-4.0	0.47-0.63 0.38-0.48 0.48-0.58 0.55-0.63	6.7 10.5-12.5	0.55-0.73 10.5-12.5

TABLE 3.5 (continued)

Manufacturer	ITT Aerospace Optical Div.	Santa Barbara Research Center	RCA	Hughes	Hughes	Honeywell
Scanner Model	Surface composition mapping radiometer	Visible IR spin scan radiometer	Very high Res. Rad.	Multispectral scanner	Multispectral scanner	Multispectral scanner (S-192)
Satellite	Nimbus V	SMS	ITOS IV	Landsat 1, 2	Landsat 3	Skylab
Method of Scanning	Single 45° mirror	Spinning spacecraft and tilting mirror	45° mirror	Pivoting mirror	Pivoting mirror	Image plane conical
Scan Lines per Second	10	1.6 Scans/s 8 lines/scan (visible) 2 lines/scan (IR)	6.67	13.6 scans/s 6 lines/scan	13.6 scans/s 6 lines/scan (refl. bands) 2 lines/scan (thermal bands)	100
Swathwidth Degrees	90	18		11.5	11.5	10
Orbit	Sun Synchronous (600 nmi)	Geostationary	790 nmi	Sun synchronous (500 nmi)	Sun synchronous (500 nmi)	235 mi
Resolution mrad	0.6	0.025 (0.55-0.75 μm) 0.25 (10.5-12.6 μm)	0.6	0.086	0.086 (refl. bands) 0.26 (thermal band)	0.182
IFOV (km)	0.66	0.9 9.0	0.88	0.079	0.079 0.237	0.08
Calibration in Flight	Black body and Outerspace	Black body, space, sun	Black body, space, sun	Lamp, sun	Lamp, sun	
Method of spectral separation	Filters and dichroics	Filters	Dichroic	Filters	Filters	Spectrometer
Spectral Bands (s) μm	8.3-9.3 10.2-11.2 0.8-1.1	0.55-0.75 10.5-12.6	0.6-0.7 10.5-12.5	0.5-0.6 0.6-0.7 0.7-0.8 0.8-1.1	0.5-0.6 0.6-0.7 0.7-0.8 0.8-1.1 10.5-12.6	0.410-0.460 0.460-0.510 0.520-0.556 0.565-0.609 0.620-0.670 0.680-0.762 0.783-0.880 0.980-1.080 1.090-1.190 1.200-1.300 1.550-1.750 2.100-2.350

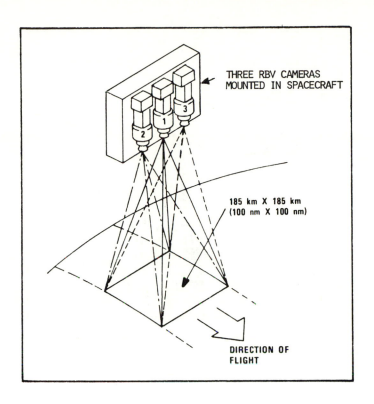

FIGURE 3.34. RBV scanning pattern (14).

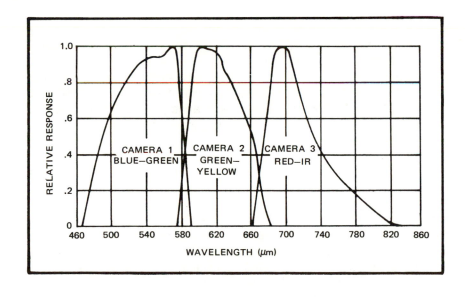

FIGURE 3.35. Spectral response, RBV camera system (14).

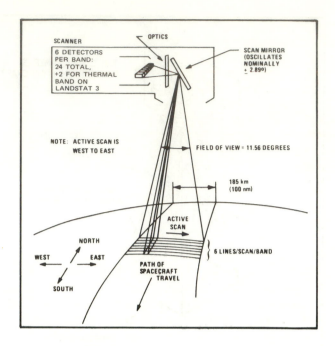

FIGURE 3.36. MSS scanning arrangement (14).

The NIMBUS passive microwave imager operates at a frequency of 19.35 GHz (a wavelength of 1.55 cm). A second imager is under development for launch on NIMBUS 6. It will operate at 37 GHz and will receive dual polarization. This change in frequency will significantly improve the detection of water droplets (rain), and the dual polarization will facilitate the distinction between water and ice as well as the estimation of water roughness.

Seasat was launched in June 1978 and carries a synthetic apertive radar. Its performance characteristics are given in Table 3.4.

3.5.3 Procurement of Data

The collection of remote sensing data is costly. Anyone wanting such data should first investigate whether or not suitable data already exist. Three federal agencies file and sell remote sensing data collected at national and international levels. These are the Department of Agriculture's Agricultural Stabilization and Conservation Service (ASCS), the Department of Interior's Earth Resources Observation system (EROS), and the National Oceanographic and Atmospheric Agency (NOAA).

ASCS photographs most of the United States every five years in black-and-white photography at scales of 1:20,000 and 1:40,000. The Western Aerial Photography Laboratory at Salt Lake City, Utah sells ASCS photographs.

The EROS Data Center at Sioux Falls, South Dakota is the outlet for Landsat and Skylab imagery and computer-compatible tapes. In addition, EROS sells all the aerial imagery taken by NASA in support of its earth observation program (largely high-altitude color and color infrared photography taken at a scale of 1:60,000 and 1:120,000) and also aerial imagery produced by the USGS in its mapping program.

NOAA is an outlet for imagery from its various meteorological satellites as well as Landsat imagery. Their address is NOAA. Satellite Data Service Branch, 5200 Auth Road, Washington, D. C. 20233. In addition to these federal agencies, aerial survey firms, state highway departments, public utility commissions, and local and regional planning commissions usually have a file of photogaphs available to interested users.

When suitable data are not available, one must procure them. Aerial photography and thermal imagery can be purchased from the many aerial survey firms throughout the country. Those procuring such data for the first time would do well to read

Chapter 6 of this volume and sections of the *Manual of Photogrammetry* (16), which discuss how to plan and procure photographic missions. Multispectral, passive microwave, and radar imagery are, in general, more specialized and costly, and only limited facilities are available. Persons desiring this type of remote sensing data would do well to read the specialized remote sensing journals and symposium proceedings to identify organizations engaged in this activity.

3.6 ADVANCED AND RESEARCH SENSORS

Advances in solid-state physics and electronics portend corresponding advances in sensor systems through new component development and miniaturization. This section discusses imaging sensors currently being developed and nonimaging sensors now in use which may someday operate in an imaging mode.

3.6.1 Imaging Sensors

SELF-SCANNED DETECTORS

The performance ability of optical mechanical scanners depends on the amount of time the scanning detector is allowed to make an observation of a scene point. As resolution decreases or scan rate increases, the dwell time and, hence, the performance, is reduced. One obvious method of improving space scanner resolution while maintaining performance and size is to increase the number of detector elements in the sensor. For example, the Landsat scanner uses six detectors per band. Large arrays of solid state detectors have been made with built-in, electronic self-scanning ability, for example charged coupled devices. Linear and two-dimensional arrays of detectors have been made which eliminate the need for mechanical scanning of the scene (6). These detectors are currently limited to operating in the visible and near-visible region, but infrared self-scanning arrays appear very promising.

LASER SCANNERS

The highly collimated radiation from lasers make them ideally suited for use in active, airborne scanning systems. In the laser scanner, the laser beam scans the scene synchronously with the receiver.

Some of the advantages of active scanning with a Laser are the following:

• Operation is not dependent on sunlight.
• No shadows occur in the image as the receiver looks down the illuminating beam.
• Operation in narrow spectral bands is possible.
• Three-dimensional imaging is possible by modulating the laser and measuring the phase or time of return of the reflected radiation.

An active/passive multispectral scanner, under development by NASA, uses two Laser bands in conjunction with 12 passive bands (20).

THEMATIC MAPPER

The next generation Landsat multispectral scanner is now under consideration. Called the Landsat-D thematic mapper, the proposed scanner will operate in seven bands in the 0.5 to 12.5-μm region, with a ground resolution of about 40 m. The output data rate from this sensor will be 150 Mbps compared with 15 Mbps for the Landsat MSS. Landsat-D will require an order of magnitude improvement in data recording and transmitting capability. The driving factors of the performance specifications for the thematic mapper are agricultural applications. The proposed wavelength bands of operation are the following:

0.45 - 0.52 μm	0.80 - 0.91 μm
0.52 - 0.60 μm	1.55 - 1.75 μm
0.63 - 0.69 μm	10.5 - 12.5 μm
0.74 - 0.80 μm	

3.6.2 Nonimaging Sensors

LASER PROFILER

In a laser profiler, the interaction of a high-powered laser pulse with matter is observed as a function of time. When operating from an aircraft, the return signal from a downward pulse contains information on atmospheric scattering, surface reflection, distributed reflectance from a textured surface, boundaries of transparent media, scattering within a semitransparent medium, and/or fluorescence. Time resolution of the return radiation gives the location and nature of the interaction. Thus, the profiler can be used to obtain data on the vertical distribution of scattering in air and water, sea state, vegetation height and crown structure, topographic relief, and fluorescence associated with chlorophyll (biomass) and pollutants (oils). When high-powered lasers with high-pulse repetition rates are developed, the above functions should be attainable with laser scanners.

FRAUNHOFFER-LINE DISCRIMINATOR

Many materials can be classified and their amounts quantitatively measured by their fluorescence properties. Short wavelength solar energy can stimulate fluorescence, but the magnitude of the broadband fluorescence is usually masked by the magnitude of sunlight reflected by the medium. One technique for observing weak, solar-induced fluorescence in the presence of reflected sunlight is to use the Fraunhoffer-line discriminator technique (21). Its principle of operation is based on the fact that sunlight is substantially reduced in the Fraunhoffer absorption lines and the spectral reflectance from terrain features is broadband. In the Fraunhoffer-line discriminator, observations of upwelling radiation of an object are made in two narrow spectral bands, one centered on a Fraunhoffer absorption line and one immediately adjacent to it. Consider the situation where the scene has a reflection of 10%, the solar radiance in the Fraunhoffer line is 5% of that outside it, and ultraviolet sunlight stimulates fluorescence equivalent to a 1% change in reflected sunlight.

In the absence of fluorescence, the ratio of signals in and outside the line is

$$\frac{5\% \times 10\%}{100\% \times 10\%} = 1/20$$

In the presence of fluorescence, the ratio of signals becomes

$$\frac{5\% \times 10\% + 1\%}{100\% \times 10\% + 1\%} = \frac{1.5\%}{11\%} = \frac{1}{7}$$

Thus, the small fluorescence changed the ratio by a factor of approximately 3. The resulting ratio is independent of the level of solar illumination if the shape of the spectral distribution is unchanged, that is, the ultraviolet stimulating radiation changes in proportion to the reflected radiation. However, the ratio does depend on both the reflectance of the material and its magnitude of fluorescence. Since the reflectance of the material can be obtained with appropriate calibration of the return signal from the band adjacent to the Fraunhoffer lines, the magnitude of the fluorescence can be calculated.

SCATTEROMETERS

A scatterometer is a device for measuring the microwave reflectance (monostatic scattering cross section) of an object or terrain element, as a function of angle. A simple pulsed radar using a downward-looking antenna can be used to carry out this measurement, as illustrated in Fig. 3.37. The beam is made wide in the "along track" dimension parallel to the aircraft's trajectory, but narrow in the "cross-track" dimension. The depression angle of a particular terrain element is close to 0° when that element is far ahead of the aircraft, and increases monotonically to 90° as the aircraft passes over the element. As this occurs, the range from the radar to the terrain element decreases monotonically, and reaches a minimum (equal to aircraft altitude) as the aircraft passes directly over the terrain element. The one to one correspondence between depression angle, ψ, and range can be used to "track" a terrain element as it passes through the beam, hence to observe its reflectance (backscatter) as a function of ψ.

An alternative to the use of a pulsed radar for scatterometer measurement is to use the Doppler shift of a continuous wave (cw) radar. In this case, there is a one-to-one correspondence between depression angle and Doppler frequency. Thus, one can use a radar that emits a constant-frequency sinusoid, and then analyze the frequency of the returns in order to determine reflectance as a function of depression angle. Each terrain element progresses through a range of Doppler frequencies as the aircraft flies toward and over it. The scatterometer has been found useful for measuring surface roughness and conditions associated with terrain, vegetation, and water.

SPECTROMETERS

Often an investigator or instrument designer wishes to measure the spectral characteristics of targets and backgrounds to a finer degree than obtainable with multispectral imaging sensors. In a spectrometer, the information bandwidth capacity

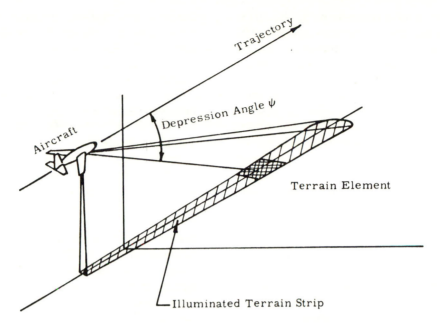

FIGURE 3.37. Geometry of a scatterometer.

of a detector is devoted to measuring the spectral distribution of radiation from an object. In contrast, the bandwidth of a detector in a scanner is devoted to measuring the spatial distribution of radiation in a scene.

The high spectral resolution of spectrometers make these sensors useful as devices for measuring atmospheric effects on remote sensing data as well as collecting spectral signature data needed for determining multispectral sensor band placement specifications and for developing spectral pattern recognition and processing techniques. The Earth Resource Experiment Package on Skylab carried a filter wheel spectrometer (S-191) having a field of view of 1 mrad. Using a dichroic beam splitter and two multisegment filter wheels, the spectrometer covers the regions of 0.4 to 2.5 μm and 5.8 to 16 μm with a spectral bandwidth of about 2% of the observation wavelength (19).

3.7 Conclusion

The widespread availability of remote sensing data for earth observations has developed a large group of users and applications. As in all emerging technological developments, we have only seen the "tip of the iceberg" as far as development of remote sensing technology and its application are concerned. Spurred by NASA's earth observation program and the development of solid-state electronics, researchers will continue to develop new and better remote sensing instrumentation. Sensors that appear impracticable today may very well be the sensors of tomorrow.

REFERENCES
1. Valley, S. L., ed., 1965, Handbook of geophysics and space environments: Mass., Air Force Cambridge Research Laboratories, 649 p.
2. LaRocca, A. J. and Turner, R. E., 1975, Atmospheric transmittance and radiance: methods of calculation: An IRIA State-of-the-art Report, ERIM Rept. 107600-10-T, Ann Arbor, Michigan, 508 p.
3. Wolfe, W. L., ed., 1965, Handbook of military infrared technology: Washington, D. C., U. S. Government Printing Office, 906 p.
4. Smith, J. T., Jr., ed., 1968, Manual of color aerial photography: Falls Church, Virginia, American Society of Photogrammetry, 550 p.
5. Reedes, R. G., ed., 1975, Manual of remote sensing, v. I and II: Falls Church, Virginia, American Society of Photogrammetry, 2144 p.
6. NASA, Advanced scanners and imaging systems for Earth observation: NASA SP-335, Washington, D. C., 604 p.

7. Biberman, L. M. and Nudelman, S., ed., 1971, Photoelectronic imaging devices, v. 1, Physical processes and methods of analysis, v. 2, Devices and their evaluation: New York, Plenum Press, 430 p. v. 1, and 577 p. v. 2.

8. Lowe, D. L., 1968, Line scan devices and why use them, *in* Proceedings of the fifth symposium on remote sensing of environment: Ann Arbor, Michigan, Environmental Research Institute of Michigan, v. 1, p. 77-101.

9. Holter, M. R., Nudelman, S., Suits, G. H., Zissis, G. J. and Wolfe, W. L., 1962, Fundamentals of infrared technology: New York, MacMillan, 442 p.

10. Brunell, D. N., Estes, J. E., Mel, M. R., Thaman, R. R. and Evanisko, F. E., 1974. The usefulness of imaging passive microwave for rural and urban analysis, *in* Proceedings of the ninth symposium on remote sensing of environment: Ann Arbor, Michigan, Environmental Institute of Michigan, v. 3, p. 1603-1620.

11. Moore, R. P. and Hooper, J. O., 1974, Microwave radiometric characteristics of snow covered terrain, *in* Proceedings of the ninth symposium on remote sensing of environment: Ann Arbor, Michigan, Environmental Institute of Michigan, v. 3, p. 1621-1632.

12. National Academy of Sciences, Remote sensing with special reference to agriculture and forestry, Washington, D. C., 424 p.

13. USAF Recon Central (prepared by Data Corporation), 1964-65, Airborne photographic equipment, v. 1, RCO13200 (1) Camera magazine and mounts, 1965, v. 2, RCO13200 (2) Controls, accessories and miscellaneous equipment, 1964, Supplement #1 for 1965 editions.

14. NASA, 1972, Earth resources technology satellite — data users handbook: General Electric Document No. 71SD4249, Prepared for NASA, Goddard Spaceflight Center, Greenbelt, Maryland, 230 p.

15. Colwell, R. N., ed., 1960, Manual of photographic interpretation: Falls Church, Virginia, American Society of Photogrammetry, 868 p.

16. Thompson, M. H., ed., 1965, Manual of photogrammetry, 3rd ed., v. 1 and 2: Falls Church, Virginia, American Society of Photogrammetry, 1199 p.

17. Slater, P. N., 1972, Multiband cameras: Photogrammetric Eng., v. 38, p. 543-555.

18. Fisher, J. J., and Stevers, E. Z., 1973, 35 mm quadricamera: Photogrammetric Eng., v. 39, p. 573-578.

19. NASA, 1973, Skylab program EREP investigator's information book: NASA, Lyndon B. Johnson Space Center, Houston, Texas, Rept. MSC-07874, 225 p.

20. Zissis, G. J., 1975, The technologies of remote sensing of the environment: *in* Proceedings of the American Congress on Surveying and Mapping, Washington, D. C., v. 1, p. 7-24.

21. Hemphill, W. R., Stoertz, G. E., and Markle, D. A., 1969, Remote sensing of luminescent materials, *in* Proceedings of the sixth international symposium for remote sensing of the environment: Ann Arbor, Michigan, Environmental Research Institute of Michigan, v. 1, p. 565-585.

CHAPTER 4

THE EVOLUTION OF GEOLOGICAL SPACE PHOTOGRAPHY

PAUL D. LOWMAN JR.

4.1 INTRODUCTION

Photographs of the Earth from space have become familiar features of geological textbooks, lectures, and technical papers. But before these pictures were available, there was practically no appreciation of their potential scientific value. The development of geological space photography is thus of interest not only for itself, but because it furnishes a striking example of the unexpected by-products of man's venture into space.

This review will cover photography of the Earth from sounding rockets, earth-orbiting satellites and manned spacecraft, and deep space missions. It will be centered on terrain photography, with meteorological orbital photography included only insofar as it has been applied to geology. The term "space photography," as distinguished from "orbital photography," is used in the chapter title because of the variety of trajectories involved in the missions discussed here, although in fact the vast majority of geologically useful images of the Earth have been returned from earth-orbiting vehicles, such as Gemini spacecraft and the Landsat earth resources technology satellites.

The literature on the geologic uses of orbital photography is already large, and will not be completely cited here. General references that will be useful for finding specific papers not separately cited include Darden (1), Derr (2), Freden et al. (3, 4), Holz (5), Kaltenbach (6), Lowman (7, 8, 9, 10, 11, 12, and 13), Reeves (14), and Rowan (15).

4.2 SOUNDING ROCKET PHOTOGRAPHY

The first pictures of the Earth from space were taken with automatic 35-mm cameras from V-2 rockets fired from White Sand Proving Ground, New Mexico, by the United States Army shortly after World War II (7). These and similar pictures from Aerobee rockets were of great inherent interest, but were put to no geologic use. However, the Viking 11 and 12 flights, which carried the much larger K-25 cameras with infrared film to altitudes of over 225 km, produced pictures of far higher quality and areal coverage (Fig. 4.1). These were used to evaluate the resolution capabilities of orbital camera systems by Amron Katz and others (7) and the geologic value of "hyperaltitude photography" by Paul Merifield. Merifield's pioneering study was the first geologic use of space photography, and was responsible for the proposal to have Project Mercury astronauts take hand-held photographs specifically for geologic purposes, as will be described later. Merifield demonstrated that even on photographs with ground resolutions of about 200 m, a wide range of geologic features could be delineated. One reason for this, since illustrated repeatedly, is that "resolution," or ability to separate objects, does not give a true idea of a system's ability to *detect* such objects. In particular, high-contrast linear features such as railroads, far smaller than the normal ground resolution, could be easily seen on Viking photographs. Since many geologic features, such as contacts and faults, are essentially linear, the geologic value of even low-resolution space photographs was evident.

Sounding rocket photography by itself did not arouse much interest among the geologic community. The pictures were all oblique, except for the point directly under the rocket, and useful coverage was limited to the area around the launch site. Orbital photography has dominated the field since 1963, and doubtless will continue to do so. However, sounding rocket photography, using the British Skylark rocket to carry cameras to orbital altitudes, has recently been revived, primarily for agricultural purposes. The advantages of sounding rockets

FIGURE 4.1. Viking 12 view of southwestern United States and northwestern Mexico, including Gulf of California at left, Salton Sea at right center, and Sierra Nevada at extreme right. Photograph taken in 1955 from an altitude of approximately 255 km with Hi-Speed Infrared film using a K-25 camera (7).

over orbiting satellites are, first, that a solid or storable liquid fuel rocket can be held in readiness for rapid launching when weather or other conditions, such as sun angle and crop growth, are just right. Second, returned film photography can be done without the elaborate tracking and recovery network needed for orbital missions. It is clear, then, that the sounding rocket as a camera platform must not be counted out; its best days may lie ahead.

4.3 WEATHER SATELLITE PHOTOGRAPHY

The advantages of meteorology of orbital surveillance—global, repetitive, synoptic coverage— were obvious long before artificial satellites were actually launched, and the first meterological satellite, Tiros 1, was put into orbit on April 1, 1960. Since high resolution was not considered necessary for most cloud pattern observations, the television images from this and succeeding weather

satellites have relatively low resolution. However, they did stimulate thinking about other applications of orbital imagery, and had applications outside meterology. Ice pack reconnaissance (7) was carried out in the early 1960s using Tiros images. As might be expected, details of snow cover are well-displayed by even low-resolution television pictures, and these sometimes indirectly outline large geologic structures. This is illustrated by Fig. 4.2(*a*), a 1964 Nimbus 1 view of Siberia showing the Sayan fault [Fig. 4.2(*b*)]. This fault, not shown on most geologic maps of the area, is of considerable interest as a possible continental transform fault, since one end is at the Baikal rift. But the picture is even more important in that it suggests the geologic value of repetitive coverage. Since major geologic features such as the Sayan fault do not change appreciably over periods of years, it is often argued that one good orbital picture of a given area is enough for geologic purposes. However, in the area of this September picture, the Sayan fault (expressed as aligned valleys)

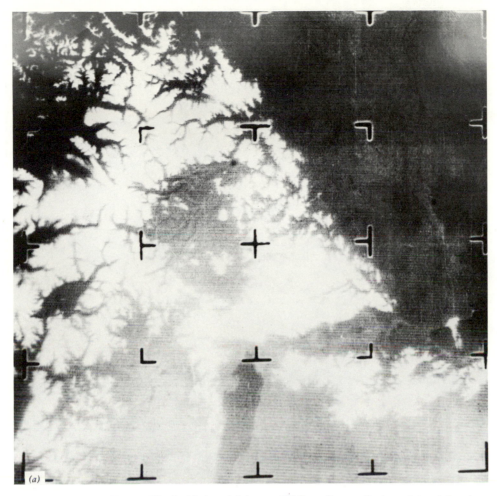

FIGURE 4.2. (a) Eastern Siberia. Nimbus 1 Advanced Vidicon Camera System picture taken in 1964, showing Sayan fault and snow-covered mountains.

might not be visible in summer if the snow was partly melted, and in the winter the terrain would be completely snow-covered, producing much the same effect. This was an early demonstration that although a particular geologic feature may not change, its *appearance* may, and that some repetition of coverage may be geologically useful.

The immense areal coverage of some meteorological satellite pictures makes them of continuing interest as supplements to high-resolution pictures such as those of Landsat. Lathram (16), for example, has produced a tectonic map of almost the entire state of Alaska from Nimbus imagery. It seems safe to suggest that, like sounding rocket photography, the imagery from weather satellites should be kept in mind by any geologist who plans to use space photography.

4.4 TERRAIN PHOTOGRAPHY

4.4.1 Mercury and Gemini Spacecraft

Orbital photography specifically taken for geologic purposes began with the later missions of the first American manned space flight program, Project Mercury. At the suggestion of P. M. Merifield, a rudimentary terrain photography experiment was proposed by P. D. Lowman, and accepted by the NASA Manned Space Flight Experiments Board, for the MA-8 and MA-9 Mercury missions. The objective of this experiment was simply to obtain high-quality color pictures of geologically interesting areas or features. Stress was put on coverage of well-mapped parts of the flight path, such as the Southwest United States, to obtain basic experience in interpreting orbital photography. The astro-

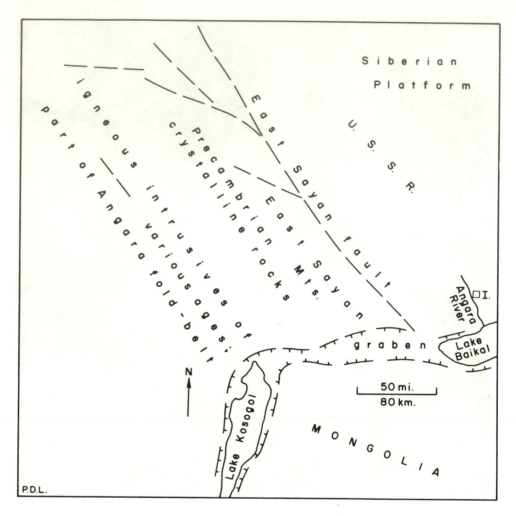

INDEX MAP
NIMBUS I AVCS PHOTOGRAPH
(b)

FIGURE 4.2. (*b*) Physiographic sketch map.

nauts were given one 3-hour briefing covering the essentials of physical geology, rationale for the experiment, targets, and photographic techniques. Coverage of these topics was something less than exhaustive. Hand-held cameras with 80-mm focal length lenses and 70-mm film were used, these being the biggest cameras practical in the confined cabins of the early spacecraft. The MA-9 mission, which at 34 hours was the longest United States space flight up to that time (1963), was photographically very successful, with 29 pictures of good to excellent quality being returned [Figs. 4.3(*a*), and 4.3(*b*)]. The areas covered, chiefly in

Southwest Asia and Tibet, were then poorly mapped, and the value of orbital photographs for cartography alone was immediately obvious. Furthermore, regional geologic structure of the Tibetan Plateau was well-displayed, showing lineaments and folds in an essentially unmapped area. When reported at the end of the Mercury Project in 1963, the results of the terrain photography stimulated considerable interest in the nonmeterological uses of orbital photography, including of course geology. A report by A. Morrison and M. C. Chown (7) on the geographic aspects of the earlier, unmanned MA-4 mission, which had returned several dozen pictures

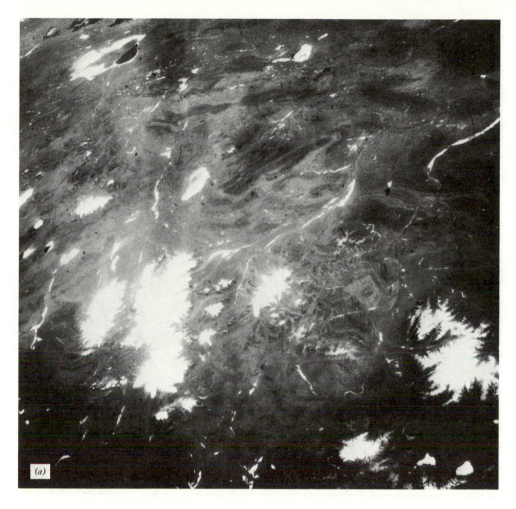

FIGURE 4.3. (a) Tibetan Plateau, Tibet. Mercury 9 photograph taken in 1963, original in color (7). Scale variable across the image, with E-W width approximately 130 km.

of North Africa, was published at the same time, providing further justification for a continuation of the terrain photography.

The low cost, simplicity, and potential value of orbital terrain photography ensured that a similar experiment (Synoptic Terrain Photography, S005) would be flown on the two-man Gemini missions, which began in 1965. Except for the first (GT-3) mission, which was only three revolutions long, all the Gemini flights lasted several days, permitting much more photography (and intriguing visual observations, such as sightings of meteor trails below the spacecraft). Furthermore, the experiment, benefiting from the Mercury experience, was more carefully planned (Fig. 4.4) and the crews extensively briefed. The initial objectives of the S005 experi-

ment were, as on Mercury, simply to obtain good color photographs, again with 70-mm hand-held cameras. The areas picked for the first long mission, GT-4, were: (1) Southwest United States (good ground truth), (2) Red Sea and adjacent parts of East African rift system (objective of the Upper Mantle Project), and (3) northern Mexico (request of the Mexican government) (Lowman and Tiedemann in Ref. 17).

The first manned Gemini mission, GT-3, although only three revolutions long, produced several good photographs of the Southwest United States. But the four-day GT-4 mission was a landmark in geologic orbital photography (17). The Gemini crew, J. A. McDivitt and E. H. White II, obtained nearly 100 pictures of all the planned areas, most of them

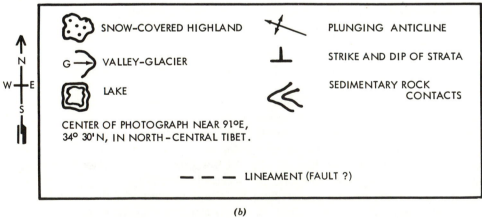

SNOW-COVERED HIGHLAND   PLUNGING ANTICLINE

G   VALLEY-GLACIER   STRIKE AND DIP OF STRATA

LAKE   SEDIMENTARY ROCK CONTACTS

CENTER OF PHOTOGRAPH NEAR 91°E, 34° 30'N, IN NORTH-CENTRAL TIBET.

– – – – LINEAMENT (FAULT ?)

(b)

FIGURE 4.3. (*b*) Regional geology sketch map (7).

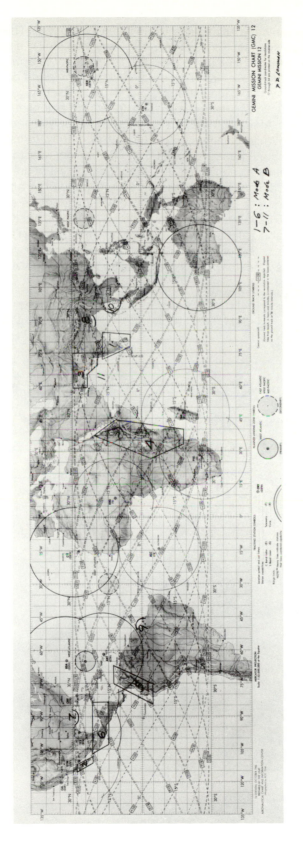

FIGURE 4.4. Typical Gemini flight path map, showing areas selected for Synoptic Terrain Photograph Experiment (17). Notice latitude restriction resulting from low inclination orbit.

good to outstanding photographically. An especially useful accomplishment was a sequence of 39 overlapping near-vertical cloud-free pictures along the flight path from Baja California to Central Texas (Fig. 4.5). Furthermore, many individual photographs showed previously unknown features in poorly-mapped parts of North Africa and the Arabian Peninsula. McDivitt and White obtained several excellent pictures showing the Bahama Bank, the Gulf of California, and other ocean areas in which bottom topography and sediment plumes were visible (9,11).

The GT-5 mission, lasting 10 days, was almost all in drifting flight because of spacecraft power problems. However, the astronauts, L. G. Cooper and C. Conrad, obtained 175 pictures, generally of excel-lent quality (Fig. 4.6). These were necessarily much less systematically distributed than those from GT-4, but covered a much wider variety of terrain, including many parts of southern Asia.

The early Gemini missions dramatically demonstrated the potential value of orbital terrain photography not only in geology but in other fields such as oceanography, forestry, agriculture, and hydrology. Beginning with the two-week GT-7 mission, therefore, the list of areas to be photographed was greatly expanded to accommodate requests from government agencies such as the Bureau of Commercial Fisheries, the Navy Oceanographic Office, and the United States Geological Survey, as well as several universities. As it happened, the GT-7 mission was plagued with window obscuration and

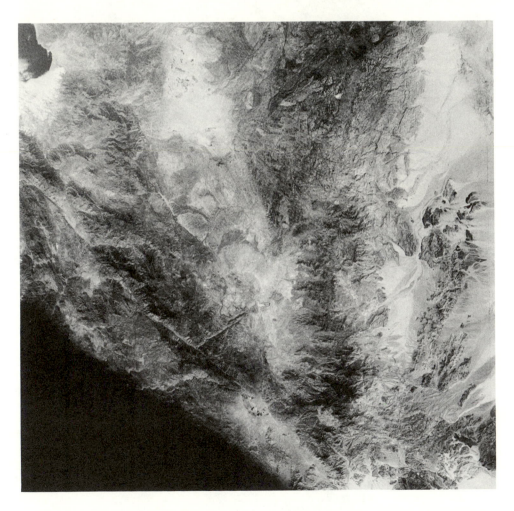

FIGURE 4.5. Gemini 4 photograph of northern Baja California, Mexico, taken in 1965; original in color (17). Ensenada, B.C. can be seen at extreme upper left. The Agua Blanca fault appears at lower left portion of the image, parallel to spacecraft window edge.

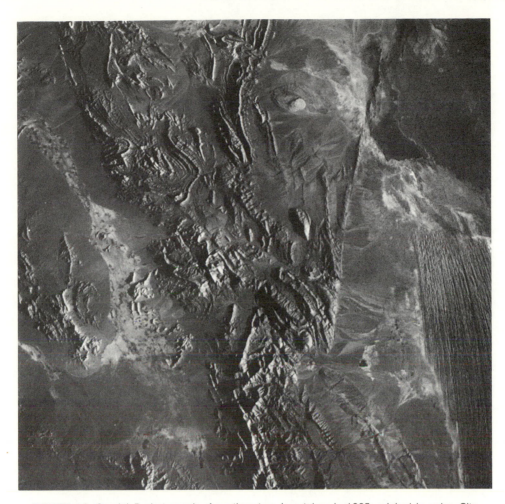

FIGURE 4.6. Gemini 5 photograph of southeastern Iran taken in 1965; original in color. City of Kerman can be seen in the valley at left center of photograph; Dasht-i-Lut at right. Adjacent linear features are yardangs (9). The white circular feature at top of the photograph is an extrusive salt dome. Notice the effect of low sun angle emphasizing relief; sun coming from right (morning).

other problems; nevertheless the crew obtained about 250 pictures usable for one purpose or another. Succeeding Gemini missions through GT-12, the last, returned still more pictures, with the eventual total of geologically useful pictures for the program reaching 1100. To summarize the geologic results of the Gemini photography in detail is impossible here, but a few major findings can be mentioned (see also Ref. 8, 9. 10, and 11).

The Texas lineament, considered by some geologists to be a dominant transcontinental fault, possibly strike-slip, was found on the Gemini 4 photos to be a broad belt of normal faults grading into, and probably controlled by, underlying fold axes [(Fig. 4.7(a) and 4.7(b)]. The same series of pictures revealed a large Quaternary volcanic field just south of the Mexican border near Palomas, Chihuahua, which was not shown on any published map. Field reconnaissance showed it to be a collection of basaltic cinder cones probably several tens of thousand years old. Some of the volcanic bombs contained dunite inclusions presumably of mantle derivation, and the Palomas volcanic field is thus of interest as a possible indicator of plate motion relative to the mantle. Photographs of North Africa revealed large areas of extensive wind erosion, suggesting that the importance of aeolian processes in deserts has been underestimated by geomorphologists. Comparisons of African and American desert landforms were made using the Gemini pictures.

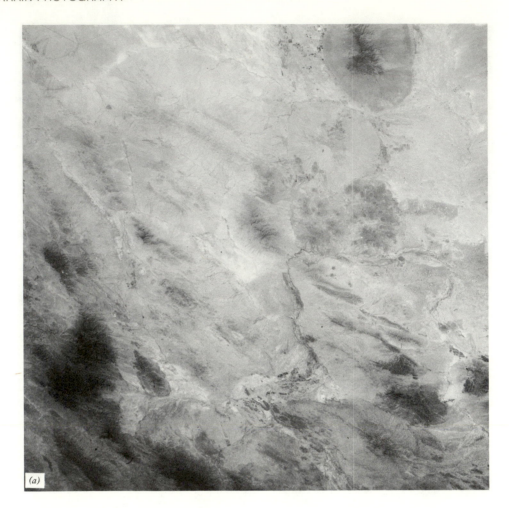

FIGURE 4.7. (a) Northern Chihuahua, Mexico, and Luna County, New Mexico (17). Gemini 4 photograph taken in 1965; original in color.

Other applications of the Gemini photographs have been reported by Wobber (18), Abdel-Gawad (19), and Pesce (20). The pictures have appeared in many geology textbooks and technical papers to illustrate particular features or areas as well. In addition, many oil and mining companies have purchased them from the Technology Applications Center at the University of New Mexico in Albuquerque for proprietary uses.

In summary, the Gemini terrain photography was highly successful, both for the pictures themselves and in opening an essentially new field of space applications (21). The pictures were also, as a permanent record of the Earth's surface features in 1965-1966, the beginning of repetitive satellite coverage of the Earth and may thus be useful for future research on long-term changes in vegetation, terrain, and land use. It may be possible, for ex-

ample, to measure the rate at which the growth of the desert is creeping southward from the Sahara by comparing the boundaries of the savannah in 1966 with those of later times. Long-term climatic changes may perhaps be studied by using the Gemini photographs to measure retreat or advance of low-latitude glaciers such as those in the Himalayas.

Before going on to the Apollo Program terrain photography, several developments that eventually had great influence on the evolution of orbital photography should be mentioned. First, NASA had organized, under the leadership of Peter C. Badgley, a Natural Resources Program. This program was an outgrowth of plans begun in the fall of 1963 for earth-orbital space stations. The successful terrain photography from the Mercury Program, then just concluded, led to the inclusion of similar

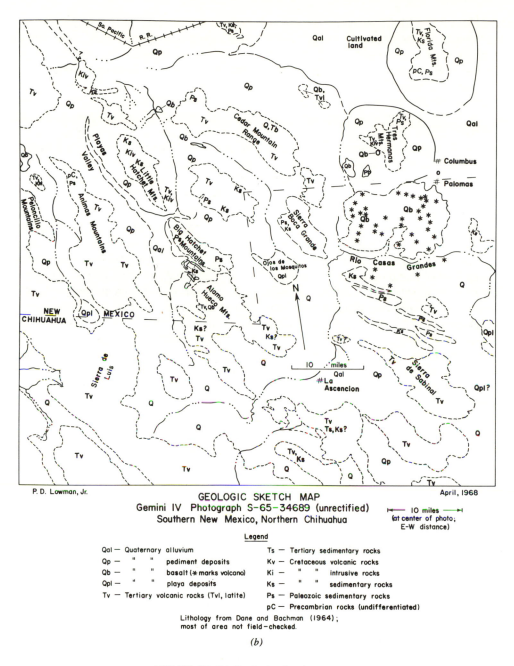

GEOLOGIC SKETCH MAP
Gemini IV Photograph S-65-34689 (unrectified)
Southern New Mexico, Northern Chihuahua

P. D. Lowman, Jr.   April, 1968

├── 10 miles ──┤
(at center of photo;
E-W distance)

Legend

Qal — Quaternary alluvium	Ts — Tertiary sedimentary rocks
Qp — " " pediment deposits	Kv — Cretaceous volcanic rocks
Qb — " " basalt (✻ marks volcano)	Ki — " " intrusive rocks
Qpl — " " playa deposits	Ks — " " sedimentary rocks
Tv — Tertiary volcanic rocks (Tvl, latite)	Ps — Paleozoic sedimentary rocks
	pC — Precambrian rocks (undifferentiated)

Lithology from Dane and Bachman (1964);
most of area not field-checked.

(b)

FIGURE 4.7. (*b*) Geologic sketch map.

photography and other types of remote sensing in space station mission discussions from that time on (culminating 10 years later in the Skylab Earth Resources Experiment Package, to be discussed later). A more immediate result of these early discussions was the formation of a multiorganization photographic team whose purpose was to help plan specific experiments for earth-orbital and lunar-orbital missions. This team functioned for several years under the guidance of Badgley and later Alden P. Colvocoresses, and contributed greatly to the planning of what was to become the NASA Earth Resources Program.

A parallel effort was begun in 1965 by the United States Geological Survey; the EROS Program (Earth Resources Observation System), directed by William A. Fischer. This was a general remote sensing project using both aircraft and spacecraft data. The EROS Program produced, among other things, specifications for an earth observation satellite which were strongly reflected in the NASA Earth Resources Technology Satellite (now Landsat), launched in 1972.

It should be pointed out here that research in remote sensing conducted for eventual space missions had great influence on aerial techniques (21), thus providing still another example of the unexpected by-products of the lunar landing program (of which the Gemini missions were an integral part).

4.4.2 Apollo Program

The Apollo Program was of course primarily directed at a lunar landing. But it produced, almost incidentally, invaluable experience in earth-orbital remote sensing. Part of this experience was the result of the close relation referred to earlier between the planning efforts for lunar and earth-orbital missions. However, after 1966 the two programs were largely separate.

Planning for the Apollo missions was of course well underway long before the end of the Gemini Program in November, 1966. A simple terrain photography experiment, using a single hand-held 70-mm camera as in the Gemini missions, had been proposed by Lowman and approved for the first earth-orbital Apollo mission (AS-204), planned for 1967. A large number of terrain photography targets, covering geologic, geographic, and oceanographic targets, was planned. This and other Apollo activities were abruptly stopped by the tragic spacecraft fire of January, 1967, in which astronauts Virgil Grissom, Edward White, and Roger Chaffee died.

The Apollo Program was delayed for nearly two years by the redesign efforts and greatly increased safety testing necessary as a result of the fire. Consequently, it was possible to advance some of the experiments intended for a later mission, the Apollo Applications Orbital Workshop, to the first Apollo flight when it was finally launched in October, 1968, or to Apollo 9 in March, 1969. It had been realized, even before the end of the Gemini Program, that multispectral orbital photography would be far more useful than single band or conventional color photography (7). At the suggestion of A. P. Colvocoresses (then at NASA Headquarters) a four-camera experiment (S065 Multispectral Terrain Photography) had been planned for the Apollo Applications Program missions. This experiment was

moved up to the Apollo 9 mission (11). Since the S065 experiment was a major step forward in orbital remote sensing, it will be described in some detail; further information is given in Refs. 2 and 6. The report by Colwell et al. (22), although concerned with agricultural uses of the photography, is of general interest as well. The objective of the S065 experiment was to obtain pictures of selected terrain and ocean areas with a four-camera array (Fig. 4.8) producing photographs simultaneously in the green, red, and near-infrared bands and with false color infrared film (Fig. 4.9). The areas picked for photography included a number of test sites selected by NASA and cooperating agencies for ground truth studies in agriculture, hydrology, geography, geology, and oceanography. Arrangements were made to make simultaneous aircraft underflights carrying various remote sensors during the Apollo 9 mission. The equipment used for the S065 experiment was a coaxially-mounted array of 70-mm cameras with 80-mm lenses, each with a different film/filter combination to give coverage in the spectral regions mentioned previously (480 to 620, 590 to 720, and 720 to 900 μm. These bands were close to those under consideration at that time for the Earth Resources Technology Satellite (ERTS-A), eventually to become Landsat-1. A general objective of the S065 experiment was in fact to conduct a simulated test, using returned film, of the ERTS-A concept. It was therefore much more important for long-range program planning than the single-camera color photography experiment carried on Gemini and Apollo earth-orbital missions, and the Manned Spacecraft Center (now Johnson Space Center) carried out an extensive program of camera development, testing, and training for the experiment. In addition, several co-investigators were brought in: R. N. Colwell, E. F. Yost, P. N. Slater, and H. A. Tiedemann. The investigators were able, for the first time in a manned mission, to provide near real-time guidance to the Apollo 9 astronauts through a Staff Support Room maintained around the clock during the mission at the Manned Spacecraft Center to provide timely information on cloud cover and flight status.

The Apollo 9 mission, whose main objective was to test the complete lunar landing combination (Saturn V, Command/Service Module, and Lunar Module), was completely successful. The last 4 days of the 10-day mission were largely available for the S065 and S005 (single camera) photog-

FIGURE 4.8. Four-camera array for SO65 experiment installed by T. Brahm, Johnson Space Center, in hatch window of Apollo 9 Command Module, 1969. Cameras are Hasselblad 500 EL model, manually triggered, but with electric film advance, with 80-mm Zeiss Planar lens.

raphy, which were both extremely successful. The crew J. A. McDivitt, R. Schweikart, and D. Scott, obtained more than 90 4-picture sets (Fig. 4.9) and several hundred individual color photographs (Fig. 4.10, color plate 1). The test areas covered were chiefly in the southern United States and northern Mexico. As summarized by Lowman (10) and Colwell et al. (22), the Apollo 9 SO65 experiment led to the following conclusions.

1. Multispectral imaging techniques are essentially mandatory for operational earth orbital photography because of the extreme range of terrain, vegetation, and atmospheric conditions encompassed, in addition to the previously-demonstrated advantages of multispectral remote sensing.

2. Sequential imaging is of great value in orbital surveys for agriculture, forestry, hydrology, and geography.

3. Multistage sampling techniques, using orbital, aerial, and ground observations, are highly desirable for the applications mentioned above.

4. The ground resolution of the S065 photographs (nominally 100 to 130 m) though adequate for many applications, such as reconnaissance geologic mapping, should be improved if the imagery is to be useful without supporting higher-resolution air photos.

5. Field checking, preferably with the use of a light

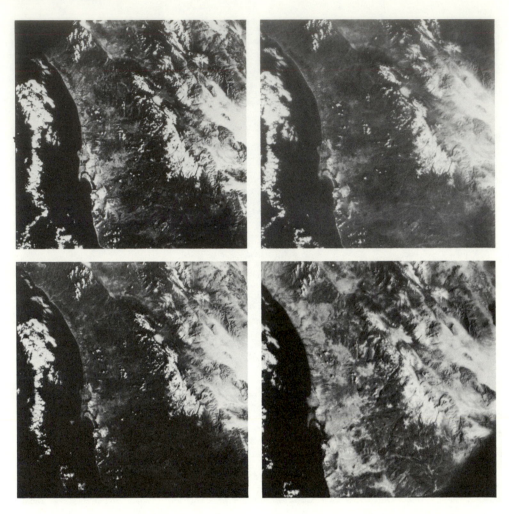

FIGURE 4.9. Set of pictures taken during Apollo 9 mission, 1969, as part of SO65 experiment, showing San Diego, California, and adjacent Peninsular Ranges of California and Baja California (11). Upper left, color infrared; upper right, green; lower left, visible red; lower right, infrared. Spacecraft altitude about 195 km.

plane, is mandatory for accurate interpretation of the photos.

The most general and important conclusion, however, was that multispectral orbital photography of the sort planned for ERTS-A was feasible and valuable in many different applications. The S065 pictures themselves were used for several years in planning image processing and interpretation techniques for ERTS, and helped stimulate interest in the Earth Resources Technology Satellite Program. The S065 cameras and technqiues were used again for the Apollo 12 mission to photograph the lunar surface (23), again producing many useful pictures.

Before concluding this review of the Apollo photography, something should be said of the single-camera color photography. Apollo 6 (unmanned) in 1967 carried an automatic 70-mm camera (24) loaded with color film, and returned one complete revolution (minus the dark side of the Earth) of extremely high-quality overlapping vertical pictures [Figs. 4.11(a), color plate 2 and 4.11 (b)]. This mission demonstrated the suitability of automated spacecraft for systematic vertical photography if there is no need to save film by avoiding cloudy or dark areas, or to photograph specific, off-track targets. Apollo 7, the first manned Apollo mission, carried in October 1968 the S005 single-camera terrain photography experiment originally intended for the ill-fated AS-204 mission (re-named Apollo 1). About 200 color pictures usable for the purposes of the experiment (geology, geography, or oceanography) were obtained [Figs. 4.12 (a), color plate 3 and 4.12 (b)], many of outstanding quality (11). Although a number of these have been published, the Apollo 7 pictures have to date not been studied in detail beyond the initial science screening (6). The reason for this was the necessarily fast pace of the Apollo missions, spaced only about three months apart and in addition the obvious fact that Apollo was above all a lunar-landing program.

The Apollo 9 crew, as mentioned earlier, took about 390 hand-held single-camera color photographs. Most of them are of outstanding quality, although frequently oblique, and are excellent for general illustrations of geologic features (11). Furthermore, the low altitude of the spacecraft (about 190 km) compared with the later Skylab and Landsat orbits resulted in unusually high ground resolution for the near-vertical pictures. Consequently, the hand-held and multispectral photographs from this mission are, like those from Gemini, of permanent archival value for comparison with future ones.

From the first lunar landing mission, Apollo 11, operations on and around the Moon naturally dominated the missions. Photography of the lunar surface was a major orbital experiment. El-Baz demonstrated the value of visual observations as a supplement to such photography (25), and later organized a similar combined experiment for the Apollo-Soyuz Test Project of 1975. Further discussion of the lunar photography would be beyond the scope of this chapter (see Chapter 20), but attention should be called to the ultrahigh altitude photographs of the Earth taken by the Apollo astronauts on the way to the Moon. Perhaps the most spectacular of these from a purely geologic viewpoint is Fig. 4.13(a), color plate 4 taken at about 11,000 km altitude from the Apollo 11 spacecraft after translunar injection. Covering as it does [Fig. 4.13(b)] the Pacific Northwest, the picture shows several geologic features, such as the Rocky Mountain trench, with unusual clarity because of their strong physiographic expression. Furthermore, this particular view was at the time the first color space picture of the northern United States, since the Gemini and earth-orbital Apollo missions had all been in low-inclination orbits confined to latitudes of 34° or less. Several other deep-space

Apollo pictures of the Earth, and some taken from the surface of the Moon, have become standard fixtures of environmental publications, testifying to the great impact they had on public consciousness. It was driven home to all that the Earth is indeed a small, fragile oasis in the desert of space, an achievement that may eventually rank as one of the most lasting monuments to the Apollo Program.

4.5 PHOTOGRAPHY FROM SKYLAB

The Apollo Applications Program Orbital Workshop mission, using a converted Saturn V SIVB stage, eventually became Skylab, a 100-ton (earth weight) structure that was the first United States space station (26) in 1973-1974. A major series of orbital remote sensing experiments was carried out during the three manned missions, the longest of which was 84 days.

These experiments included earth resource observations using three types of camera, an infrared spectrometer, a multispectral scanner, a microwave radiometer/scatterometer, and an L-band radiometer, collectively labeled the Earth Resources Experiment Package (EREP). These experiments and their products have been summarized in the Skylab Earth Resources Data Catalog (Johnson Space Center JSC 09016, 1974); only the photographic experiments will be described here.

The S190A multispectral photographic camera was an array of six high-precision cameras with 152-mm focal length lenses using 70-mm film, similar in general concept to the earlier S065 array. Film types included black-and-white infrared, black-and-white, color infrared, and color. At the nominal altitude of 435 km, a ground resolution of from 30 to 80 m was expected for low contrast targets. Several thousand pictures were returned from this experiment (Fig. 4.14, color plate 5), many taken in conjunction with simultaneous aerial remote sensing and ground measurements.

The S190B Earth Terrain Camera was aligned with the S190A array. It had a 457-mm lens and a large frame size, 11.4 by 11.4 cm. Only one film type was used at a time, but during the three Skylab missions, color, color infrared, and black-and-white film were exposed. The combination of long focal length and large format produced remarkably high ground resolutions (Fig. 4.15, color plate 6), while retaining a swathwidth (109 km) large by aerial standards although necessarily less than that of the S190A or hand-held cameras. The ground resolution with high-definition black-and-white film was estimated to be 17 m, but of course high contrast objects, such as boats, much smaller than this could be imaged although not resolved by shape.

Hand-held photography was not a formal experiment on this mission as it had been on previous earth-orbital ones. However, the Skylab astronauts had offered to do such photography on a spare-time basis. Accordingly, the crews were given several briefings and, for the last (SL-4) mission, formal classes in terrain photography and related subjects. A small manual for 70-mm photography was prepared and stored in the workshop for use during flight (Fig. 4.16) (27). The Skylab astronauts took several hundred color pictures with a hand-held camera equipped with a hand-held camera equipped with a 100-mm lens (Fig. 4.17).

The Skylab Earth Resources Experiment Package was immensely successful in terms of quantity and quality of data returned. The photographic results alone were staggering, with over 35,000 frames of imagery obtained, in addition to the data from other EREP experiments. When one considers that pictures from even the Gemini and Apollo terrain photography experiments are still under study, it is hardly surprising that only the most preliminary analyses of EREP and hand-held photography from Skylab have been published at this writing. Accordingly, conclusions reached about geologic orbital photography as carried out from Skylab must be both general and tentative, especially since comparative studies of EREP and Landsat imagery have barely begun. These conclusions follow.

1. The most general result of the Skylab experiments is that earth observations (photography and nonimaging experiments) *can* be carried out successfully from a large, multipurpose space station despite conflicting requirements on a crew time, attitude control, and other factors. This seemingly obvious conclusion is more important than it may appear, for it must be remembered that earth resource observations were only one of the four major objectives of the Skylab missions, the others being to study astronomical objects, to conduct zero-gravity materials processing, and to investigate manned space flight capabilities and basic biomedical processes. Furthermore, the performance of earth observation experiments was hampered, at least on the first manned mission (SL-2), by the repairs necessary to make the space station fully operational, such as releasing a jammed solar panel.

2. A second conclusion is in a sense the complement of the first and may seem to contradict it: the particular configuration used for the EREP experiments was not very efficient. The instruments were rigidly mounted in the Multiple Docking Adapter (MDA), which meant that the entire cluster had to be oriented for earth resources passes. This of course produced potential conflicts with other experiments, such as those involving the Apollo Telescope Mount, and when carried out with the Nitrogen Thruster Attitude Control System, involved consumption of reaction control fuel. It is therefore safe to suggest that earth photography from future space stations or large spacecraft should be done from either tethered or gimbaled platforms capable of independent pointing, or from station-keeping free-flying satellites.

3. The presence of man is a major asset to earth observation experiments, despite the tremendous success of the automated Landsat spacecraft. This conclusion is based on the overall performance of the Skylab astronauts (28), rather than just on their execution of the EREP experiments. The advantages of manned spacecraft in such experiments can be summarized thus.

A. Greater Versatility and Reliability

The Skylab astronauts made many adjustments and repairs to both the spacecraft (more properly, to the orbital workshop) and to the various instruments during the mission, far beyond those possible by ground control alone. This easily predictable advantage of manned over unmanned missions is especially impressive in view of the tremendous complexity of the scientific program carried out by the Skylab crews. No automated spacecraft has carried anything like the number of experiments flown on Skylab.

B. Acquisition of Transient or Unexpected Phenomena

The value of manned spacecraft in observing transient features was well-demonstrated by the operation of the Apollo Telescope Mount instruments on Skylab to observe solar flares (29) and other events on the Sun, and to observe and photograph with hand-held cameras unexpected oceanic features such as algae blooms. However, this

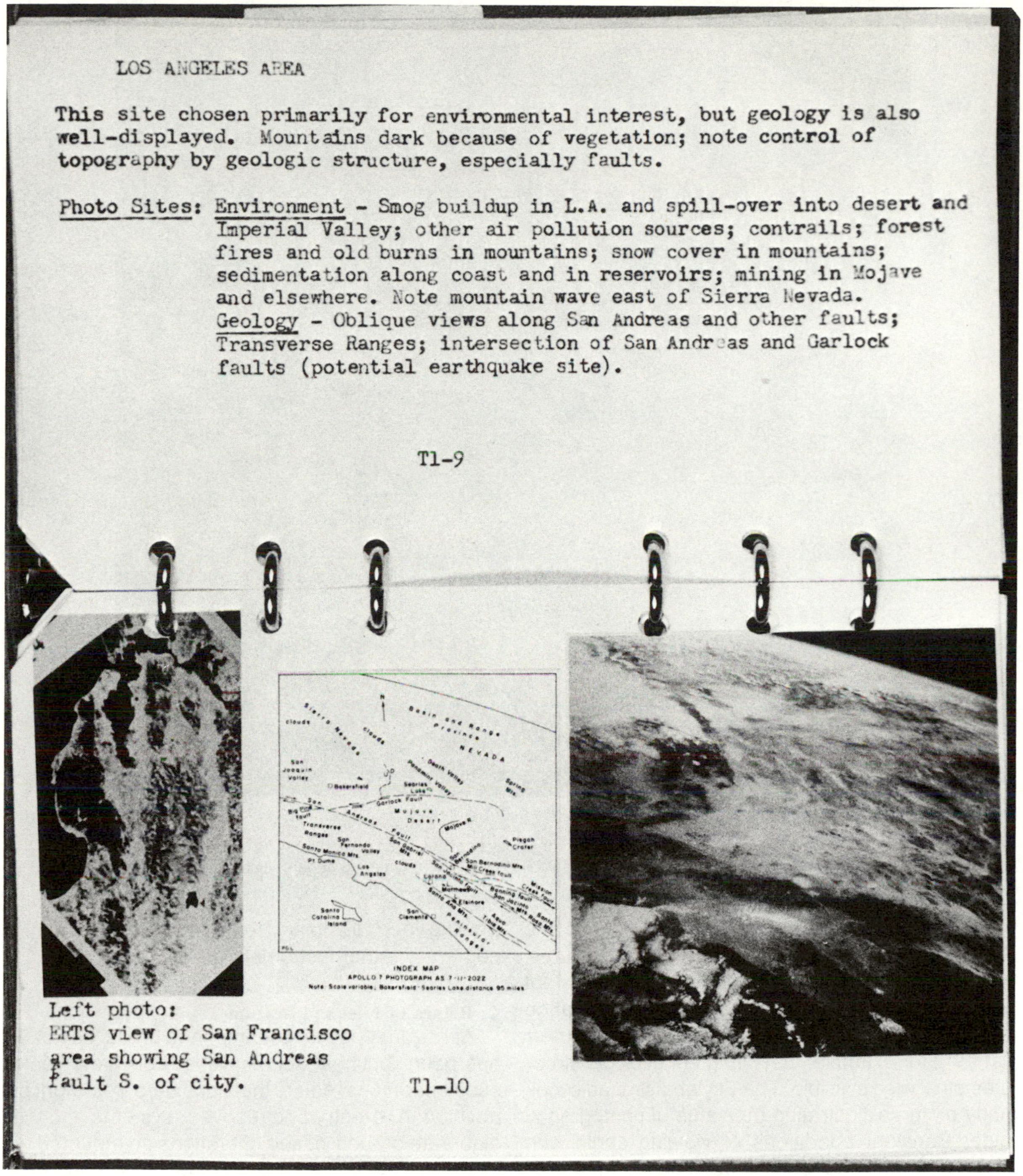

LOS ANGELES AREA

This site chosen primarily for environmental interest, but geology is also well-displayed. Mountains dark because of vegetation; note control of topography by geologic structure, especially faults.

Photo Sites: Environment - Smog buildup in L.A. and spill-over into desert and Imperial Valley; other air pollution sources; contrails; forest fires and old burns in mountains; snow cover in mountains; sedimentation along coast and in reservoirs; mining in Mojave and elsewhere. Note mountain wave east of Sierra Nevada.
Geology - Oblique views along San Andreas and other faults; Transverse Ranges; intersection of San Andreas and Garlock faults (potential earthquake site).

T1-9

INDEX MAP
APOLLO 7 PHOTOGRAPH AS 7-11-2022

Left photo:
ERTS view of San Francisco area showing San Andreas fault S. of city.

T1-10

FIGURE 4.16. Typical page from Skylab 70-mm hand-held photography manual (27). Original in color.

FIGURE 4.17. Skylab 2 hand-held 70 mm photograph of Lake Baikal, looking to southwest. Original in color. Compare with Fig. 4.2(a); note Sayan fault, terminating apparently at Lake Baikal.

capability is also useful in geology, despite the relative permanence of many geologic features, because although the features themselves may not change, as mentioned earlier, their appearance does. Low sun angle and snow cover (Fig. 4.15, color plate VI), for example, may bring out geologic structures otherwise invisible. The Skylab astronauts took many pictures illustrating the value of photography under transient conditions of viewing angle, sun angle, and snow cover. The advantage of manned over unmanned spacecraft in such photography lies in the fact that an astronaut can see (spacecraft configuration permitting) from horizon to horizon, in color, and even unaided with good enough resolution to identify many interesting features or conditions. Most important, he can, with proper training and planning, react instantly to train a camera or other instrument on the target.

C. Return of Film and Instruments

Although recovery of unmanned orbital payloads has been routinely accomplished for many years, such operations require that considerable weight be devoted to re-entry and recovery systems and that elaborate ground-based operations be mounted for the actual recovery. With manned missions, however, such return of film and cameras is almost a by-product, adding little to the cost of the mission. Since returned film photography appears to be desirable for some time to come, as will be discussed, this advantage of manned missions for photography is a significant one.

4. Despite the great success of the Landsat scanner and television systems, returned film photographs are at this time still very much in competition for geologic orbital photography. This conclusion is based partly on the work of Lee and Weimer (30), who investigated the use of S190A and S190B pictures covering the Colorado Plateau and concluded that the Skylab photographs were better than Landsat images for geologic mapping in this area. This superiority seems to rest primarily on the greater resolution of the Skylab photographs, which permitted these investigators to recognize photointerpretation keys such as texture, drainage, topography, and erosional resistance of the various stratigraphic units. Color (true color in this case) was also a major advantage, although this area is unusually well-exposed and this advantage would not necessarily apply to heavily-vegetated areas. Stereoscopic coverage was also cited as helpful. It seems safe to say that the Skylab experience provides strong arguments for the inclusion of returned film color photography in any comprehensive orbital observation program oriented toward geology.

5. Hand-held photography is still of real value in geology, although the success of Landsat has made it of distinctly lower priority than it once was. Its value lies in, first, that fact that, as described previously, trained astronauts can spot and photograph unexpected or transient geologic targets over a large area. Furthermore, oblique photographs, such as Fig. 4.17, are frequently better for illustrating large tectonic features and may bring out unexpected tectonic relationships. Obliquity may also of course lead to mistaken interpretations, and must be used with caution.

6. Sun-synchronous orbits should not be used exclusively for geologic orbital photography. This point will be discussed in relation to Landsat, which does use such orbits. However, the Skylab photographs provide a wide range of sun azimuths and sun elevation angles becuse of the different orbit (50° inclination prograde). This is a handicap in systematic repetitive coverage, but enhances the probability of getting really good pictures of, for example, lineaments with a particular orientation, or very low-relief features.

7. A firm choice on the best mix of sensors for operational geologic orbital investigations can not be made for several years. This conclusion is based on the fact that only the barest beginning has been made on, first, the combined use of Skylab sensor output (for example, combined radiometer/photograph analysis), and second, the comparison of the geologic value of different individual sensors and bands (for example, Landsat MSS versus Skylab S-192 scanner). Furthermore, the potential of existing electronic imagery recorded on computer-compatible tape has not yet been fully exploited with computer enhancement techniques; early Landsat MSS images of apparently mediocre quality have been reprocessed to produce new versions of outstanding color rendition and resolution. Finally the optimum resolution for geologic mapping has not yet been determined. Too high a ground resolution may swamp the interpreter with detail, or if attained with electrooptical sensors and telemetered may greatly increase the data load. Yet it seems clear that ground resolution two to three times the present nominal Landsat figure (79 m) is desired by most geologists.

4.6 LANDSAT IMAGERY

The concept of the Earth Resources Technology Satellite (now Landsat) was developed in the late 1960s, as mentioned previously, by joint NASA/USGS studies stimulated by the proven geologic value of orbital photographs from the Mercury and Gemini flights (31). Landsat-1 was launched on 23 July 1972 and began returning useful imagery within two days from both imaging systems, the Return Beam Vidicon (RBV) television system and the Multispectral Scanner (MSS). This satellite and its successors, Landsat-2 and Landsat-3, launched in 1975 and 1978, respectively, have been among the most successful unmanned spacecraft ever developed. The images have been used by hundreds of NASA-sponsored investigators and more recently by others at their own expense, for a wide variety of applications. The following discussion will focus on the development of geologic uses of Landsat imagery.

The first geologic use to which Landsat images was put (12) was a simple comparison with the 1:250,000 scale geologic map of California, the first state to be extensively covered (Fig. 4.18) by virtue of its generally clear weather. A structural sketch (Fig. 4.19) map was drawn from the first color composite made, showing the Monterey Bay area, and it was found immediately that there were many lineaments not shown on the state map

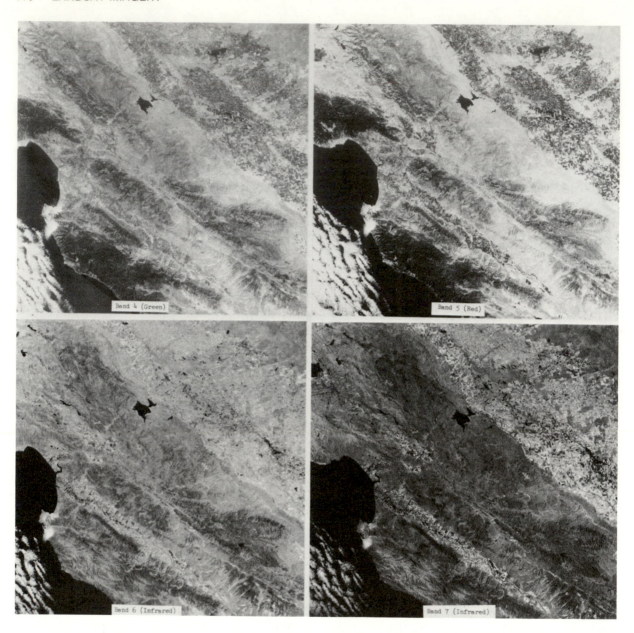

FIGURE 4.18. Landsat MSS images acquired July 25, 1972 (Image 1002-18134) showing Monterey Bay, Coast Ranges, and San Joaquin Valley.

despite its larger scale. Their nature could not be established without field checking, impossible in the time available, although it seems reasonable to interpret many of them as faults. In any event, the immediate detection of many large, unknown structures in a nominally well-mapped area gave the first indication of the great potential value of Landsat imagery for studies of regional structure. Similar studies by many other Landsat investigators,

generally in much more detail and with supporting field work, rapidly confirmed this (3, 4). Lineaments [Figs. 4.20(a), color late VII and 4.20(b)] proved to be the easiest type of structure to map with Landsat imagery, even in some areas with low relief and heavy vegetation such as Indiana (Wobber and Weir, in Ref. 4).

The value of Landsat in general geologic mapping appears substantial, although less clear-cut

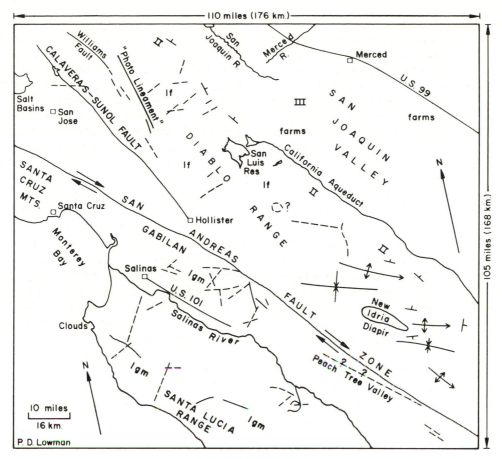

GEOLOGIC SKETCH MAP
CENTRAL COAST RANGES, CALIFORNIA

Legend:

<u>Calaveras</u>
<u>Fault</u>   Fault shown on 1:250,000 Geologic Map of California

- - - - -   Photo lineament (fault ?) not shown on Geologic Map of California

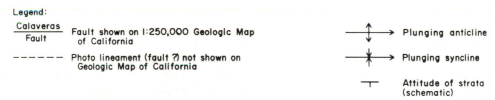

Plunging anticline

Plunging syncline

Attitude of strata (schematic)

Scale and orientation shown on map; principal point approximately 36°45'N, 121°10'W. Spacecraft altitude 560 stat. miles (900 km). Lithology after Page (1966); contacts not drawn:

   III Cenozoic terrestrial sediments

   II Late Mesozoic marine sediments (Great Valley sequence)

   Igm Granitic-metamorphic core complex with Cretaceous igneous intrusions

   If Franciscan core complex (Mesozoic eugeosynclinal rocks)

Paul D. Lowman Jr.
Goddard Space Flight Center
30 July 1972

FIGURE 4.19. First map drawn from Landsat-1 imagery (Fig. 4.18) completed 10 days after launch (12).

than in structural studies. The work of Houston et al. (in Ref. 4) provided an early demonstration that in terrain favorable for photogeology in general, in this case Wyoming, mapping at scales up to 1:250,000 could benefit from the use of Landsat pictures. This group found it possible to make many stratigraphic subdivisions at this scale, using the band 4, 5, 7 color composites despite the fact that these are not true rock colors. Their work is of particular interest since the state geologic map of Wyoming is a modern one, published in 1955.

Specialized geologic mapping in several fields has been proven to benefit from Landsat imagery. McKee and Breed (in Ref. 4) have studied 15 sand dune fields in remote areas of the eastern hemisphere (Asia, Africa, and Australia), and found it possible to identify and classify five major types of sand dune. The mapping and monitoring of seasonal changes in play as was demonstrated by Krinsley and by Reeves (in Ref. 4), although it should be noted that James Neal had demonstrated this several years earlier with Gemini photographs. Studies of coastal processes, in particular sediment transport and deposition, have been using Landsat imagery, by Pirie and Stleler (in Ref. 4).

Geologic education has already been a beneficiary of Landsat, as shown by the rapidly-increasing use of its pictures in textbooks. Orbital photography, especially in mosaics, permits students to actually see in their entirety large features such as the San Andreas fault and the Basin and Range Province, which could only be shown diagrammatically (and hence subjectively) before. Remote areas such as central Asia, which are of great tectonic interest but very poorly known, can become familar to students.

The application of Landsat imagery to exploration for minerals and hydrocarbons is at once one of the most interesting subjects and one of the hardest to discuss because of the proprietary nature of such applications. However, it is quite certain that Landsat pictures (or the original digital tape data) are in routine use by many oil and mining companies at their own expense. Economic applications by NASA-supported investigators are of course a matter of public record, and these have been summarized in Refs. 3 and 4; see also Rowan (15) and this volume. Some of these reports are indicative of the eventual economic value of Landsat. Liggett et al. (32) estimate, from a study in the Southwest United States, that in reconnaissance

exploration for mineral, groundwater, and geothermal resources, and in the study of geologic hazards, MSS imagery may provide cost savings of about 10 to 1 compared to conventional methods. Their work, and that of Robert Schmidt in Pakistan (15), show clearly that the actual application of Landsat data to mineral exploration will lie in targeting small sites for intensive investigation by airborne or surface methods; the cost savings will lie in narrowing the areas to be studied by these more expensive techniques. Exploration for oil and gas with Landsat, investigated in the Anadarko Basin by Collins et al. (in Ref. 4), can similarly benefit by, for example, reducing the cost of seismic surveys; this group estimated the possible savings at 20% to 50% of the cost of standard surveys.

In the long run, one of the most important benefits of Landsat will be increased insight into global tectonics. An excellent example of this has been provided by the work of Peter Molnar (33), who used Landsat imagery to study the tectonics of central Asia, which appears to be undergoing plate collision resulting in widespread seismic activity and crustal thickening. Kevin Burke and his colleagues have similarly used Landsat imagery to produce a photogeologic map of Tibet, finding evidence of widespread Cenozoic volcanism. A new theory for the origin of continents, holding them to be the remnants of an originally global igneous crust rather than coalesced island arcs or sialic nuclei, has been proposed by Lowman (11, 34), stimulated by studies of global geology with pre-Landsat space photography.

Several questions remain in the use of Landsat imagery similar to those discussed for Skylab previously. The optimum compromise between low and high resolution imagery is still under study. Present indications are that if only one sensing system were to be used from orbiting vehicles, ground resolution of 20 to 30 m would be best, permitting easy identification of terrain and cultural features on orbital images without the use of intermediate scale air photos, yet still retaining the synoptic view, which is the great advantage of such images. The need for repetitive coverage in geology is also under study. The dramatic change in appearance of geologic details with sun angle and season has already been noted. However, there are many kinds of geologic features that actually do change significantly over periods of a few weeks or months: glaciers, landslides, coastlines, volcanos, river channels, and sedimentation patterns. It has never been practical to monitor such changes over large areas; Landsat now makes this possible. The topic of repetitive coverage for geologic purposes can be summarized at this time by suggesting that it will take several years of repeated orbital coverage with Landsat or similar satellites to adequately record the changing *aspects* of the world's land areas; and by that time, geologic features of the sort mentioned previously will have *changed physically* since they were first imaged from space. Consequently, the geologic requirement for orbital coverage will be a permanent one, continuing into the future indefinitely. It should also be pointed out, in support of this prediction, that many scientific data are not used until years after they are collected; a picture of some remote part of Afghanistan taken in 1973 might be the final link in a study completed in 1993.

As mentioned in the discussion of Skylab, the best combination of sensors for all geologic orbital observations cannot be specified at this time, and of course continued progress in sensor development will presumably require corresponding changes in such a "best" combination. In addition, the design of satellite orbits for different applications requires further study. The sun-synchronous Landsat orbit, for example, has been shown by Goetz *et al.* (35) to bias structural interpretations by enhancing lineaments perpendicular to the sun azimuth, and a range of sun elevations and azimuths is clearly desirable. Finally, the proper combination of manned and unmanned missions for Earth observations is still to be determined; the forthcoming Shuttle/Spacelab missions will be critical for this question.

4.7 HEAT CAPACITY MAPPING MISSION (AEM-1)

The first satellite launched specifically to measure temperatures on the Earth's surface, for geological, agriculture, and other applications, was put in a circular, sun-synchronous orbit in 1978, and labeled the Heat Capacity Mapping Mission (one of the Applications Explorer Mission series). This satellite carried a two-channel scanning radiometer which produced imagery with a ground resolution of about half a kilometer along a 770 km-wide swath. One channel covers visible and near-infrared wavelengths, 0.55 to 1.1 μm, the other the thermal infrared, 10.5 to 12.5 μm.

The HCMM data were expected to be of geologic value in rock type discrimination by means of variations in thermal inertia and in the monitoring of soil moisture. Data analysis had barely begun at this is written, but the satellite had clearly achieved its nominal objectives with regard to imagery. The first data acquired permitted construction of an image of the east coast of the United States and adjacent ocean in which a wide variety of thermal features on land and water could be easily distinguished.

4.8 SEASAT-1

Another satellite of interest to geologists launched in 1978 was Seasat-1. Although intended for ocean surveillance, some of the Seasat sensors produce data of potential geologic value. Of particular importance is the first imaging radar flown in earth orbit by NASA, which despite its high depression angle proved capable of delineating geologic features, such as faults, at least in areas of high relief. The Seasat radar is discussed further in Chapter 10. An identical instrument, but with a lower depression angle for better terrain rendition, is planned for eventual launch specifically for geologic investigations. The initial success of Seasat-1 strongly supports the geologic value of orbital imaging radar.

4.9 SUMMARY

The unexpectedly sudden development of space flight is beginning to have major impact on geology in several ways. Study of the Earth's structure as revealed by its gravity field, through satellite tracking, has already undergone a revolution. Comparative planetology, made possible by the achievement of interplanetary flight, is beginning to challenge prevailing theories on the origin of continents and ocean basins. Study of the surface of the Earth itself through space photography is certainly going to have an effect at least comparable to that of aerial photography and probably a much greater one. Geologists of future generations will know the face of the Earth with a familiarity undreamed of in 1946 when space photography began.

We close this review with the eloquent words of Angus Campbell (20): "We are fortunate that Man, in his haste to explore the Universe, has taken time to look back at the planet which has succored him for so many centuries."

REFERENCES

1. Darden, L., 1974, The earth in the looking glass: Garden City, N. Y., Doubleday, 324 p.
2. Derr, A. J., 1972, Photography equipment and techniques: a survey of NASA developments: NASA SP-5099, Washington, D. C., 182 p.
3. Freden, S. C., Mercanti, E. P., and Becker, M. A., eds., 1973, Symposium on significant results obtained from the Earth Resources Technology Satellite-1: NASA SP-327, Washington, D. C.
4. Freden, S. C., Mercanti, E. P., and Becker, M. A., eds., 1974, Third Earth Resources Technology Satellite-1 symposium: NASA SP-351, Washington, D. C.
5. Holz, R. K., ed., 1973, The surveillant science: Boston, Houghton Mifflin, 390 p.
6. Kaltenbach, J. L., 1969, Science screening report of the Apollo 7 mission 70-millimeter photography and NASA earth resources aircraft mission 981 photography: NASA TM X-58029, Houston, Texas, 179 p.
7. Lowman, P. D., Jr., 1964, A review of photography of the earth from sounding rockets and satellites: NASA TN D-1868, Washington, D. C., 25 p.
8. Lowman, P. D., Jr., 1965, Space photography. a review: Photogrammetric Eng., v. 31, p. 76-86.
9. Lowman, P. D., Jr., 1967, Geologic applications of orbital photography: NASA TN D-4155, Washington, D. C., 37 p.
10. Lowman, P. D., Jr., 1969, Apollo 9 multispectral photography: geologic analysis: Goddard Space Flight Center Document, X-644-71-15, Greenbelt, Maryland, 53 p.
11. Lowman, P. D., Jr., 1972, The third planet: Zurich, WELTFLUGBILD Verlag Reinhold A. Muller, 170 p.
12. Lowman, P. D., Jr., 1969, Geologic orbital photography: experience from the Gemini Program: Photogrammetria, v. 24, p. 77-106.
13. Lowman, P. D., Jr., 1973, Chap. 21, Geologic uses of earth orbital photography in Holz, R. K., ed., The surveillant science: Boston, Houghton Mifflin, p. 170-182.
14. Reeves, R. G., ed., 1975, Manual of remote sensing: Falls Church, Virginia, American Society of Photogrammetry, 2144 p.
15. Rowan, L. C., 1975, Application of satellites to geologic exploration: American Scientist, v. 63, p. 393-403.
16. Lathram, E. H., 1972, NIMBUS IV view of the major structural features of Alaska: Science, v. 175, p. 1423-1427.
17. Lowman, P. D., Jr., McDivitt, J. A., and White, E. H., II, 1967, Terrain photography on the Gemini IV mission: preliminary report: NASA TN D-3982, Washington, D. C., 15 p.
18. Wobber, F. J., 1967, Put geology surveys in orbit to find oil: The Oil and Gas Journal, v. 59, no. 50, p. 63-67.
19. Abdel-Gawad, M., 1971, Wrench movements in the Baluchistan Arc and relation to Himalayan-Indian Ocean tectonics: Geol. Soc. America Bull., v. 82, p. 1235-1250.
20. Pesce, A., 1968, Gemini space photographs of Libya and Tibesti: in Campbell, A. S., ed., Tripoli, Libya, Petroleum Exploration Society of Libya, 81 p.
21. Lowman, P. D., Jr., 1975, The Apollo Program: was it worth it?: Forensic Quarterly, v. 59, p. 291-302.
22. Colwell, R. N., and others, 1971, Monitoring earth resources from aircraft and spacecraft: NASA SP-275, Washington, D. C., 170 p.
23. Goetz, A. F. H., Billingsley, F. C., Yost, E., and McCord, T. B., 1970, Chap. 9, Apollo 12 multispectral photography experiment, in Apollo 12 preliminary science report: NASA SP-235, Washington, D. C., 227 p.
24. Amsbury, D. L., 1973, Chap. 20, Geological comparison of spacecraft and aircraft photographs of the Potrillo Mountains, New Mexico, and Franklin Mountains, Texas, in Holz, R. K., ed., The surveillant science: Boston, Houghton Mifflin, p. 161-169.
25. Evans, R. E., and El-Baz, F., 1973, Chap. 28, Geological observations from lunar orbit, in Apollo 17 preliminary science report: NASA SP-330, Washington, D. C., p 28-1 to 28-32.
26. Canby, T. Y., 1974, Skylab, outpost on the frontier of space: National Geographic, v. 146, p. 441-493.
27. Lowman, P. D., Jr., Frey, H. V., Shenk, W. E., and Dunkelman, L., 1973, Manual for 70 mm

hand-held photography from Skylab: Goddard Space Flight Center Document, X-644-73-147, Greenbelt, Maryland, 81 p.

28. Garriott, O. K., 1974, Skylab report: man's role in space research: Science, v. 186, p. 219-226.

29. Gibson, E. G., 1974, The Sun as never seen before: National Geographic, v. 146, p. 494-503.

30. Lee, K., and Weimer, R. J., 1975, Geologic interpretation of Skylab photographs, Remote sensing report 75-6: NASA Contract NAS 9-13394, Colorado School of Mines, Golden, Colorado, 72 p.

31. Pecora, W. T., 1969, Earth resource observations from an orbiting spacecraft, *in* Singer, S. F., ed., Manned laboratories in space: New York, Springer-Verlag, p. 75-87.

32. Liggett, M. A., and research staff, 1974, A reconnaissance space sensing investigation of crustal structure for a strip from the eastern Sierra Nevada to the Colorado Plateau, Final Rept.: NASA Contract NAS 5-21809, Springfield, Virginia: National Technical Information Service, Contract No. NAFS-21809.

33. Molnar, P., and Tapponnier, P., 1975, Cenozoic tectonics of Asia: effects of a continental collision: Science, v. 189, p. 419-426.

34. Lowman, P. D., Jr., 1976, Crustal evolution in silicate planets: implications for the origin of continents: Jour. Geology, v. 84, p. 1-26.

35. Goetz, A. F. H., Billingsley, F. C., Gillespie, A. R., Abrams, M. J., Squires, R. L., Shoemaker, E. M., Lucchitta, I., and Elston, D. P., 1975, Application of ERTS images and image processing to regional geologic problems and geologic mapping in northern Arizona: Technical Rept. 32-1597, Jet Propulsion Laboratory, California Institute of Technology, Pasadena, California, 188 p.

SECTION II
IMAGE PROCESSING AND ENHANCEMENT

Most remotely sensed data used by geologists are in the form of images. Sometimes, these images are acquired in a survey designed specifically for the needs and desires of the geologist who must perform the photo-interpretation, but more often they are acquired by unmanned satellites or in aerial surveys serving many different analysts with diverse requirements. In such cases, image processing or image enhancement can be used to improve the display of the data for the individual geologist. This section of *Remote Sensing in Geology* introduces the reader to some of these methods, which range from conventional photographic darkroom procedures to complex data manipulations by digital computer.

It is important that geologists who are interested in interpreting remotely sensed data develop an understanding of the potential of image processing techniques, even if they do not intend to enhance images themselves, because only in this way can they be sure they are getting optimum use of their data.

CHAPTER 5

PHOTOOPTICAL TECHNIQUES OF IMAGE ENHANCEMENT
JAMES E. SKALEY

5.1 INTRODUCTION

Image enhancement is a general term referring to a number of operations designed to increase the amount of useful information from a scene. This term is not restricted to interpretation of imagery. It can also apply to color balance corrections or graphic-art effects that are used for illustration. Essentially, it includes all operations that modify an image to obtain an idealized output of some part of the original scene (1). It is important to ask the question: What do we modify in an image to produce the desired effect? The question is somewhat rhetorical in that one needs to know what components of a scene are most relevant to productively enhance the image. Sometimes it is not clear what factors are relevant and, hence, different techniques must be employed to produce results for comparative analysis. Eventually, a subjective judgment is made as to the most desirable approach. The point, in fact, is that there is no single approach to enhancing an image whether one works with digital or photographic processing.

Enhancement is an operation that falls between detection or recognition of the existence of a relevant feature and the interpretation and classification of that feature. We essentially enhance to improve or ease the interpretation of a scene relative to some specific information. There is the added benefit that the number of similar objects detected in a scene may also be increased. Nevertheless, one does not enhance blindly without some reference point. In many cases, this may be a priori information such as the knowledge that linear features are important clues for geological exploration. Given this reference one may use edge enhancement as a first step. (The phototechnique is discussed below. See Chapter 6 for a digital approach.) Likewise, the color in a landscape may be an important clue. On black-and-white photographs objects of varying spectral features may appear similar in tone. Recording the scene with color film instead of black-and-white would then be an enhancement. Similarly color infrared film, which records part of the visible and part of the near-infrared spectrum, provides an enhancement over standard color by increasing the contrast between land and water. It also indicates a change in reflectance between stressed and healthy vegetation due to a change in plant cell structure resulting from water loss.

The above discussion illustrates some simple enhancement concepts. We start with a monochromatic black-and-white scene on which linear contrasts can be magnified by edge enhancement. We then suggest color as a different dimension to discriminate objects having similar reflectances but different spectral characters. Finally, we shift the recorded wavelengths to take advantage of natural contrasts in a scene such as water absorbing near infrared and vegetation reflecting a large part of it. The net effect in each case is to "highlight" certain relevant information by selectively altering visual discriminants such as the contrast ratio, density, and spectral characters of a scene.

With this brief introduction, we now touch on a rationale for image enhancement and then proceed into a discussion on basic photographic film properties and concepts that are employed in enhancement. This is followed by an expanded discussion of some of the more widely used photographic techniques and then concluding with a hybrid approach to enhance and analyze multispectral imagery.

5.2 ATMOSPHERIC LIGHT EFFECTS

The requirement for enhancement of aerial photography is to compensate for some degradation of the image due to atmospheric effects. It may also be required to improve the spatial or spectral image resolution. Resolution enhancement is generally required when the recording format is less than adequate to meet the information needs. Such is the case when the objects of interest are near the limits of resolution for the photographic system or when there are problems of distortion or vignetting. Problems of the second nature are best handled by digitally reconstructing the image and making necessary corrections for distortion and exposure on a pixel by pixel basis (2) (a pixel is defined as one picture element), although resolution can be improved photographically by improving image sharpness and contrast by procedures described below.

The atmosphere affects photographic images by scattering, absorbing, and refracting light. The most important of these is light scatter. Scatter is caused by particulate matter and aerosols which are unevenly concentrated in the atmosphere. The effects of scatter are to increase the skylight and reduce the scene contrast. This results in a reduction in image resolution. Particulate matter is the principal cause of scatter in the visible portion on clear days with low relative humidity ($<30\%$); whereas, aerosols, particularly water vapor, become more important in the longer wavelengths beyond 1 to 2 μm. With clear atmosphere, more than 50% of the scatter is in the blue end of the visible spectrum. For this reason, black-and-white aerial photography is invariably recorded through a yellow filter to reduce the effects of this scatter and thereby improve image sharpness (3, 4). As the relative humidity rises above 40%, water condenses on the particulate matter forming haze conditions.

Therefore, in collecting aerial or satellite imagery in different geographical regions that have different climatic conditions, one must expect to have varying types of image enhancement problems (4). Many humid areas are continually plagued with haze or cloud cover. For some of these regions, the only type of image that can be reliably obtained is with microwave systems. Similarly, seasonal variations in both the atmosphere and spectral reflectance of vegetation have marked effects on the quality of imagery obtained.

Shadow resulting from low sun angle or rough topography can be useful in highlighting geologic structures. Many times, however, it obscures information. If the shadow is very deep, the sensor may not have recorded sufficient information to make any enhancement, but in diffuse shadow, it is possible to recover some detail especially if the shadowed area is processed separately from the rest of the scene.

5.3 IS IT WORTH IT?

Image enhancement can be costly and time consuming, so the photo-analyst must in each case consider the relative worth of the information and the difficulty involved in enhancement processes to "highlight" it. His or her decision often revolves around the economic axiom of marginal costs/marginal gains. In other words, the information derived should be of relatively large value compared to the cost of the effort to derive it. There is, of course, no a priori way of knowing in advance the worth of an added bit of information gained through enhancement. One gains insight through experience and familiarity with the techniques. In general, a rule of thumb is to keep the techniques and approach simple, using more sophisticated and costly approaches only when there is specific reason and a fair certainty that the added effort is worth the information gained.

5.4 PHOTOGRAPHIC FILM PROPERTIES

A black-and-white photograph is a continuous gradation of shades of gray (gray scales or monochromatic hues). These shades of gray reflect the continuous gradation in nature of different features intergrading with each other. In some cases, there are sharp contrasts, but in most the gradation is subtle. In classifying the information in a scene, one often desires to sharpen the contrast among features of interest, that is, define boundaries. We do this in selecting out specific gray levels, textures, or spatial patterns which help in defining features for later interpretation and classification. The following sections define several photographic properties that are employed.

5.4.1 Density

Stated mathematically, density = $\log\frac{1}{T}$, where T is the transmission of a light source through a film. Stated another way, it is a measure of the logarithm of opacity (5). As recorded on film with optimal exposure, density is a good measure of different reflectances within a scene. When related to the film exposure, it is a relative measure of luminous flux or albedo.

Within a recorded scene, solar radiance or brightness varies according to the angle of incident light, the spectral components, or the physical arrangements of surface reflectors, that is, the degree of surface roughness. The density of a scene may also be affected differentially by intervening aerosols in the atmosphere. These aerosols are of varying sizes and concentrations and scatter the light reflected off their surfaces. This has the effect of differentially increasing the amount of light recorded in shorter wavelengths due to the larger number of reflectors. The longer wavelengths, which are not subject to as much scatter, are therefore, transmitted through the atmosphere with less interference (6). As a result, the radiance values of longer wavelengths is more representative of surface albedos. The increased amount of scatter is recorded on the film as an overall lightening of tone and decreased contrast. Figure 5.1 illustrates this effect.

5.4.2 Characteristic Curve of D log E Curve

The characteristic curve or D log E curve represents the plot of density versus the log of exposure (5). The radiance from a scene can vary by as much as 10,000:1 (7). This ratio represents the variation in scene radiance expressed in lumens. By taking the log of this value, we can convert it to a reasonable scale with an upper limit of 4.0. (Film is considered to be opaque at a density of 3.0). When the density of a negative image is measured on a densitometer and plotted against the log of the film exposure time, we obtain a roughly S-shaped curve as shown in Fig. 5.2. The brightest parts of the scene are recorded toward the upper portion with the darker tones expressed below.

The curve is divided into regions: the toe, straight line portion, and the shoulder. When printed as a positive image, the "shadow" or darkest tones are derived from the toe and the "highlights" or lightest tones come from the shoulder. The straight line portion represents a region where there is a linear relationship between exposure and density. Overall the slope and placement of the curve on the axis depends on the scene brightness.

5.4.3 Gamma and Contrast Index

Gamma (γ) refers to the straight line portion of the D log E curve. It is calculated by determining the average slope of the straight line portion of the curve.

Therefore, $\gamma = \dfrac{D}{\log E}$ or $\tan \theta$

where D is measured density, $\log E$ is the logarithm of the exposure value, and θ is the angle created by the intersection of a horizontal line with the straight line portion of the D log E curve.

γ is directly related to the amount of film development. Since contrast can be altered on any film by varying the development time, it has often been used as a direct measure of contrast. However, many films have special emulsions which do not generate a straight line. In these situations, or where different film types or processing conditions are being compared, a more useful indicator is the contrast index (CI). The contrast index refers to the slope of the D log E curve between two specified points. The lower value is defined where the line intersects a circle with a radius of 0.2 density with its center on the abscissa. The second point is defined where the line intersects a concentric circle with the same center but with a radius of 2.2 density. This index has the advantage of a standard fixed reference when measuring varying shapes of curves of different films. In this sense it is a more useful replicate measure than is γ, particularly in cases where there is a long toe portion of the curve (7, 8).

5.4.4 Contrast

Contrast is often expressed in terms of γ. The higher the value of γ or the steeper the slope of the line the greater the contrast. However, expressed this way, it is only a useful measure within the straight line portion or middensity area of the D log E curve. Contrast can also refer to the total difference between extreme shadow and extreme highlight which includes both the toe and shoulder of the curve. We may also restrict its meaning to refer to toe contrast, which is a measure of the shadow detail and highlight contrast, which measures variations on the shoulder of the curve (5).

As we see it, contrast is the change in lightness between any adjacent points measured on a logarithmic scale. The eye does not perceive equal increments of intensity. It can easily discern the difference between 1- and 2-f-c of light, which represents a doubling of intensity, but it cannot distinguish between 99- and 100-f-c, which represents only a 1% increase in intensity (7). Generally, we express scene brightness differences as ratios.

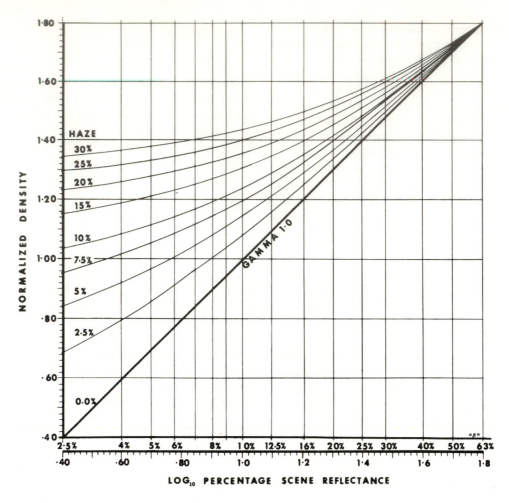

FIGURE 5.1. Curves illustrating effects of different amounts of atmospheric haze in distorting the linear reproduction scene radiances [after Ross (6)].

Taking the log of each value yields the density difference that you can expect in a target/background setting. For instance:

High contrast (100:1)   = density difference 2.0
Medium contrast (6.3:1) = density difference 0.8
Low contrast (1.6:1)   = density difference 0.2

Most features in a scene have a contrast ratio of less than 5:1 (4).

5.4.5 Color Relationships

Image enhancement is greatly facilitated through the use of color. Color is trivariant, that is, it is generated by a combination of at least three factors. This follows Grassman's laws (8), which state:

1. Three parameters are necessary and sufficient to define the color of a light.
2. All mixtures of colored lights yield continuous scales.

3. The result of additive mixture of colored lights only depends on their visual appearance and is independent of the physical origin of their colored aspect, that is, the light source may be filtered white light or a mixture of several monochromatic lights.

Color is generally considered to be a mixture of red, green, and blue light. The selection of the three primaries is favored by the fact that they are about equally separated in the visible spectrum and positioned so that different combinations can represent nearly the entire range of colors. Actually any combination of monochromatic light or hue will yield a range of color depending on the relative mix and their distance apart on the visible spectrum. Color is also described by its saturation or purity and its brightness. Saturation and purity indicate the amount of white light that decreases the color

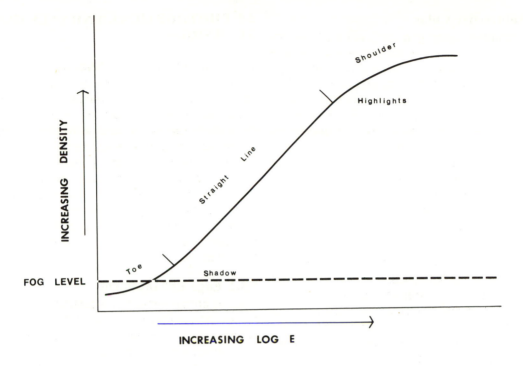

FIGURE 5.2. Characteristic curve for a standard negative. The straight line portion of the curve which shows a linear relationship between exposure and density is the region where one can systematically manipulate the density and contrast of an image so that all relevant information has a density above the fog level (0.3) but below the shoulder of the curve.

aspect; that is, as more white is added the color becomes less pure and is often referred to as a tint. Brightness refers to the intensity of the light relative to its surrounding (7, 8).

5.4.6 Additive Color

A full color scene is made up of varying mixes of red, green and blue primaries. These primary colors are reflected in proportion to the spectral characteristics of different objects in the scene. For instance, orange is a mix of red and yellow, purple is a mix of red and blue, and so forth. Similarly one can simultaneously transmit light through red, green, and blue filters and obtain a range of colors on a screen. The color varies according to the amount of overlap among the primaries and the relative proportion of light transmitted through each filter (7).

Details in multispectral images can be enhanced when combined by the additive process. When three different spectral images filtered respectively with red, green, and blue filters are projected and registered on a screen a color composite results. The shifts in color reflect differences in density in one or more of the spectral bands. A change in density on one band allows more or less colored

light to shine through, thus changing the mix of hues for that point on the screen (7). If the light transmitted through each filter is of equal intensity, the result is a white light much less intense than the source. Figure 5.3 illustrates the process.

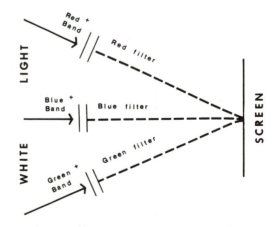

FIGURE 5.3. Additive color. Three different black-and-white images, exposed in different portions of the spectrum, can be combined by filtering each respectively with red, green, and blue filters. When projected and registered on a screen, a full color composite of the scene is rendered. Red, green, and blue light of equal intensity will produce white.

5.4.7 Subtractive Color

In subtractive color one can use the complementary colors cyan, magenta, and yellow and also obtain an approximate full color representation of a scene. This process is more appropriate to photographic methods than additive color in that it depends on the transmission through or reflection from three different dye layers as opposed to the projection and registration of three different hues. Fully saturated hues of yellow, cyan, and magenta in register produce a black. If we reduce the saturation or color density equally in each hue, we produce lighter shades of gray. If we reduce the saturation differentially, we generate a range of colors (see Fig. 5.4) (7).

Subtractive color is also more efficient in terms of the amount of white light required to obtain a color scene. In the additive process a maximum of 20% of the projected light comes through for highlight areas in contrast to nearly 100% transmission for highlight areas in the subtractive process (7). Therefore, additive color is more appropriate for electronic enhancement such as in color television where a focused beam of electrons is projected onto light emitting phosphors, which, when bombarded, emit light in a particular hue. The phosphors then take the place of the filters illustrated in Fig. 5.3. Various electronic image intensifiers are used to compensate for any loss in the signal.

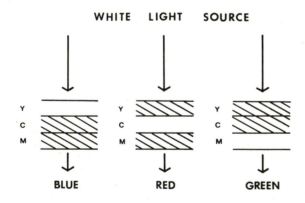

FIGURE 5.4. Subtractive color. Shining white light through different combinations of complementary hues of equal density (hatched areas) will generate different primary colors. For instance, cyan (C) and magenta (M) respectively subtract out red and green light while transmitting blue. Similarly yellow (y) subtracts out blue and transmits green and red. All three together will produce black.

5.5 PHOTOOPTICAL ENHANCEMENT TECHNIQUES

5.5.1 Improving Image Sharpness

Enlarging a portion of a scene results in a loss of definition or image sharpness unless one can compensate by exposure or processing. This is particularly true as one approaches the resolution of the film and the optics. When it is necessary to make extreme enlargements of a portion of an image the resolving power of the lens can be improved up to 2× by using a Kohler illumination apparatus attached to the enlarger. This ensures that the light entering the optics is uniformly distributed over the full aperture, thereby, improving the efficiency of the optics (9).

5.5.2 Edge Enhancement

Edge enhancement is a technique that can be achieved by using an "unsharp", that is, out of focus, negative mask sandwiched together with the positive transparency. A contact image of the two is made to achieve results pictured in Fig. 5.5. This technique is useful to increase the definition of linear features, that is, where there is a sharp density shift. The effect with directional blurring of the mask gives the illusion of topographic relief.

5.5.3 Contrast Stretching

Contrast stretching is a term more commonly used in computer enhancement where the density values in a scene are literally pulled further apart, that is, expanded over a greater range. The effect is to increase the visual contrast between two areas of different uniform densities. This enables one to discriminate more easily between areas initially having a small difference in density (2).

Somewhat analogous techniques can be accomplished photographically. Most contrast enhancement is done on the straight line portion of the D log E curve (see Fig. 5.2). The slope (γ) and length (density range) of this segment can be systematically changed by using different emulsions and developers, and by varying the development time. The total length of the line limits the range of discrete and uniform density shifts that can be printed. Therefore, it is important to select emulsion types and developers that match the contrast range required to produce the desired effect (10, 11). Figure 5.6 shows several film/developer combinations which illustrate the changes.

FIGURE 5.5. Photographic edge enhancement of linear features in a basaltic region of South Central Idaho from a Landsat I scene (scale ~ 1:330,000).

The greater the slope the more contrast per unit of line with a resulting decrease in observable levels of gray. Less slope on an equivalent straight line produces a greater number of gray levels, but with a proportionate decrease in contrast. Optimal slope for normal processing is a gamma of 1.0 so as to effect an increase in density contrast proportional to the scene contrast (5). As illustrated, some film/developer combinations produce a very long straight line while others have built-in characteristics that make them less suitable for contrast enhancements. Nevertheless, the slope and shape of the curve for each film/developer combination is variable with development time (10, 5).

This is an overly simplified explanation of how the $D \log E$ curve changes with emulsion characteristics and processing. It, however, illustrates a range of options to affect the contrast index ratio of the image scene using photo techniques.

Contrast stretching of an image photographically cannot be done with quite the same results obtained through digital processing. A digital computer typically records up to 256 shades of gray in a linear mode. Electronic Landsat images may record up to 127 levels of gray. These are discrete equidensity values directly related to the scene radiance, that is, each net increment of light is recorded as a separate shade of gray.

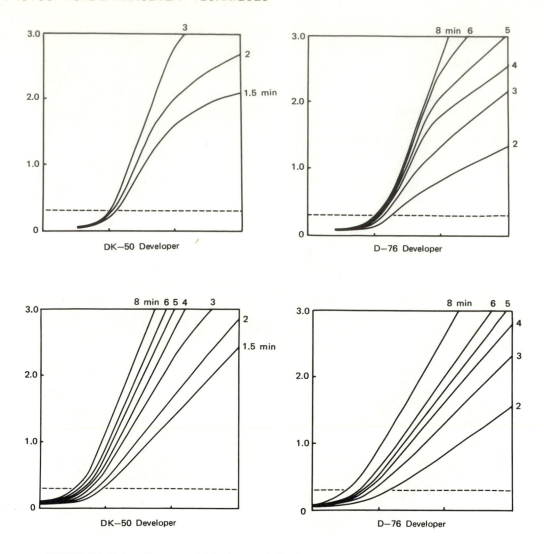

FIGURE 5.6. *D*/log *E* curves which show variation in slope (contast) produced by different emulsion/developer combinations. The top two graphs were plotted from contrast projection slides (high contrast emulsion) and the bottom two show the same developers acting on medium projection slides (medium contrast emulsion).

As discussed above, film records density values on a logarithmic scale. Therefore, equal density units on the scale do not relate to equal changes in illuminance. The eye also perceives scene brightness on a logarithmic scale. However, being less sensitive to grays than colors, it is unable to distinguish fine breakdowns on the density scale. That these more discrete values do exist in the exposed negative can be illustrated by taking densitometer readings or repeatedly enlarging a small portion of a negative until the granular structure becomes apparent.

This expansion of grays is the result of a mixture of grain sizes in the emulsion which are differentially sensitive to light intensities on exposure. The latitude or dynamic range of the film is determined by the relative mix of silver grains. If all grains are of approximately the same size, they will all tend to be developed within a very narrow range of exposures and would then yield an objectionally high gamma film (12). Varying the mix of the size of silver grains changes the sensitivity and latitude of the film.

Each enlargement increases the number of perceived shades of gray over what was recognized initially. Therefore, it is important to distinguish between absolute film density values that record varying light intensities and perceived density changes that are qualitative. Most of the scene information recorded on the negative is compressed into the middle density ranges. When compared to the digital format, the basic limitation of film is the dynamic range of various emulsions and the inability to make linear comparisons over the total range of exposures.

Depending on the scene, we can select an emulsion and generally process for an optimal density range of 1.0. For low contrast scenes this means that we can increase the contrast index by varying the emulsion, developer, or development time. Similarly, for high contrast scenes, the contrast index can be lowered. By increasing the contrast index, we create a greater separation of dark and light tones but with a simultaneous loss in gray levels. This may also be interpreted in some cases as loss in scene information. It is then important from an image enhancement perspective to weigh the question of what kinds of manipulations of the negative image make the scene most intelligible with respect to the kinds of information to be extracted.

We can compromise at this point and selectively record parts of a properly exposed negative to image only a portion of the density range of the original on the second generation image. Varying the exposure, we shift the D log E curve to right or left, selectively recording the light or dark densities, respectively. These densities can be expanded over the dynamic range of film or paper by appropriate processing.

Figure 5.7 illustrates the preceding discussion. In this case, we have chosen a single exposed film and developed three different positives under different development conditions. In each case, a prescribed point has been developed to a density of 1.0, but by varying the development time and/or developer, the contrast index is changed so that the values of other points in the scene are differentially affected. As one can see from the gray scales, the change in development time changes the density range and contrast index simultaneously. Printing each stepwedge in a different complimentary hue and registering them so that points A, B, and C are superimposed, we obtain a neutral gray. [This follows from the discussion on subtractive color process that if all three hues (cyan, magenta, and yellow) were of equal density, the resultant color would be black or a shade of gray since each acts as a minus filter for one of the other two hues.] Values for other portions of the stepwedge are not of equal density, so that there are progressive shifts in hue on either side of the neutral value.

5.5.4 Density Slicing

A second approach to image enhancement is to select out specific densities in a scene by level slicing. One method records discrete density slices in one operation with the use of Agfa contour film. This film is constructed so that it has both a negative and positive D log E curve, (see Fig. 5.8). In fact, the film acts as two films sandwiched together with one acting as a mask for the other. The amount of separation between the two curves determines the exposure latitude of the film. The high γ results in only a few density levels being printed. By using a series of increasing exposures one can shift the D log E curves from left to right in successive steps. With the addition of a yellow filter the density slice can be narrowed (13).

Nielson and other researchers have used this technique extensively for image enhancement of aerial photography and satellite imagery. The various density slices are recombined by the additive color technique. When projected on the screen, the effect is to create sharp color contrasts among the different slices. Where densities overlap the resultant color will be a mixture of the respective hues from each monochromatic light source. This follows according to the law of additivity.

Density slicing of single scenes can be a useful technique for special applications, particularly where one wants to note subtle change, but the results must be interpreted with care. One cannot necessarily classify a landscape simply by separating the various density levels that for a given set of points appear to represent discrete land/vegetation components. Such an approach fails to account for the natural variation of reflectance values expected from vegetation, soil, and so forth. In addition, other factors such as specular angles, variations in the atmosphere, and incident light result in varying angles of reflectance with a corresponding change in the energy flux recorded. At the sensor, factors, such as lens flare, vignetting, and film processing, may introduce density varia-

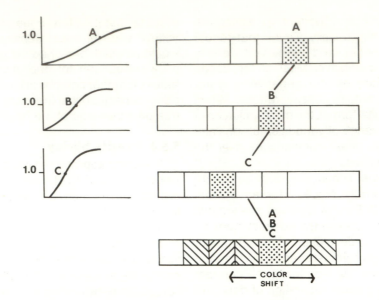

FIGURE 5.7. Varying development to isolate density by changing the contrast index ratio. In each case above a selected gray on the stepwedge is developed to a density of 1.0. Printing each transparency onto cyan, magenta, and yellow diazo film, respectively, and registering the selected value produces a neutral gray with color shifts showing changes away from the isolated density value. (For an explanation of diazo film refer to Section 5.5.7.)

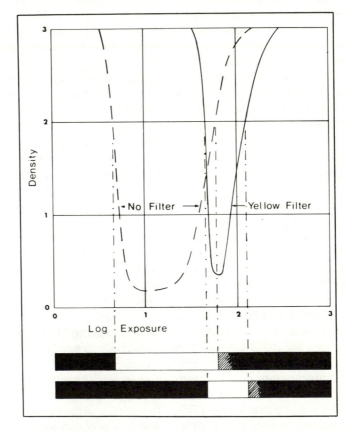

FIGURE 5.8. Characteristic density curves of Agfa contour film [modified by Nielson (13) from information supplied by Agfa-Gevaert].

tions independent of the scene reflectance. All these factors should be weighed when attempting any enhancement. In general, it is not appropriate to classify a scene by density slicing. The technique is, however, often useful in highlighting variation in low contrast scenes, such as water bodies.

5.5.5 Color Separation Techniques for Color and Color IR Film

It is often desirable to make extreme enlargements of color or color infrared films. Since these films are inherently lower in spatial resolution than black-and-white film, there may be a considerable reduction in image quality and sharpness of features. Where extreme enlargements (10 to 20 times) are necessary or where it is desirable to optically enhance a portion of the image, one can make use of standard color separation techniques to take advantage of the higher resolution of black-and-white films.

Red, blue, and green wratten filters are used to separate the cyan, magenta, and yellow layers of the color film. Information recorded on each layer is projected in the enlarger onto a high contrast black-and-white copy film to make negatives. Further enlarged black-and-white positives can now be made. The original can be simply reconstituted using complimentary diazo films (cyan, magenta, yellow) and using the subtractive color approach (11). (Diazo films are commonly used in graphic art illustrations. Further elaboration is given in Section 5.7). Figure 5.9 color plate 8 compares an original color IR enlargement with one generated by color separation techniques.

If it is desirable to make certain features stand out, one can treat the color separations as multispectral images and apply the same operations as described below in Sections 5.5.6 and 5.6. False color composites can then be generated for comparison to the enlarged color or color IR composite.

5.5.6 Photometric Corrections for Multiimage Analysis

Thus far in this discussion we have concerned ourselves with a single image of a scene. The same techniques can be applied to multiple images of a single scene, each filtered to selectively record information in different wavelengths. Of greatest importance is that the general exposure and contrast index is optimal for each spectral band. When the density values in each are translated to color

either by the additive process or the subtractive process, the result is a full color scene where each point reflects the combined spectral density value.

Generally, normal processing of each spectral image is adequate when the scene is imaged from close range under even illumination. If, however, the sensor platform is very distant, such as high altitude photography and space imagery, each spectral band may vary in density and contrast from the expected norm. This is due to aerosols that reduce the amount of radiation differentially with wavelength. The shorter the wavelength the greater the amount of light scatter with the result that the energy received by the sensor is proportionately increased in the shorter wavelengths. [There are two kinds of scatter. Rayleigh scatter refers to light reflected by particles small in size compared with the wavelength being scattered. Mie scatter results from intermediate sized spherical particles. (3)] As a result of scatter the contrast index is reduced. The number of gray levels is also reduced due to the scattering and filter effect of the aerosols. These effects are particularly noticeable on Landsat images taken over humid environments such as the northeastern United States. The green band generally suffers from overexposure and low contrast while the IR bands have in general too much contrast, particularly between land and water.

Phillips has described a photo technique to compensate for the variations of contrast and exposure in order to "normalize" or balance the contrast index and density among the different spectral bands (10). Figure 5.10 shows a Landsat scene taken in a region of Cameroon. Note the low contrast in band 4 compared with bands 5 and 7. However, even bands 5 and 7 suffer in sharpness and detail due to lower contrast when compared to the enhanced scenes in Figure 5.10. In order to compare the relative variation in contrast and density range, a series of exposures for each band are printed onto papers that are calibrated to build in a known contrast index value. In one type of paper the exposure is determined by a series of filters. Other papers are graded according to inherent exposure. The exposures are compared visually or by reflectance densitometer to obtain as close a match as possible. From the filter paper selected for each band, one can then compute from tables necessary exposure and development time required to produce a "normal print." Each print should have detail in both the "shadow" and "highlight" areas so that the

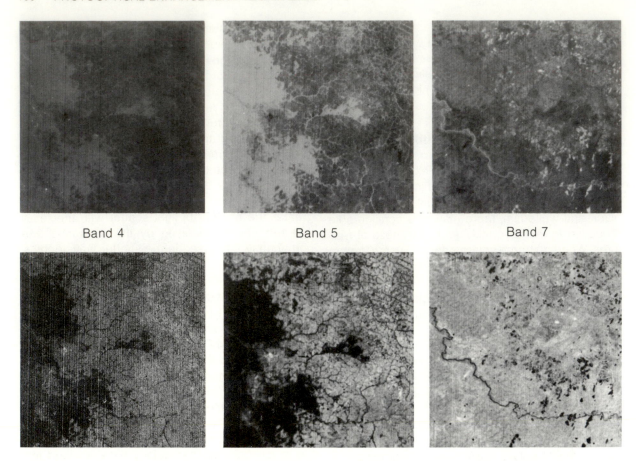

FIGURE 5.10. A set of unenhanced Landsat I images of a portion of Cameroon (negative image top) and the same set after being enhanced (positive image bottom). For each set band 5 was exposed as a "normal print" with bands 4 and 7 given identical exposure for comparison.

density range of the original is approximately equal to the log exposure of the paper. [Because Phillips processed the imagery to correspond to the specific density range of diazo materials, only the lower part of the straight line portion of the curve is used. As a result, it is important that the final product has a density no higher than about 1.3 for land details, excluding water and cloud shadows. The overall contrast index should be 1.0. This corresponds to the characteristics of the diazo materials used for making color composites (11).]

One could obtain more quanititative measures by measuring the density range of each original negative directly with a diffuse densitometer, assuming that large enough areas of equivalent density can be found in the scene. From these readings one can then compute the contrast index and bypass the use of filter paper and a somewhat more subjective result. However, in spectral images, scene reflectances for each object vary proportionally to different monochromatic radiations. Therefore, unless there is some feature that has a uniform neutral gray in the scene, it is difficult to interpret when one has equivalent quantitative measures from each spectral band. Experience has also shown that on small scale images, it is not possible to obtain equivalent density measurements within a single scene. Furthermore, the time required to collect such information is not commensurate with any improvement in the final product. With care, satisfactory and reproducible results are obtained with the method described with only a minimal added expenditure of time (11).

To match the density range in each spectral band, the proper film/developer combination must be chosen to produce equivalent D log E curves.

(See the example in Fig. 5.11). Ideally, the contrast index and density range should be equivalent in the intermediate negatives developed for each spectral band. The result should produce new images that are approximately equivalent or "in balance" among the spectral bands.

As indicated above, when the different spectral bands are printed in different hues (cyan, magenta, yellow) and registered, one can more easily distinguish visually the varying densities in a scene by the shifts in hue. Contrasts between areas of small density difference can be increased by using both positive and negative images of each spectral band in varying combinations. This is similar to using a negative and a positive mask to screen out information, such as in the case of Agfa contour film. In the case of Landsat images, one can check for color balance and processing errors by constructing a simulated color IR composite using positive images and printing the green spectral (band 4)

yellow, and red band (band 5) magenta, and one of the IR bands (band 6 or 7) cyan.

5.5.7 Color Enhancement for Black-and-White Images

The human visual system has a limited capacity to consistently pick out from a scene much more than 15 to 20 shades of gray. However, it is extremely efficient in distinguishing and comparing hundreds of different hues. In recognition of this it is often practical to convert black-and-white data to color, relating varying density values to discrete hues. This can be done electronically by devices with scanning and vidicon cameras or laser recorders with the capability to encode density slices and display them in color on cathode-ray tube (CRT) displays. More expensive models can be interfaced with computers for interactive pattern analysis (1).

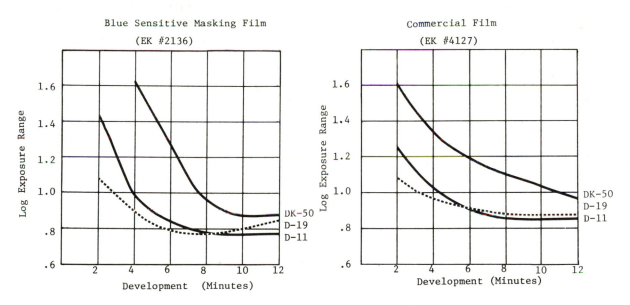

FIGURE 5.11. Shown here are plots of three different developers on two different film types. One selects the developer and film type according to the log exposure range value (see text). Using the charts, the following illustrates how to select the developer, development time, and film type in order to balance three spectral images of a hypothetical Landsat scene.

	Band 4	Band 5	Band 7
Poly contrast filter for a "normal" print	PC 3½	PC 3	PC 2
Log exposure range	0.75	0.85	1.05
Film	Blue sensitive masking film #2136		
Developer	D-11	D-11	D-11
Development time	8 min	6 min	3½ min

Conventional photo methods commonly make use of diazo graphic arts film. This film has very high resolution and is easy to use. The diazo process is a two-step process that allows one to make contact positive prints or transparency reproductions in color or black and white from positive black-and-white transparencies. The film is composed of acetate or polyester base embedded with diazo light sensitive salts and couplers. When exposed to ultraviolet (UV) light by a mercury vapor lamp, the salts are destroyed inversely proportional to the density of silver grains on the original transparency. In other words, the exposed diazo film, when processed by heated aqua ammonia vapors, results in a positive reproduction of the original (14).

Converting two or three spectrally filtered black-and-white images of a scene into different hues and registering the diazo images will produce varying color contrasts. These are proportional to the summed differences in density at each point.

Diazo film characteristically has a high contrast index with a density range of approximately 1.3. As a result, the number of shades of gray that can be printed is limited. Consequently, only a part of the range of grays on the original black-and-white image are printed at a given exposure. With decreasing exposures one can shift the lighter densities toward the shoulder of the curve, pulling the darker values off the toe. So as indicated above it is critical that the "shadow" detail on the black-and-white film transparency is printed at 0.3 density so that the density range of the transparency image closely matches that of the diazo. In effect, diazo can print only six to eight shades of gray at a given exposure due to the high γ. By increasing the exposure, the curve shifts to the right and the density values correspondingly crowd down on the toe of the D log E curve. This has the effect of bringing out detail in the "highlights," but with a loss of detail in the "shadow."

Therefore, if we consider each exposure for a hue as a density slice of six to eight shades of gray from the original image, we have the capacity to alter the contrast by change in exposure (color saturation), change in hue, or change in the ratio of densities (luminance) in a scene (by selecting different spectral bands which have recorded different albedos and/or by using combinations of a positive and negative image of each spectral band.) As can be quickly deduced, the number of possible enhancement combinations is very large. This leads us into the next topic, which deals with how to systematically manipulate the hue, saturation, and density to maximize the color contrast among points of interest in the scene.

5.6 CIE COLOR PREDICTION MODEL

In working with multispectral imagery, the photo interpreter has generally been at decided disadvantage with respect to sorting out intermediate variations in tone among several spectral bands. As a result, a great deal of the quantitative work in multispectral analysis has been accomplished by automated computer processing techniques.

The diazo methodology has been very useful as an inexpensive, straight forward way of analyzing various kinds of multispectral satellite and aircraft data. Unfortunately, it lacks any quantitative color reference to which one can relate the combined densities of two, three, or more spectral bands. Skaley et al. (11, 15) have devised a model based on a quantitative reference that permits one to produce color composites that have maximum contrast among selected points. The Commission Internationale de l'Eclairage (CIE) color coordinate system is used as a standard reference within which all colors have fixed coordinates. The colors in this system are quantitatively defined by tristimulus values for each of three primary colors (red, green, and blue). Tristimulus values can be determined directly by color matching experiments using a standard observer or by a series of equations developed by A. C. Hardy (16) and the OSA Committee on Colorimetry. These equations permit one to calculate the tristimulus values given the spectral curve of a color and the spectral energy distribution of a standard light source (17). These values then incorporate the interrelationships of hue, saturation, and brightness (reference back to Section 5.4.4). Each value for red, green, and blue represents the proportion of that primary summed over the total number of visual wavelengths represented by the color. In this way any color can then be theoretically duplicated by proportional mixing of light represented by these values. The CIE coordinate system then records the position of each color in CIE space. [See Billmeyer and Saltzman (17) for a more detailed but simplified explanation of the CIE system. Readers interested in a more quantitative description can refer to Wyszecki and Stiles (18).]

We can now calculate the tristimulus values for a given hue and exposure (density) on diazo film. These values are derived from a summation over visual wavelengths of the color matching coefficients as defined by a standard observer times the spectral energy distribution times the percent transmittance of a sample at a particular wavelength expressed in nanometers. Equations 5.1, 5.2 and 5.3 summarize these operations:

$$X = \Sigma\lambda\,(E_{D5000}\bar{X})T, \quad Y = \Sigma\lambda\,(E_{D5000}\bar{Y})T,$$

$$Z = \Sigma\lambda\,(E_{D5000}\bar{Z})T \qquad (5.1)$$

where X, Y, and Z represent the tristimulus values of red, green, and blue; E_{D5000} is the spectral energy distribution of a standard light source (19), $\bar{X}$, $\bar{Y}$, and $\bar{Z}$ represent standard observer coefficients, and T is a percent transmittance of the sample material at a particular wavelength.

Dividing each tristimulus value by the combined sum of X, Y, and Z yields the chromaticity coordinates:

$$X_{D5000} = \frac{X}{X + Y + Z}, \quad Y_{D5000} = \frac{Y}{X + Y + Z},$$

$$Z_{D5000} = \frac{Z}{X + Y + Z} \qquad (5.2)$$

X and Y values are plotted on the abscissa and ordinate axes.

Brightness is expressed as the $Y\%$ and is plotted on the Z axis, where

$$Y\% = \frac{Y}{(E_{D5000}\bar{Y})100}\,(100) \qquad (5.3)$$

These CIE coordinates form the fixed reference by which it is possible to relate different color values in Euclidean space with respect to the distance from one point to another. Figure 5.12 shows the range of values for diazo material plotted on the CIE coordinate chart.

When a selected coordinate value is translated back by substitution in Equations 5.1, 5.2, and 5.3, the relationship of the spectral band, diazo hue, and exposure value (saturation) can then be determined. This results in a selected color being assigned to a particular set of density values.

The objective of the CIE Color Prediction Model is to maximize the color contrast among two or three selected points for each composite. These points and areas having similar values are then visually enhanced. Operationally, this is accomplished by defining the greatest vector distance in Euclidean space between two points, or the greatest area most closely approximating an equilateral triangle, as represented by maximizing the distance among three points (1, 2), (1, 3), and (2, 3).

Since it is important from a perceptional reference that these coordinates lie in distinctly different color zones and since these color zones, as perceived by a human observer, vary independently from a change in hue (λ), visual sectors are imposed on the CIE chart to define and contrast color zones.

The visual sectors are defined by several factors:

1. The total color range of printing inks, (20).
2. MacAdam's ellipses illustrating the color sensitivity of the standard observer within the CIE chromaticity chart (21).
3. Judd's ellipses illustrating an equal energy distribution across the visual spectrum as it conforms to perceptibility scales (based on 100 just perceptible distances within each ellipse), (22).
4. The distribution of the total number of possible selections of coordinates for the diazo materials.

The boundaries of these associations accommodate an approximately equal number of perceived differences within each visual sector so as to obtain a reasonably equal distribution of computed coordinate selections for each sector. However, the number of possible selections using diazo materials has about a five to one bias toward the red-green line. Therefore, the distribution is divided into two groups, a light zone (red-green colors) and a dark zone (blue-green colors).

In operation, families of points representing the greatest perceptional differences are first selected. Then, from these families of points, as defined by each visual sector, coordinates which are spaced the greatest vector distance, or which define the greatest triangular area, are chosen. To select the visual sectors that have potentially the greatest visual contrast, a system of weights is used. The possible cell distances are 0, 1, 2, 3, 4, and 5. Points selected within a cell receive no weight, but the weight increases as the straight line connecting two points crosses visual sector boundaries (see Fig.

5.12). Lines that connect points crossing the light-dark line are given the equivalent weight of crossing two visual sector boundaries. This assures that no more than two points can be located on one side of the light-dark line, and that the side that has two points would have them spaced at the maximum distance as opposed to what might have occurred on the opposite side.

The criteria for assignments of bands to hues and exposure levels is represented graphically in Fig. 5.13 and summarized below.

A. Pairwise
1. Must achieve better than minimum pairwise discrimination as determined by Euclidean distance between CIE points. (See Fig. 5.13).

2. Among assignments meeting criterion for A.1, choose those with maximum cell distance (See Fig. 5.13.)
3. Among assignments with maximal cell distance, choose maximal Euclidean distance.

B. Triples
1. All pairwise distances must satisfy A.1.
2. Find maximal minimum-side-distance. Select all assignments meeting B.1 with this as their minimum-side-distance.
3. Among assignments meeting criterion for B.2, select maximal mean side length.
4. Among assignments meeting B.3, choose the one with maximum area as determined by the CIE coordinates. (See Fig. 5.13.)

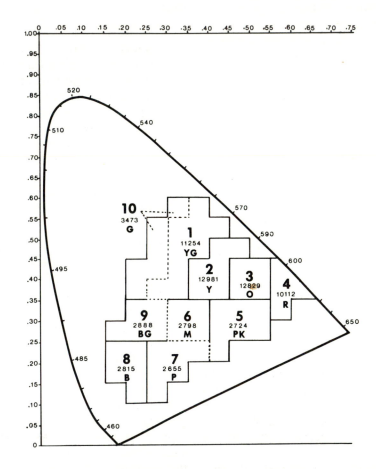

FIGURE 5.12. The maximum range of CIE coordinate values for various combinations of GAF cyan, yellow, and magenta films with respect to the theoretical total range indicated by the outer curved line are shown. Spectral shifts along this line are expressed in nanometres. The large font numbers identify the ten visual sectors of the CIE Prediction Model. The small font numbers indicate the possible CIE coordinate values computed for each sector. The dotted line represents the light-dark boundary discussed in the text. The letters under each number represent approximate color designations (YG = yellow-green, Y = yellow, O = orange, R = red, PK = red-violet, M = mauve, P = purple, B = blue, BG = blue-green, and G = green (15).

The distance from each cell to its nearest neighbors is defined as below.

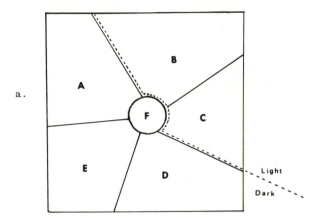

a.

The distance from "B" to "C" is 1. The distance from "A" to "B" is 2. Represented graphically.

b.

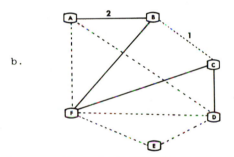

Now, no distance has been defined between "A" and "C". We define this to be the smallest distance path from "A" to "C", passing through one or more intermediate regions. Thus, from "A" to "B" is 2 units, and from "B" to "C" is 1 making a total of 3 units from "A" to "C". All paths are considered, and the shortest is selected.

This method applied to the previous graph yields:

c.

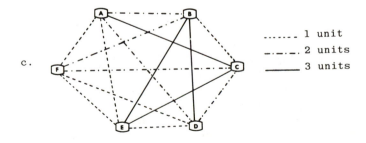

```
........  1 unit
-.-.-.-.  2 units
_____  3 units
```

As a result, cell distance is now defined between any two points.

FIGURE 5.13. Cell distance measure (15).

Upon selection of the coordinate values, assignments of hue, spectral band, and diazo exposure are computed so that all three points are represented in the composite selected. These are then similarly printed out, for example, cyan, band 4+, exposure value 6. (The exposure value corresponds to the dial setting on the diazo machine.) Figures 5.14, color plate 9, illustrates the kind of composites that can be produced. Several such composites can be used to construct a spectral map by using a quality overhead projector to project each scene onto a rear projection screen. Contrasting hues representing various classes can then be easily delineated in a cartographic operation.

The program will take density readings of up to 10 points within a scene; however, costs begin to mount appreciably if more than five or six points are run simultaneously. Using batch processing on an IBM 370 system, five points, generating 10 composites (three points compared per composite) can be run for less than $15.00, while six points generating 20 composites costs less than $20.00. It should be noted that one cannot normally expect to contrast adequately all points within a single composite. Several composites have to be examined and interpreted so as to accurately construct a map in a mosaic fashion. In this sense, the operation still requires some subjective interpretation. However, the system does afford a considerable increase in efficiency of operation and accuracy in interpretation providing the interpreter a powerful tool to analyze multispectral imagery; particularly, where more sophisticated and elaborate electronic hardware and computer software are unavailable.

It is then important to employ techniques for analysis of imagery that considers factors related to the human visual system. These factors, once identified, can then be manipulated to enhance specific information in the scene. With the exception of calculating color combinations and exposure values in the CIE model all the photooptical enhancement techniques described in this chapter are manual both in the sense of physical manipulation and in relating to the psychophysical references of the human visual system. The CIE model represents one approach to integrated photooptical/digital techniques in the analysis of imagery using the human observer as a principal interactive component. This view will be expanded upon in the following chapter.

5.7 EQUIPMENT

All of the operations described above can be accomplished at most localities that have available a well-equipped photo lab and access to a diazo machine and a small computer (minimum 250K core). An important point to note is that the quality of the photo enhancement can be sharply limited by the quality of the optics used in the enlarger. It is wise to ensure that excellent lenses and condensors are used in every case. Principal equipment required includes:

For the photo lab, a good quality enlarger with quality optics and an effective mechanism for locking the stand.

An exposure control (timer) with a capability of accurately exposing down to 0.1 s.

A photometer for measuring exposure values (optional).

A Kohler light source for improving resolution on extreme enlargements (optional).

Other equipment includes a good quality diffuse densitometer with a minimum aperature of 1 mm.

A diazo machine with a finely calibrated control for belt transport (exposure speed).

A voltage control regulator for the diazo machine to ensure continuous output of the UV light source.

A quality overhead projector, preferably one with a parabolic mirror reflector and excellent optics.

A rear projection screen. (This can simply be a frame with optical quality plate glass. Data takeoff procedures usually employ a polyester mylar film that diffuses the light when placed on the glass, allowing the interpreter to view the enlarged composite at varying scales. The image scale is manipulated by simply changing the distance from the projector to the screen.)

5.8 SUMMARY AND CONCLUDING REMARKS

In summary several more commonly used photooptical techniques have been discussed. This is not to suggest that readers should limit their imagination or not search for new techniques. Various techniques are employed only to enhance certain bits of information in a scene. Each technique is then selected on the best judgment of the experimenter as to what factors in a scene need to be modified to produce the desired result.

When extreme enlargements are made, image sharpness or definition can be improved using a Kohler light apparatus. In making extreme enlargements of color or color infrared imagery, it is desirable to use color separation techniques in order to preserve as much scene fidelity as possible.

Linear features in a scene are emphasized by using edge enhancement techniques. Often, however, scenes require more drastic modification to improve the contrast ratios within selected density ranges. Such techniques involve contrast stretching, and density slicing. Modification of both scene contrast and density range are also used in "balancing" the spectral scenes in multispectral imagery. These "balanced" black-and-white scenes can be translated to color by using diazo film. Various combinations of spectral bands, diazo hues, and exposures can be employed to construct color composites, that highlight selective bits of information.

Color composite combinations can be systematically constructed to maximize contrast among selected points in a scene by using the CIE color prediction model. This model offers the interpreter an important low-cost tool in the analysis of multispectral imagery.

As stated in the beginning of this chapter, the enhancement process should proceed as a logical progression focusing on specific bits of information. These bits of information are defined by the original objectives of the operation. The final chapter in volume I, of the *Manual of Remote Sensing*, (American Society of Photogrammetry) describes a good approach to gather data for geological investigations (23). Such information, be it empirical or a priori should be kept in mind while undertaking any enhancement of a scene.

Briefly one should proceed somewhat along the following logical sequence:

1. Determine the basic differences of importance in a scene (qualitatively). What are the properties of the features sensed and how do they differ?
2. Determine the quantitative differences within the image for the features of interest, that is, contrast, density range, and spectral character.
3. Determine what operations are required to improve the discrimination of the features of interest. This essentially involves maximizing the target to background contrast.
4. Select the simplest and most efficient technique or series of techniques required to accomplish the desired result.

REFERENCES

1. Steiner, D. and Salerna, A. E., coauthor-editor, 1975, Remote sensor data systems, processing, and management, *in*, Reeves, R. G., Anson, A., and Landen, D., eds., Manual of remote sensing, v. I: Theory, instruments and techniques: Amer. Soc. Photogrammetry, p. 611-803.
2. Goetz, A. F. H., Billingsley, F. C., Gillespie, A. R., Abrams, M. J., Squires, R. L., Shoemaker, E. M., Lucchitta, I. and Elston, D. P., 1975, Application of ERTS images and image processing to regional geologic problems and geologic mapping in northern Arizona: Jet Propulsion Laboratory Technical Rept. 32-1597, 188 p.
3. Fraser, R. S., author-editor, 1975, Interaction mechanism—within the atmosphere, *in* Reeves, R. G., Anson, A., and Landen, D., eds., Manual of remote sensing, v. I: Theory, instruments and techniques: Amer. Soc. Photogrammetry, P. 181-233.
4. Slater, P. N., 1975, Photographic systems for remote sensing, *in*, Reeves, R. G., Anson, A., and Landen, D., eds., Manual of remote sensing v. I: Theory, instruments and techniques: Amer. Soc. Photogrammetry, p. 235-323.
5. Todd, H. N. and Zakia, R. D., 1974, Photographic sensitometry, 2nd ed., Dobbs Ferry, N.Y., Morgan and Morgan, 312 p.
6. Ross, D. S. J., 1973, Atmospheric effects in multispectral photographs, Photogrammetric Eng., v. 39, p. 377-384.
7. Baines, H., 1958, The Science of photography: New York, Wiley, 319 p.
8. Kowaliski, P., 1972, Applied photographic theory: New York, Wiley, 533 p.
9. Eastman-Kodak, 1944, Photomicrography, 14th ed.: Rochester, N. Y., Eastman Kodak Co., 174 p.
10. Phillips, E. S., 1974, Photographic enhancement of ERTS imagery: Progress Rept., Dept. Commerce, Natl. Tech. Inf. Service, #E74-10259, 39 p.
11. Hardy, E. E., Skaley, J. E., Dawson, C. P., Weiner, G. D., Phillips, E. S., and Fisher, R. A., 1975, Enhancement and evaluation of Skylab

photography for potential land use inventories: Cornell Univ., Dept. Commerce, Natl. Tech. Inf. Service, #E76-10085, 214 p.

12. Rose, A., 1973, Vision: human and electronic: New York, Plenum, 197 p.

13. Nielson, U., 1974, Photographic image-enhancement: Canadian Forestry Service Technical Rept. no. 1345, 28 p.

14. Ozalid (GAF Corp.), 1950, How to get the most value with ozalid, (pamphlet): GAF Corp., 48 p.

15. Skaley, J. E., Fisher, J. R., and Hardy, E. E., 1977, A color prediction model for imagery analysis: Photogrammetric Engineering and Remote Sensing, v. 43, p. 45-52.

16. Hardy, A. C., 1936, Handbook of colorimetry; Cambridge, Mass., Technology Press, Mass. Inst. Tech., 87 p.

17. Billmeyer, F. W., and Saltzman, M., 1966, Principles of color technology: New York, Wiley, 181 p.

18. Wyszeki, G. and Stiles, W. S., 1967, Color science: New York, Wiley, 628 p.

19. Bartleson, C. J., ed., 1973, Colorimetry, in Handbook of photographic science and engineering—SPSE, New York, Wiley, 892 p.

20. Kodak, 1968, Photochemical reproduction of the visible spectrum, (pamphlet): Rochester, N. Y., Eastman-Kodak Co.

21. Wright, W. D., 1958, The measurement of colour: New York, MacMillan, 263 p.

22. Judd, D. B., 1950, Color in business, science and industry: New York, Wiley, 401 p.

23. Lee, K., ed., 1975, Ground investigation in support of remote sensing, in Reeves, R. G., Anson, A., and Landen, D., eds, Manual of remote sensing, v. I: Theory, instruments and techniques: Amer. Soc. Photogrammetry, p. 805-806.

CHAPTER 6

DIGITAL TECHNIQUES OF IMAGE ENHANCEMENT
ALAN R. GILLESPIE

6.1 INTRODUCTION

A *digital* image is a numerical representation of a sampled field. Typically the field represented is the radiance of a scene viewed in some region of the electromagnetic spectrum. However, digital images may be constructed that describe gravity or magnetic field strength, topographic relief, or even computed variables such as thermal inertia. The digital image is generated by sampling and measuring the local field strength at a number of points that are usually arranged in a rectilinear pattern. The field strength measured at each of these points is encoded as an integer. Thus the digital image is actually an array of numbers, which can be stored on magnetic tape or disk. In this form the digital image cannot be inspected visually, but can be manipulated readily by a digital computer.

Manipulation of a digital image by computer is called "digital image processing" and is performed either to prepare an image for display and interpretation, or to extract information from the image. In general usage, digital image processing consists of four procedures:

1. *Rectification*, or compensation for geometric and radiometric distortions introduced by the imaging instrument and elimination of systematic or coherent noise.
2. *Cosmetic Processing*, or removal of random noise or other imperfections in the data.
3. *Analysis*, which includes image comparison and extraction of measurement data.
4. *Display Processing*, or enhancement for visual interpretation.

The central idea behind computer image processing is quite simple. The digital image is fed into a computer one number (pixel) at a time. The computer has been programmed to insert this "data number" or "*DN*"* into an equation or series of equations. The numbers computed as a result of this are then written on magnetic tape or some other storage device. This new digital image may itself be manipulated by additional programs or it may be displayed in pictorial format, as either a television or film picture. Digital images are converted to film pictures by using the encoded radiance (*DN*) to modulate a light source that scans a photographic film. When developed the exposed picture superficially resembles a normal photographic image. However, in a normal photographic image scene detail is represented by continous changes in the distribution of opaque grains in the emulsion. In a photographic representation of a digital image scene detail is represented by changes in average brightness of small areas in the emulsion corresponding to picture elements or pixels in the digital image. Brightness of the pixels in the photographic picture corresponds to the *DN* of the pixels in the digital image.

One branch of image processing is "image enhancement," which implies that the goal is the improvement of image quality. Because image quality is a subjective measure, varying from person to person, there are no simple rules that may be followed to produce a single "best" image. Often two or more pictures made from the same "raw" image are necessary to display adequately the information required by the analyst, and even this may be only a fraction of the total usable information stored in the original image. Thus, an enhanced picture may actually contain less information than the raw image, although this is not always the case. It is always true, however, that to enhance a picture properly the information of greatest interest to the user must be optimally displayed, even at the expense of less valued information.

---

*In common usage *DN*, like *fish*, may be either singular or plural.

Digital image processing is a powerful technique largely because of its flexibility. Virtually any process or enhancement imagined by the analyst can be programmed and executed without delays caused by acquisition or modification of equipment in an optical laboratory. In practice, commonly used programs are stored in a procedure library on magnetic disk. In this way any one of several hundred algorithms is instantly accessible by computer. On the other hand, the vast amount of data contained in a typical image renders some sophisticated enhancements prohibitively expensive, and even simple procedures like contrast modification should be programmmed carefully to minimize computational expense.

The use of digital format provides a convenient way to transmit, store, retrieve, and use image data acquired by the electronic imaging devices described in Chapter 3.

Digital images, unlike their photographic counterparts, do not deteriorate with age, and the same algorithm applied to a digital image will produce identical results each time, regardless of whether the processing is performed on different machines by different analysts. This is not always true of optical processes. In addition, the copying of a digital image entails no loss of data. This simple virtue allows long, complicated series of operations, which would be impractical to perform optically because of degradation of image quality, to be performed digitally. However, photographic film is superior to digital image storage media in some ways, the most important of which are probably low cost and high information density. A single black-and-white 35 mm photographic slide can contain more than 2×10^8 bits of information and require a 12-in. reel of magnetic tape for digital storage.

Electronic imaging systems can acquire data over a much broader region of the spectrum than photographic systems, which are limited to the region from ultraviolet to near infrared (350 to 800 nm). Electronic systems can create images from thermal radiation, microwave radiation, or even sound. Often these images are stored on photographic film that is exposed directly by the sensing device, but increasingly they are stored on magnetic tape. Electronic sensors collect data in a format convenient for transmission. Partly for this reason most satellite camera systems are electronic. Both American and Russian planetologists used photographic cameras on their lunar oribters in the 1960s, but they had to digitize the automatically developed film image prior to transmission back to Earth (1). Conversion from film image to digital image and vice versa always entails a loss of data, so in general it is best to digitize early and process the acquired images completely by digital techniques. For many purposes, film images may be processed optically, but if precise geometric or radiometric control is desired and if the image area is not too large, it is often best to digitize the photograph, enhance the resulting digital image by computer, and then recreate the improved image on film for display.

In the last decade, remote sensors were generally able to process and interpret most of the image data they could acquire, but photographic systems and the electronic imaging devices discussed in Chapter 3 now have the ability to gather and are currently gathering much more data than can be assimilated. As long as geologists used only aerial photographs, the desired subset of the information available could be selected during photo interpretation, after photographic and optical processing had been completed. However, the advent of digital image acquisition and processing requires that this editing take place *before* expensive processing has been performed. The careful planning of research to minimize unnecessary processing is more important now than ever before.

6.2 CHARACTERISTICS OF DIGITAL IMAGES

No image perfectly depicts a scene. Even perfect optical lenses transmit only in limited regions of the spectrum, and represent or "image" a point source of light from the scene as a diffraction pattern. Real optical systems are even less well behaved, and introduce various geometric and radiometric errors into an image. Likewise, methods used to record an image—whether photographic or electronic—may also distort and introduce defects into the image. The very process of discretely sampling a scene to produce a digital image can distort especially fine detail (of pixel size). Random noise may be introduced by the sensing device, and periodic noise may be caused by interference from nearby electronic equipment. Finally, if the image is acquired by satellite, noise may be also introduced during transmission back to earth.

Before processing an image it is necessary to understand the source and significance of the distortions and artifacts contained in it.

6.2.1 Effects of Optical Imaging Systems

Effects of the optical system are common to all images, whether they are photographically or digitally recorded. For many purposes the most serious effect is the degradation of fine detail described by the spatial frequency response of the optical system, which is usually considered to be the response to an impulse or point source of light such as a star. The point source is blurred or spread out in the image, forming a pattern called the "point spread function" or PSF. The Fourier transform* of an *impulse* is a

---

*The Fourier transform is a function related to the image $f(x)$ (in one dimension) by the equation:

$$F(s) = \int_{-\infty}^{\infty} f(x)\, e^{-i2\pi xs}\, dx$$

where x is spatial distance (meters or pixels) and s is spatial frequency (cycles per meter, or cycles per pixel). The image is considered to be composed of sinusoidal waves of different amplitudes, phases, and frequencies s, and its Fourier transform (which is complex) is a spectrum that describes the amplitude and phase of waves comprising the image as a function of s.

Two properties of transforms are important to understand for the purpose of this chapter. First, retransformation yields the original function; that is, the Fourier transform of $F(s)$ is just $f(x)$. Second, multiplication in the Fourier (spectral) domain is equivalent to convolution in space:

$$F(s)\,G(s) = \int_{-\infty}^{\infty} [f(x) * g(x)]e^{-i2\pi xs}\, dx$$

where the symbol " * " denotes convolution. Convolution is an important concept with wide application in image processing and other fields, and examples are found in everyday life. One such example is the scanning or convolving of the signal on magentic tape by the head in a tape recorder. An example pertinent to image processing is the scanning of an image through a finite slit by a densitometer during construction of a brightness profile. In this example, the signal [$f(x)$] is scanned by a physical device [the slit, $g(x)$], which transmits some part of $f(x)$ and absorbs the rest. The resulting signal [$f(x) * g(x)$] is just the total energy passed through the slit, which changes as the slit moves over the image.

The reader should be careful to distinguish *spatial* waves, which are a mathematical device useful for representing spatial functions, from light waves.

Discussion of the Fourier transformation and its properties may be found in many standard texts, among them Bracewell (2) and Goodman (3).

---

function of constant amplitude. All spatial frequencies are present in the image to the same extent. Thus the Fourier transform of the PSF or the *image* of an impulse (the "optical transfer function," or OTF) is a good measure of the spatial frequency degradation caused by the optics and, of course, the recording device.

The Fourier transformation produces a complex function, and often for convenience only the modulus is displayed. The modulus of the OTF is called the "modulation transfer function" or MTF. An example characteristic of the NASA Landsat multispectral scanner (MSS) is shown in Fig. 6.1. Notice that the amplitude of the MTF falls off at high spatial frequencies. This reflects the fact that the high frequency waves, which make up small features and sharp edges, are the least faithfully reproduced in a blurred or defocused image.

The resolution of an imaging system is often described by the largest number of black-and-white lines per centimeter which may be seen in the image. This is a useful parameter and is an aid to the interpretation of images because we often need to know the size of the smallest features in a picture that we see or can expect to see. However, the MTF is a more complete description and may be preferable, especially if the image processing techniques described in this chapter are to be used to "restore" or "sharpen" an image by boosting high frquency information to compensate for the system MTF.

Optical systems are limited spectrally as well as spatially because the glass used to make the lenses is transparent only in certain parts of the spectrum. This means that the photosensitive film or recording device does not see all of the energy reflected from a scene. Thus the optical system cannot perfectly image a scene, although this is not necessarily a disadvantage—forming images from light in different parts of the spectrum is the basis of color photography and multispectral imaging. However, spectral transmission of the optical system and of any filters inserted into the optical path to control the region of the spectrum in which data are gathered must be considered in the interpretation of images.

Optical systems can introduce geometric errors into an image, so that a representation of a feature in a scene changes shape or scale as a function of its location in the image. With high quality lenses,

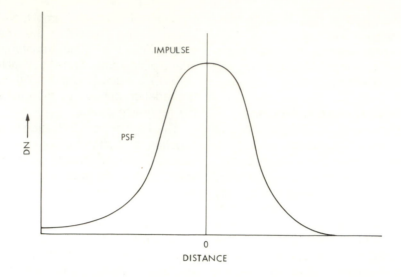

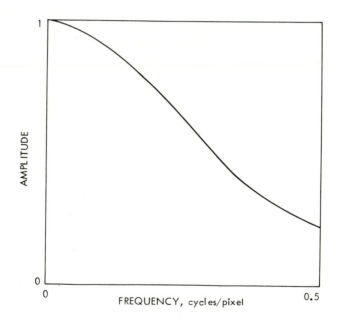

FIGURE 6.1. Impulse response of the Landsat MSS. (a) The PSF of many imaging systems resembles a Gaussian distribution. (b) The frequency response, represented here by the MTF, is found from the Fourier transform of the PSF. The MTF in this illustration was measured; the PSF is schematic.

geometric distortions are generally minimal.

Chromatic aberration can be an important effect in color or multispectral imaging. The focal length of a lens varies with wavelength, so images made in different colors will be focused differently. Chromatic aberration also results in a misregistration of images made through the same lens at the same time. It is caused by differential refraction of light of different wavelengths as the light passes through the glass of the lens, so that light from the same source is dispersed according to its color in the image. This effect can be minimized by careful construction of the lens.

A final effect of optical systems on images is called "vignetting." Vignetting is the apparent darkening radially outwards from the optic axis of an object of constant luminosity. Once this effect is described for a given camera, it can be corrected by image processing. Vignetting is most easily described by imaging uniform flat white targets under constant illumination.

6.2.2 Effects of Image Recording Systems

Limitations on image quality imposed by the recording system—whether it uses photographic film or electronic sensors—can be and often are more severe than those of the optical system. Photographic film as a recording medium was discussed in Chapter 5 and electronic cameras were discussed in Chapter 3. Two basic varieties are used in remote sensing: television or vidicon cameras and line scanners. Each contributes different kinds of distortions to an image.

Vidicon cameras basically store an image on a semiconducting plate whose conductivity is proportional to the number of photons that strike it. The image is read from the sensitive plate by scanning it with an electron beam and recording to the current, which changes as the beam scans different portions of the image. The changing current is recorded in digital form as a string of integers on magnetic tape for later reconstruction as an image.

Vidicons introduce several distortions into an image. The most important of these are gross geometric errors amounting to scale changes of several percent near the perimeter of the image. These geometric distortions change little in any local region of the image. The distortion can be described using tiny metal spots that are deposited in a grid on the vidicon plate during construction of the camera. The conducting metal spots or

"reseau marks" appear as dark spots in each image. Comparison of their measured position in the image to their known position in the vidicon defines the geometric distortion.

In addition to gross geometric distortion, the sensitivity to light changes with position across the vidicon. Uniform targets or "flat fields" have apparent brightness changes in different parts of the image, a problem that resembles vignetting.

Other local geometric errors are introduced near high contrast features because the scanning electron beam is deflected by high charge accumulations on the vidicon plate. Thus, the current actually measured does not describe the image brightness where the beam was aimed, but rather at a point a short distance away. The size of such local distortions is rarely more than one or two pixels. Nevertheless, when small scale detail is important (as in the construction of topographic maps from stereoscopic images) such geometric errors can be annoying.

A final and relatively minor geometric and radiometric effect in high contrast fine detail is caused by migration of electrons away from a local concentration. This causes a loss of high spatial frequency reponse, or loss of fine detail, and also an apparent enlargement of small features as well as loss of local contrast.

Line scanners are cameras in which a rotating mirror sweeps the image across a photosensitive diode or photomultiplier tube, which records the changing brightness as a changing voltage. The instrument is designed to be used on a moving vehicle so that every oscillation of the mirror sweeps a different part of the image across the sensor. An obvious advantage of this method is the absence of radiometric sensitivity changes in different parts of the image, because the same sensor is used everywhere. An equally obvious disadvantage is the opportunity for real changes in the scene itself between acquisition of the top line and the bottom line. However, for most remote sensing purposes the scene is essentially unchanged during the 30 s or so necessary to scan it. A far more serious consequence arises from motion of the aircraft caused by turbulence, or even the lesser motions of a satellite. These disturbances can produce severe local geometric distortions into the image, and these distortions cannot be well described without detailed comparison of the image to a map or to another accurate image of the same scene. In addition to these random distortions, in some instruments the mirror used to sweep the image across the sensor oscillates rather than rotates, thereby accelerating and decelerating during every scan and causing scale changes across the image.

The Landsat and NOAA satellites both carry line scanners. Because these satellites are in polar orbits, their velocity vectors are not in the same direction as the tangential velocity of the rotating earth and a systematic skew is introduced into the line scanner data as the earth rotates underneath the spacecraft during image acquisition.

The above list of problems encountered acquiring images is not exhaustive, but is complete enough to give a general idea of the distortions encountered in most commonly available images which must be corrected by image processing before thorough, careful interpretation can take place.

6.2.3 Construction of Digital Images From Photographs

Digital images need not be acquired directly in digital format, but may be converted from other data. Commonly the original data is recorded on photographic film. If it is desired to process the data digitally, the format conversion is done by scanning the transparency with a light beam of constant intensity. The beam should be small enough to avoid loss of data. Different amounts of light are absorbed by the transparency during scanning, and the light that is transmitted is measured by a photomultiplier tube. The signal is stored as a string of integers on magnetic tape. By using logarithmic or linear amplifiers, it is possible to record either D, the photographic density, or τ, the transmittance. For many purposes it is desirable to work with τ because transmittance is linear with scene radiance or reflected energy. However, most commercially available playback devices, which recreate pictures from digital images, expect to receive data in density units. Also, the human eye perceives energy logarithmically, so for some purposes density scans are preferred.

6.2.4 Construction of Digital Images from Nonimaging Sources

Data that are acquired by irregular sampling of a scene may be represented as a digital image. The original data need not have described reflected light. Examples shown in Fig. 6.2 depict the depth to a limestone layer as determined from well logs, and the complete Bouguer gravity anomaly in Michigan. Although the data are represented pictorially, they were acquired by irregular sampling at sites shown in Fig. 6.2(b). To convert the data to a common format, first a two-dimensional polynomial surface was fit to the original data, and then this model was resampled at the desired, regularly-spaced sample sites. By only fitting data near the immediate sample sites many small low-order polynomials were pieced together to model the entire image at great savings in computer time.

The United States Geological Survey and the Defense Mapping Agency have recently resampled the topography from 1:250,000 contour maps to create digital images. In addition, the United States Geological Survey has prepared a limited number of 1:24,500 (7½' or 7½ minute quadrangles) digital elevation maps directly from stereo air photographs, skipping the intermediate contour maps. Such "altitude images" are available on magnetic tape for most of the continental United States from the National Cartographic Information Center at the United States Geological Survey in Reston, Virginia.

The importance of these altitude images to the image analyst is that they allow computer manipulation of other image data registered to the altitude images to compensate for shading, solar heating, or other functions of topography. Some of these techniques are discussed later in this chapter, while the importance of topographic corrections for solar heating in the construction of maps of thermal properties from imagery is discussed in Chapter 8.

6.3 SAMPLING EFFECTS AND ARTIFACTS

The representation of a continuous image of a scene by a matrix of discrete DN, each representing the average local scene brightness, entails limitations and distortions of the data. These are described by sampling theory (4).

In sampling theory the image, which is a two-dimensional continuous function of scene brightness, is considered to be composed of spatial waves. The amplitude of these waves is in units of brightness, or scene radiance, and at any given point the observed scene brightness may be rep-

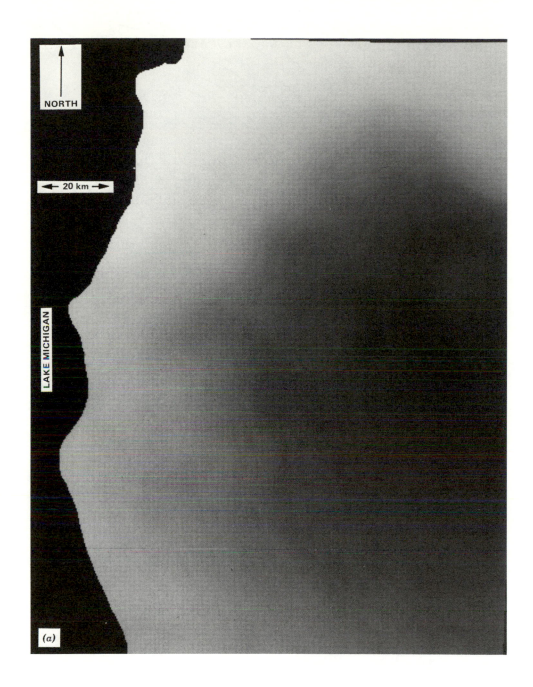

FIGURE 6.2. (a) R. J. Blackwell of the Jet Propulsion Laboratory created two registered digital images of Michigan describing data that was acquired by the Dept. of Geology at the University of Michigan from nonimaging sources. This image shows depth below ground level to a prominent limestone bed. Bright tones represent shallow depths; dark tones represent greater depths. No data was taken in Lake Michigan.

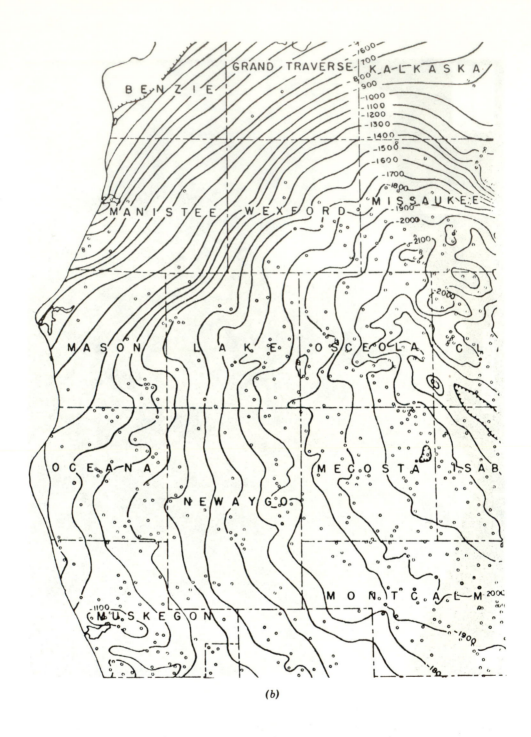

(b)

FIGURE 6.2. (b) Contour map of depth-to-bedrock image (a). Contour interval is 100 ft. Open circles are sites of wells in which depths were measured.

146

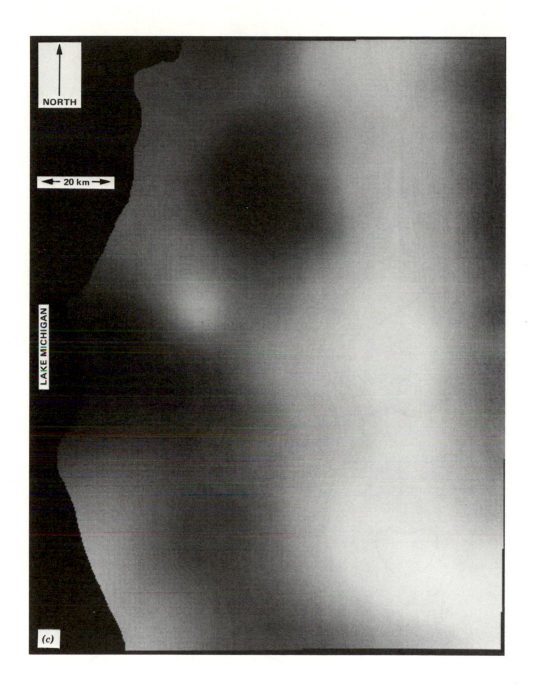

FIGURE 6.2. (c) Image shows complete Bouguer gravity for Michigan. Bright tones show gravity highs.

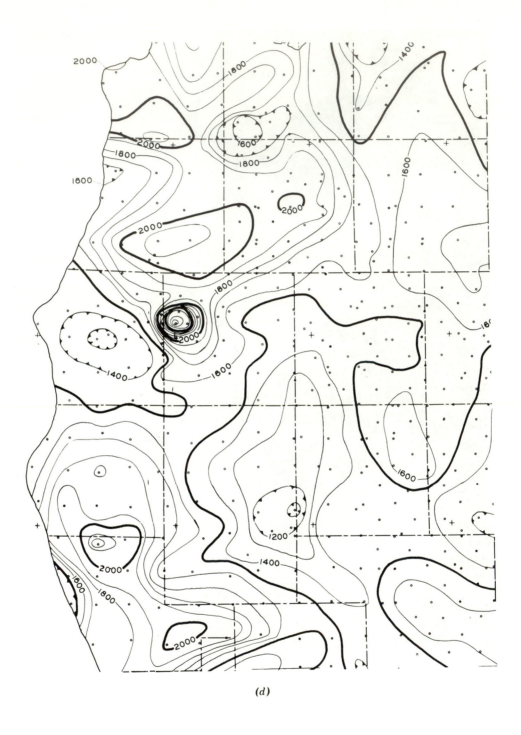

(d)

FIGURE 6.2. (d) Contour map of complete Bouguer gravity (c). Contour intervals are 1 mgal. Open circles are sample sites.

resented by the infinite sum of the amplitudes of all frequencies as shown in Figs. 6.3(a) and (b). In order to faithfully represent a sinusoidal wave it must be sampled at least twice per cycle as shown in Fig. 6.3(c). One measurement will yield an amplitude of $+a$ and the other will yield $-a$. Notice that if the measurements are not made at $\pm n \pi$ (n = 0, 1, 2, 3, . . .) the frequency will still be correctly represented although the apparent amplitude will be too small. If more than two measurements are made per cycle, the correct sine wave can still be fit to the data. In fact, the analyst may have greater confidence because of the redundancy. Because more data is taken than is required to describe the wave, this situation is called "oversampling." A wave sampled twice per cycle is "critically sampled" and the frequency of this wave is called "Nyquist frequency." A wave sampled less than twice per sample is "undersampled" and cannot be reconstructed accurately from its sample points. Effects of undersampling are important to understand because in a real scene waves of all frequencies will be encountered. Thus, no matter how small the distance between sample sites or how "frequently" the scene is sampled some components of the scene will be undersampled. Undersampling a wave results in misrepresentation of the wave's frequency, as illustrated in Fig. 6.3(d). This is called "aliasing" and the apparent frequency is always lower than the sampling frequency. In fact, if the real frequency of the wave is higher than the Nyquist frequency by some amount, its apparent frequency will be lower than the Nyquist frequency by the same amount. Thus, the Nyquist frequency is sometimes referred to as the "folding" frequency.

In image processing, it is convenient to normalize all dimensions and frequencies to units of samples or pixels, rather than meters. Frequencies are thus measured in cycles per pixel, and the Nyquist frequency is 0.5 cycles per pixel.

Another way of looking at the effects of undersampling is to consider that the smallest features in a scene, which are composed of high frequency waves, will be represented in an image, but at the wrong size. Thus, small features may be *detected* but not *resolved*.

The distinction between detection and resolution is an important one in image processing and interpretation, although the terms are often confused. Detection simply refers to the ability of the analyst to *see* that a feature is present in an image, regardless of the fidelity with which it is reproduced. To be resolved, an object must be represented faithfully.

A good illustration of the difference between detection and resolution was reported by W. Evans (5) in a Landsat image of San Francisco. Evans carefully positioned a 22-in. slightly convex mirrror so that the image of the sun was projected at the satellite as it passed overhead. A single pixel of the image represents the average brightness of a 60 by 80 m rectangle on the ground. In spite of such large "instantaneous field of view" or IFOV, Evans found that he was able to detect his small mirror, which was represented as a single bright pixel, although he might have been unable to distinguish between his mirror and an hypothetical 50-m mirror!

Obviously, in a digital image the smallest object that may be resolved is composed of waves of the Nyquist frequency (ω_n). Features in an image are composed of many frequencies, however, so ω_n does not really give a very useful measure of resolution. One reason is that the response of the imaging system decreases with frequency, as discussed in Section 6.2 and shown by the MTF of the Landsat MSS in Fig. 6.1(b). Long before ω_n is reached, the response of the MSS has fallen by over 50%. Thus, even though waves at ω_n may be correctly represented in the digital image, they may not be evident without special processing.

In general, a scene should be sampled frequently enough so that all the detail that is considered important has frequencies that fall in the high amplitude region of the MTF.

6.4 IMAGE PROCESSING

As mentioned in Section 6.1, image processing is used to achieve four different goals: rectification, removal of random defects, analysis, and display of the image data. While each of these is important, rectification and cosmetic procedures are required chiefly to allow effective analysis and display of information. Many image processing techniques and computer programs are common to all four classes of procedures. These techniques, as distinct from the more general procedures, can be considered in the following broad categories:

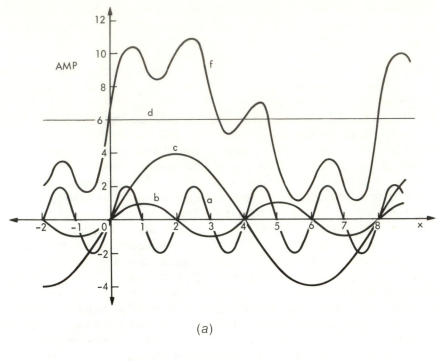

(a)

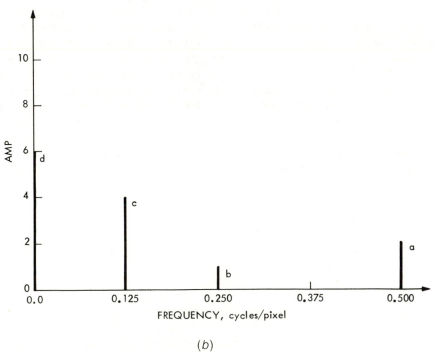

(b)

FIGURE 6.3. (*a*) Spectral decomposition and sampling of a simple function. Function *f* can be represented as the sum of a constant d and three sine waves (a, b, c) having periods of 2, 4, and 8 units and amplitudes of 2, 1, and 4 units respectively. (*b*) The spectrum of *f* shows spikes for each of the component sine waves.

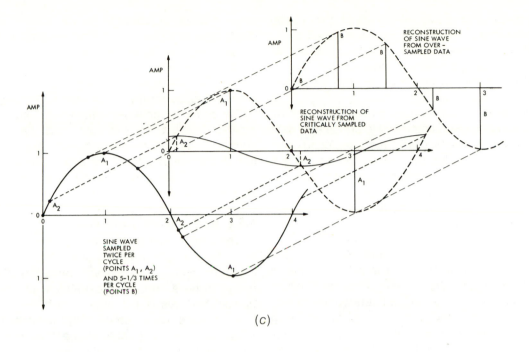

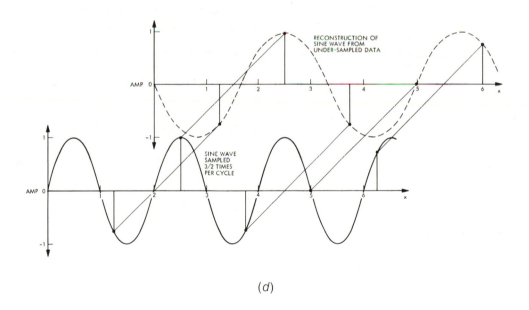

FIGURE 6.3. (c) Critical sampling and oversampling of a sine wave. (d) Undersampling of a sine wave.

1. *Geometric manipulation*, in which the spatial relationship between features is modified.
2. *Radiometric manipulation*, in which the brightness encoding scheme is modified.
3. *Spatial filtering*, in which the Fourier transform or spatial spectrum is modified.
4. *Multispectral manipulation*, in which the spectral content of images is modified to create new data formats, including thematic maps depicting computer decisions about the nature of features in a scene rather than simple pictures of those features.
5. *Image comparisons,* in which changes occurring in time are emphasized.

Of course the above list is not exhaustive, but is representative of the most important kinds of manipulations an image is subjected to. The use of the word "manipulation" is intended to emphasize the idea that the analyst really is modifying the images in an important way, one that may actually introduce artifacts into the processed images or otherwise significantly affect their interpretation. Only when a process is carefully chosen and properly executed can the "manipulation" be regarded as an "enhancement."

In the rest of this chapter, the image processing techniques listed above will be introduced as required to illustrate the four main categories of image processing procedures and the same technique will be discussed under more than one main procedure. For instance, spatial filtering used to restore or compensate for degraded frequency response is a rectification technique, but filtering used to emphasize local scene detail is a display technique.

6.5 RECTIFICATION

Rectification procedures consist of techniques designed to compensate for geometric and radiometric distortions in image data, restore the proper spatial frequency content of the image, and remove systematic noise.

In rectification it is important to distinguish between systematic and scene-dependent distortion. Systematic distortions are caused by idiosyncracies of the satellite and imaging system and are easily identified and removed. Scene-dependent distortions are more difficult to define and are thus more difficult to correct. An example of scene-dependent geometric distortion is parallax. An example of scene-dependent radiometric distortion is local charge migration in vidicon cameras as discussed in section 6.2.2. Scene-dependent distortions are not generally removed during rectification.

For some purposes all distortions must be removed. Parallax, for example, must be removed before images of the same scene taken from different camera positions may be compared. The "registration" techniques used to do this are discussed in Section 6.5.2.

6.5.1 Radiometric Corrections

The goal of radiometric correction is to rearrange the *DN* in a picture according to some desired scheme. Usually, the analyst wants the *DN* in an image to be a linear representation of scene radiance, and this is not often true in the acquired "raw" images. To accomplish this, the analyst must know how the response of the imaging system varies with radiance and also with position in the image. One advantage of the line scanners compared to photographic or vidicon cameras is that the response of a line scanner to scene radiance is spatially invariant, and radiometric correction is correspondingly simpler.

RADIOMETRIC CORRECTION OF LINE SCANNER IMAGES

Radiometric correction of line scanner images is accomplished by computer, using a "lookup table" procedure in which the correct *DN* are stored at addresses in the computer's memory which are specified by the input or acquired *DN*. The raw image data is read into the computer one pixel at a time, and the computer uses the input *DN* value to locate the correct output *DN*. The input *DN* is replaced by this new value, and in this way a radiometrically corrected picture is constructed. This method is the exact equivalent of applying some function to each input *DN*:

$$DN'_{\varrho,s} = \text{LTF}^{-1} (DN_{\varrho,s}) \qquad (6.1)$$

where *DN'* is the corrected *DN* at line ϱ and sample *s* in the image and LTF^{-1} is the specified function. The difference between the two approaches is that with the "lookup table" method, the actual amount of arithmetic performed by the computer is reduced because Equation 6.1 is used only once for each possible DN in the dynamic range instead of once per pixel. Usually there are eight bits (2^8 or 256) gray levels or DN in the dynamic range of a picture, ranging in value from 0 (black) to 255 (white). In Landsat MSS images there are only seven bits of data, or 128 DN. However, there are about 7.6 million pixels in each of four channels per image, so if the function LTF^{-1} is at all complicated the "lookup table" method is far less costly than direct computation.

The function LTF^{-1} in Equation 6.1 is the inverse of the "light transfer function" or LTF, which describes the response of the sensor to light. LTF^{-1} is therefore the function necessary to linearize the relation between image DN and scene radiance. The function LTF shown in Fig. 6.4(a) belongs to the special class of functions known as the Hurter-Driffield curves, or the D log E curves, which describe the response of a photographic emulsion to light as described in Chapter 5. In general, the LTFs of the photomultiplier tubes and photodiodes used in line scanners are much more linear than the D log E curve. Figure 6.4(b) shows the LTF^{-1} computed from the LTF in Fig. 6.4(a).

In some line scanners such as the Landsat MSS an array of sensors is used in place of a single sensor, and each sensor has its own optical filter and its own LTF. Because of this, raw images can show severe striping as seen in Fig. 6.5(a). The Fourier transform of this image clearly shows the noise spikes caused by the striping [Fig. 6.5(b)].

The Landsat MSS has a total of 24 different sensors, six in each of four different spectral channels. Six lines of image data are acquired simultaneously, one by each sensor. Six separate lookup tables must be used, therefore, to radiometrically correct the data in each channel.

Even after radiometric correction, in some Landsat images some striping remains. This may be partly caused by changes in the MSS since launch (6) and perhaps partly caused by the slight differences in the effective wavelength or color of the filters belonging to the different sensors, so that some striping may depend on scene color. This is because the LTF's are calculated only for light of a single hue, whereas the actual LTF's vary slightly with scene color.

Because not all the striping can be removed by radiometric correction, the residual must be removed by other methods. It may be treated as coherent noise and removed using spatial filtering as discussed in Section 6.5.3 under the heading "Periodic Noise Removal" or it may be regarded as an image defect and empirically removed using cosmetic techniques as discussed in Section 6.6. Cosmetic techniques include; histogram matching, in which the DN from each sensor are adjusted using different LTF^{-1}'s designed so that all six histograms describing gray level distributions in each sensor more closely resemble each other; filtering, in which an entire range of frequencies that include the noise frequencies rather than just the noise spikes themselves are suppressed; and interpolation across the worst lines.

Figure 6.5c shows the Landsat image of Fig. 6.5a after striping removal, and Fig. 6.5d shows the removed stripes.

RADIOMETRIC CORRECTION OF VIDICON IMAGES

Radiometric correction of vidicon images is more complicated than correction of line scanner images because the sensitivity of the vidicon to light varies spatially across the image, so that there is a different LTF for each pixel. These LTFs must be measured prior to use of the system by exposing the vidicon to "flat field" targets of uniform reflectance (7). Perhaps eight or nine images of the flat field are made, each at a different level of illumination. At each pixel the LTF is modeled by a polynomial fit to the eight or nine recorded DN, so that the DN corresponding to all possible light levels may be interpolated. The analyst must invert each polynomial, so that the correct exposure or scene radiance is expressed as a function of the measured DN. These functions are stored on magnetic disk for ready access by the computer. When a picture is rectified, the correct function is identified and retrieved from disk by the computer as it examines each pixel of the image. The input DN is then corrected to the value specified by the transfer function:

$$DN'_{\ell,s} = LTF^{-1}_{\ell,s} (DN_{\ell,s}) \qquad (6.2)$$

The difference between Equations 6.2 and 6.1 is

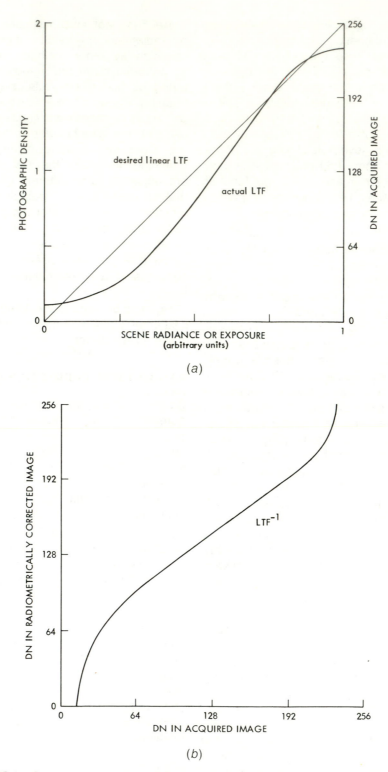

FIGURE 6.4. Radiometric correction. (*a*) Possible response of a sensor to light (LTF) and the desired linear response. (*b*) The correction function LTF^{-1} which relates measured *DN* to desired *DN'*

154

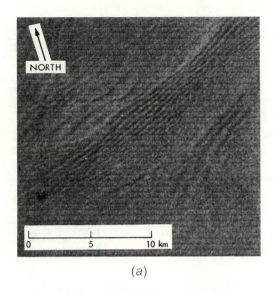

(a)

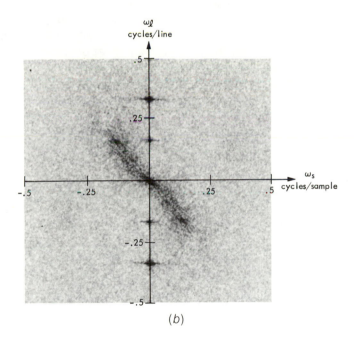

(b)

FIGURE 6.5 (a) Striping in a line-scanner image. A portion of NASA Landsat image 1024-15071 showing the ocean off New Jersey. (b) Spectrum of the same image. Dark tones represent high amplitudes in this spectrum. The six-line striping is shown by the prominent bright spikes on the vertical axis at frequencies of 0.1667, 0.3333, and 0.5000 cycles/pixel.

(c)

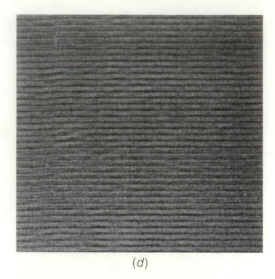

(d)

FIGURE 6.5. (c) Striping in a line-scanner image. The same image after removal of the striping. (d) Picture of noise removed from raw image.

that for a vidicon LTF^{-1} is a function of line ℓ and sample s. Figure 6.6 shows vidicon images of flat fields and raw and radiometrically corrected images.

Another characteristic of vidicon images found in neither line scanner nor photographic images is the presence of residual or "ghost" images. The charge distribution on the vidicon tube is not always completely dissipated before a new exposure is made. In this case, the new image will contain a contribution from the previous image or images. Mariner 9 pictures of Mars during the approach to the planet contained as many as four or five residual images. This effect may be reduced by flooding the vidicon with light, essentially replacing the residual image with a less obvious residual image of a flat field, which nevertheless modifies the vidicon LTF seriously enough to reduce the reliability of the radiometric correction procedure. Alternatively, residual images may be removed by computer, subtracting a small empirically determined percentage of the previous image or images from the current image (8).

RADIOMETRIC CORRECTION OF PHOTOGRAPHIC IMAGES

The correction of digitized photograpic images is like that of line scanner images in that only a single LTF is used to describe the response of the photographic emulsion to light, but similar to vidicon images in that images of flat fields are not of uniform DN. This vignetting, however, is not caused

by a spatial variation in the response of the light-sensitive element or emulsion, but rather by a decrease radially from the optic axis in the amount of light actually striking the emulsion. Vignetting, similar in appearance to the vidicon shading in Fig. 6.6(a), depends on the aperture or f-stop of the lens. It can be removed by ratioing the image of the scene to an image of a flat field taken by the same camera using the same lens settings (9). This is done by dividing the DN of the scene image at each pixel by the DN from the flat field image at the corresponding pixel and then normalizing the quotient so that the DN at the optic axis is unchanged. Because the denominator image is darker away from the center, the dark corners of of the numerator image are brightened in the corrected image, which is the desired result.

Vignetting correction is performed after correction for the nonlinear D log E light response curve in both the scene and flat field images.

Although the LTF of the photographic emulsion is nearly linear over much of the dynamic range of the emulsion, it is strongly nonlinear when there is too little or too much light. One advantage of digitizing photographic images is that it allows simple and accurate correction for these "toe" and "shoulder" regions of the LTF (Fig. 6.4). This correction is applied using lookup procedures as described in Section 6.5.1 for line scanner images.

The two-step radiometric correction procedure outlined above could just as well be done in one step, as is done for vidicon images. In this case,

FIGURE 6.6. (a) A Mariner 10 vidicon image of a flat field.

the vignetting correction would be made using a spatially dependent LTF^{-1} as in Equation 6.2. However, both the relative simplicity of the vignetting correction, which is radially symmetric about the optic axis, and the more extensive calibration required before vidicon-style corrections can be made argue for the two-step procedure. On the other hand, the more complicated spatial variation of the vidicon LTF makes the two-step approach impractical for vidicon images.

OTHER RADIOMETRIC CORRECTIONS

The radiometric corrections described in the last three sections are designed to remove systematic distortions in images which are introduced by the imaging device itself. Other radiometric effects are contained in the image which, although not distortions in the usual sense of the word, nevertheless are not intrinsic properties of the surface material in the scene. Instead, these effects are functions of illumination geometry. They occur at two different scales. If the image is taken from a great distance, it may include such a large region of the planet's surface that the position of the sun in the sky changes across the scene. This appears as gross shading in the image, following the curvature of the planet. The amount of shading at a given point on the planet changes as the planet rotates. Thus, if two pictures taken at different times are to be compared, the shading must be removed.

Images of the same scene taken at different times from aircraft or satellite are often taken from different positions, so that the viewing geometry as well as the illumination geometry is different. This effect must be removed also.

On a smaller scale the curvature of the planet is less important than topographic relief of the local surface. Topographic relief also causes changes in the illumination and viewing geometry. As an extreme example, consider a steep west-facing hillside that is shaded in the morning but fully illuminated in the afternoon.

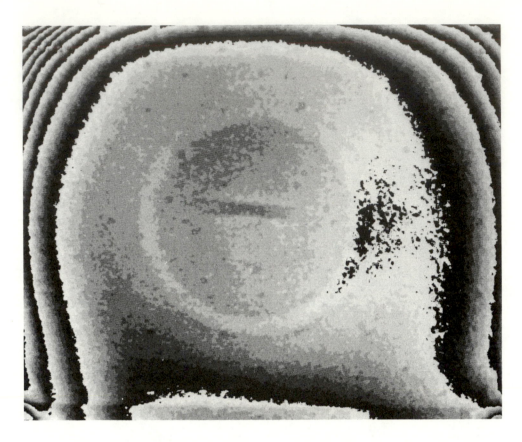

FIGURE 6.6. (b) The same flat field calibration frame contoured to show the small brightness differences more effectively. Each black to white cycle indicates a change of eight intensity levels.

Before the shading can be removed, the shape of the surface in the scene must be specified. This is done by registering the image data to a digital altitude image, or if planetary curvature is important, to a model of the planet's surface.

The relationship between radiance and illumination geometry and viewing geometry must also be known. For many materials in visible and near-infrared light this relationship is approximated by the Minnaert photometric law (10):

$$DN'_{\ell,s} = DN_{\ell,s} \cos^{-k}(i_{\ell,s}) \cos^{1-k}(e_{\ell,s}) \qquad (6.3)$$

where DN' is the radiometrically corrected brightness at line ℓ sample s of the image, i is the incident angle, e is the emergent angle, and k is an empirically determined parameter that varies with the type of surface material and texture (11). The incident angle is the angle between the local surface normal and the vector pointing from the surface to the sun, as shown in Fig. 6.7, and the emergent angle is the angle between the local surface normal and the vector pointing to the camera. DN' is the brightness that the surface should have if i and e were both zero.

Figure 6.8(b) shows the shading computed for a NASA NOAA satellite picture of Canada [Fig. 6.8(a)] made using equation 6.3 with $k = 1.0$ (Lambert surface), assuming DN was constant over the image and solving for DN', which is the brightness that would have been observed if Canada were featureless and uniformly gray. The angles i and e were computed from the known time the image was taken and the satellite tracking data. Figure 6.8(c) shows a shading picture made at the higher Landsat resolution from the USGS digital elevation image of the Salmon Mountains in northern California (Fig. 6.8d). At this larger scale shading

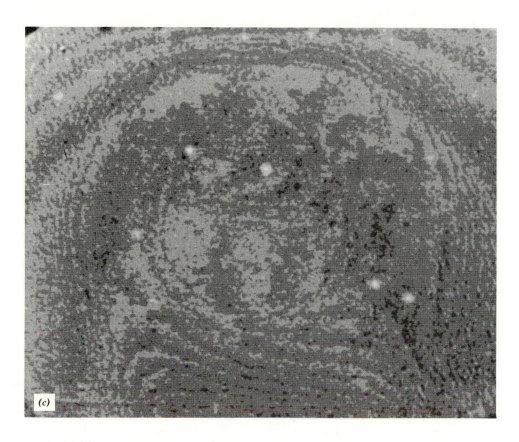

FIGURE 6.6. (c) After radiometric decalibration a contoured version of the image demonstrates that it is now virtually flat, with most of the picture consisting of only two intensity levels.

due to topographic relief rather than planetary curvature dominates the picture.

6.5.2 Geometric Rectification

The goal of geometric rectification is to rearrange the location of features in an image to agree with some desired scheme. Typically, the analyst wants to make the image conform to some standard cartographic projection used in geologic mapping, such as the Universal Transverse Mercator projection, or he may want the scene to appear as though it had been taken from directly overhead, or he may merely want to remove distortions caused by the imaging system.

Several techniques have been devised to perform geometric transformations using optical machines (12), and these are routinely used to process aerial photographs. Discussion in this chapter is of digital techniques for removing distortions which are introduced solely by the imaging system and those which are a function of the viewing geometry. Although the same computer programs are used to correct for both kinds of distortion, the corrections are usually performed separately and will be treated separately here.

The removal of geometric distortions can be accomplished in two ways:

1. The actual location of the pixels can be changed during construction of a picture.
2. The pixel grid can be retained and the individual pixels assigned new DN. This is called "resampling."

The first approach is useful primarily for simple geometric corrections such as normalization of aspect ratio, the ratio between the scales (meters per pixel) in the horizontal or sample and the vertical or line directions of an image. Because

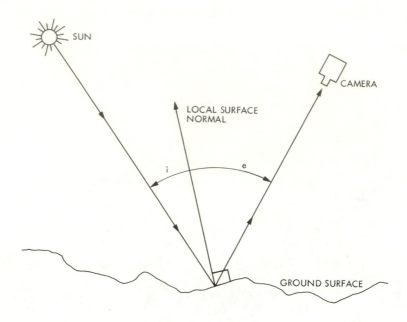

FIGURE 6.7. Illumination geometry. The incident angle, i, is the angle formed by the Sun's rays and the local surface normal. The emergent angle, e, is the angle formed by the local surface normal and the radiation reflected into the camera.

this method adjusts image geometry after all other processing is complete, it is not useful if rectification is necessary for any stage of the image processing during construction of the display picture.

RESAMPLING

Resampling is a flexible, powerful technique used to restructure the geometry in an image. There are two approaches illustrated in Fig. 6.9. In each of them some attempt is made to recreate or model the scene that was sampled to form the image. This modeled scene is then resampled to form an image with the desired geometric characteristics. Because of limitations in the size of the memory of most computers, it is necessary to model the scene only locally, in the vicinity of a desired resample site. This can still consume a large amount of time. Unfortunately, the severity of image degradation resulting from resampling increases as the size of the local area and the computation time decrease.

In resampling an image, a feature in the scene to be located at pixel (ℓ, s) in the geometrically corrected image being constructed will be found at (x, y) in the distorted image. Resampling techniques differ in the method by which the DN which should be stored at pixel (ℓ, s) of the output picture is derived from the input picture. The two principal algorithms used to accomplish this are discussed below.

NEAREST NEIGHBOR ALGORITHM

$DN_{\ell,s}$ is assumed to be the same as the DN of the pixel closest to the location (x, y) in the input picture. The closest pixel (x', y') is found by simply rounding off the values of (x, y).

Nearest neighbor algorithms are generally the fastest of the resampling algorithms. However, they may suffer from the defect that the local geometry may be inaccurate by up to $2^{-1/2}$ of the IFOV, or the size of pixel on the ground. Worse yet, the pixel from which the DN is derived shifts suddenly from the pixel just before the correct resampling location (x, y) to the pixel just after it. This problem becomes serious during digital picture comparisons because while the registration of detail in the two images may be perfect in one location, elsewhere there is misregistration. This is illustrated in Fig. 6.10, which shows a time difference picture made from Landsat images of Arizona taken one day apart. The vertical bands in Fig. 6.10 result when sudden jumps in the resampling posi-

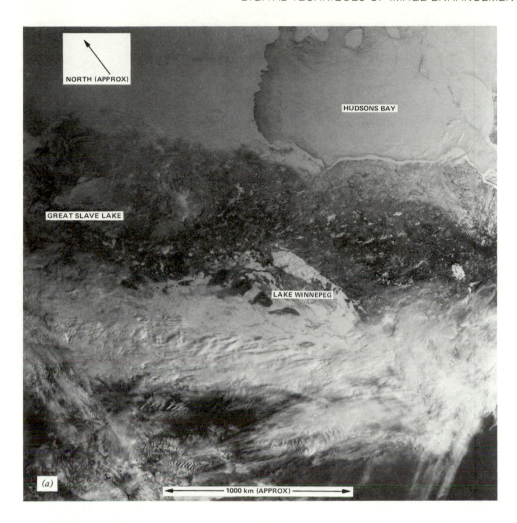

FIGURE 6.8. (a) Illumination geometry in images. A NASA NOAA-4 satellite image 76:3:9:16:25 of Canada, showing darkening because of increasing incident angle toward the morning terminator. No geometric corrections of this skewed image have been made.

tions in the two pictures from which the difference picture was made are out of phase.

Nearest neighbor resampling algorithms do not degrade the representation of fine detail, as do other resampling algorithms. This is because no interpolation, which smooths the data, takes place. On the other hand, at the discontinuities in the resample sites, the geometry of the image itself will be seriously disrupted.

INTERPOLATION ALGORITHMS

All resampling algorithms involving interpolation to find $DN_{\ell,s}$ may be regarded as convolutions of continuous functions with the discrete image. Convolution is an important basic physical concept

and is clearly discussed by both Bracewell (2) and Blackman and Tukey (4). The convolution of two functions $f(x)$ and $h(x)$ is (in one dimension)

$$g(x) = \int_{-\infty}^{\infty} f(u)h(x-u)\,du \qquad (6.4)$$

Convolution may be thought of in many ways. One useful way to regard it as a running average, with weights given by $h(x)$. For our purposes in this discussion, $f(x)$ corresponds to the discrete image. The nearest neighbor algorithm discussed above is equivalent to convolution with a rectangular kernel $h(x)$. The difference between resampling by convolution and spatial filtering by convolution is essentially that resampling requires computation at

FIGURE 6.8. (*b*) Illumination geometry in images. The shading in (*a*) caused by the earth's curvature, calculated for a Lambert surface.

irregular intervals in the input picture while digital filtering requires computation at every pixel. Regarding interpolation as convolution is useful to gain insight into the effect of resampling on the rectified image.

The most commonly used resampling method is bilinear interpolation, which is just the two-dimensional equivalent of linear interpolation. $DN'_{\varrho, s}$ in the output image is found by interpolation among the four pixels surrounding resampling site (x, y) in the input image. Because a plane fit to those four pixels' is over-determined, a higher-order surface is used instead. In practice, $DN'_{\varrho, s}$ can be found by successive linear interpolations, first along two opposite sides of the square formed by the four pixels, and then across the square, as shown in Fig.

6.9. This algorithm yields the value of the fitted surface at (x, y), which is also the desired $DN'_{\varrho, s}$.

The image degradation caused by bilinear interplation can be found in the special one-dimensional case (i.e., linear interpolation) by taking the Fourier transform H of the triangular function h as shown in Fig. 6.11(*c*):

$$h_{\text{linear}} = \Lambda(x) = \begin{cases} 1 - x; & |x| \leqslant 1 \\ 0 & ; |x| > 1 \end{cases} \quad (6.5)$$

$$H_{\text{linear}} = \int_{-\infty}^{\infty} \Lambda(x) e^{i2\pi x \omega_x} \, dx \quad (6.6)$$

FIGURE 6.8. (c) Illumination geometry in images. At a much larger scale topographic shading dominates. The "sun" has been placed 20° above the eastern horizon in this "irradiance" image of the Salmon Mountains in California.

where ωx is spatial frequency (cycles per pixel) in the x direction. Equation 6.6 may be rewritten as:

$$H_{\text{linear}} = \left(\frac{\sin \pi \omega_x}{\pi \omega_x}\right)^2 \tag{6.7}$$

This function has zeros at integer multiples of ωx cycles/sample. The weights in the convolution kernel have amplitude h at distances x of the resampling site from the nearest sample. Thus, if the resampling site is 25% of the distance from sample n to sample $n + 1$, the two weights are found at $x = -0.25$ and $x = 0.75$. The frequency degradation increases as x approaches ± 0.5 pixels [Fig. 6.11(c)].

Extension of the above argument to two dimensions is straightforward.

Resampling using a bilinear interpolation algorithm involves more computation time than resampling using a nearest neighbor algorithm, but it is geometricaly more accurate.

The amount of frequency degradation caused by interpolation can be reduced by increasing the number of pixels in the vicinity of the resampling site from the nearest four used in bilinear interpolation. For the ideal case where there is no information in the scene that is undersampled, it is possible to exactly reconstruct the scene from the digital image (13). To do this requires a special infinite convolution kernel called the sinc function. This function is just the Fourier transform of the

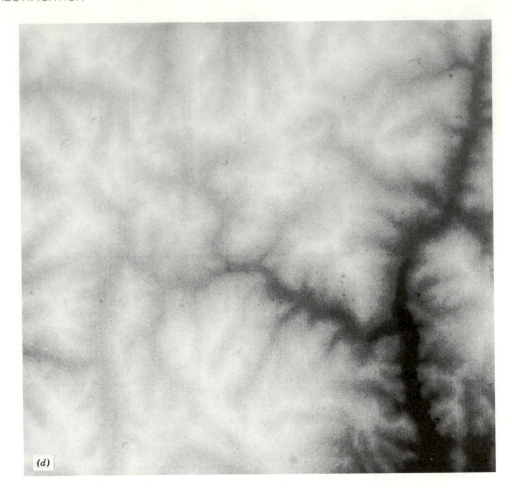

FIGURE 6.8. (d) Illumination geometry in images. The "altitude image" used to generate (c). Light tones represent high altitudes; dark tones are near sea level.

desired uniform spatial frequency response and in one dimension:

$$\text{sinc}(\omega_x) = \frac{\sin \pi \omega_x}{\pi \omega_x} \qquad (6.8)$$

where ω_x is spatial frequency in the x direction. Notice that H in Equation 6.7 is just $\text{sinc}^2(\omega_x)$. The sinc function is drawn in Fig. 6.11(a).

Reproduction of a function using a series of sinc functions results in a uniform spatial frequency response out to ω_n, the Nyquist frequency, so that the relative spatial frequency content of the image is undisturbed. The original continuous function will be reproduced exactly at all points between the samples as well as at the samples themselves. However, exact reproduction requires an infinitely large number of weights, which is not possible. One solution is to simply set to zero values of the convo-

lution kernel $h(x)$ (see Equation 6.4) for large x. In other words, image data more than some specified number of pixels from a resample site will be ignored during the interpolation.

Unfortunately as many as 19 samples in each direction are required for adequate reconstruction. Figure 6.11(a) shows the spectrum of an impulse located at the origin in an image that has been resampled using limits of $\pm\infty$, ± 19, and ± 3 pixels at $x = 0.5$ (the worst case because the site is farthest from the acquired pixels).

Clearly amplitude at higher ω in the spectrum of the resampled picture decreases as the summation limits are decreased. At limits of ± 3 this technique yields results comparable to bilinear interpolation.

Rifman (14) and others have experimented with various functions to use in place of the sinc function in Equation 6.8. These functions do not require

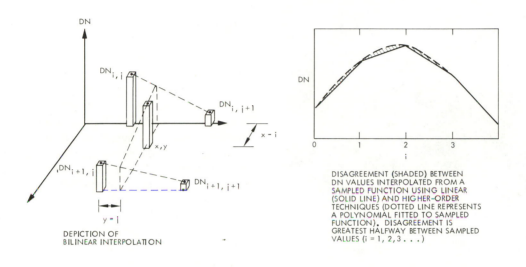

FIGURE 6.9. Nearest neighbor and bilinear interpolation resampling techniques.

as large summation limits as the sinc function, but their transforms retain a high amplitude at frequencies less than ω_n. One of these functions and the spectrum of an impulse resample using it is shown in Fig. 6.11(b).

In summary, the resampling process consists of:

1. Reconstruction of the continuum function in two dimensions from neighboring samples.
2. Resampling the reconstructed continuum at the desired points to produce a new image.

It is generally too expensive and time-consuming to compute the site in an image corresponding to each pixel in the resampled image. Because systematic distortions generally change slowly over an image, a control grid can be established for the resampled image in which the distortion is measured about every 50 pixels. Resampling sites for pixels that do not coincide with control points must be found by interpolation from neighboring points, as shown in Fig. 6.12.

It is important not to confuse loss of frequency response during resampling with degradation during acquisition of the image. Degradation in resampling varies depending on the distance from resample sites to existing sample sites, while degradation in acquisition is constant or very nearly so across the image.

SYSTEMATIC GEOMETRIC DISTORTIONS ENCOUNTERED IN IMAGES

Geometric distortions may be introduced into an image by the optical system or by the sensing apparatus as discussed in Section 6.2. In general, well-made lenses construct images with excellent fidelity, and film cameras can be made which produce images requiring no geometric rectification. However, both line scanners and vidicons introduce significant distortions into images and for most purposes these images require rectification.

In vidicon images geometric distortions caused by the vidicon tube itself are identified from the apparent location in the image of the reseau marks,

FIGURE 6.10. Difference picture showing temporal changes in the Coconino Plateau of Arizona from August 6 to August 7, 1972. Misregistration is attributed to parallax. Satellite that took images was 150 km further west on August 7 than on August 6. Vertical swaths of misregistered data (see right center) are artifacts introduced by nearest-neighbor resampling. (NASA Landsat pictures 1014-17375 and 1015-17431).

the little metal spots deposited on the vidicon tube. Measurement of the displacement between the apparent location of these marks in the images and their actual location in the vidicon defines a control grid such as the one illustrated in Fig. 6.12. For each grid intersection, or control point, the displacement necessary to undo the local geometric error is recorded, and these data are passed to the resampling program along with the image.

The reseau marks themselves are usually found automatically (15), although they may be located by inspection. The nominal or expected location of each mark is stored in the computer, and for each mark the computer searches the image for a dark spot of the right size and shape. The search is begun at the predicted location, expanding outward until the mark is found. This must be done for every image because the position of the marks may differ slightly, even though the images were made by the same vidicon.

Distortions found in line scanner images may be more complicated than those found in vidicon images. They arise from three sources: irregular motion of the oscillating mirror that sweeps the image across the sensor, irregular motion of the aircraft or satellite that carries the sensor, and rotation of the earth itself underneath the satellite-borne sensor. Thus, systematic distortions in Landsat images include skewing (about 3°) as well as scale distortions. This skewing is easily seen in the outline of the Landsat image in Fig. 6.13.

Because the Landsat program has made large numbers of digital line scanner images available to the public, the geometric correction of these images is of particular interest. The detailed corrections are discussed by Goetz et al. (16).

Line scanner distortions are essentially one dimensional. The scale in the line direction of the image is determined only by the velocity of the spacecraft in orbit, with respect to the *rotating* earth. Thus the line scale varies with the latitude of the subspacecraft point. The scale in the sample direction is determined by the period of the oscillating mirror, its deflection rate (degrees of arc per second), and the time between samples, as well as the distance of the scanner from the surface of the earth. The sample scale changes across the image because the mirror deflection rate is not constant and because the range from the camera to the earth changes as a function both of topography and of the earth's curvature.

Changes in the orientation of the line scanner with respect to the earth and to the orbit of Landsat introduce distortions into the images. Yaw, or rotation of the spacecraft about the line connecting it to the center of the earth, causes the angle between the scan direction and the orbit to deviate from 90°, and looks like a skew in the image. Roll, or rotation of the spacecraft about its velocity vector, introduces changes in the sample scale only. Pitch, or rotation of the spacecraft in the re-

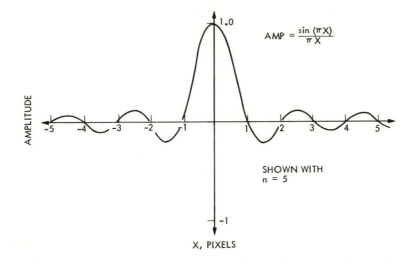

$$\text{AMP} = \frac{\sin(\pi X)}{\pi X}$$

SHOWN WITH
n = 5

X, PIXELS

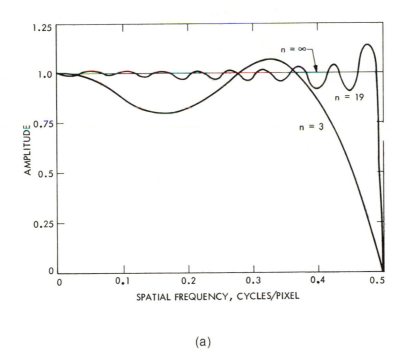

(a)

FIGURE 6.11. (a) Resampling by interpolation. An infinite sinc function (top graph) can be used to resample a band-limited image (zero amplitudes above the Nyquist frequency) with no loss of sharp detail. As summation limits are reduced to ±19 and ±3, frequency response (bottom graph) is degraded unless the resample site is identical to a sample site.

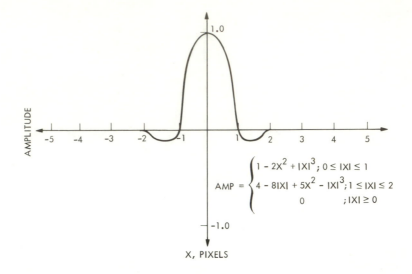

$$\text{AMP} = \begin{cases} 1 - 2X^2 + |X|^3; \ 0 \le |X| \le 1 \\ 4 - 8|X| + 5X^2 - |X|^3; \ 1 \le |X| \le 2 \\ 0 \qquad\qquad ; \ |X| \ge 0 \end{cases}$$

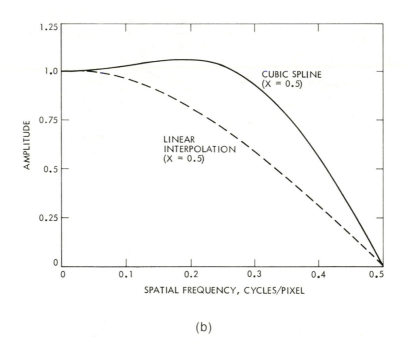

(b)

FIGURE 6.11. (*b*) Resampling by interpolation. A cubic spline function (top graph) (14) causes little image degradation even with limits of ± 2 samples (bottom graph).

168

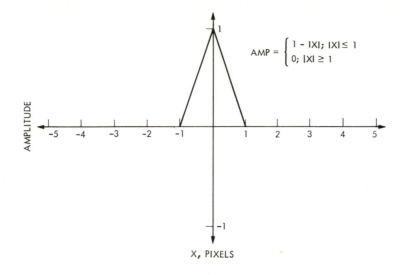

$$\text{AMP} = \begin{cases} 1 - |X|; & |X| \leq 1 \\ 0; & |X| \geq 1 \end{cases}$$

X, PIXELS

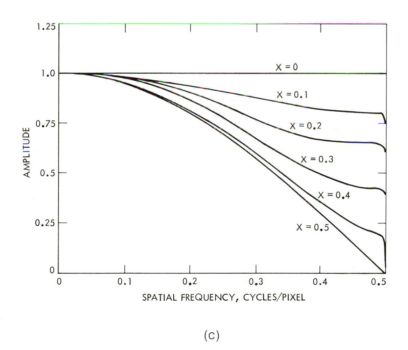

(c)

FIGURE 6.11. (c) Resampling by interpolation. Linear interpolation is performed using a Λ function (top graph). Dependence of image degradation on distance x of resample site from sample sites is shown in bottom graph.

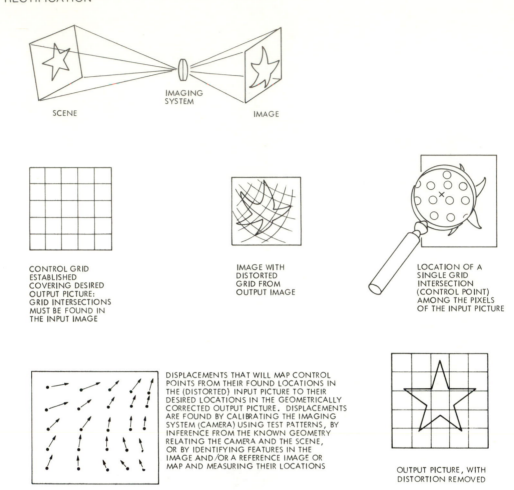

SCENE   IMAGING
                               SYSTEM   IMAGE

CONTROL GRID
ESTABLISHED
COVERING DESIRED
OUTPUT PICTURE:
GRID INTERSECTIONS
MUST BE FOUND IN
THE INPUT IMAGE

IMAGE WITH
DISTORTED
GRID FROM
OUTPUT IMAGE

LOCATION OF A
SINGLE GRID
INTERSECTION
(CONTROL POINT)
AMONG THE PIXELS
OF THE INPUT PICTURE

DISPLACEMENTS THAT WILL MAP CONTROL
POINTS FROM THEIR FOUND LOCATIONS IN
THE (DISTORTED) INPUT PICTURE TO THEIR
DESIRED LOCATIONS IN THE GEOMETRICALLY
CORRECTED OUTPUT PICTURE. DISPLACEMENTS
ARE FOUND BY CALIBRATING THE IMAGING
SYSTEM (CAMERA) USING TEST PATTERNS, BY
INFERENCE FROM THE KNOWN GEOMETRY
RELATING THE CAMERA AND THE SCENE,
OR BY IDENTIFYING FEATURES IN THE
IMAGE AND/OR A REFERENCE IMAGE OR
MAP AND MEASURING THEIR LOCATIONS

OUTPUT PICTURE, WITH
DISTORTION REMOVED

FIGURE 6.12. Identification and removal of geometric distortion from an image.

maining orthogonal direction, does not change the line scale, which depends only on the relative velocity of the spacecraft, but does change the dimensions of the IFOV and thus modifies the image spectrum.

As long as the spacecraft pitch is small, all geometric corrections may be performed in the sample direction only. These corrections include a normalization of the sample scale to the line scale. Because the correction is one-dimensional only, the resample site for each pixel in a scan line may be specified. Skewing of the image to compensate for earth rotation and spacecraft yaw is performed by adding a constant increment to the sample position of each scan line in the rectified image, as

illustrated in Fig. 6.13. The trace of the subspace-craft point over the image is thus oblique with respect to the scan lines.

CARTOGRAPHIC PROJECTION

For comparison of an image to maps or other images it is desirable to change the geometry so that features are found in the same location in both pictures. Usually the desired geometry of the projected image corresponds to a standard mapping or cartographic projection. The process of modifying the image geometry is called projection, because historically such transformations were conceived of as geometric projections of the image plane through a point onto some surface,

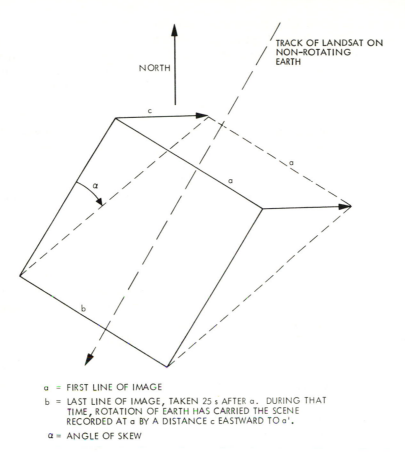

a = FIRST LINE OF IMAGE

b = LAST LINE OF IMAGE, TAKEN 25 s AFTER a. DURING THAT
    TIME, ROTATION OF EARTH HAS CARRIED THE SCENE
    RECORDED AT a BY A DISTANCE c EASTWARD TO a'.

α = ANGLE OF SKEW

FIGURE 6.13. Skewing of Landsat MSS image caused by rotation of Earth.

such as a cylinder. Descriptions of different cartographic projections and their attributes may be found in standard texts (17, 18).

Projection is usually considered to be a rectification, although it is actually a *distortion* of the image and not a *removal* of distortion.

To project an image, the analyst makes use of the same resampling programs used to correct the image for camera distortions. However, the grid of control points is conceived of as covering the projected image that has not yet been made. The analyst knows the latitude and longitude at each grid point from the equations describing the chosen cartographic projection, and he must find the location of the corresponding geographic coordinates in the input picture (19). While he may do this by visual comparison of features distinct in both the image and existing maps, usually it is more convenient to compute the locations using knowledge of the camera orientation and relation to the scene (20).

Once the analyst has found the control points in both the input image and the output projected image, he gives the displacements to the resampling program that actually constructs the projected image.

It is worthwhile to note that even photographs taken with ordinary film cameras are projected images. The scene is projected by the lens through a point on the optic axis and onto the image plane. This may be regarded as a simple perspective projection.

One reason projection is a useful tool is that it allows us to construct maps from many images pieced together in a mosaic. The map of the south polar region of Mars shown in Fig. 6.14 is a good example of such a photomosaic. In this map, five images taken by the Mariner 7 spacecraft when it passed close to Mars in 1969 were projected into orthographic format (21). The orthographic projection is the limiting case of the simple perspective projection where the observer or camera is at infinity, so that parallel rays pass through each feature in the scene and its image. The geographic coordinates and the orthographic image coordinates are simply related. The orthographic image plane is just one plane of some Cartesian coordinate system which may be related to the coordinate system containing the planet. For a spherical planet, the geographic and Cartesian coordinate systems are related by:

$$
\begin{aligned}
x &= r \cos \phi \sin \lambda \\
y &= r \cos \phi \cos \lambda \\
z &= r \sin \phi
\end{aligned}
\tag{6.9}
$$

where r is the radius of the planet, and ϕ and λ are the latitude and longitude of a point on its surface. For a *polar* orthographic projection such as the map in Fig. 6.14, the map plane coordinates are just x and y in Equation 6.9. That is, the map plane is parallel to the equatorial plane of Mars and one pole of the planet projects through the origin of the map. Figure 6.15(a) shows the original geometry of one of the high-resolution images comprising the photomosaic in Fig. 6.14.

The five projected "near-encounter" images in Fig. 6.14 together did not cover the whole polar cap, so they were superimposed on a similar map made from "far-encounter" pictures taken earlier, when the spacecraft was much farther from the planet. One of these pictures is shown before projection in Fig. 6.15(b). At the greater distance, the far-encounter pictures show much less detail than the near-encounter pictures, and that is why the background in the photomap shows less detail than the overlay. In order to approximately match *DN* across boundaries between different images in the photomap, it was necessary to remove the shading caused by changing viewing and illumination geometry.

REGISTRATION

After projection of two images to the same geometry there may still be some disagreement between the locations of the same feature in the different images. These errors arise from imprecise knowledge of the camera position or orientation with respect to the scene. The reduction of these er-

rors to a minimum so that the frames are of identical geometry is called "registration." The actual modification of image geometry is again done by the resampling program, but instead of specifying the displacement grid from prior knowledge of the camera and scene, the analyst defines it by measuring the actual location of features in one scene with respect to the other. The location of these features in the image may be found by simple inspection, or the approximate locations may be specified for the computer, which must then determine the displacement of a small region in one image that gives a "best fit" of the scene detail compared to the other "reference" image. The "best fit" is usually in the least squares sense, and the process of fitting is called cross correlation. Cross correlation is conceptually and mathematically very similar to convolution except that the function $h(x - u)$ in Equation 6.4 must be replaced by $h(u - x)$.

In registration, the goal is to find the disagreement between images at many local features and it is generally expected that these displacements will vary from feature to feature, so instead of convolving the entire images, only a small region containing each feature from one image is convolved with a small region containing the same feature in the "reference" image. The actual discrete cross correlation is

$$
c_{ij} = \sum_{m=-M}^{M} \sum_{n=-N}^{N} (DN1_{\ell_1+i+m, s_1+j+n} - \mu_1)
\tag{6.10}
$$
$$
(DN2_{\ell_2+m, s_2+n} - \mu_2)
$$

$$
\mu_1 = \frac{1}{(2M+1)(2N+1)} \sum_{m=-M}^{M} \sum_{n=-N}^{N} DN1_{\ell_1+i+m, s_1+j+n}
$$

$$
\mu_2 = \frac{1}{(2M+1)(2N+1)} \sum_{m=-M}^{M} \sum_{n=-N}^{N} DN2_{\ell_2+m, s_2+n}
$$

where c is the correlation coefficient, i and j are line and sample coordinates ranging from $(-M, -N)$ to (M, N) in the $2M + 1$ by $2N + 1$ pixel region containing the feature, $DN1$ is the DN in

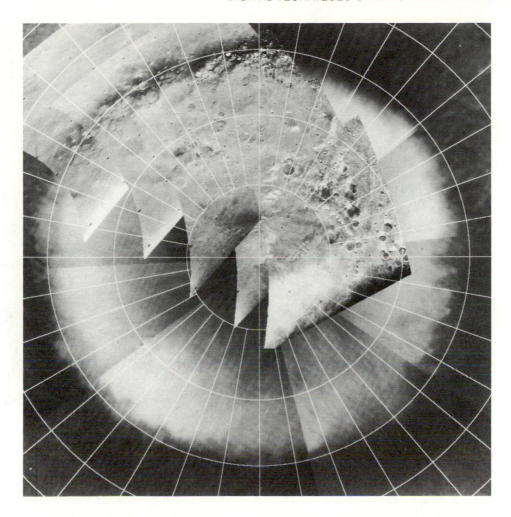

FIGURE 6.14. Orthographic photomap of the south pole of Mars. The high resolution overlay has been high-pass filtered. The black dots are reseau marks used to identify geometric distortion in vidicon images. The low resolution background was made from 10 "far encounter" or distant images made by Mariner 6 and 7 in 1969. The circles in the cartographic grid are every 10° of latitude; the lines radiating from the pole are separated by 10° of longitude. The prime meridian of Mars is up.

the first image, and $DN2$ is the DN in the reference image. The coordinates (ℓ_1, s_1) and (ℓ_2, s_2) describe the location of the center of the local region in the two images. Correlating deviations from the means μ_1 and μ_2 is done to prevent spurious registration on regional gradients that may be artifacts such as vidicon shading. When $DN1$ and $DN2$ agree perfectly, pixel by pixel throughout the region, c_{00} has a larger value than other c_{ij}. c_{00} is the correlation coefficient found with no displacement in either the line or sample direction. In general, this is not the case and the displacement required to achieve a best fit is given by (i, j), where

c_{ij} is the largest coefficient. After the maximum correlation coefficient for each feature is identified, the displacements necessary to register the images are passed to the resampling program.

The technique outlined above considers only translation errors, but in reality the images may be rotated with respect to one another as well. To accommodate rotation errors a similar but more involved approach is required.

In practice, image registration is a tedious and difficult affair. Often the correlation is poor, or there are several maxima in the correlation or c matrix. Analysts thus may have to override the com-

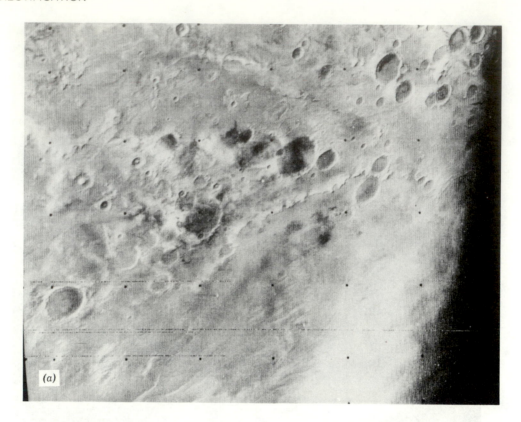

FIGURE 6.15. (*a*) A Mariner 7 near-encounter image (7N19) used in photomap, before projection to orthographic format.

puter's decision about the "best" fit. This is best accomplished interactively, with the analyst watching a television screen on which the computer displays both pictures and also the *c* matrix, with the correlation coefficients encoded as gray levels. The analyst may superimpose the two pictures, translating one picture by the amount specified by each maximum in the *c* matrix, or by an arbitrary amount. When the analyst feels the pictures register, he commands the computer to store the displacement.

An interactive user station was built at the Jet Propulsion Laboratory's Image Processing Laboratory for use in processing Mars images returned from the Viking Orbiter and Lander spacecraft in 1976. Typical correlation matrices displayed by an interactive image registration program on this display system are shown in Fig. 6.16. Fig. 6.16(*a*) shows a good fit, which is recognizable by the single bright maximum and the high degree of symmetry. The poor or ambiguous fit shown in Fig. 6.16(*b*) may be caused by noise in one or both images or by a variety of other problems. Periodic noise

presents an especially serious problem because if present in both images a very strong correlation may be found which has nothing to do with the real image data. Random noise is by definition uncorrelated and may decrease the reliability of the registration procedure. Thus, it is important to minimize both random and coherent noise before registration. This is best done by spatial filtering, as discussed in Sections 6.5.3 and 6.8.3.

6.5.3 Corrections in the Spatial Spectrum of Images

In the process of image acquisition, distortions may be introduced into the geometric and radiometric representations of a scene. In Section 6.5.1, radiometric distortions have been treated as though they were independent of the size of features in the scene, which is not strictly true. From the MTF of the camera used to create the image (see Fig. 6.1) it is clear that very small features that are composed of high frequency waves have a different LTF than very large features, in which low fre-

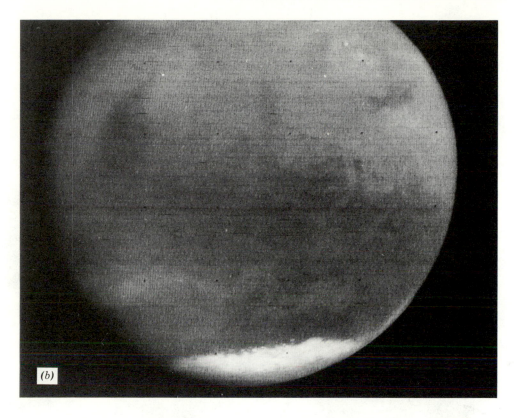

FIGURE 6.15. (b) A Mariner 6 far-encounter image (6F44) used in background of photomap, before projection to orthographic format.

quencies dominate. Small features and large features of equal brightness in the scene may be represented at different *DN* in the image. This nearly random noise may appear in the image spectrum as a constant added to the spectrum of the image as transmitted through the optical system alone before recording. Finally, interference with other nearby electronic instruments may introduce periodic noise, which appears in the image spectrum as spikes at certain frequencies.

The rectification procedure for frequency response of the camera and the addition of random or periodic noise to the image is based on modifying the image spectrum, to cause it to more closely resemble the spectrum of the scene itself. This is done by spatial filtering.

Restoration filtering may be regarded as a rectification procedure because it compensates for distortions introduced by the imaging system. On the other hand, suppression of random noise may be regarded as either a rectification or cosmetic procedure because it is usually done to make a picture *look* better, even though it does not improve

or even degrades the information content of a picture.

Periodic noise removal likewise may be regarded as a rectification or cosmetic procedure, depending on the way in which the analyst characterizes the noise.

DIGITAL FILTERING

Filtering is an extremely powerful and important tool in signal processing. In image processing, there are three important applications: restoration, enhancement, and pattern recognition. In restoration filtering, the goal is to correct for the decreased sensitivity of the camera system to signals of high spatial frequency. Thus restoration filters sharpen edges and increase the visibility of small features in the image, representing them at the correct brightness with respect to larger features. In practice, some random noise is suppressed during restoration filtering. Suppression of periodic noise is accomplished using different filters.

Filters may also be used to exaggerate or enhance certain aspects of an image for more ef-

(a)

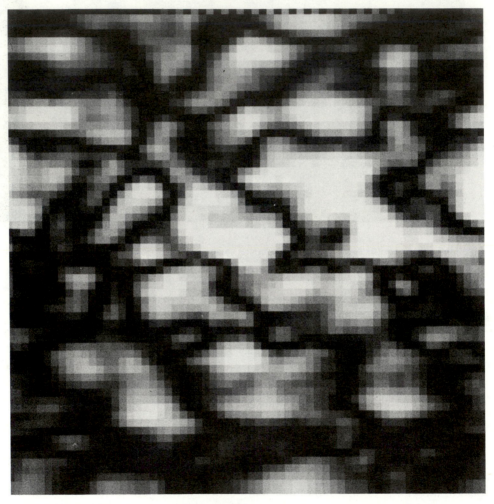

(b)

FIGURE 6.16. (a) A cross-correlation matrix depicting a good fit between two images. Registration may be achieved by translating the image 1 sample to the right with respect to the reference image. Registration occurs when correlation maximum is located at the center of the display. Bright tones show high correlation. The white axes were superimposed on the matrix to show the center. (b) A cross-correlation matrix depicting a poor fit. No single bright peak dominates the confused pattern.

fective display, as discussed in Section 6.8.3. Filters used in this way may be thought of as analogous to the filters used in amplifiers to suppress rumble or scratches in phonographic recordings. Other filters are used in image analysis, especially in pattern recognition (22). Filters of this type are not yet widely used nor have they been successfully used in processing of images for geological purposes, and are not discussed in this chapter. However, in the future they may have application to such important tasks as lineament recognition.

Filtering may be done in either the spatial domain or the frequency domain. In the spatial domain, the domain the image is acquired in, filtering of discrete functions is described by:

$$DN'_{\ell,s} = \sum_{m=-M}^{M} \sum_{n=-N}^{N} W_{m,n} DN_{\ell+m,s+n}$$ (6.11)

where m and n are integers, W is a weight matrix, $DN_{\ell,s}$ is the DN at line ℓ, sample s of the unfiltered image, and DN' is the corresponding DN in the filtered image. This equation describes convolution. The values in weight matrix W are chosen by the analyst in a way that will be discussed below.

Filtering in the frequency domain is accomplished by simply multiplying the image spectrum or the Fourier transform of the image by a chosen function F, which is the filter:

$$A'_{\omega_\ell,\omega_s} = F_{\omega_\ell,\omega_s} A_{\omega_\ell,\omega_s}$$ (6.12)

where A is the amplitude of the Fourier transform of the image, A' is the amplitude in the transform of the filtered image, and ω_ℓ and ω_s are frequencies in the line and sample directions. F may be defined in the frequency domain, but is the Fourier transform of W in the convolution equation (Equation 6.11).

Filtering in either domain is an expensive, time-consuming procedure. In general, the number of operations per pixel in convolution is proportional to the number of weights in the weight matrix W, while just one operation per pixel is required to filter in the frequency domain. However, images are acquired and displayed in the spatial domain so two Fourier transformations must be performed in addition to filtering. The number of operations in a Fourier transformation in one dimension is proportional to n

$\log(n)$, where n is the number of samples in the image.

Convolution filters that have uniform weights are significantly more economical than other filters. Because of the many redundant operations during convolution with such a filter, a short cut method may be exploited in which, as the position of W is incremented, the column of DN at the "leading edge" of W is added and the column at the "trailing edge" is subtracted from the value of the immediately previous filtered value (DN' in Equation 6.11). Using the same notation of Equation 6.11,

$$DN'_{\ell,s} = DN'_{\ell,s-1} + \frac{1}{2M+1} \sum_{m=-M}^{M} (DN_{\ell+m,s+N} - $$

$$DN_{\ell+m,s-1-N})$$ (6.13)

The cost can be further reduced by using more shortcuts, such as computing $\sum_{m=-M}^{M} DN_{\ell+m,s}$ one time only, and saving the sum in the computer's memory. In this way, the cost of convolution can be made nearly independent of the size of W. Filters of this type are often called "box filters" and their use will be discussed further as an enhancement technique in Section 6.8.3.

RESTORATION FILTERING

A real scene is composed of an infinite number of impulses, or point sources. An image can be created by the superposition of the point spread functions (PSFs) obtained from each individual impulse in the scene. Thus we can describe a continuous image prior to discretization:

$$i_{\ell,s} = \int \int s_{x,y} h_{\ell-x,s-y} \, dx \, dy + noise_{\ell,s}$$ (6.14)

where i is the image, s is the scene, and h is the PSF centered at line ℓ, sample s (subscripted).

Alternatively, we can express Equation 6.14 as the product of Fourier transforms:

$$I_{\omega_\ell,\omega_s} = S_{\omega_\ell,\omega_s} H_{\omega_\ell,\omega_s}$$ (6.15)

where I, S, and H are the Fourier transforms

of the image, the scene s, and PSF h, ω is spatial frequency, and noise has been disregarded. It follows that if H and I are known, S may be found. In other words, the Fourier transform of the scene is recoverable from an image for which the PSF is known:

$$S_{\omega_\varrho, \omega_S} = I_{\omega_\varrho, \omega_S} H^{-1}_{\omega_\varrho, \omega_S} \qquad (6.16)$$

where H^{-1} is the reciprocal of H, provided that H has no zeros in the frequency range of interest.

Frequency response can be restored either by convolving the image with a filter kernel or weight matrix created from H^{-1}, or by multiplying the transform of the image by H^{-1} and transforming the product (Fig. 6.17). In either case, it is necessary to know H, which for imaging systems is simply the MTF. The MTF is usually estimated during construction of an imaging system from the response of the camera to bar targets. In the absence of ground calibration data, the MTF can be found in flight by measuring the response of the camera to field boundaries or other features known to be sharp with respect to the resolution of the camera. Because the spectrum of the feature in the scene is known or can be estimated, the ratio of the measured spectrum of the feature in the image to the ideal spectrum is a very good approximation to the MTF. The MTF shown in Fig. 6.1 was computed this way.

Equation 6.16 states that perfect frequency response (at frequencies less than ω_n) the Nyquist frequency) can be restored to an image. However, H typically becomes small at high frequencies, and its reciprocal becomes large. Feature 6.18 shows the effect of using such a function to restore frequency response in an image with some random noise. The overall signal-to-noise ratio (S/N) for the corrected image is worse than before because the noise does not decrease in amplitude like the signal at high frequencies. To overcome this difficulty, H^{-1} is modified to reduce its amplitude frequencies. This has the effect of reducing the visual impact of random noise in the restored image. One scheme for designing H^{-1} is given by the following Wiener filter (23):

$$H^{-1'}_\omega = \frac{MTF_\omega}{MTF^2_\omega + (S/N)^{-2}} \qquad (6.17)$$

where ω is the frequency. For additive noise, S/N decreases with frequency; thus H^{-1} becomes increasingly smaller than H^{-1} as ω approaches ω_n, Figure 6.18(c) illustrates the effects of this restorative filter.

Wiener filters reduce the effect of random noise. Another way to deal with random noise is available if more than one channel of data is recorded, as is the case with multispectral line scanners. This approach, which improves the S/N by increasing the signal, is simply to average two or more image channels together. Scene data tends to be strongly correlated from channel to channel, while the random noise is not. Because of this, in averaging pictures the signal is essentially unchanged while the noise is decreased proportionately to the square root of the number of image channels being averaged (24).

PERIODIC NOISE REMOVAL

Periodic noise is composed of spatial waves having a limited range of frequencies. Consequently, periodic noise appears as spikes in the frequency domain. The six-line striping prevalent in Landsat images is a good example of simple periodic noise. Because the striping is exactly parallel to the sample or scan direction, the waves comprising it have components only in the ω_ϱ direction, and thus appear as spikes on the ω_ϱ axis of the image spectrum.

Usually, periodic noise is more complicated than the Landsat striping. The vidicon images returned from Mars in 1969 by the Mariner 6 and 7 spacecraft are examples of images with complicated periodic noise caused by the vidicon itself and also by electronic interference from other instruments on board (7). One such image and its transform are shown in Figs. 6.19 (a) and (b).

Filtering of the noise is accomplished by devising a function that has a value of unity everywhere in the frequency domain except in the vicinity of the noise spikes, where it has much lower values. The product of the filter and the image spectrum should be smooth across the location of the noise spikes. Figure 6.19(c) shows the image reconstructed from the filtered spectrum shown in Fig. 6.19(b), and Fig. 6.19(d) shows the noise that was removed in the process. This noise picture was made by subtracting the filtered image from the original image Such a picture is called a "dif-

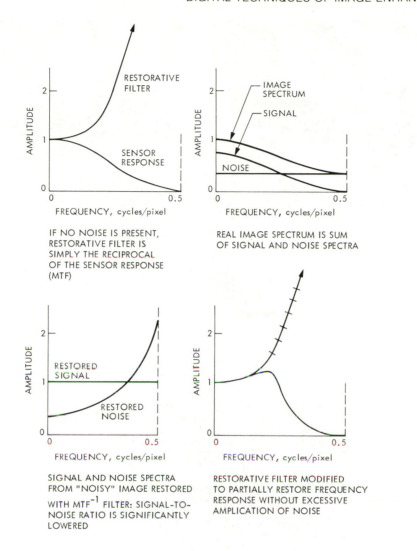

FIGURE 6.17. Design of restoration filter to correct image system frequency response.

ference picture" and is generally used to emphasize differences between two images.

It is also possible to transform the noise filter and remove periodic noise by convolution. This is actually done for some simple cases, like the Landsat images with six-line striping, but is generally impractical for the following reason: the sharper the feature in the frequency domain, the larger its transform in the spatial domain. The extreme case is the transform of an impulse, which is infinite. The periodic noise forms sharp spikes in the frequency domain, so the convolution filters required to adequately remove it are very large and expensive to use.

6.6 COSMETIC PROCEDURES

Cosmetic programs attempt to make a picture look better to a human observer, even though the information content of the picture is not increased. The removal of periodic noise, discussed in the previous section as a rectification, may also be considered to be cosmetic because the basic information content of the image is unchanged. For ex-

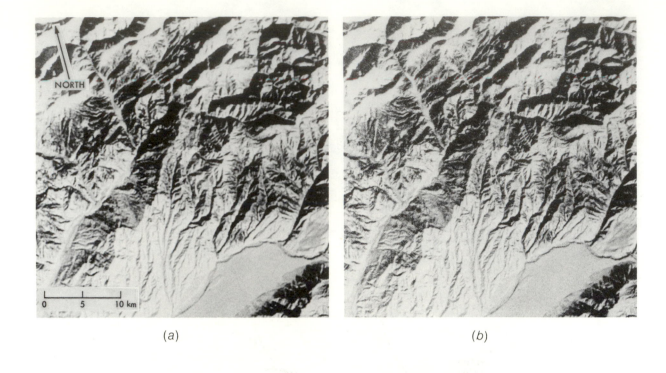

(a)   (b)

(c)

FIGURE 6.18. Restoration filtering the part of NASA Landsat image 1534-05184, near Tashkent, U.S.S.R. (a) Acquired image, with random noise added to emphasize effects of restorative filters used in (b) and (c). (b) A reciprocal MTF filter was used to restore this image. Notice graininess caused by exaggerating noise. (c) A Wiener filter was used to restore frequency response without over-emphasizing noise. Edges are sharper and more clearly defined than in (a), the unfiltered image.

180

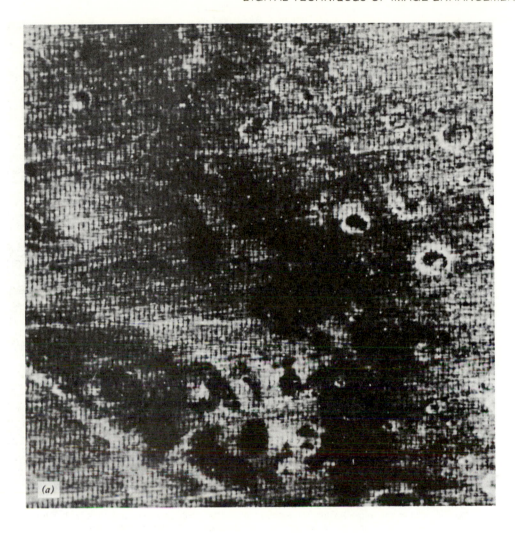

FIGURE 6.19. (a) Periodic noise removal from a Mariner 6 vidicon image (6N13) of Mars in 1969:
Contrast enhanced portion of the acquired image. Periodic noise is readily seen.

ample, striping in Landsat images can be reduced by convolution filtering, using a one-line high-pass box filter with from about 31 to 301 weights in the sample direction. Such a filter suppresses low-frequency variations in DN, displaying only local departures from average brightness. By examining only one line at a time, gross brightness differences between lines are reduced, but also radiometric fidelity is lost. Thus, the information content of the image is actually reduced.

A high-pass filtered image is essentially the original image minus the low-pass filtered image. Because addition in the frequency domain is equivalent to addition in the spatial domain, this leaves only the high frequency information. A low-pass box filter acts like a running average and has weights of unity. The high-pass box filter has weights of -1 except at the center, which has a weight of $NM - 1$ where N and M are the dimensions of the filter weight matrix. Because such a filter represents as black ($DN = 0$) a pixel whose DN is the same as the local average, a constant must be added so that pixels which are darker than average may be represented in the dynamic range also. Usually, the constant chosen is $DN = 128$, or mid-range gray. Thus, if a pixel is darker than average, it is represented as dark gray; if it is brighter than average it is represented as light gray.

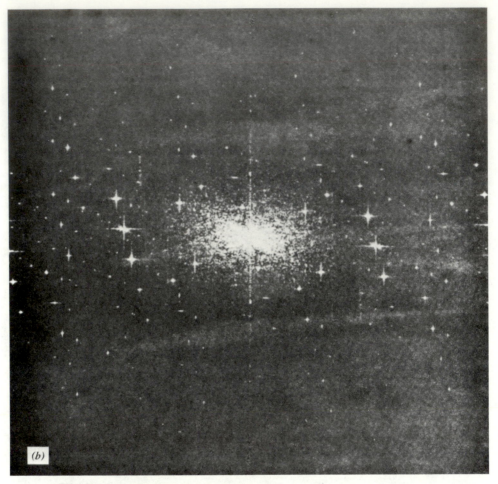

FIGURE 6.19. (*b*) Periodic noise removal from a Mariner 6 vidicon image (6N13) of Mars in 1969: Spectrum of image in (*a*). Spatial frequency zero is at the center; positive and negative horizontal spatial frequencies are displayed to the right and left of center, respectively, and positive and negative vertical frequencies are displayed above and below center. The perimeter of the display represents the Nyquist limits. Light tones depict high amplitudes. Periodic noise appears as bright spikes.

Other filtering techniques to suppress periodic random noise have been discussed in Section 6.5.3. Another approach is often used to identify and remove random noise. This approach makes use of the fact that scene brightness rarely changes by large amounts rapidly, while pixels of noise are not so constrained. The analyst may thus determine a threshold that represents the greatest anticipated change in scene brightness per pixel. If the *DN* of a pixel differs from that of its eight nearest neighbors by a value greater than this threshold, the pixel is probably noise. A noise pixel thus identified can be removed by interpolating a *DN* from adjacent pixels to replace it.

Large blemishes, several pixels in size, are occasionally found in images. Some blemishes have been attributed to dust on vidicon tubes, while others may have been caused by transmission errors. These large blemishes are usually located by visual inspection and removed by interpolation.

Reseau marks, which are the spots in vidicon images used to identify geometric distortions, are often removed before display of the image. Because the locations of the marks are already known (Section 6.5.2) they may be replaced by interpolation without intervention by the analyst, either before or after geometric rectification is complete.

Other cosmetic procedures commonly used superimpose map coordinate grids or English annotations on an image.

6.7 ANALYTIC PROCESSING

Although it may seem that most of an image analyst's effort is consumed by rectification and noise removal, these tasks are undertaken solely to prepare an image for effective analysis and dis-

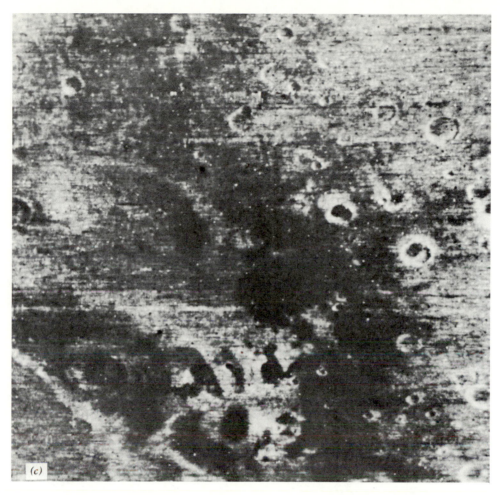

FIGURE 6.19. (*c*) Periodic noise removal from a Mariner 6 vidicon image (6N13) of Mars in 1969: Filtered picture, showing less noise than before.

play. Analysis and display of image data are the most interesting tasks performed by the image processor, especially because if they are well performed remotely sensed images can be made to yield much more information to the geologist than, say, simple aerial photographs, and they may even tell the geologist some things that he might not realize from traditional mapping or ground exploration alone.

Computer programs designed to assist in the analysis of images fall into many categories. In fact, the diversity of analytic procedures is limited only by the ingenuity and resources of the analyst himself. Some of the most important procedures are discussed below.

6.7.1 Time Difference Pictures

Detection of temporal differences occurring in a scene is most naturally accomplished by sub-

at a different time. Figure 6.10 is a time difference picture made between Landsat images of the Coconino Plateau in Arizona on August 6 and 7, 1972. The bright regions correspond to soil darkened by rain that fell in periods between the times the images were taken.

Because the images were acquired from different positions, about 150 km apart, they had to be registered before comparison. However, especially in the mountainous or hilly portions of the scene, residual misregistration is apparent. To completely remove misregistrations caused by parallax would require orders-of-magnitude more analyst and computer time.

The difference in *DN* between a pixel in one image and a corresponding pixel in the other may range from -255 to +255, although this difference is close to zero unless some change has occurred in the scene. Thus, encoded *DN* differences must

FIGURE 6.19. (*d*) Periodic noise removal from a Mariner 6 vidicon image (6N13) of Mars in 1969: Difference picture between (*a*) and (*c*), showing noise removed from image by filtering.

mitted range of 0 to 255*DN*, and there must be sufficient contrast for the analyst to see changes in the picture. This is accomplished by a linear "contrast stretch," such that:

$$DN'_{\ell,s} = a(DN1_{\ell,s} - DN2_{\ell,s}) + b \qquad (6.18)$$

where *DN*1, *DN*2, *DN'* are the brightnesses at pixel (ℓ, *s*) in the two input pictures and the difference picture and *a* and *b* are coefficients chosen to spread the actual *DN* differences throughout the permitted range. Typical values of *a* and *b* are 2 and 128, respectively, so that "no change" is represented as midscale gray (*DN* = 128), and differences of ±64*DN* are saturated white and black.

6.7.2 Stereoscopic Analysis

One of the first uses of remote sensing was to assist in the making of topographic maps (25). This is best done using stereoscopic images. When two images are made of the same three-dimensional scene from slightly different positions in space, the positions of features are slightly different in different images, and the displacement increases with proximity of the feature to the cameras and also with the separation of the two cameras. This effect is called parallax and is one way in which humans perceive depth. This may be shown by simply looking at a scene and closing first one eye and then the other. The apparent position of close objects shifts with respect to the background. By measuring the amount of this shift in a stereo pair of images, the analyst can estimate the distance from the cameras to objects in the scene, as is shown in Fig. 6.20.

In the past, the use of stereo images has been by and large restricted to making topographic contour maps from aerial photographs. However, with the advent of digitally acquired images, it became desirable to compute the elevation in a scene on a pixel by pixel basis, so that rather than a contour map a "digital elevation image" was created. The first intensive civilian use of digital stereoscopic analysis was probably begun by the Viking project to construct elevation maps (Figs. 6.21) from images taken by the landers to help guide the sampling scoop as it gathered Martian soil for chemical analysis (26). Topographic maps were also made from Viking orbiter images and earlier topographic contour maps were made from Mariner images of prominent Martian features such as the large volcano Olympus Mons and the "Grand Canyon of Mars," Valles Marineris (27).

Even before Mariner image data were used to study Martian topography, elevation maps were made of the Moon. Typically, topographic relief was estimated from shadow length at low illumination angles, but one map was made utilizing Ranger 8 images and a technique known as "photoclinometry," which requires only one image (28). Instead of computing elevation from parallax, the slope of the surface is computed using the Minnaert or some other photometric law (Equation 6.3). This requires some prior knowledge of the photometric properties (described by the parameter k in Equation 6.3) of the surface material, as well as knowledge of the viewing and illumination geometry. Young and Collins (11) measured k for different regions on Mars from Mariner 6 and 7 images, but for the Moon k was known from telescopic observation from Earth (10).

Once the slope is known for every pixel, their relative altitudes may be computed by simply integrating along an image line or, because we are discussing discretely sampled images:

$$e_{\ell,s} = x \sum_{i=1}^{s} \tan\phi_{\ell,i} \qquad (6.19)$$

where e is the elevation at line ℓ sample s, x is the sample scale in meters per sample, and ϕ is the computed slope in degrees. Because k is imperfectly known and because the Minnaert equation is only approximately correct, there is some error in the measured slope and altitudes computed by photoclinometry become progressively less accurate with distance from a reference point. In general, photoclinometric elevation maps are less accurate than those made by stereoscopic analysis.

To make elevation by stereoscopic analysis, the parallax must be measured at every pixel. To do this, the analyst may use cross correlation methods as discussed for image registration in Section 6.5.2. However, it is obviously impractical to have the analyst supervise the cross correlation because of the much greater number of computations that must be made.

It is possible to display elevation images as pictures if the elevations are encoded as gray levels. Figure 6.8(d) shows such a display where sea level is black and elevations above sea level are represented by lighter shades of gray. Such a display is not as convenient for most purposes as the more conventional contour map. Another useful method of displaying elevation information is to display the slope instead of the actual elevation. The slope is actually a vector, having magnitude and direction. A single picture is not capable of displaying both components. Of course, the component of slope in a single predetermined direction may be displayed, but a superior method is to "illuminate" the scene by a "sun" whose azimuth and angular elevation above the horizon may be specified. The DN in the displayed picture are computed using Equation 6.3 and the displayed picture looks very much like the actual scene would if all albedo variations were removed and the surface material was a uniform gray reflector. Figure 6.8(c) shows such an "irradiance map."

Irradiance maps are useful in that they allow the analyst to study the effect of lighting on his interpretation of geologic structure, and also assist him in computing heat budgets for thermal moding, as discussed in Chapter 9.

As of 1979, the only widely available digital topographic maps from which to construct irradiance maps were interpolated from 1:250,000 topographic contour maps by the Defense Mapping Agency. These digital maps are available to the public through the U.S.G.S. Pixel spacing corresponds to 0.01 in. on the original maps, or about 63.5 m on the ground. The pictures shown in Figs. 6.8(c) and (d) were made from these digital elevation maps. Unfortunately, adequate topographic contour maps do not even exist for much

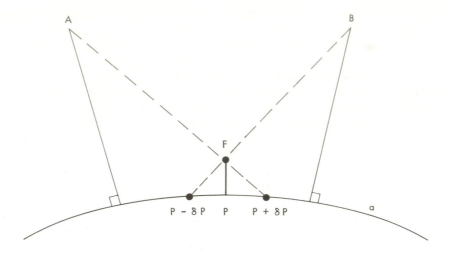

FIGURE 6.20. Parallax in stereoscopic pictures. Viewed from *A*, point *F* has an apparent displacement of δ *P*; viewed from *B*, displacement is δ *P*. Point *P* above the nominal ground surface at *a* is shown midway between *A* and *B* but in general this is not the case and the displacement in the two pictures has a different magnitude as well as direction.

FIGURE 6.21. Profiles showing altitude of the local surface of Mars in the vicinity of a trench dug by Viking Lander 1. Data for profiles were computed from these stereoscopic images taken by cameras on the Lander.

of the world. Partly to alleviate the situation, NASA may launch a satellite in 1985 whose main goal is to provide complete stereoscopic coverage of the earth.

6.7.3 Statistical Procedures

The analyst can compute any statistical measure of image data over any area contained in the image. Often he may compute statistics over a small local region, say three or five pixels on a side, and display these data for every pixel in image format, thus preserving spatial relationships. Such images give some insight into the relationship between the DN at one pixel and the DN at its neighboring pixels, or in other words the textural relationships in a scene.

TEXTURE IMAGES

Some texture images are created by convolution filtering. This approach is used to create pictures of directional spatial derivatives and of the first statistical moment, the mean. A mean image is simply a low-pass filtered image, made using a uniform weight box filter whose weights total unity [Fig. 6.22(a)]. The variance is also useful, but it must be made arithmetically using Equation 6.20 instead of by filtering:

$$DN'_{\ell,s} = \left(\frac{1}{(2M+1)(2N+1)} \sum_{m=-M}^{M} \sum_{n=-N}^{N} DN^2_{\ell+m,s+n} \right) - \mu^2$$

(6.20)

where DN' is the variance at line ℓ, sample s, DN is the input gray level, $2M+1$ and $2N+1$ are the line and sample dimensions of the "local area" for which statistics are gathered, and μ is the mean DN of the local area. In practice, DN' must be scaled to fit in the dynamic range of the display device, which is usually eight bits or 256 gray levels.

Alternatively, the standard deviation, which is the square root of the variance, may be displayed as an image. This will increase the contrast of low-variance regions in the image.

The mean picture provides insight into the gross behavior of the scene, devoid of noise and fine detail. The variance picture is like a high-pass filtered picture because it emphasizes deviations from the mean.

Directional spatial derivative pictures can be made by convolution using filter weights shown in

$\frac{1}{9}$	$\frac{1}{9}$	$\frac{1}{9}$	
$\frac{1}{9}$	$\frac{1}{9}$	$\frac{1}{9}$	(a)
$\frac{1}{9}$	$\frac{1}{9}$	$\frac{1}{9}$	
0	0	0	
0	-1	1	(b)
0	0	0	
0	0	0	
$\frac{1}{2}$	-1	$\frac{1}{2}$	(c)
0	0	0	

FIGURE 6.22. Convolution kernels used to construct: (a) A mean image with a 3-pixel square local area. (b) A first derivative image in the positive sample direction. (c) A second derivative image in the sample direction.

Figs. 6.22(b) and (c). There is rarely any reason to make pictures of any but the first and second derivatives. If the analyst wants to display the complete derivative, he requires two pictures: one to show the magnitude of the derivative and another to show the direction. In each case, the displayed variable must be encoded as a gray level. Complete derivative pictures are usually made arithmetically. The equations for the first derivative, or gradient are:

$$DNM_{\ell,s} = 0.5\Big[(DN_{\ell,s-1} - DN_{\ell,s+1})^2 +$$

$$(DN_{\ell-1,s} - DN_{\ell+1,s})^2\Big]^{1/2}$$

$$DNA_{\ell,s} = \tan^{-1}\Big[(DN_{\ell,s-1} - DN_{\ell,s+1})$$

$$(DN_{\ell-1,s} - DN_{\ell+1,s})^{-1}\Big]$$

$$(6.21)$$

where *DNM* and *DNA* are the gradient magnitude and direction or azimuth at line ℓ and sample *s*, and *DN* is the brightness in the original image. Scaling factors necessary to encode *DNM* and *DNA* in 256 gray levels have been ignored for simplicity.

Derivative pictures emphasize edges and other structural detail, just as do variance and high-pass filter pictures, but perhaps the most important use is in the analysis of topography for heat budget calculations, or for compensation for photometric slope effects in the production of albedo pictures (29). Texture pictures have also been used to improve reliability of thematic classification (30), a technique that will be discussed in Section 6.7.4. Figure 6.23 shows the different texture pictures discussed above, for the same scene.

GRAY LEVEL DISTRIBUTIONS

Not all statistical procedures are used to produce images. The most common statistical measure in image processing is used to describe the probability of finding *DN* of a given value in a picture. This probability density function, or PDF, is displayed as a histogram, with *DN* value as one axis and frequency or the number of pixels having a particular *DN* value as the other axis. The histogram thus indicates the frequency of occurrence of each gray level, or *DN*, over the area of the image examined, and is a necessary input to programs that modify the contrast within an image.

Gray level distributions reveal a great amount of information about an image. Typically the analyst looks for the number and the breadth of peaks in the PDF. An image of a homogeneous, low-contrast scene will have a PDF with a single sharp peak, which increases in breadth as the contrast is increased. Images that have more than a single type of terrain will usually have a corresponding number of peaks in the PDF. If the average brightness of pixels in two different terrains is similar, the PDF

will show a single double-peaked distribution, or even a single broad distribution instead of two distinct distributions. Figure 6.24 shows an image with a histogram of its PDF.

It is of course possible to construct histograms in more than one dimension. This is especially valuable in studying color pictures. Such an histogram will have three dimensions, one for the *DN* in each primary color blue, green, and red. It is also useful for studying multispectral images like those made by the Landsat MSS. Multispectral images are just registered images taken simultaneously, each through a filter that transmits light on a different part of the spectrum. Each image can be regarded as a single channel of a multidimensional image. By constructing two-dimensional histograms it is possible to separate features that may overlap in any single one-dimensional histogram. Thus, two-dimensional histograms allow the analyst to estimate the degree of correlation between channels of a multispectral image. Figure 6.24 shows one- and two-dimensional histograms describing PDFs of the same scene taken in red and infrared light. Correlated data, or classes of surface materials that appear the same through either filter, fall near the diagonal, which passes through the origin of the two-dimensional PDF. Uncorrelated data is located off the diagonal.

While histograms of higher dimensions are hard to depict, they may be manipulated and described statistically, and this forms the basis for image classification or multivariate analysis.

6.7.4 Image Classification

In image classification, a multichannel image is converted into a single-channel thematic map in which *DN* no longer represents any directly measurable property of the scene, but rather describes a decision assigning a pixel to a group of similar pixels, which form a cluster in a multidimensional PDF. Pixels assigned to the same cluster have a high probability of representing the same kind of surface material, for example sandstone, granodiorite, or limestone.

To understrand how classification works, it is helpful to consider a single image represented by a one-dimensional PDF. If the histogram contains two distinct, widely separated distributions it is easy to divide them, so that all *DN* smaller than some specified DN_0 are assigned to class 1 while *DN* larger than DN_0 are assigned to class 2. The

(a)

(b)

FIGURE 6.23. (a) Examples of texture images for part of NASA Landsat MSS image 1324-02433 taken of Sumatra in red light. Image after contrast stretching only. (b) Mean image, with an 11-pixel square local area or window.

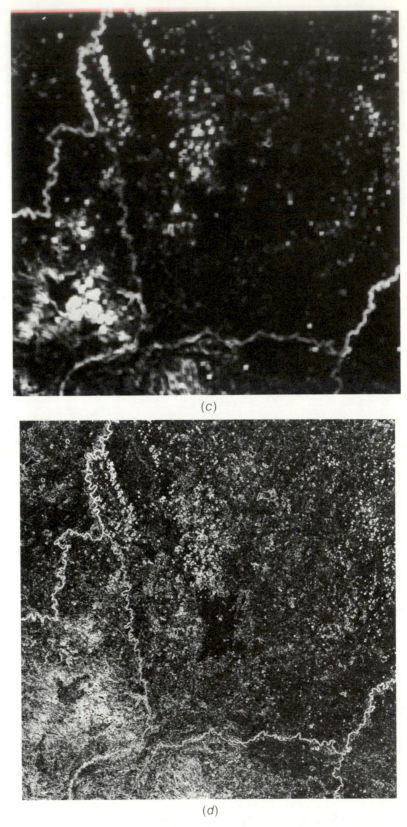

(c)

(d)

FIGURE 6.23 (c) Examples of texture images for part of NASA Landsat MSS image 1324-02433 taken of Sumatra in red light. Image showing variance of image (a). (d) Image showing magnitude of first derivative. Dark tones indicate low gradient; bright tones indicate high gradient.

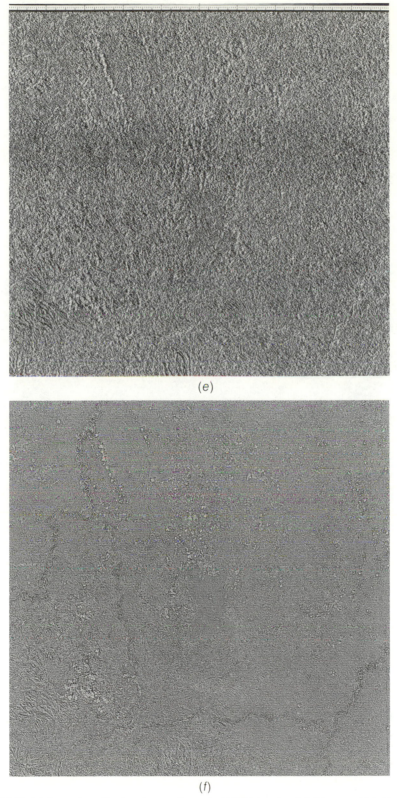

(e)

(f)

FIGURE 6.23. (e) Examples of texture images for part of NASA Landsat MSS image 1324-02433 taken of Sumatra in red light. Image showing direction of first derivative. East is dark gray, South is midgray, West is light gray, and because azimuth is a cyclic function, North is both black and white. (f) Image showing magnitude of second derivative. As in (d), low magnitude is shown by dark tones.

FIGURE 6.24. (a) Probability density functions are described by histograms. The same image as in Fig. 6.23 (a), except taken in infrared light. Both pictures have been contrast stretched for display.

output thematic map would consist of ones and twos. Pixels with DN of exactly DN_0 might be considered ambiguous and assigned an output value of zero, but such pixels would be few in number. As long as the separation of the distributions in the original PDF was high, the exact value of DN_0 is not even too important, but as the distributions come closer and overlap it becomes more critical. A small error in DN_0 has a large effect in the output thematic map. When the distributions significantly overlap so that some pixels really belonging to class 1 have the same DN as some pixels belonging to class 2, then significant errors are introduced into the thematic map regardless of the choice of DN_0. However, the ambiguity may be resolved by basing the classification on more than a single image. In two dimensions

the single value DN_0 is replaced by a rectangle or some other closed figure surrounding each cluster. Any pixel with a DN pair (one DN from each of the two channels) falling inside the boundary will be identified as a member of the associated class in the output thematic map. When the boundary is a rectangle, this procedure is called a "parallelipiped classifier." As the number of dimensions increases, the decision boundary becomes a solid or a hypersolid, but the principle remains the same.

It may not be possible to construct parallelipipeds around all clusters so as to include all parts of an actual cluster without overlapping adjacent parallelipipeds, especially at corners. One solution is to assign a pixel to the cluster whose center is nearest, that is, the center with the smallest euclidean dis-

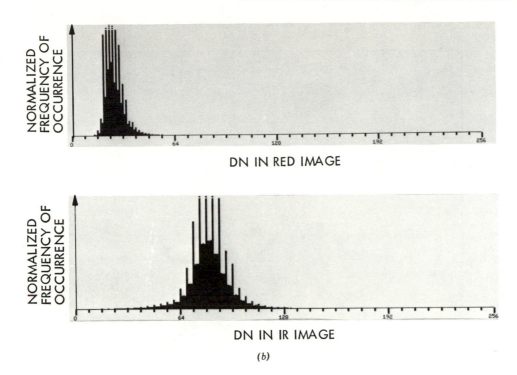

FIGURE 6.24. (b) Probability density functions are described by histograms. Histograms showing the data in the red and infrared images before stretching.

tance from the pixel in the PDF. For a decrease in ambiguity, the analyst pays a penalty in increased computation costs. Also, there is no way to allow for scatter of data around the cluster centers. Some distributions are narrow while others are wide, so that distance from cluster centers is not sufficient to assign a class. Instead, we want to know the distance in terms of the scatter. This is done with a Bayesian classifier (31), which models each cluster by an hyperellipsoid whose surface is a specified number of standard deviations from the cluster center. The eccentricity of the ellipsoid reflects the change of scatter about the cluster center in each image channel.

Perhaps the most important and difficult task in image classification is not the application of the classification algorithm, but the selection of classes of data that adequately delimit and define the themes of interest to the analyst. For example, it is purposeless to define a class describing a theme called "dolomite" if the spectral response of dolomite is not distinct from that of adjacent rocks.

There are three approaches to obtaining data from which decision boundaries can be established. Only one approach is fully developed. That approach requires the analyst to specify areas that, in his opinion, typify the list of classes or themes

he has chosen. These are referred to as "training areas." Statistics gathered by examination of the training areas are used to establish the clusters that define a class of information. In an extension of this approach, the statistics defining a very large number of themes may be stored on disk. In theory, the analyst could extract the statistics from this "signature bank" rather than from training areas in his image. In a further extension, called "unsupervised" classification, the analyst may allow the computer itself to define themes by inspection and separation of the n-dimensional PDF into clusters. Neither of the last two approaches has yet been completely successfully implemented, at least for geological applications (32).

Clearly, careless definition of themes or specification of training areas can result in a poorly classified or even misleading thematic map.

A major virtue of image classification is that it allows interpretation of a very large number of image channels. This is difficult to do for a photo interpreter studying enhanced images. He can display three at once as a color picture, but to study more he must look first at one and then the other. This is difficult to do because it is hard to remember subtle scene detail unless the image is in front of the analyst.

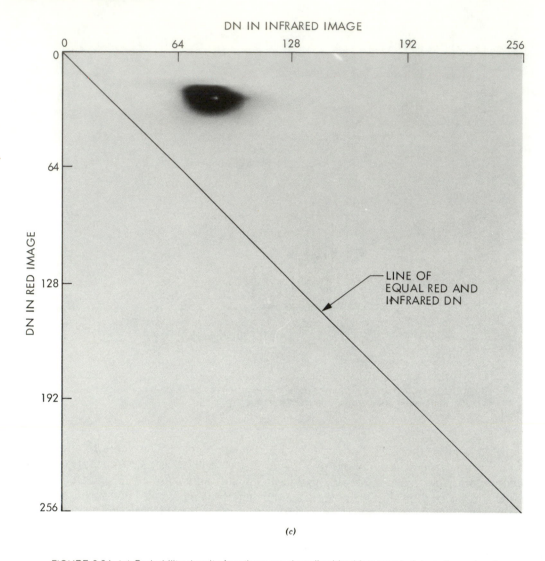

FIGURE 6.24. (c) Probability density functions are described by histograms. A two-dimensional histogram showing correlation between the red and infrared data. Dark gray tones represent high values of the PDF.

Another virtue of classification is that a thematic map can be interpreted by persons who may not be sophisticated photo interpreters. However, the process of classification is inherently exclusive. Only the actual decision and the "confidence" (a measure of the similarity between the individual *DN* vector and the mean *DN* vector chosen to represent a class) are available to the viewer, while the large amount of data from which the classification was made is discarded. Consequently, while he may be able to identify trouble spots from their low "confidence" values, the viewer may not have the necessary information to resolve ambiguities (when a pixel equally resembles more than one defined class) and uncertainties (when a pixel resembles no defined class).

In many disciplines it is sufficient to note that the thematic map has a confidence level of, for example, two standard deviations; that is, all pixels classified fell within two standard deviations of the center of the cluster defining a class. However, this is not always satisfactory. Figure 6.25, Color Plate 10, contrasts an enhanced image to a thematic map of the same scene.

Classification requires more manipulation of data than does enhancement, and thus can be significantly more expensive. The cost of a Bayesian classification is roughly proportional to $mn(n + 1)$ if m is the number of defined classes and n is the number of channels. Enhancement techniques are variable, but tend to be proportional to n alone.

6.7.5 Principal Component Transformation

Another image analysis technique that makes use of the n-dimensional PDF is called principal component analysis. We have already seen that multispectral image data is usually strongly correlated from channel to channel so that the axes of the PDF are not statistically orthogonal; that is, the variables on the axes (DN_1, DN_2, ..., DN_n) are not independent. Principal component analysis is a technique that creates new images whose PDF has orthogonal or independent axes. This is accomplished by a linear transformation of variables which corresponds to a rotation and translation of the original coordinate system of the PDF.

Because of the versatility and increasing importance of principal component analysis, the choice of the correct transformation is described below in considerable detail. First, the correlation between channels is described by the covariance matrix. The covariance matrix is defined by

$$K_{ij} = \frac{1}{(\ell_2 - \ell_1 + 1)(s_2 - s_1 + 1)} \sum_{\ell=\ell_1}^{\ell_2} \sum_{s=s_1}^{s_2}$$

$$\left[DN_{i\ell,s} - \mu_i \right]\left[DN_{j\ell,s} - \mu_j \right] \qquad (6.22)$$

where K is the covariance of image channels i and j computed in the region of the image bounded by lines ℓ_1 and ℓ_2 and samples s_1 and s_2, DN_i and DN_j are the DN in image channels i and j, and μ_i and μ are the means of the DN in the specified local region. Notice that K_{ii} is just the variance in channel i, as seen in Equation 6.20. A positive K_{ij} is found when channels i and j are positively correlated; a negative K_{ij} is found when channels i and j are negatively correlated. When channels i and j are completely independent of each other, K_{ij} is zero. To create independent images when K_{ij} is not zero, we must find a transformation that diagonalizes matrix K to K' so that for all $i \neq j$, $K'_{ij} = 0$.

We must first compute the eigenvalues of K. These numbers are actually the variance in each

channel of the transformed images associated with K'. The eigenvalues are the solutions to the "characteristic polynomial" of K. This polynomial is the determinant of the "characteristic matrix" of K, which is zero because K is symmetric ($K_{ij} = K_{ji}$). The characteristic matrix C is simply:

$$C = \begin{pmatrix} K_{11} - x & K_{12} & K_{13} & \cdots & K_{1n} \\ K_{21} & K_{22} - x & K_{23} & \cdots & K_{2n} \\ K_{31} & K_{32} & K_{33} - x & \cdots & K_{3n} \\ & & & \cdot & \\ & & & \cdot & \\ K_{n1} & K_{n2} & K_{n3} & \cdots & K_{nn} - x \end{pmatrix}$$

(6.23)

Thus in the two-dimensional case where $n = 2$:

$$[C] = (K_{11} - x)(K_{22} - x) - K_{12}K_{21} \qquad (6.24)$$

Here the eigenvalues λ_i are the solutions to the equation:

$$x^2 - x(K_{11} + K_{22}) + K_{11}K_{22} - K_{12}K_{21} = 0 \qquad (6.25)$$

Once the eigenvalues are known, the eigenvectors must be determined. These vectors are columns in a matrix R for which

$$R^t K R = K' \qquad (6.26)$$

where K' is the matrix having the eigenvalues (λ_1, λ_2, λ_3, ..., n) as diagonal elements and R^t is the transpose of R. Each eigenvector is found by substituting an eigenvalue into Equation 6.23 and finding the Hermitian normal form of C. This is the matrix for which all elements below the diagonal are zero, and for which the first nonzero element in each row is 1. If ϵ_i is an eigenvector and H is the Hermitian normal of C with λ_i substituted for x, then:

$$H\epsilon_i = 0 \qquad (6.27)$$

The matrix R formed from the column eigenvectors normalized to unit length defines the linear transformation or rotation that must be applied to the n-channel image to produce the "principal component" pictures. The eigenvalues are ranked in descending size, so that the first transformed picture will have the largest variance. Figure 6.26 shows the principal component pictures of a four-channel image.

FIGURE 6.26. (a) Principal component analysis of part of NASA Landsat image 1072-18001, showing Goldfield, Nevada. The first principal component (with an eigenvalue λ_1 of about 1500) is simply a weighted average of the green, red, and two infrared images acquired by Landsat.

FIGURE 6.26. (*b*) Principal component analysis of part of NASA Landsat image 1072-18001, showing Goldfield, Nevada Second principal component picture ($\lambda_2 \approx 30$).

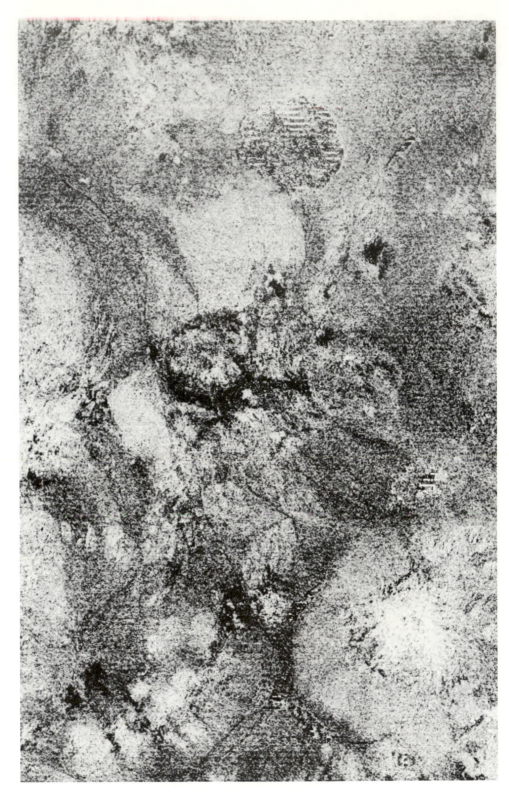

FIGURE 6.26. (*c*) Principal component analysis of part of NASA Landsat image 1072-18001, showing Goldfield, Nevada. Third principal component ($\lambda_3 \approx 10$).

198

FIGURE 6.26. (*d*) Principal component analysis of part of NASA Landsat image 1072-18001, showing Goldfield, Nevada. The fourth principal component ($\lambda_4 \approx 6$) has the lowest S/N.

It may help to understand the principal component tranformation if a time difference picture is thought of as a similar transformation or rotation:

$$\begin{bmatrix} DN'_1 \\ DN'_2 \end{bmatrix} = a \begin{pmatrix} \cos\theta & \sin\theta \\ -\sin\theta & \cos\theta \end{pmatrix} \begin{bmatrix} DN_1 \\ DN_2 \end{bmatrix} + b \qquad (6.28)$$

where DN'_1 and DN'_2 are DN in the newly created pictures, DN'_1 and DN_2 are DN in the original image channels, θ is the angle of rotation, and a and b are scaling factors necessary to locate the output DN in the permitted dynamic range. If $\theta = 45°$:

$$DN'_2 = a(DN_2 - DN_1) + b \qquad (6.29)$$

which is the difference equation, while the other image is an "average" picture:

$$DN'_1 = a(DN_1 + DN_2) + b \qquad (6.30)$$

Very often in principal component analysis, the first principal component picture is just a weighted average picture, while the remaining pictures are somewhat like pairwise differences between channels. It is useful to remember this approximate rule of thumb when trying to interpret principal component pictures.

Two other properties of principal component analysis are worth considering here. We have already seen that the variances K_{ii} are arbitrarily arranged in descending order. K_{ii} is the same parameter that is used to describe the width of the DN distribution in each one-dimensional PDF, so that the first transformed picture will have the greatest spread in data, or the greatest contrast. The sum of the variances in the transformed system is the same as the sum in the original images.

From the discussion of two-dimensional PDFs (Fig. 6.24) recall that most image data, which is correlated between channels, plots on or near the diagonal line passing through the origin of the PDF, and uncorrelated data plots off the diagonal. In three dimensions, a surface enclosing most of the non-zero values in the PDF generally resembles an eccentric ellipsoid whose long axis is nearly co-linear with the diagonal line $DN_1 = DN_2 = DN_3$. The shorter axes need not be the same length. The radius of the ellipsoid is proportional to the variance of the cluster in any direction, while the axis of the first principal component picture is colinear with the long axis of the ellipsoid. It is useful to consider an analogy with color images to help interpret principal component pictures. In color pictures, gray is produced when the primary colors blue, green, and red

are present in equal amounts. Because the first axis is usually nearly aligned with the diagonal line $DN_1 = DN_2 = DN_3 = \cdots = DN_n$, the first transformed picture is approximately an albedo picture, which depicts the average brightness of the scene devoid of color information. Just as colors other than gray are produced when $DN_1 \neq DN_2 \neq DN_3$, the other principal component pictures can be thought of as representing color or deviations from gray.

In order to achieve equal contrast in each transformed picture, the variance in each picture must be made the same; that is, in Equation 6.28 a is different for each picture so that a is really a row vector whose values a_i increase with i. This has an interesting consequence: any random noise or even periodic noise that is uncorrelated between channels will be made more conspicuous in the $(n + 1)$th than in the nth principal component picture. The principal component transformation has the effect of ranking the signal-to-noise ratio (S/N) in addition to the PDF variance. To see this, we must realize that before the gains a_i are applied the random noise is symmetrically distributed in the n-dimensional PDF, unlike the image data. Thus, if the PDF could be divided into a "noise" PDF and a "signal" PDF, under a rotation the variance along any axis through the noise PDF would remain unchanged; however, as we have already seen, the variance along an axis through the "signal" PDF changes proportionately to the shape of the ellipsoidal cluster.

It is worth emphasizing that the signal variance in the first transformed picture is always increased by the transformation while the noise variance is unchanged. It follows that principal component transformations can be exploited to produce a picture with a minimum amount of random noise. This feature of principal component transformations can be used to produce both better color pictures and also high-pass filtered pictures, as will be discussed in Sections 6.8.3 and 6.8.5.

The second important property of principal component analysis is that separation between clusters in an image can be increased. To do this the analyst computes the covariance matrix K only from regions in the image he chooses as representing the classes of data he wishes to separate. Increased separability is illustrated by Figs. 6.27 and 6.28, which depict an hypothetical situation in two dimensions. If the separation between features in one or more channels is increased, then it follows that classification will be more accurate or that fewer

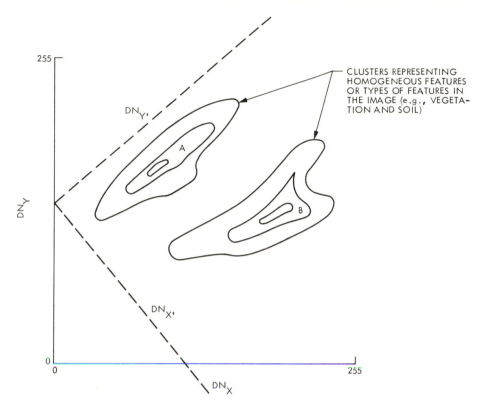

FIGURE 6.27. Contour map of two-dimensional PDF depicting frequency of occurrence of DN couplets in image channels X and Y.

channels must be used to achieve the same degree of accuracy.

By choosing training areas representative of themes the analyst wishes to discriminate in an image, he can estimate the effectiveness of an image or of a single channel within a multichannel image at separating the desired themes.

One simple measure of separability between two classes of data is given by the euclidean distance between the mean DN vectors of each class. In practice this distance is usually normalized by a factor describing the scatter of data in each class. Obviously, it is possible to have clusters of data which overlap considerably and yet have different mean DN vectors. In such a case the separability is clearly a function of the scatter in the data. One distance normalized by data scatter is the "Fisher criterion" (33):

$$d_{ijk} = \left[\frac{(M_{ik} - M_{jk})^2}{\mu_{ik} + \mu_{jk}} \right]^{1/2} \qquad (6.31)$$

where d_{ijk} is a distance separating class i from class j in channel k, M_{ik} and M_{jk} are the kth components of

the mean vectors, and μ_{ik} and μ_{jk} are the variances in each cluster. If there is a large amount of scatter in the data of either class, d will be small. The Fisher criterion or another similar measure of interclass distance can also be used to estimate the effectiveness of a single image channel at separating m classes of data:

$$D_k = \sum_{i=1}^{m-1} \sum_{j=i+1}^{m} d_{ijk} \qquad (6.32)$$

Channels of an image can thus be ranked in effectiveness at separating specified classes by ordering the channels in terms of D_k.

Equations 6.31 and 6.32 can be extended to measure the separability of data clusters in the n-dimensional space created by a multichannel image (34):

$$D = \sum_{i=1}^{m} \sum_{j=1}^{m} (M_i - M_j)^t K^{-1} (M_i - M_j) \qquad (6.33)$$

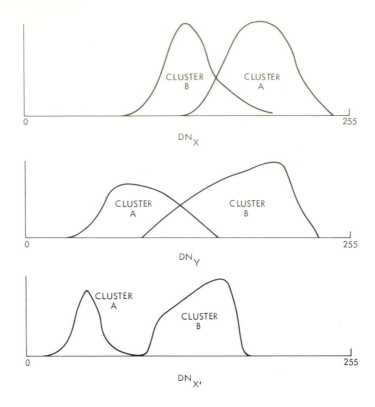

FIGURE 6.28. Single-channel PDFs. Notice that X' has better separation than either original channel X or Y. The single-channel PDF is the projection of the two-dimensional distribution onto a single axis. Vertical axes show frequency of occurrence.

where K is the covariance matrix and M is the mean vector, and the scatter is assumed to be the same in all classes.

The above techniques form the basis of one procedure for ranking and selecting those channels to be used in image classification.

6.7.6 Color Transformation

The color of an object is a function of its ability to reflect light in different parts of the spectrum. In 1853, Grassman (35) devised a trichromatic or "three color" theory to explain the way in which color is perceived, and this theory is the basis of color photography and color television. Three color theory states that mixtures of only three lights having certain colors can produce all other colors if the intensity of one or more of the lights is varied. The three colors are called primary colors and are blue, green, and red. In detail, color theory is far more complicated than this (36, 37).

In the analysis of multispectral images, it is useful to consider any three channels of an image as a color triplet, regardless of the actual portion of the spectrum which is represented. A familiar example is color infrared film, which is sensitive to green, red, and infrared light. To create a color display, green light produces a blue dye in the picture, red light is represented by green, and infrared by red. Thus the color display is not "natural" but is nevertheless useful.

RATIO PICTURES

Just as any three channels of an image can be displayed as a color image, they may also be analyzed in terms of color. In the previous section we saw that principal component pictures can be regarded as describing "color" in a scene, even though this "color" does not correspond to our human perception of color. Another way of measuring color is to create a "ratio picture" in which the DN at any pixel represents the ratio of the DN from two channels:

$$DN'_{\ell s} = a \frac{DNi_{\ell s}}{DNj_{\ell s}} + b \tag{6.34}$$

where DN' is the ratio of the DN from channels i and j at line ℓ sample s, scaled to fit in the dynamic range by constants a and b. Ratio values can range from zero to infinity, but because reflectivity in terrestrial scenes is strongly correlated across much of the spectrum, most observed ratio values lie between about 0.25 and 4.0. Typical values for a and b are 400.0 and -300.0, but these vary over about an order of magnitude with different scenes and spectral regions.

Ratio pictures are useful because they exaggerate color differences. If $DN_i > DN_j$, DN' is high or bright; if $DN_i < DN_j$, DN' is low or dark. If the albedo or the illumination changes across the scene but the color does not, the ratio picture will show no change. Thus ratio pictures have the effect of suppressing the detail in a scene which is caused by topographic effects, while emphasizing color boundaries. This property has made ratio pictures quite useful in geologic applications (38, 39, 40) because they exaggerate subtle color differences in a scene, and many geologic problems require the distinction between rock types that may appear to be quite similar. However, ratioing suppresses the ability of the analyst to discriminate between rocks with strikingly different albedos, but similar reflectance spectra (for example, between dark basalts and light marls).

Ratio techniques also enhance noise, which appears colored in multispectral images if it is uncorrelated among channels.

Atmospheric effects may also become visible in ratio pictures, because diffuse light scattered onto a scene is strongly colored blue. As the solar elevation decreases, the blue light scattered onto the scene assumes a larger share of the total illumination. To a viewer, the effect is that the color in the scene is partly a function of topography. Because the scattered light from the sky can be predicted [as a function of altitude and aerosol content (41, 42)] and because it does not change too significantly over most scenes, it can be removed by modifying Equation 6.34:.

$$DN'_{ij} = a\left[\frac{DN_{1(ij)} - atm_1}{DN_{2(ij)} - atm_2}\right] + b \qquad (6.35)$$

where atm_1 and atm_2 are the DNs that the sensors would record if the scene were colorless and were illuminated only by light scattered from the sky.

These quantities can be estimated from the DN of dark gray features shaded from the sun by large clouds or mountains. Removal of atmospheric scattering increases contrast between features in ratio pictures.

COLOR COORDINATE SYSTEMS

Over a dozen color coordinate systems have been developed, most of them in the past 40 years (43). These systems have been invented largely to help explain or predict the behavior of color photographs, as well as human perception. All are based on the tristimulus theory of color. Because most remotely sensed images are not acquired in the primary colors, many of these color coordinate systems are unnecessarily elaborate or even inappropriate for color analysis of these images. Nevertheless, it is convenient to transform image triplets into coordinate systems in which the axes represent parameters commonly used to describe color. The three physiological parameters describing color are hue, saturation, and intensity. Hue is the average wavelength of light reflected from an object. Blue, green, yellow, and orange are four hues. In addition, magentas and purples represent nonspectral hues that can be made by mixing red and blue light. Saturation is the width around the average wavelength of the spectral region in which a significant amount of light is reflected from an object. If light of a single wavelength is reflected, the color is spectrally pure or completely saturated. Most objects reflect light over a broad region of the spectrum, so that the reflected light is really a mix of different colors. Colors of such objects are pastel, or unsaturated. If light is reflected equally at all points of the spectrum, the color is completely unsaturated or gray. The third color parameter, intensity, is a measure of the total light energy reflected from the object, regardless of wavelength. Intensity is thus like albedo.

The three parameters, hue (H), saturation (S), and intensity (I), are easily approximated by a spherical coordinate system:

$$H = \tan^{-1}\left(\frac{z}{x}\right)$$

$$S = \cos^{-1}\left(\frac{y}{\sqrt{x^2 + y^2 + z^2}}\right) \qquad (6.36)$$

$$I = x^2 + y^2 + z^2$$

where x, y, z are Cartesian coordinates. H is thus like longitude, saturation like colatitude, and intensity like radius. This is appropriate because hues are often described by a "color wheel", which is the equatorial plane of the sphere. As saturation decreases to zero, a color becomes pastel and finally gray. As the colatitude decreases to zero, a vector (H, S, I) becomes colinear with the polar axis of the sphere, which passes through the gray center of the color wheel. The polar axis is thus equivalent to the diagonal line $DN_1 = DN_2 = DN_3$ of the three-dimensional PDF, which is usually represented as a Cartesian space as shown in Fig. 6.29(a). Thus the coordinates (x, y, z) in Equation 6.36 may be replaced by (DN_1'', DN_2'', DN_3'') after the DN coordinate system has been rotated so that the y axis (the polar axis of the spherical space) is colinear with $DN_1 = DN_2 = DN_3$. This is done by two successive rotations of 45° counter clockwise about the DN_2 axis and 54.7° counter clockwise about the new DN_3' axis:

$$\begin{bmatrix} x \\ y \\ z \end{bmatrix} = \begin{bmatrix} DN_1'' \\ DN_2'' \\ DN_3'' \end{bmatrix} = \begin{bmatrix} \frac{1}{\sqrt{3}} & \frac{\sqrt{2}}{\sqrt{3}} & 0 \\ \frac{\sqrt{2}}{\sqrt{3}} & \frac{1}{\sqrt{3}} & 0 \\ 0 & 0 & 1 \end{bmatrix} \begin{bmatrix} \frac{1}{\sqrt{2}} & 0 & \frac{1}{\sqrt{2}} \\ 0 & 1 & 0 \\ -\frac{1}{\sqrt{2}} & 0 & \frac{1}{\sqrt{2}} \end{bmatrix} \begin{bmatrix} DN_1 \\ DN_2 \\ DN_3 \end{bmatrix}$$

(6.37)

This transformation is drawn in Fig. 6.29(a), and if allowed values of DN_1, DN_2, and DN_3 were not restricted to the range 0 to 255, the HSI coordinate system as defined as Equations 6.36 and 6.37 would be adequate for many uses. However, because of the restricted range of DN_1, DN_2, and DN_3, no image data can fall outside of the first quadrant of the DN space. Thus no vector pointing to a color triplet in HSI space can ever lie near the equatorial plane; 100% saturation (for example, $DN_1 = 255$, $DN_2 = 0$, $DN_3 = 0$) occurs when the vector lies in one of the planes ($DN_1 = 0$, $DN_2 = 0$), ($DN_1 = 0$, $DN_3 = 0$), or as in the example above ($DN_2 = 0$, $DN_3 = 0$). Thus it is convenient to redefine S as the ratio of the actual colatitude to the maximum possible colatitude, which varies between 35.3° and 54.7° as a function of hue. In-

tensity is also defined to be the scalar sum of DN_1, DN_2, and DN_3, so that (x, y, z) are now given by $\sqrt{DN_1''}$, $\sqrt{DN_2''}$, and $\sqrt{DN_2''}$. This interpretation of intensity is appropriate because the DN themselves, as measured by the imaging system, generally describe energy fluxes. Finally, if a vector has a small colatitude, it can have a larger magnitude than if it has a large colatitude. The restriction of the magnitude of the vector arises because no DN can exceed 255. Thus, on the gray axis the intensity can be as large as 765 (or $255 + 255 + 255$) but when saturation is 100% intensity cannot be larger than 510 (or $255 + 255 + 0$). This is not an important consideration in the *analysis* of color in image triplets, but will become so in the *enhancement* for color display, in which color saturation is often increased. The modified hue, saturation, intensity color space is thus defined by:

$$H = \tan^{-1}\left(-\frac{\sqrt{DN_3''}}{\sqrt{DN_1''}}\right)$$

$$S = \cos^{-1}\left(\frac{\sqrt{DN_2''}}{\sqrt{DN_1 + DN_2 + DN_3}}\right)\phi_m^{-1}(H) \qquad (6.38)$$

$$I = (DN_1 + DN_2 + DN_3)\, I_m^{-1}(H, \phi)$$

where $\phi_m(H)$ is the maximum colatitude permitted at a given hue H, and $I_m(H, \phi)$ is the maximum intensity permitted at a given hue and colatitude. Scaling factors must be included to encode H, S, and I into the 0 to $255DN$ range. Hue is a periodic function, and the branch cut may be arbitrarily assigned to the color green, with DN_1, DN_2, DN_3 represented by blue, green, and red respectively. Thus $H = 0(0°)$ is green, $H = 85$ (120°) is red, $H = 170$ (240°) is blue, and $H = 256(360°)$ is green again.

If the transformation in Equation 6.38 is applied pixel by pixel to DN triplets from a multispectral image, pictures in which the gray levels represent H, S, and I can be constructed (Fig. 6.30). Histograms describing the frequency of occurrence of hues and saturations can be as useful as PDFs describing brightness. For instance, the PDF in Fig. 6.31 describes the hues found in the Viking Lander color picture of the Martian surface and sky. When the spacecraft first transmitted this image, there was some controversy over the actual color of the

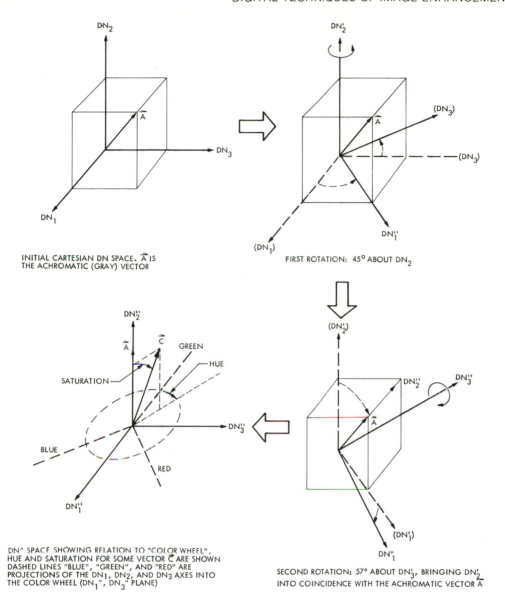

INITIAL CARTESIAN DN SPACE. $\vec{A}$ IS
THE ACHROMATIC (GRAY) VECTOR

FIRST ROTATION: 45° ABOUT DN$_2$

DN" SPACE SHOWING RELATION TO "COLOR WHEEL".
HUE AND SATURATION FOR SOME VECTOR $\vec{C}$ ARE SHOWN
DASHED LINES "BLUE", "GREEN", AND "RED" ARE
PROJECTIONS OF THE DN$_1$, DN$_2$, AND DN$_3$ AXES INTO
THE COLOR WHEEL (DN$_1$", DN$_3$" PLANE)

SECOND ROTATION: 57° ABOUT DN'$_3$, BRINGING DN'$_2$
INTO COINCIDENCE WITH THE ACHROMATIC VECTOR $\vec{A}$

FIGURE 6.29. (a) Coordinate transformation from acquired *DN* to hue, saturation, and intensity. Coordinate rotations described by Equation 6.37.

sky, which was predicted to be a dark blue before Viking. The sky was clearly not dark, and in some versions it was pastel blue but in others it was pastel red. The hue image and histogram clearly showed that the spacecraft recorded the sky at the same hue as the ground. The saturation, however, was less. The only nonred hues observed in the first pictures returned by the Viking Landers were on the spacecraft itself.

6.7.7 Thermal Inertia Pictures

Construction of geologic maps from most aerial photographs or satellite images, even of arid regions with little vegetation, is made more difficult by thin veneers of sand or soil, or weathered material or "desert varnish" covering rock exposures. It would be helpful to be able to sense below the immedate surface, even a few centimeters. One way is to use radar, as described in Chapter 11.

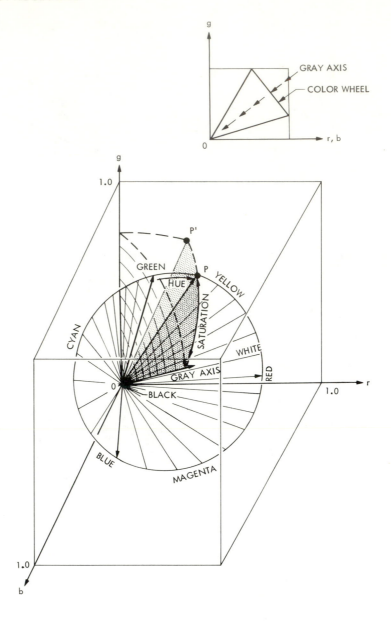

FIGURE 6.29 (*b*) Coordinate transformation from acquired *DN* to hue, saturation, and intensity. Perspective drawing of color space showing relations between hue and saturation and *DN*. The color wheel is shown as a cross section of a cone of constant saturation whose axis coincides with the gray axis.

Another way is to use thermal radiation, as described in Chapter 9. Because of daily fluctuations in solar irradiation as the earth rotates, heat is alternately conducted down into the earth and then back to the surface. This diurnal wave propagates to a depth of about 20 cm., depending on the surface material. By using albedo images registered to digital elevation maps as input to model the heat budget at each pixel, it is possible to predict the maximum change in radiant temperature which should be observed during the diurnal cycle. By comparing this value to the observed change in temperature, it is then possible to estimate a parameter called the "thermal inertia" which is simply $\sqrt{\rho\,c\,K}$, or the square root of the product of the density, the heat capacity, and the conductivity of the surface

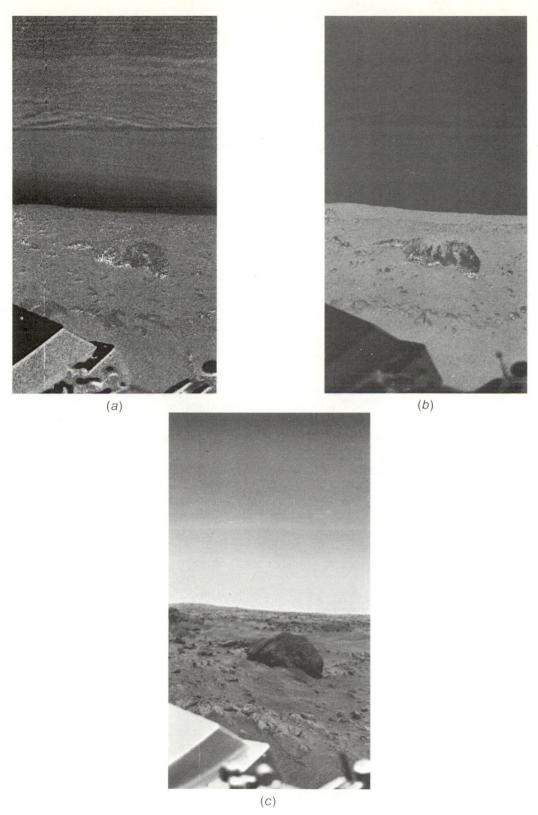

(a)   (b)

(c)

FIGURE 6.30. Hue (*a*), saturation (*b*), and intensity (*c*) pictures constructed from images taken by the Viking 1 lander on Mars in 1976. Each picture was individually stretched for display. The blue, green, and red images actually taken by the lander looked very much like the intensity picture. (*a*) Dark tones represent hues that are greener than average on Mars; light tones are redder. (*b*) Dark tones represent unsaturated or pastel colors; light tones are saturated. (*c*) Brightness is represented in the intensity picture as we would perceive it on Mars.

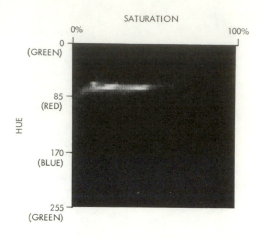

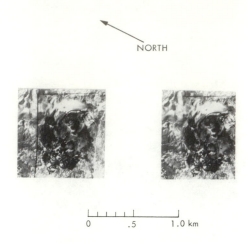

FIGURE 6.31. Two dimensional PDF of hue and saturation on Mars. Light tones denote high frequency of occurrence.

FIGURE 6.32. A simulated stereoscopic image relating thermal inertia to topography at Pisgah Crater, California. Dark tones have low thermal inertia, maximum relief is about 150 meters.

material. Thermal inertia, as the name implies, is a measure of the resistance of a material to temperature changes. Furthermore, it is a property that is measured not for the top micrometers or so of the earth but for the top few centimeters. Thus thermal inertia pictures can be used to "see through" thin veneers. In addition, many rocks that may be spectrally similar, like limestone and dolomite, have significantly different thermal inertias—in this case, by a factor of 1.7 (44). Consequently, thermal inertia pictures may be useful as an analytic aid to geologic exploration and mapping.

Figure 6.32 shows a simulated stereoscopic picture made from a thermal inertia map of Pisgah Crater in California. To create this stereo pair, parallax was calculated pixel by pixel from the digital elevation maps used in the modeling of the heat budget. The calculated parallax was used to distort the left image to create the right image. This display is particularly useful because the thermal inertia picture does not always resemble the albedo image, and correlating topography with the thermal inertia information in the same picture helps the photointerpreter relate thermal inertia anomalies to known features.

6.7.8 Lineament Recognition

Lineament recognition is listed in this discussion of image processing techniques because of its potential future importance in automating the production of geological structure maps. Lineament

recognition is a problem that has not yet been satisfactorily solved, and is one that may occupy image processing theorists for years to come. The interested reader may refer to (45) for a discussion of current technique, which consists of analysis of structure maps drawn by geologists.

One basic problem is the definition of a lineament. Lineaments are not simply straight lines on an image, but are usually inferred from discontinuous zones of small line segments or even points in an image. Even trained geologists cannot always agree on the correct structural interpretation of an image, as shown by Siegal (46). This emphasizes at the same time both the difficulty of lineament recognition and the importance of perfecting automatic procedures to reduce subjectivity in structural maps.

6.8 DISPLAY

Display procedures are designed to present to the human viewer as much information from a digital image as possible in a convenient and usable format. There are four ways of displaying image data. The DN may be printed as a matrix of numbers in a computer listing. This has the virtue that it is possible for the analyst to see the exact DN at a given pixel and also to discriminate between even the smallest possible changes in "gray level" on an image. The chief drawback is that spatial relationships between *DN*, or patterns, are not easy to recognize in listings of numbers. A second drawback is the low spatial density at which information is displayed. A single page of computer printout can list only a 50 by 30 pixel region. A Landsat MSS image contains four image channels, each of which has 7.58×10^6 pixels, so that over 20,000 pages are required to list one image.

By a technique known as "overprinting," computer listings of image data can be made in which the *DN* data is represented by a single character per pixel. Each character is formed by repetitive printing of different letters in the same location so that the ratio of ink to blank paper can be controlled. In this way, image data can be represented by gray levels, improving the ability of the analyst to recognize features but not reducing the amount of printout to a truly manageable quantity.

To achieve a usable display format for large amounts of image data, photographic representation must be made.

Photographic prints have excellent spatial resolution, but only mediocre density resolution. Transparencies, which can range from complete absorption to complete transmission of light, are somewhat better. They are, however, limited by the photographic process, which must represent the entire range of brightness (which is infinite) using a finite medium. Only a limited range of brightness can be faithfully represented; this is called "dynamic range."

Contour maps may also be used to display image data. Contour maps display large amounts of data compactly and allow pattern recognition, just as pictures do; at the same time the exact *DN* value can be read from annotation on the map or by counting contours from some known point. However, contour maps cannot depict *DN* everywhere, and most small scale detail is lost.

Photographic representation of image data is probably the most widely used technique of display. However, the limited dynamic range of a photographic image presents a problem, because a digital image does not suffer from as severe a restriction of dynamic range. To display the dynamic range of a Landsat image in a picture, for instance, requires 128 recognizable density levels. A densitometer can discriminate between gray levels only 0.02 density units apart, so the density range thus required exceeds the 1.8 unit range typical of recorded digital images.

The second problem that complicates photographic display of digital images is that the human eye is less sensitive than a densitometer. The human may be able to distinguish 15 or 20 discrete gray levels, if adjacent and sharply bounded. Thus, in order for a human eye to see subtle variations in brightness contained in a digital image, the contrast must be increased. This may be inconvenient to do when the data is acquired, as one sensitivity setting of the photosensors must be able to record the wide range or brightness encountered over the earth with a minimum of saturation. The contrast must therefore be modified after acquisition, so that gray levels actually encountered on an image are separated as much as possible in the photographic picture of the image. This process is called "contrast stretching" or simply "stretching" and is probably the most useful of all computer enhancements.

6.8.1 Contrast Stretching

Figure 6.33 shows a raw picture and three different versions of it created by applying different contrast stretches. The histograms below each picture depict the PDF of the stretched image.

The linear stretch [Fig. 6.33(b)] is created by mapping each *DN* encountered in the raw image [Fig. 6.33(a)], pixel by pixel, to the stretched picture being created using a linear equation:

$$DN'_{\ell s} = aDN_{\ell s} + b \tag{6.39}$$

where *DN'* is the *DN* in the stretched picture at line ℓ, sample *s*, *DN* is the brightness in the input image, and *a* and *b* are specified coefficients. In practice, coefficients for linear and other stretches are usually found automatically by algorithms that allow a certain amount of saturation at each end of the PDF of the stretched picture.

The Gaussian and ramp CDF ("cumulative distribution function") or equipopulation stretches shown in Figs. 6.33(c) and 6.33(d) are performed by distributing *DN* to force the envelope containing the PDF to resemble either a Gaussian distribution or a ramp. The ramp CDF stretch requires some elaboration. The CDF function is:

$$CDF(n) = \sum_{DN=0}^{n} PDF(DN) \tag{6.40}$$

where *n* is a *DN* within the allowed dynamic range. In a ramp CDF stretch *DN* are distributed to cause

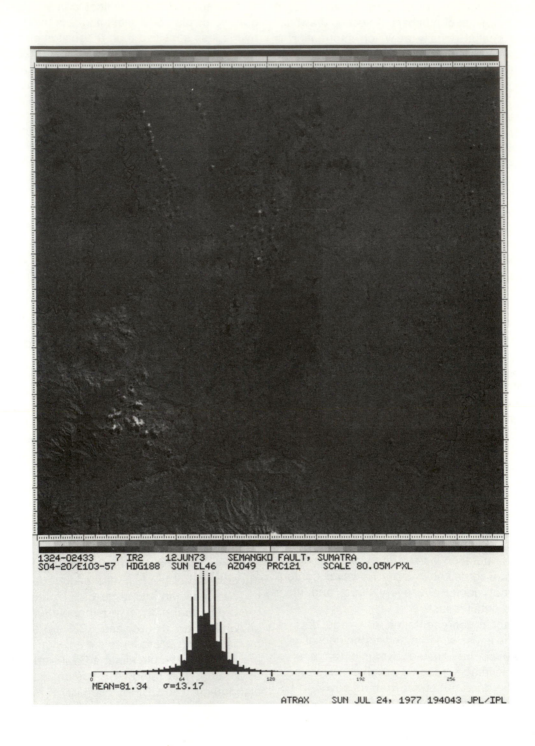

FIGURE 6.33. (a) Contrast stretches of part of NASA Landsat MSS image 1324-02433 showing Sumatra. Histograms below each picture show changes in the gray level distribution caused by contrast stretching. This image shows scene before stretching.

210

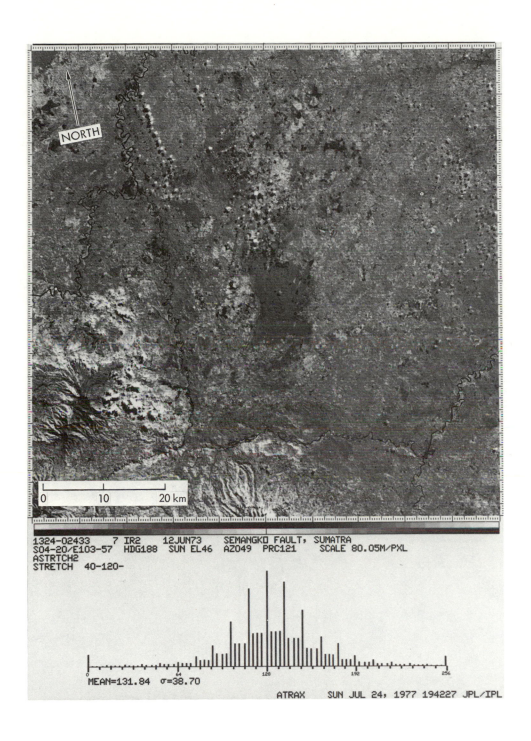

FIGURE 6.33 (*b*) Linear contrast stretch of part of NASA Landsat MSS image 1324-02433 showing Sumatra.

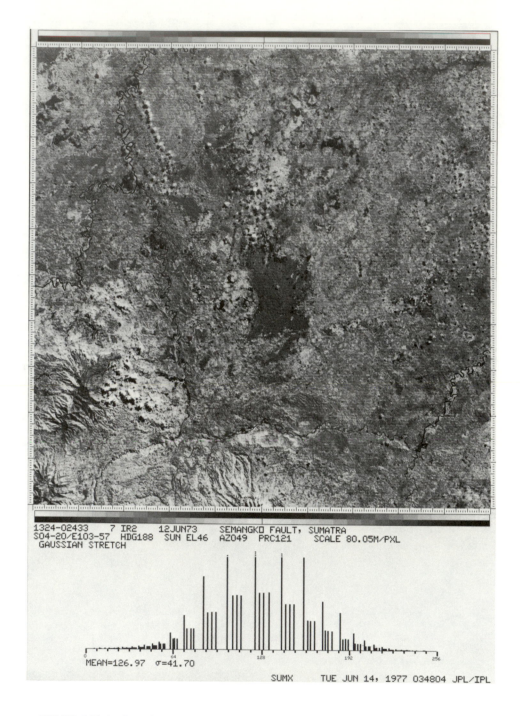

1324-02433   7 IR2   12JUN73   SEMANGKO FAULT, SUMATRA
S04-20/E103-57 HDG188 SUN EL46   AZ049   PRC121   SCALE 80.05M/PXL
GAUSSIAN STRETCH

MEAN=126.97   σ=41.70

SUMX   TUE JUN 14, 1977 034804 JPL/IPL

FIGURE 6.33 (c) Gaussian contrast stretch of part of NASA Landsat MSS image 1324-02433 showing Sumatra.

FIGURE 6.33 (*d*) Equipopulation or ramp CDF contrast stretch of part of NASA Landsat MSS image 1324-02433 showing Sumatra.

the CDF to resemble a straight line with a positive slope. Ideally, there should be equal numbers of pixels at each gray level, or the PDF of the stretched picture should be a straight line with a slope of zero.

Of course, any scheme of contrast stretching may be devised. Logarithmic stretches such as:

$$DN'_{\ell s} = a \log(DN_{\ell s} + b) + c \qquad (6.41)$$

where a, b, and c are stretch parameters will preferentially emphasize detail in dark regions, while "power" stretches such as:

$$DN'_{\ell s} = a DN_{\ell s}^{b} + c \qquad (6.42)$$

will emphasize bright detail. One particularly useful stretch applies a different linear stretch to different ranges of input DN. A variant of this technique is called "density slicing." All DN falling within a specified range are mapped to a single DN'. The result looks something like a contour map, except that the space between boundaries is occupied by pixels of the same DN. By creating three density sliced pictures from the same input image, it is possible to create a color coded display. The color can be controlled by the analyst's choice of DN' triplets representing each DN range in the input image. This technique is often used to construct interpretable thematic maps from classified images (Fig. 6.25) that resemble density sliced pictures. The only other major use of density slicing is for the display of thermal infrared images where temperatures are coded by color. Figure 10.15 shows an example of such a picture.

6.8.2 Color Display of Multispectral Data

The use of color pictures to display three channels of multispectral image data provides a dramatic increase in the amount of information that is available for ready interpretation. There are two reasons for this. First, the same gray level information that can be displayed in a black-and-white picture can also be displayed as brightness in a color picture. In addition, the extent of correlation among the three channels can be displayed as color. The human eye is very efficient at recognition and discrimination of colors. Second, in a color picture all the information contained in three black-and-white images is contained in a single picture, so that the information is easier for the analyst to interpret. He does not have to switch his attention from picture to picture.

Color pictures can be made by either additive or subtractive techniques as described in Chapter 5. In this discussion we will consider additive color pictures only. It is important to emphasize that only rarely do the colors in which image data are displayed correspond to the spectral region in which they were acquired. Thus the meticulous attention to faithful reproduction of colors, which characterizes much of the conventional literature, is largely irrelevant for the remote sensor, because the color display scheme is for him arbitrary. In fact, one particularly productive use of color is the "color ratio picture," in which three different ratio images are displayed as a color picture (39). In the color ratio picture shown in Fig. 6.34(b), color plate 11, the ratio between the acquired green (550 nm) and red (650 nm) images is displayed in blue light; the ratio of the red to the 750-nm infrared image is displayed in green; and the ratio between the 750-nm and the 900-nm infrared images is displayed as red. Color ratio pictures were invented because they are useful in revealing lithologic differences, as shown first on the Moon (47) and later on the Earth (48).

Color ratio pictures show the earth in a color scheme that is different from those to which most photointerpreters are accustomed. However, with a little thought the colors can easily be decoded in terms of the reflectance spectrum which they represent. Thus for a spectrum like that of water in which reflectivity decreases with wavelength, all three ratio pictures will have high DN and the display color will be white or bright and pastel. In the visible part of the spectrum this corresponds to blue. Similarly, if reflectivity increases with wavelength (red) then the display color will be dark. Vegetation has a more complicated spectrum because chlorophyll absorbs red light strongly. Hence vegetation in the green/red picture is bright, while it is dark in the red/infrared picture and neutral gray in the ratio between the two infrared images. Vegetation is displayed as blue or purple. The use of color ratio pictures in lithologic mapping is discussed in Chapter 13.

Regardless of what acquired or created image channels are chosen for color display, once the assignment is made subsequent enhancement is the same. Just as image data must be contrast stretched for display as a black-and-white picture, so must it be stretched for color display. Often, the three images are stretched independently before combination as a color picture. The picture shown in Fig. 6.34(a), color plate 10, was processed in this way. The special PDFs in Figure 6.35 describe the distribution of color in pictures such as this before and after stretching. The axes of the PDF are in tristimulus chromaticity coordinates:

$$r = \frac{DN_3}{\sum\limits_{i=1}^{3} DN_\ell}$$ (6.43)

$$g = \frac{DN_2}{\sum\limits_{i=1}^{3} DN_i}$$

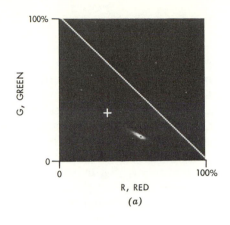

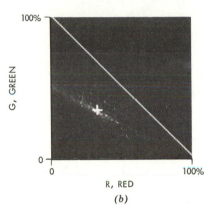

so that all colors (hues and saturations) are represented, although not intensities. Notice that only (r, g), which fall below the line $r + g = 1$, have any meaning. The goal in color enhancement is to center the cluster over the "achromatic" or gray point, $r + g = 0.333$. If this is done, all hues will be represented in the picture. If this is not done, the picture will have a monochromatic appearance, and discrimination among color differences will be lost. By increasing contrast in the three images, the cluster is generally dispersed to the corners of the tristimulus PDF. This has the effect of increasing the saturation of colors in the displayed picture.

In order to cause the cluster to center on the achromatic point, the analyst tries to stretch the images so that the means of the individual PDFs for each channel are the same, usually about $DN = 128$.

More complete control over color enhancement can be achieved if the images are first transformed either to color coordinates or to principal components. If the images are H, S, I coordinates, the hue and saturation may be stretched independently as desired by the analyst. He may, for instance, use a ramp CDF stretch on the hue picture, so that all hues are present in equal proportions. He may treat the intensity image exactly as a black-and-white picture, or he may replace it by a constant DN for all (ℓ, s) to better emphasize the hue and saturation information. Regardless of the stretches he chooses, once the H, S, and I images have been processed they must be transformed back to blue, green, and red coordinates in order to create the color picture. If principal component pictures are to be used instead of H, S, and I pictures, each is stretched as though it were a black-and-white

FIGURE 6.35. Effect of contrast stretching on color. (a) Chromaticity PDF showing distribution of color before stretching. Light tones represent high amplitudes. (b) Color distribution after stretching. Abscissa represents amount of red in picture, from 0 to 100%; ordinate represents green. Amount of blue increases from the diagonal boundary to the origin. Achromatic point is shown by cross at $g = r = 0.333$.

image. The first principal component picture, corresponding to the longest axis of the ellipsoidal DN cluster in the PDF, is similar to the intensity image. For color display, the principal component coordinate system must first be rotated and translated so that this axis becomes colinear with the gray axis ($DN_1 = DN_2 = DN_3$) of the blue, green, and red coordinate system. Two subsequent dimension rotations return the images to the blue, green, and red coordinates. These rotations are the reverse of those in Equation 6.37. In the color picture created from these images, the colors will cluster about the achromatic point in the tristimulus PDF. Figure

6.36(*a*) color plate 12, shows a picture of Sumatra, which has been enhanced by stretching each channel individually. Figure 6.36(*b*), color plate 12, shows the same scene enhanced by stretching the principal component pictures followed by the coordinate transformations described above.

Because of the nonlinear response of the human eye and brain to light, our perception of color does not entirely agree with simple mathematical descriptions such as the *HSI* system. One serious consequence of this is that moving a *DN* triplet a distance Δ during stretching has a different effect depending on original *DN* triplet. For instance, if the *DN* triplet described a saturated color, moving it Δ might not produce a noticeably different color if Δ were small. However, changing a *DN* triplet describing a pastel by the same Δ might produce a marked change. The eye is most sensitive to changes of hue in the yellow-green portion of the spectrum.

Several color spaces have been devised so that a translation of Δ produces the same perceived change in color regardless of position in the space. One such system is the Munsell Renotation System (49). This is the system typically used by geologists in the field to describe rock and soil colors. The Munsell Renotation System was designed empirically and is awkward to use in computation. A computationally simple space, which is a close approximation to the Munsell space, is the "cube root" color space (50) approximated by the transformation:

$$L^* = 25.29Y^{1/3} - 18.38$$

$$a^* = 112.49X^{1/3} - 106.00Y^{1/3} \qquad (6.44)$$

$$b^* = 42.34Y^{1/3} - 25.73Z^{1/3}$$

where L^*, a^*, and b^* are parameters related to luminance or brightness and color differences, and (X, Y, Z) are standardized chromaticity coordinates defined in 1931 by the "Commission Internationale de l'Eclairage" or CIE, X, Y, and Z are related to the previously defined tristimulus chromaticity coordinates (Equation 6.43) and hence (DN_1, DN_2, and DN_3):

$$X = \frac{0.20DN_1 + 0.31DN_2 + 0.49DN_3}{0.01DN_1 + 0.81DN_2 + 0.18DN_3}V$$

$$Y = V \qquad (6.45)$$

$$Z = \frac{0.99DN_1 + 0.01DN_2}{0.01DN_1 + 0.81DN_2 + 0.18DN_3}V$$

V is the luminance, which is proportional to the radiant flux, or the energy reflected from the scene. Because color spaces such as this are designed to model the *display* of color information, they are necessarily dependent on the characteristics of the display medium (film or TV) including the illuminant and choice of filters or phosphors representing the three primary colors. Readers interested in exploring these complexities should consult Wyzsecki and Stiles (51) or some similar text.

The most effective use of the cube root space to display data is to spread the original PDF out uniformly with a ramp CDF stretch in each channel L^*, a^*, and b^* prior to transformation back to blue, green, and red coordinates. This is effective because the data is uniformly distributed with respect to perceived hue, saturation, and intensity.

6.8.3 Spatial Filtering

Occasionally an image of a scene with a high brightness range may be encountered, so that any contrast stretch performed will saturate large portions of the original PDF. Spatial filtering to suppress gross changes in brightness over the image while exaggerating local higher frequency changes can be employed to increase the analyst's ability to discriminate detail. However, there is a loss of ability to relate brightness from one locality to another. Such a filter is called a high-pass filter because high frequency data is preserved unchanged while low frequency data is suppressed. Usually, filtering is performed by convolution because it is the most economical approach. Figure 6.37 shows the improvement a carefully chosen high-pass filter can make in a picture.

High-pass filters have three serious effects on an image. First, because most of the data in a scene is low frequency while random noise is evenly distributed, suppressing low frequency information has the effect of reducing the S/N, or making the picture more "noisy." Second, high-pass filters tend to "ring" off sharp edges or boundaries in an image. This is because the outermost weights of the filter

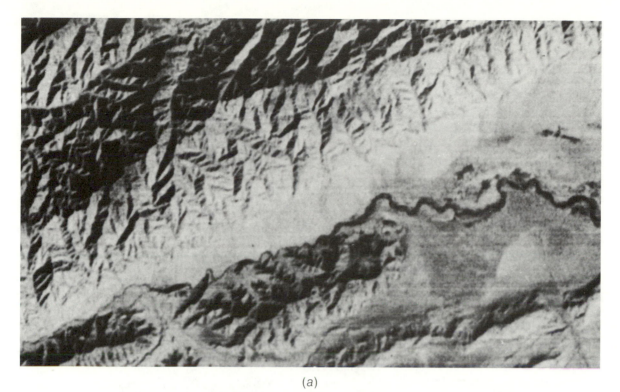

(a)

(b)

FIGURE 6.37. Part of the Astin Tagh mountains in Tibet, showing major strikeslip fault (from NASA Landsat image 1074-04253) (a) Picture as available from EROS Data Center. (b) Same picture after careful high-pass filtering. Offset streams near arrow indicate left-lateral displacement along fault.

FIGURE 6.38. (a) Directional filter artifacts in an image of glacial loess in Hai Yuan, China (NASA Landsat image 1510-03032). A 1 x 27 (horizontal) high-pass box filter was used to enhance this version. Linear features in the scene oriented about 14° from horizontal are exaggerated.

arrive at the discontinuity before the center. Thus the boundary influences the brightness of pixels in the filtered picture, which are as far away as one-half the filter size. Third, certain filters introduce undesired exaggerations of features oriented in certain directions (52). This effect is shown in Fig. 6.38 and is especially pronounced in pictures filtered with small rectangular box filters.

Random noise can be reduced by using band-pass filters that pass only spatial frequencies in a limited region of the spectrum, instead of passing all frequencies above a cutoff frequency as in a high-pass filter. Thus a bandpass filter can be chosen to minimize the S/N of the filtered picture in a way similar to that discussed in Section 6.5.3 for restoration filters.

Another approach is to filter an image made by averaging several channels, or ideally the first principal component picture, which has a maximum S/N already.

Ringing can be suppressed by using special filters such as the "variable threshold zonal filter" designed by Schwartz and Soha (53), which essentially classifies an image during filtering and then filters data belonging to different classes independently. In this way the boundaries between classes become less "visible" to the filter. However, the filter itself changes character near boundaries so interpretation of filtered pictures must be made with caution. Figure 6.39 compares the same picture filtered by an ordinary high-pass filter and by a zonal filter.

Undesired directional exaggerations are introduced even into images of random noise by some high-pass filters. This happens most obviously with one-dimensional box filters, when the number of weights is less than about 45. To explain these artifacts, we must examine the Fourier transform of the box filter.

The Fourier transform of a continuous rectan-

FIGURE 6.38. (b) Directional filter artifacts in an image of glacial loess in Hai Yuan, China (NASA Landsat image 1510-03032). A 27 x 1 (vertical) high-pass box filter was used in this version. Now features ± 14° to vertical are emphasized.

gular pulse is the sinc function, $\frac{\sin x}{x}$. The transform of a discrete box filter is very nearly a sinc function. The sinc function is periodic and has zeros at frequencies of nN^{-1} cycles per pixel, where n is any nonzero integer and N is the number of weights in the filter in the spatial domain. The side lobe at 1.5 N^{-1} has an amplitude of 0.21 [see Fig. 6.11(a)]. Frequencies in the image, which coincide with this side lobe, will be enhanced relative to neighboring frequencies. As shown in Fig. 6.40, this means that the wavelength at which enhancement occurs is a function of orientation of azimuth of the wave, with the lowest exaggerated frequencies being normal to the filter and higher frequencies increasingly oblique to it. This is exactly what is seen in filtered random noise. Not all frequencies are present in equal amounts in real scenes, and those frequencies that are present are enhanced, so the effect in filtered images can be more pronounced than in ramdom noise.

Directional exaggerations can be reduced by using filters that have simpler transforms. One possibility is to simply filter the picture twice with the same box filter. Convolution is associative so this is like convolving the filter with itself before filtering the image. The transform is $\text{sinc}^2(x)$, and the amplitude of the first side lobe is reduced to 0.04, thus essentially eliminating the problem while exploiting the advantages of box filters.

6.8.4 Compensation for Playback Devices

The film recorder can act as a low-pass filter when it produces a picture from an image. This can occur if the actual spot or pixel in the picture being created has a Gaussian cross section, so that the actual density at a pixel center is derived in some part from the density at neighboring pixels. This effect can be lessened by reducing the spot size, but this has unwanted consequences. In a region of homogeneous brightness like a flat field, individual pixels become increasingly visible, first as ripples in

FIGURE 6.38. (c) Directional filter artifacts in an image of glacial loess in Hai Yuan, China (NASA Landsat image 1510-03032). Preferential directional enhancement is minimized when a square 27 x 27 box filter is used.

the supposedly uniform surface and finally as discrete bright points set in a grid. The introduction of such a periodic pattern to the picture is all the more undesirable because the human eye is especially sensitive to periodic signals or noise.

One solution is to increase the pixel density in the display picture by resampling the image. Although this does not improve the information content of the image it effectively reduces the severity of smoothing during playback without sacrificing the flat field reproducibility.

The filtered picture shown in Fig. 6.37 has had its pixel density increased by a factor of three.

6.8.5 Filtering of Color Pictures

If the images in a color triplet are each high-pass filtered to enhance fine detail, the resulting color picture will be largely gray. This is because the filtered pictures show only the deviation of a pixel from the average DN of its neighbors. Because data are correlated among channels, they tend to deviate in the same way, and the filtered pictures are nearly identical. However, if the image is transformed to either principal component of HSI coordinates, the first principal component picture or the intensity picture may be high-pass filtered and recombined with the unmodified second and third principal component pictures or hue and saturation pictures to create a filtered picture that preserves the original color balance.

Ready and Wintz (54) pointed out the possibility of improving the S/N in a multispectral image by performing a principal component transformation of all channels available, and then discarding all except for the first three transformed pictures. The discarded pictures are the ones whose S/N's are lowered by the transformation, while the retained pictures have improved S/N's. Upon transformation back to the original coordinate system, the blue, green, and red images have less noise than before.

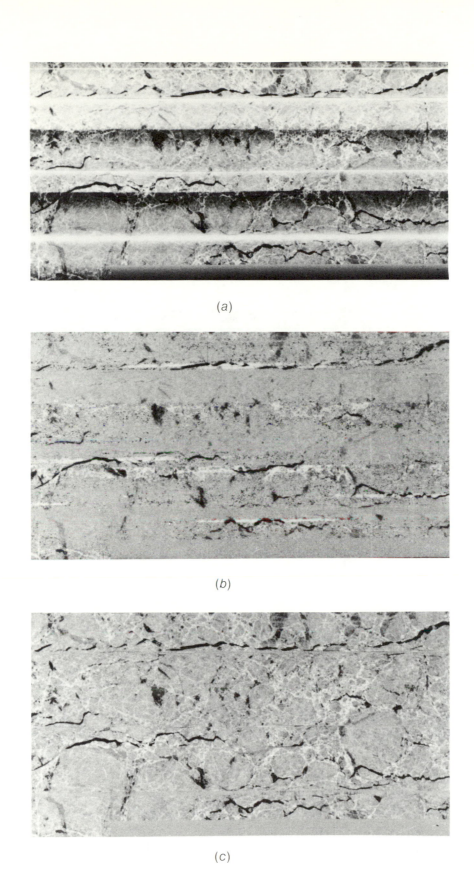

(a)

(b)

(c)

FIGURE 6.39. Supression of ringing in high-pass filtered images of pack ice. (a) A mosaic of aircraft radar images before filtering. Object of filtering is to reduce visibility of edges and other horizontal artifacts. (b) After filtering by conventional methods, ringing (horizontal bright streaks) has been introduced. (c) Ringing is suppressed by zonal filter.

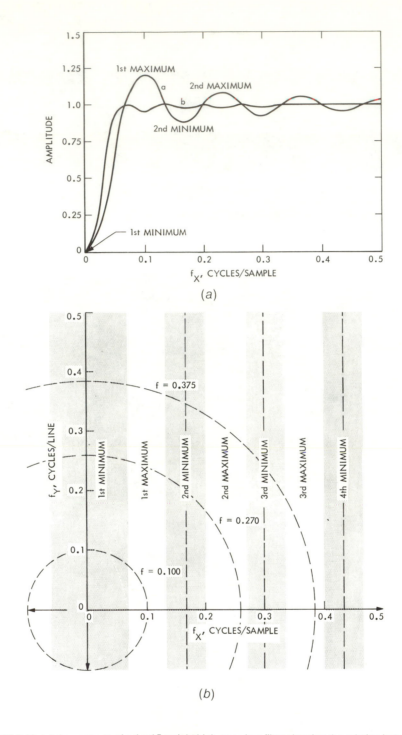

FIGURE 6.40. (*a*) A spectrum of a 1 x 15 weight high-pass box filter showing the relation between frequency and azimuth of emphasized features. Profile of the filter spectrum in the sample (f_x) direction (curve *a*). The filter spectrum is just the sinc function. For comparison, curve *b* shows $sinc^2$. (*b*)The filter spectrum in two dimensions showing maxima and minima. Regions where the amplitude is less than unity are shaded. Circles, are loci of waves of all orientations having frequencies of 0.100, 0.270, and 0.375 cycles/pixel.

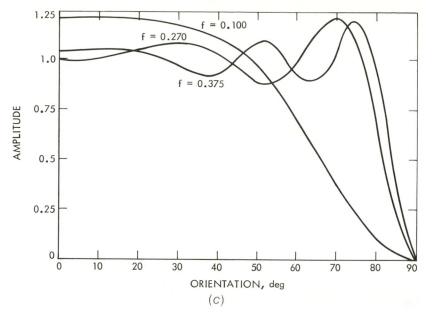

(c)

FIGURE 6.40 (c) A spectrum of a 1 x 15 weight high-pass box filter showing the relation between frequency and azimuth of emphasized features. Filter amplitude of frequencies of 0.100, 0.270, and 0.375 cycles/pixel as a function of orientation with respect to the f_x axis. At a given orientation waves having limited ranges of frequencies are exaggerated while at different orientations the same waves may be suppressed.

It is also possible to low-pass filter the third principal component picture or both the second and third principal component pictures (or the saturation and hue pictures) to reduce the amount of noise still further. Even though the color information is blurred a little by this process, the overall effect is less than if the noise were filtered out in the blue, green, and red images. Figure 6.41, color plate 13, shows the same scene as an ordinary color picture and as a filtered color picture.

6.9 SUMMARY

Digital images are readily manipulated by computer to prepare a more effective display of image data for interpretation by the geologist. Four general procedures are used: rectification, cosmetic, analytic, and display procedures. Rectification procedures remove radiometric and geometric distortions that have been introduced by the sensing instrument. The degraded frequency response of the camera may be corrected by restoration filtering, and other filtering routines may be used to remove coherent noise and to suppress random noise. Random blemishes and transmission errors may be removed by interpolation, a cosmetic procedure. Rectification and cosmetic procedures prepare an

image for analysis and display. Images may also be geometrically projected to register to other reference images or to maps.

Analytic procedures are designed to extract information from an image that may not be readily apparent by simple inspection. This information may be important in its own right, or may be used to design a more effective display for photointerpretation. The probability density function or PDF is a major analytic tool. Other analytic procedures include principal component and color coordinate transformations.

The principal tools used in displaying images are contrast stretches, spatial filtering, and color compositing. Photographic representation of digital images is generally preferred over other methods because of the immense amount of data involved and because of the need for subtle pattern recognition by the photointerpreter.

Digital processing of digitally acquired images is generally preferred to photographic or optical processing of photographic representations of digital images, primarily because digital techniques, although more expensive, are also more powerful and flexible, and there is less degradation of the image because of repeated copying. Readers who are interested in more detailed treatment of image processing techniques are referred to two recent texts: Pratt (55) and Castleman (56).

REFERENCES

1. Davies, M. E., and Murray, B. C., 1971, The view from space; New York, Columbia University Press, p. 36.
2. Bracewell, R., 1965, The Fourier transform and its applications: New York, McGraw-Hill, 381 p.
3. Goodman, J. W., 1968, Introduction to Fourier optics: San Francisco, McGraw-Hill, 286 p.
4. Blackman, R. B., and Tukey, J. W., 1959: The measurement of power spectra, New York, Dover, 190 p.
5. Evans, W. E., 1974, Marking ERTS images with a small mirror reflector: Photogrammetric Eng., v. 40, no. 6, p. 665-672.
6. Horan, J. J., Schwartz, D. S., and Love, J. D., 1974, Partial performance degradation of a remote sensor in a space environment, and some probable causes: Applied Optics, v. 13, no. 5, p. 1230-1237.
7. Rindfleisch, T. C., Dunne, J. A., Frieden, H. J., Stromberg, W. D., and Ruiz, R. M., 1971, Digital processing of the Mariner 6 and 7 pictures: Jour. Geophysical Res., v. 76, no. 10, p. 394-417.
8. Green, W. B., Jepsen, P. L., Kreznar, J. E., Ruiz, R. M., Schwartz, A. A., and Seidman, J. B., 1975, Removal of instrument signature from Mariner 9 television images of Mars: Applied Optics, v. 14, no. 1, p. 105.
9. Billingsley, F. C., 1972, Computer-generated color image display of lunar spectral reflectance ratios: Photo. Sci. and Eng., v. 16, no. 1, p. 51-57.
10. Minnaert, M., 1961, Chap. 6, Photometry of the Moon, in, Kuiper, G. P., ed., Planets and Satellites, v. 3: Chiago, Ill., University of Chicago Press, p. 213-245.
11. Young, A. T., and Collins, S. A., 1971, Photometric properties of the Mariner cameras and of selected regions on Mars: Jour. Geophysical Res., v. 76, no. 2, p. 432-437.
12. Doyle, F. J., Bernstein, R., Forrest, R. B., and Steiner, D., 1975, Cartographic presentation of remote sensor data, in, Reeves, R. G., Anson, A., and Landen, D., ed., Manual of remote sensing: Falls Church, Virginia, American Society of Photogrammetry, v. 2, p. 1077-1106.
13. Bracewell, R., 1965, op cit., p. 189-194.
14. Rifman, S. S., 1973, Digital rectification of ERTS multispectral imagery, in Symposium on significant results obtained from Earth Resources Technology Satellite-1, NASA SP-327, Washington, D.C., p. 1131-1142.
15. Kreznar, J. E., 1973, User and programmer guide to the MM'71 geometric calibration and decalibration programs: Internal Doc. 900-575, Jet Propulsion Laboratory, California Institute of Technology, Pasadena, California, 224 p.
16. Goetz, A. F. H., Billingsley, F. C., Gillespie, A. R., Abrams, M. J., Squires, R. L., Shoemaker, E. M., Luchitta, I., and Elston, D. P., 1975, Application of ERTS images and image processing to regional geologic problems and geologic mapping in northern Arizona, App. B: NASA TR 32-1597, Jet Propulsion Laboratory, California Institute of Technology, Pasadena, California, p. 131-174.
17. Thomas, P. D., 1952, Conformal projections in geodesy and cartography: U.S. Dept. Commerce Coast and Geodetic Survey Special Publication No. 251, U.S. Govt. Printing Office, Washington, D.C.
18. Richardus, P., and Adler, R. K., 1972, Map projections: New York, Elsevier, 174 p.
19. Cutts, J. A., Danielson, G. E., and Davies, M. E., 1971, Mercator photomap of Mars: Jour. Geophysical Res., v. 76, no. 2, p. 369-372.
20. Davies, M. E., and Berg, R. A., 1971, A preliminary control net of Mars: Jour. Geophysical Res., v. 76, no. 2, p. 373-393.
21. Gillespie, A. R., and Soha, J. M., 1972, An orthographic photomap of the South Pole of Mars from Mariner 7: Icarus, v. 16, no. 3, p. 522-527.
22. Rosenfeld, A., and Kak, A. C., 1976, Digital image processing: New York, Academic, p. 275-403.
23. Helstrom, C. W., 1967, Image restoration by the method of least squares: Jour. Opt. Soc. Am., v. 57, no. 3, p. 297-303.
24. Alsberg, H., and Nathan, R., 1974, Mansing using high speed digital mass storage: in Proceedings of the Human Factors Society's 18th Ann. Meeting, Huntsville, Alabama, 15, 1974.
25. Fisher, W. A., Badgley, P., Orr, P., and Zissis, G. J., 1975, History of remote sensing, in Reeves, R. G., Anson, A., and Landen, D., ed., Manual of remote sensing, v. 1: Falls Church, Virginia, American Society of Photogrammetry, p. 27-50.
26. Liebes, S., Jr., and Schwartz, A. A., 1977, Viking

'75 Mars Lander interactive computerized video stereophotogrammetry: Jour. Geophysical Res., v. 82, p. 4421-4429.

27. Wu, S. S. C., Shafer, F. J., Nakata, G. M., Jordan, R., and Blasius, K. R., 1973, Photogrammetric evaluation of Mariner 9 photography: Jour. Geophysical Res., v. 78, p. 4405-4410.

28. Nathan, R., 1966, Digital video-data handling: NASA JPL TR 32-877, Jet Propulsion Laboratory, Pasadena, California, 30 p.

29. Gillespie, A. R., and Kahle, A. B., 1977, The construction and interpretation of a digital thermal inertia image: Photogrammetry Eng. and Remote Sensing, v. 43, no. 8, p. 983-1000.

30. Haralick, R. M., and Shanmugam, K. S., 1973, Combined spectral and spatial processing of ERTS images, *in* Symposium on significant results obtained from Earth Resources Technology Satellite-1: NASA SP-327, Washington, D.C., p. 1219-1228.

31. Andrews, H. C., 1972, Introduction to mathematical techniques in pattern recognition, New York, Wiley-Interscience, p. 105-112.

32. Siegal, B. S., and Abrams, M. J., 1976, Geologic mapping using Landsat data: Photogrammetric Eng. and Remote Sensing, v. 42, no. 3, p. 325-337.

33. Fukunaka, K., 1972, Introduction to statistical pattern recognition, Chap. 4.2: New York, Academic, 369 p.

34. Fu, K. S., 1968, Sequential methods in pattern recognition and machine learning: New York, Academic, 227 p.

35. Wyszecki, G., and Stiles, W. S., 1967, Color Science: New York, Wiley, p. 228-235.

36. Committee on Colorimetry of the Optical Society of America, 1966, Chap. 7, *in* The science of color: Washington, D.C., Optical Society of America.

37. Jain, A. K., 1972, Color distance and geodesics in color 3 space: Jour. Opt. Society of America, v. 62, no. 11, p. 1287-1291.

38. Vincent, R. K., and Thompson, F. J., 1972, Rock type discrimination from ratioed infrared scanner images of Pisgah Crater, California: Science, v. 175, p. 986.

39. Rowan, L. C., Wetlaufer, P. H., Goetz, A. F. H., Billingsley, F. C., and Stewart, J. H., 1974, Discrimination of hydrothermally altered areas and of rock types using computer enhanced

40. Goetz, A. F. H., et al., 1975, op. cit. p. 97-121.

41. Rogers, R. H., and Peacock, K., 1973, A technique for correcting ERTS data for solar and atmospheric effects, *in* Symposium on significant results obtained from Earth Resources Technology Satellite-1: NASA SP-327, Washington, D.C., p. 1115-1122.

42. Crane, R. B., 1971, Preprocessing techniques to reduce atmospheric and sensor variability in multispectral scanner data, *in* Proceedings of the 7th symposium on remote sensing of the environment: Ann Arbor, Michigan, Environmental Research Institute of Michigan, p. 1345.

43. Wyszecki, G., and Stiles, W. S., 1967, op. cit., p. 450-560.

44. Janza, F. J., 1975, Interaction mechanisms, *in* Reeves, R. G., Anson, A., and Landen, D., ed., Manual of remote sensing: Falls Church, Virginia, American Society of Photogrammetry, p. 83.

45. Podwysocki, M. H., Moik, J. G., and Shoup, W. C., 1975, Quantification of geologic lineaments by manual and machine processing techniques, *in* Proceedings NASA Earth resources survey symposium: NASA TM X-58168, Houston, Texas, v. I-B, p. 885-903.

46. Siegal, B. S., 1977, Significance of operator variation and the angle of illumination in lineament analysis on synoptic images: Modern Geology, v. 6, p. 75-85.

47. Billingsley, F. C., 1972, Computer-generated color image display of lunar spectral reflectance ratios: Photogrammetric Sci. and Eng., v. 16, no. 1, p. 51-57.

48. Goetz, A. F. H., and Billingsley, F. C., 1974, Digital image enhancement techniques used in some ERTS application problems, *in* Proceedings of the third Earth Resources Technology Satellite-1 Symposium: NASA SP-351, v. 1., p. 1971-1992.

49. Newhall, S. M., Nickerson, D., and Judd, D. B., 1943, Final report of the O.S.A. Subcommittee on spacing of the Munsell colors: Jour. Opt. Society of America, v. 33, p. 385.

50. Glasser, L. G., McKinney, A. H., Reilley, C. D., and Schnelle, P. P., 1958, Cube-root color coordinate system: Jour. Opt. Society of America, v. 48, p. 730.

ERTS images, South Central Nevada: U.S. Geological Survey Prof. Paper 883, 35 p.

51. Wyzsecki, G., and Stiles, W. S., 1967, op. cit., p. 228-320.
52. Goetz, A. F. H., et al., 1975, op. cit., p. 20-21, 118-119, 165.
53. Schwartz, A. A., and Soha, J. M., 1977, Variable threshold zonal filtering: Applied Optics, v. 16, no. 7, p. 1779-1781.
54. Ready, R. J., and Wintz, P. A., 1973, Information extraction, SNR improvement, and data compression in multispectral imagery: IEEE Transaction on Communications, v. COM-21, no. 16, p. 1123-1131.
55. Pratt, W. K., 1978, Digital image processing, New York, Wiley, 750 p.
56. Castleman, K. R., 1979, Digital image processing, Englewood Cliffs, N.J., Prentice Hall, 432 p.

SECTION III
INTERPRETIVE TECHNIQUES

Remotely sensed images can be made from energy in many different parts of the electromagnetic spectrum. In this third section of *Remote Sensing in Geology*, the reader is exposed to interpretive techniques and actual examples of the use of images from the visible spectrum through thermal infrared to microwave and radar. The physics pertaining especially to each region is elaborated so that the reader does not lose sight of the physical significance during interpretation of images from different spectral regions.

CHAPTER 7

FUNDAMENTALS OF AERIAL PHOTOGRAPHY INTERPRETATION

HARMER A. WEEDEN
NANNA B. BOLLING

7.1 INTRODUCTION

Writers in the field of photographic interpretation divide aerial photographic studies into analysis and interpretation. Photo analysis is performed by all users regardless of their scientific backgrounds or goals to describe an image in terms of origin, geomorphic history, composition, and structure of separate regions within the scene. Parameters studied include topography, drainage, erosion, land use, vegetation, and photographic density, color, and texture. They are derived from shape, size, tone and color, shadow pattern, texture, and location of photographic features.

Photo interpretation, the use of photo analysis for a particular objective, is biased by a user's goals and specific discipline. Results of photographic interpretation are best represented by sketches or maps, which should be accompanied by discussions of conclusions, limitations inherent in the data, and analytical methods used. The factors affecting the quality of photographic analysis and interpretation are the first concerns of this chapter.

7.2 THE PHOTOGRAPHIC MISSION

For reasons of speed and economy, the analyst must often work with existing photographs. When a photographic mission is possible, however, careful attention must be given to the objective of the mission and the camera type, type of photography, choice of films and filters, mission control factors, and coordinated ground studies that will provide the best results. The objectives of the study determine the desired characteristics of the photography, and the mission plan. Factors requiring consideration include the characteristics of the principal features of interest; optimum scale; desirability and utility of various types of photography; and significance of season, sun angle, and moisture conditions. Consideration must also be given to multiband coverage, repeated or single coverage, and the smallest scale sufficient for feature identification for large area studies.

7.2.1 Cameras

Aerial cameras differ from hand-held cameras primarily in size, mounts, and controls. They consist of a fixed-focus lens cone assembly and a film magazine carrying up to 120 m of film. The shutter is either between the lens elements or at the focal plane; the viewfinder is usually a separate unit mounted alongside the camera.

Aerial cameras are generally classified as reconnaissance or mapping; the former has a long focal length and normal (67°) angular coverage; the latter, a short focal length and often an angular coverage exceeding 90°. Reconnaissance cameras include frame reconnaissance cameras, which have a field of view typically ranging from 10° to 40°, focal lengths ranging from a few centimeters to more than a meter, and a low degree of distortion correction; and panoramic cameras, which have a small immediate field of view but scan the terrain from side-to-side at right angles to the flight line, producing an image similar to a complex double oblique. Mapping cameras include mapping frame cameras, which have a high degree of distortion correction and a field of view of 90° to 120°; and strip cameras, which work on the principle of moving the film behind a slit in the camera's focal plane at the same velocity the image moves past the slit. Although strip cameras have a smaller field angle and less distortion than mapping frame cameras they are now less frequently used.

Two of the most common reconnaissance cameras used for quality photographs, the Wild RC 8 and the Zeiss RMK A, have a nominal focal length of 152.4 mm; focal lengths may be calibrated to six significant figures. The format (frame) size is 228.6 mm. The between-the-lens shutter has continuously varying settings, from 1/100 to 1/700 or 1/1000 s. A separate magazine carries from 60 to 120 m of film. These cameras are dimensionally stable at high altitude and low temperature. Table 7.1 outlines the specifications for several other frequently used camera systems (1).

Specialized aerial cameras are frequently used. The most common, the multiband camera, contains four to nine lenses and carries one or more rolls of film. Different film and filter combinations permit simultaneous photography in different spectral regions. Similar results can be obtained by using several cameras mounted as a unit.

Common focal lengths of aerial cameras are 15.25, 21.00, 30.50, and 61.00 cm, although focal lengths up to 244 cm have been used. The choice of focal length depends on scale and resolution require-ments, altitude restrictions, film resolution, and (in the case of color) color fidelity over the total format area. For example, by proper choice of focal length, similar photographic ground resolution can be obtained from the Skylab satellite (altitude of approximately 385 km) and U2 aircraft (altitude of approximately 19,800 m) (see Table 7.1).

7.2.2 Vertical Photography

Scale is an important consideration in photographic analysis. High altitude photographs minimize relief displacement and show a large area on a single frame, facilitating the study of regional relationships. Photographs with scales smaller than 1:40,000, however, have limited use in stereoscopic analysis; in addition, the color balance of visible color photography deteriorates with altitude because of atmospheric absorption and scattering. The most commonly used scale for detailed photo-

TABLE 7.1.   TYPICAL PHOTOGRAPHIC MISSION PARAMETERS

Camera platform:	Skylab satellite	U2 aircraft	C130 aircraft	C54 aircraft	Cessna 180 aircraft
Mission:	Reconnaissance	Reconnaissance	Reconnaissance	Reconnaissance	Mapping
Altitude:	385 km	19,800 m	4570 m	1525 m	1830 m
Camera:	Earth Terrain S190B	Vinten A/RC 10	Wild RC 8	Zeiss RMK A	Fairchild T-11
Focal length:	457.20 mm	44.45 mm	152.40 mm	152.40 mm	152.40 mm
Angular field of view:	32°	90°	93°	93°	93°
Area covered:	12,300 km^2	961 km^2	47.1 km^2	5.4 km^2	6.8 km^2
Format size:	11.5 x 11.5 cm	70 x 70 mm	23 x 23 cm	23 x 23 cm	23 x 23 cm
Scale:	1:845,000	1:445,000	1:30,000	1:10,000	1:12,000
Film:	Color SO-242	Aerocolor negative 2445	Aerochrome EK-8442	Aerochrome infrared EK-8443	Aerographic 2402
Filter:	None	2B (haze reducing)	EF (graded density) + HF3 (haze reducing)	Wratten 12 Minus Blue (deep yellow)	Wratten 25A (red)
Ground resolution: [a]	41-78 m	28-56 m	2.7-5.3 m	0.7-1.4 m	0.6-1.2 m

[a] The first number is the resolution under conditions of high contrast, the second the resolution under low contrast.

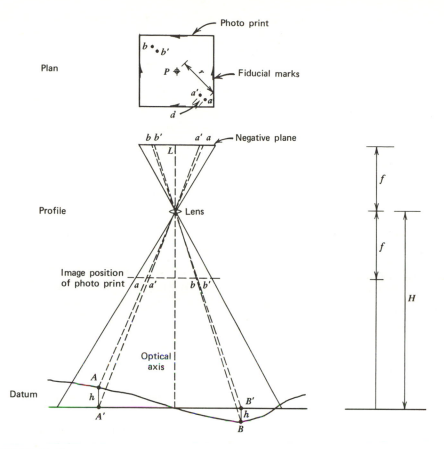

FIGURE 7.1. Geometry of vertical aerial photograph illustrating radial displacement due to relief.

graphic analysis is 1:12,000. Larger scales (1:7200 or larger) may be used in areas of flat-lying, thin-bedded sedimentary rocks.

Photographs are one-point perspectives, not orthographic projections. Consequently, scale, defined as the ratio of lens focal length to flight height (see Chapter 3), has restricted interpretation. Scale (f/H) applies only at datum and at the photograph center (see Fig. 7.1). Scale at a point h ft above (or below) datum is $f/(H - h)$. Image sizes for points above datum are larger than those at or below datum.

Images of ground features on vertical photographs are also subject to radial displacement due to relief (Fig. 7.1). If the photograph is truly vertical, the displacement, radial from the center, represents an error in map position, which must be considered in data transfer and compilation. Radial displacement due to relief is also responsible for scale differences within a photograph; the central projection characteristics of vertical aerial photographs

cause outward displacement of points above datum, and inward displacement of points below datum. The amount of radial displacement (d) of the top of an object with respect to its base is a function of its position on the photograph. It can be shown, from similar triangles, that $d = r(h/H)$, where r is the radial distance from the base of the object to the principal point[1] of the photograph. Hence, objects near the center of the photograph have minimum relief displacement, and high altitude photographs, such as those acquired by U2 aircraft, have less relief displacement than lower altitude photographs. It should be apparent that the height of objects on photographs can be determined from measurement of relief displacements.

Adjacent vertical photographs with 60% overlap (in the line of flight) are needed for stereoscopic study. The stereoscope optically straightens the

[1]The position at which the optical axis passes through the photograph is located at the intersection of straight lines joining opposite fiducial marks on the photograph.

line of sight from each eye, as shown in Fig. 7.2, recreating the condition of seeing a scene from a distance. The quality of the stereoscopic model depends on several factors, the most significant being the angular coverage of the camera lens. Wide-angle lenses create better models. Good stereoscopic models for detailed analysis can be obtained from 1:12,000 scale photographs.

7.2.3 Oblique Photography

Oblique aerial photography may be described as "high" or "low" obliques (2). High-oblique photographs (Fig. 7.3) show the horizon and a portion of the sky, whereas, low obliques (Fig. 7.4) only show the terrain. In high-oblique photographs the change in scale toward the horizon facilitates recognition of features, although it is difficult to determine their sizes. A system has been devised to determine object size on high-oblique photographs, but it is a difficult, inefficient procedure. In low-oblique photographs it is impossible to determine object size accurately. Clearly, the advantage of vertical photography lies in its uniformity of scale and angular relationships.

7.2.4 Films and Filters

Different portions of the electromagnetic spectrum can be recorded on photographic film by

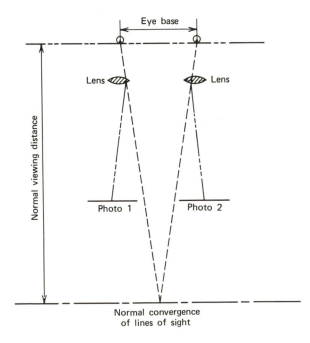

FIGURE 7.2. Ray diagram for a simple stereoscope.

proper choice of film and filters. Films are classified as orthochromatic, panchromatic, and infrared (3). Orthochromatic film is sensitive to green and blue, panchromatic responds also to red (and therefore to the entire visible spectrum), and infrared responds to the near-infrared region in addition to the visible spectrum. Orthochromatic film is rarely used in aerial photography. A variety of filters are used, depending on atmospheric conditions, the film type, and the portion of the spectrum to be registered on the film.

Three basic film types used in aerial photography are black-and-white (panchromatic or infrared), color (positive or negative), and false-color infrared (4, 5). Panchromatic black-and-white film is sensitive to green, blue, and red light, whereas black-and-white infrared film also records the near-infrared region (0.7 to 0.9 μm). On black-and-white films highly reflective surfaces, such as dry sand, appear white. With increasing moisture, they register darker, grayer tones. Organic soils appear dark gray to black, and water is black. The tonal range for vegetation is expanded on black-and-white infrared film.

Color positive film yields photographs with colors approaching those seen by the eye. It is useful for geologic interpretation, especially for the determination of sedimentary sequences (6, 7). Color negative film yields colors complementary to those on color positive film, and may be used directly to make color positive prints. False color film, originally developed for camouflage detection, now frequently used in agricultural and forestry studies (8, 9), is sensitive to green, red, and the near-infrared region (to 0.9 μm), and presents these spectral regions in the colors of blue, green, and red, respectively. Because actively growing vegetation is strongly reflective in the infrared region of the spectrum, it is seen in color IR film as bright red, whereas artificial turf, which is green on ordinary color film, is deep blue on color IR photography.

When the beginning interpreter has authority to request a photographic mission, aerial color (positive or negative) film is generally recommended because it supplies the greatest amount of information in familiar colors, and it has the widest latitude of exposure. Color may be a distinguishing characteristic of a key bed or an indicator of unusual soil conditions (4, 5). Experienced interpreters frequently prefer color IR film, which may provide more contrast for mapping drainage channels in wooded

FIGURE 7.3. High oblique photograph of Bucknell University Campus.

FIGURE 7.4. Low oblique photograph of Bucknell University Campus.

areas and identifying specific soil and rock characteristics by (enhanced) vegetation differences (10). Color IR film also improves stereoscopic model definition because it is less affected by haze than color film. As the interpreter becomes familiar with its use, color IR film will usually be considered preferable to color for most applications.

Black-and-white films are least desirable for photo analysis; they are limited in tonal range and provide weak stereoscopic models. Objects identical in tone but different in color cannot be differentiated, shadows mask features, and drainage and erosional features are difficult to identify and trace. However, convenience and economy often make black-and-white photographs acceptable, for they are available for most of the United States at a variety of scales, and often as both prints and transparencies. Much recent photography, however, is also available in color.

The major source of aerial photography is the United States Geological Survey EROS Data Center, Sioux Falls, South Dakota. Photography is also available from the Agricultural Stabilization and Conservation Service (ASCS) of the United States Department of Agriculture (USDA). In addition, many state transportation departments have their own photogrammetry divisions and commonly have photographs available at contact scale (1:12,000). Many universities also have libraries of aerial photography and other remotely sensed data (11).

Filters are usually used with aerial films (12). Yellow, Wratten 12 ("minus-blue") filters are used with black-and-white film to eliminate ultraviolet and blue wavelengths that are scattered by the atmosphere and cause hazy photographs. For color photography, the HF-3 haze filter is usually employed to remove the UV radiation; blue light is not removed because this would destroy the color balance of the film. (As a result, high-altitude color film may appear bluish and hazy due to the scattering of blue light.) Special filters may also be used to maintain the balance of hue, chroma, and saturation in color photography. The contrast and resolution of color IR film are improved by use of a yellow filter, which removes the strongly scattered UV and blue wavelengths. Recently, an IR film has been put on the market which has a yellow layer superimposed on the upper emulsion, eliminating the need for a filter on the camera (13). A filter opaque to the visible portion of the spectrum may be combined with color

IR film to confine exposure exclusively to the infrared region.

Black-and-white films sensitive to different limited portions of the spectrum, and panchromatic or infrared film with various filters, can be used in multiband cameras (commonly four or nine) or in several cameras mounted side-by-side to produce multiband photography. While the resulting unique combination of gray-tone signatures improves feature identification, the spectral characteristics of the different features and the effect of atmospheric attenuation must be known. Multiband photography is a valuable research tool in tree and crop identification and is increasingly used in soil and rock studies (14, 15). Creation of false-color images is a relatively new application of multiband photography. Two or more photographs in different bands are combined to create a color-enhanced image emphasizing particular features of interest. The choice of color for each band makes selective enhancement possible in a variety of ways, significantly improving data interpretation (7, 15). Combinations of filters and various types of film are used to confine exposure within narrow portions of the spectrum.

7.2.5 Mission Control Factors

Flight specifications vary with mission purpose (16, 17). In the northeastern United States, most geological missions should ideally be flown in the spring, before trees are in leaf but after most fields are plowed, to permit maximum exposure of soils and bedrock. Solar altitude is important in selecting the time of day for the mission; a low sun angle will result in long shadows, accentuating structural features but resulting in improper exposure of color film.

Aircraft altitude is determined by the scale desired and the camera focal length. As discussed earlier, atmospheric scattering deteriorates the color balance in color photography. Therefore, the lowest altitude consistent with mission objectives and the available aircraft range and camera focal length should be specified when using color film. For vertical photographs, the camera is mounted, with its axis perpendicular to the earth's surface. A topographic map with the intended flight lines gives the pilot a plan for coverage and shows the photographer where to change films (usually during the turn at the end of a flight line). Successive photographs (Fig. 7.5) are exposed at fixed-time intervals and correlated with ground speed to obtain 60% overlap in the line of flight (Fig. 7.6). Adjacent flight lines are planned to provide approximately 30% sidelap.

In flying a photographic mission, it is difficult to maintain straight and level flight lines because of side winds and air turbulence; these factors may also cause the camera axis to deviate from the true vertical (14). If the aircraft must head into a side wind to follow the flight path, the sides of the photographs, when assembled in a mosaic (Fig. 7.7), will not be parallel to the flight direction. The result is a strip of photos offset as shown in Fig. 7.8(a). Such photographs are said to be "crabbed," and the area of stereoscopic coverage is diminished. Because camera axis tilt results in distortions and variations in scale across the photograph [Fig. 7.8 (b)], rigid specifications concerning tilt are imposed on the photogrammetric contractor. Pho-

tography is rejected if the maximum tilt of a single photograph exceeds 3° or the average tilt for all the photographs is greater than 1°. Depending on available capital, a variety of expensive equipment, such as gyrostabilized camera mounts, auxiliary viewfinders, radar profile recorders, and inertial guidance systems for navigation, may be used to minimize tilt (1). Much available photography, however, has some degree of distortion.

Finally, the most painstakingly planned mission may be (and often is) thwarted by weather. Accurate weather reports are essential, and alternate flying dates should be included in the mission plans.

7.2.6 Coordinated Ground Studies

When possible, ground studies should be undertaken in coordination with photographic missions. Through systematic photography of various features and panoramic views of the area using both color and infrared films, accurate information on

FIGURE 7.5. Vertical photograph of the Bucknell University Campus, a portion of the area shown in Fig. 7.7.

FIGURE 7.6. Plane in flight, showing ground coverage by successive aerial photographs. (Courtesy of Wild Heerbrugg Instruments, Inc.)

shadows, moisture conditions, and the apparent color of ground features can be obtained. Field notes describing moisture, crop types, and other ground conditions can also be very helpful in interpretation.

7.3 PREPARATION FOR ANALYSIS

Preparation for analysis involves a review of the equipment available and a thorough understanding of the procedures necessary for effective stereoscopic model study. Before individual stereoscopic pairs of photographs are analyzed, a study guide of the region should be prepared using published reports and county mosaics of aerial photographs (18, 5).

7.3.1 Viewing Equipment

The type and quality of viewing equipment affect interpretive accuracy. A good stereoscope is essential. For prints, adequate room lighting is necessary. However, a good light table is needed to view transparencies.

LIGHT SOURCES

A light source that radiates energy over the entire visible spectrum is required for optimum color perception. It is possible to view transparencies with a variety of light sources and have the impression of satisfactory color balance and sufficient light. Some sources, such as sodium or mercury vapor lamps, however, are deficient in one or another of the additive primaries. Ordinary fluorescent lamps and cold-cathode lamps, often deficient in red output,

FIGURE 7.7. Rough-laid mosaic for Union County, Pennsylvania.

also affect color quality. Suitable light sources for viewing transparencies include natural daylight, incandescent tungsten light, and fluorescent light balanced to include the red portion of the spectrum (5).

Correct lighting conditions can be obtained from fluorescent tubes that provide a color-balanced source of 5000° K white light. A standard light table (normally equipped with two fluorescent tubes) can be converted to accept five such special tubes, providing excellent illumination for color transparencies. To avoid reflected glare, a brightness level of 8500 to 10,500 c/m^2 is recommended in a dimly lit room.

STEREOSCOPES

A variety of stereoscopes are available for photo analysis; each have distinct advantages and disadvantages (19, 20). In using pocket lens stereoscopes, photographs must be curled toward the center to view the complete overlap area, a procedure acceptable with prints but unsatisfactory with transparencies. Mirror stereoscopes have a longer focal distance, providing a larger, flatter field of view; the photographs need not be curled for viewing, and there is less image distortion. The optics are "folded" by two pairs of front-surfaced mirrors in order to create a compact viewing instrument.

Scanning mirror stereoscopes (19) have rotating front surface mirrors to permit scanning of the entire overlap area. Typically, an area of 23 by 46 cm can be scanned without moving either the stereoscope or the photographs. Frequently two magnifications, 1.5 and 4.5×, are available.

Zoom stereoscopes (19), give a continuous magnification range, and usually also feature optics that may be rotated to align adjacent photographs. They are ideal for viewing transparencies in roll form. Different models are available, depending on the film size to be viewed. Options include a choice of magnification range and an attachment permitting high-magnification viewing of single photographs. The zoom capability of these stereoscopes is typically 1.5 to 14×. For most efficient use, a zoom stereoscope should be mounted by means of an X - Y stage over a variable-intensity light table equipped with brackets to accommodate film reels.

A stereoscopic graphical transfer instrument (the Bausch & Lomb Stereo Zoom Transferscope) permits interpretation with direct transfer of data from a pair of photographs to a map or other base. It has a zoom capability of 0.6 to 7× for the photographs and 0.75 to 4× for the base. Two anamorphic systems, providing independent distortion correction for each photograph, permit a stretch

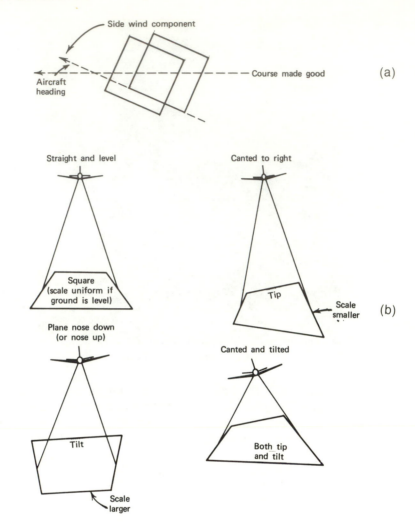

FIGURE 7.8. Distortion in aerial photographs. (a) Orientation of photographs due to "crab" (drift). (b) Variations in scale and shape of land area when the aircraft is not level at the moment of exposure (14).

range of 1:1 to 2:1 along any chosen axis.

PROJECTION VIEWERS

Projection viewers are available in single frame and stereoscopic versions. Single-frame projectors include the Saltzman projector and the Bausch & Lomb Zoom Transferscope. Simple slide and overhead projectors may also be used. The Saltzman projector is capable of -4 to +4× magnification of photographic prints, whereas the Zoom Transferscope is capable of +1 to +14× magnification of prints and transparencies and can stretch an image in any direction, making it possible to compensate for distortions.

Stereoscopic projectors include such commonly used models as the Kelsh Plotter (1), which uses glass plate positives of two paired photographs. A floating dot device (see Section 7.3.4) may be used to delineate the planimetry, trace contours, or determine individual elevations precisely. With ordinary transparencies, this instrument can be used for less precise elevation determinations or reconnaissance delineations at scales of +5 to +10×.

7.3.2 Study Guide

Before interpretation, a study guide should be prepared which contains all available information concerning the project and background information on meteorology, forest cover, cultural features, topography, geology, pedology, and hydrology; these elements have considerable influence on the gray tones and color values of the photograph (5). For example, the photographic expression of soils is directly controlled by rainfall and indirectly influenced by temperature, wind velocity, and cloud cover. Natural forest types and patterns often indicate differences in soils, geology, topography, and drainage (21). Very general differences in soils and geology may be indicated by farm practices: well-drained and poorly-drained soils are planted in different crops, whereas very steep slopes are often left in timber or used for pasture. Local customs, however, may influence individual farm practices more than natural characteristics do.

Quarries, mines, and pits are excellent clues to specific lithologies; for example, the extraction of aggregate materials frequently indicates glacial and/or river deposits. Roads usually take advantage of ridge gaps and level floodplains, whereas pipelines and powerlines are relatively independent of topography; they should not be mistaken for natural linear features. Urban and suburban areas, especially those established for some time, are generally on level, well-drained land.

Soil maps can be mosaicked and individual map units color-shaded to delineate larger, more generalized units which will reveal soil groups related to landform and internal drainage characteristics. Soil and rock characteristics influence stream flow and water storage capacity. The behavior of a stream during and after a storm can indicate the type of soil and rock in its basin; for example, streams with intermittent flow or discontinuous courses often indicate carbonate bedrock (22). Water table position distinctly influences photographic gray tone. Mottling with the soil profile indicates that the groundwater level rises to the mottled section at least part of the year, whereas gleying (a blush or grayish tone in the soil resulting from reduced iron) indicates that the section is under groundwater level almost constantly (23). Some sources of this type of information, useful in compiling a study guide, are listed in Table 7.2.

7.3.3 Comparative Analysis of a Region

Comparative analysis of a region, normally undertaken before stereoscopic analysis, can be performed using county mosaics of aerial photographs at a scale of 1:62,500, along with the study guide

TABLE 7.2. SOURCES OF INFORMATION FOR COMPILATION OF A REGIONAL STUDY GUIDE

Topic	Information Source
Meteorology	*Climatological Data and Storm Data Handbook* U.S. Department of Commerce and the Environmental Science Service Administration
Forest cover	USGS topographic maps County and larger-area aerial-photograph mosaics County agricultural soil survey reports
Farm practices	Familiarity of the interpreter with the area historical and local agricultural records
Quarries and mines	Topographic and geologic maps Historical maps and records Aerial-photographic mosaics (for patterns of distribution)
Construction	Topographic maps Detailed maps of urban/suburban areas
Topography	Topographic maps at various scales County aerial-photograph mosaics
Geology	Geologic maps and reports (older maps may be inaccurate, but they can still provide a good general idea of lithology and structure)
Pedology	County soil-survey reports, available from the U.S. Government Printing Office Preliminary pedologic maps, available from the local county office of the Soil Conversation Service, or from the state university
Hydrology	USGS water-supply papers Soil profiles in soil survey reports State groundwater reports Geologic reports

(18, 5), to delineate regional trends in topography, geology, soils, and cultural features. Major topographic features and drainage systems can be delineated, and linear features and cultivation and vegetation patterns (24) that indicate underlying geology can be noted.

Topographic and geologic maps may be compared with the mosaic to reveal possible clues to rock outcrop patterns and glacial features and to indicate how various cultural patterns may correlate with underlying structure and soils. Soil survey maps, generalized to the level of catenas, can be very useful at the county mosaic scale (5). A unique soil may serve as a key identifier; it may be consistently mottled or have a specific cultural practice associated with it (8). Cemeteries, for instance, usually indicate well-drained and easily excavated soil. Moisture retention can often be an identifying characteristic; however, because weather conditions just prior to the photographic flight can have a distinct influence, caution must be exercised.

When county mosaics are not available for study, alternate photographs of the region may be stapled together to create a rough-laid mosaic of larger scale, which should then be carefully annotated. Systematic repetition of certain features visible at this scale may reveal key beds that indicate the structural trend. In western Pennsylvania, for example, significant coal members are revealed by regional strip mining patterns. Once determined by projected profiling (5), the structural attitude of the beds can be used with the stratigraphic sequence to identify underclays or lithologic changes related to shales and sandstones. It should be remembered that the function of comparative analysis and mosaic study is to reveal regional patterns contributing to an understanding of the soils and geology, so that accurate, detailed information can be procured from stereoscopic study.

7.3.4 Stereoscopic Analysis

Stereoscopic analysis is best performed using 1:12,000 scale photographs with either glossy or semimatte finish. Although glossy finish renders fine detail somewhat better than semimatte finish, few notation instruments will register on its surface. Semimatte finish may be marked with pencil, but fine detail on such prints is harder to recognize. An alternative to prints is an emulsion image on an acetate film base, yielding a transparency with a wider range of gray scale and color tone. Although these images are easier to interpret, a light table is required. Transparencies, which are easily scratched, must be protected with frosted or clear acetate on which notation may be made (16).

STEREOSCOPIC MODEL

In undertaking stereoscopic analysis, the flight base must be established on the photographs by locating and joining the principal point and the transferred principal point on each photograph (Fig. 7.9). The photographs are placed so that the shadows fall toward the observer. The left photograph is taped to the table, under the stereoscope, and the right photograph moved left or right, maintaining the flight-base alignment, until a three-dimensional image is obtained. When an entire flight line of photographs has been recorded on a single roll of film, a stereoscope with a rotating optical system can be used to align the images.

Regardless of the type of stereoscope used, the height of objects in the stereoscopic model is exaggerated, because of lack of correspondence in angular relationships between the camera and the stereoscope. With increased experience in stereoscopic viewing, vertical exaggeration diminishes but never fully disappears (19). Because of vertical exaggeration, and its variability, the interpreter should be careful in estimating slope angles (20).

Ideal conditions for a stereoscopic model are achieved when the camera axis is truly vertical for each exposure, the azimuth of the flight base is constant, the flight altitude (hence, the scale) is constant for each exposure, and the photographs are sharply focused. The importance of these conditions can be illustrated from an understanding of the concept of *parallax*, the apparent shift in position of an object with respect to some reference system, caused by a shift in the point of observation (2). It can be shown (Fig. 7.10) that the height, h, of an object in a scene can be determined from the equation

$$h = \frac{\Delta p (H - h_1)^2}{Bf} \tag{7.1}$$

where

h = height of the object in meters,
Δp = change in parallax, in millimeters.
H = altitude of the camera lens above sea level, in meters,
h_1 = altitude of the base of the object above sea level, in meters,
B = flight base in meters,
f = focal length, in millimeters.

In Fig. 7.10 notice that the image of the tower in the two photographic prints lies on a radial line from the principal points. The X-coordinate of image

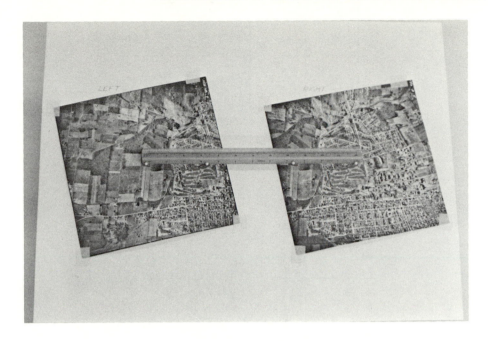

FIGURE 7.9. Alignment of a stereoscopic pair of photographs. The white-headed pins indicate the principal points and transferred principal points.

points a and b can be measured if a Y-axis is drawn through the principal points. Then $p = x_1 - x_2$, where p is the parallax of a given point and x_1 is measured in the positive direction in Print 1, and x_2 is measured in the negative direction in Print 2. Thus,

$$P_a = (x_{a_1} - x_{a_2}) \quad \text{and} \quad P_b = (x_{b_1} - x_{b_2})$$

and the differential parallax, Δp, is $p_b - p_a$

Differential parallax is most commonly determined by means of a parallax bar, or stereometer (Fig. 7.11), which operates on the "floating dot" principle (1). Two target dots, one seen with each eye, are fused stereoscopically so that the viewer sees a single dot that appears to float in space. The apparent height of this dot is related to the horizontal separation of the individual dots, which can be increased or decreased by a micrometer screw. A scale related to the screw movement reads the accurate distance between the two target dots. Differential parallax (Δp) is measured by subtracting the distance readings when the dot is set at the base and then at the top of the tower. H, h, B, and f must be known in addition to Δp. For example, assume the base of a tower is located on a slope 200 m above sea level (datum). It is imaged on 23-cm photographs (having 60% overlap in the flight direction) taken by a camera with a 15.25 cm focal length lens from height of 2000 m. The differential parallax is 5.00 mm. The height of the tower can be shown to be equal to

$$(1095)(152.5)$$

The flight base (1095) can be measured from a map or the ground, or determined by assuming a scale of 1 cm = 119 m and a 40% advance along the flight base. Then $23 \times 0.4 \times 119 = 1095$.

Any violation of the ideal conditions for a stereoscopic model creates distortion and erroneous results (19). A change in altitude or tilting of the camera axis will distort scale relationships, creating an imprecise model that tilts in space, causing the viewer to misjudge slopes and incorrectly evaluate differential parallax. The height difference of two points closely spaced in the photograph, such as in measurement of a cliff face, is not as seriously distorted as is the difference between widely separated points in the model, such as the base and summit of a hill.

PATTERN ANALYSIS

The objective of the study must be clearly defined before beginning pattern analysis to ensure realization of objective and proper study of photographic

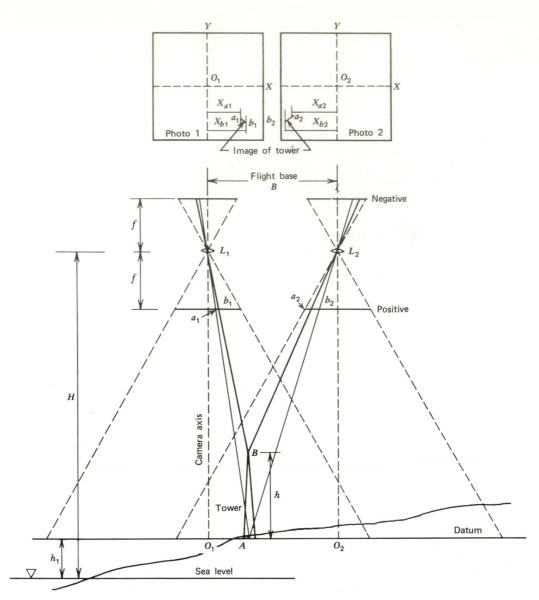

FIGURE 7.10. Photograph-to-ground relationships which create the stereoscopic model.

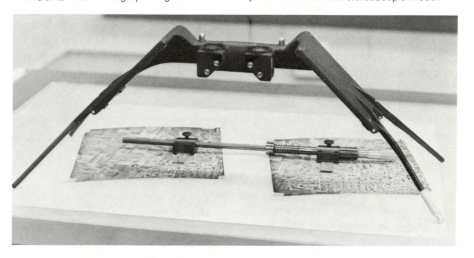

FIGURE 7.11. Parallax bar shown with a mirror stereoscope and photographs mounted for viewing.

elements. For a highway location project, the relationship of the study objectives to the photographic elements used in evaluation is presented in Table 7.3 (18, 5).

While elevation and slope may be quantitatively defined directly from the stereoscopic model, parameters such as soil texture and moisture must be qualitatively derived from analysis of gray tones (or color hue, value, and saturation), erosion scars, and supporting evidence presented by presence or absence of specific features (such as the presence of a tractor, indicating the soil was plowed on the day of the flight). Such an evaluation is admittedly subjective, and its success depends upon the skill and experience of the photo analyst. Scale, color, and photographic quality must also be considered (9, 15).

When interpreting color photography, care should be used in describing color-dependent attributes. It is not possible to assign an absolute color value as a unique signature for a specific ground condition. Film exposure and development cannot be controlled sufficiently to record colors consistently within a mission or from one mission to another, and atmospheric conditions can have a significant effect on color value and tone. Some comparisons of colors seen on IR color film with the natural color of the actual feature is shown in Table 7.4.

For pattern analysis a table of symbols should be prepared to annotate photographs and the final map (25, 26). The notation system will depend on the characteristics of the area under study; an example of such a system and its use in preparing a reconnaissance map is given in Section 7.4.

In general, only alternate photographs are annotated. Various notation schemes may be followed (27). The following method has been used successfully (5): Trace drainage in blue and sharp slope breaks in magenta using dotted lines to define ridges; outline areas of distinctive color or tonal change (such as mottled soils or linear features) in a unique color; outline in a different color areas of distinctive cultural practices; and outline areas of homogeneous vegetation in green. The resulting boundaries will not necessarily coincide with those delineated on photographs of smaller scale or on the aerial mosaic. However, the larger scale delineation should be regarded as the more accurate.

After delineation, symbols can be added to the interpretation to convey the required problem-oriented information. It is desirable, however, to conduct a field check before transferring this information to the finished map. Although time-consuming, the field check is essential and must satisfy the following needs: confirmation of questionable source data in the study guide, clarification of ambiguities arising from the interpretive process, provision of accessible subsurface data, and rationalization of inconsistent photographic elements related to season, weather, vegetation, and cultural practices. Traverses should be made across the regional strike in order to include all representative soil and rock conditions (5).

7.4 PHOTOGRAPHIC ANALYSIS

Since 1950, photo interpretive methods have been used to produce general purpose reconnaissance maps for various planning purposes. While informative, these maps may not contain sufficient detail to resolve specific problems, such as relocation of highways, placement of dams, or identification of one or more sources of construction materials. While on-site investigation will be essential, interpretation of photographs will decrease field time and give an overview of the problem unobtainable from the ground.

7.4.1 General purpose Reconnaissance Mapping

General purpose reconnaissance maps should contain a description of map units and the philosophy underlying the choice of these units. Tables, such as 7.5 and 7.6 (28) should be prepared defining symbols used for shorthand notation on the photographs and the resulting map. Symbols for a reconnaissance map intended for highway problem solving (Table 7.6) would define terms for landform, slope, soil texture, water table position, and depth to bedrock. Special symbols may be used for individual parameters, such as rock orientation or the need for ground-truth verification. The map elements used in a typical engineering geology reconnaissance mapping project are described below, followed by some examples of their grouping (18, 5).

LANDFORM

Recognition of particular soil combinations, residual or transported, is essential in describing landforms. The interpreter faces three basic landscape situations with respect to soil and rock: rock-dominated, soil-dominated, and intergrade. In

TABLE 7.3. OBJECTIVES VERSUS PHOTOGRAPHIC ELEMENTS FOR A HIGHWAY LOCATION STUDY

Evaluation Objective	Determining Characteristic	Parameter Read from the Stereoscopic Model	Photographic Elements Used for Evaluation
Depth of soil to bedrock	Topography	Elevation, slope	Landform, drainage, erosion, gray tone or color (hue, value and saturation), land use, vegetation, specific indicators
Depth to water table	Topography	Elevation, slope	
Design objectives:			
ALIGNMENT SELECTION			
Gradient	Topography	Elevation, slope	
Foundation characteristics	Soil	Texture, moisture content, organic matter	
CONSTRUCTION DRAWINGS			
Side slope safety			
Suitability for fill			
Volume change, cut to fill	Soil	Texture and moisture content	
Susceptibility to frost action			
Internal drainage			
Support value of subgrade			
Side slope safety	Rock type and structure	Coherence, jointing, bedding, faulting	
Volume change, cut to fill			
CONSTRUCTION OPERATIONS			
Equipment selection			
Trafficability	Soil and topography	Elevation, slope, moisture content, texture	
Accessibility			
Materials selection			
Equipment selection			
Accessibility	Rock type and topography	Coherence, jointing, bedding, faulting	
Materials selection			

the rock-dominated situation, landform may define rock type. In transported soils, particles are sorted or unsorted. Characteristic depositional features often assist in identification of many landforms, and characteristic erosion patterns may reveal the nature and distribution of the particle sizes present.

TEXTURE

Soil textures can be classified using the performance ratings of the American Association of State Highway Officials (AASHO). Soils are rated for subgrade performance by a series of laboratory tests. Although the resulting categories are beyond the range of accuracy possible from photo interpretation, approximations of five soil textures (based on the dominant layer—usually the C-horizon) can be made. These are presented in Section A of Table 7.6.

SLOPE

The most noticeable slope breaks are those related to landform boundaries. Precise slope measurements need not be obtained for reconnaissance mapping purposes. Arbitrarily chosen slope classes (shown in Section B of Table 7.6) can be used to estimate the dominant or overall slope, with no consideration of local steepening or flattening.

WATER TABLE POSITION

The groundwater level is controlled by various factors, including the time of year, rock permeability, topographic position, and soil texture. The date of the latest rainfall, its duration and intensity, should be known prior to analysis.

In northeastern United States, winter or spring photographs have relatively dark graytones related to wet soils, whereas summer photographs are usually lighter toned. Section C of Table 7.6 provides some suggested symbols useful for reconnaissance mapping of water table position. Summer photography, preferably at a large scale, is needed to designate low (n) watertable conditions with certainty; winter or spring photography is required for perennially wet (w) or seasonally high (h) conditions (29). Seasonal flooding (h) is indicated by meander scars or abandoned flow channels on floodplains (18).

DEPTH TO BEDROCK

While depth to bedrock can only be estimated with approximate ranges (Section D of Table 7.6),

these ranges are useful to indicate conditions that may be encountered during construction and to serve as a guide in planning drilling or seismic programs. Profiles drawn perpendicular to the main streams may be useful in estimating depth to bedrock (18, 5).

SPECIAL SYMBOLS

Symbols for geophysical surveys, drilling, sample collection, and other field verification studies for particular sites, such as those shown in Section E of Table 7.6, should be provided to coordinate ground-truth investigations.

GROUPING SYMBOLS

Symbols are grouped to designate the evaluation of particular mapping units. Three possible symbol groups are as follows:

Rock Dominated: Sh-3-S-h-0 ⤹ represents a shale landform covered by silty soil. The slope is steep and the water table is seasonally high. The soil cover is from 0 to 1 m deep. The rock dips to the southeast (assuming north is toward the top of the map).

Intergrade: C/Sh-4-G-h-1 represents a colluvial soil over shale. The colluvium consists dominantly of clay lying on a gentle slope. The water table is seasonally high and the soil is from 1 to 3 m deep.

Soil Dominated: GU-1-F-n-2 represents an unsorted glacial soil, granular in nature. The area is flat, with no foreseeable water-table problems, and the depth to bedrock (the nature of which is unknown) is from 3 to 7 m.

It is important that the defined units are independent. Care should be taken to ensure that the decision or mapping procedure for one unit has not arbitrarily restricted the evaluation of another (26).

SAMPLE RECONNAISSANCE MAP

Figure 7.12 is an example of a reconnaissance map for a portion of Erie County, Pennsylvania, produced from interpretation of color IR transparencies. Additional reconnaissance maps were prepared from interpretation of color prints and black-and-white photographs, all obtained in May, 1969. The terrain is glacial upland 300 to 900 m higher than Lake Erie.

Regional analysis from mosaics of smaller scale photographs previously revealed the overall pattern of glacial and shoreline features and was very

TABLE 7.4. INFRARED COLOR PHOTO INTERPRETATION (from Ref. 5, p. 21)

Color Viewed on IR Color Film	Color Dye Layers Operating	Real Colors in Original Scene	Example	Comments
Whitish pale blue (If IR is absorbed by subject, white will have a bluish tinge.)	All 3 layers blank	White + IR Yellow + IR Yellow + IR	Clouds Dandelion blossoms Yellow warning sign	Yellow wavelengths overlap red and green sensitive layers by 20 μ m, hence some red and green are exposed, but not as much as with white light.
Blue	Magenta and cyan			
Med. pale blue		Med. mod. red yellow	Sandstone quarry	Color difference between sandstone and shale is clearer than normal color film.
Green Blue	All 3 layers (with less yellow)			
Whitish pale green blue		Lt. pale red yellow	Bare earth	
Med. mod. green blue		Lt. mod. yellow red	Limestone soil	
Med. mod. green		Med. pale green	Water	More apparent on IR.
Blue Green	All 3 layers (with less magenta)			
Med. mod. blue green		Med. mod. red yellow	Plowed field	More green on IR where wet.
Lt. mod. blue green		Lt. pale red yellow	Channery field	Channers show up better on IR.
Med. pale blue green		Med. pale red yellow	Schist exposure	
Lt. mod. blue green		Med. pale red yellow	Dead grass	
Lt. mod. blue green		Lt. mod. yellow red	Silty soil	
Green	Yellow and cyan			
Med. pale green		Med. mod. yellow red	Dead grass	
Dk. str. green		Dk. mod. yellow red	Sandstone outcrop	
Yellow	Yellow			
Lt. mod. green yellow		Lt. mod. red	Trenching machine (painted red)	
Red	Yellow and magenta			
Whitish pale red		Whitish pale yellow	Dogwood blossoms	
Lt. pale red		Lt. pale green yellow	Forsythia blossoms	
Lt. mod. red		Lt. mod. green	Grass	Dried cornstalks show up better on IR.
Med. vivid red		Med. str. green	Broadleaf tree	
Lt. mod. violet red		Lt. mod. yellow green	Broadleaf tree	

TABLE 7.4. (continued)   INFRARED COLOR PHOTO INTERPRETATION (From Ref. 5, p. 21)

Color Viewed on IR Color Film	Color Dye Layers Operating	Real Colors in Original Scene	Example	Comments
Red Violet	Magenta and cyan			
Lt. mod. red violet		Lt. mod. yellow green	Broadleaf tree	
Dk. str. red violet		Dk. str. green	Evergreen trees	
Blue Violet	Magenta and cyan			
Med. mod. blue		Med. pale blue	Water	In turquoise, blue is essentially filtered out and green remains. In water, blue is filtered out and most of the red remains; very little IR is present.

useful in identifying specific landforms and calling attention to areas needing additional detailed study. A comparison of the reconnaissance maps produced from the different types of photographs clearly shows that mapping is most complete and accurate on the color IR transparencies, followed by color prints, then by black-and-white prints. The following observations were revealed:

1. Mapping units were similar with respect to topography, surface geology, and essential landforms.
2. The greatest difference in map units occurred in areas of peat and muck, which were identified on color IR transparencies, but not on the other photographs.
3. Differences in photographic tones and colors, and the quality of the stereoscopic models accounted for variations in boundary locations of up to 100 m on individual test profiles.

The soil maps and ground-truth information revealed the following:

1. Some soils previously designated "lake-laid" were not so interpreted because the stereoscopic model revealed that they occur on 6 to 8% slopes; it is likely that they are thin deposits over other glacial material.
2. Although there were few errors in soil depth classifications, local thin soil areas were not identified.
3. A drain-tile field, suspected to be unusually deep, was not detected.

In summary, accurate reconnaissance maps can be obtained from careful study of stereoscopic pairs of photographs, particularly color IR transparencies. Field checking, however, is essential, especially to establish the degree of accuracy obtained. In situations where an accuracy of 100 m is sufficient for the location of slope breaks, black-and-white photographs may be adequate. The degree of accuracy required must be weighed against cost (10).

7.4.2 Analysis of Specific Problems

In many instances general purpose reconnaissance maps may not be sufficient to solve specific problems, such as those encountered in planning and construction of highways. The following section describes various approaches taken in interpreting different types of photographs to address highway engineering problems in Pennsylvania. Particular attention is focused on identification and analysis of lake-bed deposits and other unconsolidated materials for stability of road and pavement structures (5).

STABILITY AND SUBSIDENCE IN LAKE BEDS

In general, study of regional mosaics of a lake bed area may reveal cultural features related to ancient shorelines, such as transportation lines following beach ridges. Low relief areas of limited size within rolling terrain indicate possible lake beds. Tonal patterns related to soil moisture often define limits of a former lake. Large flat areas of darker tone in front of moraines are often outwash plains or valley trains containing the sites of former lakes.

TABLE 7.5.   LETTER SYMBOLS FOR LANDFORMS (Derived from Ref. 26, p. 22).

Residual Soils

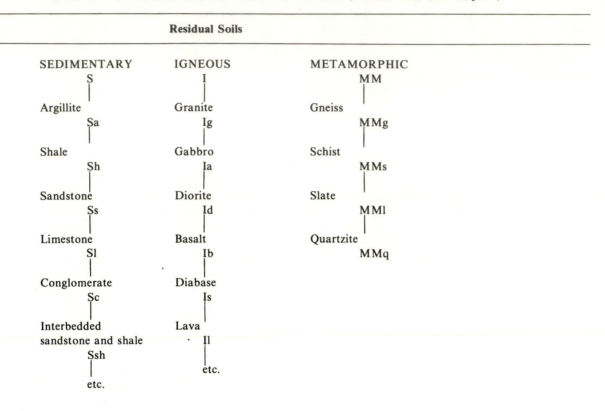

SEDIMENTARY	IGNEOUS	METAMORPHIC
S	I	MM
Argillite	Granite	Gneiss
Sa	Ig	MMg
Shale	Gabbro	Schist
Sh	Ia	MMs
Sandstone	Diorite	Slate
Ss	Id	MMl
Limestone	Basalt	Quartzite
Sl	Ib	MMq
Conglomerate	Diabase	
Sc	Is	
Interbedded sandstone and shale	Lava	
Ssh	Il	
etc.	etc.	

Transported Soils

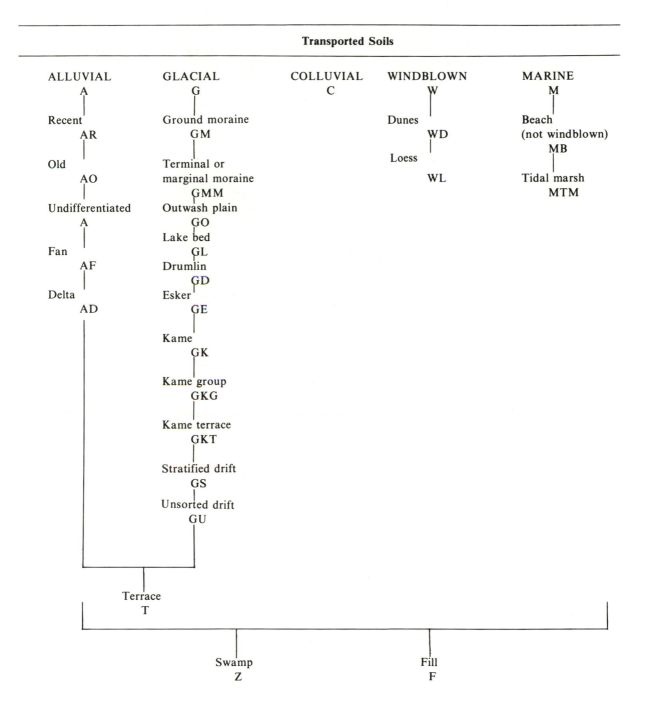

ALLUVIAL   GLACIAL   COLLUVIAL   WINDBLOWN   MARINE
A   G   C   W   M

Recent   Ground moraine   Dunes   Beach
AR   GM   WD   (not windblown)
                                                         MB
Old   Terminal or   Loess
AO   marginal moraine   WL   Tidal marsh
                 GMM   MTM
Undifferentiated Outwash plain
A   GO
Fan   Lake bed
AF   GL
Delta   Drumlin
AD   GD
                 Esker
                 GE

                 Kame
                 GK

                 Kame group
                 GKG

                 Kame terrace
                 GKT

                 Stratified drift
                 GS

                 Unsorted drift
                 GU

                 Terrace
                 T

        Swamp   Fill
        Z   F

TABLE 7.6. SYMBOLS FOR HIGHWAY-ORIENTED RECONNAISSANCE MAPPING IN PENNSYLVANIA

Element	Description	Symbol
A. SOIL TEXTURE (AASHO group and subgroups)		
A-1, A-1a, A-1b, A-2, A-2-4 to A-2-7	Gravel to clayey sand (friable)	1
A-3	Fine sand	2
A-4, A-5	Silty soil	3
A-6, A-7, A-7-5, A-7-6	Clay soil	4
Unclassified	Peat and muck	5
B. DOMINANT SLOPE		
0-4%	Flat	F
4-16%	Gentle	G
16-40%	Steep	S
Over 40%	Very steep	V
C. WATER TABLE POSITION		
Over 3 m below surface	No problem because of topographic position	n
Comes to within 1 m of surface	Seasonally high water table	h
Free water surface at, or very near, ground level	Perennially wet	w
Seasonal flooding (at least once in 4 years)	Flood plain areas related to major streams or rivers	h'
D. APPROXIMATE DEPTH TO BEDROCK		
0-1 m	Shallow	0
1-3 m	Medium	1
3-7 m	Deep	2
Over 7 m	Very deep	3
E. SPECIAL SYMBOLS		
Test for soil depth	Auger hole to bedrock	A
Test for soil depth	Seismic	S
Test for soil depth	Resistivity	R
Test for rock quality	Core drill for rock	C
Attitude of bedrock	Direction of strike and dip	⤵
Land slide area	Area of unstable condition	θ

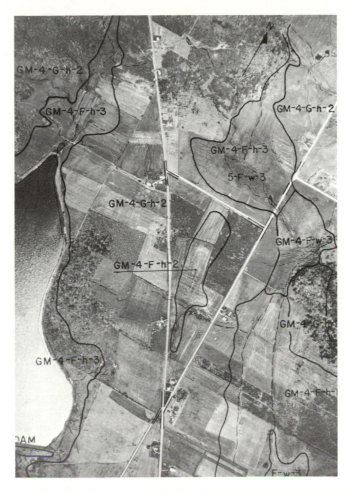

FIGURE 7.12. Reconnaissance map for a portion of Erie County, Pennsylvania, produced from interpretation of a color IR transaparency (10), printed in black and white.

Sand and gravel operations indicate coarser deposits, which are frequently far less obvious on individual photographs than on the mosaic.

For detailed interpretation in a lake-bed area, color transparencies are most desirable. Ground cover types are easily recognized and the color rendition is sufficiently close to that of "ground-truth" colors so that areas of thin ground cover or the absence of ground cover are readily identified. Tonal changes related to moisture and color changes in soil are easily seen. On the other hand, high-quality infrared color transparencies create a superior stereoscopic model. The high contrast of the color IR renditions makes it easy to trace drainage channels, and minor changes in color tone related to differences in moisture content of bare soil are readily observed. Black-and-white film records almost every discernible detail on the two types of color film except drainage in wooded areas. However, lack of color contrasts and the resulting difficulties in interpreting differences in ground cover, require more field checking than is necessary with the color films.

The use of photo analysis for highway planning and construction in lake-bed terrains has been described by Weeden and Harman (5) for an area near the town of Northeast, Erie County, Pennsylvania. A single scene, the second from the shore of Lake Erie, is presented in three formats: color, infrared color, and black-and-white panchromatic, as Fig. 7.13, Color Plate 14. The photographs were taken in May, 1969. Specific features are identified by letter, and a summary of the characteristics of these features may be found in Table 7.7.

(Because the panchromatic photographs were taken almost a month later than the color photographs, there are certain differences in vegetation and soil moisture). The following features unique to lake beds were identified on these photographs:

Landform. Flat to gently undulating terrain surrounded by beach ridges (E) or deltas; slumps and seeps where varved clays occur on slopes.

Drainage. Dendritic patterns of high density developed on clays in flat areas; parallel drainage on uniformly sloping ground; fewer tributaries on drainage channels developed in silt (B).

Erosion. Deep and V-shaped channels in first-order gullies in granular materials (D); deep and U-shaped channels in silts; and shallow and saucer-shaped channels in clays (H).

Vegetation. Alder, aspen, willow, and cattails in undeveloped swampy areas (I); shadows of vegetation noticeably influencing gray tone, especially in cover crops.

Soil. Tonal variations in bare fields related to a change in color from the A to the B soil horizon or a change in moisture content (A); darker tones indicating more moisture (J); mottled tones revealing variable internal drainage typical of silty or sandy

gravels; darker tones near a stream channel indicating that the stream is on clay (C) and lighter tones on fine sands and silts on higher ground (in this case the stream has carried away the sands and silts overlying the clay at C).

Culture. Vineyards and orchards on the better drained soils; row crops and grain on moderately drained soils; pasture on less well-drained soils; undeveloped land on very poorly drained soils (I); vegetable crops raised in muck areas (very dark soil tone); herring-bone pattern of drainage tile (these may or may not be seen, depending on the length of time since the last rain and the nature of the soil cover); highways and railroads following beach ridges.

Other features. Sudden broadening in the width of a stream cut (D) indicating dissection of a beach ridge; isolated bare spots, perhaps indicating rock near the surface; changes in gray tone of water surfaces, depending on the angle between the sun and the camera lens.

PROBLEMS IN UNCONSOLIDATED MATERIALS OTHER THAN LAKE BEDS

During construction and maintenance of highways, unconsolidated materials such as noncohesive soils and colluvium may lead to construction problems and slope stability and erosion difficulties. These difficulties are frequently aggravated by a high water table; swamp, bogs, and marshes are extreme cases of this instability.

Deposits of colluvium, consisting entirely of silt and clay or predominantly of granular materials, may be detected primarily by slope change, since they often form uniform slopes at the base of otherwise relatively steep slopes. These deposits may range in area from a local pocket to several square kilometers. Due to the gentle slope, there is usually little erosion.

Drainage density develops in inverse proportion to the permeability of the underlying material. The surface color and texture of the soil does not necessarily reflect the character of the underlying material, and vegetation and cultural practice are not strong indicators for colluvium.

PROBLEMS IN ROCK

Structure, coherence, and weathering patterns are important rock characteristics in highway en-

TABLE 7.7. CODE LETTER DESCRIPTIONS

Feature	Code Letter
Bare soil	A
Drainage detail	B
Subtle moisture variation	C
Stream cutting through beach ridge	D
Beach ridge	E
Local silty area	F
Film defect	G
U-shaped gully in clay	H
Swampy area	I
High soil moisture	J

gineering. If it must be removed, near-surface bedrock can greatly increase the expense of construction; on the other hand, deep weathering can lead to all the problems generated by unconsolidated materials. Easily weathered, highly jointed, or weak rock (such as shale) can lead to slope instability and foundation problems. Changes in lithology, alternating weak and coherent beds, and concentrations of joints (such as found in fracture traces) cause variations in foundation support which should be known before construction begins. To anticipate foundation problems, a thorough knowledge of the geologic column for the study area is essential. Lithologies in their fresh and weathered forms must be known. Regional analysis of the mosaic is necessary to determine the trends of geologic structures, lineaments, and zones of high fracture concentration (30). (See Chapter 14.)

Slope is an important key to determining structure. Equal slopes indicate horizontal bedding, whereas unequal slopes result from tilted beds. Change in slope on a hillside usually reveals a change in lithology, and trellis drainage often indicates tilted structures. Seeps along a hillside may signify an aquifer lying above a relatively impermeable bed, whereas disappearing streams usually indicate a cavernous carbonate formation.

Erosion and vegetation patterns are frequently characteristic of particular formations. Changes in color, hue, and value, or in gray tone, may indicate a change in lithology or a change in joint frequency. Field patterns are often an indication of lithology: in the Ridge and Valley Section of Pennsylvania, limestone and shale valleys are farmed, whereas sandstones are found in the ridges. A stand of scrub oak may indicate an impure carbonate that weathers to a deep sandy soil, unsuitable for farming.

SPECIAL PROBLEMS

It is impossible to list here the multitude of special problems that may be encountered. Coal mines are one problem creating difficulties for the highway engineer. Deep mines, especially when their extent is not precisely known, can result in subsidence; strip mines leave uncompacted spoil banks. Saturated, uncompacted tailings from washing and preparation of coal are another hazard, and underclays, frequently found beneath coal seams, can provide a water-lubricated stratum leading to slides.

Geologic reports indicate the approximate locations of coal seams and underclays. Stripped areas, spoil banks, and tailings deposits can be seen on USGS topographic maps, county mosaics, and full-scale mosaics. The specific location and extent of deep mines must be determined from mine maps and underground mine reconnaissance, but their presence is indicated by coal-preparation plants and associated features (such as truck facilities and/or railroad spurs). In flat-lying sediments, sandstone landforms have a blocky structure, whereas those in shale, which are usually associated with coal beds, have a softer, more rounded structure. Unless acid-resistant conifers have been planted, vegetation may be sparse or absent on highly acid soils and along streams made acid from sulfides in the coal. On infrared film, areas lacking in vegetation stand in stark contrast to the surrounding terrain and conifers appear dark red-violet rather than magenta, the typical color of other vegetation. In color photographs, some coal-mine polluted streams may appear red.

7.5 SUMMARY

The aerial photograph is the most readily interpreted remotely sensed product. Its quality and relevance to the project at hand is best when optimum mission controls are established by the interpreter. Frequently, however, the interpreter must operate under less-than-optimum conditions. Perhaps here, more than in any other remote sensing effort, the results depend on experience. Photographic interpretation is an attempt to describe the nature of the terrain surface for application to a variety of purposes. For the geologist, vegetative cover is considered as it relates to the underlying soil, whether transported or residual. The properties of a residual soil relate directly to the underlying bedrock, whereas transported soils show the effects of gravity, wind, water, or ice.

Proceeding from the general to the specific, the interpreter first obtains an overall view of the terrain from a photomosaic and prepares a study guide, incorporating information relating to the known features of the study area and their possible appearance on the photography. In study of the individual photographs, landforms are first identified, after which further refinement is obtained by study of drainage patterns, erosion, land use and land cover, cultural practices, and color and/or tonal features relating to moisture. When identifiable, individual microfeatures are noted as clues to the underlying geology and soils. The interpretation process may result in preparation of a reconnaissance map or a map designed to answer a specific set of questions. Frequently interpretation of the terrain can most effectively be shown by a set of map symbols indicating parent material, soil class, surface slope, depth to water table, and the approximate depth of soil cover. Special symbols may also be used, such as a symbol indicating need for drilling or one showing landslides.

Practicing photo analysts and interpreters are as varied in their experience and techniques as the photographs and terrains with which they deal. Some claim the world as their province, creating large-area, small-scale reconnaissance maps suitable for many general applications. Others specialize in specific terrains and applications, such as coastal processes (31), wetland analysis (32), or problems in hydrology (28). The level of mapping detail, however, is strongly correlated with the interpreter's knowledge of the area of study and his experience in interpretation of its unique features. Although the orientation adopted for this chapter has been that of the highway engineering geologist in Pennsylvania (a result of the location and bias of the authors), the principles demonstrated here may be applied to all geographic areas, with the understanding that their specific application will differ significantly from one terrain type to another.

REFERENCES

1. Thompson, M. H., ed., 1965, Manual of photogrammetry, 3rd ed.: Falls Church, Virginia, American Society of Photogrammetry, 1199 p.
2. Lattman, L. E., and Ray, R. G., 1965, Aerial photographs in field geology: New York, Holt, Rinehart and Winston, 221 p.
3. Harker, G. R., and Rouse, J. W., Jr., 1977, Flood-plain delineation using multi-spectral data analysis: Photogrammetric Eng. and Remote Sensing, v. 43, p. 81-87.
4. Stephens, P. R., 1976, Comparison of color, color infrared, and panchromatic aerial photography:Photogrammetric Eng. and Remote Sensing, v. 42, p. 1273-1277.
5. Weeden, H. A., and Harman, J. W., 1970, Manual of highway-problem-oriented photo interpretation using panchromatic, normal color, and infrared color air photos: Report TTSC 7020, Transportation and Traffic Safety Center, The Pennsylvania State University.
6. Anson, A., 1970, Color aerial photos in the reconnaissance of soils and rocks: Photogrammetric Eng., v. 36, p. 343-353.
7. Smith, P. G., Piech, K. R., and Walker, J. E., 1974, Special color analysis techniques, Photogrammetric Eng., v. 40, p. 1315-1322.
8. Committee on Remote Sensing for Agricultural Purposes, 1970, Remote sensing: with special reference to agriculture and forestry: Agricultural Board, National Research Council, National Academy of Sciences, Washington, D.C.
9. Piech, K. R., Gaucher, D. W., and Schott, J. R., 1977, Terrain classification using color imagery: Photogrammetric Eng. and Remote Sensing, v. 43, p. 507-513.
10. Weeden, H. A., and Lattman, L. H., 1971, Air photo analysis of infrared color photography: Report TTSC 7101, Transportation and Traffic Safety Center, The Pennsylvania State University.
11. Bidwell, T. C., 1975, College and university sources of remote sensing information: Photogrammetric Eng. and Remote Sensing, v. 41, p. 1273-1284.
12. Smith, J. T., Jr., ed., 1968, Manual of color aerial photography: Falls Church, Virginia, American Society of Photogrammetry, 550 p.
13. Sabins, F. F., Jr., 1978, Remote sensing principles and interpretation: San Francisco, Freeman, 426 p.
14. Strandberg, C. H., 1967, Aerial discovery manual: New York, Wiley, 249 p.
15. Turinetti, J. E., and Hoffmann, R. J., 1974, Pattern analysis equipment and techniques: Photogrammetric Eng., v. 40, p. 1323-1330.
16. Colwell, R. N., ed., 1960, Manual of photographic interpretation: Falls Church, Virginia, American Society of Photogrammetry, 868 p.
17. Highway Research Board, 1969, Remote sensing and its application to highway engineering, Special Report 102.
18. Weeden, H. A., 1962, Soil mapping for highway engineers: Engineering Research Bulletin B-82, The Pennsylvania State University.
19. Miller, V. C., 1961, Photogeology: New York, McGraw-Hill, 248 p.
20. LaPrade, G. L., 1973, Stereoscopy—will data or

dogma prevail?: Photogrammetric Eng., v. 39, p. 1271-1275.

21. Myers, B. J., 1975, Rock outcrops beneath trees:Photogrammetric Eng. and Remote Sensing, v. 41, p. 515-521.

22. Leopold, L. B., and Maddock, T., Jr., 1973, The hydraulic geometry of stream channels and some physiographic implications: Geological Survey Professional Paper 252, U.S. Government Printing Office.

23. Piech, K. R., and Walker, J. E., 1974, Interpretation of soils: Photogrammetric Eng., v. 40, p. 87-94.

24. Siegal, B. S., and Goetz, A. F. H., 1977, Effect of vegetation on rock and soil type discrimination: Photogrammetric Eng. and Remote Sensing, v. 43, p. 191-196.

25. Mollard, J. E., 1962, Photo analysis and interpretation in engineering geology investigations: a review: Review in Engineering Geology, v. 1, The Geological Society of America, New York, p. 105-128.

26. Leuder, D. R., 1950, A system of designating map-units on engineering soil maps: Highway Research Board Bulletin 28, p. 17-35.

27. Leuder, D. R., 1959, Aerial photographic interpretation, principles and applications: New York, McGraw Hill, 462 p.

28. Weeden, H. A., and Reich, B. M., 1971, Air photo analysis for runoff studies: Final Report for Federal Highway Administration Environmental Design and Control Division, Department of Civil Engineering, The Pennsylvania State University, 173 p.

29. Brown, W. F., 1973, Wetland mapping in New Jersey and New York: Photogrammetric Eng., v. 44, p. 303-314.

30. Lattman, L. H., 1958, Technique of mapping geologic fracture traces and lineaments on aerial photographs, Photogrammetric Eng., v. 24, p. 568-576.

31. Dolan, R., and Vincent, L., 1973, Coastal processes, Photogrammetric Eng., v. 39, p. 255-260.

32. Civco, D. L., Kennard, W. C., and Lefor, M. W., 1978, A technique for evaluating inland wetland photointerpretation: the cell analytical method (CAM): Photogrammetric Eng. and Remote Sensing, v. 44, p. 1045-1052.

CHAPTER 8

SURFACE THERMAL PROPERTIES
ANNE B. KAHLE

8.1 INTRODUCTION

All surfaces emit thermal radiation; the amount depends on the temperature of the body. Thermal remote sensing involves the detection and interpretation of this radiation. At the temperatures of the Earth and the planets most of this radiation lies in the infrared region of the electromagnetic spectrum; hence, we have infrared remote sensing.

Both spectral and thermal properties of the surface materials can be derived from measurements of the thermal radiation. Emission spectra of a surface, like the reflection or transmission spectra, can give diagnostic information on the type of materials present. The subject of spectral signatures has been covered thoroughly in Chapter 2 and will not be discussed further here.

This chapter will concentrate on the thermal properties of the surface material that can be determined by remote sensing. The most readily determined is thermal inertia, but some information may also be inferred about the thermal conductivity, thermal diffusivity, and density. The information is derived by remotely sensing the apparent temperature of the surface, and relating this temperature to the physical processes that have been acting to heat or cool the surface. The temperature response of the surface to the heating is a function of the thermal properties of the surface material.

This chapter first briefly describes the physics of thermal radiation. This is followed by a discussion on the natural heating processes that change the temperature of the surface, and the response of surface materials to this heating. Finally, the techniques that have been developed to interpret the relationship between remotely sensed temperature and surface thermal properties are described.

8.2 THERMAL RADIATION

8.2.1 Black Body Laws

All bodies emit electromagnetic radiation when their temperature is above absolute zero. Their temperature is a measure of the heat energy they possess, in the form of kinetic motion of their constituent molecules. As these particles collide, part of their thermal energy is converted to electromagnetic energy; the higher their temperature, the more electromagnetic radiation they emit.

The various quantities related to electromagnetic radiation are described by a number of terms, many of which are given different meanings by different authors. Those which appear to be most commonly used in remote sensing applications and which will be used in this chapter are defined as follows:

Radiant energy (Q)—energy carried by the electromagnetic radiation, in units of joules or ergs.

Radiant flux (ϕ)—amount of radiant energy which is emitted, transmitted, or received in unit time, in joules per second (watts) or ergs per second.

Radiant intensity (I)—radiant flux per unit of solid angle, in watts per steradian.

Radiant flux density—radiant flux arriving at, crossing, or leaving unit area, in watts per square meter, calories per square centimeter per second, or langleys per minute.[1]

Radiant emission (W)—radiant flux density when emitted by a surface, in watts per square meter, calories per square centimeter per second, or langleys per minute.[1]

Irradiance (E)—radiant flux density when received by a surface, in watts per square meter, calories per square centimeter per second, or langleys per minute.[1]

The quantities defined above refer to the total radiation at all wavelengths. However, all these quantities depend strongly on wavelength. The value of a given variable in a unit wavelength interval is called the "spectral" quantity, that is, the radiant emittance in a unit area in a wavelength interval centered at wavelength λ is called the "spectral radiant emittance," W_λ. In differential form, $W_\lambda = \frac{dW}{d\lambda}$. Conversely, the total radiant emittance could be found by integrating, $W = \int_0^\infty W_\lambda d\lambda$.

The rate at which energy is radiated from a body depends on both its temperature and the type of material of which it is composed. However, if the body were to be a perfect radiator, a black body, then the emitted radiation would be a function of temperature only. Although there are no true black bodies in nature, it is easiest to understand first the thermal radiation laws developed for black bodies.

A black body is an ideal absorber, absorbing all radiation impinging on it. It also is an ideal emitter, transforming kinetic energy to radiant energy at the maximum rate possible, as determined from the laws of thermodynamics. The classical example of an almost perfect black body is a cavity with a very small hole in it. Essentially all radiation falling on the hole will enter the hole and be absorbed by the inner surfaces of the cavity after numerous reflections between the inner surfaces.

If the walls of the cavity are all at a uniform temperature the hole will also emit black body radiation. Black bodies emit radiation as a function of both wavelength and temperature. Planck's law, which relates spectral radiant emittance to temperature, was given in Chapter 2, but is repeated here for convenience:

$$W_{\lambda B} = \frac{c_1}{\lambda^5} \cdot \frac{1}{e^{c_2/\lambda T} - 1} \qquad (8.1)$$

where
B = black body
T = temperature in °K
$c_1 = 2\pi hc^2 = 3.74 \times 10^{-16}$ watt m² = first radiation constant

$c_2 = \frac{ch}{k} = 0.0144$ m °K = second radiation constant

c = 2.99×10^8 m/s = speed of light
h = 6.63×10^{-34} watt sec² = Planck's constant
k = 1.38×10^{-23} watt sec/°K = Boltzmann's constant

Most of the laws describing the behavior of black body radiation were originally based on observations. Planck later derived his law from quantum mechanical principles. The earlier empirical laws were shown to be derivable from Planck's law, and the empirical constants in these laws can be expressed in terms of known physical constants.

Figure 2.3 of Chapter 2 and Fig. 8.1 of this chapter illustrate most of the important characteristics of black body radiation. Figure 2.3 shows the spectral radiant emittance for several different temperatures. Figure 8.1, the general black body radiation curve, shows the relative spectral radiant emittance as a function of λT. In this format the shape of the black body curve is the same for all temperatures, with the spectral radiant emittance being negligible at short wavelengths, rising very sharply to a peak value, and then decreasing more slowly back to a negligible value at longer wavelengths.

The position of the peak of the spectral radiant emittance, λ_{max}, can be found by differentiating Planck's law, to derive Wien's displacement law

$$\lambda_{max} = \frac{2898}{T} \ \mu m \ °K \qquad (8.2)$$

or, rewriting this, $\lambda_{max} T$ = constant. While the peak of Fig. 8.1 is always in the same location, λ_{max} is displaced toward shorter wavelengths as the temperature increases. One quarter of the total power radiated is at wavelengths less than λ_{max}. The power, on the short wavelength side of the peak varies as $T^{6.4}$. The spectral emittance varies as T^n where $n \simeq 15000/\lambda T$.

The total radiant emittance at all wavelengths is represented by the area under the curve in Fig. 8.1. Integrating Planck's law, we find the Stefan-Boltzmann law, which states

$$W = \int_0^\infty W_{\lambda B} d\lambda = \frac{2\pi^5 k^4 T^4}{15c^2 h^3} = \sigma T^4 \qquad (8.3)$$

where

$$\sigma = 5.669 \times 10^{-8} \ watt/(m^2 \cdot °K^4)$$

---

[1] One langley, abbreviated ly, equals 1 cal/cm².

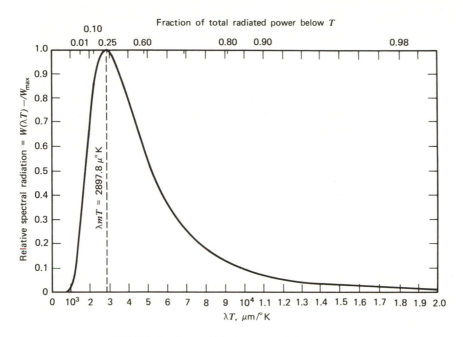

FIGURE 8.1. General black-body radiation curve.

the Stefan-Boltzmann constant. It is apparent from this equation that the total radiant emittance increases very rapidly with temperature.

Real objects are not black bodies, however. When radiation falls on the surface of a real body, only a part of the radiation is absorbed; the rest is reflected or scattered. The absorptivity, α_λ, is defined as the fraction of the incident radiation absorbed at wavelength λ. Kirchhoff, prior to the formulation of Planck's black body law, noted that all bodies emitted an amount of radiation, $\epsilon_\lambda B_{\lambda,T}$, where ϵ_λ, called the emissivity, is just equal to the absorptivity, α_λ, and $B_{\lambda,T}$ is a constant for each wavelength and temperature, *independent of the material of the body*. The quantity $B_{\lambda,T}$ is just equal to the black body spectral radiant emittance, $W_{\lambda B}$, defined in Equation 8.1. Therefore, for a black body, $\alpha_\lambda = \epsilon_\lambda = 1$, for all λ. For *real* bodies, however, Kirchhoff's law,

$$W_\lambda = \epsilon_\lambda W_{\lambda B} \qquad (8.4)$$

combined with Planck's law tells us that

$$W_\lambda = \epsilon_\lambda \frac{c_1}{\lambda^5} \cdot \frac{1}{e^{c_2/\lambda T} - 1} \qquad (8.5)$$

Good absorbers are also good emitters and vice versa. If, at a given wavelength, a body departs a certain amount from black body behavior in absorption, it departs by the same amount in emission.

Some real materials closely resemble black bodies at some wavelengths, but have strong absorption/emission features at others. Other materials, which do not show these strong absorption and emission features, but instead have an overall reduced emissivity that is practically constant over all wavelengths, are called gray bodies. The total radiant emittance from an ideal gray body is $\epsilon \sigma T^4$.

8.2.2 Temperature

When determining the temperature of a body by measuring the radiation emitted by it, some assumptions must be made about its emission properties. The most common technique is to assume that the surface is radiating like a black body. The "apparent brightness," or "equivalent black body" temperature of an object is the temperature of a black body that would emit the same total amount of flux as that measured by the sensor. Most detectors are only sensitive to a portion of the spectrum (see Chapter 3). Thus, for a sensor measuring the flux between 8 and 14 μm, for example, the assumption is made that the quantity measured, W_{8-14}, is equal to

$$\int_8^{14} \frac{c_1 d\lambda}{\lambda^5 \left(e^{c_2/\lambda T} - 1\right)}$$

and the instrument is calibrated to read the appropriate apparent temperature. Because real surfaces emit less flux than a black body at the same temperature, this apparent temperature will always be less than the kinetic temperature of the surface.

Another method of remotely determining a temperature of the body is to determine the radiation at two or more wavelengths and to fit the *shape* of this measured spectral curve to the black body curve. The temperature determined this way is called the "color temperature." If a body were an ideal gray body, the color temperature would be the same as the black body temperature since the spectral curves have the same shape. Also, if a body has strong absorption/emission lines, but the radiation has been measured at wavelengths other than where these spectral features appear, the color temperature will be a fairly good approximation to the kinetic temperature.

The concept of surface temperature is itself somewhat vague, because no real surface material will have a totally uniform temperature. The temperature not only will vary horizontally, but will vary with depth, and some of the radiant flux can originate at depth if the material is partially transparent. Comparisons of the apparent radiation temperature with contact temperature measurements can be used in an attempt to determine emissivity of the surface. However, contact temperature measurements are difficult to make without disturbing the surface and changing its temperature. For this reason, and because the emissivity of natural materials is usually quite high (0.92 to 0.98), the apparent

radiation temperature is usually the best, easily measured approximation to the surface temperature.

Most infrared temperature measurements of the Earth's surface are made in the 8-14 μm region. This is partly because at terrestrial temperatures this spectral region contains most of the emitted flux, while reflected solar radiation in this wavelength region can be ignored. Figure 8.2 shows the black body emission curves for 250° K, 275° K, and 300° K, typical values of the Earth's surface temperature. Also shown are typical values of solar radiation reflected from the surface for two values of albedo and solar elevation. It can be seen that infrared measurements made between 3 and 4 μm could contain significant radiation from both sources. Infrared measurements made at wavelengths shorter than 3 μm during daylight hours would be sensing primarily reflected solar radiation, and no attempt should be made to relate this type of infrared measurement to temperature.

The 8-14 μm spectral region is a very good "window" through the atmosphere. Only minimal corrections need to be made for absorption of the emitted radiation as it passes through the atmosphere to the sensor. These corrections are discussed in Chapters 2 and 3.

The basis of thermal remote sensing is thus the determination of surface temperature by measurement of emitted radiation, taking into account the appropriate corrections for emissivity and atmospheric effects. The remainder of this chapter discusses how one can use this temperature to derive information on the nature of the surface material.

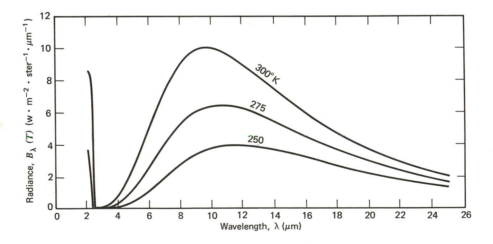

FIGURE 8.2. Blackbody spectral radiance for terrestrial temperatures, and reflected solar radiance for: *A*, Reflectivity = 0.4, solar elevation = 90°. *B*, Reflectivity = 0.2, solar elevation = 30°.

8.3 SURFACE TEMPERATURE VARIATIONS

An infrared remote sensing mission typically produces a profile, an image, or a map of the Earth's or another planet's surface radiant emittance. If instrument calibration, emissivity, and atmospheric corrections are possible, the resulting product can represent the surface temperature. Boundaries between materials of different temperature may be discerned. These differences may be due to topographic effects on surface heating, which may enable very subtle structural features to be seen on infrared images that cannot be seen in conventional aerial photographs (1, 2). The temperature differences can also be due to differences in the thermal properties of the surface materials. To quantify the information contained in such infrared images, the physics of surface heating must be understood, including the various mechanisms acting to heat and cool the surface, and the reaction of the surface material to these heating processes. Surface heating will be discussed here in the context of the Earth. Appropriate modifications can make the discussion equally applicable to the moon and planets.

8.3.1 Surface Heating

A surface will increase or decrease in temperature depending upon whether there is a net gain or loss of heat energy in the surface material. Heat balance equations are based on the law of conservation of energy. Let us consider a column of surface material extending from the surface to a depth where vertical heat transport is negligible or constant. Consider the flux of heat across all the boundaries of this column. Because the temperature of most surface materials varies much more rapidly vertically than horizontally, the lateral transport of heat across the sides of the column is usually negligible. The heat transport across the lower boundary will depend on the depth of the boundary. Diurnal heating usually penetrates to a depth of 1 m or less, while seasonal heating variations penetrate roughly to 20 m. Terrestrial heat flow from depth (including geothermal heat) may add a small constant heat source across the bottom surface of the column. If we consider the bottom of the column to be below the dirunal heating limit, then seasonal heat flow will be added to or subtracted from the terrestrial heat flow. The largest and most variable heat sources are those transferring heat across the

Earth-atmosphere interface. These include solar radiation, atmospheric radiation, surface thermal radiation, sensible heat transfer between the ground and the atmosphere, and latent heat transfer, to be defined below.

The principal source of heat for the Earth's surface is the sun. The amount of solar radiation falling on the surface depends upon the solar elevation (a function of latitude, time of day, and time of year), the topography, the cloudiness, and (to a lesser extent) other atmospheric conditions. The amount of solar radiation absorbed by the surface also depends upon the reflectivity of the surface. The topography effects the amount of sunlight reaching the surface in three ways: The slopes facing toward the sun intercept more radiation than those facing away from the sun; topographic relief combined with low sun angle can cause some surfaces to be shaded, resulting in only atmospherically scattered radiation reaching the surface; and finally, the elevation of a site will be a factor in the amount of atmospheric attenuation of the incoming solar radiation. The incoming solar radiation can be computed (3), including the effects due to sloping surfaces if topographic information is available. However, the presence of clouds makes an accurate calculation of the solar radiation much more difficult, if not impossible.

The long wave radiation at the surface has two components, the upward radiation emitted from the ground and the downward long wave radiation emitted by the atmosphere. This upward long wave radiation removes much of the heat from the surface, particularly in the afternoon when the ground is hot. The upward radiation from the surface can be approximated by $\epsilon \sigma T^4$ if ϵ can be assumed to be constant with wavelength. Otherwise the radiant emittance can be calculated from integration of Equation 8.5.

The long wave radiation downward from the atmosphere is primarily discrete line and band emission from the various atmospheric constituents, particularly water vapor, CO_2, and O_3. Although this radiation is in discrete wavelength bands it often is expressed as "effective" sky temperature, the temperature of a black body that would emit the same flux as is observed. This effective temperature can be on the order of 260°K. It is interesting that a black body at this temperature would emit maximum radiation λ_{max} at 11.15 μm. However, 11 μm is directly in the middle of a very good atmospheric window where there is very little absorption, and consequently also very little emission. Also, 260°K is not a reasonable value for the average sky temperature. This disparity illustrates the danger of trying to interpret an effective or apparent temperature as anything but a substitute for the amount of radiative flux present. The sky radiation depends strongly upon the water vapo present in the lowest level of the atmosphere. Several empirical formulas are available incorporating this moisture into the sky radiation calculation. A typical expression for the nocturnal sky radiation is Brunt's formula (4), $R_{sky} = \sigma T_0^4 (a + b\sqrt{e})$, where the temperature used is the *ground* temperature, e is the atmospheric water-vapor pressure at the surface, and a and b are empirical constants.

The sensible heat, H, is the heat removed from or added to the Earth's surface by the atmosphere through conduction and convection. The heat flow can go in either direction, taking heat from the hotter medium to the cooler. Transfer of sensible heat is the principal mechanism for removing heat from the surface under dry windy conditions. Sensible heat transfer depends upon the ground temperature, the vertical temperature gradient in the lowest few hundred meters of the atmosphere, the wind speed, the atmospheric stability, the elevation, and a ground "roughness" parameter.

The latent heat is the heat removed from the surface by evaporation, or added in the process of dew formation. Latent heat exchange between the surface and the atmosphere is a very important factor in the energy balance of the Earth-atmosphere system. Not only can it be the principal process removing heat from the surface under moist conditions, but surprisingly, the release of this heat in the atmosphere during the formation of rain clouds is the largest single heat source for the atmosphere. Latent heat exchange depends upon all the variables involved in the sensible heat exchange plus the ground moisture and the vertical moisture gradient in the lowest level of the atmosphere. Empirical expressions exist (4, 5, 6) for the calculation of both sensible and latent heat transfer.

The above processes act to heat or cool the Earth's surface. Figures 8.3(a) and 8.3(b), after Sasamori (5), show measured values of the heating terms under wet and dry conditions, respectively. Positive terms indicate heat being transferred away from the surface. Under dry conditions [Fig. 8.3(a)], the transfer of sensible heat, H, is the principal

mechanism for removing heat from the surface, almost balancing, though lagging slightly behind, the solar heating of the surface, which is the major component of the net radiation, R_n, during daylight hours. Under wet conditions [Fig. 8.3(b)], the transfer of latent heat, LW, by evaporation from the surface removes most of the heat. We can infer that the surface remained cooler than the air throughout the day, because sensible heat was transferred to the surface during the entire period of observation.

Figure 8.4 presents some calculated values of these same terms under dry conditions prevailing at Pisgah Crater in the Mojave Desert of Southern California (6). In this figure the signs have been reversed from those in Fig. 8.3 so that now heat transferred toward the surface is positive and heat removed is negative. Also the net radiation, R_n, has been subdivided into solar radiation, S, and long

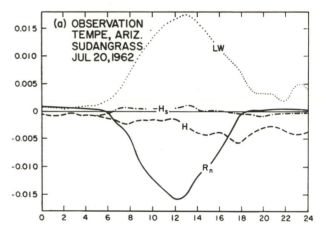

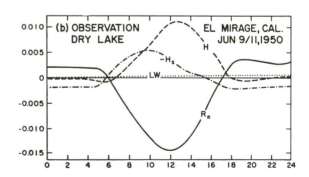

FIGURE 8.3. Energy balance for (a) a wet surface and (b) a dry surface: R_n = net radiative flux, H_s = sensible heat flux, H^n = heat conducted in the soil, LW = latent heat flux [after Sasamori (5)].

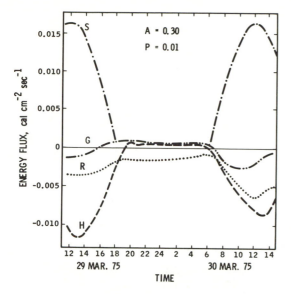

FIGURE 8.4. Energy balance at Pisgah Crater under dry conditions: S = solar flux, R = long wave radiative flux, G = heat conducted in the soil, H = sensible heat (6).

wave radiation, R. The calculation of the heating terms in Fig. 8.4 was based on observed meteorological conditions, with considerably more wind the first day than the second. During the first, windier day, much of the heat was removed by transfer of sensible heat. On the second, less windy day, the surface became hotter and a greater percentage of the heat was removed by transport into the ground, G, and long wave radiation from the surface.

8.3.2 Thermal Properties of Surface Material

The thermal properties of the surface materials determine how the heating from these various heat sources is distributed in the surface layers and how the temperature varies as a function of time and depth in response to the heating. The various thermal properties that are of importance are defined as follows:

Thermal conductivity (K) is the rate at which heat will flow through unit area in unit time in a uniform medium having a temperature gradient of one degree per unit distance. Rocks and soils in general are relatively poor conductors of heat, with values ranging from 0.0012 cal•cm^{-1}•s^{-1}•°K^{-1} for sandy soils to 0.012 for quartzite and dolomite. This compares with 0.0014 for water and 0.00006 for air. The conductivity can be quite variable for a given type of material, and also varies with amount of soil moisture present, with particle size, and as a function of temperature.

The *specific heat (c)* of a substance is the ratio of the amount of heat required to raise the temperature of some mass of the substance by one degree, to the amount of heat required to raise the same mass of water at 15°C by one degree. The specific heat of most surface materials are quite similar: quartzite, 0.17 cal/(gm •°K), and sandy soil, 0.24, being near the extremes of range.

Another physical property of materials that is important to their heating characteristics is the *density (ρ)* with units of grams per cubic centimeter.

Additional thermal properties of interest, which can be derived from the terms already defined, are the *heat capacity (C)*, the *thermal diffusivity (κ)*, and the *thermal inertia (P)*. Heat capacity, defined as $C = \rho c$, is the ratio of the amount of heat required to raise the temperature of a volume of the substance by one degree to the amount of heat required to raise the same volume of water by one degree, again at 15°C. Units are calories per cubic centimeter per degree Kelvin. Thermal diffusivity is defined by $\kappa = K/c\rho$ in units of square centimeters per second. Thermal inertia defined by $P = \sqrt{K \rho c}$, can be thought of as a measure of the resistance of a material to a change of temperature. Thermal inertia is the thermal property most amenable to determination by remote sensing, as will be demonstrated shortly. Typical values of thermal inertia are for quartzite, 0.074 cal•cm^{-2}•°K^{-1}•s$^{-\frac{1}{2}}$, and for sandy soils, 0.024 cal •cm^{-2}•°K^{-1}•s$^{-\frac{1}{2}}$. Typical values of these thermal properties for several common rocks are reproduced in Table 8.1 from Table 4-1, "Thermal Properties of Common Rocks-Handbook Values," in the *Handbook of Remote Sensing* (7).

If the surface layer of the Earth consists of a dry material, either soil or rock, then the temperature anywhere in the layer can be found from the one-dimensional heat conduction equation,

$$\frac{\partial G}{\partial z} = -\rho c \frac{\partial T}{\partial t} = -\frac{\partial}{\partial z}\left(K \frac{\partial T}{\partial z}\right) \qquad (8.6)$$

where

G = heat flux in the ground
z = depth
T = temperature
t = time

TABLE 8.1. THERMAL PROPERTIES OF COMMON ROCKS—HANDBOOK VALUES. VALUES ARE GIVEN IN CENTIMETERS-GRAM-SECOND UNITS FOR 20° C

	K	ρ	c	κ	P
Basalt	0.0050	2.8	0.20	0.009	0.053
Clay soil (moist)	0.0030	1.7	0.35	0.005	0.042
Dolomite	0.012	2.6	0.18	0.026	0.075
Gabbro	0.0060	3.0	0.17	0.012	0.055
Granite (granite rocks)	0.0075 0.0065	2.6	0.16	0.016	0.052
Gravel	0.0030	2.0	0.18	0.008	0.033
Limestone	0.0048	2.5	0.17	0.011	0.045
Marble	0.0055	2.7	0.21	0.010	0.036
Obsidian	0.0030	2.4	0.17	0.007	0.035
Periodotite	0.011	3.2	0.20	0.017	0.084
Pumice, loose (dry)	0.0006	1.0	0.16	0.004	0.009
Quartzite	0.012	2.7	0.17	0.026	0.074
Rhyolite	0.0055	2.5	0.16	0.014	0.047
Sandy gravel	0.0060	2.1	0.20	0.014	0.050
Sandy soil	0.0014	1.8	0.24	0.003	0.024
Sandstone, quartz	0.0120 0.0062	2.5	0.19	0.013	0.054
Serpentine	0.0063 0.0072	2.4	0.23	0.013	0.063
Shale	0.0042 0.0030	2.3	0.17	0.008	0.034
Slate	0.0050	2.8	0.17	0.011	0.049
Syenite	0.0077 0.0044	2.2	0.23	0.009	0.047
Tuff, welded	0.0028	1.8	0.20	0.008	0.032

For a homogeneous material, the conductivity will not vary with depth, so Equation 8.6 can be rewritten

$$\frac{\partial T}{\partial t} = \frac{K}{\rho c}\frac{\partial^2 T}{\partial z^2} = \kappa\frac{\partial^2 T}{\partial z^2}$$ (8.7)

Also from Equation 8.6 it can be seen that the heat flow,

$$G = -K\frac{\partial T}{\partial z}$$ (8.8)

Thus, heat will flow in the soil from the hotter to the colder region at a rate proportional to the thermal conductivity. As the top surface heats up during the day, heat will be conducted downward into the cooler material at depth. When the top surface cools in the afternoon, this heat stored at depth will flow back up toward the surface. The rate of flow of heat, and the temperature distribution in the soil as a function of time (Equation 8.7) will depend on the surface heating, which is the driving mechanism, and also upon the thermal properties of the material.

Equation 8.7 can be solved for the temperature at any time and depth subject to knowledge of the initial conditions, the boundary conditions, and the diffusivity. This is the basis for the thermal models described in the next section.

These calculations have been made for typical initial and boundary conditions. Results of one such calculation are shown in Fig. 8.5, illustrating the diurnal heating wave. Each curve in the figure shows the temperature as a function of depth for a given time of day. Comparison of the curves for successive times throughout the day illustrates how the diurnal heating wave propagates downward and is attenuated. The annual heating wave is similar, but it penetrates to a depth of perhaps 20 m with a comparably larger magnitude. The dirunal temperature wave will be superimposed on the upper meter of the annual wave.

The temperature as a function of depth, at 6:00 A.M. and 2:00 P.M., is shown in Figs. 8.6 through 8.10 for various values of the thermal properties. The initial and boundary conditions were held constant, and the thermal properties were varied one at a time. The effect on the temperature distribution of a change in conductivity is shown in Fig. 8.6, and the effect of a change in specific heat in Fig. 8.7. An increase in conductivity allows the heating wave to penetrate the surface further, with less rise in temperature at the surface. An increase in either the specific heat or the density (the latter not illustrated) will cause more of the heat to be used to raise the local temperature so less is left to penetrate to depth. The dependence of diurnal temperature patterns on the derived quantity diffusivity ($\kappa = K/\rho c$), can similarly be calculated. If diffusivity increases due to an increase in conductivity, then more heat will be conducted to depth, and the diurnal surface temperature variation will decrease as in Fig. 8.6. On the other hand, if the diffusivity increases due to a decrease in specific heat (or density) then the surface temperature range will increase as in Fig. 8.7.

Figures 8.8 through 8.10 show the effect of thermal inertia on the diurnal temperature distribution. In Fig. 8.8 the thermal inertia is varied by changing the conductivity. As thermal inertia increases, the diurnal change in surface temperature decreases. In Fig. 8.9 the thermal inertia is varied by changing the density. Again, for an increase of thermal inertia the diurnal surface temperature change decreases. However, comparing Fig. 8.8 with Fig. 8.9 we see that when thermal inertia increases due to increasing conductivity, more heat is conducted to depth, but when thermal inertia increases due to increasing density, more heat stays near the surface. Of particular interest is Fig. 8.10, which illustrates the effect of keeping the thermal inertia constant while changing the component physical properties. The density was doubled from 3.5 g/m^3 to an unrealistically high value of 7 g/cm^3, while the conductivity was halved from 0.010 cal•s^{-1}•$^\circ$K^{-1}•cm^{-1} to 0.005 cal•s^{-1}•$^\circ$K^{-1}•cm^{-1} More heat is being stored near the surface in the more dense rocks, rather than being conducted to depth. The slight difference between the two values of surface temperature at 6:00 A.M. is caused by

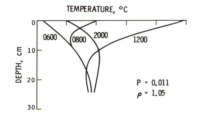

FIGURE 8.5. Temperature as a function of depth for various times of day.

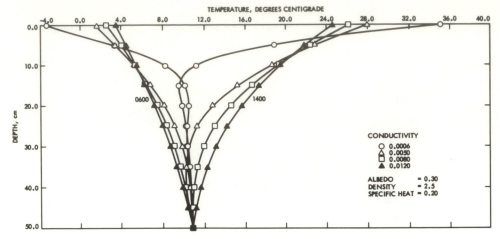

FIGURE 8.6. Temperature as a function of depth and time, for various values of conductivity.

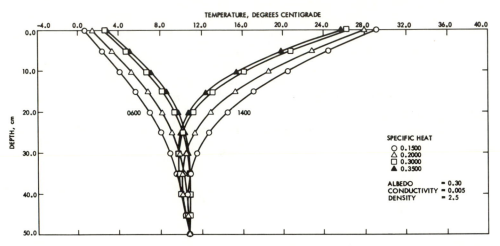

FIGURE 8.7. Temperature as a function of depth and time, for various values of specific heat.

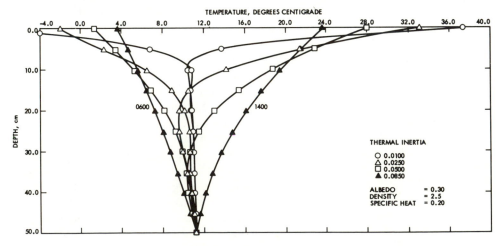

FIGURE 8.8. Temperature as a function of depth and time, for various values of thermal inertia, derived by varying conductivity.

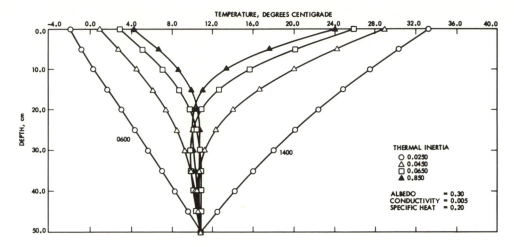

FIGURE 8.9. Temperature as a function of depth and time, for various values of thermal inertia, derived by varying density.

numerical inaccuracies in the calculations. The surface temperatures throughout the day remain essentially equal. Although this effect is not unexpected from the theory, it helps to illustrate graphically the physical significance of thermal inertia. It also illustrates why thermal inertia is the physical property most amenable to remote sensing, since only the *surface* temperature can be determined remotely. To determine any other thermal property of the material, additional information must be available.

The quantitative thermal remote sensing problem is to relate the measured surface temperatures to the thermal properties of the surface material. The

final section of this chapter will deal with the numerical models that have been developed for this purpose.

8.4 THERMAL MODELS

Several numerical models have been developed to relate the observed diurnal surface temperature variations to thermal properties of the surface materials. Most models are based on the same principles. The heat flow equations for the surface materials are solved for temperature as a function of time for a range of values of the thermal properties of the surface materials. The surface heating terms are included as boundary conditions. The models differ in the number of simplifying assumptions made, the form of the boundary conditions and the method of solution. This type of analysis was successfully employed to determine the thermal inertia of the lunar surface (8, 9, 10), giving insight into the nature of lunar surface materials prior to spacecraft landings. These studies have recently been extended to geologic applications on the Earth (6, 11-20). There are also numerous surface heating models or studies, many very sophisticated, which have been developed for meteorological problems (5, 21-33). Some of the more advanced models (5, 33) include the simulation of both heat and moisture transport in the surface layers and across the air-Earth boundary layer. Some of the different types of models that have been applied to geologic problems will be described here.

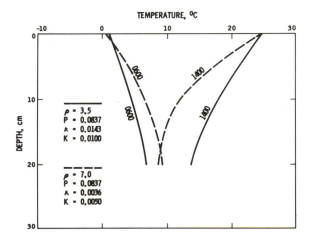

FIGURE 8.10. Temperature as a function of depth and time, for constant thermal inertia, but varying density and conductivity.

Watson (11, 12, 13) assumed one-dimensional periodic heating of a uniform half-space (a region bounded by a plane on its upper side and extending downward to infinity) of constant thermal properties. The temperature obeys the heat conduction equation, Equation 8.7. The solution to this equation for periodic heating of angular frequency ω is

$$T(z,t) = \sum_{n=0}^{\infty} D_n e^{-k\sqrt{nz}} \cos(n\omega t - \epsilon_n - k\sqrt{nz}) \quad (8.9)$$

where D_n and ϵ_n are arbitrary coefficients and $k = \sqrt{\omega/2\kappa}$ is the wave number of the first harmonic (34). The arbitrary coefficients were evaluated by Watson by expressing the surface boundary condition as an energy balance between incoming solar and sky radiation, outgoing ground radiation and conduction into the ground, ignoring other heat transfer mechanisms, including sensible heat and latent heat. He also introduced the effect of a geothermal heat flux into his model.

Figure 8.11 illustrates the diurnal temperature variation of the surface predicted by this model for variations in four important parameters that influence surface temperature: thermal inertia, albedo, geothermal flux and surface emissivity. These curves illustrate another reason why thermal inertia is a significant variable for remote sensing. The day-night crossover in the temperature curves for variations in thermal inertia implies that the day-night change in temperature, ΔT, will be very sensitive to thermal inertia. Surface materials with low thermal

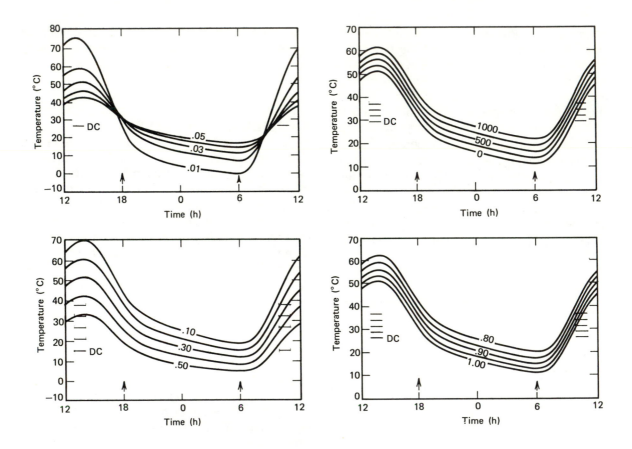

FIGURE 8.11. Diurnal temperature curves for varying (a) thermal inertia [(in cal/(cm²•s½)], (b) albedo (in fractions 1), (c) geothermal heat flux (in HFU), and (d) emissivity (in fractions 1). Times are local solar vlaues. The mean diurnal temperature DC is indicated by a horizontal line for each curve. The vertical arrows near the time axis mark the sunset and sunrise times. Fixed parameter values used in the computation of these curves are: inertia 0.03 cal/(cm² • s½), albedo 0.3, emissivity 1.0, latitude 30°, solar declination 0°, sky radiant temperature 260°, dip 0°, cloud cover 0.2. [After Watson (13)].

inertia will have a large ΔT and materials with high thermal inertia will have a small ΔT. The effect of systematic errors in the measurement of T will be minimized. None of the other parameters demonstrates this crossover effect.

Using Watson's model, Pohn and others (15) generated parametric families of curves of ΔT as a function of thermal inertia for various albedo values. These were combined with data acquired from the Nimbus meteorological satellites (8 km resolution) for part of Oman in the Arabian peninsula to produce a thermal inertia contour map. This map showed fair agreement with the major units on a reconnaissance geologic map. Contous shapes corresponded in some places to geologic contacts. The high thermal inertia of an ophiolite suite, the intermediate thermal inertia of limestone units, and the low thermal inertia of wadi deposits and gravel units made it possible to infer the location of these materials even at 8 km resolution. Pohn later found that some anomalies in the thermal inertia map, which did not correspond to features on the reconnaissance map, did match features on a more detailed geologic map and in a more detailed image obtained by the Landsat spacecraft.

Watson's model requires that the heating terms can be described as periodic functions. The version described here does not explicitly include sensible or latent heat, but values of these heat exchange mechanisms could be included as part of the "effective" radiation terms.

A different model has been developed by Kahle (6), again based on solution of the heat conduction equation subject to the surface heat balance boundary condition. In this model, numerical solution is accomplished by difference equations, and the meteorological heating terms are included. Nonperiodic transient heating can be included. The boundary condition across the Earth's surface used in this model is the surface heat balance equation,

$$S + R + H + L + G = 0 \qquad (8.10)$$

where

S = net solar radiation absorbed by the surface
R = net thermal radiation downward from atmosphere and upward from the surface
H = sensible heat flux between the atmosphere and the ground
L = latent heat flux between the atmosphere and the ground

G = heat flux in the soil

All fluxes are considered to be positive if directed toward the Earth-atmosphere boundary. The lower boundary condition is usually considered to be T = constant, $G = 0$ at some appropriate depth below the level reached by the diurnal heating wave. This boundary condition can be modified to include a constant flux representing a seasonal trend or geothermal heat flow.

Simple formulations for the various heating terms are used, based on various theoretical and empirical expressions available in the literature. A complete description of the calculation of these terms is given in (6).

The surface material is assumed to be homogeneous down through the depth to which the diurnal heating wave penetrates. The lower boundary is chosen to be a depth $z = 50$ cm. This surface layer is subdivided into 1-cm thick layers. At an initial time, $t = 0$, the temperature $T(0, z)$ is defined for each layer based on measured values or theoretical estimates. The temperature at the next time step, $t = 1$, is then computed for each layer from the finite difference form of the heat conduction equation,

$$T(t + 1, z) = T(t, z) + \kappa \frac{\Delta t}{\Delta z^2} \left[T(t, z + 1) - 2T(t, z) + T(t, z - 1) \right] \qquad (8.11)$$

The updated value of the surface temperature $T(t + 1, 0)$, is calculated by solving for the temperature, T_g, that will satisfy the surface heat balance equation (Equation 8.10). This equation can be rewritten in the form

$$T_g^4 + C_1 T_g + C_2 = 0 \qquad (8.12)$$

where C_1 and C_2 are constants determined from prevailing meteorological conditions and site parameters, including albedo, topography, the effective sky temperature, the wind speed, the air temperature, the saturation water mixing ratio, the atmospheric water mixing ratio, a ground moisture parameter, site elevation, and a surface roughness parameter. These terms must all be measured or estimated, and are assumed to be known as a function of time. Details of their computation are given in Ref. (6). Equation 8.12 can be solved numerically for T_g by use of the Newton-Raphson iteration technique (35).

The vertical step size, Δz, is chosen to be 1 cm, because the ground temperature does not vary rapidly over this distance, except perhaps in the top centimeter. Once the vertical step size has been fixed, constraints are placed on the time step size. If the stability criterion

$$\kappa \frac{\Delta t}{\Delta z^2} < 0.5 \qquad (8.13)$$

is not met, the computation rapidly becomes unstable resulting in extremely unrealistic temperature values. For most surface materials, the diffusivity, κ, is in the range of approximately 0.001 to 0.05 cm^2 s^{-1}. Thus, with Δz = 1 cm, Δt must be in the range of about 50 s or less. The model time step is changed to meet this stability criterion as the diffusivity is changed. Because the time steps are small (50 s), satisfactory convergence is achieved after only one or two iterations.

Having thus computed the new value of the surface temperature at t = 1, the subsurface temperatures are computed for the next time step, t = 2, using Equation 8.11. This procedure is repeated throughout the diurnal cycle. The end result of running the model is a complete set of temperature values in the ground, as a function of depth and time.

For a given set of environmental parameters (including the thermal properties of the surface), the model will produce the diurnal variation in the surface temperature. However, in the geological remote sensing problem, one measures the diurnal temperature range of the surface, ΔT, and seeks to determine the surface thermal properties. The inversion of this solution is accomplished by 1) running the model for a sufficiently complete range of values of the thermal properties, thus 2) creating a lookup table that gives the diurnal surface temperature range, ΔT, as a function of thermal inertia, then 3) inverting this table to give thermal inertia as a function of ΔT.

In practice, the meteorological conditions have been assumed to be uniform over the remotely sensed site. The parameters included in Equation 8.12 that are allowed to vary over the site are albedo and topography (which affect the solar heating) and the thermal inertia. The computer model is run, allowing these quantities to vary separately, generating a four-dimensional table of ΔT as a function of albedo, slope, slope azimuth, and thermal inertia. This is inverted to give a four-dimensional lookup table of thermal inertia as a function of albedo,

slope, slope azimuth, and ΔT. The use of ΔT rather than the actual values of the surface temperature minimized errors introduced by ignoring surface emissivity and atmospheric transmissivity.

This model has been applied to a remotely acquired data set from the Pisgah Crater-Lavic Lake area of the Mojave Desert in California. The albedo and ground surface temperature were measured by an 11-channel multispectral scanner on board the NASA NP3A aircraft in a series of predawn and afternoon flights, with concurrent field measurements of the required meteorological variables. A quantitatively reasonable thermal inertia image was created (19) from this data set along with an albedo image (Fig. 8.12).

The dominant feature in both the thermal inertia image [Fig. 8.12(a)] and the albedo image [Fig. 8.12(b)] is the Pisgah basalt flow. This flow, which extends from Pisgah Crater (feature A in the illustrations) south to Lavic Lake (feature B) has a low albedo (0.06-0.13) but a relatively high thermal inertia (0.04-0.08 cal•cm^{-2}•°K^{-1}•s$^{-1/2}$). It is thus dark in the albedo picture and light in the thermal inertia picture. Lavic Lake, a clay playa, has a high albedo (0.35-0.50) but a lower thermal inertia (0.01-0.03 cal•cm^{-2}•°K^{-1}•s$^{-1/2}$), so it is light in the albedo picture and dark in the thermal inertia picture. These computed values of thermal inertia are typical of the measured values for these materials.

A strong correlation exists in these images between areas with low albedo (dark) and high thermal inertia (light). The correlation is not an artifact of the model, but a coincidence at this particular site because of the type of rock and soil present. Had the site contained high-albedo granite instead of dark basalt, for instance, the thermal inertia picture would be essentially unchanged (the thermal inertia for granite is about 0.06 cal•cm^{-2}•°K^{-1}•s$^{-1/2}$) but the albedo picture would be much different.

Deviation from the negative correlation between albedo and thermal inertia is best illustrated at location D, where there was an area of basalt cinders used as road fill. These cinders have both a low albedo and a low thermal inertia. Other cinders, especially in the flat area immediately northwest of the cinder cone, likewise have both low albedo and low thermal inertia, and can be readily distinguished from the basalt flows, which have low albedo but high thermal inertia. In this image, which is not topographically compensated, the topographic effect on the solar heating gives incorrect values for

Thus thermal insertia information can be used in distinguishing between materials of similar reflective properties but different thermal properties, and in assisting in the identification of surface materials covered with a thin layer of foreign matter, especially wind-blown sand or dust or weathering products such as desert varnish. It is clear that, by itself, thermal inertia information is not sufficient to allow identification of surface materials, but when interpreted in conjunction with other remotely sensed data it can be of considerable value in resolving ambiguities.

The models discussed so far have a thermally homogeneous surface medium. However, there is often a great deal of variability of thermal properties in the top meter or so of the soil or rock. The cause can be either actual variation in soil or rock material or a variation in the moisture content with depth. Both the heat flux equation and the thermal conductivity equations must be modified when considering temperature changes in the presence of such inhomogeneities. In addition, the moisture flux in the soil in both liquid and vapor form must be considered. It is possible, however, to model the combined heat and moisture fluxes in a consistent way (5). One such model has been constructed by Rosema (33). He used a difference equation formulation to solve the coupled one-dimensional heat conduction and soil moisture transport equations. The equations define how the temperature and water content of the surface layers change with time when water is present in the soil in both liquid and vapor form and can move up and down and evaporate and recondense elsewhere, with resultant heat transport.

Rosema's model has been used to increase the understanding of the physical processes occurring under wet soil conditions. It has yet to be applied to a remotely acquired data set to determine soil thermal properties.

8.5 SUMMARY

Thermal remote sensing can be used to determine various thermal properties of surface materials. These include thermal conductivity, thermal diffusivity, density, and, most readily, thermal inertia. Data is gathered by remotely sensing the apparent surface temperature and relating it to the surface, the temperature response of which is a function of the thermal properties of the surface material.

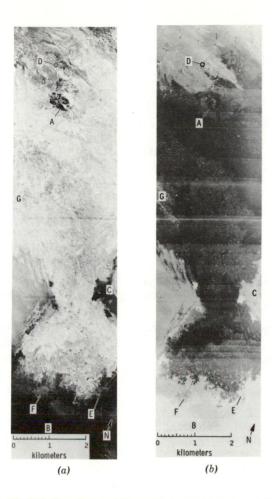

FIGURE 8.12. Pisgah basalt flow, (a) thermal inertia and (b) albedo.

thermal inertia at the cinder cone itself. In subsequent work, Gillespie and Kahle (20) have used digitized topographic information to obtain reasonable values of thermal inertia for the cinder cone.

Examination of the two images at several locations where one material is overlain by a thin layer of a different material (locations E, F, G) shows that the resultant value of the thermal inertia is either intermediate between the two or, for a very thin surface layer, resembles the thermal inertia of the underlying material. Locations E and F show some examples of intermediate values of thermal inertia at the boundary between the lava flow and the playa, where an outwash of basalt a few centimeters thick overlies the playa. At location G, where a very thin layer of sand has blown across the basalt, the thermal inertia is representative of the underlying basalt.

Numerical modeling techniques have been developed to relate measured surface temperatures to the thermal properties of surface materials and to interpret these relationships. Of particular importance is the application of such techniques in using thermal inertia information to distinguish between materials of similar reflective properties (albedo) but different thermal properties, and to assist in identifying surface materials covered with a veneer of foreign matter, particularly wind-blown sand or dust or weathering products such as desert varnish. Thermal inertia information alone is insufficient to allow identification of surface materials, but when supported by other remotely sensed data it can be of great value in resolving ambiguities.

REFERENCES

1. Rowan, L. C., Offield, T. W., Watson, K., Cannon, P. J., and Watson, R. D., 1970, Thermal infrared investigations, Arbuckle Mountains, Oklahoma: Geol. Soc. Amer. Bull., v. 81, p. 3549-3561.
2. Offield, T. W., 1975, Thermal-infrared images as a basis for structural mapping, Front Range and adjacent plains in Colorado: Geol. Soc. Amer. Bull., v. 86, p. 495-502.
3. Kondratyev, K. Ya., 1969, Radiation in the Atmosphere: New York, Academic Press, 912 p.
4. Sellers, W. D., 1965, Physical climatology: Chicago, Illinois, Univ. of Chicago Press, 272 p.
5. Sasamori, T., 1970, A numerical study of atmospheric and soil boundary layers: Jour. Atmos. Sci., v. 27, p. 1122-1137.
6. Kahle, A. B., 1977, A simple thermal model of the Earth's surface for geologic mapping by remote sensing: Jour. Geophys. Res., v. 82, p. 1673-1680.
7. Janza, F. J., 1975, Chap. 4, Interaction mechanisms, in Reeves, R. G., ed., Manual of remote sensing: Falls Church, Virginia, American Society of Photogrammetry, p. 75-180.
8. Wesselink, A. F., 1948, Heat conductivity and nature of the lunar surface material: Bull. Astron. Institutes Netherlands, v. X, p. 351-363.
9. Jaeger, J. C., 1953, Conduction of heat in a solid with periodic boundary conditions, with an application to the surface temperature of the moon: Proc. Cambridge Phil. Soc., v. 49, no. 2, p. 355-359.
10. Sinton, W. M., 1962, Chap. 11, Temperatures on the lunar surface, in Kopal, Z., ed., Physics and astronomy of the Moon: New York, Academic Press, p. 407-428.
11. Watson, K., 1971, A computer program of thermal modeling for interpretation of infrared images: Rep. PB 203578, U.S. Geological Survey, Washington, D.C., 33 p.
12. Watson, K., 1973, Periodic heating of a layer over a semi-infinite solid: Jour. Geophys. Res., v. 78, p. 5904-5910.
13. Watson, K., 1975, Geologic applications of thermal infrared images: Proc. IEEE, v. 63, p. 128-137.
14. Watson, K., Rowan, L. C., and Offield, T. W., 1971, Application of thermal modeling in the geologic interpretation of IR images, in Proceedings of the seventh international symposium on remote sensing of environment: Ann Arbor, Michican, Environmental Research Institute of Michigan, v. 3, p. 2017-2041.
15. Pohn, H. A., Offield, T. W., and Watson, K., 1974, Thermal inertia mapping from satellite—discrimination of geologic units in Oman: Jour. Res. U.S. Geological Survey, v. 2, p. 147-158.
16. Lyon, R. J. P., 1974, Field mapping determinations: ground support for airborne thermal surveys with application to search for geothermal resources: Stanford Remote Sensing Lab. Technical Rept. 74-8, 50 p.
17. Marsh, S. E., 1975, The feasibility of satellite thermal infrared sensing for geothermal resources: Stanford Remote Sensing Lab., Tech. Rept. No. 75-12, 71 p.
18. Kahle, A. B., Gillespie, A. R., Goetz, A. F. H., and Addington, J. D., 1975, Thermal inertia mapping, in Proceedings tenth international symposium on remote sensing of environment: Ann Arbor, Michigan, Environmental Research Institute of Michigan, p. 985-994.
19. Kahle, A. B., Gillespie, A. R., and Goetz, A. F. H., 1976, Thermal inertia imaging: a new geologic mapping tool: Geophys. Res. Let., v. 3, p. 26-28.
20. Gillespie, A. R. and Kahle, A. B., 1977, The construction and interpretation of a digital thermal inertia image: Photogrammetric Eng. and Remote Sensing, v. 43, No. 8, p. 983-1000.
21. Estoque, M. A., 1963, A numerical model of the atmospheric boundary layer: Jour. Geophys. Res., v. 68, p. 1103-1113.
22. Manabe, S., Smagorinsky, J., and Strickler, R. F., 1965, Simulated climatology of a general circulation model with a hydrologic cycle: Mon.

Weather Rev., v. 93, p. 769-798.

23. Myrup, L. O., 1969, A numerical model of the urban heat island: Jour. Applied Meteor., v. 8, p. 908-918.

24. Gadd, A. J., and Keers, J. F., 1970, Surface exchanges of sensible and latent heat in a 10-level model atmosphere: Quart. Jour. Royal Meteor. Soc., v. 96, p. 297-308.

25. Delsol, F., Miyakoda, K., and Clarke, R. H., 1971, Parameterized processes in the surface boundary layer of an atmospheric circulation model: Quart. Jour. Royal Meteor. Soc., v. 97, p. 181-208.

26. Gates, W. L., Batten, E. S., Kahle, A. B., and Nelson, A. B., 1971, A documentation of the Mintz-Arakawa two-level atmospheric general circulation model: The Rand Corp., Santa Monica, CA, R-877-ARPA, 408 p.

27. Zdunkowski, W. G., and Trask, D. C., 1971, Application of a radioactive-conductive model to the simulation of nocturnal temperature changes over different soil types: Jour. Applied Meteor., v. 10, p. 937-948.

28. Deardorff, J. W., 1972, Parameterization of the planetary boundary layer for use in general circulation models: Mon. Weather Rev., v. 100, p. 93-106.

29. Outcalt, S. I., 1972, The development and application of a simple digital surface-climate simulation: Jour. Applied Meteor., v. 11, p. 629-636.

30. Bhumralkar, C. M., 1974, Numerical experiments on the computation of ground surface temperature in an atmospheric circulation model: The Rand Corp., Santa Monica, California, R-1511-ARPA, 52 p.

31. Jacobs, C. A., Randolfo, J. P., and Atwater, M. A., 1974, A description of a general three-dimensional numerical simulation model of a coupled air-water and/or air-land boundary layer: Center for the Environment and Man, Inc., Hartford, Connecticut. Rept. No. 4131-509A, 85 p.

32. Rosema, A., 1975, Chapter 8, Simulation of the thermal behavior of bare soils for remote sensing purposes, in de Vries, D. A., and Afgan, N. H., eds., Heat and mass transfer in the biosphere: Washington, D.C., Scripta Book Co., p. 109-123.

33. Rosema, A., 1975, A mathematical model for simulation and the thermal behavior of bare soils, based on heat and moisture transfer: NIWARS-publication No. 11, 3 Kanaalweg, Delft, The Netherlands.

34. Carslaw, H. S., and Jaeger, J. C., 1959, Conduction of heat in solids, 2nd ed.: New York, Oxford Univ. Press, 510 p.

35. Henrici, P., 1964, Elements of numerical analysis: New York, Wiley, 336 p.

CHAPTER 9

INTERPRETATION OF THERMAL INFRARED IMAGES

FLOYD F. SABINS, JR.

9.1 INTRODUCTION

This chapter describes the acquisition and interpretation of data in the thermal infrared portion of the electromagnetic spectrum which ranges from 3 to 14 μm in wavelength. For geologic purposes, the 8- to 14-μm band is generally employed because this spans the radiant power peak of the Earth at its ambient temperature of 300°K. Radiant temperature is recorded as spot data or as profiles along aircraft flight lines by nonimaging devices called IR radiometers. Imagery is acquired by airborne optical-mechanical scanners.

This chapter describes the interpretation of thermal IR images for geologic mapping, monitoring of volcanism, and recognition of geothermal activity.

9.2 ACQUIRING THERMAL INFRARED DATA

9.2.1 Radiometers

Infrared radiometers measure radiant temperature by means of an IR-sensitive detector and a filter. In a typical radiometer (Fig. 9.1), a thermistor-controlled, electrically heated black body provides an internal calibration source (1). Radiation from the target and from the black body are alternately directed onto the detector by a rotating chopper blade. The signal difference between the target and the internal calibration source is converted into target temperature for display, with a sensitivity of approximately 0.2°C. The field of view, or spot size, of a typical radiometer is determined by the optical system and typically is either 2° or 20°. Radiant temperatures may be measured in the field by portable radiometers equipped with battery packs.

9.2.2 Airborne IR Scanners

Airborne IR scanners (2) consist of three basic components: (1) optical-mechanical scanning system, (2) thermal IR detector, and (3) image re-

cording system. The scanning system consists of an electric motor mounted in the aircraft with the rotating shaft oriented parallel with the aircraft fuselage and the flight direction (Fig. 9.2). A front-surface mirror with a 45° facet is mounted on the end of the shaft and sweeps the terrain at a right angle to the flight path. Infrared radiation from the terrain is reflected onto a detector, producing an electrical signal that varies proportionally with the intensity of IR radiation. In the simplest scanners of the mid-1960's the amplified signal modulates the intensity of a small light source. The image of the light source is swept across a strip of recording film by a second mirror rotating synchronously with the scanning mirror. The recording film advances at a rate proportional to the aircraft ground speed and altitude, so that each scan line on the ground is represented by a scan line on the film. In the late 1960s, direct film recording was largely replaced by recording on magnetic tapes, which were played back in the laboratory to produce optimum images (3). Recently developed scanner systems provide a supplemental paper copy of the image continuously during the flight to provide a valuable in-flight monitor of system performance and a check on aircraft navigation.

In some scanners the faceted mirror rotates about a vertical axis to generate a conical sweep of the terrain. A segment of the circular ground path in advance of the aircraft is recorded as an image, which is called "Forward-Looking IR (FLIR)" imagery. This pattern provides a constant path length, resulting in a constant spot size on the ground, in contrast to the horizontally rotating scanner in which the path length at either edge of the scan line is longer than the vertical path directly beneath the aircraft.

For geologic investigations the 8- to 14-μm wavelength region is generally preferred because it includes the radiant power peak of the Earth. The

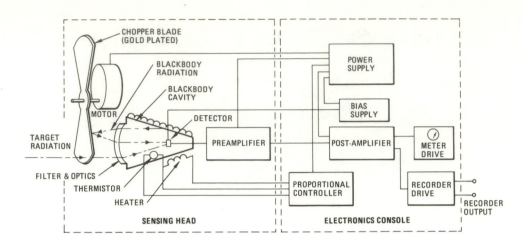

FIGURE 9.1. Schematic diagram of an IR radiometer. As the chopper blade rotates, the detector alternately receives radiation from the target and from the black-body temperature standard From Moxham (1). Reproduced by courtesy of Barnes Engineering Company.

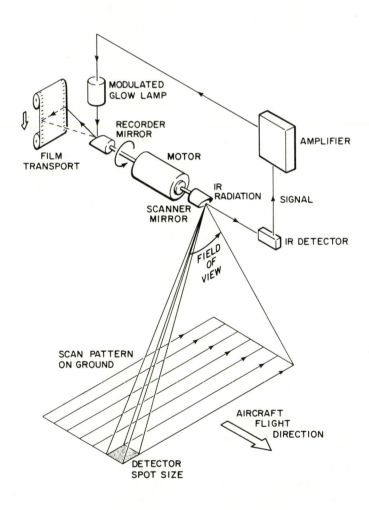

FIGURE 9.2. Airborne IR scanner system. After Sabins (2).

radiation is detected with a mercury-cadmium-telluride detector. At night this detector is generally unfiltered, because its response range closely matches the 8- to 14-μm atmospheric window and a filter attenuates incoming IR radiation. The spot size, or instantaneous field of view, of a detector is typically 2 to 3 milliradians (mrad), which corresponds to a circle 2 to 3 m in diameter at a distance of 1000 m. Thermal sensitivity is typically 0.1°C.

9.3 FACTORS INFLUENCING IMAGERY

The appearance of ground features on thermal IR images is determined by numerous system and physical factors, some of which were described in the preceding section and in Chapters 3 and 8. Other factors, such as moisture content and the time at which images are acquired also influence the appearance of ground features on images. Before discussing these factors, it is emphasized that

IR scanners and radiometers detect radiant temperature. Recall that the radiant temperature of a surface is proportional to the product of its emissivity to the ¼ power times the kinetic temperature (in degrees Kelvin). It is impossible to distinguish radiant temperature differences caused by emissivity variations from those caused by differences in kinetic temperature of the target.

The effect of emissivity differences on radiant temperature is illustrated in Fig. 9.3 for an aluminum block with a uniform internal, or kinetic, temperature of 10°C (283°K). The half of the block that is painted dull black has an emissivity of 0.97, resulting in a radiant temperature of approximately 8°C, whereas the polished surface, with an emissivity of 0.06, has a radiant temperature of -133°C. Metallic objects, such as aircraft and buildings with metal roofs, have low radiant temperatures because of their low emissivity. For typical geologic materials the value of emissivity has an approximate range from 0.80 to 0.95 [See Ref. (4)].

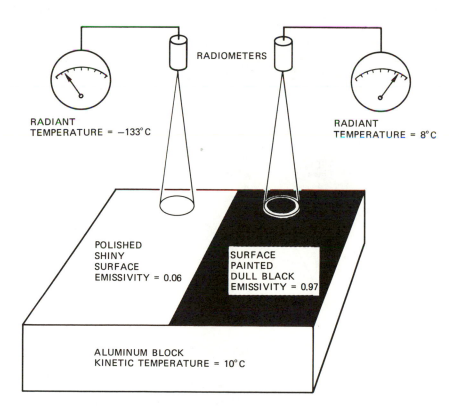

FIGURE 9.3. Effect of emissivity differences on radiant temperature (5). Kinetic temperature of aluminum block is uniformly 10°C. Different emissivities cause different radiant temperatures. From Remote Sensing: Principles And Interpretation by Floyd F. Sabins. W. H. Freeman and Company. Copyright© 1978.

9.3.1 Water and Vegetation Influence

On nightime thermal IR images standing water has a warm signature relative to the adjacent ground, whereas on daytime imagery water is cool and the ground is relatively warm. Sabins (5) pointed out that the thermal inertia of water (0.037 cal •cm^{-2}•s$^{-1/2}$•°C^{-1}) is lower than that of most geologic materials; therefore these day and night signatures cannot be explained by differences in thermal inertia. K. Watson (quoted in Sabins, 5) noted that convection in water bodies can maintain a relatively warm surface temperature at night. In contrast to standing water, damp terrain appears cool on images because of evaporative cooling.

Green deciduous vegetation appears warm on nighttime imagery because of the high water content. During the day transpiration of water vapor lowers leaf temperature, causing vegetation to appear cool relative to the surrounding soil. The relatively high nighttime and low daytime radiant temperatures of conifers, however, do not appear to be related to their water content; the composite emissivity of the needle clusters of such trees approaches that of a black body. Dry vegetation, such as crop stubble in agricultural areas, appears warm on nighttime imagery in contrast to bare soil which is cool. Dry vegetation insulates the ground to retain heat, resulting in its warm nighttime signature.

9.3.2 Diurnal Temperature Variations

Typical diurnal, or daily, variations in radiant temperature for different materials are shown in Fig. 9.4. The most rapid temperature changes occur near dawn and sunset. At the times when the radiant temperature curves of different materials intersect (thermal crossover) there is no temperature difference between them. For geologic reconnaissance of large areas, the predawn hours are optimum because of the relatively stable temperatures. To determine the thermal properties of local areas, repeated flights during periods of maximum temperature difference are used.

9.3.3 Time of Day

As a result of differential solar heating and shadowing, topographic features are emphasized on thermal IR images acquired during daylight hours. The effect of topography is reduced or absent on images acquired during predawn hours, as illustrated in Fig. 9.5 showing the Caliente Range and Carrizo Plains, California (6). On the daytime image, differential solar heating and shadowing clearly define ridges and canyons in the Caliente Range and the small hill (Feature 9) in the Carrizo Plains. On the predawn image, topographic features, other than stream channels, are obscure and

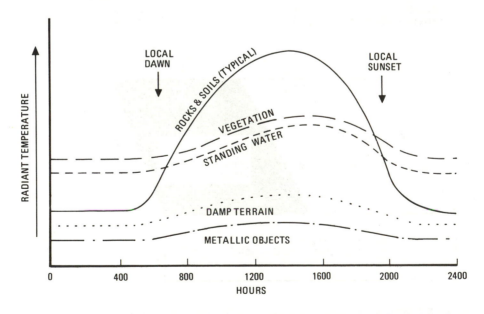

FIGURE 9.4. Diurnal radiant temperature curves (diagrammatic) for typical materials. From Remote Sensing: Principles and Interpretation by Floyd F. Sabins. W. H. Freeman and Company. Copyright © 1978.

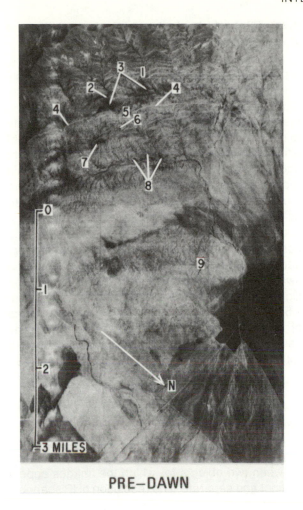

PRE–DAWN   **POST–SUNRISE**

FIGURE 9.5. Comparison of day and night thermal IR images (8 toμ m) of caliente Range and Carrizo Plain, California (6). See text for explanation of annotation.

geologic features are emphasized as illustrated at localities 1, 4, and 6, which correspond to sandstone and basalt outcrops ("warm" expression), and localities 7 and 8, which correspond to outcrops of mudstone with abundant expanding clay ("cool" expression). Predawn and postsunrise images of the Carrizo Plains are similar, because there is little topographic relief to cause differential solar effects.

In the Arbuckle Mountains of Oklahoma, Rowan and others (7) have also shown that predawn imagery is preferable for identifying rock types and mapping fracture zones. On high-altitude thermal IR imagery of the Colorado Front Range, flown two hours after sunrise, Offield (8) noted that the differential heating and shadowing effects aided recognition of structural features having subtle topographic expressions. This application is analogous to low-sun-angle photography.

Maximum thermal contrasts between materials generally occur near sunset, but radiant temperatures and temperature contrast decrease as the night progresses. As discussed in Section 9.3.2, the most stable radiant temperatures occur in predawn hours. If a number of images are to be flown, more uniform results will be obtained from predawn flights. The mapping of thermal inertia is an operation requiring daytime, as well as nighttime, imagery for the same area.

9.4 CONDUCTING SUCCESSFUL IR SURVEYS

Thermal IR imagery is generally not available for most regions but must be acquired from aerial contractors who can provide these specialized services. The following general guidelines are intended to aid in planning, flying, and evaluating an IR survey for geologic applications.

9.4.1 Type of Survey and Aircraft

Infrared surveys may be single flight-line coverage, or mosaic coverage with parallel sidelapping image strips. Single flight-line coverage over a small test site is simple to plan and conduct and may be repeated at different times of the day to evaluate diurnal thermal variations. Mosaic coverage, employed for reconnaissance surveys, requires proper positioning of flight lines to obtain parallel sidelapping image strips, a difficult task for nighttime surveys.

Successful IR surveys can be conducted using aircraft as small as a single engine, 6-passenger Cessna 206, or as large as a twin turboprop, 55-passenger Hawker Sidley 748. The minimum requirement is that the aircraft accommodate the pilot, navigator, equipment operator, and several hundred pounds of equipment. An open camera port in the belly of the aircraft is essential for mounting the scanner.

9.4.2 Flight Line Orientation and Length

For geologic projects, flight lines should be oriented parallel or at an acute angle to the regional structural strike. This orientation avoids masking of linear features by scan line patterns.

The maximum length of flight lines for tape recording systems is determined by recording time per tape. One commonly used system has a 15-min recording time which, at an aircraft ground speed of 300 kph, permits a maximum line length of 75 km. Aircraft speed and effects of possible headwinds must be taken into account in estimating line length. New tape recorder systems provide an hour of continuous recording and allow for longer flight lines.

9.4.3 Flight Altitude, Image Scale, and Ground Coverage

Flight altitude determines image scale and lateral ground coverage. Flight altitude is meaningful only when expressed as height above average terrain. Although terrain elevation generally varies along the flight line, it is impractical to alter aircraft altitude to maintain a constant height above terrain. A flight altitude is selected that provides optimum height above the average terrain elevation. For geologic reconnaissance, imagery acquired 2000 m above terrain is generally optimum. This is not advocated as a standard elevation of all imagery, but it is a good tradeoff among the many factors that must be considered. At 2000 m above terrain, imagery acquired at a 120° scan angle covers a ground swath 7 km wide, or 3.5 km on either side of the flight path. By spacing flight lines 5 km apart, 1.5 km of sidelap is obtained for adjacent strips of imagery. This allows a margin of error for any navigation problems and covers the project area without flying an excessive number of lines (9).

9.4.4 Resolution and Information Content of Imagery

Image quality is often expressed in terms of resolution, which is the minimum ground separation between two objects for which the images appear distinct and separate. The resolution of IR scanners is determined by spot size of the detector and typically ranges from 2 to 3 mrad. At a height of 1000 m, 1 mrad amounts to 1 m of separation. For example, a 3-mrad detector at 2000 m above terrain produces imagery with ground resolution of 6 m along the center line. Resolution becomes lower (poorer) toward the margins of the imagery because the greater slant distance results in a larger spot size. Resolution alone may not be a valid measure of the quality or useful information content of imagery; detectability and recognizability must also be taken into account (10).

9.4.5 Nighttime Navigation

The problem of navigating a pattern of parallel flight lines in the dead of night can only be appreciated after one has actually attempted such a mission. The two phases of the mission are to fly from the base air strip to a starting point in the survey area and then fly the mosaic pattern. A typical nighttime survey at 2000 m above terrain requires a series of parallel lines 70 to 80 km long and spaced 3 to 4 km apart.

Different aircraft navigation methods are available for nighttime IR surveys and have been described by Sabins (9). Inertial navigation systems are probably best, but are very expensive and not available on most survey aircraft. Very low frequency (VLF) radio navigation systems are less expensive and have been successfully employed on a number of thermal IR surveys (9). The capability of viewing imagery in-flight is a valuable check on the navigation.

9.4.6 Field Measurements

Information on weather, soil moisture, and vegetation conditions at the time of an IR survey may facilitate interpretation even though ground crews can make only a limited number of measurements during the four to six hours of a survey. The measurements rarely coincide with anomalous areas on images, where such measurements would be most useful. Field excursions after images have been interpreted are essential, and useful measurements may be after the airborne surveys. Radiometer measurements provide information that may be useful for understanding image signatures.

9.5 IMAGE INTERPRETATION FOR GEOLOGIC MAPPING

Thermal IR imagery can be interpreted using conventional photogeology techniques with an understanding of the physical properties of earth materials, primarily surface thermal properties. Consideration must also be given to the instrumental and environmental factors that affect data acquisition and display. The following section presents three case examples of thermal IR image interpretation, and calls attention to specific characteristics of geologic features on IR images.

9.5.1 Indio Hills, California

The Indio Hills, located in the eastern Coachella Valley, are a ridge of deformed clastic sedimentary rocks of late Tertiary age that trends southeast parallel with the San Andreas fault zone (Fig. 9.6). Bedrock is well exposed and vegetation is sparse (11).

Two types of bedrock can be distinguished on the thermal IR image [Fig. 9.7(b)]. One type consists of poorly stratified, moderately to poorly consolidated conglomerate, with minor sandstone and silt-

stone that has a relatively warm and uniform signature, as illustrated in the lower center of the image on the far outlying hill (location 1.2, D.5 to F.5). The second type of bedrock is the Palm Spring formation which consists of well-stratified alternating beds of resistant conglomeratic sandstone and nonresistant siltstone up to 12 m thick that weather to a ridge and slope topography. On the nighttime image the sandstone beds have a warm signature and the siltstone beds have a cool signature. Alluvium appears cool and featureless.

Geologic structure is also well expressed. Plunging anticlines and synclines in the Palm Spring formation are indicated by arcuate patterns of the alternating warm and cool bands. The trough of one syncline [locality 3.0, D.5 of Fig. 9.7(b)] is marked by a distinct tonal anomaly, cool on the south flank and warm on the north. The linear interface coincides with the axis of the syncline. Careful field checks show no faulting or fracturing, nor is there any apparent stratigraphic or topographic cause of the anomaly. The cause remains enigmatic.

The San Andreas fault borders the west side of the Indio Hills; to the south it passes along the east side of the warm-appearing outlier of poorly stratified Ocotillo conglomerate. Farther south the fault trace is concealed by alluvium and has no topographic expression. On the aerial photograph [Fig. 9.7(a)] the alluvium-covered trace of the fault is marked on the east side by denser vegetation that abruptly terminates at the fault trace (1.5, D.5 to 1.2, G.5). On the IR image the fault trace is clearly expressed by an alignment of very cool (dark) anomalies along the east side (1.4, F.7 to 1.5, I.0). The cool anomalies are not related to vegetation distribution, for vegetation appears warm on nighttime imagery. The cool anomalies are probably related to the barrier effect of the San Andreas fault on groundwater movement. In the spring of 1961, a few months before the infrared survey, Cummings (12) reported a 15 m difference in elevation of the water table across the fault in this vicinity. The shallower water table and higher moisture content on the east side of the fault may cause evaporative cooling Similar anomalies appear on nighttime imagery of the San Andreas fault in the Carrizo Plains 320 km to the northwest, and in the northern Indio Hills (11) where the Mission Creek fault is indicated by cool anomalies at Thousand Palms Oasis.

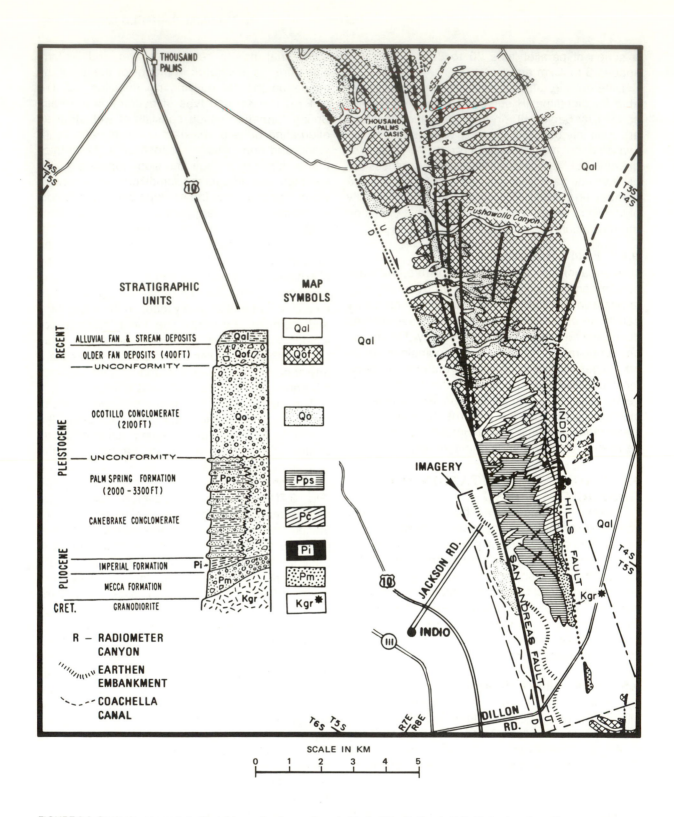

FIGURE 9.6. Geologic map and stratigraphic section for south part of Indio Hills, California (11). Notice location of imagery coverage.

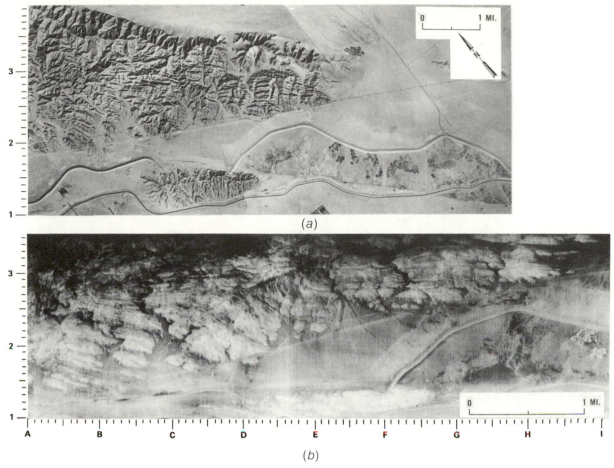

FIGURE 9.7. Indio Hills, Riverside County, California (11) (a) Aerial photograph. (b) Nighttime IR image (8 to 14 μm).

In January 1969 a portable radiometer was used in the field to obtain temperature measurements on eight pairs of sandstone and siltstone outcrops [locality 2.9, E.2 in Fig. 9.7(*b*)] at midnight, predawn, immediate postdawn, and daytime. These measurements are plotted in Fig. 9.8. From midnight through postdawn the sandstones are warmer than siltstones; when the canyon is fully illuminated (0835 h) the siltstones are considerably warmer than the sandstones. The field localities were selected for uniformity of exposure, but the association of sandstones with ridges, and siltstones with slopes suggested that topography might control the radiant temperature. However, diurnal radiometer measurements of large slabs of sandstone and siltstone placed on the ground matched those made at the outcrops and correlated with the nighttime image signatures. This experiment established that rock type, rather than topography, is respon-

sible for the thermal expression of the Palm Spring formation.

The sparse vegetation has a conspicuous warm signature on the night time imagery as shown by large patches of mesquite, green at the margin but dead in the center that appear as irregular bright rings at locality 2.0, H.8. A diurnal sequence of radiometer measurements were made at locality 2.6, H.0 where three salt cedar trees are distinctly warmer than the surrounding soil. Beginning at noon, radiometric temperatures were measured for the salt cedars and three smaller shrubs and the values were averaged. For each observation period radiometric temperature measurements of six soil exposures were also averaged. The results are plotted in Fig. 9.9, together with air temperature readings. The vegetation is consistently warmer than the soil at night, with the maximum differential being 4°C. Relative temperatures are re-

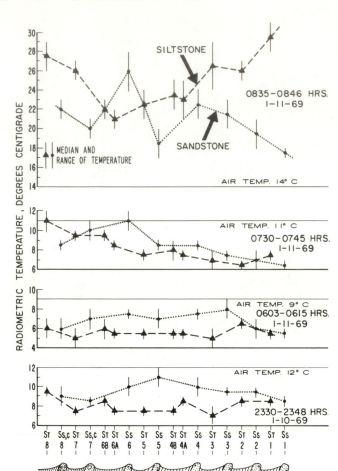

FIGURE 9.8. Radiometric temperatures (8 to 14 μm) of sandstones (Ss) and siltsones (St) of Palm Spring formation. Temperatures were measured at four different times of day and illustrate the thermal crossover of sandstone (dotted curve) and siltstone (dashed curve) after sunrise. Measurement stations are shown on the diagrammatic stratigraphic section that represents approximately 122m of beds. Locality 2.9, E.2 of IR image in Fig. 9.7b. From Remote Sensing: Principles and Interpretation by Floyd F. Sabins. W. H. Freeman and Company. Copyright© 1978.

versed during the day when the soil is much warmer than vegetation. Notice that the thermal crossovers occur within less than an hour, both in the evening and morning. These diurnal temperature relationships of soil and vegetation have since been confirmed on day and night IR imagery at many localities.

9.5.2 Imler Road Anticline and Superstition Hills Fault, California

The Imler Road anticline and Superstition Hills fault are located at the southwest margin of the Imperial Valley. Aerial photography and thermal IR imagery were obtained in 1953 and 1961, respectively, and are shown in Fig. 9.10. Bedrock is the Borrego formation, which consists of brownish gray lacustrine siltstone with thin interbeds of well-cemented brown sandstone (Fig. 9.11). Flaggy pieces and concretions of sandstone litter the surface where the sandstone beds crop out. Light-colored, nodular, thin layers within the siltstone help define bedding trends within this monotonous unit. The siltstone bedrock appears relatively cool and the sandstones relatively warm on the imagery [Fig. 9.10(a)], corresponding to thermal signatures of similar rock types in the Indio Hills.

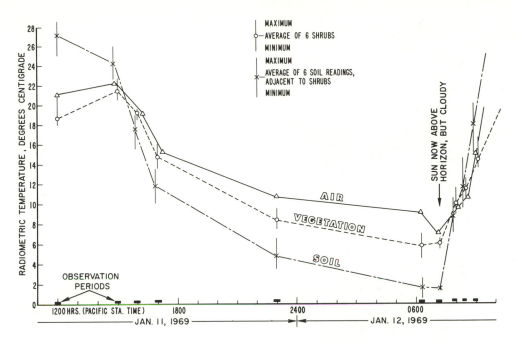

FIGURE 9.9. Diurnal radiometric temperatures (8 to 14 μm) of vegetation and soil at Indio Hills. Locality 2.6, H.O of Fig. 9.7b. From Remote Sensing: Principles and Interpretation by Floyd F. Sabins. W. H. Freeman and Company. Copyright© 1978.

In addition to the dunes and patches of sand, there is a sheet of windblown sand that covers much of the area and appears warmer than the bedrock exposures. During Quaternary time, pluvial Lake Coahuila covered this area. Except for travertine coatings on large boulders, Coahuila deposits have been reworked by the wind.

Most of the area is very low-relief desert with sparse clumps of vegetation that localize patches of windblown sand on their east margin, downwind from the prevailing westerly winds. Larger sand dunes stabilized by mesquite bushes (Fig. 9.11) appear warm (bright) on the imagery and dark on the photography. The very warm-appearing, Y-shaped feature at locality 3.2, B.1 in Fig. 9.10(a) is a thick accumulation of windblown sand lodged against an earthen embankment. Imler Road in the northern part of the area is actually straight, but appears curved because of distortion by the IR scanner. The warm-appearing road was surfaced with hard-packed sand at the time the imagery was acquired, but today is paved with asphalt.

The irrigated fields in the southern part of the area are part of the Imperial Valley agricultural area. Water in the Fillaree irrigation canal at the south edge of the area appears relatively warm

on the nighttime image, for the reason given earlier. At the time the imagery was acquired, the very warm (bright) field area north of the canal [1.8, C.0 on Fig. 9.10(a)] was probably flooded with standing water to leach salt from the soil. The very cool (dark) fields were probably damp from recent irrigation, resulting in evaporative cooling.

Two structural features, the Imler Road anticline and the southeast extension of the Superstition Hills fault, arc well expressed on the imagery but are invisible or obscure on aerial photographs. The Imler Road anticline at locality 4.2, B.7 in Fig. 9.10(a), is 1.2 km long and 0.4 km wide, and was first mapped by Sabins (2) on thermal IR imagery. There are no conspicuous lithologic or topographic patterns associated with the anticline. Without the imagery it would be possible to walk across the anticline in the field without recognizing its presence. In the field the plunge of the anticline has been defined by walking out light-colored nodular beds and the resistant sandstone units. Structural attitudes are obscure in the siltstone, but dips up to 45° were measured in the sandstones and the dip-reversal across the fold axis was located. A broad, low ridge with up to 9 m of relief coincides with the general area of the anticline.

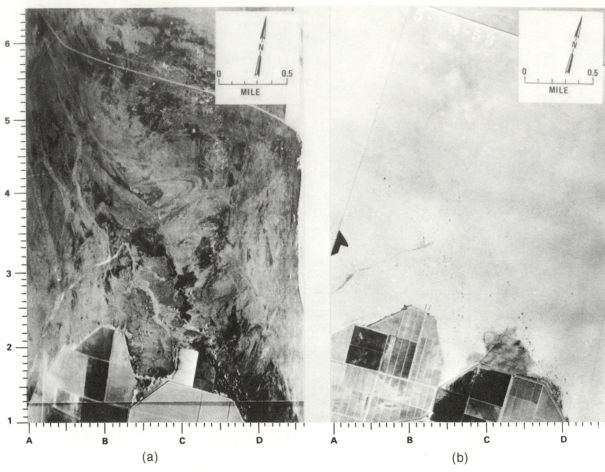

FIGURE 9.10. Superstition Hills anticline and fault, Imperial County, California. From Sabins (2). (a) Nighttime thermal IR image (8 to 14 μm). (b) Aerial photograph.

In Fig. 9.11 the anticline is shown as solid bedrock, but there are numerous thin patches of windblown sand that account for the gray tones on the imagery. The core of the anticline consists of contorted siltstone with a characteristic cool signature. The alternating light and dark pattern outlining the limbs of the fold appears to correlate with outcrops of sandstone (warm) and siltstone (cool), respectively. The west end of the anticline is abruptly truncated at the southeastward projection of the Superstition Hills fault. The inferred trace of the fault is obscured by windblown sand, but outcrops of siltstone in the immediate vicinity of the anticline are strongly deformed, suggesting drag folding.

South of the anticline there is a small arcuate pattern of alternating cool and warm bands [3.6, B.5 in Fig. 9.10(a)] that correspond to an exposure of gently dipping siltstone and sandstone bedrock with an outcrop pattern generally resembling that of the imagery. The pattern is more pronounced on the imagery than on the ground.

The Superstition Hills fault is a right-lateral strike-slip fault that was originally named in the Superstition Hills, 14.5 km to the northwest of Imler Road, and projected into this area on the El Centro sheet of the State Geologic Map of California. The fault alignment shown in Fig. 9.11 differs from that on the El Centro sheet. In addition to truncating the anticline, the fault is marked in the southeast part of the image by a subtle, but definite, southeast-trending linear feature marked by the sharp contact between cooler-appearing areas on the east and warmer on the west (2.2, C.5 to 1.2, D.0). The trend of the linear feature is parallel with, and about 160 m to the east of, the row of prominent sand dunes with warm signatures. On April 9, 1968, the Borrego Mountain earthquake caused surface

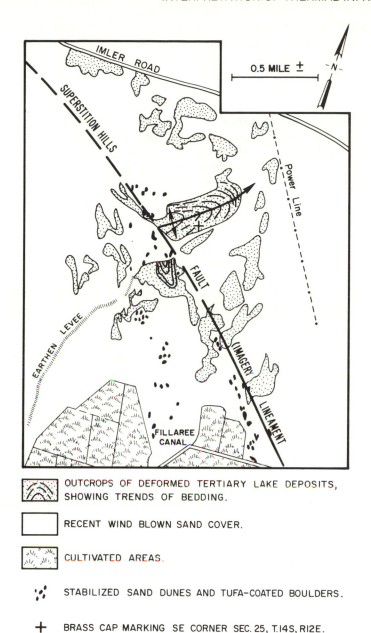

OUTCROPS OF DEFORMED TERTIARY LAKE DEPOSITS, SHOWING TRENDS OF BEDDING.

RECENT WIND BLOWN SAND COVER.

CULTIVATED AREAS.

STABILIZED SAND DUNES AND TUFA-COATED BOULDERS.

BRASS CAP MARKING SE CORNER SEC. 25, T.14S, R12E.

FIGURE 9.11. Geologic interpretation of IR image of Imler Road anticline and Superstition Hills fault. From Sabins (2).

breaks along the trace of the Superstition Hills fault, which were mapped by A. A. Grantz and M. Wyss. In this area their map (13) shows a break with less than 2 cm of right-lateral displacement that closely coincides with the linear feature on the imagery. The imagery, flown four years prior to the earthquake, located an important structural feature that is obscure on photography and in the field.

The greater information content in the thermal IR image relative to the aerial photograph has been analyzed by Sabins (2). He noted that the difference is not due to the time lapse between the 1953 photography and 1963 imagery, because of the relative stability of the desert and the absence of noticeable changes observed during annual field trips. Sabins (2) concluded that the rocks in this area have greater variations in thermal radiance than in visible reflectance. Furthermore, subsequent attempts at aerial photography of the anticline with color and IR color film have not yielded as much information as the thermal IR image.

9.5.3 Pisgah Crater and Basalt, California

Pisgah Crater and its associated basalt flows of Recent age are located in the Mojave Desert 50 km east of Barstow, California as shown in the geologic index map [Fig. 9.12(a)]. The elevation of the flow is approximately 600 m, with the Pisgah cinder cone rising about 75 m above the flow surface. The predominant vegetation is creosote bush, occurring as scattered clumps and local concentrations in the dry washes. The fine-grained, porphyritic, vesicular Pisgah basalt consists of pahoehoe which has a smooth ropy surface with flow structures, and aa which has a rough clinkery scoriaceous surface. The basalt types are readily distinguished in the visible spectral region by the medium gray signature of the pahoehoe and the black signature of the aa.

Pisgah Crater is a symmetrical cinder cone of scoriaceous pebble- and cobble-size fragments of basalt pumice, with minor ash and volcanic bombs. The central depression is floored with pahoehoe

basalt. Roads and quarrying operations have modified the shape of the cone and are visible on the imagery. A low cinder ridge extends northward from the cone. The younger alluvium consists of fan deposits of gravel eroded from the Pisgah basalt and older rocks. Fine- to medium-grained quartz sand occurs in patches deposited by the prevailing westerly winds.

Pisgah Crater and vicinity has been a remote sensing test site of NASA and the United States Geological Survey since the mid 1960s (14). In 1975 NASA acquired multichannel imagery at predawn and postnoon times, using a 10-channel scanner in the visible and photographic IR spectral regions (0.4 to 1.1 μm), and one thermal IR channel (8 to 14 μm). Data were processed by the Jet Propulsion Laboratory to produce daytime and nighttime thermal IR images and an albedo (reflected sunlight) image of the visible channels. These images are shown in Fig. 9.13; Fig. 9.12(b) is an interpretation map.

Because of differential solar heating and shadowing, topography is clearly expressed on the daytime thermal IR image, as illustrated at Pisgah Crater and the gullies in the northern part (left side) of the image [Fig. 9.13(b)]. The warmest appearing areas are the cinders of Pisgah Crater

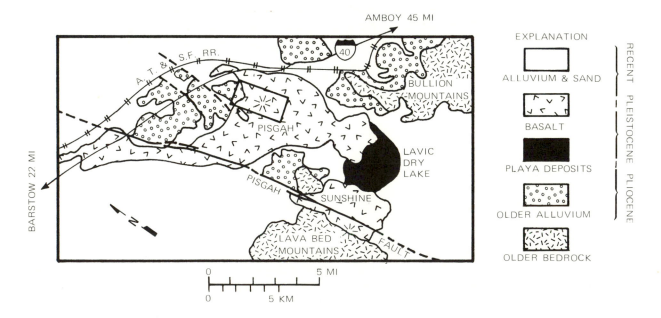

FIGURE 9.12. Maps of Pisgah Crater and vicinity, California. A. Geologic map. Rectangle is location of thermal IR images and interpretation map. From Remote Sensing: Principles and Interpretation by Floyd F. Sabins. W.H. Freeman and Company. Copyright© 1978.

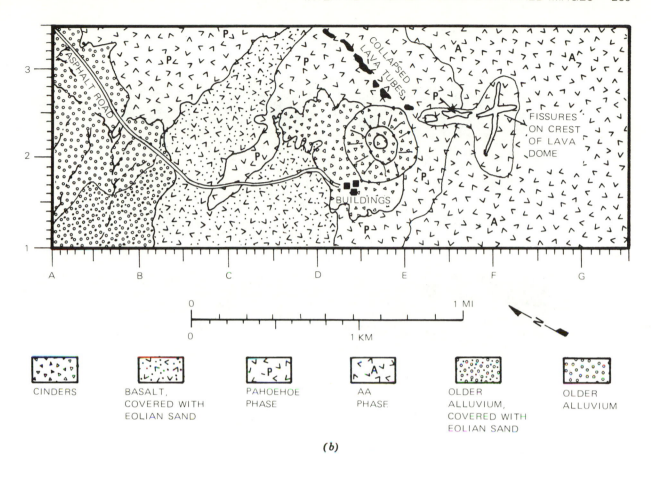

CINDERS

BASALT,
COVERED WITH
EOLIAN SAND

PAHOEHOE
PHASE

AA
PHASE

OLDER
ALLUVIUM,
COVERED WITH
EOLIAN SAND

OLDER
ALLUVIUM

(b)

FIGURE 9.12. Map of Pisgah Crater and vicinity, California. B. Geologic interpretation of thermal IR images. From Remote Sensing: Principles and Interpretation by Floyd F. Sabins, W. H. Freeman and Company. Copyright© 1978.

and the reworked cinders and basalt dust of the quarry operation immediately west of the crater. South of the road at locality, 1.7, B.9, an isolated patch of cinders surrounded by eolian sand is used as a parking area. Here the warm daytime signature of the cinders is not influenced by topography. South of Pisgah Crater the contact between pahoehoe (bright) and aa (dark) is prominent on the daytime thermal IR image. Alluvium and eolian sand have similar signatures on the daytime image, being darker (cooler) than the cinders and pahoehoe, but approximately the same temperature as the aa.

Topographic effects are negligible on the nighttime thermal IR image [Fig. 9.13(c)]. The alluvium and sand have about the same temperature and are the coolest materials; the cinders are only slightly warmer. There is a tendency for the aa to

be warmer than the pahoehoe, representing a reversal of the daytime signatures, but this trend is not as pronounced as it is on the day and night IR images acquired in 1965 and illustrated in Sabins (5).

Several features have distinctly different day and night signatures. On the crest of the large lava dome are intersecting fissures (2.5, F.1 in Fig. 9.13). On the daytime image the sunlit portions of the fissures are warm and the shaded portions are cool. On the nighttime image the fissures are warm. A string of collapsed lava tubes east of Pisgah Crater [see Fig. 9.12(b)] are similarly cool on the daytime image and warm on the nighttime image. The heat absorbed in the fissures and tubes during the day is radiated at night from the walls and absorbed and reradiated from the opposite walls, thus forming an approximation of a black-body cavity.

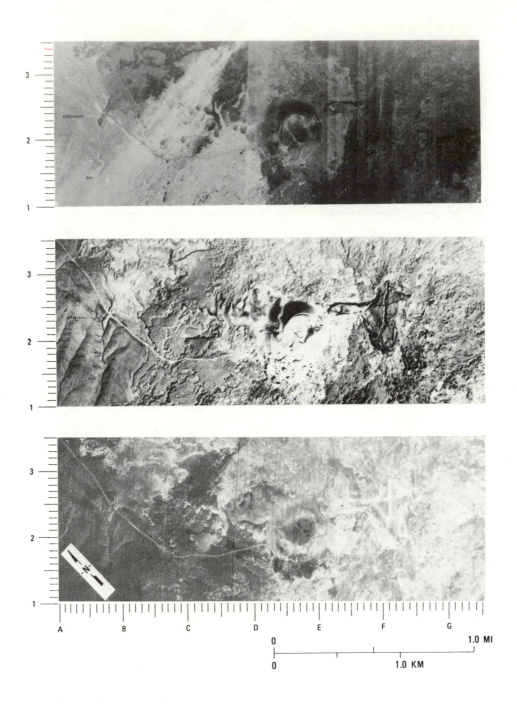

FIGURE 9.13. Pisgah Crater, California. Images acquired May 30, 1975, by NASA. Processed by Jet Propulsion Laboratory. Courtesy Ann B. Kahle, Jet Propulsion Laboratory. (a) Albedo image, (b) Daytime thermal IR image, (c) Nighttime thermal IR image.

9.6 SURVEILLANCE OF ACTIVE VOLCANOES

In 1971 the history and status of thermal surveillance of volcanoes with remote sensing and contact temperature devices was summarized by Moxham (1) who concluded that ". . . results to date are more tantalizing than elucidating. At some volcanoes thermal forerunners (of eruptions) have been well documented; elsewhere, temperature measurements have been inconclusive or negative." Few systematic, repetitive thermal IR surveys have been made; most of these used older scanners without the magnetic tape recording and quantitative imagery described earlier. Systematic thermal studies should be made in conjunction with geophysical and geochemical investigations at sites selected to yield significant results about volcanic processes. Friedman and Williams (15) listed all IR surveys of volcanoes reported to 1968. More than 23 volcanoes throughout the world have been surveyed but generally only by a single flight or, at best, several flights within a few weeks. Surface data are correspondingly sparse.

Moxham (16) analyzed three IR surveys of Taal Volcano, in the Philippines, during the nine-month quiescent period between the 1965 and 1966 eruptions. Taal is well suited for IR surveys because most of the volcano is below the water table and there is both a crater lake and a surrounding water body. Because of the uniform and high emissivity of water, small temperature changes are readily detected and there are no masking effects of vegetation. The first IR survey showed the two known fumarole fields and several previously unknown fumaroles on the north flank of the central cone. Both IR and ground investigations showed new hot springs, one of which enlarged prior to the 1966 eruption, along the flanks of the 1965 explosion crater. Hydrothermal activity persisted around the rim of the 1965 cinder cone, which was the site of the 1966 eruption midway between the two areas of maximum thermal water discharge.

9.6.1 Halemaumau, Hawaii

Figure 9.14 is a thermal IR image and accompanying geologic sketch of Halemaumau, a collapse crater on the floor of Kilauea caldera. Halemaumau has been the center of surface activity at Kilauea for the past 150 years. The historic flows within Kilauea caldera are pahoehoe lava. The older lavas beyond the Kilauea escarpment are

mantled by ash deposits up to 2 m thick. The faults in the vicinity of the Crater Rim Road are generally downdropped toward the caldera and bound topographic benches. The southwest-trending swarm of cracks lack appreciable displacement, although many are open at the surface and range from a few centimeters to 5 m in width (17).

As shown in Fig. 9.14(a), the highest radiant temperatures at Kilauea occur at three localities: (1) The entire outer escarpment of Halemaumau is warm, probably because the dense interior portions of lava flows are exposed there; (2) discontinuous arcuate warm zones on the floor of Halemaumau partially coincide with a low interior escarpment and may represent convective transfer of heat along an unmapped fracture zone; (3) irregular reticulate warm zones occur on the floor of Kilauea Crater. The zone intersecting the north rim of Halemaumau is especially pronounced. On imagery acquired in 1963, Fischer and others (18) correlated these in part with steaming fissues on the caldera floor.

There is no consistent correlation of radiant temperature with the different basalt flows. The fractures and faults in the vicinity of Crater Rim Road are distinctly warmer than their surroundings and resemble the fissures at Pisgah Crater, which are also warm on nighttime imagery. Both the Hawaii and California fractures are natural approximations of black-body cavities that retain solar heat from daytime exposure.

Despite differences in scale, the 1973 image (Fig. 9.14) can be compared with the 1963 image that was presented and interpreted by Fischer and others (18). The reticulate warm pattern of steaming fissures on the floor of the main crater is similar in both images. In 1963 the fissure that would intersect the north rim of Halemaumau 10 years later was not so extensive. The 1961 vent and spatter cone inside the southwest rim of Halemaumau (Fig. 9.14) show the highest thermal anomaly on the 1963 image, which correlates with rock temperatures of 100°C measured one m below the surface (18). On the 1973 image, no warm anomaly occurs at this vent, apparently because of cooling over the intervening 10 years. The warm anomalies at this general location on the 1973 image appear to be steaming fissures on the main caldera floor, outside Halemaumau Crater. On November 29, 1975, an earthquake triggered new eruptions at Kilauea. At the time of this writing,

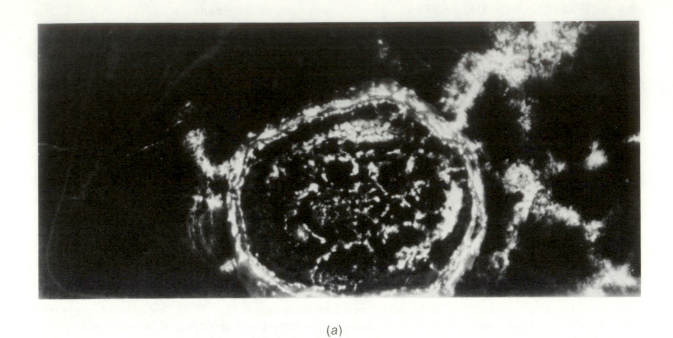

(a)

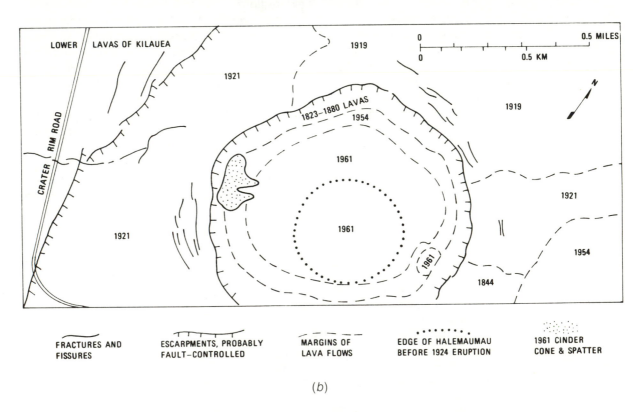

LOWER LAVAS OF KILAUEA

1921

1919

1919

CRATER RIM ROAD

1823–1880 LAVAS

1954

1961

1961

1961

1921

1921

1954

1844

0   0.5 MILES

0   0.5 KM

N

| FRACTURES AND FISSURES | ESCARPMENTS, PROBABLY FAULT-CONTROLLED | MARGINS OF LAVA FLOWS | EDGE OF HALEMAUMAU BEFORE 1924 ERUPTION | 1961 CINDER CONE & SPATTER |

(b)

FIGURE 9.14. Halemaumau Caldera and Kilauea Crater, Hawaii. (a) Nighttime thermal IR image (8 to 14 μm), courtesy Daedalus Enterprises, Inc., Ann Arbor, Michigan, (b) Geologic map after Peterson (17).

the relationship of the new volcanism to thermal anomalies at Halemaumau is not known. Calibrated thermal IR images of Kilauea and Mauna Loa are illustrated in color in Sabins (5).

9.7 GEOTHERMAL APPLICATIONS

Thermal IR imagery is very useful for detecting such surface expressions of geothermal activity as fumaroles, steaming ground, and hot springs. The Hawaiian imagery illustrated the application to active volcanic areas. Infrared imagery was used in Mexico to locate thermal springs and zones of hydrothermal alteration in a geothermal area (19). Nonquantitative IR imagery of the Geysers geothermal area in northern California, flown in 1966, was interpreted by Moxham (20). Hot springs and fumaroles were readily detected on the imagery. Less intense thermal features associated with the steam field are present in the imagery but are not readily detectable on the ground. There was little evidence however, of a regional thermal IR anomaly associated with the known geothermal reservoirs.

9.7.1 Low Intensity Geothermal Anomalies

In contrast to fumaroles, steaming ground, and hot springs, the slightly elevated ground temperatures associated with some geothermal areas are much more difficult to detect on IR imagery or radiometer profiles. Watson (21) pointed out that temperature differences associated with variations in geology and topography can readily overwhelm geothermal anomalies of several hundred heat flow units [1 HFU = 1 X 10^{-6} cal/(cm•s)]. Mathematical model studies were made to evaluate relative effects of various factors on surface radiant temperature. These studies suggest that both thermal and reflectance imagery should be acquired at least three times in the diurnal cycle (21). A midmorning 8- to 14- μm image in Idaho revealed a weak thermal anomaly, which was confirmed by ground measurements.

9.7.2 Mount Rainier, Washington

This area could also be included in the section on volcanoes; however, because Mount Rainier has been inactive for the past 550 to 600 years (22), it is treated here as a possible geothermal area. The stratovolcano rises to an elevation of 4392 m, more than 3.2 km above its base. The summit is a cone of andesite lying within a broad depression at the top of the volcano. The depression was probably created about 5000 years ago by phreatic explosions and avalanching of the former summit (23). Two small summit craters are nearly filled with snow and ice, but steam jets at the eastern crater have melted caverns beneath the ice. The west crater is about 300 m in diameter and is partly overlapped by rocks of the younger east crater, 390 m in diameter. The only rock exposed on the summit is the andesite around the crater rims, which is swept bare of snow by the winds. Eleven glaciers radiate from the mountain.

Infrared images are acquired for the United States Geological Survey in 1964, 1966, and 1969 to evaluate increased seismic and thermal activity at the summit of Mount Rainier. Moxham (24) reported that fumaroles lining the rims of the two summit craters were similar in extent and apparent thermal intensity on the 1964 and 1966 imagery. Warm thermal anomalies occurred along the north rims of the east and west craters and at an isolated spot farther west.

The Environmental Research Institute of Michigan (ERIM) acquired 8- to 14- μm imagery of Mount Rainier on the night of July 18, 1970. The tape-recorded data were processed into 23 discrete thermal intensity levels. These were displayed in 11 colors and in black (violet was repeated), and separated by white intervals to produce the image shown in fig. 9.15(a) color plate 15 (25). Temperature calibration data were not available; therefore, the colors representing thermal intensity are ranked from warmest to coolest. An interpretation map is shown in Fig. 9.16(a) color plate 16. The glaciers are marked by yellow and sepia in Fig. 9.15(a) and are separated by narrow rock ridges, shown in red, that are warmer. The concentric temperature zones on the flanks of the mountain are parallel with the topographic contours and become colder upward. The two summit calderas, which are filled with ice and snow, show the coldest temperatures (black) on the crest of the central cone (violet). The other cold (black) areas on the summit are not explained. The geothermal anomalies are the arcuate blue areas coinciding with some of the rock outcrops at the summit. Despite the smaller scale of the ERIM image, the location and configuration of these 1970 anomalies correlate very well with the thermal anomaly map prepared by Moxham (24) from 1966 imagery.

9.7.3 Crater Lake, Oregon

Crater Lake was formed approximately 5000 years ago when the volcano Mount Mazama collapsed, creating a caldera 8 to 10 km in width with a depth of almost 600 m. The andesite bedrock is mantled on the north with glowing-avalanche deposits of pumice formed by the explosive activity. Wizard Island is a small volcanic cone in the lake that formed long after the caldera subsidence. Although volcanic activity at Crater Lake is now dormant or inactive, geothermal activity (25) may be indicated on thermal IR imagery.

Nighttime IR imagery was acquired by ERIM on July 19, 1970, and processed into the nine uncalibrated temperature intervals displayed in color in Fig. 9.15(b) color plate 15. Figure 9.16(b) color plate 16 is an interpretation map. One of the coolest areas is Lao Rock on the northwest rim [Fig. 9.16(b)], which forms on a high cliff of dacite lava capped by pumice deposits. Areas covered by pumice north and south of the lake are generally cool because of the low thermal inertia of this porous material. The barren cone of Wizard Island is cool in contrast to the warm signature of the vegetated lower slopes. Other warm areas, such as Grouse Hill, appear to correlate with vegetation cover. The radiometer studies at Indio Hills noted the relatively high nighttime temperatures of vegetation. The park headquarters on the south side of the lake has the maximum temperature of the land areas.

Within the lake, the western margin is distinctly cooler. Minimum temperatures occur along the shore north of Wizard Island, where water depths are 30 m or less; this contrasts with the northeast margin, where depths exceed 570 m. Very warm plumes of water originate at bays and coves along the eastern margin, with a pattern suggesting clockwise circulation in the lake. The IR energy is radiated from the few surface millimeters of water; therefore the depth of the warm plumes cannot be inferred from IR imagery. A geothermal origin for the warm plumes is a distinct possibility, although the plumes may represent runoff into the lake.

9.8 SUMMARY

Thermal IR remote sensing, generally in the 8- to 14-μm spectral region, provides information about radiant temperature, which is determined by kinetic temperature and emissivity of surface materials. Infrared radiometers record temperature as profiles along flight lines or as point measurements. Imagery is acquired by airborne scanners in either a qualitative or quantitative mode. Interpretation of imagery can provide information on geologic structure, lithology, volcanism, soil moisture, and steaming ground.

REFERENCES

1. Moxham, R. M., 1971, Thermal surveillance of volcanoes, in The surveillance and prediction of volcanic activity: Paris, UNESCO, p. 103-124.
2. Sabins, F. F., 1969, Thermal infrared imagery and its application to structural mapping in Southern California: Geol. Soc. America Bull., v. 80, p. 397-404.
3. Sabins, F. F., 1973, Recording and processing thermal IR imagery: Photogrammetric Eng., v. 39, p. 839-844.
4. Buettner, K. J. K., and Kern, C. D., 1965, Determination of infrared emissivities of terrestrial surfaces: Journal Geophys. Res., v. 70, p. 1329-1337.
5. Sabins, F. F., 1978, Remote sensing-principles and interpretation: San Francisco, Freeman, 426 p.
6. Wolfe, E. W., 1971, Thermal IR for geology: Photogrammetric Eng., v. 37, p. 43-52.
7. Rowan, L. C., and others, 1970, Thermal infrared investigations, Arbuckle Mountains, Oklahoma; Geol. Soc. America Bull., v. 81, p. 3549-3562.
8. Offield, T. W., 1975, Thermal-infrared images as a basis for structure mapping, Front Range and adjacent plains in Colorado: Geol. Soc. America Bull., v. 86, p. 495-502.
9. Sabins, F. F., 1973, Flight planning and navigation for thermal IR surveys: Photogrammetric Eng., v. 39, p. 49-58.
10. Rosenberg, P., 1971, Resolution, detectability and recognizability: Photogrammetric Eng., v. 37, p. 1255-1258.
11. Sabins, F. F., 1967, Infrared imagery and geologic aspects: Photogrammertric Eng., v. 29, p. 83-87.

12. Cummings, J. R., 1964, Coachella Valley investigation: Calif. Dept. Water Resources Bull. 108.

13. Allen, C. A., and others, 1972, Displacements on the Imperial, Superstition Hills and San Andreas faults triggered by the Borrego Mountain earthquake, *in* the Borrego Mountain earthquake of April 9, 1968: U. S. Geological Survey Prof. Paper 787, 207 p.

14. Sabins, F. F., 1975, Remote sensing field trip guide to Los Angeles Basin, Antelope Valley and Pisgah Crater, unpublished, 19 p.

15. Friedman, J. D., and Williams, R. S., 1968, Infrared sensing of active geologic processes, *in* Proceedings of the fifth international symposium on remote sensing of the environment: Ann Arbor, Michigan, Environmental Research Institute of Michigan, p. 787-815.

16. Moxham, R. M., 1967, Changes in surface temperature at Taal Volcano, Philippines 1965-1966: Bull. Volcan., v. 31, p. 215-234.

17. Peterson, D. W., 1967, Geologic map of the Kilauea Crater Quadrangle, Hawaii: U.S. Geological Survey Geol. Quad. Map GQ-667.

18. Fischer, W. A., Moxham, R. M., Polycn, F., and Landis, G. H., 1964, Infrared surveys of Hawaiian volcanoes: Science, v. 146, n. 3645, p. 733-742.

19. Valle, R. G., Friedman, J. D., Gawarecki, S. J., Banwell, C. J., 1970, Photogeologic and thermal infrared reconnaissance surveys of the Los Negritos-Ixtlan de Los Hervores geothermal area, Michoacan, Mexico: Geothermics Spec. Issue 2, p. 381-398.

20. Moxham, R. M., 1969, Aerial infrared surveys at the Geysers geothermal steam field, California: U. S. Geological Survey Prof. Paper 630-C, p. C106-122.

21. Watson, K., 1975, Geologic applications of thermal infrared images: IEEE Proceedings Annals n. 501, p. 128-137.

22. Fiske, R. S., Hopson, C. A., and Waters, A. C., 1963, Geology of Mount Rainier National Park, Washington: U. S. Geological Survey Prof. Paper 444, 93 p.

23. Crandell, D. R., 1963, Paradise debris flow at Mount Rainier, Washington: U. S. Geological Survey Prof. Paper 475-B, p. B135-B139.

24. Moxham, R. M., 1970, Thermal features at volcanoes in the Cascade Range, as observed by aerial infrared surveys: Bull. Volcan. v. 34, p. 77-106.

25. Wagner, T. W., 1971, Processing of thermal remote sensor data from geothermal areas: Univ. of Michigan Willow Run Lab. Rept. No. 31069-5-T, 16 p.

CHAPTER 10

TECHNIQUES AND APPLICATIONS OF IMAGING RADARS

HAROLD C. MACDONALD

10.1 INTRODUCTION

Prior to World War II the concept of using reflected radio waves for remote target location purposes evolved in several countries with the development of systems that transmitted from ground to airborne targets. By the end of the war, the techniques (collectively called radar, an acronym for *radio detection and ranging*) and their basic principles had been developed to a high degree. Since then, radar technology has been continuously refined, and in fact radar was one of a host of devices originally developed for military use, which in the early 1960s spawned the field of remote sensing.

Radar is generally described as mapping radar, although the direct product of the data collection is not a map but a visual output called imagery, from which maps can be constructed. Although the resolutions of most *side-looking airborne radar* (SLAR) systems do not match those of photography, these same systems have the unique ability to delineate physical characteristics of the Earth's surface providing an image of "photographic" quality independent of visibility or weather conditions. This operational advantage is related to the wavelength region of radars, which can penetrate clouds, smoke, fog, and most precipitation. Radar mapping surveys can thus be conducted virtually every day or night.

A task shared by almost every nation in the world today is the identification, development, and conservation of natural resources. From the geologist's point of view, it is essential that older exploration methods be improved, and newer exploration techniques be developed. The unique operational capabilities and information content available in radar remote sensing establish it as a prime candidate for continued evaluation and development as an exploration tool.

Microwave sensing shows two distinct areas of sensor utilization, active and passive. The active systems, which will be explained in detail in this chapter, have been employed to generate radar images for a variety of geoscience studies for which cameras were exclusively used in the past. In contrast, passive systems have not yet proven to be practical for most geological studies. The passive microwave radiometer, for example, when directed toward the ground, provides composite signals resulting in a brightness temperature of the terrain. Applicability of passive systems appears to have value for sea state determination, sea ice identification, soil moisture studies, and basic investigations for microwave-terrain interaction mechanisms.

The design of imaging radars for geological investigations requires an interdisciplinary effort, which includes both a theoretical (provided by the radar engineer) and applications-oriented (provided by the geologist) understanding of the sensor's operation. Certainly the geologist utilizing the imagery need not be capable of system design; however, certain fundamentals of radar operation should be understood in order to provide optimum data retrieval for the interpretation.

10.2 RADAR OPERATION

10.2.1 Operational Frequencies

Conventional radars utilize the frequency range from 230 to 40,000 MHz although neither end of this range is truly definitive of the frequency limitation for radar operation. A letter code of frequency-wavelength bands, *K*, *X*, *L*, and so forth, was arbitrarily selected to ensure military security in the early developmental stages of radar and has continued in use for convenience.

10.2.2 Antenna Configuration

Side-looking airborne radars can produce imagery of the terrain on one or both sides of the aircraft flight path. SLAR differs from other scanning radar systems in that the antenna is fixed to the aircraft and its radiation pattern is directed perpendicular to the ground track. Scanning of the area is accomplished only by the movement of the aircraft in flight. Discussions of imaging radar antenna configurations have separated the systems into two primary methods of operation: (1) real aperture or brute force [Fig. 10.1(a)], and (2) synthetic aperture [Fig. 10.1(b)]. Advantages of one mode of operation over the other are related to achieving improved azimuth resolution (parallel to ground track.) Real-aperture systems provide improved azimuth resolution by increasing the length of the antenna mounted on the airborne platform. The Westinghouse AN/APQ-97 (*Ka*-band system, no longer operational) and the APS/94D Motorola

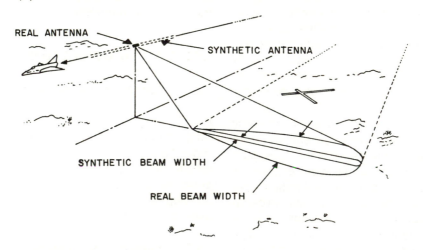

FIGURE 10.1 Sketch diagrams, side-looking radar systems. (A) Real-aperture system. (B) Synthetic aperture system.

MARS (*X*-band) real-aperture systems use an antenna approximately 5 m in length. Synthetic-aperture radars (SAR) such as those of Goodyear-Aero Service AN/APQ-102 (*X*-band), Environmental Research Institute of Michigan (*X*- and *L*-bands) and the Jet Propulsion Lab (*L*-band) provide synthetic antenna lengths many times larger than real-aperture systems. With synthetic-aperture systems, relatively good resolution is possible at great distances, even from satellites. Details of both real-aperture and synthetic-aperture antennas will be provided in Section 10.4.

10.2.3 Basic Case—Ground to Air

Radar is unique among the commonly used remote sensors in that it provides its own illumination, electromagnetic energy of radio or microwave frequencies. For the simplest mode of operation, that is, ground to air, a transmitter on the ground emits electromagnetic radiation in the form of RF (radio frequency) energy, and when the RF energy is interrupted by any object such as an aircraft, part of the energy is reflected back to a receiver. The reflection or reradiation of energy sent back to the receiver is referred to as an "echo," and the object reflecting or reradiating the signal is called the target. If, however, reradiation is returned from the surrounding terrain which makes it difficult to select the desired target, the unwanted return signals are termed "clutter." It is exactly this "clutter" that the geologist would like to analyze.

10.2.4 Complex Operation—Air to Ground

Figure 10.1(*a*) shows an area being imaged by a typical real-aperture SLAR system. The transmitter generates short bursts of pulses of RF energy. These pulses are propagated into space by means of a directional antenna (A), and radiate from the antenna as a block of energy (B) at the velocity of light (3 × 10⁸ m/s).

The RF energy is confined to a narrow path, as shown in Fig. 10.1(*a*). At any one instant, the terrain illuminated by the transmitted pulse is limited in the range direction by the pulse duration and in the azimuth direction by the beamwidth of the directional antenna (a more detailed discussion of pulse duration and beamwidth will be given later in Section 10.4). Thus, the size of the illuminated patch on the terrain as indicated by the crosshatched area in Fig. 10.1(*a*) is determined by the radar's resolving power.

If a terrain feature capable of intercepting RF energy is irradiated at point (a), a fraction of the transmitted energy will be reradiated (backscattered) in a direction toward the antenna (A). An object at (b) will also reradiate energy back to the antenna at a later time when object b is illuminated by the pulse packet of RF energy. The same is true for features at (c) and (d). The portion of energy returned to the antenna from the terrain features (a), (b), (c), and (d) is converted to a video signal by the receiver. The signal return from these features is displaced from the origin (c) as a function of the range from the antenna to the target, whereas the amplitude of each return is a function of the backscattering properties of the terrain.

The video signal is displayed by intensity modulating the beam of a cathode-ray tube (CRT) as the beam is swept across the face of the CRT. If the antenna is repositioned by aircraft forward motion to "look" at a new strip of terrain adjacent to the one just imaged, and each resultant line is displayed on the CRT in the same relative position, an image of the terrain can be generated by sequentially exposing a continuous strip of film to the face of the CRT.

In summary, the antenna (A) is repositioned laterally at the velocity of the aircraft (*Va*). Each RF pulse transmitted (B) returns signals from the targets within the beamwidth. These target returns are converted to a time-amplitude video signal (C), which is imaged as a single line (E) on photographic film (F). Returns from subsequently transmitted pulses are displayed on the CRT at the same position (D) as previous scan lines; however, with each subsequent pulse transmitted, the photographic film is moved past the CRT display line at a velocity (*Vf*) proportional to the velocity of the aircraft (*Va*), and an image of the terrain is recorded on the film (*F*) as a continuous strip of radar imagery.

10.3 TERRAIN-SIGNAL INTERACTION

Radar return is that portion of the transmitted radar energy that returns to the radar receiver and is directly related to the nature of the terrain illuminated by the transmitted electromagnetic energy as well as certain system parameters. The terrain-signal interaction is complicated because the terrain parameters affecting the radar return are complex. For example, return signal amplitude is influenced by the overall composition of the illuminated area, including such parameters as moisture content, vegetation extent and type, and surface configuration. The relationship between return and terrain becomes even more difficult to determine if system parameters are considered such as angle of incidence, transmitted and received polarizations, and radar frequency.

10.3.1 Radar Equation

That phase of geological interpretation of radar imagery consisting of inferring significant characteristics of the terrain from observed tonal variations on the image requires that the interpreter be familiar with the parameters that influence radar backscatter or signal return. The signal strength (or intensity of terrain return) received at the antenna determines the relative brightness of a resolution cell on the imagery. The performance of a radar that relates return signal (return power) to both the target parameters and radar system characteristics is conveniently described by the radar equation:

$$Pr = \frac{P_t G^2 \lambda^2}{(4\pi)^3 R^4} \sigma \qquad (10.1)$$

where

Pr = received power
Pt = transmitted power
G = antenna gain
λ = wavelength of system
R = slant range distance between radar and target
σ = radar cross section; effective backscatter of the target

The equation for received power indicates that Pt, G, and λ are radar system parameters, and R is determined by the antenna location with respect to the ground. Radar cross section σ, then, includes all of the geoscience information about the illuminated terrain (except for location) that the radar is capable of sensing (1).

The radar equation as shown above (10.1) is for a single scatterer, explicitly, the area that would intercept sufficient energy from the incident field to produce the observed echo intensity by isotropic reradiation. However, in practice the radar returns from many scatterers (extended targets usually much larger than a resolution cell) are averaged, and average differential cross section (radar cross

section per unit area) is used in conjunction with average return power. The average differential cross section, which is also known as the scattering coefficient σ^0, is a dimensionless real number whose magnitude is a function of incidence angle, surface roughness, dielectric constant, frequency or wavelength, and polarization. In a moment we shall examine each of these parameters in detail, but here it is important to note that when system parameters (such as frequency, polarization, transmitted power, and antenna gain) are held constant, the gray tone on a radar image is proportional to average return power strength and hence to σ^0. The reader is referred to Long (2) for an analytical discussion of the radar equation.

10.3.2 Incidence Angle

The angle of incidence, θ, is the angle formed by an impinging beam of radar energy (propagational vector that is a perpendicular to radar wavefront) and a perpendicular to the incident surface at the point of incidence [Figs. 10.2, (a) and (b)]. The angle between a line from the transmitter to a point on the terrain and a horizontal plane passing through the transmitter is the depression angle. The geometric parameters of SLAR imaging systems are such that along the swathwidth of an area imaged (near to far range), there is a continuous change in the angle of incidence [Fig. 10.2(b)]. When imaging homogeneous flat terrain, for any constant depression angle along the flight path, the angle of incidence will remain constant. Under more typical natural terrain conditions, however, local variations in terrain slope can change the effective angle of incidence. The consequence of terrain slope on both the incidence angle and aspect angle (complement of incidence angle), at a constant depression angle, is shown in Fig. 10.2(c). Notice on Fig. 10.2(c) that if the terrain is flat, the aspect angle equals the depression angle. However, if terrain slopes are oriented at an angle toward the imaging radar, the effective angle of incidence decreases (with increasing terrain slope angle) to a point where the aspect angle equals 90°, the angle of incidence is 0° (vertical incidence), and maximum reradiation is the result. Conversely, if terrain slopes are oriented away from the imaging radar, the angle of incidence increases (with increasing terrain slope angle) to a point where grazing (minimum reradiation) is the result. The relationship between minimum and maximum return can also be understood if we consider just

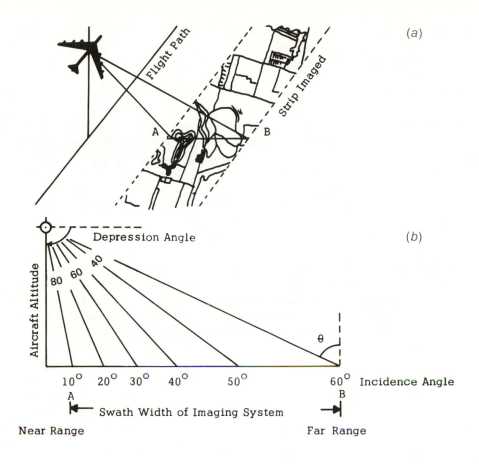

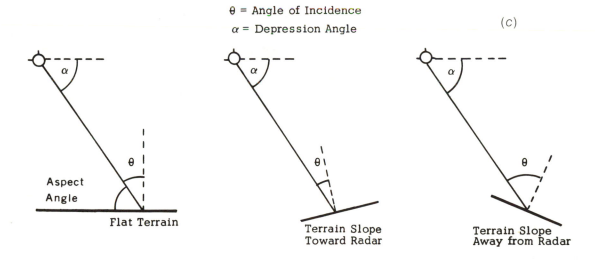

FIGURE 10.2. SLAR ground swath. (A) Ground coverage. (B) Swathwidth for flat terrain, change in incidence and depression angles near to far range. (C) Effect of terrain slope (with constant depression angle) on incidence and aspect angles.

terrain slope and depression angle. When terrain foreslopes are oriented toward the radar, and the terrain slope angle equals the depression angle, maximum return will result. Conversely, when terrain backslopes are oriented away from the radar, and the terrain slope equals the depression angle, minimum reradiation will occur. Between these two extremes there will be variations in reradiation. These contrasting angles of terrain illumination are somewhat analogous to the situations encountered when gathering aerial photography during times of high and low solar illumination (elevation). For example, when the sun is essentially at the zenith (vertical incidence) shadowing is minimized; however, during sunrise or sunset hours, as the sun approaches the horizon, shadowing of terrain features correspondingly increases. However, unlike photography where shadows can have partial illumination because of short wavelength visible light scattering and reflection, radar shadows are totally black (no return).

10.3.3 Surface Roughness

Surface roughness, a geometric property of the terrain, generally has the most important influence upon the return signal amplitude (imagery tone), as it dictates the extent of scattering that forms the reradiation pattern. Surface roughness that is dependent upon terrain textural features such as tree leaves and branches, surface vegetation, soil and regolith particles, and so forth, is not an absolute roughness, but the relative roughness expressed in wavelength units. Terrain surfaces may be divided into two major categories according to surface roughness, depending upon the relationship of the root-mean-square (rms) surface roughness to the signal wavelength, λ. For rms surface roughness much less than a wavelength ($\lambda/10$) the surface appears "smooth" or mirror-like to the imaging radar (under these conditions, reflection follows Snell's law where the angle of incidence equals the angle of reflectance), while for surface roughness of the order of a wavelength or more, the surface appears "rough." For a more general example, a terrain surface can be classed according to roughness between the extremes of a perfect specular reflector, where backscatter exists only near vertical incidence and reflected energy is contained in a small angular region about the angle of reflection conforming to Snell's law, and a diffuse or isotropic

scatterer, where the scattering coefficient (σ^0, previously discussed with the radar equation) is essentially independent of the angle of incidence.

10.3.4 Complex Dielectric Constant

The complex dielectric constant (the electrical properties of the surface) will influence reflectivity of the radar return from terrain surfaces. The microwave reflectivity of the terrain surface, for example, is a function of the magnitude of the complex dielectric constant, and observed changes in reflectivity from natural surfaces are primarily due to changes in the moisture content of either the vegetation or soil surface. In the microwave region of the spectrum, the dielectric constant of most naturally occurring materials, when dry, is in the range of 3 to 8. Here, the radar energy would travel through a relatively large volume of material and reflectivity would be comparatively low. However, in the same frequency region, the dielectric constant of water may be near 80. An increase in the dielectric constant increases the reflectivity of a surface and the radar energy would be reflected back without significant travel through the material. The effect of water content on the return from a dense volume of foliage has been observed where variations in return were attributable principally to the moisture content of the vegetation itself (3).

We may also see that the loss of attenuation of the microwave energy is a function of the conductivity of the material and the frequency of the energy. In general, the higher the frequency, the greater is the attenuation in the material and hence the effective penetration is less. This may have a marked effect on the return from vegetated surfaces, as at the higher frequencies the return is essentially from the top of the vegetation canopy, while at lower frequencies, with greater penetration capability, the return may be primarily a volume return contributed by leaves, branches, trunks, and perhaps even the ground. The complex dielectric constant varies almost linearly in response to contained moisture per unit volume, and penetration of microwave energy is greatest and reflection least with low moisture contents. Conversely, penetration is least and reflection is greatest when moisture contents are high (4).

In summary then, significant changes in dielectric properties of natural materials usually result from changes in moisture content. An increase in the dielectric coefficient increases the reflectivity, and therefore the level of return signal. For two comparably rough surfaces, the difference in their scattering coefficients and thus the amplitude of the return signal is indicative of the difference in their dielectric properties.

10.3.5 Frequency

The variations attributable to frequency are directly related to the two parameters previously discussed, surface roughness and complex dielectric constant. In general, the rougher the surface in terms of wavelength, the more diffuse the return. Consequently, a given surface will normally appear rougher at a higher frequency (shorter wavelength) than at a lower frequency. Varying the frequency of the wave incident upon a natural terrain surface of any roughness produces an effect similar to variations in surface roughness.

10.3.6 Polarization

The polarization of an electromagnetic wave describes the orientation of the electric field strength vector at a given point in space during one period of oscillation. With traditional, single polarization SLAR configurations, a horizontal electric field vector is radiated and upon striking the terrain most of the energy is returned to the antenna with the same polarization as the transmitted pulse. Thus, only the horizontal component of the return signal from the terrain will be displayed on the imagery; however, independent of the transmitted polarization, the return signals can also contain a depolarized component, that is, energy depolarized by the terrain surface and vibrating in various directions. Depolarization refers to the change in polarization an electromagnetic wave undergoes as a consequence of interaction with the terrain. Radar systems that provide a traditional horizontally polarized antenna and an additional vertically polarized antenna allow simultaneous display of return signals of both polarizations. The **HH** (transmit and receive horizontal electric-field vector) return is known as the like-polarized return; the **HV** (transmit horizontal and receive vertical electric-field vector) return is known as the cross-polarized return.

Radar images produced by like-polarized return signal may differ from those produced by the cross-polarized return, because of the differences between the physical processes responsible for the two types of return. Although the mechanism responsible for recording a significant cross-polarized component has not been clearly demonstrated, several theoretical models support depolarization by double bounce, volume scatter, or multiple reflections (2). For example, at short wavelengths such as Ka- and X-band, when the terrain is exceedingly rough in comparison to the radar wavelength (such as a forested area), there is little difference between like- and cross-polarized images; however, for relatively smooth terrain there is a definite polarization dependency (5). In grassland and marshland regions where the configuration of specific plant communities (especially aquatic) provides marked contrasts in plant density and geometry, polarization differences are marked (6, 7).

Limited geological utility has been provided using the multiple polarized imagery. Dellwig and Moore (8), and McCauley (9) have provided two published geologic reports where marked differences between like- and cross-polarized imagery appear to have geologic application. Figure 10.3 provides **HH** and **HV** imagery of the nonvegetated terrain of the Pisgah Crater area in the California Mojave Desert, which is dominated by Tertiary and Quaternary lava flows and Quaternary alluvium. The nearly black area of low return extending in nearly all directions from the letter A is Lavic Lake, a playa. The most obvious differences between the **HH** and **HV** images are along the contacts between Sunshine Flow lava field (area C) and alluvium (contact shown at B). The alluvial fan in area A is readily apparent on the cross-polarized image. To the northwest of Sunshine Flow, contrasts between images also can be noted in the alluvial materials along the alluvial apron (areas E). Pisgah Flow (area D) shows no significant difference in return on the dual images.

The Pisgah Crater area has been selected by several authors (5, 8) as an example to illustrate that volume scatter, as a result of nonhomogeneities due to surface roughness, actually does contribute more to the cross-polarized return than does multiple scattering. For example, Sunshine Flow (area C) has a relatively homogeneous rough surface in comparison with the surrounding alluvium. The multiple signal return (scatter) from Sunshine Flow is relatively insignificant if one considers the roughness or volume scatter from the nonhomogeneous alluvium. For this reason, the return on the **HV** image is less than that of the **HH** image, and Sunshine Flow

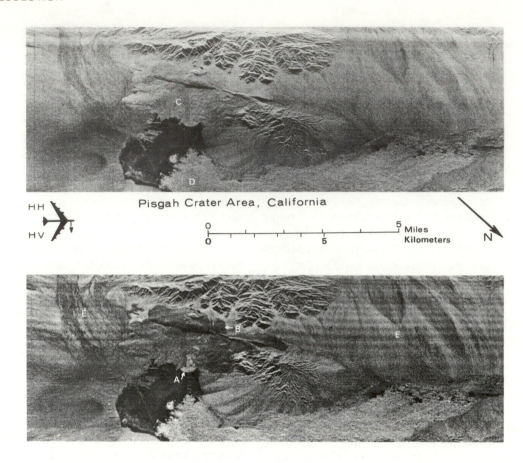

FIGURE 10.3. Pisgah Crater Area, California. Top image like-polarized (*HH*); bottom image cross-polarized (*HV*).

appears dark on the cross-polarized image. The same area on the like-polarized imagery has a return similar to that of the alluvium, that is, high return from both Sunshine Flow and the alluvium. Here direct surface scatter, which provides little contribution to the cross-polarized return, contributes significantly to the like-polarized return. In contrast, Pisgah Flow (area D) is a nonhomogeneous basaltic body with many vesicles (approximately one-third the volume), which provides a volume of scatterers (for depolarization) similar to the alluvium (5). Consequently, the returns recorded on both images for Pisgah Flow are similar. Future research and collection of multiple polarization data will undoubtedly prove the full geological value of such imagery in selected situations.

10.4 RESOLUTION

One of the key parameters often used to judge the quality of an imaging radar is resolution; however, one should be careful to distinguish the difference between detection and resolution. Depending on the resolution of a particular radar system, it may not be possible to resolve (that is, separate on the imagery) two objects only 50 ft apart. Yet, one of the objects may be detectable even though it has dimensions of only several feet across. For example, a power line pole may be clearly detectable on the radar imagery although it is impossible to distinguish between poles spaced closely enough to be below (or within) the resolution distance of the radar system. If the target has sufficient reflectivity, the radar will detect targets such as power line poles, vehicles, corner reflectors, and so forth. This is in spite of the fact that they may be physically much smaller than the resolution cell of the imager, and we will see in a moment why.

While the ultimate system resolution is a function of several parameters, the single most important criterion is the radar pulse rectangle size as projected on the ground. At a given instant of time, the

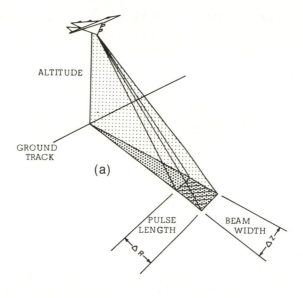

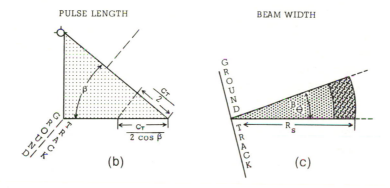

FIGURE 10.4. SLAR imaging system resolution. (a) Real-aperture resolution cell. (b) Range resolution for real- or synthetic-aperture systems. (c) Azimuth resolution for real-aperture system.

ground area [ΔR by ΔZ, Fig. 10.4(a)] is simultaneously superimposed in the radar receiver and this area is called the pulse rectangle of ground resolvable area (similar to instantaneous field of view for Landsat MSS systems). Objects within the pulse rectangle are recorded as if they were all located at the center of the rectangle, and two objects that occupy the same pulse rectangle appear as a single return on the imagery. The integrated reflectance of all objects within the ground resolvable area at any given instant is, in essence, presented as one value of reflectance on the imagery (pixel of Landsat MSS systems). When this composite reflectance value differs from those in adjacent pulse rectangles, the sensor is said to be able to discriminate between targets. It must be realized that if a single "hard" target (such as a small

truck) is in flat, open surroundings and is illuminated by the radar, it most likely will be the dominant reflector in one pulse rectangle. The radar accordingly records, as a single bright target, the composite reflectance of the truck and the surrounding area within the pulse rectangle. To the radar, the truck is as large as the ground resolvable area and is distinguishable from the surrounding, less bright pulse rectangles. If objects are separated by a distance greater than the corresponding dimension of the pulse rectangle, they will be imaged in separate, resolvable pulse rectangles.

10.4.1 Real-Aperture Systems

Radar resolution dimensions for real-aperture SLAR systems vary in both range [normal or perpendicular to ground track, ΔR of Fig. 10.4(a)] and azimuth [parallel to the ground track, ΔZ of Fig. 10.4(a)] directions. The slant range resolution (normal to ground track) of either real- or synthetic-aperture systems is determined by the pulse duration (pulse length) of the transmitted energy, that is, $c\tau/2$ (where $c = 3 \times 10^8$ m/s and τ = pulse length). Short pulses lasting 10^{-7} s may be readily achieved and these provide a slant range resolution of 30 m. If finer resolution is required, the length of the pulse can be reduced. The equivalent ground range resolution is the projection of this distance onto the ground plane, $c\tau/2 \sin\theta$ where θ is the angle of incidence. This relationship may be alternatively expressed in terms of aspect angle β, where $c\tau/2 \cos\beta$ [Fig. 10.4(b)]. Thus objects separated by a distance equal to or less than $c\tau/2 \cos\beta$ will not be resolved as individual targets and will appear on the radar as signal return from a single resolution cell. If the separation distance between two objects exceeds $c\tau/2 \cos\beta$ then the two targets will be resolved. Examining the simplest situation where terrain is flat, depression angle and aspect angle are equal, we may increase the altitude (slant range distance) of our radar (keeping depression angle to target constant) to orbital heights and see that the range resolution stays the same. Resolution in range then is independent of distance, being a function of viewing angle and pulse length.

Real-aperture radars obtain azimuthal resolution by employing a relatively narrow beam of transmitted energy such as the beamwidth shown in Fig. 10.1(a) and 10.4(a). The width of the beam [B_θ, Fig. 10.4(c)] on the ground determines the azimuthal size (ΔZ) of the recorded resolution cell where

$$\Delta Z = B_\theta R_s \qquad (10.2)$$

where
ΔZ = azimuth (or along track) resolution, in meters
B_θ = beam width or radar antenna, in radians
R_s = slant range to target, in meters.

Further, antenna beam width is, to the first order:

$$B_\theta = \frac{\lambda}{D} \qquad (10.3)$$

where λ is the wavelength of imaging system and D is the azimuth aperture of the illuminating antenna (antenna length in the case of SLAR). Substituting 10.3 in 10.2, real-aperture azimuth resolution can be written

$$\Delta Z = \frac{R_s \lambda}{D} \qquad (10.4)$$

Azimuth resolution then, is directly proportional to range and wavelength and inversely proportional to antenna length. Although the beamwidth is a constant angle, the distance subtended at the target by this angle depends upon the range. Thus, the width across the beam is a function of slant range. Examination of Fig. 10.4(a) will provide evidence that azimuth resolution is variable and improves at the near ranges where the width of the fan shaped beam is at a minimum, but deteriorates in the far range. Thus, azimuth resolution can be improved by reducing range (distance to target) or wavelength, or by increasing antenna length.

The operational wavelength and altitude are usually selected becuse of desirable terrain reradiation characteristics or desirable weather-penetration capabilities, and consequently optimum azimuthal resolution is generally achieved by antenna design. A longer antenna focuses the beam into a narrower angle, which improves the resolution.

Imaging systems in which pulse ranging is used to provide resolution in one coordinate of a radar image, while the angular beam that is scanned provides the second coordinate, are referred to as noncoherent radars. Systems in this class rely only to a very small extent, if at all, on the preservation of phase relationships between transmitted and received waveforms, nor are signal-processing techniques used to generate an image. Coherent systems, however, provide a high degree of phase stability; they preserve phase relationships between transmitted and received waveforms, and employ signal processing to achieve resolution finer than that permitted by noncoherent radars. The synthetic-aperture radar is the most important example of a coherent radar system and will be discussed next.

10.4.2 Synthetic-Aperture Radars
Synthetic-aperture radars (SAR) use signal storage and processing techniques to simulate the

performance of an antenna that is electronically much longer than the actual physical antenna employed. A synthetic antenna is produced by having an aircraft with a relatively short real antenna transmit and receive pulses at regular intervals along the flight path. When these individual signals are stored and then processed, an antenna of long effective length is synthesized in space. This synthetic antenna is many times longer than the physical length of the real antenna and, therefore, can provide a very narrow beam of constant width that results in improved azimuth resolution [Fig. 10.1(b)]. There is an important difference between return signal processing of real-aperture antennas [Fig. 10.1(a)] and that of the synthetic antenna [Fig. 10.1(b)], because with the real antenna, individual pulses are transmitted, received, and displayed sequentially as a line on the image. Typically, the image is simply a photographic presentation of the reradiated signal intensity that is recorded as a function of antenna scan and slant range. For the synthetic antenna, however, each target will produce a large number of pulses that must be stored and then combined in a special way so that the synthetic antenna can simulate a physical antenna of a length equal to time of target illumination. A simple analogy described in a Goodyear Aerospace brochure (10) illustrates this.

An object in the center of the concentric circles of Fig. 10.5(a) is bobbing up and down in the water to produce waves at a frequency of 10 cycles per minute. A passenger in a boat at position B would count 10 waves passing in 1 min. However, if a boat were moving somewhat toward the waves as in position A, the passenger would count a higher number of waves passing his position, perhaps 12 per minute. At position C the boat is moving away from the wave source and the apparent frequency would be less than at positions A and B, perhaps only 8 cycles per minute. The difference between the received and transmitted waves per minute or frequencies (a target's radial component of movement relative to the transmitter) is called the doppler frequency.

If the observer were to keep a record of the level of the water as the boat moved from position A to B, and then to C, it would look like the curve recorded in Fig. 10.5(b). Now consider the wave source to be an aircraft transmitting a radar signal. The boat would correspond to a target or reflecting object

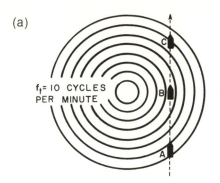

(a)

$f_t = 10$ CYCLES PER MINUTE

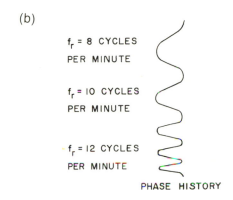

(b)

$f_r = 8$ CYCLES PER MINUTE

$f_r = 10$ CYCLES PER MINUTE

$f_r = 12$ CYCLES PER MINUTE

PHASE HISTORY

f_t = TRANSMITTED FREQUENCY

f_r = RECEIVED FREQUENCY

FIGURE 10.5. Doppler frequency effect and phase history.

moving through the antenna beam as a consequence of the forward motion of the aircraft. The signals received by the antenna would have a shift in frequency because of the apparent motion of the target relative to the aircraft. A record kept of the received signals would be similar to that obtained by the observer in the boat. Such a record is called the phase history of the return.

By means of some rather sophisticated electronic and optical devices, the amplitude and phase history are usually stored on data film by using a cathode-ray tube to display the incoming signals and a lens to focus this display on a continuously moving strip of film (11). The radar receiver detects the radar waves scattered from each point on the terrain. In the equipment aboard the sensor platform, the signals generated in the receiver by the series of reflected microwave pulses are combined with a sequence of reference signals to create interference patterns. The interference signal from each scatterer emerges as a voltage which controls the spot brightness on a cathode-ray tube and is recorded as a broken line parallel to the edge of the data film. During times of constructive interference when the received and reference pulses coincide, the spot will be bright. When the phase of the return and reference pulse are destructive, the voltage will be low and the moving spot will be dim. Line by line each interference pattern is displayed on the cathode-ray tube and photographed on a moving strip of data film, the velocity of which is proportional to the velocity of the aircraft. The series of opaque and transparent dashes on the film is actually a one-dimensional interference, and the data film can be regarded as a radar hologram. On the ground the holographic film is illuminated with a source of coherent light in an optical correlator to produce an image.

We have now seen that synthetic-aperture radars can artificially simulate very long antennas by coherently processing the returns; however, two subtypes of synthetic-aperture radars should be mentioned—focused and unfocused. In the focused radars, phase corrections are added to the recorded phase histories to compensate for the fact that the reflected wavefront from a distant target is not a plane wave, but is actually a spherical surface. The unfocused radars operate on the assumption that the reflected wavefront is a plane; therefore, their azimuth resolution is spoiled in proportion to the extent that this assumption is not true. Theoretically, fully focused synthetic-aperture radars have their azimuth resolution limited by $\Delta Z = D/2$, where D is antenna length, while for unfocused synthetic-aperture radars, the finest possible theoretically obtainable resolution is $\Delta Z \simeq 1/2 \sqrt{\lambda R}$.

The azimuth resolution of focused synthetic-aperture radars is simply limited by the length of the synthetic aperture. This manner of operation obtains the azimuth dimensions independent of range to the target; however, an unfocused synthetic-aperture radar has resolution that is still dependent on wavelength and range. The reader is referred to Moore (12) for a comprehensive theoretical analysis of SLAR resolution.

In summary then, significant azimuth resolution improvement has been achieved by employing signal processing called synthetic aperture. A synthetic-aperture SLAR uses signal storage and processing techniques to simulate the performance of an antenna aperture that is much longer (in the direction of flight) than the actual physical antenna employed. This results in improved azimuth resolution beyond that available from the beamwidth of the physical antenna. The physical antenna of the synthetic aperture scans the terrain in the same way as the real-aperture SLAR. However, at intervals along the flight path as a signal is transmitted, the phase and amplitude of the returns are measured and stored. After the physical antenna has traversed a distance L, the stored signals can be combined to produce a return similar to that obtained from a real antenna having a physical length L. Thus, although the physical or real antenna may only be a meter long, the signals stored and processed over a flight 400 m long result in a synthetic antenna 400 m long [Fig. 10.1(b)]. Theoretically, the azimuth resolution of the synthetic-aperture SLAR is half the actual aperture of the real antenna in the configuration used.

10.5 IMAGERY FORMAT AND RADAR GEOMETRY

10.5.1 Imagery Format

SLAR systems use either a slant-range or ground-range presentation on their video recorders. Slant-range sweeps are linear, so the spacing between return signals on the image is directly proportional to the time interval between the terrain features being recorded. For ground-range sweeps, the spacing is modified (hyperbolic waveform applied to video display circuitry) to equate the image scale to that which the terrain features would actually have on the ground if the latter were both flat and at a fixed altitude beneath the aircraft. The distinction between ground range and slant range is illustrated in Fig. 10.6(a), (13) where H is the altitude, Sr is the slant range, and Gr is the ground range; where radar altitude and slant range are known, the ground range can be calculated by

$$Gr = \sqrt{Sr^2 - H^2} \tag{10.5}$$

The smaller triangle of Fig. 10.6(a) illustrates the relationship between slant-range and ground-range distance or dimension, where Ds is the slant-range distance and *Dg* is the ground-range distance from the wavefront to the same point. Small increments of the wavefront are considered to be a straight line over small distances, thus perpendicular to *Sr*.

Figure 10.6(*b*) further illustrates the geometry of an imaging radar in the range direction. A pulse transmitted from the aircraft produces a wavefront that strikes in sequence, terrain points *a*, *b*, *c*, and *d*. The slant-range distance [*Ds* of Fig. 10.6(*a*)] between the return signals received at the aircraft will be proportional to the distances *a'b'*, and *c'd'*; even though the ground-range distance (*Dg*) between *ab* and *cd* are equal. On slant-range imagery

then, the spacings of features at near range are compressed (near-range compression) while in the far range, slant-range and ground-range distances (*Dg/Ds*) approach unity as the depression angle decreases. Near-range compression reaches an extreme directly below the aircraft, because all features in this region produce returns that are received at essentially the same time and cannot be separated.

10.5.2 Radar Shadow

Shadows on aerial photographs are a function of the position of both the camera and the sun. Radar

(a)

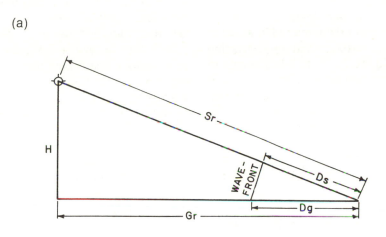

(b)

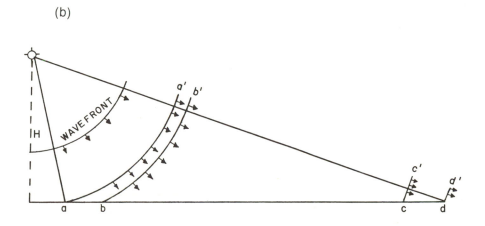

FIGURE 10.6. (a) Measurement of slant range and ground range. (b) Wavefront geometry, slant range and ground range [from LaPrade (13)].

provides its own "illumination" and the parameters that determine whether or not a terrain feature will produce radar shadow are depression angle and the angle of terrain slope (slope facing away from the impinging radar beam). Radar shadows will always occur whenever terrain back slope exceeds depression angle. The relationship between depression angle and terrain back slope is such that three cases are possible; (1) back slope fully illuminated and no shadow results (Fig. 10.7, terrain features 1 and 2), (2) the back slope is partially illuminated and partially in shadow (grazing), and (3) the back slope is not illuminated by the impinging beam resulting in a no return or shadow area (Fig. 10.7, terrain features 3 and 4).

The shadowing of terrain features (of equal terrain slope and height) being imaged in a swath from near to far range will increase as depression angle decreases (Fig. 10.7). Thus it is possible to image relatively homogeneous topography with SLAR imaging systems, and have equivalent slopes along the swathwidth fully "illuminated" in the near range, and in partial or complete shadow in the far range.

Side-looking radars image the terrain at oblique angles of illumination and because of this, the shadows formed on the imagery are somewhat analogous to those shadows formed on aerial photographs taken at low sun angle. This oblique shadowing has aided in defining topographically expressed geologic features on both photography (14) and radar imagery (15). In low relief areas for example, shadowing is desired for the geological interpretation of subtle drainage or fracture patterns. To illustrate the effect of varying depression angles on terrain interpretation, Fig. 10.8 shows a plaster terrain model, photographed at various angles of illumination (14). Here we can equate the angles of illumination to radar depression angles. At relatively shallow depression angles (10°-40°) subtle terrain texture and topographic detail become enhanced, while at steeper depression angles terrain discontinuities become less obvious. Especially in the low relief areas, the oblique illumination and resultant shadowing have aided geological interpretation, but in mountainous terrain, extensive shadowing can prove to be a distinct interpretive disadvantage. The interpretive disadvantage of extensive radar shadowing will be examined in detail later in Section 10.9; however, it is also important to mention that under the constraints for using shadows on photographs for height measurement, radar

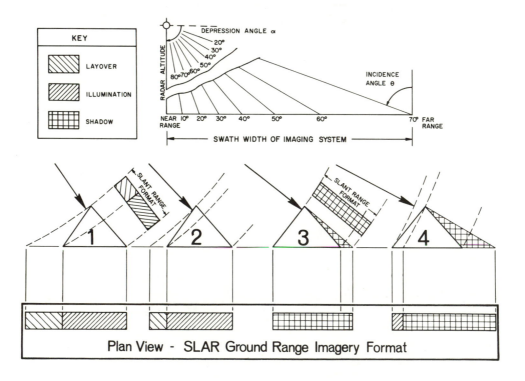

FIGURE 10.7. Radar layover, foreshortening and shadowing, a function of terrain slope and radar depression angle.

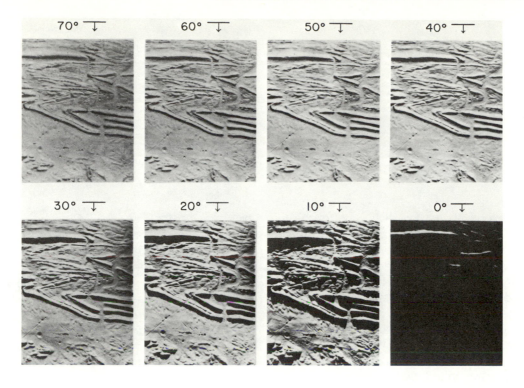

FIGURE 10.8. Illumination of a three-dimensional plaster terrain model from depression angles (illumination) of 70°-0° [courtesy R. J. Hackman (14)].

shadows can be similarly utilized.

To calculate the height of elevated terrain, one can measure the length of the radar shadow that is very distinctive on the imagery. Figure 10.9 includes the slant range length of the radar shadow Ss, the projection of the slant range shadow onto the ground Sg, and the height of the terrain feature, h. The height then is equal to

$$h = \frac{H(Ss)}{Sr} \qquad (10.6)$$

Sr is the slant range distance from the aircraft to the end of the shadow.

10.5.3 Radar Geometry

The imagery format of most SLAR systems can display errors of two types, geometric displacement and distortion. Distortion errors arise from atmospheric effects, platform motion, improper adjustment, and so forth. Geometric displacement errors are inherent to any SLAR imaging system and will be of primary concern for our examination.

RADAR FORESHORTENING

Characteristic of all radars imaging irregular terrain surfaces is the variation in the length of equal terrain slopes when the slope measurements are taken at different incidence angles, that is, radar foreshortening. Foreshortening results in the shortening of a terrain slope on radar imagery in all cases except when the incidence angle is equal to grazing or 90°, at which time the terrain slope and slope length on radar imagery will be equal (assuming that the scale factor between radar and terrain is accounted for). For example in Fig. 10.7 the fore (toward radar) and back (away from radar) slope lengths of terrain features 1, 2, 3, and 4 are equal; however, these same slope lengths as they are seen and recorded on the imagery, vary considerably over the total range of incidence angles. More simply, just as the slant range distance that is measured as a function of time, determines the sequence in which targets are displayed, the length of a terrain slope on imagery is dependent on the time it is illuminated. The fore slope length of terrain feature 2, recorded on the imagery, is less than half of the the back slope, because the radar illuminates

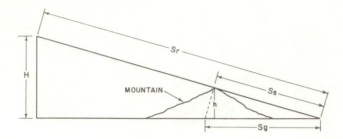

FIGURE 10.9. Radar shadow geometry.

the back slope almost twice as long as the fore slope. The same fore slope length on terrain feature 3 is imaged as a single line, when the top, middle, and bottom of the slope are "seen" at the same time. Where the incidence angle approaches vertical incidence as it has with terrain feature 3, radar foreshortening is maximum, but where the incidence angle approaches grazing (largest incidence angles, such as the far range of Fig. 10.7) the effects of foreshortening are minimal. Only at grazing are terrain slopes seen in their true length.

We now can visualize that with imaging radars, the way in which terrain slopes are recorded on imagery is determined by the relative time it takes the reradiation to travel from the radar transmitter in the airborne platform to the top, middle, and base of the feature, and back to the radar receiver. Because the speed of the transmitted and received radar signal is a constant (the speed of light), the distance in slant range can replace time. We now describe the terrain fore slope as a function of slant-range distance to the top, middle, and base of the feature (Fig. 10.10), which provides a more graphic presentation of radar foreshortening. What we might consider as a normal view of a terrain feature is shown in Fig. 10.10(a), where the slant-range distance (Sr) increases progressively from the bottom to the top of the feature, that is, $Srb < Srm < Srt$. Although the slope is foreshortened, the facets of the slope are displayed as a direct function of their height above the surrounding datum plane, the lowest point being imaged first and the highest point being imaged last. This is designated as "normal" because it is closely akin to the visual perception the observer has of the terrain feature.

Figure 10.10(b) provides the illumination of a terrain fore slope similar to that previously discussed for Fig. 10.7, feature 3. Here we see that the

slope facing the aircraft is concave upward and describes an arc of a circle whose center is located at the aircraft. When this condition is met, the slant-range distances from base to top of the feature are equal, that is, $Srb = Srm = Srt$ and the complete terrain slope is imaged as a point. Fortunately, when imaging natural terrain slopes the conditions for imaging an entire terrain slope of any magnitude as a point are not generally met. However, as we shall see in a moment, the imaging of individual slope

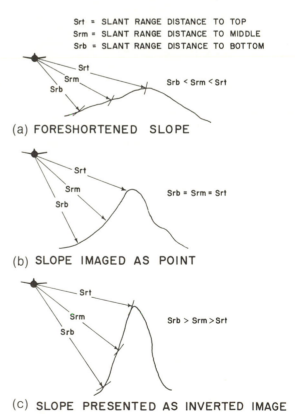

Srt = SLANT RANGE DISTANCE TO TOP
Srm = SLANT RANGE DISTANCE TO MIDDLE
Srb = SLANT RANGE DISTANCE TO BOTTOM

$Srb < Srm < Srt$

(a) FORESHORTENED SLOPE

$Srb = Srm = Srt$

(b) SLOPE IMAGED AS POINT

$Srb > Srm > Srt$

(c) SLOPE PRESENTED AS INVERTED IMAGE

FIGURE 10.10. Radar foreshortening and layover.

facets as a single point is a much more common occurrence.

Included also in Fig. 10.10 is a terrain feature illustrated in situation (c), where the slant-range distance from the aircraft is greater at the base (Srb) than it is in the middle (Srm) or top (Srt), that is, $Srb > Srm > Srt$. The resultant radar distortion known as radar layover, to be discussed in the next section, is a phenomenon where the top of a feature is imaged before the bottom, thus providing an inverted image as viewed by the interpreter.

RADAR LAYOVER

In the previous section we have again seen that a radar imaging system is time dependent, that is, the return signals are directly proportional to the time interval (range) between the transmitter and the terrain feature imaged. Because of this inherent characteristic, one geometric displacement that usually can be recognized on the imagery format in rugged mountainous areas, is the "leaning" of terrain features toward the nadir of the aircraft. This relief displacement is called layover effect, and is somewhat similar to the relief displacement of aerial photography, but it should be remembered that while a camera forms an entire image from one point in space, radar forms its image one line at a time as it moves along its flight path. As an example, a tree located on a vertical or oblique photograph some distance from the photo's perspective center will be imaged as if it were laid over away from the nadir. The same tree top as seen by SLAR would be displaced toward the nadir, because the radar imaged the top (being closer to the radar) before it did the bottom (Fig. 10.11).

Whenever terrain fore slope exceeds the complement of the depression angle, radar layover will

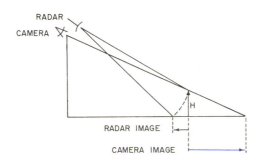

FIGURE 10.11. Relief displacement, photography and radar.

occur. Illustrated in Fig. 10.10(c), the top of the terrain feature is recorded on the imagery prior to the base because the slant-range distance from transmitter to the top of the feature is less than the slant range distance to the base. With terrain feature at b, the top, middle, and base are imaged as a single, point (maximum radar foreshortening). The top feature is imaged normally, that is, with the base portrayed closer to the nadir of the aircraft because the slant-range distance to the base of the feature at A is less than the slant-range distance to the top. Obviously, the effect of radar layover is most pronounced for large depression angles (near range), coincident with relatively steep sloping terrain elements.

Recognition of both radar foreshortening and layover is possible in the canyon country of Arizona. Figure 10.12 illustrates the gross topographic features of the Coconino Plateau, Arizona, and Fig. 10.13 provides slant-range imagery of the same area. The average slope of the canyon walls is approximately 50° (Table 10.1) and the relative relief of canyons is approximately 450 m. Although the canyon topography (that is, slope and relief) is quite uniform across the area, shadowing, layover, and foreshortening on the radar imagery are not. This contrast between actual terrain configuration and that which is displayed on the imagery provides a practical example of the effect of changing depression angle across the swathwidth of the imagery, even though terrain slope and relief may be relatively uniform.

The inherent disadvantages of a SLAR imagery format which can sometimes hinder geologic interpretation are extensive foreshortening, shadowing, and layover. For example, the fore slopes illustrated in Fig. 10.13 vary in length from a single line (south of point b) in near range, to approximately 35% of the actual length in the far range. Assuming constant fore slopes of 50° the combined effects of radar layover, foreshortening, and shadowing show the marginal utility of the imagery for geologic interpretation in the vicinity of such slopes perpendicular to flight path.

Especially in mountainous regions having extremely rugged relief, SLAR's inherent geometric distortions and shadowing may cause the same area to look quite different, if the images are obtained at different depression angles. Figure 10.14 illustrates slant-range imagery recorded near

COCONINO PLATEAU, ARIZONA

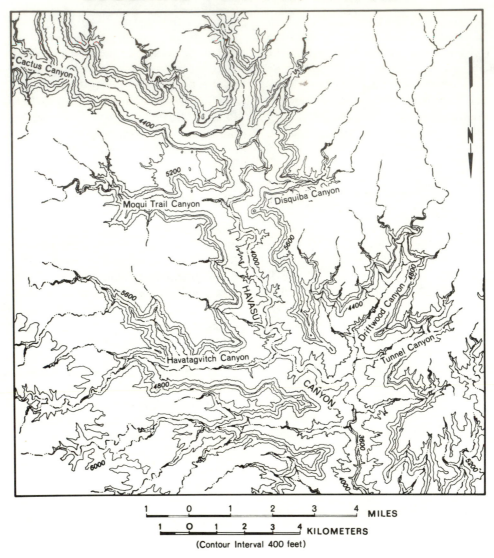

FIGURE 10.12. Generalized topographic map, Coconino Plateau, Arizona (from USGS Supai, Arizona quadrangle 1:62,500).

Cascade Glacier in Washington. The fore slope of ridge *a-b* [Fig. 10.14(*b*)] is foreshortened considerably, but is not in layover. Figure 10.14(*a*) shows the same ridge imaged at a steeper depression angle than Fig. 10.14(*b*) and consequently illustrates layover of ridge *a-b*. The bright band along the top of the ridge represents the total extent of layover. Careful comparison between the portrayal of the back slope of *a-b* as recorded on both images reveals that where layover has occurred [Fig. 10.14

(*a*)], the length of the back slope has been decreased accordingly. This apparent shortening of the back slope can be seen graphically in Fig. 10.7, when comparing the slant-range imagery presentation of terrain features 1 and 3. For example, Fig. 10.7 shows that part of the back and all of the fore slope of terrain feature 1 is recorded in the layover region, and recording of the true back slope length has been shortened accordingly. Contrasting other landforms of Fig. 10.14 (points *c*, *d*, and *e*, for

FIGURE 10.13. Radar foreshortening shows variation of foreslope length across swathwidth of imagery from near to far range.

TABLE 10.1. TERRAIN SLOPE DATA COCONINO PLATEAU, ARIZONA

Area	Average Terrain Slope	Average Depression Angle	Terrain Slope Required for Layover	Layover
Tunnel Canyon	50.5°±	19°	71°	NO
Driftwood Canyon	51.0°±	22°	68°	NO
Havatagvitch Canyon	51.5°±	21°	69°	NO
Disquiba Canyon	50.5°±	33°	57°	NO
Moqui Trail Canyon	51.0°±	34°	56°	NO
Cactus Canyon	53.0°±	45°	45°	YES

CASCADE GLACIER AREA, WASHINGTON

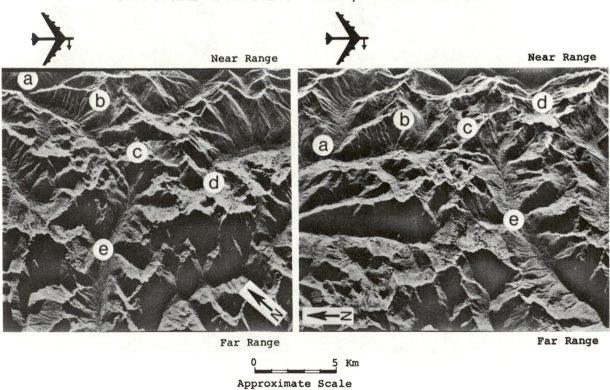

FIGURE 10.14. Contrasting terrain appearance, two different images of same area.

316

example) will reveal that images of the same area do indeed appear different as portrayed on the SLAR format. Extensive shadowing on these images is especially troublesome for the interpreter.

10.5.4 Look-Direction

SLAR imaging systems, because of their large areal coverage, enable the eye to integrate subtle topographic differences over long distances. This synoptic presentation in combination with an oblique angle of incident "illumination" provides enhancement of topographic features that are not readily apparent on conventional aerial photography even when viewed stereoscopically. More recently the advantages of radar have been compared. SLAR systems can image the terrain at comparatively low incidence angles, and shadows formed on the imagery are analogous to those of aerial photographs taken at low sun angle. Because of radar's oblique angle of illumination, features of geologic continuity have been observed which were previously undetected. Thus, the geologist is able to see the terrain portrayed in a configuration that is normally not available to him. Many examples have been cited in the literature where radar has actually defined structural features such as lineaments and faults that had not been previously detected using normal geological reconnaissance methods (15). However, as we have just seen, there are limitations, and while radar shadowing has proven particularly helpful to the geologist if the topographically expressed geologic features have relatively little relief, large relief features such as mountains provide extensive shadowing which can deter geologic interpretation. In addition, the orientation of the imaging radar (look-direction) in relation to structural trends in an area can be both an advantage and a disadvantage.

The Humboldt Range of western Nevada characterizes Basin and Range fault-block mountains, that is, subparallel block-faulted mountain ranges separated by alluvial valleys of approximately equal size. The core of the central range illustrated in Fig. 10.15 is composed predominantly of Triassic volcanics, whereas limestones comprise the younger strata flanking the mountain range on the west. The limestone-alluvium front-fault along the western front of the range is particularly pronounced in Fig. 10.15(a) between (B) and (C). The northward-trending, westward dipping, high-angle normal fault

is conspicuous because of the triangular-faceted spur ends, flatirons, and steep, V-shaped canyons. Figure 10.15(b) was imaged from nearly the opposite look-direction and the fault scarp is not as conspicuous because of a slightly different radar look-direction with respect to the diagnostic landforms.

The effect of shadowing in the delineation of subtle stream channels is illustrated at location (D). In the far range of Fig. 10.15(b) the delicate dissection of topography resulting in numerous distributaries in the alluvial apron is well defined, but the same pattern cannot be detected as easily in the near range of Fig. 10.15(a). The delineation of distributaries in the far range is totally attributable to alternate highlighting and shadowing which is nearly absent in the near range.

Orthogonal look-directions provide marked contrasts in terrain extractable information and this is particularly evident in the Trinity Mountains area of Nevada (Fig. 10.16). An obvious north-trending joint system (resulting in a serrated topography of igneous rocks) is visible in Fig. 10.16(a) between locations (A) and (A'); however, the same linear alignments are conspicuously absent in Fig. 10.16(b). The look-direction for Fig. 10.16(a) is perpendicular to the linear trends, whereas the look-direction for Fig. 10.16(b) is parallel to the linears. The differences in scale on these two images (obvious between similar locations A'-B) are related to the slant-range geometry of an earlier version of the AN/APQ-97, Ka-band system. The scale in the along-track (azimuth) direction is constant if we assume synchronization between aircraft ground speed and film speed (Fig. 10.1A); however, the scale in the range direction is variable on this slant-range format. Ground-range imagery formats available with present-day commercial systems provide uniform scale in both range and azimuth.

Multiple flight coverage over the southern portion of the Boston Mountains, Arkansas, has provided radar imagery over a geographic area that has received considerable attention by geologists involved in radar-remote sensor studies. The detection of a pronounced system of north-south trending lineaments on radar imagery was the primary reason for concentration of effort in this particular area (16). Initial field investigations provide evidence that the lineaments, originally delineated on low-resolution radar imagery were in fact, structurally controlled stream valleys which appear to have

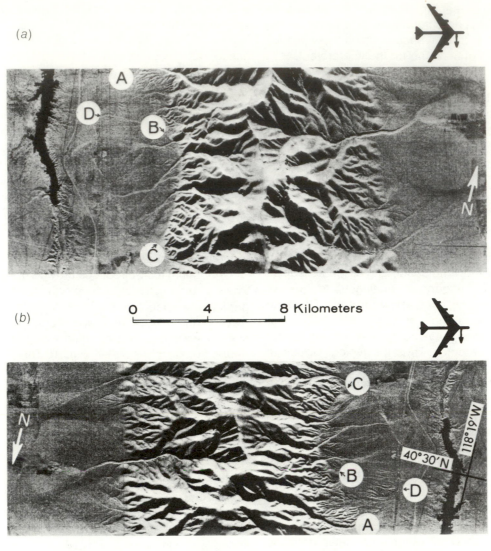

CENTRAL HUMBOLDT RANGE, NEVADA

FIGURE 10.15. Central Humboldt Range, Nevada.

the same general trend as one of the dominant joint-sets recorded at the outcrop. Part of the original study area is illustrated in Fig. 10.17, and the north-south linearity of many stream valleys is obvious on Fig. 10.17(a). Three different directions have been provided, (1) look-direction perpendicular to the linear stream valleys [Fig. 10.17(a)], (2) look-direction oblique to the stream valleys [Fig. 10.17(c)], and (3) look-direction parallel to the trend of the stream valleys [Fig. 10.17(b)]. On the imagery of the flight perpendicular to these linear trends [Fig. 10.17(a)] the enhancement of valleys (A)-(A'), (D)-(D') is

particularly striking. These same valleys, however, do not express the distinctive linear parallelism in Fig. 10.17(b). Similarly, the linearity of (F)-(F') is well defined in Fig. 10.17(b) (look-direction is now orthogonal to stream valley), whereas in Fig. 10.17(a) no such distinction can be made. Numerous other examples can be isolated when comparing Figs. 10.17(a) and (b), especially when the minor tributaries are examined [see for example, Area (C)].

Figure 10.17(c) provides us with an intermediate look-direction, that is, between the look-directions of Figs. 10.17(a) and (b). The imagery produced at

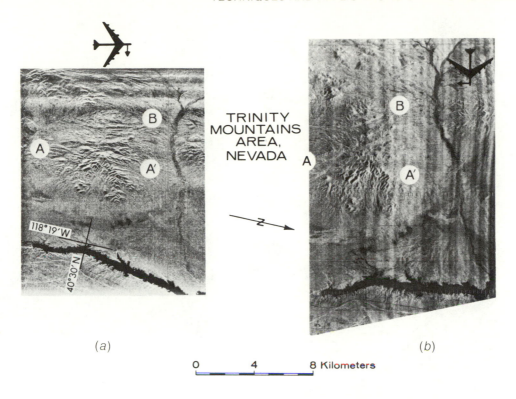

FIGURE 10.16. Orthogonal look-directions of the same area, Trinity Mountains area, Nevada.

this intermediate look-direction appears to have compromising qualities for the detection of linear topographic features. This is to say that Fig. 10.17(c) displays most of the features discussed above at an intermediate stage of definition between suppression and accentuation.

Depending on the relative topographic relief, effective incidence angle, and look-direction, geologic features can be advantageously enhanced or can be completely suppressed. Even where the structural grain of an area is known, opposite-look coverage for all flight lines has improved the geologic information extractable from the imagery (17). Maximum data retrieval from radar geological reconnaissance in poorly mapped areas may necessitate imaging the specific region from two orthogonal look-directions.

10.6 PENETRATION

10.6.1 Atmosphere

Propagation losses through the atmosphere (absorption by water vapor and oxygen) generally increase with increased radar frequency. This is to say that atmospheric effects on radar are confined

to shorter wavelength systems; however, absorption and reflection from precipitation are somewhat more of a problem. The reflection from water droplets is used by weather radars to distinguish regions of precipitation. Heavy rain (between airborne radars and ground) can provide an area of high return on radars having wavelengths of 1 cm or less, which will effectively "shadow" the terrain making interpretation impossible. The effect of moderate rain on radar return is illustrated in Fig. 10.18. Only about 15% of the energy from the terrain is returned to the 1-cm radar through moderate rainfall, whereas the same amount of precipitation is essentially negligible for 3-cm radar wavelengths (18).

10.6.2 Vegetation and Surface

Vegetation may obscure significant geologic relations or it may be of considerable help in geologic mapping. Although vegetative differences may be related to topographic, soil, rock, or moisture factors or various combinations of these, vegetation can sometimes exhibit a definite geologic preference or selectivity. Field geologists, for example, frequently use selective vegetation growth as a key to the recognition of certain rock units. Aerial

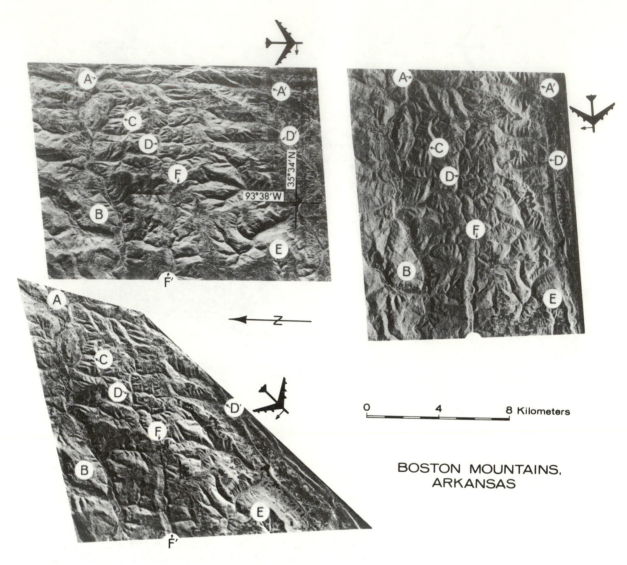

FIGURE 10.17. Multiple flight coverage, Boston Mountains, Arkansas.

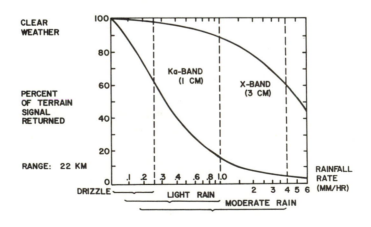

FIGURE 10.18. Effect of rain on radar return at *Ka*- and *X*-band wavelengths. [from Aero-Service Goodyear Brochure (18)].

320

cameras and scanners in the visible- and near-IR regions depend on color and textural contrasts for detection of terrain characteristics in vegetated terrains. Recently numerous studies have shown the seasonal variation in data content of Landsat images, and some of the differences were attributable to contrast in vegetative cover (19).

The wavelengths used in microwave sensing are considerably longer than those of aerial cameras or Landsat-MSS systems. Therefore, the microwave return from vegetation is primarily influenced by roughness (plant morphology) and dielectric properties rather than the cellular structure of vegetative types. In humid areas where vegetation conceals the ground surface, the average radar image tone of short wavelength systems (Ka-X-band) depends on such variables as spacing of individual trees, spacing of branches, size and density of leaves, and orientation or slope configuration of the plant community. Although SLAR systems have successfully provided mapping data for many of the world's most remote jungle regions, geologic interpretation was performed on imagery that basically portrayed the top of a continuous canopy of vegetation. The relative degree of penetration or the depth of material to which the return is sensitive varies directly with the sensing wavelength. At Ka- and X-band wavelengths, the relatively high moisture content and consequent high conductivity of vegetation preclude substantial penetration beneath the vegetative canopy. In contrast with shorter wavelength systems, long wavelength radars hold promise for improving penetration and thereby providing additional geologic information; however, only a very limited amount of long wavelength imagery has been available for geologic analysis. The Environmental Research Institute of Michigan (ERIM) radar system operates two-wavelength synthetic-aperture radar systems that simultaneously image the terrain at X-band (3.2 cm) and L-band (23.0 cm) wavelengths. Alternate X- and L-band horizontally polarized radar pulses are transmitted; however, both like- and cross-polarized energy are recorded for each wavelength resulting in four channels or imagery strips. Shuchman, Davis, and Jackson (20) have published what can be considered some of the highest quality multifrequency SLAR imagery available to the nonmilitary, and comparisons between dissimilar wavelength and polarization images of the same area in southeastern Kentucky are provided in Fig. 10.19. The range of returns from vegetated surfaces at the L-band frequency appears to be substantially less than at X-band. This would suggest that many contrasting terrain conditions defined primarily by subtle vegetation changes will not be provided with relatively long wavelength systems.

L-band imagery obtained in forested areas of central Arkansas during time of full leaf (summer) and partial defoliation (late spring) has provided evidence that at L-band wavelengths, a significant amount of leafy matter penetration occurs in deciduous forests, and the resulting roughness (return) comes from a combination of branches, trunks, and perhaps even the ground (21). In forested areas, little difference could be detected between the imagery obtained in summer and when the trees were almost defoliated. Forested areas provided relatively uniform gray tones regardless of the amount of leaf volume; however, cultivated fields, pasture, and bare fields were recorded as relatively dark areas.

Although the amount of L-band imagery has been limited, it is apparent that long wavelength systems with improved penetration capability have the potential for minimizing the vegetation contribution of radar return signal, perhaps resulting in the enhancement of certain surface return variations. Where vegetation can be eliminated as a factor influencing return, the surface reradiation may be analyzed for variations in roughness or composition (primarily moisture content).

The success of imaging radars to provide evidence of penetration of surface materials where vegetation is absent has not been well documented. Schaber et al. (22) have examined L-band imagery in the Death Valley area of California and infer some penetration in carbonate- and sulfate-cemented sands and silts. Very shallow (few centimeters) penetration is also suggested in the floodplain deposits where the influence of the water table is believed to be reflected in the return signal.

10.6.3 Glacial Ice and Lunar Soils

Glacial ice, in contrast to vegetation and soil, can be penetrated to significant depths with radar energy. Glacial ice attenuates microwave energy much less than soil or rocks, and lack of moisture allows for low conductivity and low attenuation. Porcello and others (23) have provided interesting penetration data with the Apollo 17 Lunar Sounder

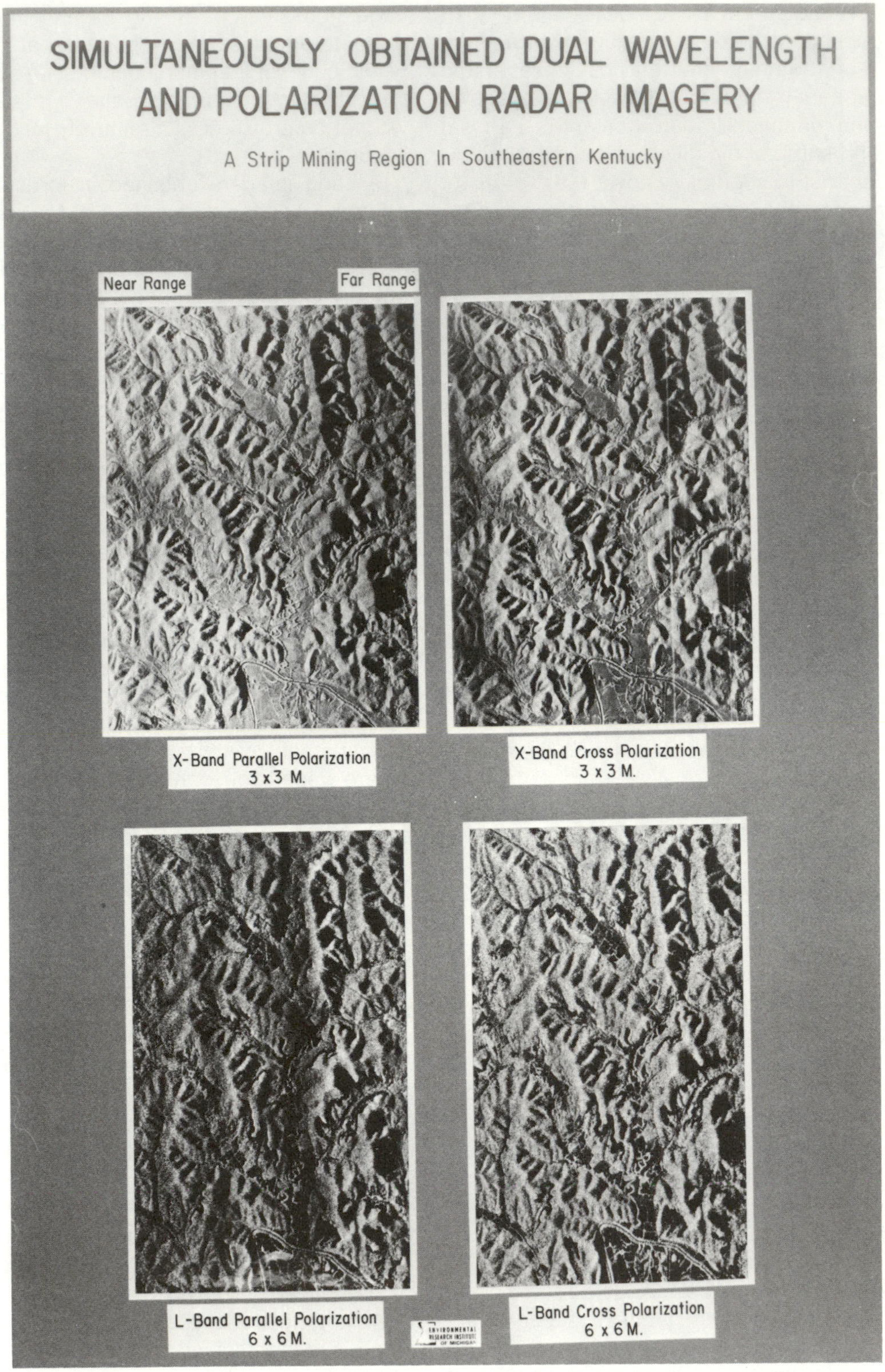

FIGURE 10.19. Simultaneously obtained dual wavelength and polarization radar imagery of a strip mining region in southeastern Kentucky, resolution of X-band imagery 3 m and L-band 6 m (courtesy Environmental Institute of Michigan, ERIM).

Experiment (ALSE). A three-wavelength synthetic-aperture radar operating at 60 m, 20 m, and 2 m (5, 15, and 150 MHz) was designed to detect subsurface geologic structure. Lunar soils, because of their very low microwave attenuation (absence of free water) allow significant penetration depths at radar wavelengths, and detection and location of subsurface discontinuities (sounding) has proven feasible.

10.7 IMAGERY INTERPRETATION

When using remote sensors that cover any part of the electromagnetic spectrum other than the visible, the geologist is initially handicapped because he normally cannot use interpretive techniques that were developed through experience. Through their long association with vertical aerial photographs, geologists tend to be less familiar with nonimaging data formats and prefer a presentation that provides a reproduction of the terrain. When using data from newer remote sensing imaging systems, the geologist must discover new "signatures" that may yield clues for identifying or inferring a particular terrain feature. With the output array of SLAR systems (radar imagery) however, basic interpretive techniques developed for photogeologic interpretation are quite applicable. While radar imagery may differ from aerial photographs in geometry, spatial resolution, and terrain phenomena sensed, most competent photogeologists find an ease of transition from interpreting aerial photographs to interpreting radar imagery.

10.7.1 Imagery Recognition Elements

For geologic interpretation of radar imagery the tone, texture, shape, pattern, size, and shadow are recognition elements on the imagery which contribute to the extraction of geologic data.

TONE

Radar imagery tone can be defined as the intensity of signal backscatter, converted to a video signal and recorded on photographic film as shades of gray ranging from black to white. When a transmitted radar signal interacts with the ground, reradiation signal strength will vary according to the character of the terrain or target area. The strength of the return signal determines the intensity of a point on the cathode-ray tube. High intensity returns representing a large intercept of reradiation appear as light tones on positive imagery prints, whereas

low signal returns appear as dark tones on the imagery.

Both the brightest (high return) and darkest (other than shadow) radar imagery tones are generally caused by the influence of specular reflection. Strongest returns from this smooth surface occur when the target surface is oriented to allow a major part of the energy to be returned to the receiver. Conversely, weakest returns will occur when this specular surface is oriented away from the receiver allowing only minor parts of the transmitted signal to be returned.

If the surface is rough to the impinging radar energy, reradiation will occur in many directions, depending on the orientation of smaller reflecting surfaces. This type of reflection, diffuse reflection, allows a broad range of tones to be recorded on the imagery. In the natural terrain environment, where vegetation forms often cover the surface, the average tone on radar imagery is dependent on such variables as spacing of individual trees, spacing of branches, size and density of leaves, and orientation and slope configuration of the plant community. At Ka- and X-band wavelengths, most vegetation surfaces are "rough" except for some leaf surface composites which are relatively flat and act as facets perpendicular to the imaging direction. However, where vegetation is sparse (such as in arid and semiarid environments), the radar imagery tone is influenced by the scattering characteristics of the terrain, that is, surface roughness and geometric orientation. Terrain imaging radars have operated at a variety of wavelengths, from several meters to a few millimeters. In general, the shorter wavelengths are preferable for terrain sensing because they scatter diffusely from a greater variety of surfaces allowing greater discrimination of terrain composition, as reflected in the imagery tone.

TEXTURE

The distribution and frequency of tone changes of individual resolution elements can be considered image texture. Photographic texture has been described as the frequency of tone change within the image and is produced by an aggregate of unit features too small to be clearly discerned individually on the photographs. The scale of the photographs would obviously have an important bearing on this definition of texture. Radar imagery texture similar to photographic texture then is a comparative feature within any one general scale.

Imagery texture can be evaluated as fine, smooth, coarse, grainy, speckled, mottled, irregular, and so forth. Radar image texture is generally used to permit identification and delineation of unit areas contained within boundaries of homogeneity. This concept of texture contrasts with the recognition of imagery pattern (to be discussed later) which refers to an orderly spatial arrangement of unit features. Because of the dependence of image texture on the distribution of tones, this recognition element is less affected by the lack of image calibration than is tone (20). This is to say that regardless of the radar system's gain settings, the image texture remains relatively constant from one image to another.

SHAPE

The shape of objects depends on their genesis. Shape can be defined as a spatial form with respect to a relatively constant contour or periphery or more simply, the telltale outlines or general configuration of surface expressed features. Because cultural features generally have regular geometric shapes, the radar image interpreter can usually distinguish natural from cultural. Even though many landforms usually have irregular outlines, numerous geologic features can be interpreted by their shape alone, that is, alluvial fans, volcanic cones, river terraces, many glacial features, and folded strata are but a few examples.

PATTERN

Pattern can be defined as the arrangement of geologic, topographic, or vegetation features, that is, areas or groups of areas throughout a region with recurring configuration. Patterns resulting from particular distributions of gently curved or straight lines are common and are frequently of structural significance; they may represent faults, joints, bedding, and so forth.

Drainage pattern analysis provides the geologist with an extremely important geomorphic and structural interpretive technique. Drainage patterns (spatial relationships) generally reflect at the surface the underlying structure and rock type. In addition to drainage patterns, vegetation and soil patterns commonly reflect structural conditions, lithologic character of rock type, and distribution of surficial materials.

SIZE

The surface or volume dimensions of an object have been used as a qualitative recognition element on radar imagery. The size of known features on the imagery provides a relative evaluation of scale and dimensions of other terrain features.

SHADOW

The interpretation of subtle terrain configurations is often improved by radar shadow. Shadows on radar, in contrast to those on photography, are black (no return), and while radar shadowing may aid in determining the shapes of certain terrain features, it also limits stereoscopic interpretation. The length of shadows will, under the similar constraints of shadow measurement on single aerial photos, allow for the determination of the height of certain features. Shadow frequency can also be used to infer relative relief, and this aspect of terrain interpretation is discussed under the section dealing with radar geomorphology.

The six imagery recognition elements (previously discussed) should be used in concert by the geologist when interpreting radar imagery. Remembering the general aspects as to why imagery "looks" the way it does, the photo-like rendition of volcanic terrain on the Island of Bali can be readily appreciated (Fig. 10.20).

10.8 RADAR MOSAICS AND MAPS

Nearly all geological surveys require a basis of general geologic mapping available at the commencement of operations and all geologists need some form of base map or image mosaic to record the ground position of sampling sites, and so forth. Where large areas are to be surveyed, it is common to construct map mosaics from adjacent strips of radar imagery. Good quality mosaics require that the overlap of each strip be properly controlled and that the amount and direction of the image shadows be the same for each strip. Most operational SLAR systems provide imagery that conforms to certain mapping standards established by the United States Geological Survey. The results of two geometric accuracy studies involving the use of *Ka-* and *X-* band SLAR operational systems were published by Derenyi (24), and Van Roessel and De Godoy (25), respectively.

FIGURE 10.20. Radar image of volcanic terrain (courtesy Aero-Service Corporation, and Goodyear Aerospace Corporation).

10.9 STEREO RADAR

Generally, SLAR imagery has been interpreted using monoscopic viewing. For examining positive prints or positive image transparencies, standard roll-type light tables equipped with a zoom binocular microscope are quite adequate. For stereoscopic viewing, however, paper prints are much preferable to image transparencies. If a radar mosaic is available for a particular study area, regional geology is first interpreted directly from the mosaic. Those regions on the mosaic that necessitate more detailed examination are then viewed stereoscopically with a standard 2× pocket type stereoscope or a binocular mirror stereoscope. Stereoscopic interpretation is facilitated by using the combination of radar mosaic as the base map, and paper positive prints of overlapping strips from adjacent flight lines. When the problem area has been resolved positive prints are removed from the work table and interpretation is continued on the mosaic. Monoscopic viewing of individual imagery strips can be facilitated by a table-mounted, rear-projection viewer having a variable light source and variable magnification.

Even in areas of subtle terrain relief, radar shadows on the SLAR imagery provide the interpreter a clear relief impression of three-dimensionality. This relief impression is similar to that observed on low sun angle aerial photography. This "pseudo-stereo," which is responsible for the enhancement of terrain features on radar, does not, however, provide the advantage of true stereoviewing.

The geometric fidelity of current operational systems combined with stereoviewing of overlapping radar strips offers considerable interpretive advantages over monoscopic viewing. Koopmans' (26) 1973 radar study for example has provided a comparison of drainage patterns derived from monoscopic radar, stereoscopic radar, stereoscopic aerial photography, and topographic maps in an area near Baudo, Choco, Colombia. In the low relief areas with a canopy of tropical rain forest it was found that drainage information obtained from stereo radar interpretation was quantitatively better than that shown on 1:25,000 topographic maps of the area.

Although geological interpretation can be accomplished with improved confidence on stereo images, there are problems. For example, the shifting of the source of illumination (radar provides its own illumination source) can cause varying intensities to result from similar slopes (refer to Section 10.3.2). Fore slopes may be illuminated and foreshortened, or in layover, while back slopes may be illuminated and foreshortened, or in shadow, being near their true length only where they are imaged at an incidence angle of 90° (grazing). The psychophysical difficulties encountered by the interpreter when viewing stereo radar taken from opposite look-directions are somewhat similar to the topographic inversion problems that result when one aerial photograph is taken early in the morning while another is taken late in the afternoon. Under these conditions, adequate fusion into a three-dimensional model is somewhat difficult with photography, but extremely so with radar imagery because radar shadows are absolutely black (no return). Similarly, totally foreshortened slopes on one radar image can be represented by a single line on the imagery, while the same slopes imaged from opposite look-direction may be near their true length (grazing). Whereas same-side stereo coverage has provided more satisfactory stereo pairs for the geologist than opposite-side, right-angle flight paths are considered to be superior to parallel flight paths (27). Figure 10.21 provides same-side stereo coverage of an area in southeastern Venezuela. The spacing between flight paths used is normally chosen to provide about 60% sidelap, that is, just enough to ensure duplicate coverage of terrain with allowance for navigation errors and areas of near-range layover. In addition to stereo interpretation, available radar systems are adequate for medium-scale stereomapping with proper selection of flight paths (27).

10.10 RADAR GEOLOGY

Detailed observation of the terrain is required for all phases of applied geology which include exploration for minerals, petroleum, groundwater, and construction materials. Certain remote sensing techniques provide the geologist with a valuable supplement to field geology and for exploration because greater areas and inaccessible terrain can be observed in a relatively shorter time period. Photogeologists, for example, have made use of aerial photographs for several decades, and expert interpretation has provided the identification of geologic features reflected by a characteristic terrain configuration and faithfully reproduced in the photo image. In some areas of the world where the collection of photographic data is constrained by adverse weather conditions, imaging radar systems are proving to be a major remote sensor tool for geological reconnaissance studies. Thus radar geology enables us to obtain geological information through the study and analysis of radar imagery, even though the microwave scattering characteristics of terrain cover and terrain objects are largely unrelated to optical reflectivity.

Determining surface geology from radar imagery has found more practicality than perhaps any other area. The geologist who utilizes radar imagery for geologic syntheses usually has four ultimate objectives: (1) correlation of outcrops from one location to another, (2) determination of the stratigraphic sequence, (3) delimitation of rock types or lithologic units, and (4) determination of geologic structure. Radar imagery provides the geologist with a terrain format approximating a three-dimensional strip map that can reveal varying amounts of geologic information depending on the terrain environment and stage of erosional development. For example, in areas where rock type and structure are reflected in the topography, it may be possible to recognize certain geological features quite unequivocally from the evidence provided by the radar imagery. However, in many areas of the world, where bedrock is obscured by surficial deposits or vegetation, and where the underlying geologic conditions have little direct influence on the surface, the interpretive skill of the geologist to analyze landforms becomes the critical link for structural and even lithologic evaluation. The extent to which radar geological interpretation methods can be successful varies considerably depending primarily upon the geological and geomorphological character of the region and, of lesser importance, the vegetation cover.

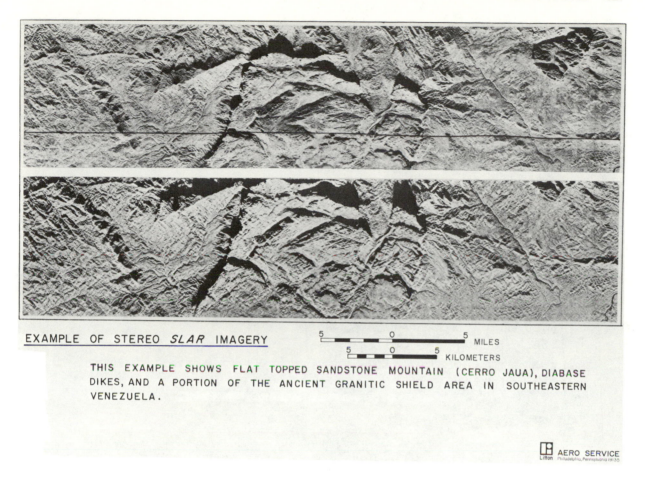

EXAMPLE OF STEREO *SLAR* IMAGERY

5   0   5 MILES

5   0   5 KILOMETERS

THIS EXAMPLE SHOWS FLAT TOPPED SANDSTONE MOUNTAIN (CERRO JAUA), DIABASE DIKES, AND A PORTION OF THE ANCIENT GRANITIC SHIELD AREA IN SOUTHEASTERN VENEZUELA.

AERO SERVICE
Litton Philadelphia, Pennsylvania 19135

FIGURE 10.21. SLAR stereo coverage, showing flat-topped sandstone mountain (Cerro Jaua), diabase dikes, and part of the ancient granitic shield area in southeastern Venezuela. Scale 1:250,000 (courtesy Aero-Service Corporation, and Goodyear Aerospace Corporation).

Geologic maps can be compiled from detailed mapping on the ground, from analyses of aerial and satellite photography and imagery, and from radar imagery. When possible, the geologist uses all these techniques, generally starting with small-scale satellite data and eventually employing ground surveys. Landsat and Skylab results indicate that much of the geologic information also can be obtained through geomorphic analysis, primarily of topography and drainage patterns. Though satellite imagery provides an excellent base map because of lack of distortion, optimum data extraction is often restricted because of relatively low resolution, poor weather, single look-direction, and lack of terrain definition in low relief areas. Radar has demonstrated its capability for providing detailed terrain information regardless of weather or lighting conditions plus a choice of multiple look-directions and illumination angles.

10.10.1 Panama

The first extensive geological reconnaissance mapping using SLAR imagery was completed in eastern Panama (15, 28). The radar imagery, covering all of Darien Province, Panama, part of northwestern Colombia, and much of east-central Panama, was obtained in 1967 and 1969 for the United States Army Engineer Topographic Lab in less than 20 h of actual imaging time. During the preceding 30-year period, there had been minimal success in obtaining aerial photographic coverage because of almost perpetual cloud cover, especially over mountains and foothills.

The 1967 data provided a slant-range display that made mosaicking very difficult. The right half of Fig. 10.22 is composed of slant-range imagery. In contrast, the 1969 imagery was a ground-range display, which made mosaicking less difficult. The left half of the uncontrolled radar mosaic (Fig. 10.22) was constructed from ground-range imagery. With the exception of data provided by scattered field investigations, geologic information interpreted from the radar imagery of eastern Panama far exceeded that which was previously available. In this first geologic "test" of imaging radars it was found that radar imagery facilitated physiographic differentia-

tion on a regional scale. At the minimum, quick subdivision of igneous and sedimentary rocks could generally be made. Large scale structures were synoptically studied, and the single strip format used in conjunction with a radar mosaic helped familiarize geologists with features of structural provinces. On a more detailed scale, a relative stratigraphic sequence could be determined if lithic units were expressed in the terrain configuration and the structure not too complicated. Here, stereo-viewing of large-scale opaque prints was a great help. A summary sketch map of the final interpretive product from the Panama study is provided in Fig.

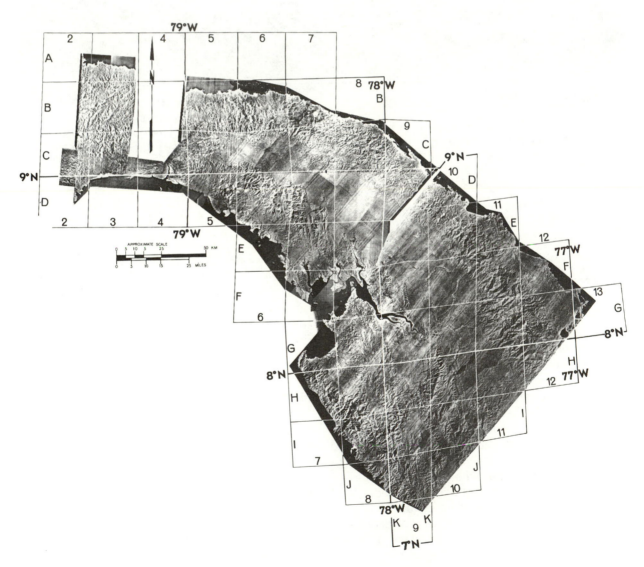

FIGURE 10.22. Radar mosaic, Eastern Panamanian Isthmus.

10.23; however, the reader is referred to the 1971 study published by Wing (29) for an insight to the geological significance of this particular investigation.

The Panama radar mapping project proved to be the catalyst that ignited radar remote sensing interest throughout the world. Since that time many worldwide radar mapping programs have been conducted by mining and petroleum companies, as well as by the foreign governments themselves. One radar mapping company for example, Goodyear-Aero Service, has radar mapped some 12 million km² in 11 countries on four continents from 1971 to 1977. Areas of extensive radar mapping include: Australia, Bolivia, Brazil, Colombia, Indonesia, Guatemala, New Guinea, Nicaragua, Nigeria, Peru, Philippines, Togo, and Venezuela. Unfortunately, because of the proprietary nature of much of this information, specifics relating to the success of such programs have not yet been widely publicized. There are exceptions and they are related as follows.

10.10.2 Venezuela

The first major radar survey in South America was contracted for by the Venezuelan government in the spring of 1971. This program was significant for a number of reasons. First, it provided the

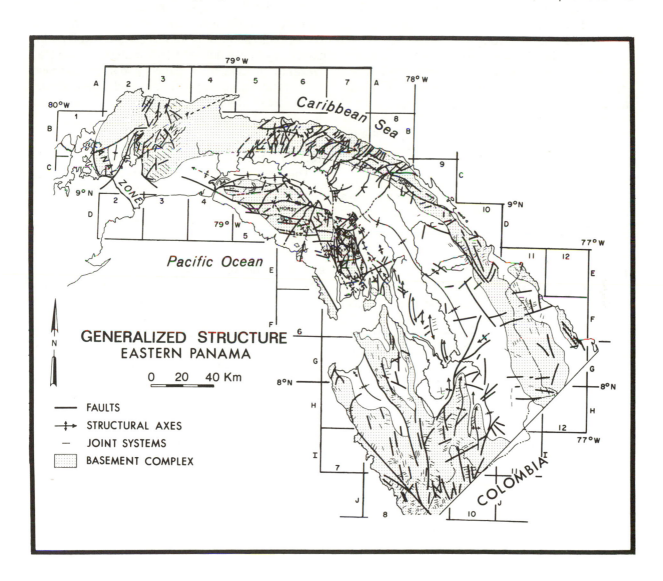

FIGURE 10.23. Generalized structure map, Eastern Panamanian Isthmus. Compare with Fig. 10.22.

introduction of synthetic-aperture, side-looking radar in its first unclassified nonmilitary application. Second, it represented at that time the largest single, commercial radar mapping program in history, ultimately covering more than 344,000 km². Last, but far from least, it was to produce positive results that would lead to programs that have provided more than 10 million km² of radar mapping in South America.

The Venezuelan radar project produced a 10-fold increase in accuracy of the known location of boundaries of the territory with neighboring countries, and systematic mapping of the territory's water resources, including the discovery of the source of several major rivers. The geologic interpretations have not only given the first true picture of the general distribution of rocks in Amazonas but also pinpointed a highly anomalous location, later announced as a major mineral discovery (iron and rare earth elements). Radar mosaics of the quality illustrated in Fig. 10.24 were one of the products of this study. Here a 160 km wide swath of imagery of the Guayana shield area of Venezuela is illustrated, with the Orinoco River in lower right. This is largely a granitic complex containing a large number of granitic intrusions of various ages. The Marahuaca Mountains, which rise to 2538 m, are formed from a diamond bearing massive quartzite known as the Roraima Formation.

10.10.3 Brazil

Until the organization of project RADAM (*RAD*ar of the *AM*azon) in 1971, the Brazilian Amazon Basin was one of the largest poorly mapped areas in the world. The reconnaissance survey of the Amazon and the adjacent Brazilian Northeast initiated the largest SAR project ever undertaken. More than 160 semicontrolled radar mosaic sheets (1:250,000) have been prepared as an information base on mineral resources, soils, hydrology, vegetation, land use, and to obtain a standard base map product. The survey conducted by Goodyear Aero-Service Corporation provided radar imagery and mosaics for an area of 8.5 million km², now being used routinely in all phases of the massive program to conquer the Amazon (18). Worthy of note is the fact that a large part of RADAM project lies outside the cloud-covered equatorial rain forest. Airborne radar was selected over other sensors such as photography because of the distinct advantages to be gained in both time and quality of information

available in producing meaningful maps and interpretation data.

10.11 GEOMORPHOLOGY

Geomorphologists are generally interested in describing or identifying landform features or regions and understanding the processes responsible for shaping the landscape. The inherent characteristics and distortions of radar imaging systems make them especially valuable in terrain analysis study based on topographic relief, slope, or texture. This can be approached qualitatively or quantitatively, and radar imagery can be helpful in either approach. Qualitatively radar imagery can be used for regionalization of landform units as well as for identifying individual landform features. Quantitative landform data, relative relief and slope, can also be determined using inherent radar distortions, radar foreshortening, shadowing, and to a lesser degree layover and parallax.

10.11.1 Qualitative Regional Analysis

The texture, pattern, and shapes of the radar shadows allow the delineation of discrete landform units accurately and easily. Landform units mapped on radar have been found to correlate well with units derived from topographic maps. In fact along a proposed sea-level canal route in Panama the agreement between map-derived and radar-derived units was remarkable, and when determining only three landform units, plains, hills, and mountains, over 90% agreement was found in all cases (30, 31). Geomorphological maps were derived by project RADAM personnel from interpretation of radar imagery and available multispectral photography combined with field work for ground truth checking (18).

10.11.2 Quantitative Regional Analysis

The accurate description of landforms is the first step in any geomorphic study. The three most important vertical dimensions used in landform analysis are elevation, relative relief, and slope. Although radar imagery can be theoretically used to determine individual slope values, its real value focuses on the determination of slope angle distribution on a regional scale. Radar foreshortening, radar shadows, and radar power return have all been evaluated as a source of slope data by comparing the radar-derived data with map-derived

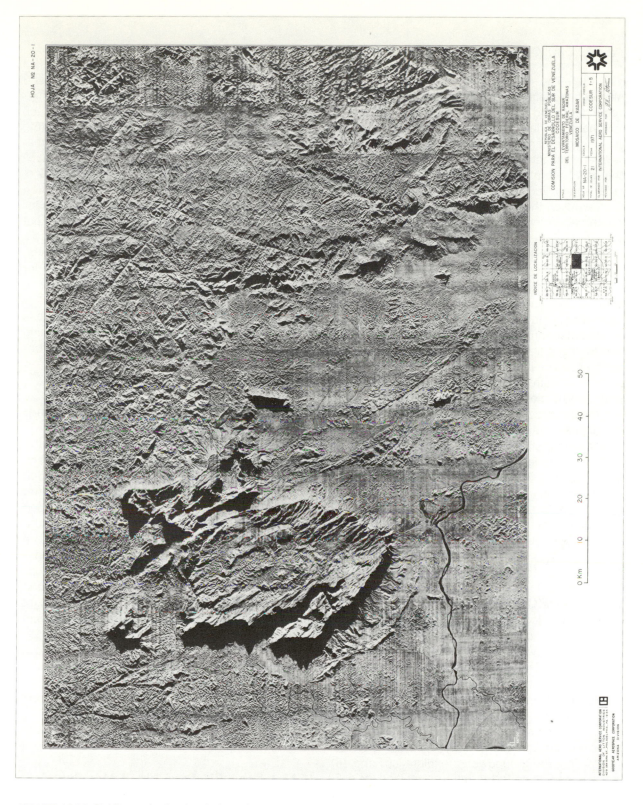

FIGURE 10.24. SLAR mosaic showing Archean basement complex overlain by Roraima formation quartzites in south Venezuela. Numerous linear features indicate extensive faulting and shearing in the granitic complex on top half of mosaic (courtesy Aero-Service Corporation and Goodyear Aerospace Corporation.)

data (32). The studies met with varying success. Although radar layover and parallax are in part slope dependent and therefore a potential source of morphometric data, the difficulty in accurately measuring layover and parallax on radar imagery is a severe limitation to their use. The most practical source of morphometric data is the frequency of radar shadowing (33). For simplicity in determining relative relief, the measurements of radar shadow length can be obtained.

10.11.3 Coastal Analysis

Imaging systems are of special interest to the coastal and wetland geomorphologist. Continuous strip presentation of the land-water interface many kilometers wide and hundreds of kilometers long is advantageous for the study of the relatively narrow coastal zone. In addition, the near all-weather, 24-h imaging capability is a particular asset in coastal and wetland environs commonly obscured by cloud cover.

Change detection is an obvious and documented use of radar imagery. Radar imaging systems are of special importance if a short-term phenomenon or process is being monitored and time of imaging is paramount (34, 35). Lack of water depth penetration makes an imaging radar of limited value for most detailed coastal studies such as turbidity and sedimentation monitoring. For inland water pollution surveillance, SLAR systems appear to have no realistic utility.

10.12 HYDROLOGY

Most hydrologists have primary interest in the occurrence and variation of both surface and groundwater. Within the entire field of hydrology SLAR systems have considerable limitation; however, the following applications are generally compatible with radar's capabilities: (1) estimating stream flow through basin geometry analysis, (2) mapping surface water bodies and marshlands, (3) mapping of snow fields and flood areas, (4) obtaining terrain information for groundwater exploration, and (5) perhaps sometime in the future a method for estimating runoff based on soil moisture changes.

10.12.1 Basin Geometry Analysis

Operational SLAR systems have produced considerable information for drainage analyses, delineation of stream patterns, and water body detec-

tion. However, extensive radar shadowing and relatively crude resolutions as compared with photography make regional studies more feasible. Such applicability has been documented by McCoy (32, 36) and Lewis (30). These studies have shown that each different radar system (depending on wavelength and resolution) yields different amounts of detail, but that the consistency of information content provided by most radar systems allows for the extrapolation of data to the level of detail that would be available on a 1:24,000 scale topographic map. This potential exists for each of the geometric variables of drainage basins, but is strongest for basin area, total network length, total number of stream segments, and basin perimeter.

As a drainage mapping tool, radar imagery has proven to be very useful in compiling and updating drainage maps in inaccessible, cloudy environments such as the Amazon Basin.

10.12.2 Surface Water Bodies and Marshlands

One of the most significant water-related measurements possible with Landsat imagery is that of the areal extent of surface water: lakes, rivers, streams, flooding, and so forth. However, flooded areas are often overcast for long periods after flooding, and estimating the total extent of a flood could be obtained with radar. The high ratio of radar return from land surfaces, compared to its return from open water aids in detecting land-water boundaries. However, there are problems related to the diffuse boundary that exists between lakes and swamp/marsh zones, dark tones (low return) of open sand, covering of the water surface with aquatic vegetation, and in areas of high relief, distinguishing water bodies from radar shadow. Regardless of the qualifications, determining the extent of water surfaces is a task well suited for radar, because of water's specular surface and resultant lack of return on the imagery. Water quality and depth parameters cannot be obtained with imaging radars.

10.12.3 Snow Field and Flood Mapping

The role of radar in snow mapping is a dual one. For freshly fallen, dry snow, radar signals at almost any radar wavelength can penetrate to rather sizable depths. Thus, the radar can disclose the underlying terrain in such cases. On the other hand, hydrologists are strongly interested in forecasting runoff and landslides; and to do so with radar means that the hydrologist must be able to see the snow, not see through it. Old snow (firm or névé) provides a very high return on imaging radars, while glacier ice provides a low return (37). At least at *Ka*-band frequencies, the areal extent of snow fields and glaciers can be accurately determined. Consequently the use of radar for mapping snow and forecasting runoff may be promising, but quantitative experiments have not yet been conducted.

For flood mapping two types of information are of primary concern: (1) timely information on character and extent of flooding, and (2) accurate floodplain delineation. SLAR's capability to provide timely data is well documented; however, the utility of radar for mapping flood areas has not been tested. Less sensitivity to vegetative cover by long wavelength radar systems might provide a valuable sensor for delineating flooded areas in forested regions. Remotely sensed data of floods offers the element of timeliness, synoptic coverage of large areas at low costs, and when active microwave sensors are employed, the advantage of all-weather observations.

10.12.4 Soil Moisture

Perhaps the most critical area in the whole field of hydrology is soil moisture. It is an important factor in runoff and flood prediction, and also of vital importance to plant and crop growth and health. Although theoretically radar appears to have great potential for many hydrologic purposes because of the sensitivity of the dielectric constant to changes in moisture content, little has been done quantitatively to apply radar to hydrology.

10.13 MERGING RADAR WITH OTHER MULTISPECTRAL DATA

There seems to be a growing awareness among remote sensing specialists that ultimately remote sensor data from many sources have to be combined for optimum interpretability. A major radargrammetric application is the merging of radar image data with imagery from other sensors. Graham (38) has shown that a radar-Landsat synergism appears to retain all the information available from each sensor while providing additional detailed data resulting from the simultaneous viewing of the two in superposition. The problems of geometric alignment can be lessened if radar

swathwidths and look-directions are chosen to approximate Landsat's track and direction of solar illumination (azimuth).

10.14 SEASAT

Seasat-A launched on June 26, 1978 provided a five-sensor complement: altimeter, scatterometer, microwave radiometer, visible/infrared radiometer, and an imaging radar. The satellite was designed to provide precision altimetry for marine geoid and sea surface topography studies, measurement of global wind speeds and directions by means of radar scatterometry, all-weather global monitoring of sea surface temperatures with radiometers, and an accurate determination of wave directional spectra using the first synthetic-aperture imaging radar flown in space by NASA.

All of the instruments (except the synthetic aperture radar-SAR) were expected to be operated continuously during most of the mission. The SAR was to operate in real time only when it was over appropriate high-data-rate ground stations, and provided land and ocean images with a resolution of 25 m and a swath of approximately 100 km (39).

Seasat SAR depression angles, which are centered on 70° (20° incidence angle for flat terrain), will provide minimal shadowing and consequently, landform discrimination in high-relief terrains will prove difficult because of excessive layover. However, there may be considerable geologic value in the first space radar (40). Unfortunately, a power failure aboard Seasat-A terminated SAR imagery acquisition after mid-October, 1978.

Figure 10.25 provides one of the first Seasat SAR images to be released by the Jet Propulsion Lab, California Institute of Technology. The 90-km swath (centered near Knoxville, Tennessee) was obtained at 6:07 A.M. local time on July 8, 1978. Three physiographic regions can be distinguished: the Great Smokey Mountains of North Carolina to the south; the Valley and Ridge province of the folded Appalachians in the center of the image; and the Cumberland Plateau province of Kentucky to the north. Whereas most major structural trends (such as the Pine Mountain fault at E, Fig. 10.25) are distinguishable in the northern two-thirds of the image, excessive layover in the Smokey Mountains prevents optimum geologic interpretation. Land-water boundaries are clearly recognizable such as Douglas Lake, Tennessee (A), Tennessee River (B),

Lake Cherokee, Tennessee (C), and Norris Lake Tennessee (D).

10.15 SPACE SHUTTLE

A Spaceborne Imaging Radar (SIR-A) geologic mapping experiment is tentatively planned for the shuttle OFT-2 flight in the mid to latter part of 1980. The prime objective of this experiment is to demonstrate the applicability and potential of spaceborne imaging radars as tools for geologic mapping, mineral exploration, petroleum exploration, and fault detection. The experiment will also develop and test techniques to evaluate the utility of extending the spectral coverage of remote sensors, and in particular the complementary roles of Landsat and radar for resource studies (41).

The instrument to be used for the SIR-A, five-day mission will consist primarily of available hardware from previous space programs. The nucleus of the instrument will be the Seasat engineering model, a synthetic-aperture L-band imaging radar. Except for modular additions, the system is essentially unmodified, and the optical recorder is a castoff from the Apollo Program. During the five-day mission the recorder will allow up to six hours of coverage over a variety of predetermined geologic test sites. The 50-km swath will be centered at approximately 47° incidence angle, providing imagery with 40-m resolution in range and azimuth. Only a single polarization mode (**HH**) will be provided; however, the output image will be digitized (with full dynamic range) and recorded on computer compatible tape.

10.16 SUMMARY

Several unique characteristics of microwave sensors make them valuable for geologically oriented investigations. The feasibility of an all-weather, day-night sensor is certainly an obvious advantage in many poorly mapped regions of the world. However, the geologist must keep in mind that where vegetation cover masks the ground surface, the radar signal may be influenced by the combination of the vegetation and the terrain surface beneath the vegetation, and consequently, the extent to which geological parameters can be mapped from microwave terrain data varies considerably depending on the geologic and geomorphic characteristics of a region, on the density of vegetation, and the amount of surficial debris. The low angles of oblique illumination have been demonstrated to emphasize subtle terrain features that normally would have been overlooked using common photographic techniques. Where vegetation is sparse or absent, however, an imaging radar becomes extremely sensitive to the actual surface roughness, and surface particle size and texture dominate the microwave return signal.

The last two decades have provided rapid technological progress in radar remote sensing. Imaging radars that were relatively unknown to geologists only a few years ago are now the geological reconnaissance tools responsible for extensive

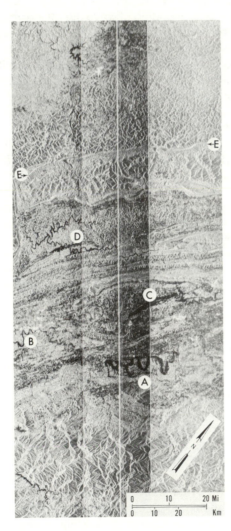

FIGURE 10.25. Seasat SAR image centered near Knoxville, Tennessee. Douglas Lake, Tennessee (A); Tennessee River (B); Lake Cherokee, Tennessee (C); Norris Lake, Tennessee (D); and the Pine Mountain Fault (E).

mapping in the cloud-shrouded tropics. Many geological applications that appear to be theoretically feasible have not been tried, and future active microwave research will provide additional utility for imaging radars. Multifrequency and multipolarization systems from satellite altitudes will ultimately have a tremendous impact on geologic exploration.

REFERENCES

1. Rouse, J. W., Jr., MacDonald, H. C., and Waite, W. P., 1969, Geoscience applications of radar sensors: IEEE Trans. Geoscience Electronics GE-7, no. 1, p. 2-19.
2. Long, N. W., 1975, Radar reflectivity of land and sea: Lexington, Mass., Heath, p. 31-62.
3. Cosgriff, R. L., Peake, W. H., and Taylor, R. C., 1960, Terrain scattering properties for sensor system design (Terrain handbook II): Engr. Exp. Sta. Bull., no. 181.
4. Estes, J. E., and Simonett, D. S., 1975, Fundamentals of image interpretation, in Bowden, L. W., and Pruitt, E. L., eds., Manual of remote sensing, v. II: interpretations and applications: Falls Church, Virginia, American Society of Photogrammetry, p. 995-998.
5. Janza, F. J., 1975, Interaction mechanisms, in Janza, F. J., ed., Manual of remote sensing, v. I: theory, instruments, and techniques: Falls Church, Virginia, American Society of Photogrammetry, p. 144-146.
6. Hanson, B. C., and Moore, R. K., 1976, Polarization and depression angle constraints in the utilization of SLAR for identifying and mapping surface water, marsh, and wetlands: Seattle, Washington, Proceedings American Society of Photogrammetry, p. 499-505.
7. Morain, S. A., 1976, Use of radar for vegetation analysis, in Lewis, A. J., ed., Remote sensing of the electromagnetic spectrum: Assoc. Amer. Geographers, v. 3, no. 3, p. 61-78.
8. Dellwig, L. F., and Moore, R. K., 1966, The geological value of simultaneously produced like- and cross-polarized radar imagery: Jour. Geophysical Res., v. 71, no. 14, p. 3597-3601.
9. McCauley, J. R., 1972, Surface configuration as an explanation for lithology-related cross-polarized radar image anomalies: Fourth Ann. Earth Resources Program Review, NASA/JSC, v. 2, p. 36-1 to 36-6.
10. Goodyear Aerospace Corporation, 1971, Basic concepts of synthetic-aperture, side-looking radar G1B-9167, Litchfield Park, Arizona, 11 p.
11. Jensen, H., Graham, L. C., Porcello, L. J., and Leith, E. N., 1977, Side-looking airborne radar: Sci. Amer., v. 237, no. 4, p. 84-95.
12. Moore, R. K., 1975, Microwave remote sensors, in Janza, F. J., ed., Manual of remote sensing, v. I, theory, instruments, and techniques: Falls Church, Virginia, American Society of Photogrammetry, p. 446-458.
13. LaPrade, G. L., 1970, Subjective considerations for stereo radar: Goodyear Aerospace Corp. Rept. G1B-9169, 13 p.
14. Hackman, R. J., 1967, Time, shadows, terrain, and photointerpretation: U.S. Geological Survey Prof. Paper 575-B, 155-B160.
15. MacDonald, H. C., 1969, Geologic evaluation of radar imagery from Darien Province Panama: Modern Geology, v. 1, no. 1, p. 1-63.
16. Dellwig, L. F., MacDonald, H. C., and Kirk, J. N., 1968, The potential of radar in geological exploration in Proceedings Fifth Symposium Remote Sensing of Environment: Ann Arbor, Michigan, Environment Research Institute of Michigan, p. 747-764.
17. Gelnett, R. H., 1978, Importance of look direction and depression angles in geologic applications of SLAR: Motorola MARS Tech. Rept. TR-04823, Phoenix, Ariz., 15 p.
18. Aero-Service—Goodyear, 1976, Applications and scope of terrain imaging radar: Aero-Service Goodyear Brochure—Litchfield Park, Arizona, 44 p.
19. Short, N. M., Lowman, P. D., Jr., and Freden, S. C., 1976, Mission to earth: Landsat views the world: NASA SP-360, Washington, D.C., 459 p.
20. Shuchman, R. A., Davis, C. F., and Jackson, P. L., 1975, Contour strip-mine detection and identification with imaging radar: Bulletin Association of Engineering Geologists, v. 12, no. 2, p. 99-118.
21. MacDonald, H. C., Waite, W. P., Tolman, D. N., and Borengasser, M., 1978, Long wavelength radar for geological analysis in vegetated terrain: Proceedings of the Amer. Soc. Photogrammetry Fall Technical Meeting, Albuquerque, New Mexico, p. 386-394.
22. Schaber, G. G., Berlin, G. L., and Brown, W. E., Jr., 1976, Variations in surface roughness within

Death Valley, California: geologic evaluation of 25-cm-wavelength radar images: Geol. Soc. Amer. Bull., v. 87, no. 1, p. 29-41.

23. Porcello, L. J., Jordon, R. L., Zelenka, J. S., Adams, G. F., Phillips, R. J., Brown, W. E., Jr., Ward, S. H., and Jackson, P. L., 1974, The Apollo lunar sounder radar system: Proc. IEEE, v. 62, p. 769-783.

24. Derenyi, E. E., 1974, SLAR geometric test, Photogrammetric Eng., v. 40, no. 5, p. 597-604.

25. Van Roessel, J. W., and DeGodoy, R. C., 1974, SLAR mosaics for project RADAM: Photogrammetric Eng., v. 40, no. 5, p. 583-595.

26. Koopmans, B. N., 1973, Drainage analysis on radar images: ITC Journal, v. 3, p. 464-479.

27. Graham, L. C., 1975, Flight planning for stereo radar mapping: Photogrammetric Eng., v. 41, no. 9, p. 1131-1138.

28. Viksne, A., Liston, T. C., and Sapp, C. D., 1969, SLAR reconnaissance of Panama: Geophysics, v. 34, no. 1, p. 54-64.

29. Wing, R. S., 1971, Structural analysis from radar imagery of Eastern Panama Isthmus: Modern Geology, v. 2, no. 1, p. 1-21, p. 75-127.

30. Lewis, A. J., 1971, Geomorphic evaluation of radar imagery of Southeastern Panama and Northwestern Colombia: CRES Tech. Rept. 133-18, Univ. Kansas, Lawrence.

31. Lewis, A. J., 1974, Geomorphic-geologic mapping from remote sensors, in Estes, J. E., and Senger, W., eds., Remote sensing techniques in environmental analysis: Santa Barbara, California, Hamilton Publishing Co., p. 105-126.

32. McCoy, R. M., and Lewis, A. J., 1976, Use of radar in hydrology and geomorphology, in Lewis, A. J., ed., Geoscience applications of radar systems: Assoc. Amer. Geographers, v. 3, no. 3, 152 p.

33. Lewis, A. J., and Waite, W. P., 1973, Radar shadow frequency: Photogrammetric Eng., v. 38, no. 2, p. 189-196.

34. Lewis, A. J., and MacDonald, H. C., 1973, Radar geomorphology of coastal and wetland environments: Proceedings American Society Photogrammetry, Part II, Fall Convention, Lake Buena Vista, Florida, p. 992-1003.

35. MacDonald, H. C., Lewis, A. J., and Wing, R. S., 1971, Mapping and landform analysis of coastal regions with radar: Geol. Soc. Amer. Bull., v. 82, no. 2, p. 345-358.

36. McCoy, T. M., 1969, Drainage network and analysis with K-band radar imagery: Geographical Review, v. 39, no. 3, p. 493-512.

37. Waite, W. P., and MacDonald, H. C., 1970, Snowfield mapping with K-band radar: Remote Sensing of Environment, v. 1, no. 2, p. 143-150.

38. Graham, L. C., 1977, Synthetic aperture radar geologic interpretation techniques: in Microwave Remote Sensing Symposium Proc., NASA/JSC Houston, Texas, Dec. 1977, p. 129-140.

39. Dunne, J. A., 1976, SEASAT-A data acquisition and distribution: Second Annual Pecora Symposium Amer. Soc. Photogrammetry and USGS, Oct. 25-29, p. 71-84.

40. Estes, J. E., and others, 1976, SEASAT land experiments, in Matthews, R. E., ed., Active microwave workshop report: NASA Conference Publications 2030, NASA/JSC, Houston, Texas, 300 p.

41. Simonett, P. S., 1976, Space shuttle radar and its applications to mapping, charting, and geology: applications review for a space imaging radar, Santa Barbara: Remote Sensing Unit Tech. Rept. 1, UCSB, Santa Barbara, California, p. 6-1 to 5-27.

ADDITIONAL REFERENCES

Bryan, M. L., 1973, Radar remote sensing for geoscience: an annotated and tutorial bibliography: ERIM Rept. 193500, Environmental Research Institute Michigan, Ann Arbor.

Dellwig, L. F., and others, 1975, Use of radar images in terrain analysis: an annotated bibliography: ETL Rept. 0024, U.S. Army Eng. Topo. Lab., Fort Belvair, Virginia, 318 p.

ADDITIONAL READING

Long, M. W., 1975, Radar reflectivity of land and sea: Lexington, Massachusetts, Heath.

Matthews, R. E., ed., 1975, Active microwave workshop report: NASA SP-376, Houston, Texas.

CHAPTER 11

TECHNIQUES AND APPLICATIONS OF MICROWAVE RADIOMETRY

THOMAS SCHMUGGE

11.1 INTRODUCTION

By means of sophisticated radio telescopes carried aboard artificial satellites, a new and striking view of our world as a planet can be obtained, using a technique called passive microwave remote sensing, where the microwave region of the spectrum is taken to be the wavelength range from 1 mm to approximately 30 cm (1 to 300 GHz).

In studying the microwave emission from other planets, radio astronomers have used an important relationship between the observed radio brightness temperature, T_B, and the surface conditions. This relation states that brightness temperature is proportional to the product of the surface temperature and a quantity called the emissivity, & , which depends on the nature and composition of the surface. The presence of a planetary atmosphere adds some smaller contributions. Surface temperature does not vary much locally, but the emissivity does, so small changes in surface properties can dominate the observed pattern of brightness temperature. Thus, a radio image of the Earth can differ considerably in appearance from visible-light or infrared images.

Observations of this kind are made with a microwave imager carried on the Nimbus 5 meterological satellite, launched on December 11, 1972 (Fig. 11.1). This device is known as an electrically scanning microwave radiometer (ESMR). It consists of a stationary slotted wave-guide array about a meter square, with a sensitive receiver operating at a frequency of 19.35 GHz (1.55-cm wavelength). Electronic techniques are used to sweep the beam from side to side across the satellite's track over the earth's surface. Thus, no moving parts are needed. In each such sweep, the beam is halted (by an onboard computer) at 78 positions for 0.047s each, while a brightness temperature is observed with an accuracy of 1.5°K. A total of 4 s is required for each sweep.

Nimbus 5 is traveling around the earth in an almost circular retrograde orbit, about 1100 km (nearly 700 miles) high and inclined 81° to the equator. This orbit was designed to be sun-synchronous, with the satellite crossing the equator northward at local noon. One revolution or 107 min later, it crosses the equator at a longitude 27° further west, again very near noon. Hence, the lighting conditions and therefore the surface temperatures at a given place are approximately the same on consecutive revolutions.

Other earth viewing sensors on board the Nimbus 5 spacecraft include a five-channel microwave radiometer (NEMS) for atmospheric soundings and several infrared radiometers. Two of these, the Selective Chopper Radiometer (SCR) and the Infrared Temperature Profile Radiometer (ITPR) were for atmospheric soundings. The remaining two, the Surface Composition Mapping Radiometer (SCMR) and the Temperature Humidity Infrared Radiometer (THIR), were for surface temperature observations in the 8-14μm band. The remaining components indicated in Fig. 11.1 were for data transmission and satellite operations and control.

Figure 11.2, color plate 17 is a radio image of the entire earth, prepared from Nimbus microwave-image data taken January 12-16, 1973. To show T_B differences, the charts are color coded at intervals of 10°K. Occasional black squares indicate gaps in the data. The most striking feature of the radio map is the strong contrast between land and water, with differences of up to 160° in T_B. Central South America, central Australia, and central Africa show the highest values, due mainly to the very high emissivity of vegetation over hot ground. The lowest values occur over oceans in the temperate and subarctic or subantarctic zones. Soil moisture effects can be seen in the Mississippi Valley, where damp ground covered by dead or dormant vegetation has a lower T_B than its sur-

NIMBUS 5

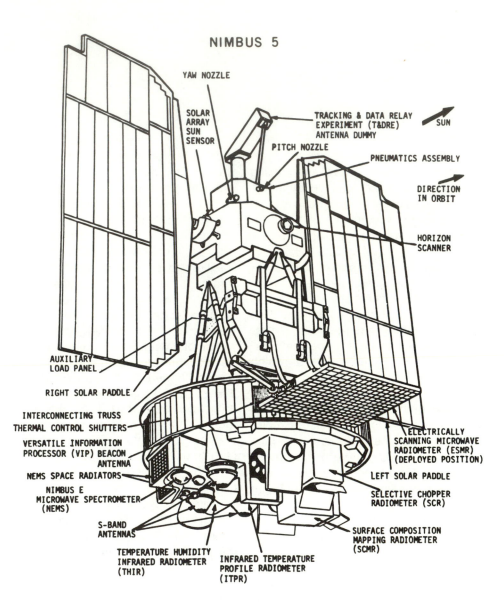

FIGURE 11.1. Nimbus 5 spacecraft with associated experiments.

roundings. The sharp boundaries between the Sahara and central Africa, also the boundary between South American coastal desert and the Andean plateau, reveal differences in soil moisture and soil fertility through the change in vegetation cover. The low T_B over the oceans can be strongly increased by water vapor, clouds, or rain in the intervening atmosphere. The lighter blue band along the equator is associated with the intertropical convergence zone, long known as a region of extensive bad weather and high humidity. Other bands extending off toward higher latitudes correspond to extratropical weather fronts. The series of alternating cold and warm bands from Japan to the west coast of the United States represent the successive positions of a fast-moving storm which nearly crossed the Pacific Ocean during the four days of observation (1).

The microwave radiometers on the Nimbus 5 spacecraft were not the first in space. Radiometers on board the Mariner 2 spacecraft made observations of Mars at 1.35- and 1.9-cm wavelengths and the first observations of the Earth were made from the Russian satellite, Cosmos 243. However, this satellite, which carried four radiometers in the wavelength range from 0.8 to 8.5 cm was in space for only a short period, that is, a few weeks. Nimbus 5 thus was the first spacecraft to obtain substantial amounts of microwave radiometric data, as we shall see in the applications section of this chapter.

Microwave radiometry involves the measurements of the thermal emission from the surface at the microwave wavelength and should not be confused with the active microwave or radar measurements discussed in the previous chapter. In radar, a pulse of microwave energy is sent out from the sensor and the return or reflected signal is measured. The strength of the return depends on the surface roughness and dielectric properties of the terrain being studied but not directly on the temperature of the material. The strength of the thermal emission depends on the temperature and dielectric properties of the material and to a lesser extention on the surface roughness.

So far, radar systems have been used primarily for obtaining imagery and have not been used for quantitative interpretation of surface dielectric properties. This is because of the difficulty in obtaining an absolute measurement of the returned signal. Radiometric measurements have been used quantitatively for making estimates of surface and atmospheric properties. Because of the noncoherent nature of thermal emission it is not possible to use the signal handling techniques used in radar systems to acquire the high spatial resolution obtained by these systems.

11.2 BASIC PHYSICS

11.2.1 Microwave Brightness Temperature

What is this quantity T_B that is measured by a microwave radiometer and how is it related to the physical temperature of what is observed? Figure 11.2 shows that T_B was as low as 130°K over the ocean, which has a temperature of about 290°K, while over land with about the same physical temperature it was approximatly 280°K. This difference is a result of the difference in the radiating capacity (emissivity) of the two materials.

Recall that by definition a black body is an idealized body that absorbs all the radiation incident on it. In order to maintain a constant temperature, it must therefore emit an equal amount of energy. For a given temperature T and frequency f the brightness of such a black body is given by the Planck equation

$$W_{fB} = \frac{2hf^3}{c^3} \cdot \frac{1}{\exp(hf/kT) - 1} \qquad (11.1)$$

where h is Planck's constant, k is Boltzman's constant and c is the velocity of light. This equation differs from Equation 8.1 in that it gives the spectral radiant emittance per unit area in a unit frequency interval centered at the frequency, f, instead of the wavelength interval specified in Equation 8.1. In the microwave region the frequency of the wave is used more commonly than its wavelength. This is the reverse of the convention used in the infrared and visible portions of the spectrum and probably results from the fact that microwave techniques evolved from radio technology. Also recall that the frequency of the wave remains constant as the wave travels from one medium to another. In this chapter we will describe the wave by either its frequency, f, or wavelength, λ, which are related by

$$\lambda = \frac{c}{f} \cong \frac{30 \text{ cm}}{f(\text{GHz})} \qquad (11.2)$$

taking the velocity of light, c, to be 3×10^{10} cm/s.

At microwave wavelengths and at all temperatures occurring in terrestrial environments hf is much less than kT and the exponential in Equation 11.1 can be expressed in a series expansion that reduces to

$$W_{fB} = \frac{2f^2}{c^2}kT = \frac{2kT}{\lambda^2} \qquad (11.3)$$

This expression is called the Rayleigh-Jeans approximation and it is accurate to within a few percent if the frequency in giga hertz is less than the temperature in degrees K. Thus at microwave wavelength the brightness of a black body will be linearly proportional to its temperature. For natural objects that are not perfect emitters (gray bodies) the power radiated when that body is at a temperature, T, will be less than that emitted by a black body at that temperature. However, the power emitted by a gray body may be considered as equal to the power emitted by a black body having a cooler temperature. The lower temperature is referred to as the brightness temperature, T_B, of the object. The ratio of T_B to T is the emissivity, $\&$, of the object, that is

$$\& = \frac{T_B}{T}$$

It is the variation of this quantity, the emissivity, that accounts for the wide range of brightness temperature observed in Figure 11.2.

As noted in the discussion of Figure 11.2, the variations in T_B over the ocean are caused by atmospheric effects. To understand this behavior it is necessary to consider all the contributors to the observed T_B.

The radiation received by an airborne or spaceborne radiometer comes not only from the surface being observed but also from the atmosphere as well. This is represented schematically in Fig. 11.3. The atmospheric contribution can be split into two components; the first is the reflection at the surface of the downwelling radiation from the atmosphere and the second is the upwelling radiation of atmosphere directly incident on the radiometer. Mathematically the observed T_B by a radiometer at an altitude H is given by

$$T_B = \tau(H, \theta)[RT_{\text{sky}}(\theta) + (1 - R)T_{\text{surf}}] + T_{\text{atm}}(H, \theta)$$
(11.4)

where τ is the atmospheric transmission and R is the surface reflectivity. The first two terms represent the radiation coming up from the Earth's surface and will be reduced by the factor τ (<1). The factor T_{sky} is the brightness temperature that would be observed by an upward-looking radiometer at the Earth's surface. The third term is the upwelling radiation produced by the atmosphere. The factors τ, T_{sky}, and T_{atm} are interrelated in that a large atmospheric absorption, small τ, implies large values for T_{sky} and T_{atm}.

The quantity $(1 - R)$ is the emissivity for the surface. The large variation at microwave wavelengths of this quantity for terrestrial surfaces causes the major variations in the T_B values observed in Fig. 11.2, color plate 16. These variations in R are caused by dielectric constant differences of the surface materials, which will be discussed next.

11.2.2 Dielectric Constants

The dielectric properties of a material are characterized by its dielectric constant, ϵ, which is a measure of the response of the material to an applied field, such as an electromagnetic wave. This response can be split into two factors. The first determines the propagation characteristic of the wave in the material, that is, velocity and wavelength, and the second is a measure of the energy losses in the media. These two factors are represented by the real (ϵ') and imaginary (ϵ'') parts of a complex dielectric constant. Thus

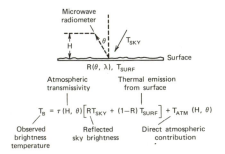

FIGURE 11.3. Schematic representation of the sources of radiation received by a radiometer at an altitude H.

$$\epsilon = \epsilon' - j\epsilon'' \tag{11.5}$$

where

$$j = \sqrt{-1}$$

The ratio of ϵ''/ϵ' is called the loss tangent for the material. In general the values of ϵ' and ϵ'' will be functions of temperature and frequency. Table 11.1 presents values of ϵ' and ϵ'' for a few selected materials, to illustrate the range that may be observed in nature.

It is the large contrast between the dielectric properties of water and those of most solids that make the use of microwave radiometric techniques important for water-resources-related problems. Because of this it is useful to study the dielectric properties of water in more detail. The large dielectric constant for water results from the ability of the dipole moment of the water molecule to align itself along the direction of an applied electric field. The degree of alignment, and thus the magnitude of ϵ, is a function of frequency and temperature. The frequency dependence is given by the Debye formulae for the real and imaginary parts of ϵ.

They are

$$\epsilon' = \epsilon_\infty + \frac{(\epsilon_S - \epsilon_\infty)}{1 + (\omega\tau)^2} \tag{11.6}$$

and

$$\epsilon'' = (\omega\tau)\frac{\epsilon_S - \epsilon_\infty}{1 + (\omega\tau)^2} + \frac{4\pi\sigma}{\omega} \tag{11.7}$$

where
ϵ_S and ϵ_∞ = low and high frequency values of ϵ, respectively
τ = molecular relaxation time resulting from the damping effects in the liquid
$\omega = 2\pi f$
σ = ionic conductivity of the solution in centimeter-gram second units.

The factors ϵ_S, τ and σ are functions of temperature and salinity. The values for pure water and sea water at 20° C are illustrated in Fig. 11.4, where the dielectric constant is plotted versus frequency.

In the microwave region the dielectric constant is changing rapidly from ϵ_S at low frequencies to a much smaller value (ϵ_∞) at high frequencies. The decrease of the dielectric constant with temperature is the result of decreased dipole alignment resulting from thermal agitation. The effect of salinity, for example, in sea water is to introduce an ionic conductivity factor that increase the imaginary part of ϵ at low frequencies but has little effect on the real part.

TABLE 11.1   DIELECTRIC CONSTANTS FOR SELECTED MATERIALS.

	Frequency	ϵ'	ϵ''	&
Water @ 20° C	1 GHz	80	4.5	0.36
Water @ 20° C	20 GHz	35	40	0.39
Dry soil	1.4 GHz	2.8	~0	0.94
Wet soil	1.4 GHz	19.6	4.8	0.59
Granite (ρ = 2.6)	10 GHz	4.4	0.03	0.88
Limestone (ρ = 2.6)	14 GHz	8.2-8.6	0.02-0.04	0.77
Ice (pure)		3.2	< 0.01	

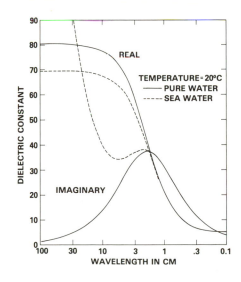

FIGURE 11.4. Real and imaginary parts of the dielectric constant of both pure water and sea water at a temperature of 20° C.

When the water molecule is in ice the dielectric constants have the same Debye relaxation behavior as those of water, with the important difference that the relaxation time in ice is about six orders of magnitude larger than it is in water. Thus the transition from the large static value ($\epsilon' \cong 100$) to the high frequency value of 3.2 occurs in the kilohertz frequency range for pure ice. The imaginary part, ϵ'', is very small for pure ice (less than 0.01 at -1°C) and decreases with decreasing temperature.

For sea ice the situation becomes more involved because of the presence of brine pockets (mixtures of salt and liquid water) in the ice. The average salinity of new sea ice may be as high as 20 parts per thousand, but it decreases with age as the brine pockets drain. The effect of these brine pockets is to increase both real and imaginary parts of ϵ. The resultant values are complicated functions of both salinity and temperature and will not be discussed further here.

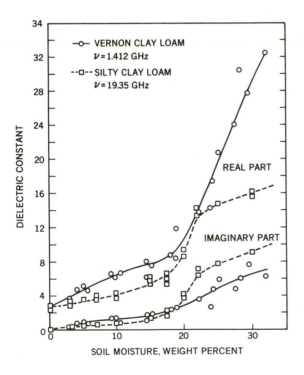

FIGURE 11.5. Real and imaginary parts of the dielectric constant for clay loam soils. The 21-cm data are from the U.S. Army Waterways Equipment Station (2).

The dielectric constants for soils are very strongly dependent on the amount of water in the soil. The results of laboratory measurements of ϵ for soils at wavelengths of 21 cm and 1.55 cm are presented in Fig. 11.5(2). For soil moistures below a certain level, there is a slow increase of ϵ with soil moisture, but above this level there is a sharp increase in the slope. This difference results from the behavior of water in soils. When water is first added to a soil it is tightly bound to the soil particles; in this state the water molecules are not free to become aligned, and the dielectric properties of this water are similar to those of ice. As the layer of water around the soil particle becomes larger, the binding to the particle decreases and the water molecules behave as they do in the liquid; hence, the greater increase of ϵ as water is added at the higher soil-moisture values. The break in slope depends on the soil texture or particle size distribution and occurs at lower soil moisture levels for a sand than for a clay.

The dielectric constants for vegetation have not been measured to the same extent as those for soils, and as a result their behavior is not as well understood. Water is the main contributor to the dielectric properties of plant matter, and it appears that there is an approximate linear dependence of dielectric constants for plants on their water content.

The dielectric constants for rocks and dry minerals are generally independent of frequency in the microwave region of the spectrum. The real part, ϵ', varies from 2.5 for low density rocks to 9.5 for high density basaltic rocks; however for most rocks ϵ' is between 5 and 7. The ratio ϵ''/ϵ', called the loss tangent, is generally between 0.01 and 0.1 for most rocks, indicating that rocks are not very lossy. Because the dielectric constants of most rocks fall within a relatively narrow range, it will be difficult to distinguish rock types on the basis of radiometric observations of dielectric constant differences.

11.2.3 Emissivity

How do we determine the emissivity of a surface? Kirchkoff's law states that the emissivity equals the absorptivity, A, of a surface. For a semi-infinite media, one which permits no transmission through the medium, this yields

$$R + A = 1 \qquad (11.8a)$$

where R is the reflectivity of the surface. Thus the emissivity (ε) is given by

$$\varepsilon = 1 - R \tag{11.8b}$$

The reflectivity R is determined from the Fresnel relations, which describe the behavior of an electromagnetic wave at a smooth dielectric boundary. The relations, derived in any standard textbook on electromagnetic theory, are

$$R_v = \left|\frac{\cos\theta - \sqrt{\epsilon - \sin^2\theta}}{\cos\theta + \sqrt{\epsilon - \sin^2\theta}}\right|^2 \tag{11.9a}$$

$$R_h = \left|\frac{\epsilon\cos\theta - \sqrt{\epsilon - \sin^2\theta}}{\epsilon\cos\theta + \sqrt{\epsilon - \sin^2\theta}}\right|^2 \tag{11.9b}$$

where the subscripts v and h refer to the vertical and horizontal polarizations of the wave. The horizontal polarization is that in which the electric field of the wave is oriented parallel to the reflecting surface and perpendicular to the direction of propagation. The vertical polarization is that in which the electric field has a component perpendicular to the surface. At $\theta = 0°$ these polarizations are equivalent and Equation 11.9 reduces to

$$R = \left|\frac{\sqrt{\epsilon} - 1}{\sqrt{\epsilon} + 1}\right|^2 \tag{11.10}$$

This equation was used to calculate the emissivities presented in Table 11.1. In Fig. 11.6 the reflectivities and emissivities resulting from Equation 11.9 are plotted as functions of nadir angle, θ. It is important to note the difference in behavior between the vertical and horizontal polarizations. As the nadir angle increases from $0°$ the vertical emissivity increases until it reaches 1 at what is called Brewster's angle, θ_E, which is determined by

$$\tan\theta_B = \sqrt{\epsilon} \tag{11.11}$$

An excellent demonstration of this behavior is presented in Fig. 11.7, where measured values of of T_B for three smooth surfaces asphalt ($\epsilon = 4.5$), glacially smoothed limestone ($\epsilon = 6.0$), and a smooth coal bed ($\epsilon = 2.9$) are plotted along with the

values expected from Equation 11.9a for the vertical polarization (3). These results demonstrate the predicted angular behavior and indicate that a radiometer can be used as a means of remotely sensing the value of the dielectric constant for exposed surfaces. This capability is particularly useful when the dielectric constant depends on a parameter that is of geophysical interest, such as soil moisture.

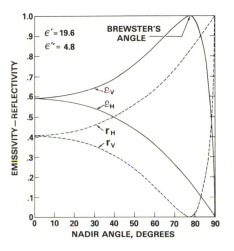

FIGURE 11.6. Emissivity and reflectivity calculated using the Fresnel relations (Equation 11.9) for a wet soil.

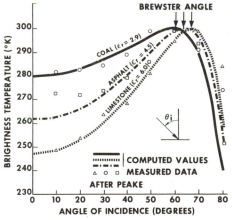

COMPUTED AND MEASURED BRIGHTNESS TEMPERATURE OF COAL (NEAR CADIZ, OHIO), LIMESTONE (NEAR MARBLEHEAD, OHIO), AND ASPHALT SURFACES AT 10 GHz, VERTICAL POLARIZATION. (DATA NORMALIZED TO 300°K AT THE BREWSTER ANGLE)

FIGURE 11.7. Computed and measured values of T_B for smooth surfaces of limestone, asphalt, and coal at 10 GHz and vertical polarization (3).

11.2.4 Atmospheric Effects

The radiation from the Earth's surface interacts with the atmosphere as it travels to the radiometer. This interaction will degrade the information of the surface carried by the radiation. This effect can be seen by considering a homogeneous atmospheric layer at a temperature T_o, absorptivity γ, and thickness d, with the incident radiation characterized by a brightness temperature T_B incident at an angle θ as shown in Fig. 11.8. The outgoing radiation will be given by

$$T'_B = T_B \exp(-\gamma d \sec \theta) + T_o[1 - \exp(\gamma d \sec \theta)]$$ (11.12)

where the first term is attenuated incident radiation and the second is the reradiation by the layer itself. If γ is small, then $T'_B \sim T_B$, and conversely, if γ is large, $T'_B \sim T_o$. Note that if $T_B = T_o$, there is little or no change in the intensity of the outgoing radiation. In this case the effect of the atmospheric layer would be to reduce the contrasts between the adjacent sources, that is,

$$\triangle T'_B = \triangle T_B \exp(-\gamma d \sec \theta)$$ (11.13)

thus degrading the information concerning the surface.

The terms τ, T_{sky}, and T_{atm} in Equation 11.4 can be obtained by summing or integrating the effects of many such layers. The transmissivity for an atmospheric layer of thickness H is

$$\tau(H) = \exp(-\int_0^H \gamma(z) \sec(\theta)\, dz)$$ (11.14)

and for the sky brightness

$$T_{sky} = \int_\infty^0 T(z)\gamma(z)\sec \theta \exp\left[-\int_z^0 \gamma(z') \sec(\theta)\, dz'\right]dz + T_{cb}\tau(\infty)$$ (11.15)

where T_{cb} is the cosmic background radiation, it is usually taken to be 3° K. The direct atmospheric contribution is

$$T_{atm}(H) = \int_0^H T(z)\gamma(z) \sec(\theta) \exp\left[-\int_z^H \gamma(z') \sec(\theta)\, dz'\right]dz$$ (11.16)

The magnitude of absorptivity, γ, depends on the composition of the atmosphere and the frequency of the radiation. The two causes of the atmospheric interaction are the absorption arising from transitions between molecular states of atmospheric gases and the scattering and absorption by cloud and rain particles. For frequencies below

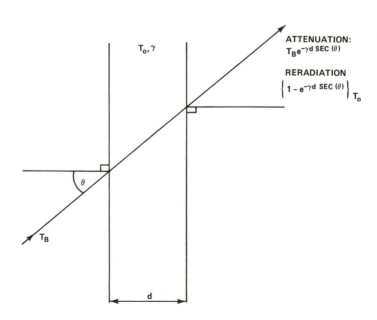

FIGURE 11.8. The effect of an absorbing slab of thickness d on the radiation passing through it.

about 15 GHz both of these interactions become rather weak.

The atmospheric gases that primarily interact with microwave radiation are water vapor and oxygen .The H_2O absorption lines are at frequencies 22.235 and 183 GHz and higher, while the O_2 absorption is due to a band of lines centered about the frequency of 60 GHz and a single line at 118 GHz. At sea level the individual absorption lines are broadened so that the band of lines appears to be one broad line. At lower pressures it is possible to resolve the lines. The effect of these two gases can be calculated by using their known absorption coefficients and by approximating the atmosphere by a large number of thin layers. The result is demonstrated in Fig. 11.9 in which the sky brightness temperature that would be observed by a radiometer looking straight up is plotted versus frequency for two atmospheric conditions. The temperature and pressure profiles used were those of the Standard United States Atmosphere. In one case the relative humidity was set to zero so that only the effects of O_2 are indicated, that is, the absorption band at 60 GHz. In the second case the relative humidity levels of the standard atmosphere are included, and the H_2O line at 22.2 GHz is shown. Included in this brightness temperature is a 3°K cosmic background brightness temperature.

The effects of clouds will depend on the phase of the cloud particles, ice or liquid, and the size of these particles in relation to the wavelength. The effect of an electromagnetic wave on a material particle is to produce a motion of the charges contained within the particle. These charges in motion will absorb some of the incident energy and also reradiate some producing a scattered field. Thus some of the incident energy is scattered out of the beam and some is absorbed by the particle. A more detailed discussion of the interaction with particles involves the theory of scattering of electromagnetic waves and is beyond the scope of this book. However, we can make a few general comments about the interaction:

1. Clouds of ice particles, that is, cirrus clouds, have negligible effect on the microwave signal because of the small particle sizes and the low dielectric constant of ice.
2. For nonraining clouds of water particles, for example, stratus and fair weather cumulus, there is negligible effect for frequencies below about 15 GHz. The absorptivity for clouds of this type is proportional to the square of the frequency.
3. For raining clouds, which have large particle sizes, the effects of clouds are considerable for frequencies above 3 GHz.

For a more detailed discussion of atmospheric effects, the reader is referred to the section by Gaut and Reifenstein (4) in Chapter 5 of the *Manual of Remote Sensing*.

11.3 INSTRUMENTATION

The microwave radiometer is an ultrasensitive electromagnetic receiver that responds to thermal radiation at microwave frequencies. The output of the receiver is proportional to the brightness temperature of the radiating source.

Basically the radiometer consists of an antenna for collecting the thermal radiation and a radio receiver for measuring this radiation. A block diagram of a common type radiometer is presented in Fig. 11.10.

The components of the radiometer, in addition to the antenna, are the microwave switches, which determine the sources of the energy that will be inputed to the receiver. The loads, which are es-

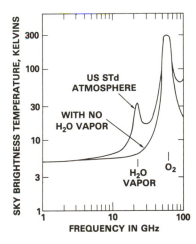

FIGURE 11.9. Brightness temperature observed by an upward looking radiometer as a function of frequency for the U.S. Standard Atmosphere, and for the same atmosphere with no water vapor.

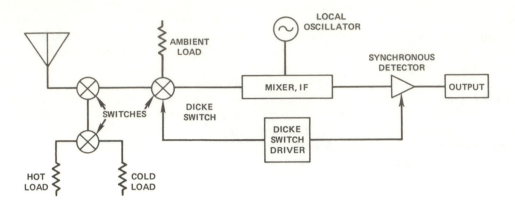

FIGURE 11.10. Block diagram of a radiometer operating at frequency f_0.

sentially black-body sources at known temperatures, are used for calibration and stability purposes. The components of the receiver, for example, mixer and IF, will be described later.

At microwave wavelengths the angular resolution and thus the spatial resolution is limited by the antenna size. The limit is approximated by

$$\Delta\theta \cong \lambda/d \qquad (11.17)$$

in radians, where d is the size of the antenna. Thus a 1 × 1m antenna at a wavelength of 1.5 cm would have an angular resolution of

$$\Delta\theta \cong 1.5 \text{ cm}/100 = 15\text{mrad}$$

The plot of the relative sensitivity for an antenna as a function of angle is called the antenna pattern. Knowledge of this pattern is necessary for the proper interpretation of the observed signal.

A typical pattern is shown in Fig. 11.11(5). The value of $\Delta\theta$ given above is the full beam width at the half-power points. The angular beamwidth out to the first null or zero is about 2.5 times the half power beamwidth. Another complication is the existence of side lobes, that is, the ability of the antenna to receive power from directions other than that of the main beam. If the side lobe level is too high, the interpretation of received signal would be difficult due to the uncertainty of the direction of the source. Therefore, a radiometer should have a narrow main beam with low sidelobe levels. Unfortunately, for a given size the antenna with the narrowest main beam will have the highest side lobe levels; conversely reducing the side lobe levels increases the beamwidth. For radiometers, beamwidth is sacrificed, and the antenna is designed so that 90% or more of the received power comes in through the main beam. Thus the 1 × 1 m antenna of the Nimbus 5 ESMR (λ = 1.55 cm) has a spatial resolution of 25 mrad instead of 15 as obtained from Equation 11.15.

The function of the radiometer receiver is to detect and measure the radiation collected by the antenna. The power level of the signal in radiometer receivers is very small, so that both high sensitivity and stability of the receiver are important requirements. The most common type is a superheterodyne receiver, which includes the mixer and IF or intermediate frequency amplifier. In the mixer, the incoming radiation, which covers a broad frequency range, is mixed with the radiation from a local oscillator at the measurement frequency. When two waves are added together, that is, mixed, the output has components at the sum of the two frequencies and at their difference. The sum component is thrown away and the difference com-

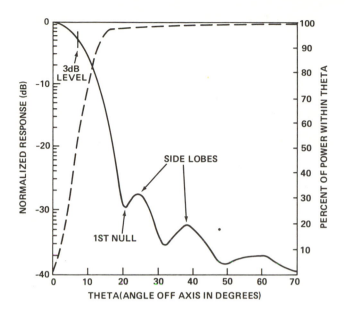

FIGURE 11.11. Antenna pattern for the Skylab 21-cm radiometer. This figure also shows the fraction of the power received within a given angle (5).

ponent is amplified in the intermediate frequency (IF) amplifier. The frequency range of the IF amplifier determines the bandwidth of the radiometer. The term intermediate is used because the frequency of this amplifier, usually tens or hundreds of mega hertz, is between the microwave frequencies of the incoming radiation and the low frequency variations of the T_B signal. Conventional radio and television receivers have these same types of components.

The amplifiers of a highly sensitive receiver always exhibit some drift and gain variations, especially in response to temperature changes, which would prevent the absolute determination of the input radiation. The periodic comparison of the unknown input power with the radiation from a controlled source, the Dicke load, essentially eliminates the effect of drift and gain variations. If the switching is done at a high enough frequency the gain has no time to change during one cycle. Typical frequencies range from 10 to 1000 Hz. This is known as a Dicke radiometer after the scientist who first used the technique for radio astronomical observations in 1946 (6).

The output of the IF amplifier will be amplitude modulated in a square wave at the switching frequency. The amplitude of modulation will be propor-

tional to the difference between the antenna and reference temperatures. This modulated signal is measured in the synchronous detector, which is a switching detector that follows the pattern of the Dicke switch in changing its polarity. Thus the output is tuned to the switching frequency, and will average out amplitude variations at other frequencies.

To calibrate the radiometer, that is, to relate the voltage output to the antenna temperature, the input can be switched to calibration targets or loads at known temperatures for the hot and cold references indicated in Fig. 11.8. For example, the ESMR on the Nimbus 5 satellite uses cold space ($T_B \cong 3°K$) as one calibration target and an internal target at the ambient temperature of the instrument as a second. Thus the voltage obtained with the antenna as input can be compared with the voltage obtained from the known temperature targets to determine the antenna temperature.

This section is an introduction to the basic instrumentation techniques of microwave radiometry. For more details the reader is referred to texts on radio astronomy (7) or the section in the *Manual of Remote Sensing* (8) on microwave radiometry.

11.4 APPLICATIONS OF MICROWAVE RADIOMETRY

Most applications of microwave radiometry have been concerned with water and water-related phenomena because of the strong dielectric properties of water. For example, the difference in the dielectric properties between wet and dry soils permits the remote detection of moisture content in soil. The dielectric difference between ice and water makes possible the determination of the ice water boundary. In atmospheric studies, the strong absorption of the liquid water in rain makes possible the measurement of rain rates over the ocean. These and few other applications of microwave radiometry will be discussed in detail in this section.

11.4.1 Soil Moisture

In Fig. 11.7 it was observed that materials with different dielectric constants would produce different T_B's. Since the dielectric constant for soils are strongly dependent on their moisture content (Fig. 11.5) it should be possible to determine moisture content with T_B observations. Indeed, results similar to those presented in Fig. 11.7 were obtained for soils (Fig. 11.12) by a radiometer mounted on a tower (9). The results for the smooth field in Fig. 11.12(a) indicate the range of T_B produced by soil moisture changes. Notice the difference in behavior of the two polarizations. The vertical values increase as the angle increases, while the horizontal values decreases. If the measurements were extended out to 90° the vertical values would reach 300°K at the Brewster's angle (~70° to 80°) and then drop rapidly to zero at 90°, while the horizontal values would decrease monotonically to zero, as predicted in Fig. 11.6. When the surface is roughened (for example, by plowing) the differences between the polarizations are reduced [Fig. 11.12(b)] and the sensitivity of T_B to soil moisture is also reduced.

This sensitivity to soil moisture has been observed by radiometers on board aircraft flown over test sites at which the soil moisture was measured on the ground. Figure 11.13 presents the results at the 21-cm wavelength from a flight in February 1973 by the NASA Convair 990 aircraft over the north end of the Imperial Valley in southern California (10). The flight line started over the Salton Sea, having at that time the salinity of

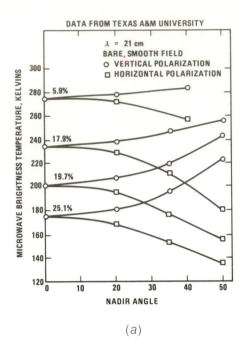

(a)

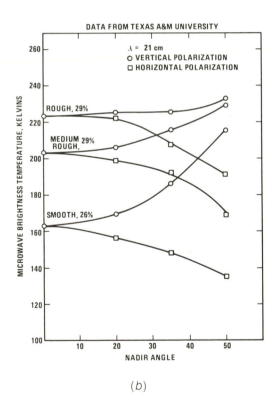

(b)

FIGURE 11.12. Results from field measurements performed at Texas A&M University of T_B versus angle: (a) for a smooth field at different moisture levels; (b) for fields with different surface roughness but essentially the same moisture levels (9).

ocean water, continued east over irrigated agricultural fields, and ended over the desert. The measured T_B ranges from 96°K over the water to approximately 280°K over the dry fields and desert resulting from changes in the surface emissivities. The surface temperatures observed with a 10-12 μm thermal IR radiometer are also plotted and indicate that there were only small variations ($\sim 10°$K) in soil temperature compared to the large differences in ($\sim 80°$K) in T_B resulting from the soil moisture differences in the agricultural fields. Ground measurements of soil moisture were made for a number of fields along the line. The values presented in Fig. 11.13 are for the 0-2.5 cm layer of the soil and are the averages obtained from four samples taken in each 40-acre field. The results from a number of such fields in both the Imperial Valley and Phoenix areas from flights in 1973 and 1975 are presented in Fig. 11.14. The straight line is the linear regression result for the data and is given by

the equation presented in the figure. The scatter in the data results from a number of sources: variations in soil temperature, variations in surface roughness, and uncertainty of the ground measurements. Two fields with the same moisture content can have different soil temperatures which will produce proportional changes in T_B. As noted in Fig. 11.12(b), roughness differences can cause a 60°K spread in T_B for wet fields. The aircraft data does not have a spread this large and indications are that most of the fields are in the medium rough category represented in that figure. The third contributor to the scatter is the uncertainty of the ground measurements of soil moisture. A problem arises when one tries to obtain the average of a highly variable quantity, such as soil moisture, with a small number of point samples. Intensive sampling has indicated that the uncertainty of the average may be as high as 15 to 20% for soil layers near the surface (0- to 2- or 2- to 5-cm layers).

INFRARED (10-12 μm) AMD 21 CM
BRIGHTNESS TEMPERATURES
OVER
IMPERIAL VALLEY, CALIFORNIA

AIRCRAFT ALTITUDE = 0.6 KM

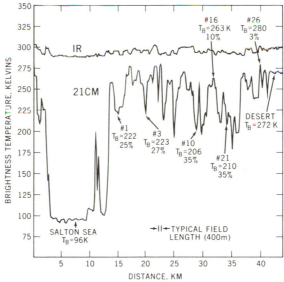

FIGURE 11.13. Thermal infrared (10 to 12 μm) and microwave (21 cm) brightness temperatures versus distance, for a flight path over the north end of the Imperial Valley in Southern California. The 21-cm brightness temperatures and measured soil moisture are indicated for several of the fields (10).

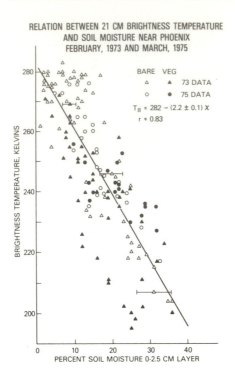

RELATION BETWEEN 21 CM BRIGHTNESS TEMPERATURE
AND SOIL MOISTURE NEAR PHOENIX
FEBRUARY, 1973 AND MARCH, 1975

FIGURE 11.14. 21 cm brightness temperatures versus the soil moisture in the 0-2.5 cm layer of soil.

An important conclusion that can be drawn from the data in Fig. 11.14 is that the effect of moderate amounts of vegetation on the relationship between T_B and soil moisture may be minimal for the 21-cm radiometer. The vegetation encountered in these experiments was wheat 50-60 cm tall. This ability to penetrate vegetation along with clouds greatly extends the utility of this microwave approach for soil-moisture sensing.

The ability of microwave sensors to see through clouds is apparent in the data that is available from satellite, especially in the 1.55-cm Nimbus 5 ESMR and the 2.2- and 21-cm data from Skylab. Figure 11.15 is a comparison of the data obtained by these three sensors on the same day from a Skylab pass across Texas (11). There is substantial change in T_B at 21 cm, but there is little or no change at the shorter wavelengths. While there is considerable difficulty in comparing the space observations that have a large field of view on the ground (25 km at 1.55 cm to 120 km at 21 cm) with ground measurements of soil moisture, it was possible to compare these data with

the precipitation history and a number of ground soil-moisture measurements along the track. From this it was clear that the 21-cm radiometer was responding to soil moisture differences, which were not observed by the shorter wavelength sensors, presumably due to the presence of vegetation obscuring the ground. These results from both aircraft and spacecraft platforms indicate that a 21-cm radiometer has potential for making large area soil-moisture measurements.

11.4.2 Sea Ice

In Section 11.2.2 we noted the large difference in the dielectric properties of liquid and solid water. This is manifested in large brightness temperature difference between the open ocean (140-150°K) and the sea ice (200-220°K) in the polar areas represented in Fig. 11.2. This difference can be used to monitor the areal extent of sea ice for climate studies and as an aid to shipping. Figure 11.16 illustrates this capability using the Nimbus ESMR for the Antarctic, comparing the boundaries in the early springs of 1973 and 1974 (12). The cross-hatched area is the region that was covered with ice in 1973 but not in 1974. The most prominent open area is the 200 × 1000 km region just off the Antarctic coast at 10° longitude and completely surrounded by ice. The 1974 observation was the first of persistent open water on such a large scale well within the sea ice boundary. The opening formed again in 1975 and 1976 but further to the west and is believed to result from the upwelling of relatively warm water.

The dielectric properties of sea ice depend upon the age of the ice. Brine pockets that are formed as the ice freezes drain with time, leaving little air holes in the ice. The magnitude of the effect these air holes or cavities have on the emission from the sea ice depends on the wavelength. As the wavelength approaches the size of the air holes the interaction of the wave with them becomes stronger and more of the radiation is scattered out of the beam, thus lowering the observed T_B for sea ice. This phenomena is called volume scattering and it also occurs in snow. This lowered T_B has been observed in a series of aircraft flights over the sea ice in the Beaufort Sea north of Alaska and Canada. The multispectral response for a 90-km track covering a large multiyear ice floe in an area of predominantly first-year sea ice is shown in Fig. 11.17 for several wavelengths between

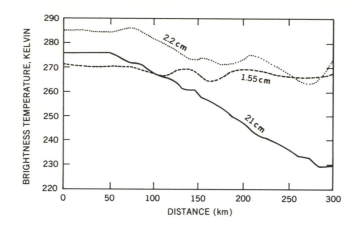

FIGURE 11.15. Variations of the Skylab 2.2-cm and 21-cm and the Nimbus 1.55 cm brightness temperatures along a 300 km track across north central Texas on June 5, 1973 (11).

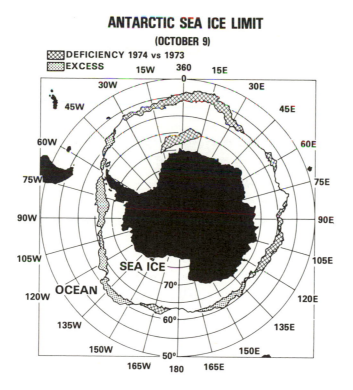

FIGURE 11.16. Comparison Antarctic sea ice maps derived from Nimbus 5 ESMR for 1973 and 1974 (12).

21 and 0.3 cm (13). The flight was over a camp of the Arctic Ice Dynamics Joint Experiment (AIDJEX) so the age characterizations of the ice were verified by surface observations. For wavelengths 2.8 cm and shorter there was a significant decrease in T_B when crossing from the first-year ice to the older ice floe. The magnitude of the difference increases inversely with wavelength. There is a 40° to 50° difference at 0.81 cm, 20° to 30°K at 1.55 cm and only about 15°K at the 2.8-cm wavelength. The refrozen leads or cracks in the multiyear ice floe show up as warm spikes in the data. This ability to distinguish between first-year and multiyear sea ice is important for navigation purposes because the first-year ice will be thinner and softer than the old ice and thus easier to break through.

11.4.3 Meteorological Applications

In Section 11.2.4 is was noted that the role of the atmosphere is that of an absorbing layer between the Earth's surface and the radiometer and as such it degrades the information content of the radiation received by the radiometer. However, these effects can be used to obtain information about the atmosphere. From Equation 11.12 it was observed that if the atmospheric temperature were the same as the T_B of the surface then little information could be obtained about the atmosphere. Therefore, the brightness temperature of the background surface should be significantly different from the atmospheric temperature. This is the case over the ocean where T_B is in the 100° to 150°K range. Therefore, observations of such quantities as rainfall, liquid-water content, and water-vapor content are easy to do over the oceans, but much more difficult over land with its variable T_B background.

In Fig. 11.2, the microwave brightness temperature map of the Earth, the regions of elevated T_B over the oceans correspond to areas of increased moisture content, liquid or vapor. A particular case is observed over the North Pacific where a storm front is seen to move to the east during the four-day period used to average the data. The front is denoted by the dark areas, higher T_B, against the lighter, 130°K, background. The lighter blue band along the equator results from the increased rainfall and humidity associated with the Inter Tropical

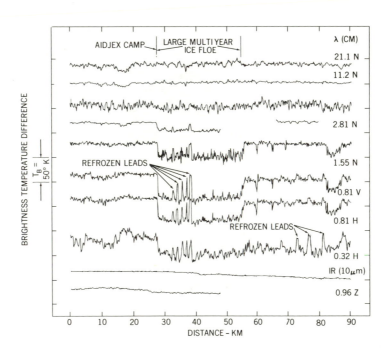

FIGURE 11.17. Multispectral data obtained in March 1971 during a low-level pass at 300-m altitude along a track over the Beaufort Sea. Here, Z = zenith-viewing, H = 45°-viewing in horizontal polarization, and V = 45°-viewing in vertical polarization. The data dropouts in the 2.81- and 0.96-cm radiometers correpond to calibration cycles at those times (13).

Convergence Zone (ITCZ), where the trade winds from the northern and southern hemisphere meet.

Scientists at the NASA Goddard Space Flight Center have used these data from the Nimbus 5 ESMR to determine rainfall rates over the oceans, thus providing information for regions of the earth for which there were very few conventional data available. The first step in this process was the determination of the relationship between T_B and rainfall. This was done by comparing the satellite T_B's with radar observations of rainfall off the coast of Florida. The results are shown in Fig. 11.18 (14). In this figure a curve of the calculated values of T_B versus rain rate is also plotted and it is seen that the radar data are generally within a factor of 2 of the curve. The crosses in the figure are data that were inferred from ground-based radiometer measurements made by looking at the sky during rains.

With the relationship between T_B and rainfall established it was then possible to make the rainfall determinations routinely, and an atlas of the monthly averages of rainfall over the oceans has been prepared (15). The zonally averaged results for the Atlantic ocean are presented in Fig. 11.19. The peak near the equator is the ITCZ which is seen to move north and south seasonally. There are a number of

differences between the two hemispheres that are apparent in these data. For example, the polar fronts near 40° North latitude are more active than their southern hemisphere counterparts.

The more detailed analysis of these monthly maps has yielded some interesting discoveries. In the Atlantic, southeast of South America, an extensive area of rainfall was discovered. This rainy region was not known before, and does not appear in any existing map of global rainfall, probably because few ships traverse the area. Another example is the interannual variation of rainfall over the Pacific in the months of January 1973 and 1974. In January 1973, intense rainfall occurred over a wide region all along the Equator (between 0° and 8° N). This was the time of the El Niño phenomenon (warm ocean current due to relaxation of upwelling along the coasts of Ecuador and Peru), with its disastrous effect on the plankton and fish in the waters of the Pacific off the west coast of South America. For January 1974 (a non-El Niño year), this region was relatively dry. The ratio of rainfall in the period December 1973 to February 1974 to the rainfall in the period December 1972 to February 1973 is 1 to 6, a significant difference. It will require the analysis of many additional years of rainfall to show that this rainfall variation is connected with the changes in the atmospheric circulation over the South Pacific that cause the El Niño phenomenon. In any event, this is a distinctive example of a rainfall anomaly. Investigations of similar anomalies will be very valuable in weather and climate studies (15).

In addition to the ESMR, the Nimbus 5 spacecraft carried the Nimbus E Microwave spectrometer (NEMS) which is a five-channel radiometer designed to make atmospheric temperature soundings and measurements of atmospheric liquid-water content and water vapor. The temperature soundings are done with three channels (53.65, 54.90, and 58.80 GHz) on the low frequency side of the oxygen absorption band, (see Fig. 11.9) and two channels (22.235 and 31.4 GHz) in the water-vapor region for determining the water content (liquid or vapor) for the atmosphere. The spatial resolution of each of these channels is approximately 200 km.

The three oxygen channels were chosen at different distances from the center of the oxygen band to sample the temperature at different levels in the atmosphere. The weighting functions give the frac-

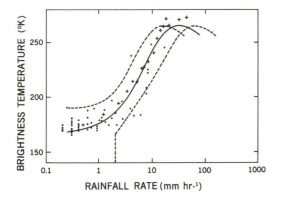

FIGURE 11.18. Brightness Temperature as a function of rain rate: ●Nimbus-5 ESMR vs WSR-57 Radar; +Inferred from ground based measurements of brightness temperature and direct measurements of rain rate. The solid line is the calculated brightness temperature for a 4-km freezing level. The dashed lines represent departure of 1 mm/h or a factor of two in rain rate (whichever is greater) from the calculated curve (14).

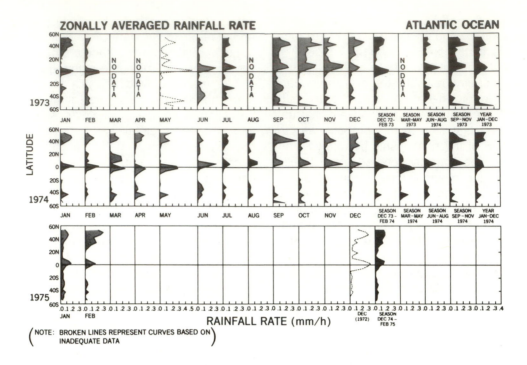

FIGURE 11.19. Monthly zonally averaged rain rates from Nimbus 5 ESMR (15).

tion of radiation received by the radiometer coming from each level of the atmosphere. The channel nearest the center of the oxygen band (58.8 GHz) samples the highest altitude (~ 17 km) due to the stronger absorption at this frequency. The other channels further from the center of the band have their peak responses lower in the atmosphere, at 10 and 4 km, due to the weaker atmospheric absorption at these frequencies. While these channels primarily contain information about the temperatures near these altitudes, they also have some information about the entire profile. Thus it is possible to retrieve information about the entire profile from these data. The mathematical process is complicated and will not be described here. The results from these retrievals are generally within 2° or 3°K of the conventional temperature measurements (16). Microwave temperature sounders are planned for the next generation of operational meteorological satellites so that they will provide temperature profile for even the cloud-covered areas obscured from the present infrared sounders.

As was noted earlier the 22 and 31 GHz channels were selected for the purposes of measuring atmospheric water-vapor and liquid-water content.

Both species would appear as warm emitters against a cold background, such as the ocean, and therefore would be undetectable over a warm background such as land. The 22 GHz channel on the peak of the water vapor line is approximately 2.5 times more sensitive to water vapor than the 31 GHz channel, but it is only half as sensitive to liquid water (17). Thus with these differences in sensitivities, observations at these two frequencies should make it possible to infer the water vapor, W, and liquid water, Q, contents of the atmosphere.

A theoretical analysis has been performed to relate the 22 and 31 GHz T_B's to these quantities W and Q through a pair of coupled linear equations (18). for water vapor

$$W = w_0 + w_1 T_B(22 \text{ GHZ}) + w_2 T_B(31 \text{ GHz}) \qquad (11.18)$$

and for liquid water

$$Q = q_0 + q_1 T_B(22 \text{ GHz}) + q_2 T_B(31 \text{ GHz}) \qquad (11.19)$$

where the quantities w_1 and q_1 depend on the background conditions, that is, surface temperature and emissivity. The constants were evaluated by calculating values of T_B for 60 temperature and water-vapor profiles with and without clouds covering the expected range of these variables. These calculated values were compared with water-vapor and liquid-water content of these atmospheres to determine the constants. An example showing the global variations of water content derived from the NEMS 22 and 31 GHz channel over the Pacific Ocean is presented in Fig. 11.20. Cloud images

from the thermal IR channel of the THIR instrument on Nimbus 5 compare well with the microwave determinations of liquid water. Note particularly the liquid water peak in the region of cloudiness at 50° South. This region also shows the increased sensitivity of the 31 GHz channel to liquid water compared to that of the 22 GHz channel. The high levels of water vapor around the equator produce the elevated T_B's observed at 22 GHz in this region. The atmosphere above 40°N contains little

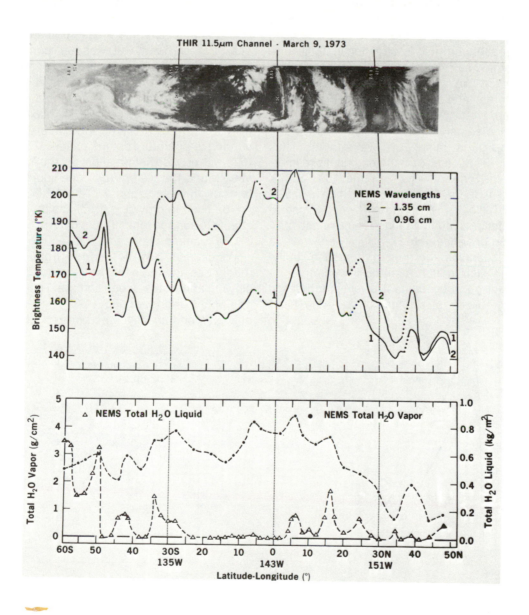

FIGURE 11.20. Distribution of liquid water content and water vapor content over the Pacific as derived from the 22 (1.35 cm) and 31 (0.96 cm) GHz channels of NEMS (18).

vapor and no appreciable liquid water; therefore, both channels are primarily responding to the surface emission.

The predicted ability of the NEMS instrument to probe the atmosphere for temperature profiles and water-content has been demonstrated by the Nimbus 5 data. The Nimbus 6 satellite, launched in June of 1975, carried a scanning version of the NEMS to produce these data over larger areas (17).

11.4.4 Oceanographic Applications

By a suitable choice of wavelengths radiometric observations over the ocean may be used to obtain oceanographic information in addition to the meteorological information described in the previous section. The oceanographic parameters that are of interest are sea surface temperature and wind speed.

As was noted earlier, the emissivity of sea water is less than 0.5 at microwave wavelengths. Thus Equation 11.3 indicates that the brightness temperatures of the ocean will be in the 100°-150° K range and that there will be less than a 0.5° K change in T_B for a 1° K change in ocean temperature. An additional complicating feature is the temperature dependence of the emissivity for water. At certain frequencies there is a decrease in the emissivity which counteracts a temperature increase resulting in a constant or even decreased value of T_B. This effect is demonstrated in Fig. 11.21

where the sensitivity (the change in T_B for a 1° K change in water temperature) is plotted as a function of frequency (19). There is a broad maximum around 5 GHz and nulls near 1.5 GHz and 20 GHz. At the null points, the emissivity is inversely proportional to the absolute temperature and thus the product is independent of temperature. Above and below these null points T_B actually decreases with increasing water temperature.

The scanning multifrequency microwave radiometer (SMMR), which will be on board the Nimbus G and SeaSat-A satellites to be launched by NASA in 1978, is designed to measure sea surface temperature for nearly all weather conditions (20). This instrument will measure T_B at five frequencies (see Table 11.2) in both horizontal and vertical polarizations at an incidence angle of 50°. The 6.6 GHz channel, which has the maximum sensitivity in Fig. 11.21, will be the primary one for sea surface temperature determinations. The remaining four will provide information on the surface wind speed, and the water-vapor and liquid-water content of the atmosphere. This latter information will be used to provide corrections for atmospheric effects to surface temperature determinations of the 6.6 GHz channel.

Windspeed may be inferred from the surface roughness and foam that are produced by the wind blowing across the ocean surface. The effect of the roughness and foam (Fig. 11.22) is to increase the observed brightness temperature for winds above a

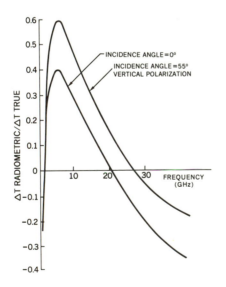

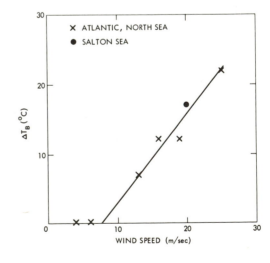

FIGURE 11.21. Sensitivity of T_B of sea surface thermodynamic temperature over the range 0 to 30°C (19).

FIGURE 11.22. Increase of T_B at 19.35 GHz caused by wind at the ocean surface (19).

threshold value of 7 m/s at the rate of about 1 K/(m/s). Subsequent experiments have indicated that the sensitivity is essentially independent of freqency above 10 GHz but that the sensitivity decreases rather sharply at lower frequencies.

Thus in the next few years it should be possible with the SMMR data to make observations of sea surface temperature and surface wind speed on a global basis, along with some properties of the overlying atmosphere. These data will be very useful in studying the air-sea interaction and its possible influence on global weather patterns.

Another oceanographic application of microwave radiometry is the detection of oil slicks. Scientists at the Naval Research Laboratory have studied the change in T_B that is produced by a thin film of oil on the ocean surface (21). The oil film acts as a matching layer between the air and

the sea, thus reducing the reflectivity at this surface. This behavior is much the same as that of an optical coating on a glass lens. As the thickness of the oil film increases, the observed T_B first increases and then passes through alternating maxima and minima due to the standing wave pattern set up by the reflections from the oil-air and the oil-water surfaces. This pattern is illustrated in Fig. 11.23 where the change in T_B is plotted for two frequencies, 19.4 and 31 GHz (free space wavelengths 1.55 and 0.96 cm), as a function of oil thickness. The maxima and minima occur at successive integral multiples of a quarter of the wavelength in the oil, which are 1.06 cm and 0.67 cm for the type of oil used. Thus, the first maxima occur at thickness of 2.7 and 1.6 mm for the two frequencies. Because of the oscillatory behavior of the change in T_B with oil thickness, measure-

Table 11.2. ANTICIPATED SMMR OPERATING CHARACTERISTICS (20)

Frequency (GHz)	6.6		10.69		18.0		21.0		37.0	
Wavelength (cm)	4.55		2.81		1.67		1.43		0.81	
3-dB beamwidth	4.2°		2.6°		1.6°		1.4°		0.8°	
Integration time (msec)	126		62		62		62		30	
Instrument noise and stability (K)	0.4		0.5		0.7		0.7		1.1	
IF range			10-110 MHz							
RF bandwidth			250 MHz							
Polarization isolation			>18 dB							
Dynamic range			10-330 K							
Antenna beam efficiency			> 90%							
Projected antenna aperture			79 cm							
Absolute accuracy			2 K							
Scan period (back and forth)			4.096 s							
View angle			42°							
IFOV, Nimbus G[1]	148	95	91	59	55	41	46	30	27	18
(km) Seasat A[2]	121	79	74	49	44	29	38	25	21	14
			(Nimbus G)				(Seasat A)			
Scan range			-25 to +25°				-3 to 47° to right of track			
Earth incidence angle			50°				49°			
Satellite altitude			955 km				794 km			
Swath			822 km				595 km			
Subsatellite velocity			6.414 km/s				6.623 km/s			

[1]Launched October 23, 1978.
[2]Launched June 26, 1978; failed October 6, 1978.

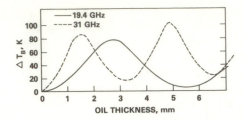

FIGURE 11.23. Increase of T_B at 19 and 31 GHz as a function of oil slick thickness (21).

ments at only one frequency can lead to ambiguous thickness determinations. For example, a ΔT_B at 19.4 GHz of 40°K can be produced by oil thicknesses of 1.6 or 3.9 mm (Fig. 11.23). However, if simultaneous measurements are made at 31 GHz the ΔT_B's will be 90° and 35° K, respectively for these two thicknesses. Therefore, by using two frequencies the thickness ambiguities introduced by the oscillations can be removed.

This approach was tested over a series of small controlled oil spills. The results of a typical experiment are given in Fig. 11.24, which shows (1) the visible slick with contours representing changes in the optical reflectance, (2) isotherms of ΔT_B at 31 GHz, (3) isotherms of ΔT_B at 19.4 GHz, and (4) the inferred value of oil thickness. The latter contour maps are superimposed on the outline of the visible slick. Integration of the thickness contours gives a volume of 2460 liters ± 10%, which is in good agreement with the actual volume of oil spilled, 2380 liters. The conclusions based on a number of such experiments are that the thick region of oil, (≈ 0.1 to 1.0 mm thickness), contains 90% of the oil in less than 10% of the area of the visible slick and that the volume of oil as

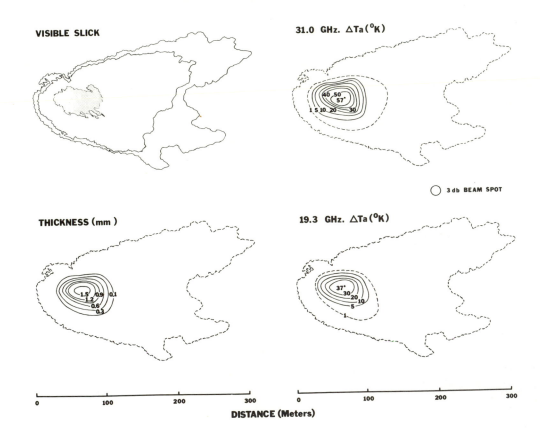

FIGURE 11.24. Sketch of visible oil slick, radiometric isotherms at 31 GHz and 19 GHz, and inferred oil thickness (21).

derived from the microwave measurements was within 25% of the actual volume spilled. Thus multi-frequency passive microwave radiometry offers the potential to measure the distribution and volume of sea surface oil slicks on an all-weather basis.

11.4.5 Geologic Applications

Up to the present time not much work has been done using microwave radiometry for geologic purposes. This is due to the lack of large T_B contrasts between the different rock types. Also the spatial resolution has generally been too coarse to distinguish geologic features. Recently there has been a study using the Nimbus 5 ESMR data over the deserts of North Africa in a particular region where there is an extensive amount of exposed limestone (22). From Table 11.1 it is seen that the dielectric constant of limestone is significantly higher than that of most other rock types. This will cause a lower T_B for those areas where the predominant rock type is limestone. Figure 11.25 is a T_B contour map for eastern North Africa showing a region of lowered T_B's in the middle of the map west of the Red Sea, with the lowest

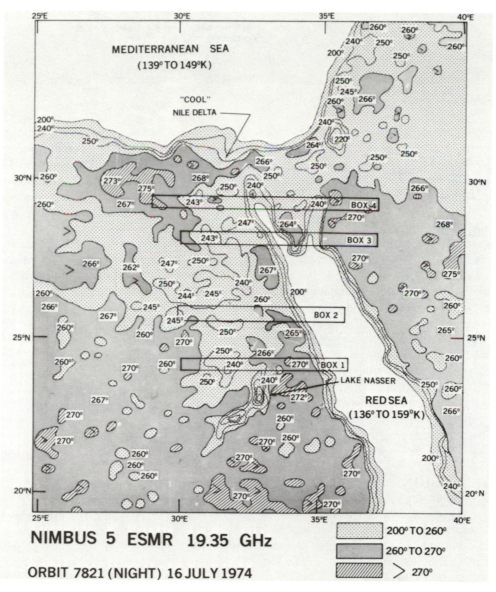

FIGURE 11.25. Nimbus 5 ESMR T_B contour map for North Africa for the night of July 16, 1974 (22).

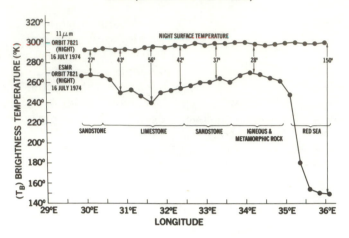

FIGURE 11.26. Comparison of THIR thermal infrared and ESMR 1.55-cm brightness temperatures for box 1 of Fig. 11.25 (22).

value, below 240° K, just northwest of Lake Nasser. The T_B of most of the surrounding area is 260°K or greater. Figure 11.26 is a comparison of the THIR thermal IR channel and microwave responses approximately along the 24° N parallel. The IR surface temperatures were essentially constant, with less than a 10° K warming from west to east, while there was a 30° K change in the microwave T_B along this line. The region of lowest T_B was identified as Eocene limestone, while the areas on either side were identified as Nubian sandstone, which has a lower dielectric constant. This is an example of the radiometer response to rather large scale geologic features. With improved spatial resolutions made possible by large antennas, it should be possible to do further mapping of this nature. However the approach is inherently limited by the lack of large contrasts between different rock types and the fact that soil cover will mask the rock signature in most situations.

11.5 CONCLUSIONS

In this chapter an introduction to microwave radiometry and its application to various fields of geophysics has been presented. The applications have been primarily limited to meteorological or oceanographic studies because of the low spatial resolutions currently available in satellite data. These applications have included temperature and water content soundings of the atmosphere, global rain-rate observations over the oceans, and sea ice mapping. The SMMR on Nimbus-G and Seasat-A will extend these capabilities to include observations of oceanographic parameters, such as sea surface temperature and surface wind speed. In the future, spatial resolutions will be improved when larger antennas become possible through the use of the space shuttle. With the better spatial resolution more land applications will be developed, especially soil moisture or geologic mapping. At that time the capabilities of microwave radiometry will be more fully exploited.

REFERENCES

1. Webster, W. J., Wilheit, T. T., Chang, T. C., Gloersen, P., and Schmugge, T. J., 1975, A radio picture of the earth: Sky and Telescope, v. 49, p. 14-16.
2. Lundien, J. R., 1971, Terrain analysis by electromagnetic means: Technical Rept. No. 3-693, Rept. 5, U.S. Army Waterways Experiment Station, Vicksburg, Mississippi.
3. Peake, W. H., 1969, The microwave radiometer as a remote sensing instrument: Technical Rept. 1903-8, Electrosciences Laboratory, Ohio State University, January, 1969.
4. Gaut, N. E., and Reifenstein, T., 1975, Atmospheric effects on remote sensing for the micro-

wave spectrum, *in* The manual of remote sensing: American Society of Photogrammetry, v. 1, p. 206-216.

5. Lerner, R. M., and Hollinger, J. P., 1975, Analysis of microwave radiometric measurements from Skylab: Naval Research Laboratory, NASA Contract Rept. CR-147442, December, 1975.

6. Dicke, R. H., 1946, The measurement of thermal radiations at microwave frequencies: Rev. Sci. Instruments, v. 17, p. 268-275.

7. Tiuri, M. E., 1965, Radio telescope receivers, *in* J. D. Kraus, Radio Astronomy, Chap. 7: New York, McGraw-Hill, 1965. Also G. E. Evans and C. W. McLeish, RF radiometer handbook: Artech House, Inc., Dedham, Massachusetts, 1977.

8. Ulaby, F. T., 1975, Microwave radiometry, *in* The manual of remote sensing: American Society of Photogrammetry, v. 1, p. 499-534.

9. Newton, R. W., 1977, Microwave remote sensing and its application to soil moisture detection: Technical Rept. RSC-81 of the Remote Sensing Center at Texas A&M University, College Station, Texas, January, 1977. [Available through University Microfilms No 77-20,398]

10. Schmugge, T., Wilheit, T., Webster, W., and Gloersen, P., 1976, Remote sensing of soil moisture with microwave radiometers-II: NASA Technical Note TN D-8321, September, 1976.

11. Eagleman, J. R., and Lin, W. C., 1976, Remote sensing of soil moisture by a 21-cm passive radiometer: Journal of Geophysical Research, v. 1, 81, p. 3600 and J. R. Eagleman, Detection of soil moisture and snow characteristics from Skylab: Final Rept. 237-23, NASA Contract 9-13273.

12. Zwally, H. J., and Gloersen, P., 1977, Passive microwave images of the Polar Regions and research applications: Polar Record, v. 18, p. 431-450.

13. Gloersen, P., Nordberg, W., Schmugge, T. J., and Wilheit, T. T., 1973, Microwave signatures of first year and multiyear sea ice: Jour. Geophysical Res., v. 78, p. 3564-3572.

14. Wilheit, T. T., Chang, A. T. C., Rao, M. S. V., Rodgers, E. B., and Theon, J. S., 1977, A satellite technique for quantitatively mapping rainfall rates over the oceans: Jour. Applied Meteor., v. 16, p. 551-560.

15. Rao, M. S. V., and Theron, J. S., 1977, New features of global climatology revealed by satellite-derived oceanic rainfall maps, Bulletin of the American Meterological Society, v. 158, 1285-1288, and Rao, M. S. V., Abbott, W. V., and Theron, J. S., Satellite-derived global oceanic rainfall atlas (1973 & 1974): NASA SP—410, 1976.

16. Waters, J. W., Kunzi, K. F., Pettyjon, R. L., Poon, R. K. L., and Staelin, D. H., 1975, Remote sensing of atmospheric temperature profiles with the Nimbus-5 Microwave Spectrometer: Journal of the Atmospheric Sciences, v. 32, p. 1953-1969.

17. Staelin, D. H., Rosenkranz, P. W., Barath, F. T., Johnston, E. J., and Waters, J. W., 1977, Microwave spectroscopic imagery of the Earth: Science, v. 197, p. 991-993.

18. Grody, N. C., 1976, Remote sensing of atmospheric water content from satellite using microwave radiometry: IEEE Transactions on Antennas and Propogation, v. AP-24, p. 155-161.

19. Wilheit, T. T. Jr., 1978, A review of applications of microwave radiometry to oceanography: Boundary Layer Meteorology, 13, p. 277-293.

20. Gloersen, P., and Barath, F. T., 1977, A scanning multichannel microwave radiometer for for Nimbus-G and Seasat-A: IEEE Journal of Oceanic Engineering, v. OE-2, p. 172-178.

21. Hollinger, J. P., and Mennella, R. A., 1973, Oil spills: measurements of their distributions and volumes by multifrequency microwave radiometry: Science, v. 181, p. 54-56.

22. Allison, L. J., 1977, Geological applications of Nimbus radiometer data in the Middle East: NASA Technical Note TN D-8469, 1977.

SECTION IV
APPLICATIONS OF REMOTE SENSING TECHNIQUES TO GEOLOGIC PROBLEMS

The 10 chapters in this section describing the actual use of remotely sensed data in different geologic disciplines together comprise the heart of *Remote Sensing in Geology*. Most geologists who use remote sensing during their careers will be involved in interpreting data rather than acquiring or processing it. Accordingly, the greatest emphasis in *Remote Sensing in Geology* is on presenting the reader with examples of the role of remote sensing in problems ranging from mineral exploration to extraterrestrial exploration during the decade ending in 1979.

The section ends with a brief look at new trends, evident today, which will set the pace of remote sensing in the 1980s. These trends will help shape the careers of at least some of the readers of *Remote Sensing in Geology*, and they in turn will influence the course and scope of this developing science.

CHAPTER 12

VEGETATION AND GEOLOGY

GARY L. RAINES
FRANK C. CANNEY

12.1 INTRODUCTION

In most geologic studies vegetation is considered to be a hindrance. However, two-thirds of the land area of the world is moderately to heavily vegetated (1) and most of the world's mineral production comes from the remaining areas of low to moderate vegetation cover. Thus, areas of heavy vegetation, such as the belt of tropical forests, probably contain the greatest reserves of undiscovered ore deposits on land. With the world's increasing needs for raw materials, geologists are now required to explore these heavily vegetated areas.

The purpose of this chapter is to provide an overview of the physical properties of vegetation that are relevant to geologic remote sensing and of the established remote sensing techniques that can be used to obtain geologic information from vegetation studies. Our view of new techniques that are being developed is presented. In order to understand the use of vegetation in geologic studies, an understanding of geobotany is required. Therefore, a brief review of geobotany is given; a more complete review is available in Hawkes and Webb (2), and Brooks (3).

12.2 GEOBOTANY

Geobotany is generally defined as the *visual* survey of vegetation in order to define geologic differences in the landscape. With the advent of remote sensing applications to geobotany and, therefore, the additional capability of using the nonvisible parts of the electromagnetic spectrum, geobotany can now be more broadly defined. Geobotany is the analysis of electromagnetic radiation reflected and emitted from vegetation in order to define geologic differences of the landscape.

The basis of geobotany is the relationship between plant's nutrient requirements and two interrelated factors, the availability of nutrients in the soils and the physical properties of the soils, including the availability of soil moisture. Implicit to the application of geobotany is the assumption that there is some relationship between the soils and the underlying rock.

The nutrients required by plants can be grouped into two categories, macronutrients and micronutrients. Macronutrients are the elements that are needed in large quantities. Some macronutrients are, of course, derived from the atmosphere, and thus are not of geologic interest. Micronutrients are trace elements that are needed only in very small amounts, yet are essential; the list of identified micronutrients continues to increase as analytical procedures and nutritional studies become more sophisticated. The list of macro- and micronutrients recognized as of 1972 is shown in Table 12.1 (3, 4).

All plants have an optimum range of requirements for macro- and micronutrients. Also, certain plants require elements not shown in Table 12.1, such as *Astragalus spp.* that requires selenium.

TABLE 12.1. NUTRIENT REQUIREMENTS OF PLANTS. THE QUERIED ELEMENTS HAVE NOT DEFINITELY BEEN ESTABLISHED AS NUTRIENTS. DATA IS FROM SUTCLIFF (4) AND BROOKS (3).

Macronutrients	Micronutrients
Carbon	Boron
Hydrogen	Chlorine
Oxygen	Copper
Nitrogen	Iron
Phosphorous	Manganese
Sulphur	Molybdenum
Potassium	Zinc
Calcium	? Cobalt
Magnesium	? Silicon
	? Sodium
	? Vanadium

However, the optimum range of nutrients required is not static; plants appear to have variable requirements and tolerances that are dependent on the full range of environmental factors. For example, in greenhouse experiments where all parameters can be controlled, plants can tolerate more adverse conditions in one parameter, such as a high metal content in the soil, without the adverse effects that are known to occur in natural environments where few factors are ideal. This does not imply that greenhouse experiments are not valid, but great care is required in applying concepts developed in greenhouse experiments to natural environments.

The amounts of nutrients in the soil are a function of the effect of the soil-forming processes and the quantity in the parent material of the soil. However, all of the contained nutrients are not necessarily available to the plants. Availability is a function of the physical and chemical properties of the soil, such as soil pH, redox potential, soil texture, soil moisture, exchange capacity, and the presence or absence of complexing agents. Furthermore, the climate and topographic relief greatly effect the availability of nutrients.

The plant community at a particular place is a function of the whole environmental complex, only one factor of which is the local geology. That the influence of geology can be observed in such a complex situation may seem impossible, but it is a fact that has been demonstrated many times.

The application of geobotany can be categorized into three conceptual approaches: the study of plant communities including characteristic florae and specific indicator species, the study of vegetation density including the extreme case of complete absence of vegetation, and the study of plant morphology. Each approach involves studies of the various ways that individual plants, or plant communities, respond to environmental stress. Stress can also affect plants in a fashion that is not detectable by the human eye. This nonvisual aspect of response to stress will be discussed in a later section of this chapter.

The application of remote sensing techniques to these geobotanical concepts requires consideration of the size or aspects of the plants or vegetation canopy to be observed, and therefore the scale of the images to be acquired. Foresters generally prefer images at a scale greater than 1:10,000 in order to accurately identify tree species. Therefore, because of the very large scale images required,

the plant communities approach, where particular isolated indicator species are looked for, will not be applicable except in unusual situations; similarly, most of the diagnostic morphological changes will not be observable. However, plant community information such as gross plant community changes and some morphological changes such as changes in the leaf orientation or size are often most easily observed by remote sensing techniques. Thus, if specific geobotanical phenomenon are being investigated the scale has to be considered.

12.3 SPECTRAL PROPERTIES OF VEGETATION

In what part of the electromagnetic spectrum is the effect on vegetation of a particular geologic factor most likely to be observed? The complete answer to this question is not really known; however, in this section we will review what is known of the spectral properties of vegetation.

A plant is a living organism that uses energy from the sun in order to incorporate the inorganic nutrients into the living organism. This energy from the sun changes in quality and quantity with the seasons; and also a plant's requirements for solar energy change with the seasons. Associated with a plant's changing requirements are changes in the spectral properties of the vegetation. For example, deciduous and coniferous trees have strikingly different seasonal patterns of growth. Consequently, the reflectance spectra of coniferous and deciduous trees are most different in the fall, when deciduous trees have no leaves and the trees are not covered with snow. Clearly this seasonal aspect of the spectral properties may be exploited with infrared photography, even at small scales.

The spectral properties also vary with time of day. Because the angle of incidence of sunlight varies through the day, different mixtures of the sides and tops of the vegetation-canopy components will be illuminated.

The spectral properties can be divided into four categories on the basis of wavelength and the different physical phenomena involved. These categories are reflectance properties in the 0.4 to 2.5 micrometer (μm) region, luminescence properties in the 0.4 to 0.7 μm region, thermal properties in the 3 to 5 and 8 to 14 μm regions, and radar reflectance properties in the 1 cm to 3 m region. These categories will be considered in detail in this section.

COLOR PLATE 1 FIGURE 4.10. Sequence of consecutive
70-mm hand-held pictures taken during Apollo 9 mission by
R. Schweikart. Cape Hatteras at upper left corner is first of
sequence, followed by Delmarva Peninsula, New Jersey, then by
Cape Verde Islands. Last seven frames start at Sierra Nevada
and cover flight path looking north; final frame shows Grand
Canyon and San Francisco Mountains, Arizona (snow-covered).
Taken March, 1969.

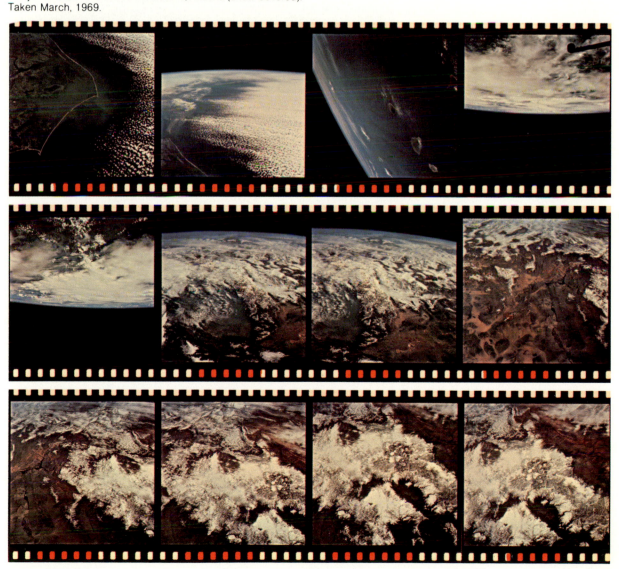

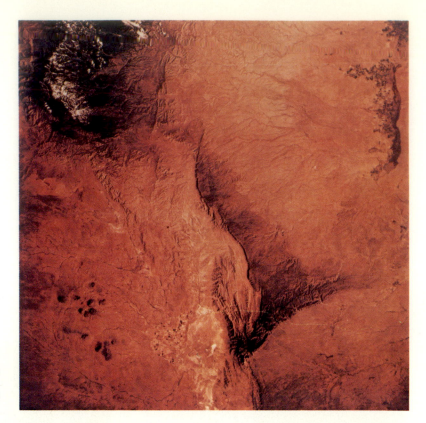

COLOR PLATE 2 FIGURE 4.11. Guadalupe
Mountains, New Mexico and Texas.
(a) Apollo 6 photograph taken by automatic
70-mm camera in 1967.

(b) Geologic sketch map (11).

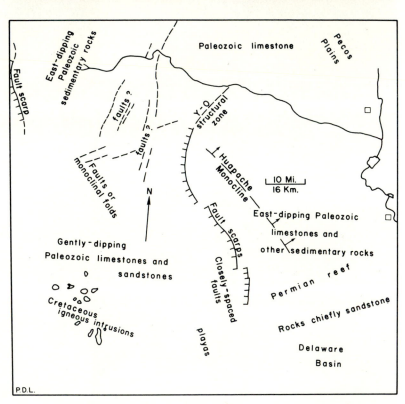

GEOLOGY
APOLLO 6 PHOTO AS 6-2-1450

COLOR PLATE 3 FIGURE 4.12. Los Angeles and surrounding area.
(*a*) Apollo 7 hand-held photograph.

(*b*) Geologic sketch map. Scale variable; Bakersfield-Searles Lake distance 95 miles.

INDEX MAP
APOLLO 7 PHOTOGRAPH AS 7-11-2022
Note: Scale variable; Bakersfield-Searles Lake distance 95 miles.

Map labels:

N

Sierra Nevada
clouds
clouds
Basin and Range Province
NEVADA
San Joaquin Valley
□ Bakersfield
U D
Death Valley
Panamint Valley
Searles Lake
Spring Mts.
Garlock Fault
Mojave Desert
San Andreas Fault
Big Pine fault
Transverse Ranges
San Fernando Valley
Santa Monica Mts.
Pt. Dume
San Gabriel Mts.
clouds
Mojave R.
San Bernadino
San Bernadino Mts.
Mill Creek fault
Pisgah Crater
Los Angeles
Corona
L. Matthews
San Jacinto fault
Banning fault
San Jacinto Mts.
Mission Creek fault
Santa Rosa Mts.
L. Elsinore
Santa Ana Mts.
Aqua Tibia Mts.
Santa Catalina Island
San Clemente □
Peninsular Ranges
P.D.L.

COLOR PLATE 4 FIGURE 4.13. Pacific Northwest.
(a) Apollo 11 view from about 12,000 km altitude.

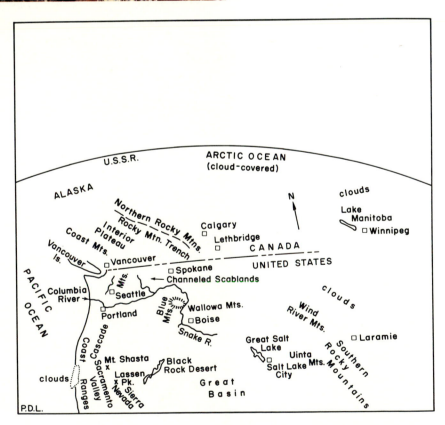

(b) Geologic sketch map.

GEOGRAPHY

APOLLO II PHOTOGRAPH AS II-36-5302

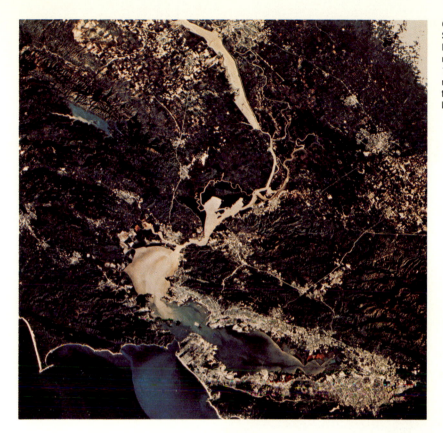

COLOR PLATE 5 FIGURE 4.14. Skylab 4 S190A photograph (SL 4-92-336) of San Francisco taken 1974 by one of six multispectral cameras, each with 152-mm focal length lens, using high-resolution color film (SO-356). Estimated ground resolution 40-46 m. Skylab altitude 435 km.

COLOR PLATE 6 FIGURE 4.15. Skylab 4 S190B photograph (SL 4-93-067) of Flagstaff, Arizona, and surrounding area (north at top right), taken 1974 by Earth Terrrain Camera, with 457-mm focal length lens, using high-resolution color infrared film (SO-131). Estimated resolution 23 m. San Francisco volcanic field at upper left; Meteor Crater at lower right.

COLOR PLATE 7 FIGURE 4.20. Salton
Sea and Peninsular Ranges, California.
(a) Landsat false color composite
image taken Nov. 6, 1972 (Image
1106-17504).

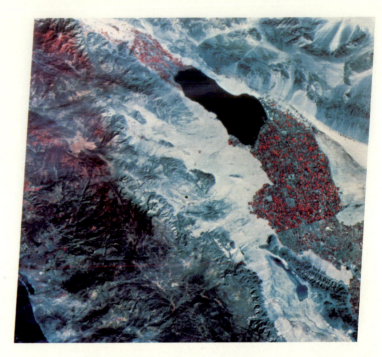

(b) Structural geology sketch map (12).

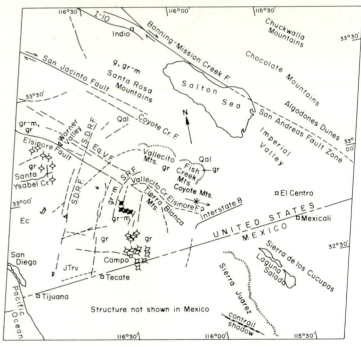

STRUCTURE SKETCH MAP

Peninsular Ranges, San Diego County, California

ERTS-I Image 1106-17504 (6 Nov. 72)

STRUCTURE	LITHOLOGY*
⎯ ⎯ ⎯ Fault (solid where confirmed, dashed where inferred or nature not certain).	Qal Quaternary alluvium.
	Ec Eocene nonmarine sediments.
	gr Mesozoic granite rocks.
◆ Foliation in metamorphic rocks.	gr-m Pre-Cenozoic granite and
◇ Flow structure in intrusive igneous rocks (inclusions, crystals, etc.).	metamorphic rocks.
	*Lithology from Geologic Map
✳ Plunging Syncline	of California (1:250,000 sheets)

Paul D. Lowman, Jr.
Goddard Space Flight Center
October, 1973

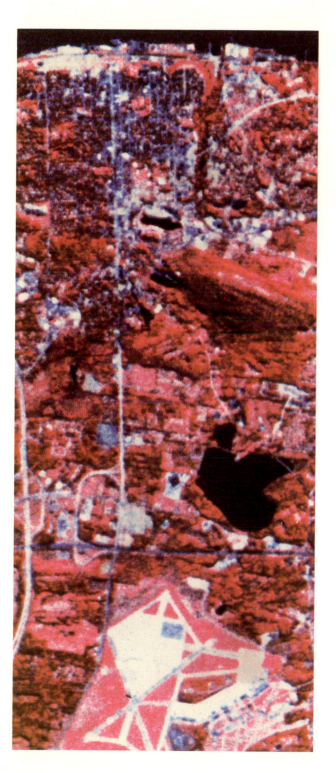

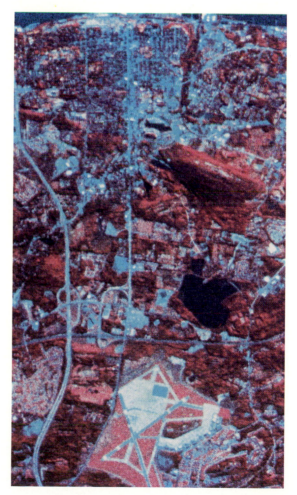

COLOR PLATE 8 FIGURE 5.9. Color separation tech-
ique applied to a Skylab S190B color infrared image. Left
view shows a standard enlargement of the Newburgh, New
York area (scale 1:80,000). Right view is a diazo composite
from the separation negatives of the same image (scale
1:48,000). Note the improved image fidelity and resolution
in the diazo composite even at this larger scale.

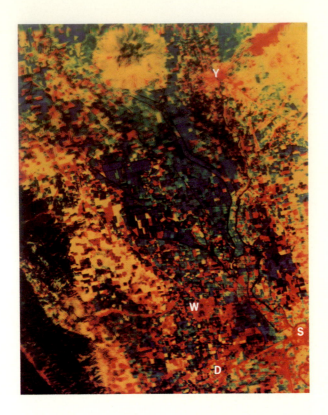

COLOR PLATE 9 FIGURE 6.11. Color composite generated by the CIE color prediction model of the Sacramento Valley, California. Geologic features of interest are the shallow marine sediments (late Cretaceous) of sandstone and shale dipping to the east and a circular volcanic intrusion (Pliocene), which is prominent near the top center. The remainder of the area is alluvium. The cities of Sacramento (S), Woodland (W), Davis (D), and Yuba City (Y) are visible in the scene.

COLOR PLATE 10 FIGURE 6.25. Comparison between an enhanced image and a thematic map of the same region created using Bayesian classification. The legend on the right of each picture relates types of surface materials to colors in the display. The colors in the thematic map were chosen by the analyst. The colors on the enhanced picture were found by inspection. The legend identifies two playa surfaces, felsic rocks, basalt flows with and without sagebrush, altered volcanic rocks showing a limonitic stain, altered silicified rocks, and vegetation. Images show Goldfield, Nevada from NASA Landsat image 1072-18001.

HYDROTHERMAL ALTERATION DETECTION
IN THE GOLDFIELD TEST AREA (NEVADA)

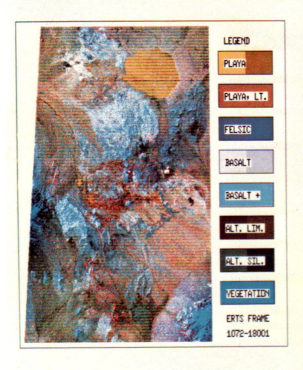

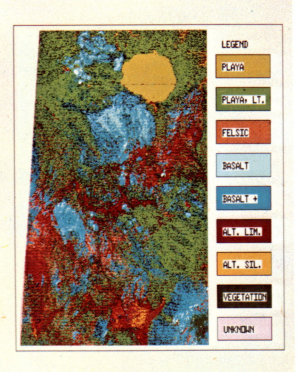

COLOR RATIO ENHANCED PICTURE
MSS 4/5 = ■ MSS 5/6 = ■ MSS 6/7 = ■

CLASSIFICATION MAP
LARSYS USED MSS 4, 5, 6, AND 7

COLOR PLATE 11 FIGURE 6.34. Ethiopia (a) Color Additive composite picture showing part of Ethiopia (NASA Landsat image 1319-07185). MSS 4 is represented by blue; MSS 5 by green; MSS 7 by red picture.

(b) Color ratio composite image. MSS4/MSS5 is represented by blue; MSS5/MSS6 by green; and MSS6/MSS7 by red. Images supplied courtesy of GeoImages, Inc.

COLOR PLATE 12 FIGURE 6.36. (*a*) Color additive composite of Sumatra (see (Fig. 6.23).

(*b*) Same scene enhanced by contrast stretching of individual principal component pictures before transformation back to blue, green, and red pictures. Blue and cyan colors in (*a*) are here represented by blue and green, with increased discriminability. NASA Landsat image 1324-02433.

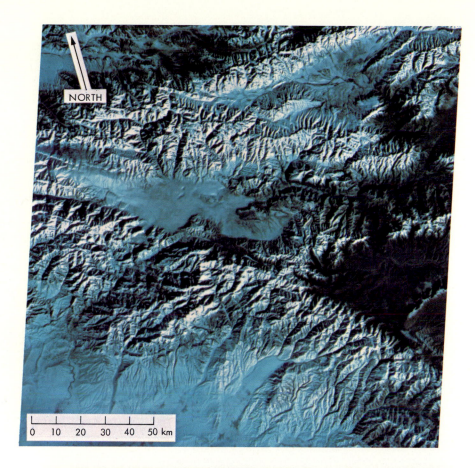

COLOR PLATE 13 FIGURE
6.41. High-pass filtering in
the color domain. (a) In-
trinsic brightness range in
the scene exceeds dynam-
ic range of the display unless local contrast is sacrificed.

(b) High-pass filtering only
the intensity information
without loss of data in
bright and dark regions.
Image shows Talaso-
Ferghana fault in the U.S.S.R.
(NASA Landsat image 1534-
05184).

COLOR PLATE 14 FIGURE 7.13. Analysis of aerial photographs (near Northeast, Pennsylvania) of a lake bed terrain.

(a) Color photograph.

(b) Infrared color photograph.

(c) Black-and-white photograph. See Table 7.6 for description of units.

(b)

(c)

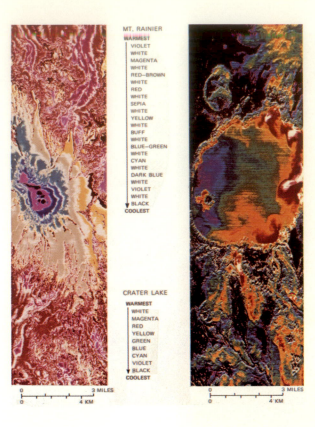

MT. RAINIER

WARMEST
VIOLET
WHITE
MAGENTA
WHITE
RED-BROWN
WHITE
RED
WHITE
SEPIA
WHITE
YELLOW
WHITE
BUFF
WHITE
BLUE-GREEN
WHITE
CYAN
WHITE
DARK BLUE
WHITE
VIOLET
WHITE
BLACK
COOLEST

CRATER LAKE

WARMEST
WHITE
MAGENTA
RED
YELLOW
GREEN
BLUE
CYAN
VIOLET
BLACK
COOLEST

0   3 MILES
0   4 KM

0   3 MILES
0   4 KM

COLOR PLATE 15 FIGURE 9.15. Nighttime thermal IR images (4.5 to 5.5 μm) of Mont Rainier (left) and Crater Lake (right). Images acquired July 1970, and processed by Environmental Research Institute of Michigan (24).

COLOR PLATE 16 FIGURE 9.16. Interpretation maps of images in Fig. 9.15.

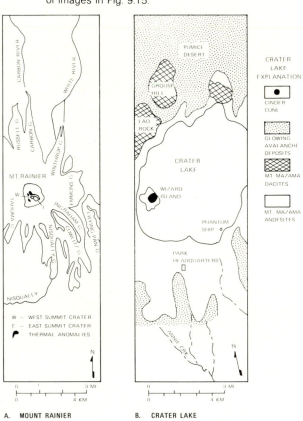

A. MOUNT RAINIER

B. CRATER LAKE

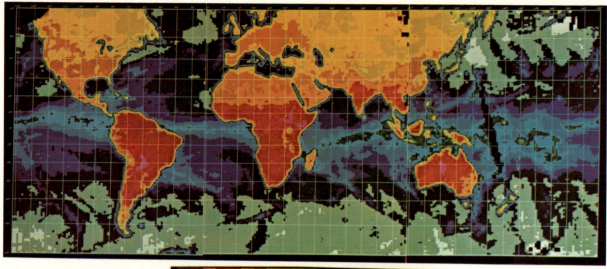

COLOR PLATE 17 FIGURE 11.2. Color coded brightness temperature map of the Earth using the Nimbus 5 ESMR data for the period 12-16 January 1973.

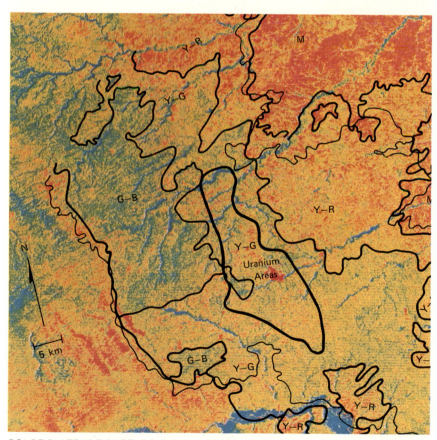

COLOR PLATE 18 FIGURE 12.5. Color-coded ratio image of Landsat bands 5 and 6 showing variation in vegetation density that is related to bedrock lithology. Cyan is the lowest ratio, highest vegetation density, and magenta is the highest ratio, lowest vegetation density. The colors are arranged in normal order, cyan, green, yellow, red, and magenta. Vegetation density is related to lithology as explained in the text.

Explanation of Symbols

Contact of upper uranium-containing formation with non-uranium-containing formation, from Denson and Horn (23).

Outline of uranium producing area. X marks major mines.

Color boundaries as visually interpreted from this ratio image. G-B means green to blue and is a coarser-grained facies, Y-G means yellow to green and is an intermediate-grained facies, Y-R means yellow to red and is a finer-grained facies, and M means magenta and is the lower formation that is very fine-grained.

COLOR PLATE 19 FIGURE 12.6. Clearing associated with the Vazante zinc deposit, Brazil. Photography by G. L. Raines, U.S. Geological Survey.

COLOR PLATE 20 FIGURE 12.8. Effect of different concentrations of lead on bean plants grown in standard nutrient solutions. Top row, left to right, 0, 10, and 25 ppm lead. Bottom row, left to right, 50, 100, and 200 ppm lead. Chlorotic leaves, inhibition in overall growth, and reduction in root size are clearly related to lead content. The lower most green leaves in all six plants represent growth produced by stored nutrients in the seeds and thus are little affected by the lead in the nutrient solution.

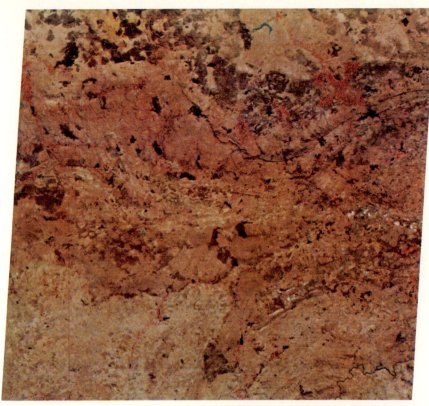

(a) Dry season.

(b) Wet season. Courtesy Mineral Science and Engineering.

COLOR PLATE 21 FIGURE 13.8. Portion of Landsat scenes showing the effect of vegetation on lithological discriminability in a sedimentary terrain, South Africa.

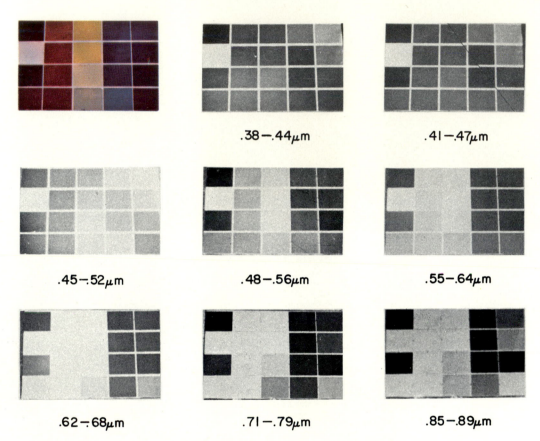

.38 –.44μm

.41 –.47μm

.45 –.52μm

.48 –.56μm

.55 –.64μm

.62 –.68μm

.71 –.79μm

.85 –.89μm

COLOR PLATE 22 FIGURE 13.14. Wavelength dependence of relative photographic tone.
Courtesy LARS.

COLOR PLATE 23 FIGURE 13.17. Stratigraphic components of the sedimentary Nama system in South West Africa.

(a) Landsat scene #1147-08163.

(b) Interpretive geologic map. Courtesy Mineral Science and Engineering.

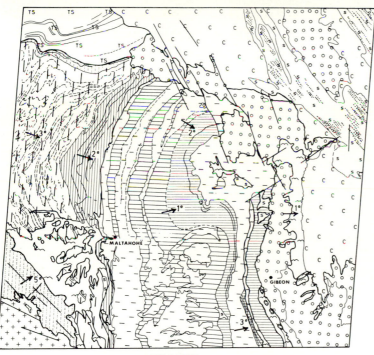

EXPLANATION

SCALE 1:1 000 000

WINDBLOWN SAND INCLUDING DUNES	RED ARGILLACEOUS SANDSTONE		
CALCRETE	RED QUARTZITIC MUDSTONE	FISH RIVER FM.	
ALLUVIUM	RED MUDSTONE		NAMA GROUP
SEDIMENTS AND VOLCANICS – KARROO	GREEN SANDSTONE AND SHALE		
TS TSUMIS FORMATION	GREEN SHALE INTERBEDDED WITH GREY QUARTZITE	SCHWARZ RAND FM.	
BASEMENT	BLACK LIMESTONE	KUIBIS FM.	
	TRENDS		
	ERTS LINEARS		

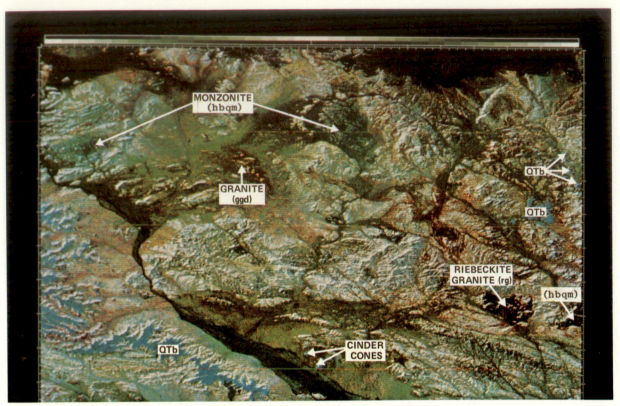

COLOR PLATE 24 FIGURE 13.22. Landsat false color ratio composite images in northwest Saudi Arabia.

(a) Image produced from MSS ratios 4/5, 5/6, 6/7 displayed in blue, green, red, respectively.

(b) Image produced from MSS ratios 5/4, 6/5, 7/6 displayed in blue, green, red, respectively. Courtesy GSFC.

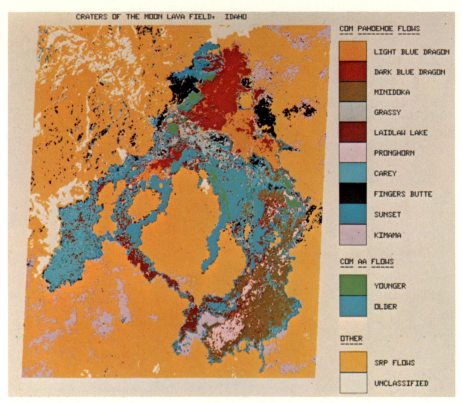

COLOR PLATE 25 FIGURE 13.24. Computer classification map of Craters of the Moon Volcanic Field, Idaho. Courtesy JPL.

COLOR PLATE 26 FIGURE 13.25. Unsupervised multispectral classification of Landsat data for an area in southern France. The six classes, from white to black, are (1) sea water, (2) fresh water, (3) recent alluvium, (4) ancient alluvian, (5) undifferentiated terrains, (6) outcrops. Courtesy Institute Francais du Petrole.

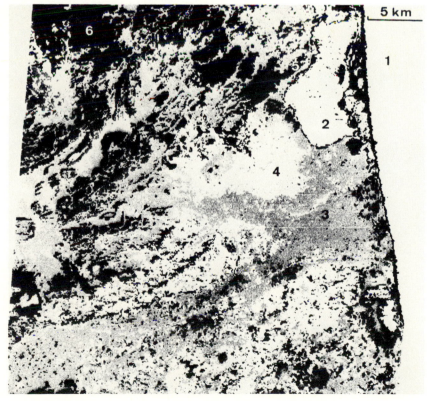

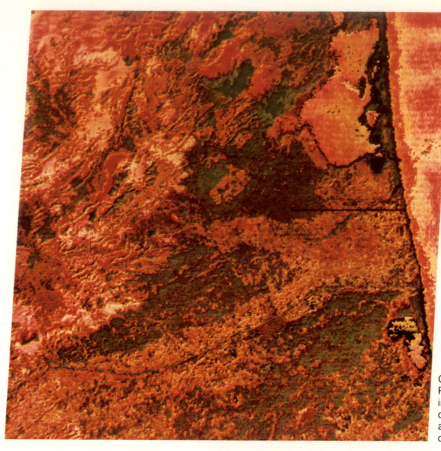

COLOR PLATE 27 FIGURE 13.26.
Principal component image of an area
in southern France. First component is
displayed as red, second component
as yellow. Courtesy Institute Francais
du Petrole.

COLOR PLATE 28 FIGURE 13.27.
Photo interpretation map of fig. 13.26.

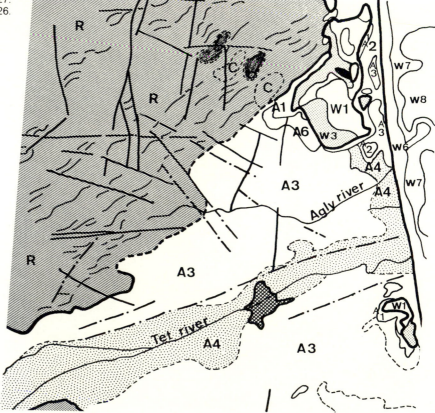

COLOR PLATE 29 FIGURE 15.9. Apollo 6 automatic 70-mm photograph taken in 1967, showing southwest New Mexico and northern Chihuauhua. Note dominance of fluvial erosion features, such as pediments surrounding Florida and Tres Hermanas Mountains at center. See Lowman (4) for discussion and maps of photo.

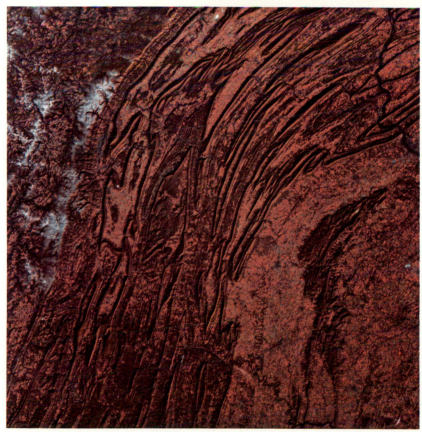

COLOR PLATE 30 FIGURE 15.10. Landsat image of the Ridge and Valley section of the Appalachians; Potomac River at bottom, Harrisburg at upper right. See Fig. 15.3, also Short et al. (5). Image no. 1495-15222, 30 Nov. 73.

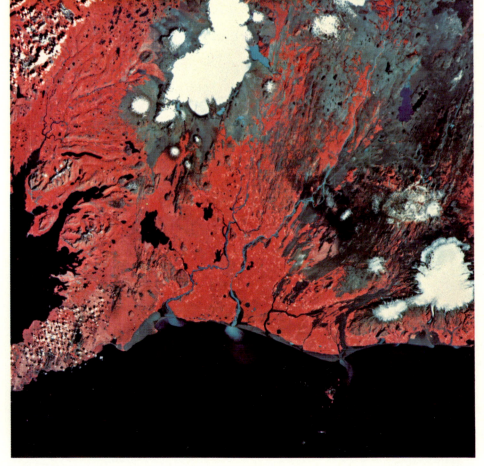

COLOR PLATE 31 FIGURE 15.11. Apollo 7 photograph of southern Zagros Mountains, showing anticlinal mountains; Queshm Island at lower right, Persian Gulf at bottom. Dark features are salt domes. See Lowman (4) for discussion and map.

COLOR PLATE 32 FIGURE 15.17. Landsat image of southern Iceland, showing central rift, which is an emergent segment of the Mid-Atlantic Ridge, a spreading center. Movement is normal to the trend of fissures. See Short, et al, (2), for explanation, also Williams and Carter (13).

COLOR PLATE 33 FIG-
URE 15.18. Apollo 6 auto-
70 mm photograph taken
in 1967, showing north-
west Sonora, Gulf of Cal-
ifornia, and extreme southwest Arizona. Dark feature
at top center is Pinacate
volcanic field. Great Sonora Desert at upper left.

COLOR PLATE 34 FIG-
GURE 15.19. Apollo 9
S065 experiment 70 mm
photograph taken in 1969,
showing Mississippi River
valley; Vicksburg at lower
right. Broad light north-
south-trending swath at
left center is the Teche
Mississippi, the course of
the river about 1000
A.D. See Lowman (4) for
discussion.

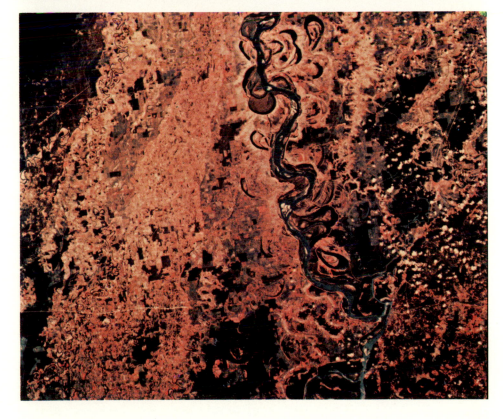

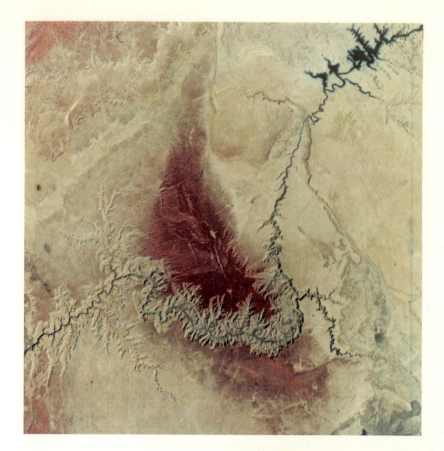

COLOR PLATE 35 FIGURE 15.20. Landsat image of Grand Canyon. Kaibab Plateau is dark because of heavier vegetation resulting from increased precipitation. Lake Powell at upper right.

COLOR PLATE 36 FIGURE 17.15. Skylab S190A color photograph of south-central Nevada.

COLOR PLATE 37 FIGURE 17.19. Color-infrared composite of south-central Nevada consisting of linearly stretched MSS band 4, 5, and 7 images using blue, green, and red filtered light. A, yellow to yellow-orange color of vividly colored limonitic areas. Image E1072-18001, acquired October 3, 1972.

COLOR PLATE 38 FIGURE 17.21. Color-ratio composite of south-central Nevada.

COLOR PLATE 39 FIGURE 17.23. Simulated natural color composite image of the Nabesna quadrangle, Alaska.

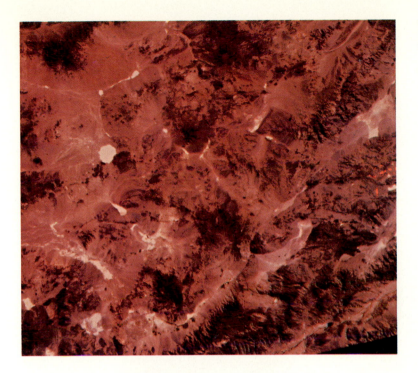

COLOR PLATE 40 FIGURE 17.24. Simulated natural color composite image of south-central Nevada.

COLOR PLATE 41 FIGURE 17.27. Simulated natural color composite and color-ratio composite images of the Cameron, Arizona distric. (Image courtesy of Gordon Swann, U.S. Geological Survey, Flagstaff, Arizona).

N

4   0 1 2 3 4 MILES
4   0 1 2 3 4 KILOMETERS

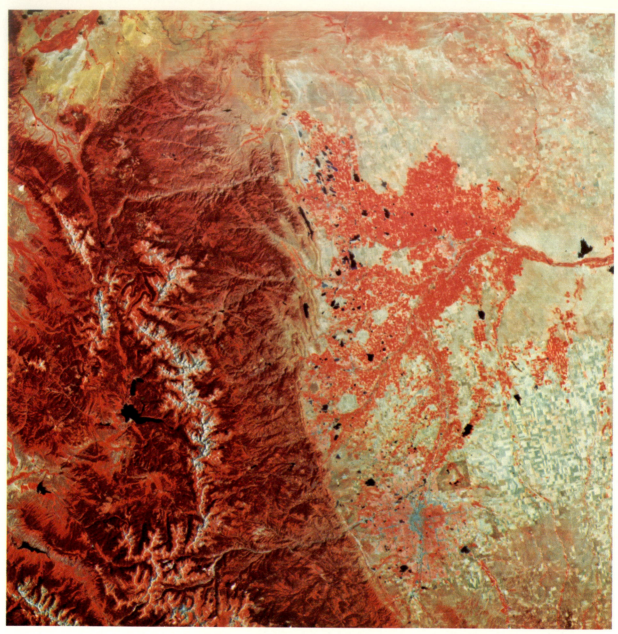

COLOR PLATE 42 FIGURE 18.4.
Landsat-1 false color image of
eastern Colorado including the
Denver, Rocky Mountain National
Park, Fort Collins, and Greeley areas.

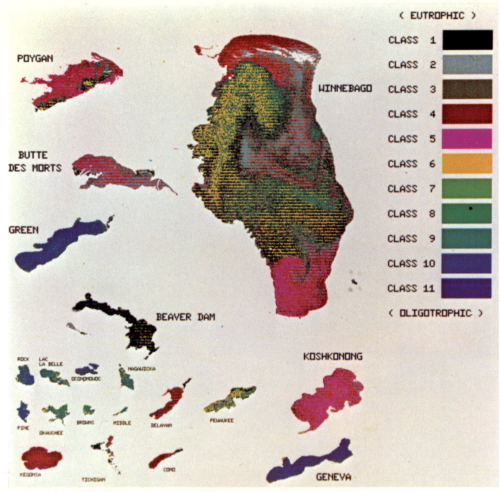

COLOR PLATE 43 FIGURE 19.1.
Color coded trophic indices of
several lakes in Wisconsin based
on satellite data.

COLOR PLATE 44 FIGURE 19.2.
Color infrared photograph of a
segment of Alum Creek, central Ohio.
Trees and other vegetation stressed
or killed by oil-field brine appear as
light red, pink, or gray.

COLOR PLATE 45 FIGURE 19.15. Color coded category map of east-central Ohio produced directly from CCT computer processing. Much of this area has been strip mined for coal and reclaimed. Landsat-1 scene 1407-15352. Blue is water, red is strip mined areas, brown is reclaimed strip mine, orange is grassland.

COLOR PLATE 46 FIGURE 19.16. Stripped earth category map derived from Landsat data (21 August 1972) overlaying an aerial photograph (9 September 1973) of the same mine shows close agreement between aircraft and satellite data.

12.3.1 Reflectance Properties

The simplest component of vegetation is a leaf; before discussing the complexities of assemblages of leaves, bushes, trees, and forests, the spectral properties of an individual leaf will be discussed. The spectral reflectance properties of a leaf in the 0.4 to 2.5 μm region are a function of the leaf pigments, primarily the chlorophyll pigments, the leaf cell morphology, internal refractive index discontinuities, and the water content. The pigments of the chlorophyll group are the primary control from 0.4 to 0.7 μm, the cell morphology and internal refractive index discontinuities are the primary controls from 0.7 to 1.3 μm, and the water content is the major control in the 1.3 to 2.5 μm region. A detailed discussion of leaf spectral reflectance properties is beyond the scope of this chapter; so only a brief review will be presented. Gates and others (5), Myers (6), and Gausman (7) give more detailed reviews.

Figure 12.1 gives a typical spectral reflectance curve for a green leaf. The minima near 0.45 and 0.68 μm are due primarily to absorption by chlorophyll a, but these absorption bands are further broadened by the presence of other chlorophyll bands. Various other leaf pigments also contribute to broadening the absorption band near 0.45 μm. Absorption of radiation in these bands supplies the energy required for photosynthesis. The high reflectance region from 0.7 to 1.3 μm is due to refractive index discontinuities within the leaf. The primary type of discontinuity is the internal air-cell interfaces of a leaf, the cell morphology (5). However, leaf components, stomata, nuclei, cell walls, crystals, and cytoplasm, also produce significant refractive index discontinuities that contribute to the reflectance of a leaf (7). This high reflectance is believed to be a functional requirement of the leaf to maintain a proper energy balance and not overheat and destroy the chlorophyll. In the region from 1.3 to 2.5 μm

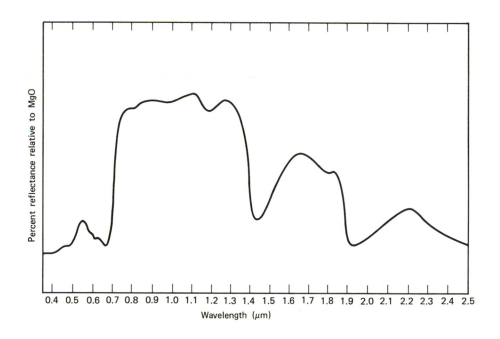

FIGURE 12.1. Typical spectral reflectance curve for a leaf.

the spectral reflectance curve of a leaf is essentially the same as that of pure water, and the spectral curve can be matched to that of an equivalent thickness of water.

Spectral properties are not static because they are intimately related to the life processes of the plant, and so they follow a cycle from youth to maturity and death that can be a useful, but complicating, factor in applications. The precise cycle is not well documented; however, a general pattern seems to exist (5). The cycle for a deciduous leaf consists of three phases to maturity. In the initial phase the near-infrared reflectance is well developed, very high, and coupled with a high yellow visible reflectance. Second, as chlorophyll begins to be manufactured, the blue-absorption band appears. Finally, with an increasing amount of chlorophyll, the red-absorption band becomes progressively stronger resulting in a general overall decrease in visible reflectance, and simultaneously an increase in near-infrared reflectance. After a period of maturity, and until death of the leaf, the red-absorption band disappears as the chlorophyll pigments disintegrate and the blue-absorbing pigments remain. This produces the yellow and red colored leaves of fall. There is also a decrease in near-infrared reflectance as the cells shrink and dehydrate. In contrast to deciduous trees, coniferous trees always contain a large amount of chlorophyll; however the young leaves tend to be brighter, because the chlorophyll absorption bands for a collection of conifer leaves are subdued early in the growing season. Also there is no period when all the leaves die so there is no drastic change in reflectance throughout a year.

Because of the many factors influencing the spectral properties of leaves and the different arrangements of chlorophyll pigments, other pigments, cell morphology, internal leaf components and water content, different types of leaves have slightly different spectral reflectance properties. Even for a single leaf the upper surface generally has a different and lower reflectance than the lower surface, because a different arrangement of the control factors is encountered by radiation as it enters the leaf from the upper or lower surface.

A plant is a collection of leaves and numerous voids between leaves. Radiation is reflected, absorbed, emitted, and transmitted by each of these leaves, and the reflected, emitted, and transmitted light may be retransmitted and re-reflected numerous times before leaving the plant. Thus, the spectral reflectance and emittance properties of a plant are much more variable and more complex that that of a leaf. Colwell (8) considered the following parameters as well as leaf reflectance properties to be important when considering a vegetation canopy; (a) transmittance of the leaves; (b) number and arrangement of leaves; (c) spectral characteristics of the stalks, trunks, limbs, and other components of the plants; (d) spectral properties of the background including soil, rocks, vegetation litter, understory shrubs, and so forth; (e) solar zenith angle; (f) look angle; and (g) azimuth angle. Also, individual plants generally are not resolved in most remote sensing applications, so that the vegetation canopy appears as a combination of all the variables mentioned, plus variation due to different species and different mixtures of plants.

Detecting plants under stress, especially stress related to an excess concentration of a metal ion, is geologically interesting. The chlorophyll pigments are generally believed to be the part of a plant most sensitive to the various types of stress. Therefore, the chlorophyll absorption bands are thought to be sensitive indicators of stress. However, adverse affects on chlorophyll also tend to decrease the vigor of growth; thus, in a stressful environment, leaves may be smaller, there may be fewer leaves, the leaves may be oriented differently, or other growth-related phenomena may occur. All of these factors greatly affect the leaf area index which is defined as the ratio of total leaf area of the plant canopy to the ground area in the field of view (9). The 0.7 to 1.3 μm region is the most sensitive to changes in the leaf area index (10); thus, in practice the 0.7 to 1.3 μm region seems to be more sensitive to stress when whole plants, not just a leaf, are considered.

12.3.2 Luminescence Properties

Vegetation is naturally luminescent under solar radiation due to the luminescence of some plant pigments, especially the chlorophyll pigments. In biological studies wide use is made of luminescence spectra of plants to assess the vigor of photosynthesis and metabolism. Any adverse environmental parameter that stresses vegetation by affecting photosynthesis might be detectable by monitoring plant luminescence.

Plant luminescence is basically ultraviolet, visible, and near-infrared radiation that is naturally emitted by plants that are irradiated by solar radiation.

The amount of naturally emitted radiation is very small and has been observed by making spectral measurements in the Fraunhofer lines of the solar spectrum at 396.8, 442.1, 486.1, 518.4, 589.0, and 656.3 nm (11). The Fraunhofer lines are absorption lines in the spectrum of the sun due to gases in the sun's atmosphere. The line widths are generally less than a few tenths of a nanometer, and the central intensity is less than 10% of the intensity of an adjacent part of the spectrum.

Vegetation may show anomalous luminescence because of stress from excess metal concentrations in the soil. One example comes from a study in which bean plants were grown in the laboratory in nutrient solutions containing different concentrations of molybdenum, copper, zinc, and lead (12). The results obtained with 10 plants, 5 nonstressed and 5 stressed with 10 parts per million (ppm) sodium molybdate, show that the differences observed in luminescence after two weeks are reversed after five weeks, and this reversal not only occurs between stressed and nonstressed plants, but also between young and old plants. Thus, the luminescence properties of vegetation are affected by geologically important parameters, but not in a simple way.

12.3.3 Thermal Infrared Properties

It is the radiometric temperature of a vegetation canopy that is detected by a remote sensing system. The radiometric temperature is related in a complex and poorly understood fashion to the temperature of individual leaves. Because of this lack of understanding, this section will deal only with what is known about the thermal infrared properties of individual leaves. The parts of the spectrum considered here are the 3 to 5 μm and the 8 to 14 μm regions. The 3 to 5 μm region involves both reflectance and thermal emittance; thus, data analysis in this region is very complex. The 8 to 14 μm region involves only thermal emittance related properties.

A plant absorbs large amounts of solar radiation, especially at visible wavelengths, and re-emits most of this energy at thermal infrared wavelengths in order to maintain a proper energy balance. In this process a plant generally maintains a leaf temperature that is approximately 10° to 15°C above air temperature in full sunlight and is approximately 5°C below air temperature at the coldest part of the night. During the warming and cooling parts of the day, leaf temperature follows a cycle between these

extremes that is dependent upon local environmental factors such as the local radiation flux at the ground, the air temperature, the relative humidity, wind velocity, and the active plant factor of transpiration capabilities of the plant (13, 14). The contribution of metabolic processes to the energy budget and thus to plant temperature is not significant (13).

Transpiration is the process by which moisture is moved out of the plant and is the active mechanism a plant uses to control the temperature of its leaves. Plant factors and environmental factors (15) affect the rate of transpiration. The plant factors are: (a) root-shoot ratio; (b) leaf area; and (c) leaf structure. These plant factors are influenced by the availability of nutrients, which is influenced by geologic parameters as discussed in Section 12.2 of this chapter. Environmental factors are: (a) availability of light—an important controlling parameter as it has a dominating effect on stomatal opening and closure (transpiration nearly ceases during the nighttime when the stomata are closed); (b) humidity—controls the magnitude of the vapor pressure gradient between the internal and external atmospheres [the greater the gradient (low relative humidity), the higher the rate of transpiration]; (c) temperature—other things being equal, an increase in temperature increases the transpiration rate due to its effect on stomatal movements and vapor pressure gradients; (d) wind—a rather complex effect that normally increases transpiration by dispersing the water vapor in the immediate leaf vicinity with consequent increases in the vapor pressure gradient; and (e) a geologic factor—availability of soil moisture to the roots. Thus, factors that are influenced by geologic parameters affect the rate of transpiration of plants, and, therefore, geologic parameters affect the temperature of leaves.

12.3.4 Radar Properties

There has been very little published concerning the physical properties of vegetation in the radar region of the spectrum, especially as a function of geologic parameters. Therefore, in this section only the primary physical properties of vegetation involved in forming the radar signal will be reviewed. A more extensive review and list of references is available in Matthews (16) from which the following discussion is summarized.

The radar return from vegetation is primarily influenced by the roughness of the plants or vegetation canopy and by the dielectric properties of the vegetation. The roughness controls the geometry of the scattering of the signal, and is determined by the size, shape, orientation, and number of leaves on a plant. The dielectric properties of vegetation are determined primarily by the water content. Thus, water stress and vegetation species having differing water content are factors determining dielectric properties.

The roughness and dielectric properties of vegetation vary with wavelength, angle of observation, and polarization. Wavelength and angle of observation determine which surfaces reflect the radar signal. The wavelength of the signal controls the penetration into or even through the vegetation. As the wavelength of the radar signal increases, penetration into the vegetation increases. Thus, it is theoretically possible with an appropriate selection of wavelength to see through vegetation to the soil and rocks below. The polarization differences observed in radar images are very poorly understood for vegetation; however, different species of vegetation apparently depolarize the radar signal to varying degrees; polarization changes are helpful in discriminating species.

12.4 ESTABLISHED APPLICATIONS

The ancients were well aware of geobotanical indications of mineral deposits as indicated in the writings of Georgius Agricola in his *De Re Metallica*, first published in 1556. Agricola (17) certainly was referring to indicator species and perhaps to bare spots as indicators of mineralization when he wrote:

Likewise along a course where a vein extends there grows a certain herb or fungus which is absent from the adjacent space, or sometimes even from the neighborhood of the veins. By these signs of nature a vein can be discovered.

Likewise, it is even more startling to read that the ancients recognized and used thermal anomalies as guides to ore, for Agricola also writes:

Further, we search for the veins by observing the hoarfrosts which whiten all herbage except that growing over the veins, because the veins emit a warm and dry exhalation which hinders the freezing of the moisture, for which reason such plants appear rather wet than whitened by the frost.

Numerous other examples could be quoted from the writings of observant individuals between the time of Agricola and the present.

The established geobotanical techniques that can be considered applications of remote sensing are primarily those that involve the interpretation of photographs and Landsat images by inspection by an experienced interpreter. Thus, only the 0.4 to 1.1 μm region has been used. The primary considerations are tone or color, texture, pattern, shape, size, relationship to other features, and experience of the interpreter. This approach is basically airborne geobotany and, when knowingly applied, it has been productive. The basic underlying principle is recognition of the norm for a given area and then recognition of deviations from this norm.

12.4.1 Discrimination of Geologic Materials

Two basic geobotanical techniques have been applied to discrimination of geologic materials, indicator flora, and vegetation density. However, today only those species that are large enough and/or constitute a significant part of the vegetation canopy are practical for study using remote sensing techniques. With this limitation in mind, applications have been made in detection of groundwater (18), saline deposits (19), hydrothermal alteration (20), limestone (3), and serpentine (3). Also in localized areas particular types of vegetation grow only on particular lithologies. For example, trees grow only on the nonshale units at Canon City, Colorado, as shown in Figure 12.2.

Vegetation density is also a useful factor that can be detected and quantified on photographs (21) and Landsat images. Limestones and ultramafic rocks, especially serpentines, commonly cause the vegetation density to decrease in normally vegetated areas. Figure 12.3 shows the effect of limestone on vegetation density. The cause here is probably related to soil pH factors both in the soil formation process and preference for particular soil pH of the pinons and junipers growing in this area. However, in other environments limestones support the densest growths of vegetation and this emphasizes the complexity of the problem and our lack of understanding of the cause-and-effect relationship.

Vegetation density differences affect the tone or color of photographs even when particular shrubs involved are not completely resolvable. An example is the shale area of Fig. 12.2. These residual soils

FIGURE 12.2. The trees in this semiarid environment only grow on nonshaley lithologies. The sandstone and the limestone hogbacks in the background are both tree covered, while the shale units in the foreground are treeless. The changes in tone in the black shale units in the foreground are due primarily to changes in nontree vegetation density.

FIGURE 12.3. The trees are noticeably thinner on the area of limestone than on the arkose to the right and the granitic rocks to the left. The area is near Canon City, Colorado.

are sufficiently saline that a halophyte vegetation community has developed. Halophytes are plants that are tolerant of saline soils. The tonal differences seen in Fig. 12.2 are primarily due to changes in the vegetation density shown in Fig. 12.4.

An understanding of the spectral properties of vegetation, soils, and rocks suggests that a ratio of Landsat multispectral scanner bands 5 (0.6 to 0.7 μm) and 6 (0.7 to 0.8 μm) should be sensitive to vegetation density (8, 22) and, in some cases, to species differences. In addition, because the scanner records images of an area once every 18 days, it is possible to detect subtle temporal changes in the dynamic vegetation canopy. The concept used for mapping vegetation density with this Landsat 5/6 ratio is that as the vegetation density increases there will be a corresponding decrease in the radiance of Landsat band 5 due to the chlorophyll absorption near 0.68 μm and an increase in Landsat band 6 due to the very high reflectance of vegetation in the near infrared. Therefore, the ratio is inversely proportional to vegetation density.

Figure 12.5, color plate 18 shows a Landsat 5/6 ratio image of part of the Powder River Basin, Wyoming. Cyan is the lowest ratio, highest vegetation density, and magenta is the highest ratio, lowest vegetation density. Variation in vegetation ground cover without regard to the species of vegetation can be correlated here with subtle changes in the amount of clay in the two formations involved, which are basin-fill deposits. The vegetation is a sage-grass assemblage. Sage prefers a loose, open soil to areas of higher clay contents, and the grasses prefer more sandy soils. These two formations are very difficult to differentiate in the field; however, they are easily differentiated in this ratio image. Also, subtle variations can be seen within the upper formation that are related to coarse- and fine-grained facies changes of the formation. Uranium appears to be only associated with the intermediate facies, denoted Y-G. Knowledge of these subtle variations is contributing to an understanding of the roll-front uranium deposits of the Powder River Basin as discussed in Raines et al (24).

Radar can also be used to detect changes in vegetation density and species differences, especially under those circumstances where photography cannot be acquired. A prime example of this is the work in Panama where radar images were acquired in a short time after photographic coverage was not obtained in 30 years (25). Side-looking radar can make a unique contribution when subtle differences in tree height might be enhanced by radar's low illumination angle. Also, radar can enhance subtle differences in the vegetation canopy because of differences in the scattering characteristics of such factors as shapes of individual plants and leaves, and differing orientation of the leaves.

12.4.2 Mineral Exploration

Observation of the distribution of indicator plants, vegetation density changes, especially bare spots, and morphological changes in plants are the three principal applications of geobotanical techniques in mineral exploration. Due to the small size of most individual indicator species and the small areal extent of most patches of such species, the chances of detecting significant concentrations of indicator species with any current remote sensing technique appears to be small. Some indicator species that produce large and colorful blooms, such as the California poppy, which can be a copper indicator in the southwestern United States, might, however, be detected if observed precisely at the time of maximum blooming.

Deviations from normal vegetative cover in a heavily forested region, generally bare spots, have been useful in prospecting. The general environment of mineral deposits often is unfavorable for vegetation growth so that bare spots or grassy clearings with only a few stunted shrubs and perhaps a few long-growing species result. Such clearings are probably the most widely used geobotanical indicator in prospecting. An example of such a clearing is shown in Fig. 12.6, color plate 19.

This significance of bare spots and clearings was recognized and used successfully in copper exploration nearly 50 years ago in the Rhodesian Copper Belt where systematic searches were made using panchromatic aerial photography. More recently, Lag and Bolviken (26) have reported the occurrence of chlorotic vegetation and clearings due to lead-poisoned soils and vegetation in five different areas in Norway where galena mineralization was present in the underlying rock.

Morphological and physiological changes in plants, and toxicity symptoms are also useful guides in geobotanical prospecting surveys and have long been used in ground-based geobotanical surveys. Brooks (3) states that such changes include among

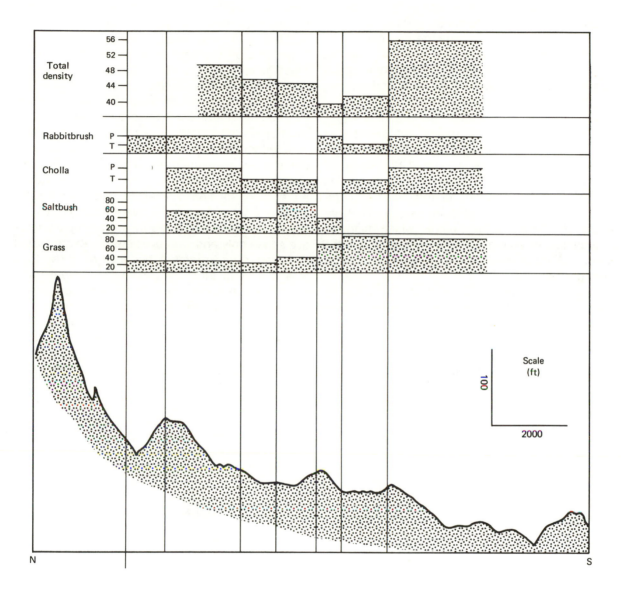

FIGURE 12.4. Vegetation density and topographic profiles of the black shale area near Canon City, Colorado. Total density is percent of ground coverage. For saltbush and the grasses, the units are percent of the total vegetation community. For rabbitbrush and cholla, P means the plant is present but so dispersed that it is not well-recorded, and T means the plant is present but there are fewer individuals than P. The names for the plants considered are rabbitbrush [*Chrysothamnus nauseosus* (Pall.) Britt.], cholla [*Opuntia imbricata* (Haw.) DC.], and saltbush (*Atriplex confertifolia*). The location of this profile is shown in Fig. 12.2.

others: dwarfism, gigantism, mottling and chlorosis of leaves, abnormally shaped fruits, changes of color in the flowers, disturbances in the rhythm of the flowering period, and changes in growth form. An example of a growth-form change is shown in Fig. 12.7 from the Pine Nut area of Nevada (Cannon, written communication, 1975). The abnormal plant was growing in molybdenum-rich soil and the normal plant in soil of background metal content. Several other morphological changes related to mineralized ground were also noted. For example, a shrub, squawapple, growing in a mineralized area had small white flowers compared to the normal pink flower, and juniper trees known to contain anomalous concentrations of molybdenum commonly exhibited abnormal branching. Another example of a flower-color change, which might be detectable by remote sensing techniques, is the change of red fireweed to a white variety over uranium deposits (27).

Detection of morphological changes usually requires a trained botanist and most changes can only be observed during specific and usually short intervals in their growth cycle; this limits the usefulness of the technique. It also seems unlikely that the majority of such changes will ever be detectable by remote sensing techniques.

One observed toxicity symptom is chlorosis, a yellowing of the normally green leaf. Chlorosis is caused by any factor that interferes with chlorophyll production. Other plant pigments are usually present, but in different and normally lesser amounts depending on the species; however, the colors of these other pigments are not observed because of the overpowering absorption of visible light by the chlorophyll pigments. When chlorosis occurs, the decrease in chlorophyll content allows the yellow of the carotenoid pigments to become visible. This toxic effect is clearly illustrated by the results of a greenhouse experiment shown in Fig. 12.8, color plate 20 wherein bean plants were grown in standard nutrient solutions containing different amounts of lead. The relationship between concentrations used and the resulting effects are useful in a relative sense only and cannot be directly extrapolated to a natural environment, because all of the lead is available to the plants in these nutrient solutions. Obvious mineral deposit-related chlorosis appears to be rather uncommon, again this is a subtle effect that has to be knowingly looked for. This is to be expected because natural selection processes would undoubtedly eliminate those species that were unable to develop some degree of tolerance.

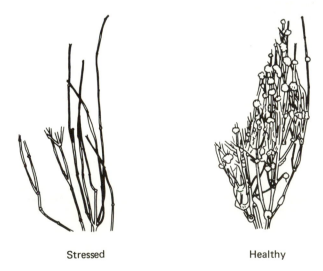

Stressed   Healthy

FIGURE 12.7. Normal (right) and abnormal (left) *Ephedra viridis*. The length of the plant stem between joints in the abnormal plant greatly exceeds the length in the normal plant. Sketches by Daniel Michalski, U.S. Geological Survey.

Such an example of natural selection is the distribution of Ponderosa pine in the Virginia Range of Nevada (20). The vegetation in the area consists of a sage-pinon-juniper assemblage except where hydrothermal alteration has occured. Many of these hydrothermally altered areas are covered with pure stands of Ponderosa pine that are outside their normal environmental range. The Ponderosa are able to survive due to lack of competition from the dominant sage assemblage and because of the reduced nutrient requirements of Ponderosa (20). These large trees can clearly be mapped with aerial photography and thus some, but not all, of the hydrothermal alteration can be mapped.

12.4.3 Soil Moisture

Soil moisture is probably the geologic factor that most significantly influences vegetation. Plants take up nutrients in water-soluble forms, and water is needed in certain quantities by any plant for growth. One example of probable lack of soil moisture has already been mentioned, the lack of trees on the shale lithologies in Colorado (Fig. 12.2). Geologic factors that cause soil moisture concentrations can result in concentrations of vegetation in normally sparsely vegetated areas. Figure 12.9 shows a concentration of trees and larger shrubs along a fault that dams groundwater in a normally grassy and sage-covered area. Similarly, plants often concentrate along a particular bedding plane in sedimentary rocks or along joints because there is more water available at such locations.

Phreatophytes are plants that require their roots be in the water table or the capillary fringe above the water table (18). In addition, there are other plants that often behave as if they were phreatophytes. Robinson (18) gives an extensive list of such plants. Many of the plants are large and can be easily observed on aerial photographs, and if the photographs are of sufficiently large scale, these plants can be identified. However, in some environments, such as semiarid, these plants are sufficiently differ-

FIGURE 12.9. Large shrubs and trees (dark areas in foreground) concentrated along a fault due to damming of the ground water.

ent that they can be recognized as anomalous even on Landsat images. Additional useful information can be obtained if the species is identified and its rooting characteristics are known. With this information, the depth to the water table can be estimated, and with information on climate and depth to groundwater, evapotranspiration rates can be estimated for determining the water requirements of an area.

12.5 CURRENT RESEARCH AREAS

This section provides an overview of the general areas of research. The areas of current research all deal with defining the spectral properties of natural vegetation in attempts to enhance either aspects of the vegetation canopy that are very subtle, such as a change in the leaf area index, or a change in a spectral property such as chlorophyll absorption, which can be related to a geologic factor. Three research areas will be discussed, the luminescence properties, the thermal properties, and the reflectance properties.

12.5.1 Luminescence of Stressed Vegetation

Extensive studies of the luminescence of geochemically stressed and nonstressed Ponderosa pine trees were made in a relatively undisturbed area containing a good selection of trees growing both in normal soils and in soils containing highly anomalous contents of copper and zinc (11, 12). All environmental parameters affecting the health and vigor of the stressed and nonstressed trees appeared to be similar except for the metal content. Chemical analyses of the two groups revealed anomalous amounts of copper and zinc in the needles of the pines growing in the metal-rich soils. The luminescence of selected groups of anomalous and background trees were measured, both on a diurnal and seasonal basis. Figure 12.10 is an example of the type of differences observed. From these data and studies in other mineralized areas, luminescence contrast between geochemically stressed and nonstressed trees seems to be independent of air temperature, relative humidity, and wind speed; but, a correlation with cloud cover is probable (11, 12).

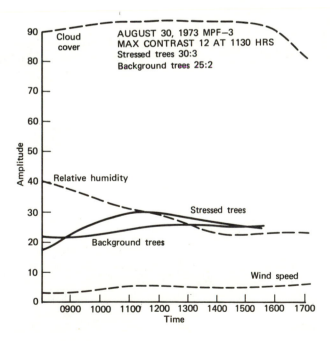

FIGURE 12.10. An example of the changes in luminescence of stressed and nonstressed *Pinus ponderosa* (11, 12). Important environmental parameters are also shown. Relative humidity and cloud cover are in percent and wind speed is in miles per hour. The curves for stressed trees and background trees are in units of relative luminescence, relative to Rhodanine WT dye. Data from Watson and Hemphill (11).

12.5.2 Thermal Infrared

Little research is presently being done on the relationship between the thermal infrared properties of vegetation and geologic parameters primarily because of the complexity of this problem. When any remote sensing technique is applied to a particular problem, it is necessary to have an appreciation for the relative magnitudes of the effects that all operative parameters have on a particular measurement. As an illustration, an attempt was made by Canney to test whether or not tree temperatures could be correlated with anomalously high metal contents in the tree and supporting soil. In thermal surveys performed in 1971 over a mineralized site in Maine, it was found that areas that were warmer than their surroundings correlated crudely with areas where the high-metal-content trees and soils were located. However, when these surveys were repeated a year later (1972), probably under less than ideal weather conditions, the anticipated correlation was not observed; the temperature changed primarily with elevation. Thus, there may be a correlation between the temperature of vegetation and geologic parameters, but from a consideration of the complexity of the temperature control mechanism of vegetation and the results of Canney's experiment, the differences are probably small and very difficult to detect in a consistent fashion.

12.5.3 Reflectance Properties

Most current research is concerned with the 0.4 to 2.5 μm spectral reflectance properties of vegetation because more is known about them, and, in our opinion, there is more potential for useful geologic information in this region. There are two interrelated areas of research, one concerned with identifying the spectral properties of vegetation in different geologic environments, and the second concerned with analysis of remote sensing images.

Two approaches currently used to analyze reflectance data are the computer classification approach and the human-interpreter approach. The computer classification approach relies on the use of spectral and textural properties of the image. The human-interpreter approach relies on the interpretative skills of an analyst; thus, the spatial and spectral aspects of an image as seen by the analyst are interpreted. Digital image processing techniques are being used to present new types and improved versions of images for human interpretation. These digital techniques are discussed in Chapter 5. The computer classification and human-interpreter approaches should not be thought of as competing approaches but rather they should be considered to be complementary.

Photographic techniques can be used to study approximately 0.1 μm-wide spectral bands in the 0.4 to 0.9 μm region, and multispectral scanners can be used to study the 0.4 to 2.5 μm region in the 0.05 to 0.1 μ bands. The advantages of photographic techniques when gross spectral differences exist are that very high spatial resolution is achieved and the data are simple and inexpensive to acquire, which is why photographic techniques have been most commonly used for the established geobotanical remote sensing applications. The disadvantages are that the photographs are never exactly reproducible, they are generally nonquantitative, and spectral band widths of less than approximately 0.075 μm are not obtainable. These disadvantages of photographic techniques, coupled with the complexity of vegetation canopies and the apparently very subtle aspects of spectral reflectance changes that may occur from stress, are probably responsible for the consistently negative results obtained in attempts to detect subtle stress phenomenon related to metals in soils.

Commonly, spectral differences can be easily detected under optimized conditions, such as when homogeneous samples are examined with high performance instruments in a laboratory or field situation. However, these spectral differences cannot be detected when spatial and spectral resolutions and sample homogeneity are degraded in a remote sensing situation. An example of this is found in the work of Canney who performed careful spectral studies in the field on balsam fir and red spruce growing in metal-rich and background soils. Large spectral differences between metal-rich and background trees from *in situ* spectral measurements are shown in Fig. 12.11. When the same trees were investigated by remote sensing systems using color infrared photography, multiband photography, and multispectral scanners, these spectral differences could not be observed.

The human-interpreter approach using digitally processed multispectral scanner images appears to offer the greatest potential for discriminating very subtle reflectance differences. The major advantages of this approach are very narrow spectral bands, very high quality images, precise and repeatable image manipulation, and, most important, the power of a human interpreter.

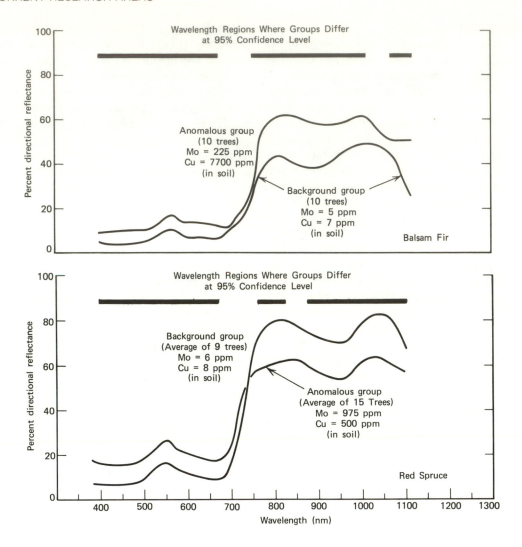

FIGURE 12.11. Reflectance of balsam fir and red spruce. The black bands at the top show where the differences are significant.

The digital Landsat multispectral scanner data has been used successfully with this approach to discriminate very subtle changes in rock reflectance properties. The Landsat 5/6 ratio (Fig. 12.5) is an example of this approach. Detection of change in a vegetation canopy as a function of season is another technique that might allow for discrimination of geological factors. One possible seasonal aspect is that water stress effects should be more pronounced at the end of the dry part of the year. Also, if the vegetation is being stressed by other geologic factors, such as excess metal concentra-tion, then possibly the stress effects, such as changes in spectral reflectance properties, might be greater at the end of a dry, more stressful, period. Lyon (28) has made spectral reflectance measure-ments of pinons and junipers growing on an area of anomalous molybdenum concentration in the Pine Nut Range of Nevada and has shown that there are spectral differences in certain ratios of thes spec-tral data. By using Landsat digital data and digital image processing techniques, specifically band ra-tioning, these areas with the anomalous trees can be discriminated from the surrounding areas. However, in this case, since the vegetation cover is only 10 to 15%, it is not clear whether vegetation spectral reflectance differences are being observed on Landsat data, or if soil differences or a combination of soil and vegetation differences are being observed.

As digital multispectral scanner data in narrower spectral bands become available, new research avenues are opened. For example, Collins (written communication, 1977) has used 10 nanometer band-width spectral bands on the long wavelength edge of the 0.68-μm absorption band of chlorophyll to make a very subtle discrimination of the state of maturation of wheat and to discriminate wheat from other crops. The position of this absorption edge is a function of the degree of maturation of the wheat. Collins (oral communication, 1977) has suggested that similar shifts of this absorption edge may occur in natural vegetation under geochemical stress. This relationship has not been established as yet, but it appears promising.

12.6 SUMMARY

The application of remote sensing to geobotany is a complex problem requiring an understanding of the pertinent physical properties of vegetation plus an understanding of the complex relationships between the vegetation and the underlying rocks and soils. This chapter has attempted to review the pertinent physical properties and to suggest the basis for the relationships between vegetation and the associated rocks and soils. It is clear that much more research is needed. However, as discussed in Section 12.4, there are many cases where useful geologic information can be obtained by the application of geobotany combined with remote sensing techniques. The underlying principle in these applications is recognition of the norm and then recognition of deviation from this norm.

REFERENCES

1. Draeger, W. C., and Lauer, D. T., 1967, Present and future forestry applications of remote sensing from space: Fourth Meeting Amer. Institute of Aeronautics and Astronautics, Anaheim, paper 67-765.
2. Hawkes, H. E., and Webb, J. S., 1962, Geochemistry in mineral exploration: New York, Harper & Row, 415 p.
3. Brooks, R. R., 1972, Geobotany and biogeochemistry in mineral exploration: New York, Harper & Row, 290 p.
4. Sutcliffe, J. F., 1962, Mineral salts absorption in plants: New York, Pergamon Press, 194 p.
5. Gates, D. M., Keegan, H. J., Schleter, J. C., and Weidner, V. R., 1965, Spectral properties of plants: Applied Optics, v. 4, p. 11-20.
6. Myers, V. I., 1975, Crops and soils, in Reeves, R. G., Ansom, Abraham, and Landen, David, eds., Manual of remote sensing: Falls Church, Virginia, American Society of Photogrammetry, p. 1715-1813.
7. Gausman, H. W., 1977, Reflectance of leaf components: Remote Sensing of Environment, v. 6, p. 1-9.
8. Colwell, J. E., 1974, Vegetation canopy reflectance: Remote Sensing of Environment, v. 3, p. 175-183.
9. Tucker, C. J., 1977, Asymmetric nature of grass canopy spectral reflectance: Applied Optics, v. 16, p. 1151-1156.
10. Knipling, E. B., 1970, Physical and physiological basis of the reflectance of visible and near-infrared radiation from vegetation: Remote Sensing of Environment, v. 1, p. 155-159.
11. Watson, R. D., and Hemphill, W. R., 1976, Use of an airborne Fraunhofer Line Discriminator for the detection of solar-stimulated luminescence: U.S. Geological Survey Open-File Rept. 76-202.
12. Watson, R. D., Hemphill, W. R., Hessin, T. D., and Bigelow, R. C., 1974, Prediction of the Fraunhofer line detectivity of luminescent materials, in Proceedings of the ninth international symposium on remote sensing of environment: Ann Arbor, Michigan, Environmental Research Institute of Michigan, v. 3, p. 1959-1980.
13. Gates, D. M., 1963, The energy environment in which we live: American Scientist, v. 51, p. 327-348.
14. _____ 1970, Physical and physiological properties of plants in Remote sensing with special reference to agriculture and forestry: National Academy Science, Washington, D. C., p. 224-252.
15. Devlin, R. M., 1967, Plant physiology: New York, Van Nostrand Reinhold, 446 p.
16. Matthews, R. E., 1975, Active microwave workshop report: NASA SP-376, 502 p.
17. Hoover, H. C., and Hoover, L. H., 1950, De Re Metallica: New York, Dover, 638 p.

18. Robinson, T. W., 1958, Phreatophytes: U.S. Geological Survey Water-Supply Paper 1423, 84 p.

19. Gates, D. M., Stoddart, L. A., and Cook, C. W., 1956, Soil as a factor influencing plant distribution on the salt desert of Utah: Ecol. Mon., v. 26, p. 165-175.

20. Billings, W. D., 1950, Vegetation and plant growth as affected by chemically altered rocks in the western Great Basin: Ecology, v. 31, p. 62-74.

21. Reeves, R. G., Anson, A., and Landen, D., eds., 1975, Manual of remote sensing: Falls Church, Virginia, American Society of Photogrammetry, 2144 p.

22. Rouse, J. W., Jr., Hess, R. W., Schell, J. A., Deering, D. W., Harlan, J. C., 1974, Monitoring the vernal advancement and retrogradation (green wave effect) of natural vegetation: Final Rept., Sept. 1972-Nov. 1974; Natl. Tech. Information Service, N75-28492, 371 p.

23. Denson, N. M., and Horn, G. H., 1975, Geologic and structure map of the southern part of the Powder River Basin, Converse, Niobrara, and Natrona counties, Wyoming: U.S. Geological Survey Map I-877.

24. Raines, G. L., Offield, T. W., and Santos, E. S., 1978, Remote sensing and subsurface definition of facies and structure related to uranium deposits, Powder River Basin, Wyoming, Economic Geology, v. 73, no. 8, p. 1706-1723.

25. Wing, R. S., and MacDonald, H. C., 1973, Radar geology—petroleum exploration technique, eastern Panama and northwestern colombia: Amer. Assoc. Petroleum Geologists, v. 57, P. 825-840.

26. Låg, J., and Bølviken, B., 1974, Some naturally heavy-metal poisoned areas of interest in prospecting, soil chemistry, and geomedicine: Norges geologiskl undersøkelse, v. 304, p. 73-96.

27. Shacklette, H. T., 1964, Flower variation of *Epilobium angustiofoium* L. growing over uranium deposits: The Canadian Field-Naturalist, v. 78, p. 32-42.

28. Lyon, R. J. P., 1975, Correlation between ground metal analysis, vegetation reflectance, and ERTS brightness over a molybdenum skarn deposit, Pine Nut Mountains, western Nevada, *in* Proceedings of tenth international symposium on remote sensing of environment: Ann Arbor, Michigan, Environmental Research Institute of Michigan, p. 1031-1044.

CHAPTER 13

LITHOLOGIC MAPPING
MICHAEL J. ABRAMS

13.1 INTRODUCTION

Geologists are often surprised to learn that much of the Earth's surface is inadequately and sometimes poorly mapped, generally at scales of 1:250,000 or smaller. Even after 100 years of persistent mapping activities in the United States, much of the 50 states is not depicted properly on up-to-date, accurate maps. In other areas, existing geologic maps represent the results of only a few reconnaissance traverses where even regional relationships may not be depicted accurately. Maps produced from remotely sensed images can alleviate this problem by serving as adequate reconnaissance geologic maps in areas where base maps are lacking, access is difficult, or ground truth is minimal.

Reconnaissance mapping by remote sensing is cost-effective and time-efficient. Lithologic contacts can be extended over large areas with a minimum of ground control, and identification of rock types can be extrapolated on the basis of spectral and geomorphic information. In addition, the regional perspective provided enables the interpretation of small-scale features in relation to a more regional context.

In general though, maps made from remotely sensed images are not as accurate as those produced by conventional field mapping. In the field, geologists closely examine individual bedrock outcrops and extrapolate bedrock exposures under soil and vegetation cover in producing the final map; the result is a bedrock map, depicting the distribution of various rock types as if colluvium and vegetation were stripped away. In contrast, a reconnaissance photo-interpretation map depicts "telegeologic" units which are similar-appearing units in the images, representing bedrock and associated soil and vegetation cover. These telegeologic units do not necessarily correspond to geologic units which might be mapped in the field.

The accuracy of photo-interpretation maps thus relies heavily on the amount of ground control available, the type of images examined, and the skill of the photo interpreter.

Interpretation of remotely sensed images is best accomplished by conventional aerial photographic analysis techniques used with an understanding of the spectral properties of rocks and other Earth materials and their appearance on images. For lithologic mapping, consideration must be given to analysis of drainage patterns, vegetation, landforms, climatic influences, and most importantly, the spectral character of the rocks represented by photographic tone or color on the images. However, the criteria used often vary from one region to another as a result of climatic influences on weathering, vegetation cover, and mass movement.

This chapter describes the application of geologic interpretation techniques for reconnaissance lithologic mapping. The emphasis is placed upon the use of images acquired in the visible and reflected infrared spectral region; lithologic mapping from thermal IR, and SLAR has already been discussed in Chapters 9 and 10, respectively. Consideration is given to some of the problems that affect the interpretation of these images. Only a brief discussion is presented on conventional photogeology interpretation techniques; the reader is referred to Chapter 7 of this volume and standard texts (1, 2) for additional details.

While reconnaissnce mapping from images will never replace field mapping, it may be a desirable preliminary to any lithologic mapping program. In areas of the world where access is difficult and topographic maps are nonexistent, photo-interpretation maps are invaluable in characterizing regional relationships of broad geologic patterns, serving to define specific areas for detailed mapping, and serving as base maps for compilation of geologic data.

13.2 INFORMATION AVAILABLE FROM IMAGES

Lithologic mapping from images can often be accomplished by analysis and interpretation of the spectral and spatial information within the images, as modified by the type of terrain present, its climatic environment and history, and the prevailing geomorphic processes and their stage of development. Consideration must also be given to the environmental and operational factors affecting instrument response (see Chapter 3), the spectral region being recorded (Chapter 2), and the characteristics of the data display (Chapter 5 and 6). The spectral and spatial information available to the interpreter for lithologic identification is expressed by landform development, drainage density and pattern, vegetation differences, and spectral reflectivity, all interpreted in the context of climatic effects.

13.2.1 Climatic Effects

Significant differences in landforms result from widely different climatic conditions. Under a given set of climatic conditions some geomorphic processes will dominate over others, and impart to the landscape diagnostic characteristics. In regions of abundant water and moderate to high temperatures, the predominant process is chemical weathering and material transport. In the humid tropics, for example, deep chemical decay of the rocks often produces an extremely thick soil mantle, totally obscuring the underlying bedrock. The resultant topography is characterized by smooth, flowing slopes which are often further obscured by a thick vegetative cover. In areas of moderate vegetation, during periods of prolonged rainfall, soil and clay masses produced by weathering become saturated and highly mobile, moving downslope by slumping, mudflows, or creep. This further obscures the underlying bedrock and its topographic expression. Accordingly, these areas yield the least amount of geologic information to photo interpreters.

In more temperate regions, chemical weathering is not as dominant. Landforms tend to be more angular and topography less subdued, reflecting the greater importance of mechanical weathering and the diminished importance of soil development. Mass movement of surface material is not as pronounced as in humid regions, resulting in a closer association of residual soil and parent rock. Additional consequences of a more moderate climate are the reduced vegetative cover and greater areal exposure of rock outcrops. Consequently, more lithologic information is discernible from images of temperate regions than from tropic regions.

Under both climatic regimes, however, there is a tendency for soils from differing rock formations to become more similar at maturity due to leaching by rainfall of salts from these soils (3). This diminishes apparent differences on images and makes identification of rock types more difficult. In arid and semiarid regions, where precipitation is scant, the dominant geomorphic process is physical weathering. Mass movement of surface material is minimal and therefore the varying resistance of rock units beneath slopes is often sharply reflected in the topography. Outcrops of bedrock in these areas are further enhanced by the exceptionally slow rate of soil formation and the sparsity or even lack of vegetation. Owing to the almost total absence of water, the weathering mantle is not excessively leached, and closely resembles the composition of the underlying bedrock. These areas therefore yield the greatest amount of geologic information to remote sensing.

The differences in landscapes produced under widely different climatic conditions can be well illustrated by the diverse landforms of limestone terrains. In humid regions, limestone is very susceptible to weathering by solution, and is considered a rock of low weathering resistance. It typically underlies topographic lows and often develops characteristic solution cavities called karsts, which are readily discernible in images. In arid regions, however, limestone is a resistant rock and commonly forms cliffs or ridges.

13.2.2 Landform Analysis

Analysis of landforms provides valuable information on the composition and mode of origin of different terrains. Coupled with an understanding of the climatic influences on geomorphic processes, landform analysis is one of the most important techniques for lithologic identification from images.

In general, sedimentary terrains are more readily recognizable in images and yield more lithologic information than either igneous or metamorphic terrains. Sedimentary terrains have a greater diversity of rock types, often displaying pronounced differential resistance to erosion. As a result of what Rich (4) termed the "etching concept," resistant rock units are brought into relief and less resistant beds are lowered by weathering and erosional processes. Consequently, sedimentary terrains are often readily distinguished in images by their distinct banded appearance. The intensity of the banding reflects many factors, including the thickness, diversity, and dip of the rock units, as well as the extent and amount of surficial cover by unconsolidated materials.

Flat-lying or horizontal bedding can be distinguished by banding along topographic contours, and by changes in slope characteristics due to the presence of both resistant and nonresistant rocks. This is particularly evident in arid regions, where the parallel retreat of slopes produces distinctive landforms such as mesas and buttes. Where the beds have been severely folded or warped, the recognition of sedimentary terrains is particularly easy, as illustrated by the Landsat image of the Zagros Mountains, Iran (Fig. 13.1).

Extrusive igneous terrains are also easy to recognize because of their characteristic landforms, structural features, structural relations to surrounding rocks, and their topographic and erosional characteristics.

In geologically young or undeformed areas, lava flows are distinguished by their lobate patterns, expressed topographically or by vegetation, and their frequent association with volcanoes and their relation to local tectonic features (Fig. 13.2). In high resolution images, flow structures such as channels and surface characteristics aid in interpretation, and sometimes provide sufficient information on the viscosity of the lava to infer their composition.

In older or deformed areas, recognition of extrusive rocks can be difficult. Where lava flows have been folded or tilted, they may be similar in appearance to resistant sedimentary beds. Where erosion has been dominant, cinder cones and com-

FIGURE 13.1. Folding of sedimentary rocks in the Zagros Mountains, Iran. Dark circular features are salt diapirs. Landsat scene #1076-06232, MSS band 5. Courtesy EROS Data Center.

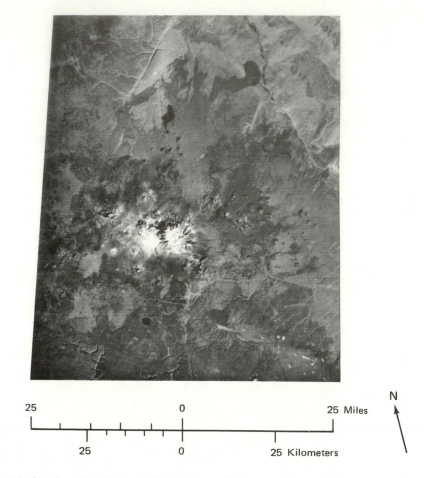

25 ————————————— 0 ————————————— 25 Miles

25 ————————————— 0 ————————————— 25 Kilometers

N

FIGURE 13.2. San Francisco Volcanic Field, Arizona. Note young flows, cinder cones, and related graben. Landsat scene #1103-17323, MSS band 5. Courtesy EROS Data Center.

posite volcanoes weather relatively rapidly, losing their distinctive morphological appearance.

While intrusive igneous rocks seldom have distinctive landforms, they can sometimes be recognized by their structural relations to the surrounding rocks. Dikes are often readily identifiable by their linear or curvilinear topographic expression. They may form spine-like ridges or depressions, as can occur in a humid climate where a basalt dike is intruded into sandstone or quartzite. In contrast, sills are often not distinct for they produce landforms similar to bedded sedimentary rocks. Large intrusive bodies such as plugs, stocks, laccoliths, and lopoliths often have circular shapes which can be expressed by the topography (Fig. 13.3).

The presence of these features can sometimes be inferred by the tilting or doming of the over-

lying rocks. Because tectonism and salt diapirs can also cause tilting and doming, consideration must be given to the regional geology. Where exposed, these intrusives can be recognized by their circular or elliptical outcrop, which may be accompanied by tilting of the country rock as in the Black Hills of South Dakota.

Where intrusive rocks cover large areas, distinctive landforms are absent although they have a massive appearance and lack bedding. This homogeneous nature is itself a diagnostic characteristic. Nearly all large intrusive bodies also display a crisscross pattern of jointing, which is expressed tonally, topographically, or by vegetation (Fig. 13.4).

The identification of metamorphic rocks is often difficult, with the greatest amount of information obtained in terrains where metamorphism and de-

FIGURE 13.3. Diapiric granitic intrusions, East Pilbara region, Western Australia. Landsat scene #1148-01282, MSS band 5. Courtesy EROS Data Center.

formation have not been intense and the primary structures of the original rocks are still preserved. Banding or bedding, and relict folding, can best be recognized where the metamorphic rocks are derived from thin-bedded sedimentary rocks of contrasting lithologies (5). However, a similar appearance can also be due to foliation rather than bedding. Foliation can impart a topographic grain to the region, which can be expressed by parallel alignment of ridges and low areas due to differential weathering.

13.2.3 Drainage Analysis

Drainage analysis is an established and widely used interpretative technique. Drainage density, stream frequency, and drainage pattern must be interpreted in context of climate, relative relief, and ground condition. For example, climate influences the type and amount of vegetation and the capacity of the soil to absorb water, which in turn affects the amount and rate of surface runoff. As a consequence of these factors, in the presence of similar rock types, a humid region would be expected to show a coarser drainage texture than a semiarid region (6).

Permeability of the ground is probably the single most important factor in controlling drainage texture. Many factors determine permeability, including: the nature and degree of soil development, its texture and structure, moisture content, and surface condition; vegetation cover; and degree of cracks and fractures in bedrock.

In general, impermeable materials will have more numerous drainage lines than will permeable ones, as illustrated in Fig. 13.5, an aerial photograph of a previously glaciated area in Indiana. The near absence of drainage lines on the permeable sand and gravel terrace is well displayed, in contrast to the large number of drainage lines on the impermeable silt and clay till plain.

Approximately 1 mile

FIGURE 13.4. Jointing in igneous terrain, Wyoming. Major fault (A) is offset by younger fault (B). Note alignment of streams on opposite side of divide (C). Courtesy of R. Ray and U.S. Navy.

FIGURE 13.5. Difference in drainage texture upon a sand and gravel terrace and adjacent till upland near Clinton, Indiana. After Thornbury (6).

The same phenomenon can be observed in sedimentary terrains where sandstones exhibit coarse drainage because of their high porosity and permeability, and shales exhibit a fine drainage texture because of their low permeability.

The amount of initial relief influences drainage texture, for a larger number of drainage lines will be developed on an irregular surface compared to one lacking significant relief. The degree of irregularity controlled by the massiveness of the rocks, jointing, and faulting, further influences the drainage texture. High relief also provides for the cutting of deep valleys with resultant development of headward erosion.

The drainage pattern primarily reflects the influence of structure, and secondarily the influence of lithology. Many drainage patterns have been recognized (Fig. 13.6); complete discussion may be found in standard references (6, 7). The significance of the most frequently occurring drainage patterns, dendritic, trellis, rectangular, radial, and deranged, is summarized in Table 13.1.

DENDRITIC

TRELLIS

RECTANGULAR

Mt. Katahdin

RADIAL

Scale

0   1   2   3   4 Miles

FIGURE 13.6. Types of drainage patterns. After Thornbury (6).

TABLE 13.1. SIGNIFICANCE OF DRAINAGE PATTERNS

Pattern	Characteristics	Significance
Dendritic	Irregular branching of streams, leaf appearance	Develops in flat-lying areas of uniform composition with little or no structural control. Suggests near-horizontal sedimentary rocks, massive intrusive rocks, or strongly metamorphosed rocks.
Trellis	Subparallel streams aligned along nearly parallel topographic features with short perpendicular inter-connected channels	Most likely to develop in areas of sedimentary rocks with marked structural control due to alternating resistant and less resistant beds.
Rectangular	Abrupt right-angle deflections of stream directions	May indicate structural control of instrusive igneous rocks or tilted sedimentary rocks.
Radial	Streams diverge from a centrally elevated point	Develops on domes, volcanic cones, and other types of isolated conical or subconical hills.
Deranged	Irregular stream courses that flow into and out of lakes or swamps	Develops on glacial deposits where the postglacial drainage has not yet been integrated into a well-defined pattern.

13.2.4 Vegetation

Although vegetative cover often obscures the land surface, it can still provide information useful for direct lithologic discrimination and recognition of structures. Under uniform climatic conditions, plant communities are often specific to rock type or soil. This results primarily from differences in the chemical and mineral composition of the rock or soil and differences in the physical characteristics of the medium.

Vegetation specificity in an igneous-metamorphic terrain in Alaska is illustrated in Fig. 13.7. The area underlain by massive marble (A) and other carbonate rocks supports a dense cover of conifers, whereas the areas underlain by diorite (B) have only a patchy vegetative cover, which in places enhances the typical crisscross joint pattern common to intrusive rocks.

The appearance on images of vegetation species associated with residual soils or parent rock materials is seasonally dependent, as has been dramatically illustrated in a study conducted by Grootenboer (8). Figure 13.8, color plate 21 shows an area in the Western Transvaal, South Africa, composed primarily of dolomite. The top image (*a*) is a late winter Landsat scene, acquired in September 1972 when most of the vegetation was brown and the trees were leafless. Slight tonal variations are evident in the dolomite series. The bottom image (*b*) was acquired in December 1972 at the height of the rainy season with the vegetation in full leaf. Strong tonal contrasts are due to the association of characteristic vegetation with underlying geology and soil moisture differences. The variations appear to be related to stratigraphic components in the dolomite series; such units had not been previously recognized or mapped (8).

Differences in vegetation health can sometimes be recognized in images, and may reflect stressful growth conditions induced by abnormal chemical or mineral concentrations in the soil, as described in the preceding chapter.

FIGURE 13.7. Vegetation contrast between igneous and metamorphic complex (A) and younger volcanic flows (B). Note that certian faults (C) may be traced up to the lava flows and continuations may be picked up at the opposite side of the flow area. Courtesy of R. Ray and U.S. Geological Survey.

Even in areas where the vegetative cover completely blankets the surface, geologic structures can sometimes be recognized by their topographic expression alone. In areas of more modest vegetative cover, geologic structures such as faults and joints are often enhanced by anomalous vegetation patterns. These patterns reflect the higher moisture content generally found along fractures and in fault zones, due to structural control of groundwater movement. This phenomenon is illustrated in Fig. 13.9, a low altitude aerial photograph of rolling hills in northern California. Annual grasses are dry but deeper-rooted weeds stay greeen by drawing moisture from clays in gouge along northeast-striking faults and northwest-striking shale interbeds in the sandstones. Small offsets in the shale are particularly well-defined.

13.2.5 Spectral Reflectivity

The spectral reflectance of earth materials is often the most useful and diagnostic criterion for lithologic discrimination. Spectral reflectance is a measure of the amount of light reflected by a material, and is expressed in images by photographic tone or color. Reflectance is a consequence of the chemistry and structure of the material modified by environmental factors and the physical condition of the material.

In the visible and near-infrared spectrum, information is a consequence of reflectance variations due to electronic and vibrational processes. The principal constituents of igneous rock-forming minerals, and hence all rocks, are silicon-oxygen, aluminum-oxygen, and magnesium-oxygen structures, which have neither electronic nor vibrational transitions in this energy level. The spectral information that appears as bands or wings of bands is due to the presence of other minor components that are present constitutionally, substitutionally, or as impurities. The majority of discernible features in the spectra of igneous rocks occurs as a result of the presence of iron, its oxidation state, and water. The same is true for sedimentary and metamorphic rocks, with the exception of carbonates, which display strong absorptions caused by vibrational processes due to the CO_3^{-2} ion and Al-O-H deformation in clay materials (see Chapter 2).

Interpretation of spectral reflectivity data can be aided by analysis of laboratory spectra or spectra acquired *in situ*. This section describes the information available in spectral reflectance data, and how it can be used for lithologic mapping.

LABORATORY SPECTRA

Analysis of laboratory spectra provides a preliminary basis for determining the wavelength regions that show diagnostic spectral features, and for determining the separability of various materials. Spectral reflectance data from laboratory samples minimize the complications resulting from environmental factors such as atmospheric absorption and scattering, soil cover, vegetation cover, and impurities. However, the absence of these environmental factors can make extrapolation to naturally occurring materials risky, and the limited size of each sample may preclude adequate representation of the inherent variability in earth materials. Nevertheless, study of laboratory spectra is necessary for understanding and interpreting remotely sensed data.

Representative laboratory spectra of igneous rocks are presented in Fig. 13.10(9). The spectra are arranged in order of high acidic composition at the top to mafic rocks at the bottom. Graphic granite displays absorption bands at 1.4, 1.9 and about 2.2 μm. These are due to vibrational overtones and combinations of OH$^-$ and H_2O. Biotite granite and granite have much less water, so that the absorption bands diminish or disappear.

Intermediate and basic (mafic) rocks contain less water and have an increased content of the opaque mineral magnetite. The effect of magnetite is to reduce the overall reflectance and decrease the size of any spectral absorption features. However, the increased presence of ferric and ferrous iron produces diagnostic spectral absorptions near 0.7 and 1.0 μm, respectively.

Ultrabasic rocks contain much more opaque materials and Fe^{+2} bearing pyroxenes and olivine, and consequently the ferrous iron bands are more pronounced. Pyroxenite, for example, has a strong absorption at 1.0 μm and near 2.0 μm from Fe^{+2} in octrahedral coordination with oxygen in pyroxene, amphibole, or olivine.

Representative laboratory spectra of sedimentary rocks are illustrated in Fig. 13.11. In general, the dominating features are due to the additional presence of the carbonate radical, which produces absorption bands between 1.9 and 2.3 μm. The most intense band is located near 2.3 μm, and

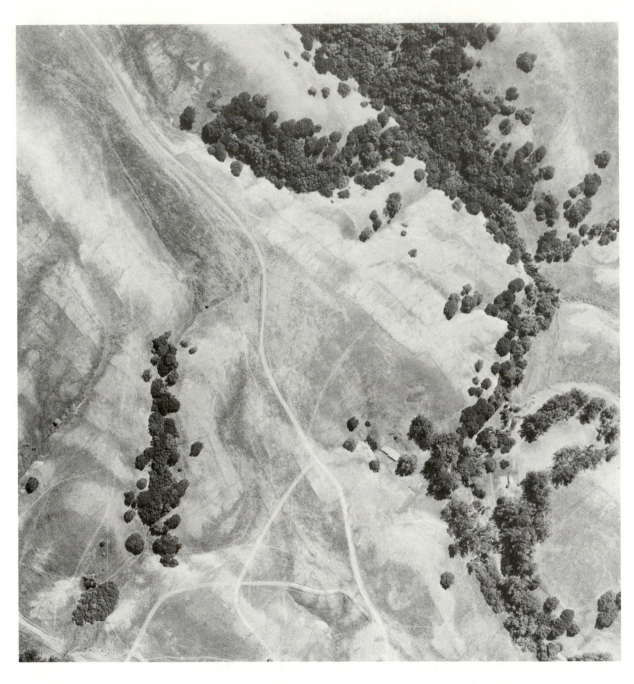

FIGURE 13.9. Preferential vegetation growth along fault zones in interbedded shale and sandstone. Courtesy of Harding-Lawson Associates.

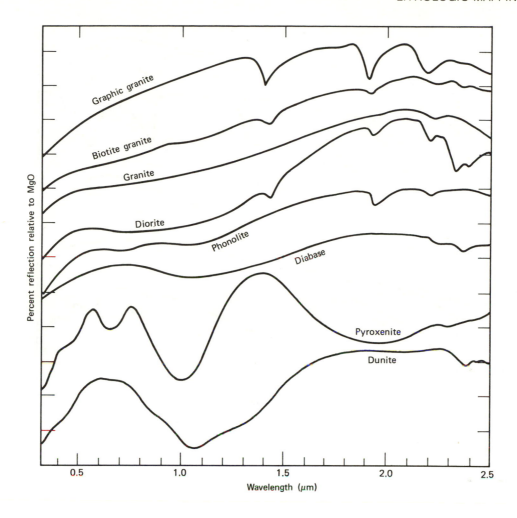

FIGURE 13.10. Bidirectional reflectance spectra of igneous rocks, separated vertically for clarity. Reflectance divisions are 10%. After Salisbury and Hunt (9).

is prominently displayed by fossiliferous limestone. Most limestones also display water bands at 1.4 and 1.9 μm; some show a broad absorption at 1.0 μm due to ferrous iron. Sandstones typically show a band near 0.85 μm due to ferric oxide stain on the grains. In addition, the presence of fluid inclusions in the grains, and calcareous cement cause absorption bands at 1.4, 1.9, 2.2, and 2.3 μm.

Shales often contain sufficient carbonaceous material to mask spectral features. When features are retained, they are usually the H_2O and OH^- features of clay minerals, and the broad absorption bands due to the presence of iron. Both features are illustrated in the spectrum of the illite-bearing shale.

Representative laboratory spectra of metamorphic rocks are presented in Fig. 13.12. The features displayed by marbles are strong carbonate absorptions and high reflectivity in the near infrared. The other metamorphic rocks display hydroxyl or water bands in their spectra, one of which is particularly strong at 1.4 μm. The features produced by ferrous iron are evident in several of the spectra; the broad absorption at 1.0 μm appears in the spectra of hornblende schist and tremolite schist.

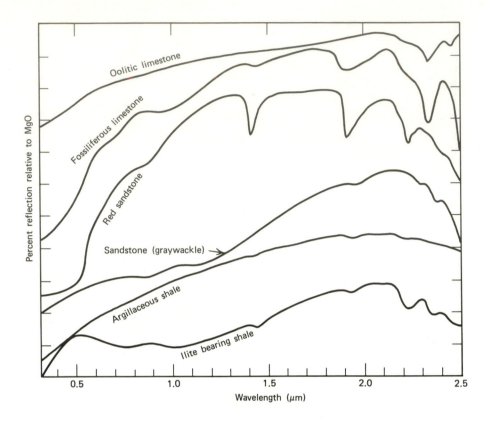

FIGURE 13.11. Bidirectional reflectance spectra of sedimentary rocks, separated vertically for clarity. Reflectance divisions are 10%. After Salisbury and Hunt (9).

FIELD SPECTRA

When dealing with *in situ* materials, environmental factors complicate the spectrum. The presence of vegetation, water, organic compounds, and human-made objects contribute to the observed spectral response due to limits imposed by spatial resolution, and make determination of the lithologic composition difficult. In addition, atmospheric absorption, turbulence, and scattering restrict the wavelength ranges that may be investigated to the amtospheric "windows," which may not be optimal for lithologic discrimination. Consequently, laboratory spectra should only be used as a guide in interpreting spectral information acquired in the natural environment, and not as a definitive tool. A more practical approach is to use *in situ* spectral reflectance data for interpreting multispectral data on imagery. Reflectance spectra acquired in the field are more representative of the spectral response recorded in images, for the sample size allows inclusion of the inherent variability in the unit as well as including some of the physical and environmental factors that modify the spectral response of the observed material.

The differences between laboratory spectra and spectra acquired *in situ* have been noted by Longshaw (10). Comparison of the two kinds of spectra revealed that major spectral features are observed for both, but that the environmental effects can modify the field data to the extent that they no longer resemble the spectra of laboratory materials. This is due to the fact that the natural environment introduces surficial weathering effects which can cause ambiguity in the relationship between the *in situ* reflectance of a rock and its internal minerology.

This point is well depicted in Fig. 13.13, which shows two laboratory spectra of andesite, one of the

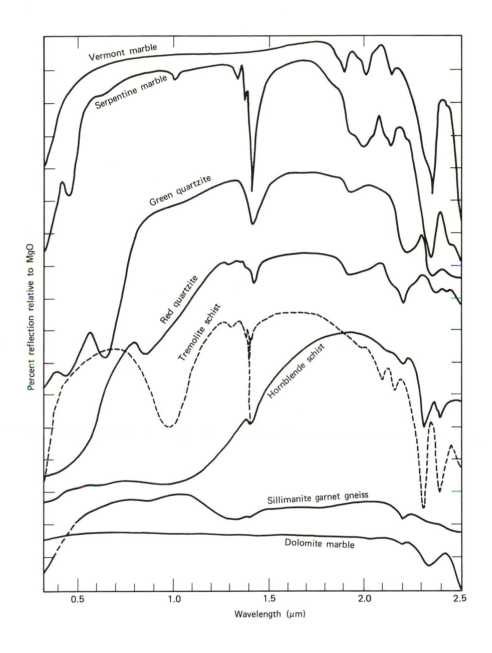

FIGURE 13.12. Bidirecitonal reflectance spectra of metamorphic rocks, separated vertically for clarity. Reflectance divisions are 10%. After Salisbury and Hunt (9).

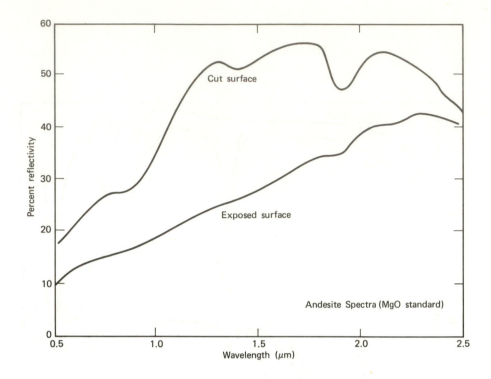

FIGURE 13.13. Laboratory reflectance spectra of andesite for cut and exposed surfaces. Courtesy JPL.

exposed surface, the other of a freshly cut surface. The 0.9- μm absorption band is practically gone in the exposed surface, and the reflectance curve is greatly depressed.

EXPRESSION OF SPECTRAL REFLECTIVITY IN IMAGES

Spectral reflectivity is expressed in images as either photographic tone or color. Photographic tone is a measure of the relative amount of light reflected by an object in a given wavelength region, and displayed as shades of gray on a black-and-white photograph. The wavelength dependence of photographic tone can be illustrated by examining the appearance of a color target photographed in different wavelength regions (Fig. 13.14, color plate 22).

Photographic tone is modified by a material's geometry relative to the sun, water content, vegetative cover, atmospheric effects, sensor response, and photographic processing. The effect of atmospheric scattering, particularly at shorter wavelengths, is partially to obscure the spectral re-

sponse of the material by adding unwanted signal noise, which diminishes the relative contrast between different materials. A material's geometry relative to the sun determines the amount of incident radiation, and also the amount of reflected radiation. On an image, similar rock types facing toward or away from the sun, such as on either side of a hill, will show major brightness differences, which are purely due to topography. Moisture reduces the reflectance in the visible and near-infrared wavelength regions, so that the same rock type may show different tones due to moisture content. Photographic processing can also affect the relative tones on an image. Without careful processing, the brightness of a material may vary from image to image, making extrapolation of lithologic identification difficult. On a single image, the relative separability of two materials may also be reduced due to improper processing.

Examination of a material's reflectance spectrum can provide an *a priori* indication of its relative photographic tone in images. Basalt, for example, has a low reflectance in all wavelength regions;

its appearance in images in any wavelength region would therefore be uniform and dark. Playa material has a very high and uniform reflectance in the wavelengths greater than 0.6 μm, and it would appear relatively bright in images in that wavelength region. Red shale with abundant clay shows variable reflectance in different wavelength regions. The spectrum is dominated by the effect of oxidized iron, which is responsible for the red color in the visible region. It also causes the dip between 0.7 and 0.9 μm. The sharp dip at 2.2 μm is due to OH⁻ stretching from the clay component. The relative photographic tone of this shale would depend upon which wavelengths of light were being sensed, but in general it would be bright.

Color is probably the single most useful recognition element in interpreting imagery because the human eye is capable of distinguishing nearly 1000 times as many tints and shades of color as shades of gray. Color provides a means of simultaneously displaying data from more than one wavelength region. In a color photograph, the spectral response of a scene is portrayed by assigning a color to each of three distinct wavelength regions, with the intensity of that color corresponding to the relative reflectance of the components. This can be accomplished directly, with the use of color film, or indirectly, by color reconstruction from three black-and-white transparencies.

True color pictures represent the spectral response of a scene in the visible wavelength region as sensed by the human eye. The hues and intensities in a true color picture are a function of the film characteristics, photographic processing, and the type of illumination used in viewing. Under optimum conditions the picture very closely approximates the interpreter's perception of the natural scene and is therefore the easiest to interpret.

Spectral information in the near reflected infrared wavelength region is routinely recorded and displayed on color infrared film (see Chapters 3 and 7). The film is sensitive to light from 0.36 μm to about 0.9 μm. The blue, green, and red components of the picture represent the amount of reflected light in the green, red, and near-infrared wavelength regions, respectively. On color infrared images, vegetation is red due to the strong chlorophyll reflection in the infrared; red rocks are yellow; gray materials tend to remain unchanged. Because of the lack of sensitivity in the blue wavelength regions, color infrared film readily penetrates haze.

Spectral information in the wavelength region beyond 1 μm cannot be recorded directly on photographic film. Instead, optical-mechanical scanners are used as discussed in Chapter 2, to measure simultaneously information in several infrared wavelength regions; the information is then recorded on magnetic tape for processing or displayed directly on black-and-white film. Information from three wavelength regions can be displayed simultaneously by color reconstruction from black-and-white transparencies. This can be readily accomplished by use of an additive color viewer, in which three transparencies may be registered and projected on a screen in blue, green, red, or white light, forming a false color image. The analyst has the flexibility to choose color schemes for the display that maximizes the separability of features of interest, without regard to the scene's true color appearances. Color reconstruction can also be accomplished by conventional photographic techniques using color film.

SELECTION OF SPECTRAL BANDS

The separability of materials in images depends on relative photographic tone, a measure of the reflectivity in the recorded wavelength region (except for shadowed areas). Maximum contrast, hence separability, occurs when two materials have large reflectivity differences over the sensed spectral region. In general, spectral differences are evidenced as narrow absorption bands and therefore the width of the sensed wavelength region is crucial. Conventional black-and-white aerial photographs, for example, record spectral information in the broad wavelength region of 0.5 to 0.7 μm. The relative photographic tones represent the average reflectance of the material. Difficulties in separating materials due to the broad wavelength sensed can be illustrated by examining the behavior of limestone and shale. Figure 13.15 illustrates the relative reflectance of weathered samples of the Bernal formation (a red thin-bedded shale and shaly siltstone) and the San Andres limestone (a grayish limestone). Pronounced differences exist in the relative spectral intensity in the blue and red parts of the spectrum, but the integrated light reflectance of those materials between 0.5 and 0.7 μm is very similar and little or no photographic distinction could be expected (5).

By narrowing the sensed wavelength regions, better separation of materials can be expected. However, instrumental responses limit the narrow-

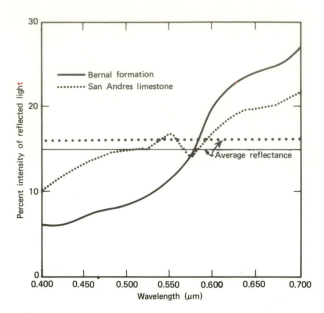

FIGURE 13.15. Spectra reflectance curves of limestone and shale, depicting separability as a function of wavelength, and similarity of average integrated reflectance. After Ray (5).

ness of the spectral bands that can be sensed (see Chapter 3).

In addition to using narrow wavelength bands, the optimal wavelength regions for separating different materials must be determined. Qualitative or quantitative analysis of the spectral features associated with the materials of interest can provide the necessary information to select these bands.

In one study to determine the optimal bands and ratios for separating various materials using Skylab S-192 images, Vincent and Pillars (11) performed linear discriminant analysis on 211 laboratory spectra of soils, minerals, and rocks to rank 12 wavelength regions and the best 12 of a possible 66 wavelength ratios between 0.4 μm and 2.34 μm. Linear discriminant analysis is a multivariate statistical technique that can be used to determine the separability of given classes or groups of materials. The ranking for the wavelength regions and ratios is shown in Table 13.2. The first three single wavelength regions correspond to spectral features due to the presence of carbonate, OH^-, water, and iron, as previously described. The highest-ranked ratios correspond to spectral regions that have maximum slopes or have components in absorption regions that are also due to the key components.

Their study indicates that the data analyzed are most useful for discriminating minerals deposited by hydrothermal alteration, weathering, and evaporation; discrimination among igneous silicates appears to be more difficult.

In an unpublished study using field acquired spectra of various materials, Siegal used linear discriminant analysis to determine the separability of different materials and the best wavelength regions to be used. Table 13.3 presents the classification accuracy and best wavelength regions for separating one material from all the others. Between 86% and 100% of each material was correctly identified as that material; up to 25% of the other materials were misclassified. The large variation in the best wavelength regions indicates that no single wavelength region can be used to separate accurately all given materials in a scene. Rather, greater accuracy can be achieved by carefully selecting wavelength regions to separate one class or a few classes of materials from all the others.

These classifications are based upon variations in reflectance as a function of wavelength. Other schemes of discrimination may be possible, for example, ones based upon slope rather than ratios. The advantage to such a method would lie in discrimination of particle size variations between targets of chemically and petrologically similar materials.

Analysis of field-acquired spectral data can be used to determine the separability of lithologic units, and to select images in the proper spectral regions to maximize lithologic differences. The application of this technique for reconnaissance lithologic mapping has been described by Goetz et al. (12) in a study comparing the usefulness of computer enhanced Landsat and Skylab data for mapping in northern Arizona. By using discriminant analysis of spectral reflectance measurements in the wavelength region of 0.4 to 2.4 μm, they were able to determine the best wavelength regions for separating the lithologies exposed in the study area. Skylab S-192 imagery was then used to approximate the three best regions (1.3, 1.0, and 0.5 μm) and a false color composite image was produced and interpreted. For comparison, computer enhanced Landsat ratio composites were also interpreted for lithologic information. The photo-interpretation lithologic maps are presented in Fig. 13.16. The Skylab S-192 image allowed the most accurate separation and identification of lithologic

TABLE 13.2   RANKING OF SINGLE WAVELENGTH REGIONS AND RATIOS AFTER VINCENT AND PILLARS (11)

RANK	WAVELENGTH REGION (μm)	WAVELENGTH RATIO (μm)
		0.77-0.89
1	2.10-2.34	0.60-0.65
		0.50-0.55
2	0.93-1.05	0.45-0.50
		0.93-1.05
3	0.45-0.50	0.54-0.60
		1.15-1.28
4	1.55-1.73	1.03-1.19
		2.10-2.34
5	0.60-0.65	1.55-1.73
		0.77-0.89
6	0.54-0.60	0.50-0.55
		0.54-0.60
7	0.77-0.89	0.45-0.50
		0.54-0.60
8	1.03-1.19	0.50-0.55
		0.77-0.89
9	1.15-1.28	0.45-0.50
		0.77-0.89
10	0.65-0.73	0.54-0.60
		0.93-1.05
11	0.50-0.55	0.50-0.55
		0.93-1.05
12	0.42-0.45	0.77-0.89

TABLE 13.3. CLASSIFICATION ACCURACY AND WAVELENGTH REGIONS FOR SEPARATING VARIOUS MATERIALS [AFTER SIEGAL (UNPUBLISHED)].

| Sample | Best Wavelength Regions (μm) | | | | Classification Accuracy | |
	1	2	3	4	% of Sample Correctly Classified	% of Other Material in Column 1 Classified as Samples
Basalt	0.75	0.40	1.1	2.3	100	16
Andesite	2.3	1.2	2.4	2.0	100	19
Rhyodacite	1.7	2.4	0.40	0.45	89	25
Quartz latite	1.2	2.1	0.55	1.3	86	15
Rhyolitic tuff	1.6	2.4	0.80	0.60	98	14
Limestone	0.70	2.2	2.4	2.3	100	6
Shale	2.1	1.1	2.4	2.3	91	16
Playa	0.45	0.55	0.40	0.50	95	0
Tailings	2.0	1.3	1.7	1.6	94	0
Altered rocks	1.6	2.2	0.65	2.1	87	2
Salic limestone	1.0	0.40	0.50	0.55	100	22
Limonitic marble	2.2	2.4	1.7	2.3	94	0

units in the study area; however both types of images could be used to produce adequate reconnaissance geologic maps.

13.3 LITHOLOGIC INTERPRETATION

13.3.1 Interpretation of Standard Images

The previous sections have examined the information available in images that can be used to discriminate and map lithologic units. This discrimination can be most readily accomplished when the materials have relatively large spectral differences, display significant differences in landform and drainage pattern and density, and are not obscured by nonspecific vegetative cover. These conditions exist in arid or semiarid regions of the world, where standard black-and-white or color images can provide adquate information to produce excellent reconnaissance maps. With the availability of sufficient ground truth data, the inferred composition of the various lithologies can be confirmed and a true geologic map can be prepared. This point has been demonstrated by Viljoen et al. (13) who studied an arid region in South West Africa using a standard Landsat false color infrared

image [Fig. 13.17(a) color plate 23] to produce an excellent geologic map [Fig. 13.17(b) color plate 23]. Geological formations present in the area include Archaean basement, folded Proterozoic sediments, undeformed Paleozoic Nama Group sediments and Karoo System volcanics, and young surficial deposits. The Nama Group consists of a basal limestone sequence (Kuibis formation) overlain by greenish shales with sandstone interbeds (Schwarz Rand formation), overlain by reddish argillaceous sandstone and mudstones (Fish River formation).

The basal limestones are distinguished by their grey color and dendritic drainage pattern in the southwestern part of the image where they are relatively flat-lying. In the northwest, where they are more steeply dipping, they are no longer as distinct.

The Schwarzrand formation reveals a two-fold division: a lower unit of interbanded red sandstone and green shales, represented on the image as interbanded lighter and darker olive colors; and an upper unit of greenish shales, represented by dark bluish and brown colors.

The Fish River formation shows broad stratigraphic subdivisions, characterized by color variations on the image. The color is directly related to the lithology, becoming progressively darker with increasing content of argillaceous material. In the image, volcanics are characterized by a vivid bluish green color, the result of their dark brown color and excellent exposure.

Surficial deposits can be readily distinguished on the basis of distribution pattern, texture, and color. Alluvial deposits show as a very white tone; calcrete is characterized by a light brown tone, often with a characteristic pitted texture; wind-blown desert sand shows up yellow with the larger dunes distinguishable as linear features.

In less favorable environments, standard small-scale images may not provide adequate information for producing accurate reconnaissance maps. Drainage patterns and densities, jointing, vegetation type, and landform development may not be discernible because of the low resolution of the image. Furthermore, spectral differences may not be large enough to be distinctly portrayed. In such cases, the use of larger scale images such as conventional aerial photographs, U-2 aircraft photographs, and high resolution satellite images must be used. Even the difference in resolution between Skylab S190A images (73-m resolution) and S190B images (20-m resolution) may be significant in improving the quality of the map that can be produced (14). Conventional aerial photographs provide even higher resolution images, and correspondingly allow mapping in much greater detail, often to scales as large as 1:10,000.

The greater accuracy and detail provided by large-scale images, however, have a tradeoff with areal coverage. Photo interpretation of a 10,000 mile^2 area, for example, could be done using a single Landsat image; the same area would require interpretation of 62 U-2 photographs, at a correspondingly larger increase of invested time. Although the resolution of U-2 images (commonly 3 m) is nearly two orders of magnitude better than Landsat images, the added effort to compile and mosaic data from the large number of individual scenes may not justify the improved detail. In addition, vignetting problems and tonal variations from image to image can make extrapolation of lithologic information difficult.

Besides problems of scale, it is necessary to consider the effects of temporal change in images. Temporal changes include differences in vegetation health, density, and community; variation in shadowing due to sun elevation and azimuth; enhancement of landforms and structural features

with light snow cover; change in relative tone of materials due to variable moisture retention; enhancement of small drainage features due to heavy precipitation. The time of year when an observation is made, therefore, has a strong control over the amount and type of information that can be derived by an interpreter. An example of this type of seasonal control was reported by Moore and Gregory (15) for an area in the Northwest Territories, Canada. Studying Landsat scenes they found that an extensive hummocky morainal deposit and much bedrock structure could be seen on winter images. On the summer images a vegetation pattern could be seen which was related to the distribution of lacustrine and marine clay deposits.

Snow cover may enhance topographic features, particularly under conditions of low sun angles. A heavy blanket of snow (that is, > 23 cm) accentuates major structural features, and a light dusting of snow (that is, < 2.5 cm) accentuates more subtle topographic expressions (such as jointing and drainage patterns) (16). By monitoring snow cover under varying states of melting and accumulation, subtle topographic features may be accentuated, thus increasing their detectability (Fig. 13.18). This is of particular value in interpreting drainage patterns, jointing, and landform development, which are used for lithologic identification.

13.3.2 Interpretation of Processed Images

Multispectral data can be acquired in analog and digital form. In both cases, more data is available than can be displayed or analyzed in a single picture; in addition, the analyst is usually interested in only a subset of the data relevant to the problem being considered. Techniques for extracting and displaying various subsets of data have been described at length in Chapters 4 and 5. This section describes the use of some digital processing techniques to maximize the display of lithologic information. Although many of these techniques can also be implemented by photographic and optical procedures, greater control and flexibility is afforded by digital processing.

CONTRAST ENHANCEMENT

Contrast enhancement is one of the most widely used image processing techniques for lithologic mapping. In its simplest form, it is the process of redistributing the brightness levels in a picture

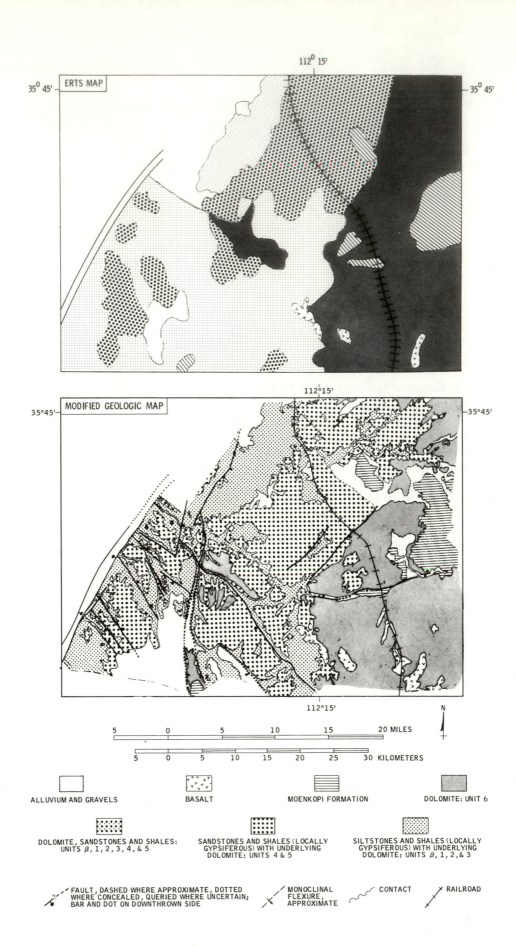

ERTS MAP

MODIFIED GEOLOGIC MAP

112° 15'

35° 45'   35° 45'

112°15'

35°45'   35°45'

112°15'

N

5   0   5   10   15   20 MILES

5   0   5   10   15   20   25   30 KILOMETERS

ALLUVIUM AND GRAVELS   BASALT   MOENKOPI FORMATION   DOLOMITE: UNIT 6

DOLOMITE, SANDSTONES AND SHALES: SANDSTONES AND SHALES (LOCALLY   SILTSTONES AND SHALES (LOCALLY
UNITS β, 1, 2, 3, 4, & 5   GYPSIFEROUS) WITH UNDERLYING   GYPSIFEROUS) WITH UNDERLYING
          DOLOMITE: UNITS 4 & 5   DOLOMITE: UNITS β, 1, 2, & 3

FAULT, DASHED WHERE APPROXIMATE, DOTTED   MONOCLINAL   CONTACT   RAILROAD
WHERE CONCEALED, QUERIED WHERE UNCERTAIN;   FLEXURE,
BAR AND DOT ON DOWNTHROWN SIDE   APPROXIMATE

402

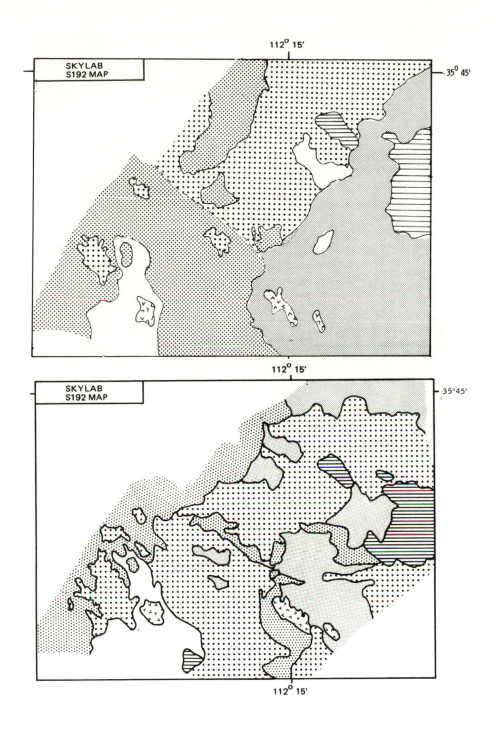

FIGURE 13.16. Geologic and photo interpretation maps of a southeast section of the Coconino Plateau, Arizona. Courtesy JPL.

403

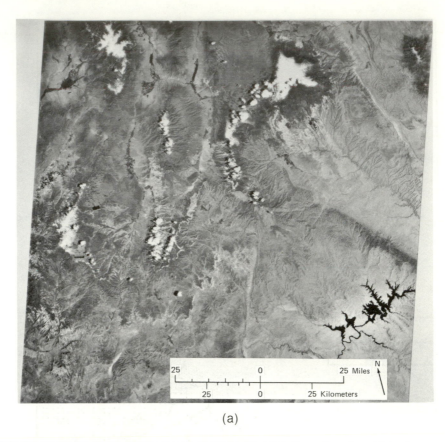

(a)

FIGURE 13.18. (a) Snow enhancement of structural detail in southwestern Utah. Summer Landsat scene #1320-17375, MSS band 5

to utilize the entire dynamic range of the display device. Many linear and nonlinear algorithms are available and can be applied to the data. The algorithm chosen depends on the subset of data which is of interest to the analyst. In applying contrast enhancement for lithologic mapping, it is necessary to examine the probability density function (PDF) of the scene (see Chapter 5) and determine the region of the histogram that contains the areas of interest. A suitable algorithm is chosen to maximize the contrast of those areas. If the analyst recognizes that the most interesting or significant information in an image is contained in bright regions, he might increase its contrast at the expense of contrast in darker, less important regions. Figure 13.19 illustrates this technique applied to the study of Qom Playa, a dry lake in Iran (17). The original image was processed to maximize

the display of information for bedrock and alluvium. As a result, the brightness levels representing the playa are packed at the high end of the PDF, and are displayed as white. By redistributing the few brightness levels of the playa over the entire dynamic range, subtle differences attributable to variations in moisture content are made evident.

To maximize the display of information for each component in an entire scene requires more sophisticated contrast stretching. A scene that contains land, snow, and water has a trimodal PDF. Simple linear stretching would only increase contrast in the center of the distribution, and would force the high and low peaks further toward saturation. To display such a scene adequately, each mode could be stretched independently to fill the dynamic range. This is equivalent to producing and combining three pictures, one of only water,

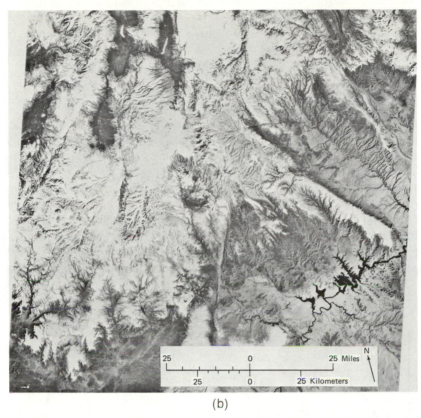

(b)

FIGURE 13.18. (b) Snow enhancement of structural detail in southwestern Utah. Winter, Landsat scene #1230-17382, Courtesy EROS Data Center.

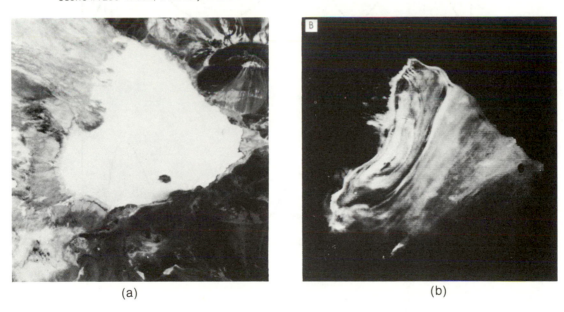

(a)   (b)

FIGURE 13.19. Qom Playa, Iran. (a) Computer processed digital image, processing designed to portray general geologic information. (b) Computer processed to maximize information in playa. Courtesy JPL.

one of land, and one of snow (Fig. 13.20). When the three modes in the original data do not overlap, the relative photographic tones of each region are preserved. When the modes overlap, the overlap regions will be included in the wrong mode and incorrectly stretched. To avoid the problem of a black-white boundary at the dividing point between two modes, the output gray levels of adjoining modes may be reversed. Figure 13.21 illustrates this technique applied to a Landsat scene of Iceland. In the original picture [Fig. 13.21(a)], water and snow-ice are saturated. In the processed picture, details in both the water and ice are evident, in addition to information in the land.

With any type of contrast enhancement, the relative photographic tone of different materials is modified. Simple linear stretches have the least effect on relative tone, and brightness differences can still be related to differences in reflectivity. In extreme cases, the relative tone can no longer be meaningfully related to the reflectance of the materials. An analyst must therefore be fully cognizant of the processing techniques that have been applied to the data.

RATIOING

Ratioing is a processing technique in which two spatially registered spectral images are divided, pixed by pixel (see Chapter 6). The resultant data are rescaled to fill the dynamic range of the display medium by contrast enhancement techniques.

Ratioing can be thought of as a method of enhancing minor differences between materials by defining the slope of the spectral curve between two bands. Ratio images also minimize the first-order effects of brightness variations due to topographic

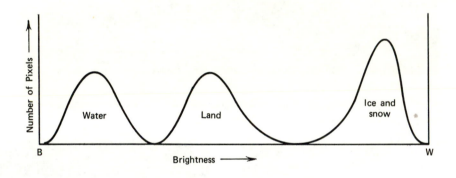

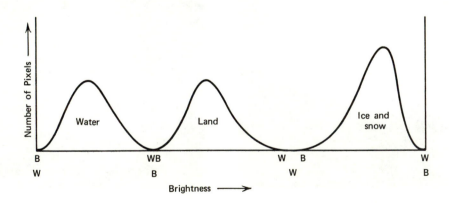

FIGURE 13.20. Probability density function of a scene with water, land, and ice before (a) and after (b) processing.

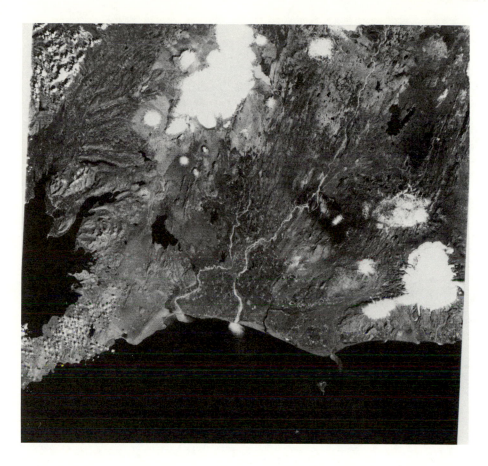

FIGURE 13.21. (a) Landsat image of southeast Iceland. Courtesy JPL. Linear contrast stretch.

slope, and they permit the display of differences between slopes of reflectance spectra of the bands in one image. At the same time, however, ratio images accentuate noise, making interpretation more difficult. In addition, dissimilar materials having similar spectral slopes but different albedos, which are easily separable on standard images, may become inseparable on ratio images.

Ratio images can be meaningfully interpreted because they can be directly related to the spectral properties of materials. Increased information can be obtained when the analyst uses those ratios that maximize the differences in the spectral slopes of materials in the scene. The presence of the oxidized-iron absorption band at about 0.85 μm lends itself well to this technique (see Chapter 17). The detectability of limonitic areas in a scene would be enhanced by using a ratio image that includes this wavelength region. Even better separation could be achieved by combining several ratio images, each displaying information about a different part of the spectrum where key absorptions or reflectances occur. It is therefore important that the interpreter be cognizant of the general types of materials found in the scene and their spectral properties in order to select the best ratio images for interpretation. Because it is impossible to optimize the simultaneous display of all materials in a scene, it is often necessary to examine several different combinations of ratio images, each displaying a different subset of the data.

Ratio images have been used in many geological investigations to recognize and map areas of mineral alteration (18, 19) and for lithologic mapping (20, 21). To date most studies have used Landsat data, and consequently success has been limited

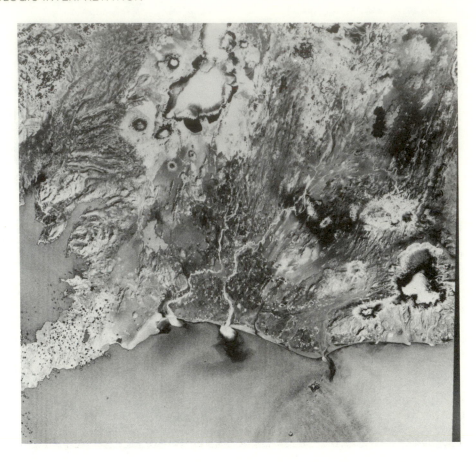

FIGURE 13.21. (*b*) Landsat image of southeast Iceland. Courtesy of JPL. Trimodal contrast stretch.

by the narrowness of the spectral region and the relatively broad wavelength bands of the MSS. One representative study illustrating the advantages and disadvantages of using Landsat data for geologic mapping was reported by Blodget *et al.* (21) in Saudi Arabia. The study area included complex pre-Cambrian metamorphic assemblages intruded by plutonic rocks varying in composition from gabbro to granite. Cambrian sandstones and Tertiary-Quaternary flood basalts overlie the pre-Cambrian metamorphic assemblages in places; locally small volcanic flows are also found. Young aeolian and alluvial sediments fill the topographic depressions and drainages. Two of the ratio composites used for interpretation are presented in Fig. 13.22, color plate 24. Distinctive color differences allow recognition of three kinds of plutonic rocks: monzonite, granite, and riebeckite granite. In addition, cinder cones, basalt, sandstone, and alluvial materials are distinct and recognizable; basalt flows of different ages can also be distinguished. The greatest amount of discriminative data was provided by the Landsat band ratio composite, using 4/5, 5/6, and 6/7. Other ratio combinations were necessary to extract additional lithologic information. Rowan *et al.* (18) also found that the 4/5, 5/6, 6/7 composite provided the greatest amount of information for discriminating between hydrothermally altered and unaltered rocks, as well as separating various types of igneous rocks in and near the Goldfield, Nevada mining district. A complete discussion of their techniques and findings is presented in Chapter 17.

CANONICAL ANALYSIS

Canonical analysis, which is also referred to as multiple discriminant analysis, produces a data transformation based on spectral signatures of a set of targets (22, 23) to obtain the maximum separability among the targets (groups). Mathematically, canonical analysis computes a set of transformed variables based on maximizing the among-category and minimizing the within-category covariance matrix. Geometrically, it produces a set of mutually orthogonal (independent) axes which are fitted to the data by rotation, translation, and scaling such that the first new axis (transformed variates) accounts for the greatest amount of variance with succeeding axes containing lesser variance.

Canonical analysis is particularly useful for dimension reduction and is similar to principal component analysis (see Chapter 6) in that an orthogonal linear transformation is computed from the data. The techniques differ however, in the method of computing the transformation (23). Principal component analysis uses the total covariance matrix, whereas, canonical analysis is based on the two components of the total covariance matrix.

The principal advantage of canonical analysis is the increased discrimination between rock materials with small spectral differences because the transformation is determined from the chosen target areas only, rather than from the entire image.

In application, canonical analysis can be used to classify and enhance multispectral data. At least one "training area" in the image must be defined to characterize each of the materials of interest (groups). The groups must be spectrally homogeneous and representative of the classes to be investigated. The training areas are used to compute the transformed variables that can be applied to elements of the scene (pixels) to produce a new display, or to classify pixels by assigning them to the group for which their probability is the greatest.

Canonical analysis has been used by several investigators to enhance or classify Landsat data (24, 25, 26, 27). Podwysocki *et al.* (27) for example, used canonical analysis to analyze a section of the scene in Saudi Arabia described by Blodget et al. (21). The four canonically transformed Landsat images of the Sahl al Matron test site are presented in Fig. 13.23. The first canonical variable (Axis 1) accounts for 78% of the total variance contained in the Landsat scene, and the second, third, and fourth variables account for 18%, 3%, and 1%, respectively. Over 99% of the total variation is explained by the first three canonical variables. The first variable predominantly displays albedo differences of the different materials. Lithologic identification is limited to basalt (A) and alluvial materials (B), which are distinctively dark and bright. The second variable primarily displays spectral information related to lithologic variability. The distinct bright oval area (A) corresponds to outcrops of granite and granodiorite. Exposures of basalt (B) are very distinctive. Several small cinder cones (C) can also be recognized. The gray area (D) represents outcrops of monazite. Its appearance on the first variable (C) is similar to alluvial material (B), but is relatively darker on the second variable. The third variable, which contains 3% of the total variation, displays no interpretable additional lithologic or albedo information. Many of the lithologies which are separable on the second variable can also be recognized on the third, although they are not as distinct. The fourth axis displays no additional information and consists mainly of the unassigned sources of variation (noise).

Interpretation of canonically transformed data can be improved by combining and registering three of the images to produce a false color composite picture. In such a display the inter- and intraimage tonal variations can be more easily assessed. Although canonically transformed images may contain more geologic information than any one individual processed image, they are more difficult to interpret because the original wavelength regions are no longer displayed as a single color.

AUTOMATIC CLASSIFICATION

Classification is a data analysis technique that can be used to define and recognize classes or groups whose members have certain characteristics in common. The classes should be mutually exclusive and exhaustive; that is, there should be one and only one class to which an element is assigned, and all elements in the domain of interest may be so assigned. In practice, however, these requirements are difficult to fulfill and often are not achieved (28).

As discussed in Chapter 6, two approaches of classification are used in the analysis of multispectral data. The classification can be supervised, in which case training areas for each of the classes of interest are established by the analyst, or unsupervised, in which case the classes are objectively determined from an algorithm.

Sahl al Matran Test Site Saudi Arabia
Canonical Transformed LANDSAT Data
(PSU ORSER)

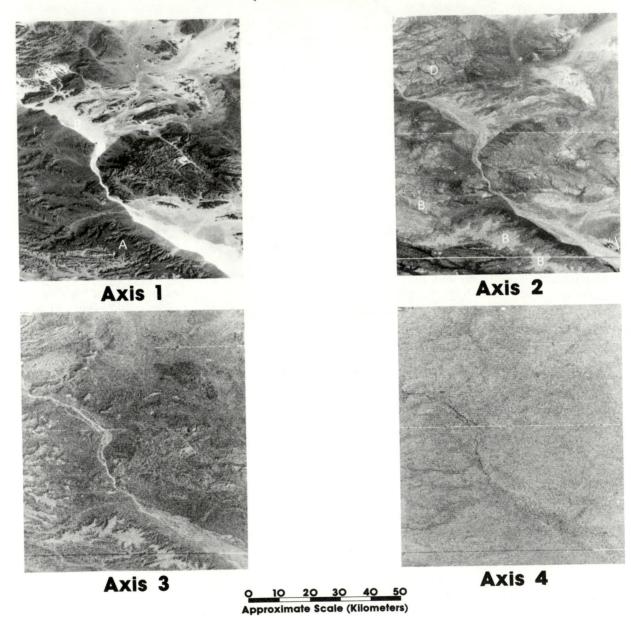

Axis 1

Axis 2

Axis 3

Axis 4

Approximate Scale (Kilometers)
0 10 20 30 40 50

FIGURE 13.23. Canonically transformed Landsat image of Sahl al Matran, Saudi Arabia. See text for annotation information. Courtesy GSFC.

The training areas for supervised classification schemes must be based on *a priori* knowledge, that is "ground truth." Statistics are computed for each class and are used in various algorithms to identify other areas within the image which have similar spectral characteristics. In unsupervised classification, no *a priori* knowledge is assumed and the classes are based on the actual relations among the variables. In such classification schemes, the number of classes can be established by the analyst, or more objectively, through the use of an algorithm in which the data themselves are used to suggest the number of natural categories (26).

Automatic classification techniques have been applied successfully in the fields of land-use mapping, agriculture, and even water quality monitoring. In general, geologic applications have met with less success. Lithologic units are rarely homogeneous, and are masked by various amounts of soil, vegetation, colluvium, and organic debris. These factors make selection of representative training areas extremely difficult, especially in areas of highly dissected terrain. Even with carefully chosen training areas, classification accuracy under these conditions often does not exceed 50% (26). This is due to the influence of soils, vegetation cover, and transported surficial materials, which mask and alter the spectral characteristics of the underlying rocks.

In addition, the spectral features of most rock types and derived soils are not very distinctive in the wavelength region sensed by most existing instruments; that is, the visible and near-infrared. The few differences that do exist are, however, difficult or impossible to use in a generalized remote sensing effort to map the composition of all rocks within a given scene. Instead, this spectral region is best suited to precise and particular applications, such as enhancing the separability of a given material with a known and distinctive spectral feature (9). This provides one rationale for using classification techniques for lithologic mapping. In addition, classification of multispectral data (1) requires less sophistication on the part of the analyst, as long as the classified image is well described by the training areas; (2) is relatively fast to use once the thematic map is made; (3) provides results that are easier to communicate to others.

An example of the use of classification of Landsat data for geologic mapping has been presented by Lefebvre and Abrams (29) for the Craters of the Moon Volcanic Field, Idaho. Based on limited field reconnaissance, training areas were chosen representing five pahoehoe units and two aa units. A Hybrid Classifier (30), which incorporates a parallelpiped classifier, and the Bayesian maximum likelihood algorithm (see Chapter 6, Section 6.7.4), was used to produce a thematic map (Fig. 13.24 color plate 25). The resulting map was later checked in the field and found to be accurate. The spectral response of the units for the Landsat MSS bands were examined and revealed a systematic increase in reflectance that could be correlated with increased vegetative cover and degree of weathering. This increase was interpreted to be indicative of the relative age of the units.

Another successful application of automatic classification for geologic mapping has been presented by Melhorn and Sinnock (31). For the test area in Colorado, training areas were selected representing the three major geological constituents, sandstone, shale, and alluvium. An automatic pattern recognition procedure using the Bayesian maximum likelihood classifier was then used to classify multispectral Landsat data for the test area. The geologic map produced showed fairly good correspondence with existing geologic maps.

Fontanel et al. (32) have taken the classification of Landsat imagery one step further by combining the techniques of multispectral classification with principal components analysis (PCA). PCA is a variance maximizing transformation technique in which orthogonal linear combinations of the observed variables are produced to represent the important information with a smaller number of dimensions (see Chapter 6). The motivation for combining these two techniques is to produce an optimal enhancement of the data by maximizing the display of information in different subsets of the total data. In their approach, Fontanel *et al.* (32) first use an unsupervised classification technique to establish groups within a scene. PCA is then performed on the spectral data for each of the groups and the eigenvalues for each principal component are used to transform the multispectral data. Using PCA alone, only an average enhancement is obtained which is not optimum for any part of the image.

Using this technique, Fontanel et al. (32) examined Landsat data for an area in southern France. Six identifiable groups were delineated by the classification algorithm (Fig. 13.25) color plate

26. After application of PCA, the first two principal components, which account for 96% of the total information, were combined to make a color picture (Fig. 13.26, color plate 27) from which an interpretive map was produced (Fig. 13.27) color plate 28. Unit R corresponds to Jurassic and Cretaceous limestones and sandstones with several marker beds. The pinkish areas within R are scrub covered areas; dark green areas are clouds. The A units are six separable alluvial deposits, ranging in age from Quaternary to Pliocene. The W units are different qualities of water. Coastal waters (WC) can be followed all along the shoreline. The outflow of River Agly can also be detected.

13.4 DIFFICULTIES IN INTERPRETING MULTISPECTRAL DATA

As with any interpretive science, geologic photo interpretation has both intrinsic and extrinsic problems which can adversely affect the quality and amount of available information. In addition to the instrumentational and operational factors that affect the data quality (see Chapter 3), some of the same factors that are useful for lithologic identification, such as vegetation and soil cover, can obscure the surface and lead to misidentification of rock types.

13.4.1 Vegetation Cover

While natural vegetation may be specific to a particular lithology, as described previously, the type and extent of vegetation cover more often reflects only the influence of topography, relative rainfall, insolation, prevailing wind direction and velocity, and the maturation stage of the floral community.

Nonspecific vegetation may be present in all but the most arid regions and its presence may severely hinder or limit computer-automated and photointerpretive studies of multispectral data for soil mapping and lithologic discrimination. In large-scale, high resolution images the effect of vegetation may be greatly reduced because of the relatively small instantaneous field of view of the scanner. With high resolution systems it is possible, in all but completely covered areas, to look between the vegetation clumps and obtain a reasonably accurate measurement of the spectral response of the ground. However, in small-scale, low resolution images such as Landsat, the spec-

tral response at each picture element represents the area-weighted average of the reflectance characteristics of all components of the surface in the measured wavelength region. Consequently, the measured brightness represents a composite spectrum composed of soil, rock, vegetation, and other organic materials.

The effect of varying amounts of natural vegetation on rock and soil discrimination has been described by Siegal and Goetz (33). Using *in-situ* spectral reflectance data (0.45 to 2.4 μm) for several common rock types and natural vegetation, they used a computer to perform simple geometric weighting to produce composite spectra to simulate the spectral effect of varying amounts of vegetation cover on natural materials.

The spectral reflectance contribution of varying amounts of green grass and dry sage cover on andesite, limestone, and limonitic argillized fragments and soil, for example, is illustrated in Figs. 13.28 and 13.29.

As illustrated, the effect of vegetation on the reflectance spectrum is most pronounced for rocks with low albedos. With only 10% green grass cover, the spectral characteristics of andesite and limestone are masked and identification is difficult. The spectrum showing the addition of 30% green vegetation is totally dominated by the effect of the vegetation, although in the case of limestone, the characteristic reflectance fall-off beginning at 2.3 μm is still evident. For limonitic argillized fragments and soil, the effect of green grass cover is not as pronounced. With 30% vegetation cover, the strong and broad absorption band at 0.85 μm due to ferric iron is still recognizable although greatly subdued. The absorption band at 2.2 μm due to the OH^- stretching is still distinct. With 60% cover, the green grass spectrum completely dominates the composite spectrum. In all three cases, the slopes of the spectral curves are more strongly affected at shorter wavelengths because of the steep rise of reflectance in the vegetation beyond 0.68 μm.

Dead or dry vegetation does not have as large an effect on the underlying materials as green vegetation because of the increased proportion of stem material and the pronounced differences in pigments, internal leaf structure, and water concentration. Figure 13.29 illustrates the effect of varying amounts of dry sage on andesite, limestone, and limonitic argillized fragments and soil. With

EFFECT OF GREEN GRASS COVER

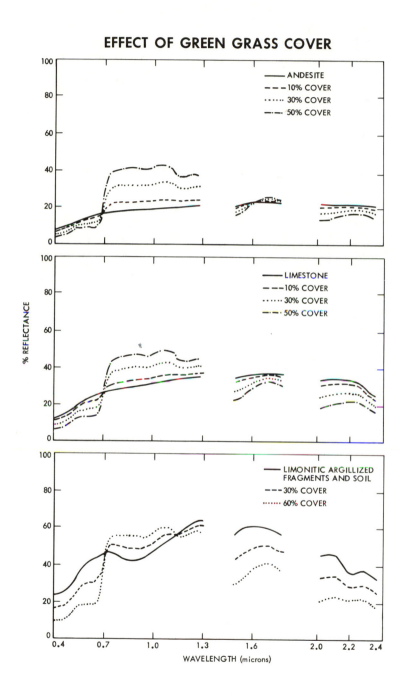

FIGURE 13.28. Vegetation effects of green grass cover on spectral reflectance of various materials. Courtesy JPL and Photogrammetric Engineering and Remote Sensing.

EFFECT OF DRY BUSH COVER

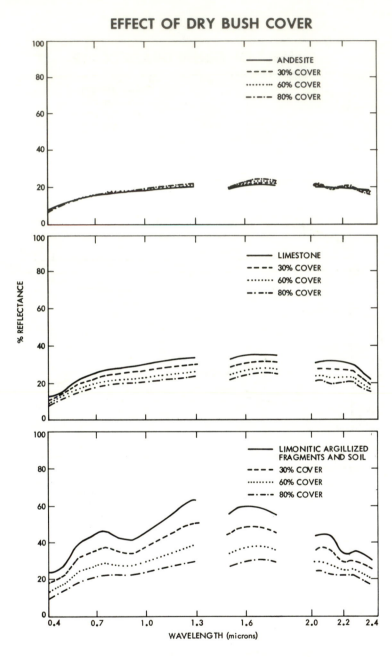

FIGURE 13.29. Vegetation effects of dry sage cover on spectral reflectance of various materials. Courtesy JPL and Photogrammetric Engineering and Remote Sensing:

414

even 60% vegetation, the shape of the composite spectrum for each of the different rock types still resembles that of the lithic materials, although the average albedo has been changed.

The effect of vegetation on the spectral response of earth materials must also be considered when using ratio images of data. The effect of increasing green grass cover, for example, on the spectral response of andesite, limestone, and limonitic argillized fragments and soil for the Landsat MSS band ratios 4/5, 4/6, 4/7, 5/6, 5/7, and 6/7 is illustrated in Fig. 13.30. With increasing grass cover the MSS ratios, 4/6, 4/7, 5/6, and 5/7 for andesite can vary from approximately 0.90 to 0.10. MSS ratio 4/5 is the least afffected, because the wavelength region of 0.5 to 0.7 μm does not include the pronounced green vegetation reflectance zone in the near infrared. Manzanita cover and other types of green vegetation will produce similar results. Dead or dry vegetation cover will have little effect on Landsat MSS band ratios, because the shapes of their spectral curves are similar to those of rocks. The severity of the effect thus depends upon the wavelength region considered, and the spectral reflectance characteristics of the vegetation and the parent material (33).

13.4.2 Surficial Material Cover

Soil cover and other surficial deposits can greatly affect the spectral appearance of the ground and can mask structural features and other indicators useful for lithologic identification. By far the most significant problem in dealing with surficial materials is caused by the phenomenon of material transport. Water, ice, and wind are active transporting agents, capable of moving objects and materials of all sizes for great distances. Therefore, surficial deposits of this nature do not reflect the composition of the underlying material or bedrock. Surficial deposits of these types are sometimes recognizable in images by their distinct morphological appearance, such as stream terrace deposits, glacial deposits, or sand dunes, and are mapped accordingly. Where morphological appearance of the surficial deposits is not distinct, it is often difficult to determine if the soils are *in-situ* or are derived from a foreign source.

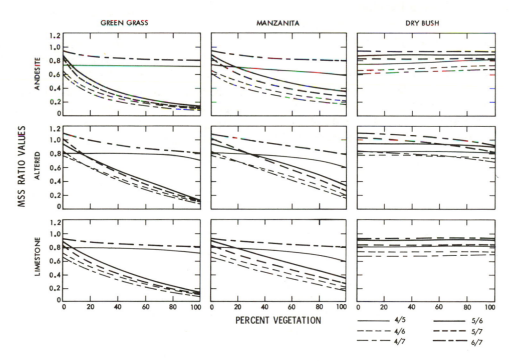

FIGURE 13.30. Vegetation effects on Landsat MSS ratio values. Courtesy JPL and Photogrammetric Engineering and Remote Sensing.

The recognition of *in situ* material does not necessarily allow unambiguous identification of the underlying source rock. The physical and chemical processes that produce soils can modify their composition to such an extent that their spectral signature no longer reflects that of the source. This is particularly prevalent in areas of extreme chemical weathering, where leaching of salts is dominant. In addition, the presence of organic material in the soil can further modify its spectral characteristics. Even in the areas where soil cover is minimal or absent, bedrock exposure does not guarantee that the observed spectral response is indicative of the parent material. Chemical modification of the surface by weathering can significantly alter the material's spectrum. An extreme example of this effect is desert varnish, which produces a coating of manganese and iron oxides on exposed surfaces. This phenomenon affects only reflected wavelengths and does not alter emitted radiation; therefore, measurements taken in thermal infrared wavelength regions are more representative of the material's body properties, and do not reflect surficial properties (see Chapters 8 and 9).

13.4.3 Quality of Image Products

A problem that the interpreter must be aware of is the variable quality of image products. This is primarily encountered when dealing with products supplied by a commercial source; where the analyst has no interaction with the photoprocessor, and must rely on the judgment of an untrained technician to produce a satisfactory product. Although the image may be processed for "optimal" appearance, it may not display the maximum amount of geologic information for a particular task.

13.5 SUMMARY AND CONCLUDING REMARKS

In this chapter an attempt has been made to describe the applications of geologic interpretation techniques for reconnaissance lithologic mapping. Emphasis was placed on the use of images acquired in the visible and near-infrared reflected spectral regions, as these data are the most commonly used and easily available at the time of writing.

Interpretation of remotely sensed images requires an understanding of the spectral properties of earth materials, and consideration of conventional analyses of drainage patterns, vegetation, landforms, and climatic influence.

The availability of images in digital form allows the use of digital processing techniques to enhance the display of information. When properly chosen and interpreted, these techniques can enable the geologist to produce an accurate lithologic reconnaissance map. As an integral part of almost any geologic project, this map is an invaluable tool, and can serve as the base for further work.

REFERENCES

1. Lattman, L. H., and Ray, R. G., 1965, Aerial photographs in field geology: New York, Holt, Rinehart and Winston, 221 p.
2. Avery, T. E., 1977, Interpretation of aerial photographs, 3rd ed.: Minneapolis, Minnesota, Burgess Publishing Company, 392 p.
3. Murray, A. N., 1955, Growing vegetation identifies formations: World Oil, v. 141, no. 1, p. 102-104.
4. Rich, J. L., 1951, Geomorphology as a tool for the interpretation of geology and earth history: New York Academy Sci., Trans., Ser. 2, v. 13, no. 6, p. 188-192.
5. Ray, R. G., 1960, Aerial photographs in geologic interpretation and mapping: Geol. Soc. Amer. Prof. Paper 373, 230 p.
6. Thornbury, W. D., 1969, Principles of geomorphology, 2nd ed.: New York, Wiley, 594 p.
7. Small, R. J., 1972, The study of landforms: Cambridge, England, Cambridge University Press, 486 p.
8. Grootenboer, J., 1973, The influence of seasonal factors on the recognition of surface lithologies from ERTS-1 imagery of the Western Transvaal, *in* Proceedings of the third Earth Resources Technology Satellite-1 symposium: NASA SP-351, v. 1, sec. A, p. 643-655.
9. Salisbury, J. W., and Hunt, G. R., 1974, Remote sensing of rock type in the visible and near infrared, *in* Proceedings ninth international symposium on remote sensing of the environment: Ann Arbor, Michigan, Environmental Research Institute of Michigan, v. III, p. 1953-1958.
10. Longshaw, T. G., 1974, Field spectroscopy of multispectral remote sensing: an analytical approach: Applied Optics, v. 13, p. 1487-1493.

11. Vincent, R. K., and Pillars, W. W., 1974, Skylab S-192 ratio code of soil, mineral, and rock spectra for ratio image selection and interpretation, *in* Proceedings ninth international symposium on remote sensing of the environment: Ann Arbor, Michigan, Environmental Research Institute of Michigan, v. II, p. 887-896.

12. Goetz, A. F. H. *et al.*, 1976, Comparison of Skylab and Landsat images for geologic mapping in northern Arizona: Final Rept. NASA 7-100, Task Order, RD-161, Johnson Space Center, Houston, Texas.

13. Viljoen, R. P., Viljoen, M. J., Grootenboer, J., and Longshaw, T. G., 1975, ERTS-1 imagery: an appraisal of applications in geology and mineral exploration: Minerals Sci. Eng., v. 7, no. 2, p. 132-168.

14. Lee, K., and Weimer, R., 1975, Geologic interpretation of Skylab photographs: Remote Sensing Rept., 1975-76, Colorado School of Mines, Golden, Colorado, 77 p.

15. Moore, H. D., and Gregory, A. F., 1973, A study of the temporal changes recorded by ERTS and their geological significance, *in* Third Earth Resources Technology Satellite-1 symposium: NASA SP-351, v. I, sec. A, p. 845-855.

16. Wobber, F. J., and Martin, K. R., 1973, Exploitation of ERTS-1 imagery utilizing snow enhancement techniques, *in* Proceedings symposium on significant results obtained from Earth Resources Technology Satellite-1: NASA SP-327, v. I, sec. A, p. 345-351.

17. Krinsley, D. B., 1974, The utilization of ERTS-1 generated images in the evaluation of playa sites for economic and engineering development: U. S. Geological Survey Final Rept.—Contract S-70243-AG, MMC 195A.

18. Rowan, L., Wetlaufer, P., Goetz, A. F. H., Billingsly, F., Stewart, F., 1974, Discrimination of rock types and detection of hydrothermally altered areas in South-Central Nevada by the use of computer-enhanced ERTS images: U. S. Geological Survey Prof. Paper 883, U. S. Government Printing Office, Washington, D. C., 35 p.

19. Vincent, R. K., 1975, Commercial applications of geologic remote sensing: in Proceedings of the 1975 IEEE Conference on Decision and Control, Houston, Texas, Dec. 10-12, 1975, p. 258-263.

20. Goetz, A. F. H., Billingsley, F. C., Gillepsie, A. R., Abrams, M. J., Squires, R. L., Shoemaker, E. M., Lucchitta, I., and Elston, D. P., 1975, Application of ERTS images and image processing to regional geologic problems and geologic mapping in Northern Arizona: JPL Tech. Rept. 32-1597, California Inst. of Tech., Pasadena, California, 188 p.

21. Blodget, H. W., Brown, G. F., Mock, J. G., 1975, Geological mapping in Northwestern Saudi Arabia using Landsat multispectral techniques: Proceedings of the American Society of Photogrammetry, 41st Annual Meeting, Washington, D. C., p. 971-990.

22. Seale, H. L., 1964, Multivariate statistical analysis for biologists: London, Methuen & Co., 209 p.

23. Merembeck, B. F., Borden, F. Y., Podwysocki, M. H., and Applegate, D. N., 1977, Application of canonical analysis to multispectral scanner data, *in* Ramani, R. V., ed.: 14th Annual Symposium on Computer Applications in the Mineral Industries, Society of Mining Engineers of American Institute of Mining, Metallurgical, and Petroleum Engineers, New York, p. 867-879.

24. Lachowski, H. M., 1973, Canonical analysis applied to the interpretation of multispectral scanner data, unpub., M. S. Thesis, The Pennsylvania State University, University Park, Pennsylvania.

25. Lachowski, H. M., and Borden, F. Y., 1973, Classification of ERTS-1 MSS data by canonical analysis, *in* Proceedings symposium on significant results obtained from Earth Resources Technology Satellite-1: NASA SP-327, v. I, sec. B, p. 1243-1251.

26. Siegal, B. S., and Abrams, M., 1976, Geologic mapping using Landsat data; Photogrammetric Eng. and Remote Sensing, v. 42, no. 3, p. 325-337.

27. Podwysocki, M., Gunther, F., Blodgett, A., 1977, Discrimination of rock and soil types by digital analysis of Landsat data: Goddard Space Flight Center, Document 923-77-17, Greenbelt, Maryland, 37 p.

28. Griffiths, J. C., 1968, Geological data for classification, *in* Proceedings of a symposium on decision making in mineral exploration Vancouver, Canada: Reprinted in Western Miner, February, April, 1968. Contribution No. 67-37 College of Earth and Mineral Sciences, The Pennsylvania State University, University Park, Pennsylvania.

29. Lefebvre, R., and Abrams, A., 1977, Relative

age-dating of craters of the Moon Volcanic Field, Idaho: Abstract, GSA meeting, Missoula, Montana, May, 1977.

30. Addington, J., 1975, A hybird clasifier using the parallelpiped and Bayesian techniques: Proceedings of the American Society of Photogrammetry, 41st Annual Meeting, Washington, D. C., p. 772-784.

31. Melhorn, W. N., and Sinnock, S., 1973, Recognition of surface lithologic and topographic patterns in Southwest Colorado with ADP techniques, *in* Proceedings symposium on significant results obtained from Earth Resources Technology Satellite-1: NASA SP-237, v. I, sec. A, p. 473-481.

32. Fontanel, A. C., Blanchat, C., and Lallemand, C., 1975, Enhancement of Landsat imagery by combination of multispectral classification and principal component analysis, *in* Proceedings of Earth resources survey symposium, Houston, Texas: NASA TM X-58168, v. I-B, p. 991-1012.

33. Siegal, B. S., and Goetz, A. F. H., 1977, Effect of vegetation on rock and soil type discrimination: Photogrammetric Eng. and Remote Sensing, v. 43, no. 2, p. 191-196.

CHAPTER 14

STRUCTURAL GEOLOGY
DAVID P. GOLD

14.1 INTRODUCTION

This chapter is organized around basic principles and concepts, and draws heavily on the techniques developed in structural analysis. It is aimed at the reader who has a background knowledge in structural geology, geomorphology, and photogeolgy, and attempts to systematize some of the methodology used by the field geologist. Gathering data on a scale consistent with the size of the feature is the main link in the application of remote sensing systems for mapping structures. By emphasizing the *principles* and *concepts* that are "glossed over" in most structural geology texts, it is hoped the reader will be able to extrapolate the remote sensing responses into a meaningful three-dimensional model suitable for kinematic and dynamic analysis on the appropriate scale. Thus the scheme of presentation differs from the conventional "description of structures" approach, but this aspect is integrated into the text as examples to illustrate or clarify a principle or concept. As such the chapter is neither comprehensive nor exhaustive, because only those situations amenable for analysis by remote sensing techniques are discussed.

Gathering data on a scale consistent with the size of the feature being analyzed is the main key to the application of remote sensing to structural studies (see Table 14.1), (1, 2, 3). The size relationships of structures, structural domains and the limits of structural homogeneity, analogy with known terranes and models of deformation should provide the correct framework for scaled extrapolations of structures. Large structures require the structural geologist to change the framework of his observations from direct measurement of primary and secondary discontinuities to an interpretation of bedrock and landforms. Although the prime tool is the geological map, topographic maps and remotely sensed images are now used extensively in gathering the field data. In the same way that aerial photography revolutionized field geology during the late 1940s and early 1950s by portraying position, configuration, and to a limited extent, attitudinal data on a macroscopic scale (see Table 14.1), so have the remote sensing systems launched in the late 1960s and early 1970s expanded the scope of direct structural observations on a megascopic and gigascopic scale.

14.2 THEORY AND PRINCIPLES

14.2.1 Morphotectonic Principles

Morphotectonics is the study of landforms of regional or tectonic significance, and as such it is the basis for the interpretation of geologic structure from remotely sensed images. The interpretation of bedrock geology from the topography also is used in the allied disciplines of *photogeology* and *structural geomorphology* (4), and is based on the geomorphic principles involving structure, process, and stage that were formulated by Davis in his classical paper *Erklarende Beschreibung der Landformen*. Hills (5) has used the term morphotectonics for "something broader and applicable to regional study on a large scale, using topography as the primary criterion . . . . It involves a study of the external form and outlines of major topographic units—mountain massifs, plateaux, and river basins—as well as their internal structures . . . , and permits interferences relative to areas as large as continents." Many of these inferences involve tectonic processes and are dynamic in nature. They require an understanding of the mechanism(s) and scales of deformation(s) that produced the structural features and their expression in landforms—conditions that are rarely fulfilled.

In addition to an accurate geological map, a high quality topographic base map is needed for any quantitative studies of the relationship between landscape development by erosion of bedrock lithol-

TABLE 14.1   OPTIMUM SCALES AND CORRESPONDING SENSING PLATFORMS FOR MAPPING DEFORMATION AND ª GEOTECTONIC PHENOMMENA

Structural Scale and Order	Natural Scale and Size of Feature	Types of Phenomena	Types of Maps
Gigascopic 1	1:50 million > 1000 km	Continental or oceanic plates, seismic plates	Globes; world-wide and continental geologic, tectonic and seismic maps
Megascopic 2	1:1 million 10 to 1000 km	Mountain belts, basins, island arcs, rift valleys structural provinces, plutons, megalineaments, craters	Large globes; continental, national, state geologic, tectonic and seismic maps
Macroscopic 3	1:1000 10 m to 10 km	Folds, faults, lineaments, craters volcanoes, dikes, fracture traces, etc.	Regional, geological and structural maps; fabric diagrams
Mesoscopic 4	1:1 1 cm to 10 m	Folds, faults, cleavage, joints, bedding, geologic contacts	Detailed geologic and structural maps; fabric diagrams
Microscopic 5	1000:1 10 μm to 1 cm	Micro-fractures, deformation lamellae, grain size	Microfabric orientation diagrams
Submicroscopic 6	100 million:1 1 Å - to 10 μm	Lattice defects	Crystal structure charts

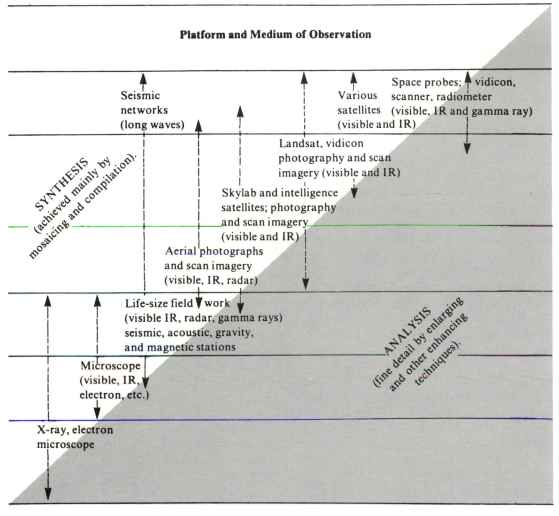

Platform and Medium of Observation

SYNTHESIS (achieved mainly by mosaicing and compilation).

Seismic networks (long waves)

Various satellites (visible and IR)

Space probes; vidicon, scanner, radiometer (visible, IR and gamma ray)

Landsat, vidicon photography and scan imagery (visible and IR)

Skylab and intelligence satellites; photography and scan imagery (visible and IR)

Aerial photographs and scan imagery (visible, IR, radar)

Life-size field work (visible IR, radar, gamma rays) seismic, acoustic, gravity, and magnetic stations

ANALYSIS (fine detail by enlarging and other enhancing techniques).

Microscope (visible, IR, electron, etc.)

X-ray, electron microscope

[a] Larger-scale maps are produced from the higher orders of observation and smaller-scale maps from the lower orders. The arrows on the right indicate the range in scale: (a) up into the unshaded region by synthesis (integration and mosaicing) of the primary observations, and (b) downward into the shaded region by analysis (enlarging and enhancing to the limit of resolution). Compiled from various sources (Ref. 1, 2, 3).

ologies in different structural configurations and states. Although accuracy and quality generally are reduced by compilation to regional scale maps, orthoimages sensed from satellite platforms are now available that provide a direct view of the landforms at appropriate scales for most land areas of the world. The utility of a suitable regional base map is well illustrated in the Appalachian fold belt of central Pennsylvania and West Virginia (Fig. 14.1) where Davis showed the development of trellis drainage in folded and dissected stratified rocks with different erosional properties, and a direct correlation of long straight stream segments to strike valleys. On a regional scale the folds are manifest as sinuous ridges, underlain by Tuscarora quartzite (Silurian), and as subsidiary ridges underlain by the less resistant Oswego sandstone (Ordovician). Superposition of strata in the Appalachian fold belt reveals an inversion of structural relief and topography (Fig. 14.2), so that synclinal ridges prevail among anticlinal valleys. These are the normal geomorphic expressions of lithology and structure in orogenic belts, and it is only these structural elements that commonly are represented on tectonic maps. Not all of the cross-strike topographic features apparent in Fig. 14.1 can be explained by the conventional "lithology and primary structure" control. This has led some geologists (5, 6, 7, 8) to conclude these features are associated with transgressive, secondary structures. Most of these transgressive features are relatively linear in aspect, and are referred to as lineaments (6) or fracture traces (8) depending on their length (greater or less than 1.6 km, respectively). There is no apparent displacement of strata across the lineaments, and some have been demonstrated to represent zones of increased fracture density between rigid basement blocks, (7) hinge zones between basins (9), the surface expression of dikes, strain boundaries associated with stepped tear faults (10), zones of stress concentration associated with regmatic shear zones, or the propogation of basement fractures to the surface. Those lineaments with associated displacements undoubtedly represent the trace of faults or fault zones. Although these linear features may be discontinuously expressed as alignment of wind and water gaps or straight stream segments, and tend to be masked by the dominant lithostructural elements in folded terranes, they are the main morphotectonic element of the shield and platform areas (see Figs. 14.1, 14.30).

Some of the subtle features of the cratons were recognized long ago on small scale topographic maps (6) and more recently on relief models (5, 11). While building a relief model of Australia on a scale of approximately 1:500,000 Hills (12) suggested on morphotectonic grounds the "existence of crustal blocks, of longer and narrower belts expressed as ridges and troughs, and of linear elements expressed chiefly as block edges and in the pattern of streams and interfluves." Observations such as these have been greatly facilitated in the past decade by the widespread use of satellite imagery. Morphotectonic studies have made us aware of the contrast in the structural data available between the orogenic belts and the shield and platform areas. The uneven coverage is a sampling problem as well as a general lack of awareness for the different styles of deformation resulting from the same event in different domains. In addition, structural expressions in the cratons tend to be subtle because they generally are (a) regions of low relief, and (b) metamorphic terranes of low geologic diversity, and hence with little or no differential erosion of the lithologies in the crystalline rocks.

Direct observations from remotely sensed images on a megasopic scale have led to revisions of some of our morphotectonic concepts. We are aware now, for example that;

1. The cratons are less stable than originally supposed.
2. Lineaments and semirigid crustal blocks are common in and to all cratons.
3. Many continental basins are bounded by lineaments.
4. Lineaments are not limited by tectonic province boundaries nor geologic age and appear to be rejuvenated fractures along ancient zones of weakness.
5. Long linear fractures and semirigid crustal blocks probably are manifestations of intraplate deformations, and they should be considered along with fold mountain belts, plateaux, basins, crustal arches, major faults and rift systems, suture zones, and tectonic province boundaries in any tectonic analysis.

14.2.2 Scales of Observation and Structural Phenomena

We are concerned here with phenomena related to the size of objects and the scale of observa-

FIGURE 14.1. Landsat-1 image of central Pennsylvania, with superimposed trends of the Tyrone-Mount Union lineament, *A*, and the Everett-Bedford lineament, *B*. The black arrowheads mark the position of the McAlevys Fort-Port Matilda lineament. State College is located under the star symbol. Note the physiographic provinces traversed by these lineaments. The sinuous pattern in the Valley and Ridge Physiographic Province is formed by the resistant beds in denuded folds in the Appalachian fold belt. The course of the Juniata River (center) follows strike valleys and breaks through the ridges along lineament traces to join the Susquehanna River (top right) about 15 km northwest of Harrisburg (*H*). Strip mines for coal in the Allegheny Plateau Province are visible (dark tones) as the vermicular patterned terrain in the top left corner. The small "cotton-ball" features are clouds. The axis of the Appalachian orocline extends from the Maryland border, south of Harrisburg, northwestward through State College. Northeast of this axis, folds plunge to the northeast; south of the axis, the folds plunge to the south-southwest. Landsat image E-1045-15243-7, September 6, 1972.

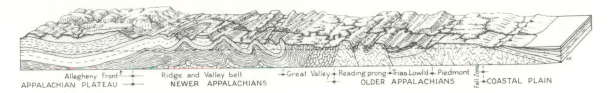

FIGURE 14.2. Morphotectonic section across the Appalachian orogenic belt. This diagram is not to scale; it should be viewed in conjunction with Figure 14.27, a Landsat mosaic of Pennsylvania, with the section extending from southeast to northwest. [After Strahler, 1963(13)]

tion, and possible mechanical relationships between structures of different size. Although large geological features have been characterized from regional maps, there have been all too few attempts to study structural features on different scales (1, 2, 3, 14). (see Table 14.1), or the duration of deformational events as a function fo size (2) (Fig. 14.3). Until recently the lack of suitable observational platforms for mapping large features accounted, in part, for the paucity of quantitative studies involving scale. An adequate theory for predicting deformations on a mega- or gigascopic scale is presently lacking.

The means for extrapolating field observations to deduce the configuration of larger structures are (a) compilation from large scale maps, (b) direct mapping on a mosaic base prepared from large scale topographic maps or aerial photographs, (c) use of Pumpelli's rule and the enveloping surface principle, and (d) use of scale models and mechanical similitude. Each scale contains its own information; some information may be unique to a given scale, and there is the danger that all, or some of this information may not be preserved or transferred faithfully in mosaics or compilations. For example, contour lines, small streams, and

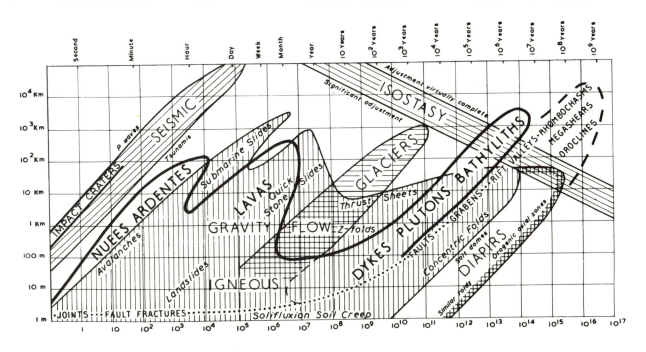

FIGURE 14.3. Geological events and geotectonic phenomena as a function of time (duration) and size. Note the positive slope of all the phenomena listed, except for isostatic rebound. [After Carey, 1962(2); reproduced with the author's permission.]

minor features commonly are omitted during a compilation of a regional map of sufficiently small scale for mesoscopic viewing. Only on a relief model made with careful attention to detail can one fully comprehend the realities of the topography of large areas (5). On such relief maps subtle linear patterns may be enhanced by side-lighting at a low angle (11). Fortunately, the resolution of a photograph is much greater than can be achieved on a topographic map, and subtleties of the relief can be studied particularly in stereo pairs.

One important aspect of remote sensing at different altitudes is that it provides directly different views of the same feature. A study of small objects involves *analyzing* the subject in progressively finer detail at increasing scales, for example, at greater than natural size (1:1) by using a microscope, electron microscope, or x-ray techniques. Conversely, a *synthesis* or compilation of data observed in one scale is the vehicle for studying large objects. Table 14.1 illustrates the link between scale and structural phenomena, incorporating remote sensing platforms appropriate to each scale. Clearly, these same scale considerations apply to many other classes of problems that involve spatial relationships.

Other factors related to scale are the sample interval and resolution. Resolution is, in part, a function of the wavelength of the sensing medium: A geological body or structure could remain undetected if the wavelength of the transmission medium or the sampling interval is much larger than the size of the target. While the optimum wavelength for characterizing the shape of an object varies inversely with increasing scale the resolving power of a remote sensing system generally varies directly with scale.

By using folding optics in cameras, long focal lengths can be achieved in small systems, and when long local lengths are combined with a high resolving power lens, images of remarkably high resolution can be procured from a satellite. Active microwave systems hold great promise for structural studies because (a) good resolution can be achieved by increasing the effective focal length through synthetic-aperture systems, (b) it has a superior spatial resolution compared to the shorter wavelengths, and (c) the low angle, oblique imaging mode yields more information on the subvertical joints and fractures and eroded contacts than the near-vertical optical systems.

14.2.3 Structural Analysis Principles

Structural geologists observe and map the present condition of the rocks exposed. Because tectonic strain rates are of the order of 10^{-14} per second, the present condition of most rock bodies can be considered as the final stage in a deformational event. The stages are:

$$\text{Initial condition} \longrightarrow \text{Process (Deformation mechanisms)} \longrightarrow \text{Final condition (Observed structure)}$$

The deformation or strain is the change in attitude (rotation), position (translation), shape (distortion), or volume (dilation) that the rock has sustained from its initial condition. Without an initial "fabric" in a known condition to serve as a reference, strain cannot be determined. Conditions of deposition, consolidation, and crystallization that produce the primary structural or fabric elements must be known if the kinematic and dynamic stages of analysis are to be attempted. By observing or mapping the structural geometry, and the sequence and number of imposed "fabric" elements and/or structures, the geologist attempts to deduce the process(es) (motions and stresses) from an estimate of the nature of displacements (movement picture) and the amount of deformation (strain).

Some of the vehicles of structural analysis and the morphotectonic analysis of large structures are contained in the following principles and rules:

1. Structural homogeneity is inversely related to scale.
2. The geometric order of the structure reflects the geometric order of the movement picture, that is, the rock structure is related to the deformation that produced it.
3. The symmetry of the structure is related to the symmetry of the deformative movements, that is, there is a relationship between the symmetry of cause and effect and by measuring the former we can deduce the latter.
4. The influence of structure on geomorphology is profound, but the expression differs with the condition of internal and external variables, that is, the main cause of differential erosion in rocks are variations in (a) composition, (b) state (grain size, texture, attitude, porosity, and so forth), and (c) climatic conditions.
5. The state of a rock body can be altered by super-

imposed structures, (for example, increased density of fractures, differential cementation, strain) and these "conditions" can have an over-riding effect locally on the rate of erosion.

6. The most accurate information on the form and shape of large structures is gathered on a scale appropriate to their size.

A corollary to the last principle is that the use of remote sensing techniques for recording morphological expressions is the most viable means of analyzing progressively larger structures. In addition, remote sensing can be used to map lithology in terranes where conditions are favorable, that is, abundant outcrop and spectrally distinct rock types.

It should be realized that quantitative dynamic analysis requires control over the boundary conditions, the stress-strain relationships, the strain-displacement relationships, and other geologic parameters (see Fig. 14.4) that seldom are known with any degree of accuracy. Geometric extrapolations from two-dimensional remote sensing data introduce additional uncertainties so that the dynamic analysis of large structural and tectonic features are both controversial and speculative. The current "state of the art" is interpretation by either structural simulitude to well-known terranes of similar configuration, or by analogy to conceptual models based on plate tectonics theories.

14.3 STRUCTURAL CONCEPTS

14.3.1 Introduction

The concept of a rock structure and the definition of structural geology varies with the nature and the scale of the investigation. In this chapter structural geology will be considered in the broadest sense to include not only the geometry of the primary and secondary discontinuities, and their outcrop pattern on geological maps, but also the possible processes and mechanism(s) that were responsible for their development. The emphasis will be on the influence of the intrinsic inhomogeneities and imposed discontinuities that define the structural elements on the sculpturing of the surface by weathering and erosion.

14.3.2 Discontinuity Concept

Unlike body force sensors and seismic recording arrays, which can portray features in space, the conventional remote sensing systems that utilize

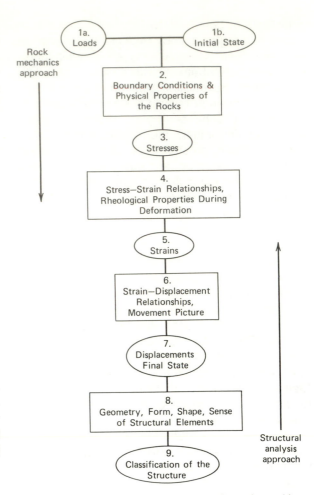

FIGURE 14.4. General solution of a rock deformation problem, showing the stress-strain-displacement relationships of tectonic forces and the resultant structures. [Modified after Hardy, 1971(15).]

electromagnetic radiation portray rock structures as strictly two-dimensional configurations of some unspecified discontinuity. The expression of the discontinuity varies with the contrasting physical and chemical properties of the host rocks and the signal recorded by the sensing medium. The interpreter may use the outcrop pattern and topography such as the break in slope, or unique radiation values (that is, spectral signature) to map the discontinuity.

In mapping structures, geologists are concerned mainly with discontinuities that are readily perceived by the naked eye. Primary discontinuities in sedimentary rocks are expressed in the Griffithsian formula, $D = f(c, o, sh, s, p)$ (16) as a change in composition, orientation, shape, size, or packing.

Perturbations within or initiation of a new energy cycle in the depositional environment create the discontinuities (formation contacts, bedding surfaces, and unconformities) so useful in geological mapping. It is the differential erosion of a layer or bed that has the most profound effect on the landscape. Thus the initiation of a new cycle generally can be mapped at or near the base of resistant ridges. The effect of dip of a resistant bed on the topographic expression is illustrated in Fig. 14.5. Notice that the streams developed along the axis of the valley commonly are strike streams—these morphological features provide a basis for mapping primary structural elements in sedimentary terranes.

The primary discontinuities of most interest to the structural geomorphologist are the unconformity, bedding surface, or formational boundary that separates rocks of differential degradational rates, fracture orientation and frequency, or over which are developed drainage systems of contrasting types of density. Using optical and microwave range remote sensing systems, discontinuities between bedrock and/or their soils may be detected by differences in color or tone, or in the type and abundance of the vegetation.

The secondary structural discontinuities of interest involve changes in the state of the host rocks, imposed during deformation (Figs. 14.5 and 14.6). The difference in chemical and physical properties may be induced by comminution, differential strains in fault zones, or by the creation of voids (joints) or interconnecting fractures to yield zones of high porosity and permeability. Commonly these discontinuities weather more readily than the country rocks and are characterized by an align-

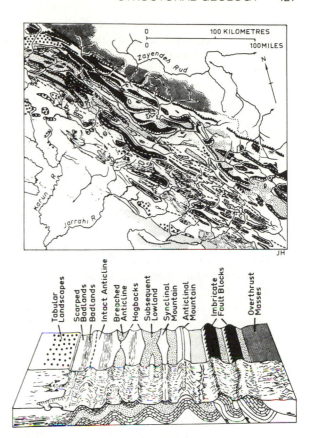

FIGURE 14.6. The Tertiary and Mesozoic rocks exposed in the Zagros Mountains of western Iran offer a good example of the relationship of specific landforms to structure. An arid climate, relatively recent deformation (Pliocene), a simple fold style, and the presence of several massive and resistant formations have preserved table lands, hogsbacks, cuestas, intact and breached anticlines, anticlinal valleys and synclinal mountains, elongate domes and basins, and fault block mountains. To the southeast, these rocks have been pierced by extrusive salt domes. Diagrams after Twidale (1971) (17) as modified from Oberlander (1965) (18). (Published with permission of the Australian National University Press.)

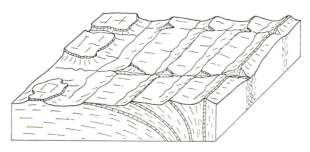

FIGURE 14.5. Varied topographic expressions of primary and secondary discontinuities is a region of stratified rocks. Note the effect of the change in dip of the resistant beds on the landforms. The cross-strike lineaments are shown conceptually to be controlled by zones of greater fracture density.

ment of negative topographic features, such as straight valley segments, swales, and wind and water gaps—some cemented fracture zones may be expressed as resistant ridges.

The boundary represented on a geological map should have physical meaning in time and space. For primary discontinuities to be meaningful geological marker horizons they should (a) represent an identifiable setting in an energy cycle, (b) be a distinctive interface of wide areal extent, and (c) be amenable to mapping by normal field techniques. The importance of the discontinuities from a remote sensing point of view lies not only in their use as geological boundaries, but also in their genetic connotation for determining the nature and amount of deformation. The principles of "Superposition" and "Original Horizontality" expounded by the Danish physician, Steno (1638-1687) serve to specify the initial conditions of primary discontinuities. The secondary discontinuities are imposed structural features such as faults, fractures, joints, cleavage, and other foliations, dikes, and so forth, that have played an active role during deformation. If the nature of imposed structures are characteristic of a rock type or formation, then any areal change in density, attitude, or length, and so forth, may reflect a primary discontinuity. For example, the position of the Tertiary unconformity in relatively flat-lying rocks in West Texas is apparent in the fourth order trend surface map (see Fig. 14.7) of fracture trace frequency and orientation pattern (19). In summary, remote sensing systems can be used to (a) portray the geometry of major discontinuities, (b) help distinguish domains of "fracture" patterns, and (c) to map discontinuities whose subtle surface expression would be overlooked by other techniques.

Other discontinuities may be defined by artificial and variable socioeconomic factors such as the extraction and recovery costs of some economic commodity. The magnetic reversal is potentially the most useful geological boundary for it is global in extent, short in duration, and the magnetic stratigraphy is well-known over parts of the geological time scale. Unfortunately the distribution of reversals through time is not uniform, nor do reversals necessarily coincide with natural breaks in the depositional energy cycle.

14.3.3 Fabric and Structural Elements

We have seen from the previous section that without chemical, physical, or spatial inhomogeneities or discontinuities, no deformation would be apparent. Fabric and structural elements are specific forms of planar and linear discontinuities. Structural elements have lateral (plane or surface) or linear (lineation) continuity and a fixed position that can be mapped respectively along strike or down plunge. Some planar structural elements are sufficiently widespread to serve as a marker horizon or a geological boundary. Fabric elements are "inhomogeneities" or dislocations that are discontinuous laterally or linearly, yet the aligned platy or linear components can be used to define planes

(or surfaces) or lines in a continuum. The general orientation of the family of planar (for example, foliation) or linear (for example, alignment of elongate mineral grains) fabric elements is used to define their attitude. Attitudes are constant within any structurally homogeneous volume, but there is no unique position and hence no mappable marker horizon.

In addition we need to distinguish between the structural/fabric elements that are confined to a specific zone (nonpenetrative), and those that are distributed throughout (penetrative) the rock body. The *nonpenetrative* features, which include faults, igneous contacts, and unique beds or layers, contrast with the pervasiveness of some folds, foliations, or alternating beds in the rock body. The degree to which a particular structural or fabric element is *penetrative* and preferentially oriented (structurally homogeneous) or random (strictly homogeneous) within a rock body, determines whether a banded or uniform surface texture will develop as a result of weathering and erosion.

The relative size of a structural or fabric element depends on the scale of the observation. Generally they are thought of as mesoscopic scale features because most structural/fabric measurements are made in the field with a compass and clinometer. There is nothing unique or sacred about this scale—microstructural elements are measured with a universal stage on a microscope; macro- and megastructural elements can be measured from photographs and orthoimages acquired from aircraft and satellite platforms. The former is the basis for petrofabric analysis, the latter, for modern fracture trace and lineament analysis. The physical nature of discontinuities, their influence on morphology development, and their response to various remote sensing systems will be discussed in the next section.

14.3.4 Structure and Texture

A *structure* is the alignment of structural/fabric elements into a definite form or configuration. It is a geological entity, independent of size, defined by some unique shape (for example, plane, curved surface and planar wave, cyclinder, funnel, trough, dome, bowl/basin) or characteristic such as a zone of fractures or joints, or a zone of intense seismic activity (subduction zone), or transverse troughs across the midocean ridges (transform faults).

Primary structures result from depositional, consolidational, or crystallization processes, and in their

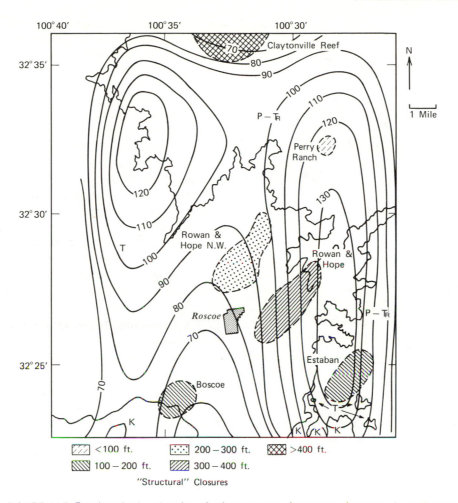

FIGURE 14.7. Fourth-order trend surface for fracture trace frequency of an area in northwest Texas. This surface answers 89 percent of the variation between cells, and shows higher frequency values over the Permo-Triassic rocks than over the Cretaceous rocks (*K*). The Cretaceous unconformity is covered by a thin veneer (not more than 6-8 feet thick) of Tertiary sands, gravels, and caliche (*T*). [Published with permission of Melvin Podwysocki (19).]

original state, tend to be planar and nearly horizontal. They are the boundaries commonly represented on geological maps, and are the contacts most likely to show in "remotely sensed" images on a macroscopic or mesoscopic scale. Under favorable conditions some of these are apparent on a megascopic scale, for example, formations of the Transvaal Supergroup and the layered members of the Bushveld Igneous Complex, in Fig. 13.17.

Second structures are deformed primary structures, and may also include imposed features generated on or from the active components. The latter include the mechanically induced fractures, in-

trusive contacts, and the transposed and recrystallized or redistributed primary structures in metamorphic rocks.

A whole spectrum of sizes of structures may be generated from a single deformational event as the stress is redistributed and attenuated through inhomogeneities to act on progressively smaller units. The size relationship of structures is important in deciding what is a structural element on one scale and a structure on the next order of scale (see Fig. 14.8). The hierarchy is demonstrated in folds by Pumpelli's rule, which states that the "degree and direction of a fold is indicated by those minor plications on the sides" if they are formed

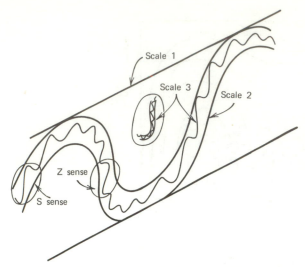

FIGURE 14.8. Diagram to illustrate enveloping surfaces on different scales. The sense of the "minor or drag" folds can be used to predict the sense of the next-order flexures. Note that the sense of the "minor" folds across each axial plane is the same on each scale, and that they are enantiomorphous forms.

by the same event. The concept of smoothing out the "plications" with an imaginary surface tangent to their crests is used extensively for conventional field work (mesoscopic scale), and can be applied equally as well to remote sensing situations (macro- and megascopic scales) for predicting the sense and closure of major folds from the associated "drag folds," and by eliminating some of the background noise to show the "big picture." Note the three scales (orders) of enveloping surfaces in Fig. 14.8, and the right- and left-handed senses in the asymmetry of the "minor" folds. The attitude of the enveloping surface measured around a fold form can be used for generation of π or β diagrams (3, 20, 21)* for the corresponding scale.

From a remote sensing point of view, *texture* refers to the morphological pattern developed on

---

*The π and β diagrams are spherical projections (usually a Schmidt or a Wulff net), generated by plotting the attitude respectively of the pole (π) and the tangent plane (β) measured on both limbs of a fold. For cylindrical folds the poles will lie on a great circle (the π girdle), whose pole represents the attitude of the axial line of the fold. Likewise, all the planes on the β diagram should intersect at a point coincident with the attitude of the axial line of the fold. If the sample points are adequately distributed and sufficient, then the nature of the distribution of the poles in the diagram will yield additional information on the geometry of the hinge zone and the interlimb angle. For additional information refer to Ref. 3, 20 and 21.

the surface. Textures generally reflect the spatial relationships (orientation and density) of those structural elements with morphological expression. Some patterns are characteristic of specific rock types and their variation with attitude and climate can be used in terrain analysis (see Chapter 13), for example, where secondary structures superimposed on the natural rock fabric dominate to exhibit the fracture-textured pattern so characteristic of shield terranes (see Fig. 15.4). Surficial deposits also may have a characteristic morphology and texture (for example, washboard textures from recessional moraines, transverse to ice-flow features such as fluting scours, roche mutonne, and eskers), particularly if the degradation agencies act independently of bedrock inhomogeneities (note the northeast glacial fluting transverse to the south-southeast structural trend in Fig. 14.9).

14.3.5 Structural Style, Domains, and Domain Boundaries

Structural style refers to the shape and form of penetrative structures such as folds. If the physical properties and strain rates in the rocks are similar, then terranes exhibiting similar structural style imply a similarity in the deformation process and stress regimes. Structural styles on the megascopic scale could be characterized using remote sensing techniques and the dynamic considerations worked out in a well-known tectonic setting might be extrapolated to lesser known regions.

A *structural domain* is a rock body that is structurally homogeneous with respect to a penetrative structural/fabric element(s), for example, fold axes associated with cylindrical folds, or folds of similar structural style. The nature of large, noncylindrical folds, especially those involved in polyphase deformations, can be deduced by integrating data from adjacent structural domains. This technique should be useful in extrapolating mesoscopic (ground-based data) into lower order scale structures interpreted from remote sensing systems. The *domain boundary* apparent in remote sensing data is likely to be a significant component in a plate tectonics model. Strain discontinuities such as the arch or hinge zone between basins or a discontinuity within a structurally homogeneous terrane separating regions of different structural style are subtle domain boundaries. They may coincide with growth faults, ramped subsurface décollement slices and tear faults, basement faults, and frac-

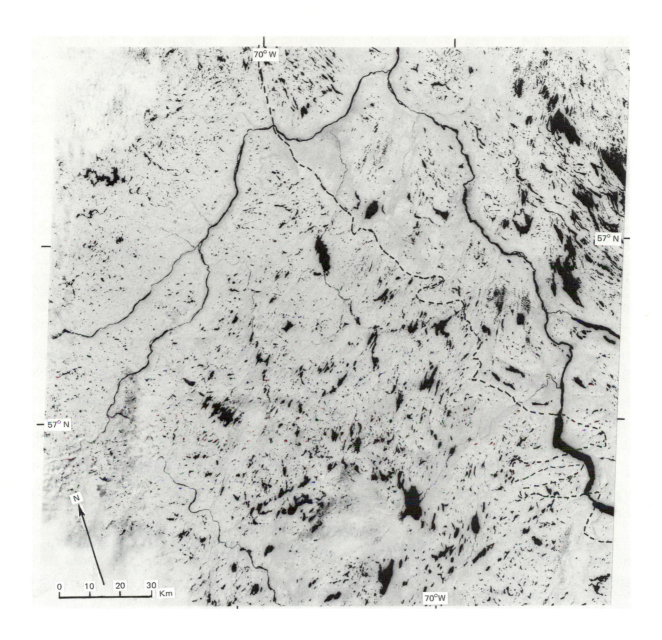

FIGURE 14.9. Landsat 1 scene of northern Quebec showing the junction of the Larch River (to southwest) and the Caniapiscau River (from south) to form the Koksoak River, about 100 km south of Ungava Bay (22). The Larch River, which drains a granite gneiss terrane of the Superior Province (Kenoran orogeny, 2480 m.y.) to the southwest, is controlled mainly by recent ice movement features and transgresses rocks of the "Labrador Trough" about 30 km from its junction with the Caniapiscau River. In contrast, the course of the Caniapiscau River exhibits a strong structural control by bedrock lithology of the folded and faulted Proterozoic rocks (Hudsonian orogeny 1735 m.y.) of the "Labrador Trough." Glacial lakes (dark) are formed either by ice scouring (NE trends), by damming behind moraines (arcuate trends to SE), or by differential erosion along fractures (fracture traces and lineaments—E-W, N-S, and NW trends), and or by bedding (general WNW sinuous trends of folds in the northeast segment of the image). The prominent tonal lineaments that converge to the northeast are ice-flow features such as fluted and ribbed moraines, crag and tail hills, eskers and ice scoured valleys that are independent of bedrock lithology and regional structure.* The dashed line marks approximately the western boundary of the "Labrador Trough." Landsat image E-1027-15193-7, August 19, 1972. [*A more detailed annotation of this scene is given in Short and others, 1976 (22).]

tures that have exerted a curtaining effect in a narrow zone through the overlying sedimentary rocks. Others may represent the locus of facies changes that have persisted through long intervals of geological time. Preferential erosion along these boundaries yield the linear troughs and valley segments apparent on some topographic maps and the lineaments evident on remotely sensed images. A conceptual view of the nature of some of these boundaries in the central Appalachian region (Fig. 14.10) was developed by Kowalik and Gold (10) from subsurface drill-hole data, and seismic, gravity, and surface geological mapping surveys.

Dislocations and other inhomogeneities in the body may redistribute stress and cause local concentrations that could influence the nature and position of any deformation. Even without the inhomogeneities inherent in any rock body, the limbs and hinge regions in folds represent vastly different strain levels. This inhomogeneous response is fortunate for structural geologists because some of the initial conditions of the rock are preserved and provide the means of determining the nature and intensity of deformation. Only for deformation associated with high grade metamorphism is it likely that the initial structural/fabric elements will be obliterated by "overprinting" or recrystallization. This concept of "channeled" strain boundaries is important to the kinematic and dynamic analysis of large fractures.

14.3.6 Depth Zone Concept for Superimposed Structures

Because many parts of the cratons that are amenable to analysis are deeply eroded, we should be aware of some of the structural characteristics and style associated with macro- and mega-scopic structural features in different depth zones (see Fig. 14.11 for example of epi-, meso-, and catazonal plutons from part of the Front Range in Colorado). The large size of most plutons, the contrast in composition, grain size and fabric, erosional resistance, and morphological texture between most intrusions and their host rocks, render these sorts of terranes readily interpretable in remote sensing studies.

In a study on the emplacement of granitic plutons, Buddington (23) characterized the epi-, meso-, and cata-depth zones by the nature of the intrusive contacts, and the width and mineral assemblages in the contact metamorphic aureole. A depth of 10-20 km and temperatures of 450°-700°C (23) ensures the country rocks are in amphibolite and higher grades of regional metamorphism. Adjacent to the plutons, migmatites are common, and contacts are gradational and generally conformable with the gneissic foliations developed in the country rocks (Fig. 14.11). The diapiric emplacement of mantled gneiss domes of tonalitic and granitic composition in a granite-greenstone terrane is illustrated in Fig. 13.3 of the Pilbara region of Western Australia (24). The light colored, elliptical bodies in Fig. 14..13 also represent granitic batholiths surrounded by darker colored greenstone belts in the Rhodesian shield.

Mesozonal intrusions over a depth range of 6-16 km and temperatures of 250°-500°C (23), which imparts a regional metamorphic regime of greenschist to epidote-amphibolite facies. These are characterized by locally concordant and discordant contacts in complex contact zones, with well-developed contact metamorphic aureoles. Foliations are developed in the plutons, mainly near the contacts. The shape commonly is irregular, and pendants of country rock may be present (see Fig. 14.11 and 14.23b).

The epizonal intrusions represent a depth of 0-10 km in an ambient temperature range of 0°-300°C (23). Their discordant habit (except for concordant intrusions), distinctive shapes, sharp contacts with chilled margins and thin metamorphic aureoles, as well as contrast in composition, structure, and landforms ensures optimum discrimination. The ring dikes and plugs of the Pilanesburg alkaline complex (see Fig. 13.17) and of the Air Plateau, in Niger (Fig. 14.12) are readily apparent by their characteristic tone and shape. The Great Dyke of Rhodesia averages about 6 km wide and extends for 480 km (see Fig. 14.13). The ultramatic and gabbroic rocks flare outward at the surface in a series of interconnecting basins and troughs. The faults, apparent in some of these scenes by their sharp separations, typify the brittle behavior of the rocks in the epizone.

The eruptive rock bodies are equally easy to discriminate by their characteristic size, shapes and habits, and their composition (color) and structural contrast with the country rocks. In addition there may be vegetational enhancement in eroded or exhumed volcanic bodies (for example, the greenbelts portrayed in Figs. 13.3; 14.13, and in the Triassic Basin of eastern Pennsylvania and New Jersey),

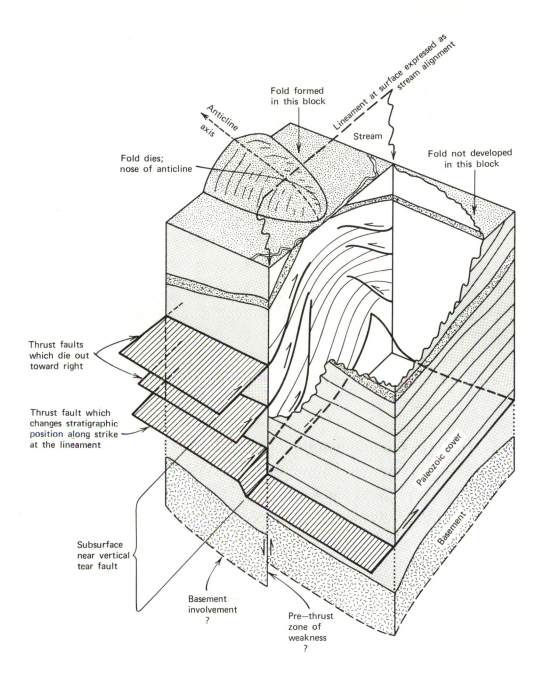

Fold formed
in this block

Lineament at surface expressed as
stream alignment

Anticline
axis

Stream

Fold not developed
in this block

Fold dies;
nose of anticline

Thrust faults
which die out
toward right

Thrust fault which
changes stratigraphic
position along strike
at the lineament

Paleozoic cover

Basement

Subsurface
near vertical
tear fault

Basement
involvement
?

Pre—thrust
zone of
weakness
?

FIGURE 14.10. Idealized block diagram showing the postulated three-dimensional structure of a "Gwinn-type" lineament. The lineament is the topographic expression, usually a linear valley, of a strain discontinuity between two semi-independent thrust blocks that may be linked by a stepped tear fault. The effect is to create a discontinuity that is not necessarily a fault everywhere along the plane, but that enables the strain in adjacent blocks to be manifest in different ways, for example, folds of different form and non-coincidence of fold axes. This strain boundary is envisaged as an essentially vertical zone, as much as 2 km wide, of anomalously high fracture density (joints and minor faults). Decoupling of regional stress across this fracture zone may be a "thin skinned" manifestation above a decollement, or the rejuvenation of a deep-seated fault, for example, the basement fault illustrated in the diagram. [After Kowalik and Gold, 1976 (10).]

Epizone — Tertiary stocks
Mesozone — Silver Plume batholith
Transitional mesozone—catazone —
    Boulder Creek batholith
Catazone — quartz monzonite
    gneiss stocks & phacoliths

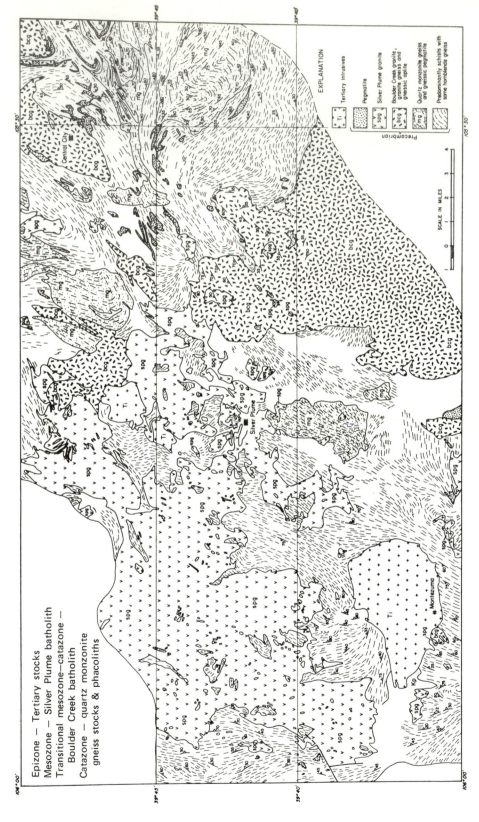

FIGURE 14.11. Geological map of part of the Front Range in Colorado, illustrating the structure and contact relationships of plutons emplaced in the epi-, meso-, and cata-depth zones. Note the regular and transgressive contacts for the Tertiary plutons, and the progressively more irregular and less discordant contacts for the Silver Plume granite bodies, the Boulder Creek granite batholith, and the quartz monzonite gneissic stocks and phacoliths. Modified after a map by Lovering and Goddard, (1950) (25) and published by Buddington, (1959) (23). This area is shown as the dashed rectangle on the Geological Map of the Front Range (Fig. 14.23a), and on the accompanying Landsat images (Fig. 14.23b, c, and d).

EXPLANATION

Ti   Tertiary Intrusives

spg   Pegmatite

spg   Silver Plume granite

bcg   Boulder Creek granite,
      granite gneiss and
      gneissic aplite

mg   Quartz monzonite gneiss
      and gneissic pegmatite

      Predominantly schists with
      some hornblende gneiss

Precambrian

SCALE IN MILES

0   1   2   3   4

434

or its absence over recent flows (for example, the irregular, narrow dark bands following drainage in Fig. 14.12). Rocks extruded from central vents generally produce conical shaped volcanic piles (see Fig. 14.12) of varying size, peakedness, and completeness, depending on the composition and viscosity of the lavas and the amount of ash. Lava flows of wide extent are inferred to be extruded by fissure eruptions, the dikes of the epizone.

Craters, vents, and calderas are discriminated by their association with volcanic terrains and their unusual quasicircular shape. The unique form of vents (craters) at the summit of volcanic cones has been used to characterize the volcanoes of Mars (26) as observed by Mariner 9 orbiter images. Spectacular summit craters can be identified (see Fig. 14.14) on Olympus Mons, a Martian stratovolcano of enormous proportions. Other types of circular structures include the isolated domes, cryptoexplosion structures, and meteorite impact sites and craters. Attention on some of these features was generated from the observations of astronauts, for example, the large Richat structure (see Fig. 14.15) photographed in 1965 by McDivitt and White from the Gemini IV space capsule. Inward facing cuestas of quartzite and annular valleys in shale and limestone attest to a domal structure of the Lower Paleozoic rocks as a result of magma emplacement at depth (22).

14.3.7 Limits for Structural Extrapolations

The form of large folds commonly is determined by mapping a marker horizon on a base map of appropriate scale, or by integrating the orientation data in a spherical projection such as the π diagram. Not only is the form of the fold determined by the geometric arrangement of the structural/fabric elements on the next larger scale, but also deductions can be made from the minor plications on the limbs as to its shape and symmetry if conditions are favorable for the use of Pumpelli's rule (see Fig. 14.8). The down-plunge projection of cylindrical folds to determine subsurface form can be extended only a limited distance with any confidence. These limits depend on scale and the homogeneity of folds within a domain, that is, not beyond the domain boundaries established for the minor fold elements.

It is unlikely that a structural domain would extend over an area as large as most remote sensing systems cover, for example, approximately 34,000 km² (13,000 miles²) in a Landsat scene. Neither would a structure persist indefinitely with depth, for the deformation style would change, even in a uniform stress field, with the change in physical properties of the rock in response to the increase in temperature, vertical load, and confining pressure. Discontinuities of this type are recognized in the "basement" rocks in shield areas by a change in either metamorphic grade or structural style, and are referred to in the relative terms of infra- and supracrustal units. In addition, the Conrad and Mohorovicic seismic discontinuities are likely to be domain boundaries. The presence of these subhorizontal crustal discontinuities suggest that the megascopic scale rock masses on the continents should be considered as thin curved shells rather than rectangular slabs, a geometric consideration that would introduce inhomogenous responses to a widespread deformational event. Besides the "horizontal" discontinuities, subvertical discontinuities in the form of faults, fracture systems, dikes, domain and block boundaries, and so forth, probably are more common in the crust than has been previously realized. Some geologists have proposed that these crustal "fractures" are part of a continental (28), or even global fracture system (29, 9) caused by body tides and the rotation of the Earth (30). It is these major impressed discontinuities in the crust that will create additional lateral inhomogeneities and effect subsequent deformations. This has led to the concept of tectonic blocks (5) bounded by discontinuities (fractures), which become more diffuse in the ductile rocks at depth. Some of these discontinuities may behave as block, domain, or strain boundaries during later deformation(s), or even as strain relief zones for earth body tides, and would be expressed at the surface as lineaments.

Constrained in depth by horizontal discontinuities and a room problem in displacing denser materials beneath, crustal (infra and supra) folds and other structures should not persist down plunge to depths much greater than the catazone. Culminations and depressions are to be expected in supracrustal fold systems caused by horizontal compression of a curved shell segment, especially where vertical discontinuities might influence the deformation patterns. Many noncylindrical plane folds [Fig. 14.16(a)] may form in a unique event, particularly if the deviation in axial line curvature is relatively small, or if they are zig-zag type *en echelon*

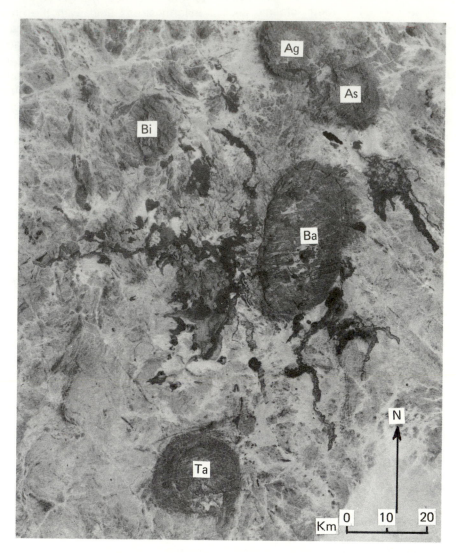

FIGURE 14.12.(a) Landsat-1 image of the Air Plateau, a Precambrian massif in Niger, North Africa. The dark colored circular and elliptical mountains of Ashkout (As), Agaluk (Ag), Bilete (Bi), Baquezans (Ba), Drayan (Dr), and Taraouadji (Ta) are Late Paleozoic and Mesozoic peralkaline granitic intrusions, with economic potential fortin and niobium minerals. Dark, irregularly shaped flows of olivine basalt are associated with a number of recent volcanic vents, and contrast well with the Precambrian "basement" rocks. The dominant fractures are northwest trending strike faults, with an apparently conjugate set in a north easterly direction (27). The igneous activity is thought to represent a mantle plume. Landsat image E-1286-09191, May 5, 1973.

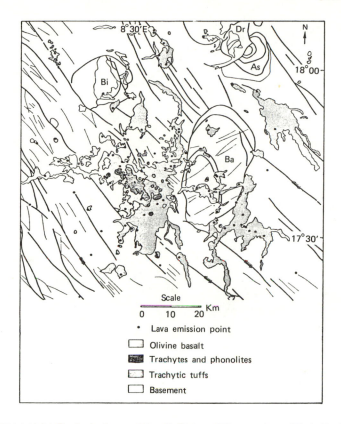

FIGURE 14.12.(*b*) Geological map of the Air Plateau. [The map is modified after Black and Girod, 1970 (27).]

folds with opposing plunge directions [see Fig. 14.16(*b*)]. Aligned depressions may coincide with plunging terminations of macroscopic scale folds and the hinge trough between zig-zag type *en echelon* folds in parts of the Appalachian fold belt [Fig. 14.17 (*a*)]. Some of the cross-strike lineaments mapped in Pennsylvania (14) from Landsat images, are based on the alignment of these unusual fold features, and straight stream and valley segments. The Zagros fold belt of southwestern Iran affords an example of intricate patterns (see Fig. 15.11) from an apparently simple folding event (see Fig. 14.6 for map view and cross section showing the relationship between the morphological expression and the structural style and setting).

14.3.8 Overprinting and Superimposed Structures

Successive or episodic compressional events would impress new forms on the earlier structures and produce the interference fold patterns apparent in many orogenic belts and in most of the shield areas of the world. *Polyphase deformations* may result in complex outcrop patterns of marker horizons or seemingly irregular patterns of fabric elements. Fortunately, the dominant events will generate the dominant structures, with local perturbations in form to attest to the other episodes. The interested reader should consult Refs. 3, 21, 31, 32, and 33 for additional information on polyphase folds. With variation in size, form (concentric, similar, chevron, and so forth), shape (flexure curvature and degree of appressedness of the limbs), attitude and shape of the axial plane and axial line, superposition angle, order of events and magnitude, and the mechanics of folding, many different interference patterns could result. Only two simple, but fairly common patterns of the periclinal folds (basins and domes) and the geoflex (curved fold belt or orocline) will be discussed in greater detail here, because most of these show up well in aerial photographs and satellite imagery.

438

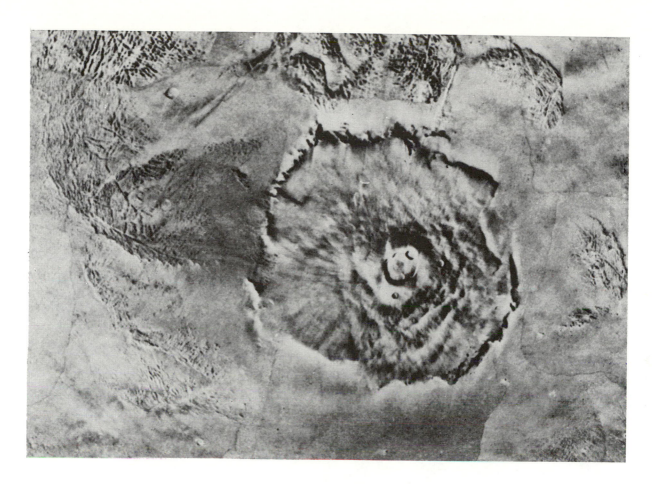

FIGURE 14.14. Olympus Mons on Mars is a relatively recent, large shield volcano with a complex summit calders (26). It is enormous by terrestrial standards, with a shield 500 to 600 km across and a summit 23 km above the surrounding plain. A summit caldera some 70 km across contains subsidiary craters, and the entire shield drops to the plain in a cliff that ranges up to 4 km in height. Concentric terraces on the slopes, and scalloped aureoles of distinctive textured terrain that extend outward from the base of the cliff as far as 500 km, attest to a complex origin and some unusual extrusive phenomena for this Martian volcano. Using an impact crater frequency model, Olympus Mons is dated as a 200 m.y. old feature (26). The mosaic was prepared from Mariner 9 Orbiter images (26).

FIGURE 14.13. Landsat images showing part of the Great Dyke of Rhodesia (Zimbabwe) transgressing an Archean terrane of gneisses, granitic batholiths, and greenstone belts. The dike, which averages 6 km wide and extends for 480 km, is unique among layered intrusive bodies for its large size, linear continuity, and economic potential (chromite). The ultramafic and gabbroic rocks flare outward near the earth's surface in a series of basins and troughs (see arrows). Among the host rocks are tonalitic batholiths (T), (light toned elliptical masses) intruded into ancient gneissic belts, and separated by irregularly shaped greenstone belts (G) (dark tones). Metamorphic rocks of the superimposed Limpopo mobile belt (L) underlie the northeast trending foliated terrain in the southeast corner of the mosaic. Right lateral offsets of the dike are apparent on faults near its southern end. The regular shaped light areas represent overgrazed range lands. Landsat images E-1103-07285-7 and E-1103-07291-7, November 3, 1972.

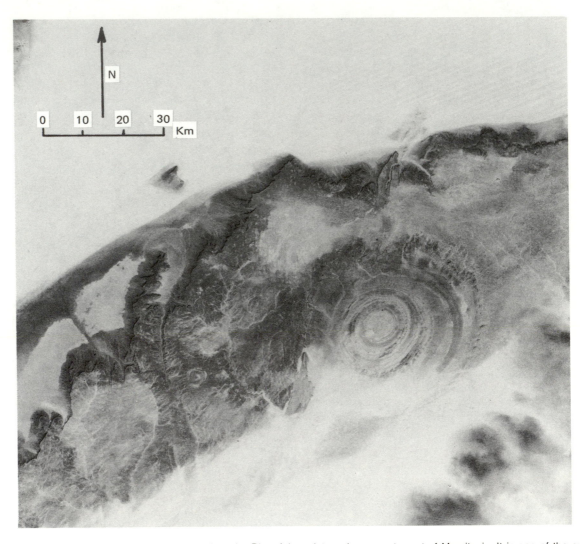

FIGURE 14.15. The Richat structure is located on the Dhar Adrar plateau in a remote part of Mauritania. It is one of the eye-catching sites from near space and overshadows its smaller (4 km diameter) companion, the Semsiyyat dome about 30 km to the west southwest. The concentric rings at Richat are as much as 30 by 40 km across, and represent inward facing cuestas of quartzite and annular valleys in shale and limestone differentially eroded in Ordovician sedimentary rocks (22). The prevailing wind, which blows from upper right to lower left (22), has swept much of the plateau (dark tones) free of desert sands (light tones), and has developed the spectacular seif dunes (linear wave pattern) of the Mekteir dune field to the north. Landsat-1 image E-1103-10434-7, November 3, 1972.

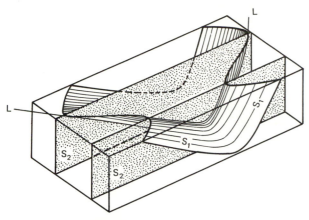

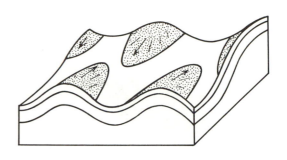

FIGURE 14.16. (a) Down plunge projections of a noncylindrical plane folds in S_1 with axial planes S_2 to illustrate the geometry of culminations and depressions. Although a depression of the appressedness depicted above is unlikely to form without a superimposed cross-fold, those of relatively small deviation in plunge angle may form from a unique event.

FIGURE 14.16 (b) The surface pattern of some zig-zag type of *en echelon* folds. In the Appalachian fold belt some of the transition or saddle zones portrayed above coincide with major cross-strike lineaments.

The geometric analysis of polyphase deformation structures is complicated by the new structural/fabric elements and structures induced by each successive deformational event. In Fig. 14.18(a) an upright fold with horizontal plunge is refolded by another upright fold with a vertical plunge to produce a simple geoflex. The later F_2 fold with a vertical axial plane S_3, deforms the earlier, close, upright fold, F_1, and generates three new fold axis lineations. One is associated with the right limb domain of the F_1 fold ($L_{S_{1R}\text{-}S_3}$), another with the F_1 axial surface ($L_{S_2\text{-}S_3}$), and the other with the left limb domain ($L_{S_{1L}\text{-}S_3}$). If S_3 is planar then the three lineation directions will lie in this plane. Note that the S_{1L} domain produces an antiform plunging northwest, the S_{1R} domain, a synform plunging southeast, and S_2, a vertically plunging fold.

Some geoflexes may result from simple shear [Fig. 14.18(b)], a mechanism proposed by Knowles et al. (34) for the Appalachian geoflex in central Pennsylvania [see Fig. 14.17(a)]. This geoflex hinge coincides with a major culmination; folds plunge respectively to the northeast and the south south-west, east and south of the hinge line, which extends from north of State College towards Harrisburg, Pennsylvania.

Basins and domes commonly result from two orthogonal intersecting, quasiupright fold systems with essentially horizontal plunges and approximately equal magnitude [see Fig. 14.19(a)]. More complex patterns result if the axial planes of the folds intersect obliquely [Fig. 14.19(b)], or are inclined. Intersecting synclines are represented by *basins*, anticlines by *domes*, and the syncline-anticline interaction produces a *saddle*. Excellent examples of folds of this type are exposed in the Flinders Ranges of South Australia (Fig. 14.20), and the eastern Grenville structural province in Canada. The remarkably symmetrical Mecatina Crater [Figs. 14.21(a) and (b)] near the Labrador border in eastern Quebec, is formed by the nearly perpendicular intersection of two fold sets of almost equal wavelength and amplitude (35).

14.4 SOME APPLICATIONS

14.4.1 Introduction

Remote sensing techniques are perhaps the most important tools to the geologist for mapping geology and structure on the macroscopic and smaller scales. Direct lithostructural mapping is possible only if the different rock types are sufficiently well exposed and distributed to render a diagnostic spectral signature of characteristic texture. This works best in terrains with little or no

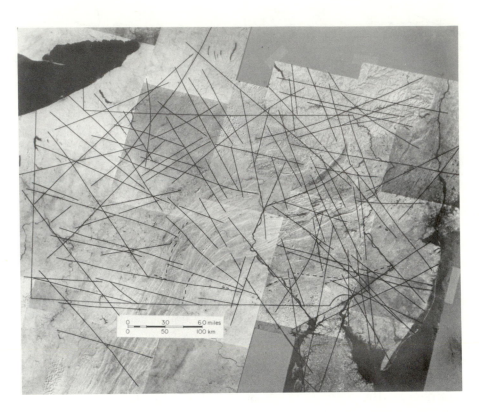

FIGURE 14.17. (a) This Landsat-1 mosaic of Pennsylvania was used as a base for mapping inter-mediate and megascopic lineaments (solid lines), and also provides a synoptic view of the major physiographic provinces. Some of the notable geographic features are the Hudson River (upper right), New York City and Long Island (center right), the barrier islands off the coast of New Jersey (lower right), Lake Erie (dark tone, upper left), the Finger Lakes of New York (top center), and the crescent-shaped area (dark tone) of the Wyoming Valley "anthracite basin" of northeast Pennsylvania. The Pennsylvania orocline is apparent from the curvature of the Valley and Ridge physiographic province (curvilinear tonal pattern) of the Appalachian fold belt (from lower center to top right corner of the state). The Allegheny Plateau province is the relatively flat-toned terrain to the west and north; to the south and east of the fold belt and narrow belts representing successively the Great Valley, Blue Ridge, Triassic Basin, Piedmont, and Coastal Plain physio-graphic provinces. Note the discordant attitude of many lineaments to the regional structural grain, and their independence of either the major physiographic and tectonic province boundaries, or the age of the bedrocks. The variable age of ore deposits along some lineaments (Pre-cambrian to Mesozoic) suggest that they are long-lived, periodically rejuvenated features. Some of the lineaments are discernable in the Tertiary coastal plain sediments and through Recent glacial deposits. The dashed lines represent the trace of mapped faults. The photograph is reproduced with the author's permission [Gold et al., 1974 (14).]

442

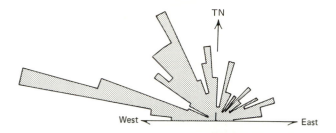

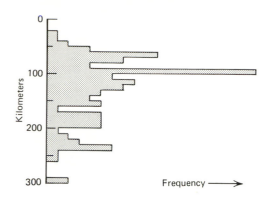

FIGURE 14.17. (b) Rose diagram showing the strike frequency of the intermediate and long lineaments mapped on the Landsat-1 mosaic of Pennsylvania and adjacent areas. The longer cross-strike lineaments account for the dominant west-northwest and north-northwesterly trends: some of these strike across physiographic province boundaries and are still discernible in recent glacial and coastal plain sediments. The shorter lineaments tend to "fan" with the Pennsylvania orocline. (Published with permission of D. P. Gold.)

FIGURE 14.17. (c) Histogram of the lengths of the intermediate and long lineaments mapped on the Landsat-1 mosaic of Pennsylvania and adjacent areas. The uneven frequency distribution probably is due to the small population ($n = 84$). (Published with permission of D. P. Gold.)

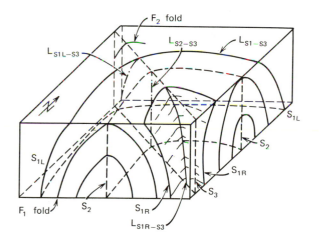

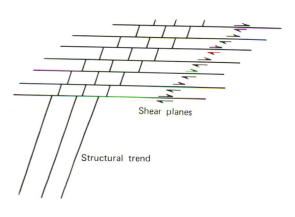

FIGURE 14.18. (a) Block diagram to show and identify the structures generated by multiple phases of deformation. An upright fold with an horizontal axial line Ls_1-s is refolded by another upright fold F_2, whose kinematic transport direction is horizontal. Note, the F_2 fold generates a northwesterly plunging antiform on S_{1l} (left limb), a vertically plunging fold on S_2, and a southeasterly plunging synform on S_{1R} (right limb). Their respective fold axes are Ls_1L-s_1, Ls -s_1, and Ls_1R-s_1 will lie in the same plane, if the axial surface of the F_2 fold is planar.

FIGURE 14.18. (b) Plan view of an orocline formed by simple shear. Note that the individual segments are translational but not rotational, a condition Knowles and Opdyke (34) sought to test from the direction of paleomagnetic vectors.

443

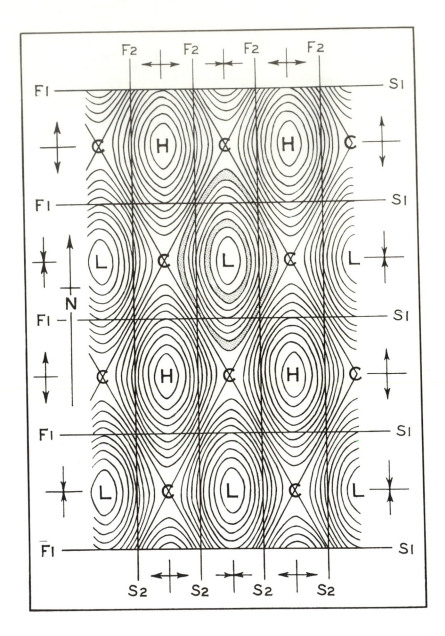

FIGURE 14.19. (a) Contoured relief plan of interference surface(s) for intersecting upright folds, showing the development of domes (H), basins (L), and saddles or cols (C). Primary fold axes are orthogonal. [After O'Driscoll, 1962 (32), reproduced with the author's permission.]

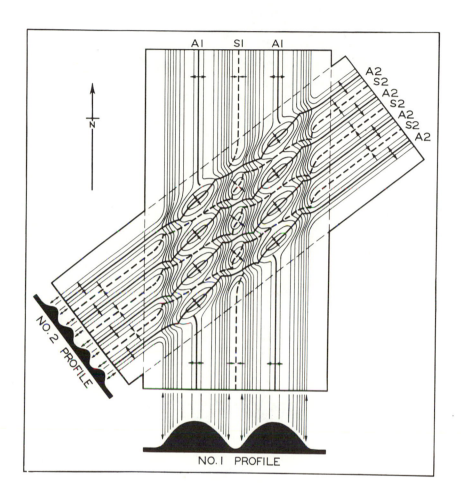

FIGURE 14.19. (*b*) Contoured relief plan of interference for intersecting upright folds. Primary fold axes are inclined. [After O'Driscoll, 1962 (32), reproduced with the author's permission.]

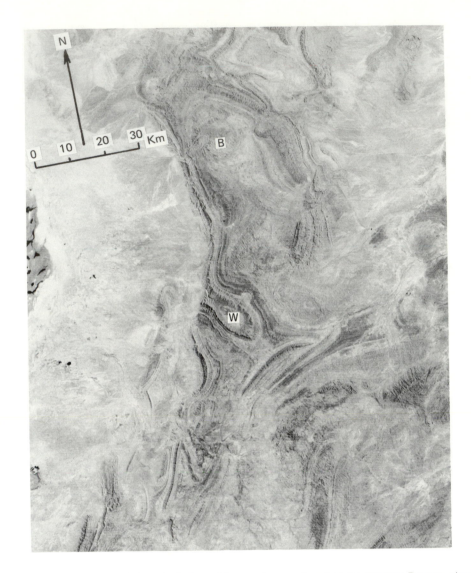

FIGURE 14.20. (a) Landsat-1 image of the principal sandstone outcrops of part of the Flinders Ranges, about 650 km north of Adelaide in South Australia. The Flingers Ranges block of Proterozoic and Lower Paleozoic sedimentary rocks lies to the east of the structural trough (Sunklands) that extend from Spencer Gulf to Lake Torrens. Although the folds are generally simple and open, the cross-folded pattern apparent in the figures is due to the intersection of trends at the junction of two divergent fold belts (28). Lake Torrens (L), Blinman Dome (B), and Wilpena Pond (W) are marked in the Landsat image for reference. Landsat image E-1565-00003-7, February 8, 1974.

446

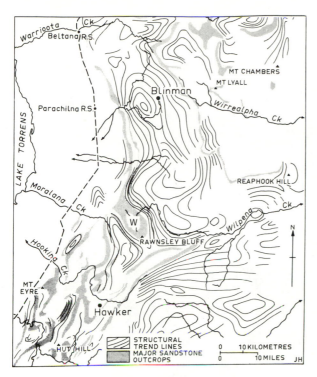

FIGURE 14.20. (b) Abridged structural map (17) of the principal sandstone outcrops of part of the Flinders Ranges about 650 km north of Adelaide in South Australia. (Map reproduced with the permission of the Australia National University Press).

soil cover and sparse vegetation, such as desert and arctic regions. Excellent examples have been worked up by NASA scientists for parts of Saudi Arabia (37). Fortunately, primary and secondary "structures" can be mapped indirectly from an analysis of their morphological expression in most terrains where direct lithological observations are masked by soil and vegetation. Some of the most obvious features are secondary structures, particularly if they are discordant or transgressive. However, not all the linear features (lineaments, fracture traces, joint traces) so apparent on remotely sensed images can be attributed to underlying faults, dikes, fissures, fractures or joint sets, or zones of fracture or joint concentrations, and care must be exercised in their structural interpretation.

14.4.2 Linear Features—A Discussion on Lineaments and Fracture Traces

There is a great deal of misleading terminology on linear features in the literature, and this has proliferated since the advent of spacecraft and high-altitude imagery. O'Leary et al. (38) have addressed this problem and have defined "lineament" *senso stricto* in a geomorphological sense as a "mappable, simple or composite linear feature of a surface, whose parts are aligned in a rectilinear or slightly curvilinear relationship and which differs distinctly from the patterns of adjacent features and presumably reflects a subsurface phenomenon." This is consistent with the original definition of Hobbs (39) who considered them as "nothing more than a generally rectilinear earth feature manifest in the landscape by (1) crests of ridges of the boundaries of elevated areas, (2) the drainage lines, (3) coast lines, and (4) boundary lines of formations, of petrographic rock types, or of lines of outcrops." Although he pointed out the nonequivalency to "tektonische linien" of German geologists because the latter "are mostly regarded as lines of displacements . . .", he later (40) included seismotectonic lines, which introduced tectonic implications to the term. Sonder (30, 41) expanded the concept of a "lineament" to a general regional designation to " . . . denote a definite direction which is contained in the tectonics, the joints, and the relief." This was unfortunate because the corollary has not been proven and our ability to map "lineaments" on images far outpaces our ability to characterize them physically in three dimensions. In addition, the similarity in scale of tectonic features and "lineaments" on satellite images has led to a host of speculative dynamic interpretations.

More linear features will be defined on derived data as the physical field surfaces become more refined. For example, linear features are apparent in the second derivative gravity map of the Sudbury area, Ontario (42), and the Nye-Bowler lineament in Montana is seen best in color composite, ratioed Landsat images. Linear features and patterns are common to all types of maps, and to restrict the term "lineaments" solely to topographic features (6, 43) fails to accommodate the allied and derived forms. Among the three main categories of linear features are: (a) those defined by point and/or segment alignments, such as mud volcanoes in the Caspian Sea (6), hot-springs in Italy (44), plutons (see Fig. 14.13), fissures and pipes (45), dikes and volcanic centers (see Fig. 14.12), oil and gas fields (Nye-Bowler lineament in Montana), wind and water gaps (see Fig. 14.1), and topographic depressions and stream or valley

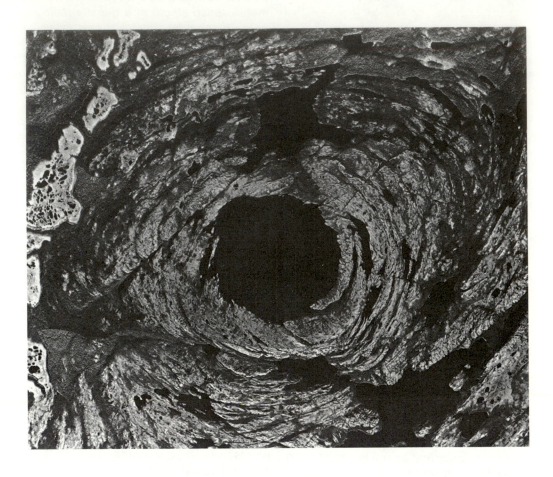

FIGURE 14.21. (a) Mosaic of aerial photographs of the Mecatina Crater, Quebec. The circular lake overlies the "eye" of concentric saucerlike shells that define a structural basin, caused by the interference of cross-folds (two synclines) of nearly equal wavelength and amplitude. Saddles or cols (anticline-syncline interaction) are apparent in the lithostructural trends in the upper right and left corners. Mecatina basin is one of a regional group of basins, domes, and cols that occur in a regular pattern in the Grenville rocks of northeastern Quebec, Labrador, and Newfoundland. Most lakes (dark tones) serve to define the lithostructural grain; others are preferentially developed along fracture zones (lineaments). Most of the saucer-shaped lithostructural units have outward dipping cuestas, some as high as 100 ft (30 m). Bedrock is abundantly exposed in the terrain around the crater, and it is covered sparsely with a low arctic type of flora (light tones) of cariboo moss and heather. Stunted pines and spruce trees in the valleys exhibit a matte texture and darker tone. The honeycomb textured, very light toned areas to the west are identified as muskeg swamps. The mosaic was prepared for the Dominion Observatory by the Air Photo Division of the Department of Energy, Mines and Resources, Ottawa, Canada.

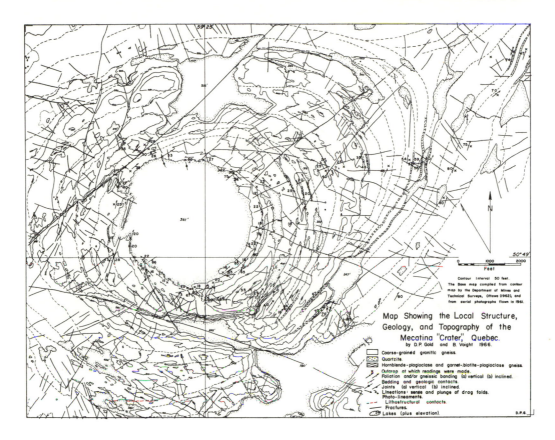

Map Showing the Local Structure,
Geology, and Topography of the
Mecatina "Crater", Quebec.
by D.P. Gold and B. Voight 1966.

FIGURE 14.21. (b) Structural map (36) of the Mecatina Crater, Quebec.

segments (see Fig. 14.22); (b) by linear trunca-
tions and offsets of contours in topographic maps
or in gravity, magnetic, isopach, dip angle, or
structure contour maps; and (c) linear boundaries
in derived maps showing facies, structural domains,
or strain zones. Gwinn (46, 47) drew "lineaments"
on fold axis culminations and transverse sedimentary facies changes deduced from subsurface data,
and Rogers (48) drew "lineaments" from structural
data on a geological map.

"Lineaments" on various geophysical measurements are important components in plate tectonics.
Another class of first order linear feature is the
alignment of volcanic centers or their roots. Some
are associated with rift valley settings. Two examples are the Oligocene alkaline intrusions that
extend some 1400 km from northeast Mexico to
southern New Mexico (49) and the carbonatites,
kimberlite, and alkalic rock plutons of the Monteregian Petrographic Province that extend for some

500 km from near Ottawa, Canada, eastward across
the St. Lawrence lowlands and into the Applachian
fold belt (50). Other first order linear features
represent the intrusive or extrusive manifestations
of a mantle plume, irregularly erupting through a
moving lithospheric plate, for example, the "alkaline
igneous rock lineaments" of Angola, South West
Africa (Namibia), Brazil, and Uruguay (51). The
implications of these settings will be discussed in
Section 14.4.

All too few lineament maps specify the nature
(10) of the expression mapped or its quality. Some
features such as the 38th Parallel Lineament (52)
have a combination of expressions that include
seismic zones and other deep-seated indicators
in aligned kimberlite and alkalic rock bodies and
offsets in aeromagnetic and gravity maps along
parts of its length, interspersed with near surface
features of faults and rift zones and epizonal ore
deposits of varying age. Perhaps the solution lies

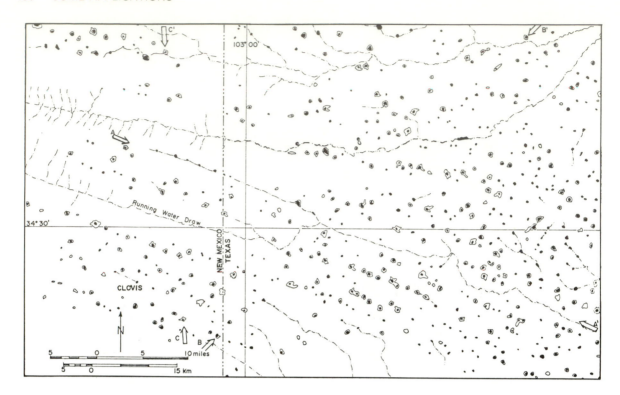

FIGURE 14.22. Distribution of depressions and streams on part of the USGS 1:250,000 scale topographic quadrangle map No. NI 13-6 of the Clovis Area, New Mexico-Texas. Depressions are shown by the 50-ft contour interval, and the playa lakes and swamps are marked in black. The regular geometric alignments for most of the depressions suggests a structural control by subsurface fractures. The main lineament directions defined by the point alignments are given by *AA'* (N70°W), *BB'* (N45°E), and *CC'* (N5°W) in order of decreasing dominance.

in using appropriate qualifying adjectives to indicate the mode of the expression (38). Lattman (8) defined *photogeologic lineament* as "a natural linear feature consisting of topographic (including straight stream segments), vegetation, or soil tonal alignments, visible primarily on aerial photographs or mosaics, and expressed continuously or discontinuously for many miles." However, these are explicitly transgressive or secondary features "... not obviously related to outcrop pattern of tilted beds, lineation and foliation, and stratigraphic contacts ..." Werner (53) has suggested that these latter types be designated *photolineaments* to distinguish them from *topographic lineaments*, but the former term specifies the sensor and not the form of the expression. Other types of lineaments include *magnetic* (or aeromagnetic) and *gravity lineaments*, *structural* and *stratigraphic lineaments* [the Gwinn-type of Kowalik and Gold, (10)], *volcanic* and *alkaline igneous rock lineaments* (51), and *geological lineaments* for the composite types such as the 38th Parallel Lineament (53). This usage is con-

sistent with the revised geomorphological definition of O'Leary et al. (38) as they specify neither the mapping base nor the surface—and the qualifying adjective(s) give the nature of any derived data.

The concept that a lineament "reflects a subsurface phenomenon" behooves the interpreter to distinguish among primary structures, that is, *strike lineaments* from *cross lineaments* (54). In addition the geological significance of a topographic lineament will vary depending on whether it is degradational or aggradational in character. Most of the lineaments *senso stricto* are a result of differential erosion but, the streams of the Murray River system in Australia express several pronounced lineaments in an interior basin of alluviation: an anastomosing system of the Darling River and some of the aligned headwater tributaries define a *riverine lineament* more than 640 km (400 miles) long (5) over mainly Quaternary sediments.

The Running Water Draw-White River lineament system in Texas extends for some 270 km (170 miles) in a belt 40 km wide (25 miles) striking

N70°W. It is characterized by families of closely spaced depressions, linear streams, and ridges developed on the upper flat caliche caprock surface of the Ogallala formation (Pliocene) in eastern New Mexico and West Texas (see Fig. 14.22). The caliche, formed most likely by an eolian aggrading soil process, is covered by soil and windblown sand, locally in sand dunes (55). Playa lake deposits are developed in the depressions, whose origin is ascribed mainly to wind deflation (56), and to a lesser extent by differential solution in the underlying Permian evaporite beds (55). An inflection of the post-Ogallala contours in the lineament zone indicate a dip separation of about 8 m (25 ft) at the east end and 30 m (100 ft) at the west end. The subparallel lineaments of this system reflect a zone of structural weakness that has influenced geologic events since the Cretaceous and perhaps much longer (55). Reeves (57) identified three dominant directions, N70°W, N45°E, and N5°W (see Fig. 14.22) and suggested they represent a post-Pliocene-pre-Late Pleistocene rejuvenation of an ancient deep-seated regmatic shear pattern.

Current research on lineaments is focused on some perceptive concepts and observations by Hobbs, namely: (a) in recognizing that " . . . many lineaments are identical with seismotectonic lines and they therefore afford a means of to some extent determining in advance the lines of greatest danger from earthquake shock . . ." (40); (b) in suggesting that even their composite nature" . . . is some surface expression of a buried feature . . ." (40); and (c) in observing their relationship to scale ". . . lineaments which may appear rectilinear on maps, may be so only in proportion as the scale of the map is small." (39). Neglecting scale may be tantamount to discarding data on the mechnical relationships. Perhaps by reporting size, nature, and quality of expression a meaningful genetic classification of "lineaments" (senso lato) will be developed. More small scale geophysical studies are needed to determine the three-dimensional nature of "lineaments." The suggested scale for linear traces in Table 14.2 has been applied to lineament mapping in Pennsylvania (10) and it could be adapted to include all natural linear expressions regardless of the mapping base.

The thickness (width) of lineaments is a parameter seldom addressed—a coarse pencil line on a Landsat image may represent a band on the ground 1-2 km wide! A general relationship of increasing width with length is suggested from joints, measured in millimeters to centimeters, and widths of 1 to 33 m reported (58) for fracture traces from a variety of terrains. In a study of the ridge crest expression of 15 intermediate to long lineaments across Bald Eagle Mountain in central Pennsylvania, Krohn (59) mapped anomalous zones, 0.5 to 2 km wide, asymmetrically coincident with the lineaments mapped on Landsat imagery. These zones are characterized by an increase in fracture and joint density, the presence of gossan or sulfide mineralization, and the abundance of faults of small displacement. More work is needed to define the ground characteristics of lineaments and their zone of influence.

Additional problems have arisen because most of the current lineament mapping is done on remotely sensed images, which contain problematic linear features and patterns expressed by vegetation distribution and vigor, soil color and tone, and textural patterns, besides the more geologically evident features shown by aligned topographic features of valleys, troughs, scarps, ridges, and straight stream segments. Nor are these latter features consistent in their expression, but they may be highlighted by shadow and seasonal effects, to give a low mapping reproducibility between operators, especially if conventional photogeologic mapping techniques are used on single images. The monoscopic mapping of linear features on images is highly subjective (60), and only in orientation frequency is there any close accord between different operators.

Operator variability in monoscopic mapping of linear features may be reduced by means of a Ronchi grid, a diffraction grating, commonly ruled with 200 lines per inch (78 lines/cm),*which preferentially enhances or suppresses linear features on an image (61). Because the grid spacing is many times larger than the wavelength of light, a Fresnel diffraction pattern is produced if the grid is placed near (inside the near-field focus) the eye. Light is diffracted by the grid to produce bright and dark bands (or color fringes) parallel to the line edges or slits in the grating. Linear features subparallel to the grid will appear diffused and suppressed,

---

*Ronchi grids of different spacing are available from various scientific supply houses.

TABLE 14.2   SCALE OF SOME LINEAR TRACES WITHOUT OBVIOUS DISPLACEMENTS [a]

Surface Feature	Size	Mapping Base	Possible Structures
Joint traces	Centimeters to tens of meters	Outcrop maps, orientation diagrams, large scale aerial photographs	Bedding
Fracture traces	Approx. 100 m to 1.6 km	Aerial photographs, large scale topographic maps	Narrow, steeply dipping zones of joint concentrations up to 33 m wide
Lineaments	(a) Short: 1.6 to 10 km	Topographic maps, small scale aerial photographs	Broad zones of up to a few km wide of disrupted rocks, including concentrations of narrow fracture zones
	(b) Intermediate: 10 to 100 km	Topographic maps and relief models (1:250,000), high altitude imagery	
	(c) Long: 100 to 500 km	Satellite imagery, small scale relief models	Petrographic provinces and aligned volcanic centers, rift valleys, aulacogens, and continental sutures as much as 100 km wide
	(d) Megalineaments: 500 km	Satellite imagery, and mosaics of Landsat imagery	

[a] Some of these features may be dikes or faults with little or no strike separation.

as will all other lines except those perpendicular to the grid [see Figs. 14.23(c) and (d)]. Not only is the grid section highly selective in enhancing the orthogonal direction, but it also efficiently links aligned linear segments in this direction in a continuous line of different brightness. Unfortunately, this efficiency also brings up many spurious linear features such as tree shadows and sun angle effects and care must be taken to discriminate them. However, the main value of the grid is in visually integrating aligned pattern elements of landscape; spurious features must be distinguished on an individual basis.

Another way of improving the significance of observations of different operators is by counting for orientation [the orientation-frequency histogram or rose diagram, for example, Fig. 14.17(b)] or density (total length) within a cell of appropriate size † (10). In Fig. 14.24 only those features whose midpoints lay within a cell were counted for frequency orientation data, and the total length of the lineaments in each azimuth class were summed within each cell to give a measure of their density. Not only is the histogram format for individual cells amenable to statistical tests (for example, Chi-square) for similarity, but also to computer proces-

†The cell should contain at least 100 observations in order to make valid statistical comparisons.

sing and presentation as thematic rose diagrams (see Fig. 14.24) on a CALCOMP plotter (63). Additional filters of sun azimuth, scan line, and regional bedding strike directions were applied in this Pennsylvanian study (see Fig. 14.24) in order to reduce operator subjectivity in mapping subtle features whose expression may depend on seasonal and sun angle conditions: The local nature of these variations impose limits on the cell size.

Mapping reproducibility of length and position of linear features is improved greatly if stereoscopic analysis is undertaken. The stereoscopic portrayal of gravity and magnetic contours reveal geophysical lineaments on aligned anomalies, offset contours, and steep gradients that are not apparent in the monoscopic data (64).

In an analysis of lineaments it should be realized that the orientations and density measured on one scale are not necessarily the same as those mapped on another scale. The rose diagram [Fig. 14.17(b)] of the long lineaments mapped on the Landsat mosaic of Pennsylvania (65) [Fig. 14.17(a)] should be compared with the short lineaments mapped on individual Landsat images and plotted in cells of comparable size (see Fig. 14.24). The former tend to be through-going features with a consistent west-northwest trend and cross-strike lineaments (66) that fan with the Appalachian orocline. The latter diagrams reveal the variations in orientation coincident with the major structural and physiographic provinces in Pennsylvania (10). An integration of the small cell data into a *synoptic diagram* generally reveals an oblique relationship of frature data with scale that has been rationalized in terms of a second order shear model (65). In a detailed study of linear features on three scales long the Tyrone-Mt. Union lineament in central Pennsylvania (see Fig. 14.1), an orthogonal pattern, coincident in directions is apparent in the mesoscopic (joint traces) and megasopic (lineaments) scale data but not in the macroscopic (fracture traces and short lineaments) scale features (67). The latter exhibit preferred orientations conjugate to the dominant lineament direction of 135°. These geometric relationships are portrayed in hierarchial scale form in Fig. 14.25.

14.4.3 Groundwater Exploration in Fracture Traces and Lineaments

The mapping of fracture traces and short lineaments for the purpose of locating weathered zones with relatively higher porosity and permeability within the bedrock, was developed by Parizek, who showed that the bedrock beneath fracture traces in the Lower Paleozoic carbonate rocks of central Pennsylvania yielded 10 to 1000 times more water than wells in the same rocks at adjacent "off fracture trace" sites (68). Fracture traces and lineaments in central Pennsylvania are expressed as: aligned surface sags and depressions; gaps in ridges; soil tonal changes revealing variations in soil moisture; aligned springs; seeps in perched surface ponds; natural vegetation alignments due to vigor, stress, or species; straight stream and valley segments; abrupt changes in valley alignment; and the presence of gullies and waterfalls. Some of these settings in central Pennsylvania are illustrated in Fig. 14.26. The fracture traces here are underlain by narrow zones, from 2-20 m wide of fractured country rock, and are similar in size to the nearly vertical (87°-88°) zones of fracture concentration measured in the siltstone and shale beds for northeastern Pennsylvania (58). With an average width of 13 m, the Pennsylvanian fracture traces are comparable to the 3-33 m widths documented in a variety of terranes in the United States (69). From 5 to more than 60 closely spaced, subparallel joints (fractures) may transgress all beds with little variation in orientation or density, or the joints may

concentrate in the competent beds (see Fig. 14.5). As few as five distinct high angle joints have been observed underlying fracture traces in some settings; elsewhere these zones comprise hundreds of closely-spaced, intersecting joints so as to impart a shattered appearance to the rock.

Many factors combine to influence well yields at a particular location in terranes where fracture permeability predominates. Some of the more important factors are: well radius and depth; casing depth; well efficiency; drilling method; degree of fracture development; depth to water table; variation in rock type; topographic setting; structural setting and dip of beds; and type and number of joints and faults and other zones of fracture concentration. Similar geological variables influence the depth and extent of weathering, a factor of importance in highway and engineering foundation studies, and in mine and tunnel roof stability analyses (70). If these variables are normalized or held constant, then water yields from wells on and off fracture traces and lineaments will give a measure of the permeability and porosity of the adjacent rocks.

Well yields data from the Upper Sandy dolomite member of the Gatesburg Formation (Late Cambrian) in the Nittany Valley, Pennsylvania, are plotted in Fig. 14.27. These data, adjusted for depth of saturated rock penetration or saturated rock exposed below the well casing, are given as "productivity" in units of liters per second per meter of drawdown per meter of saturated rock penetrated $1\ell \cdot s^{-1} \cdot m^{-1} \cdot m^{-1} = 1$ gal$\cdot$min$^{-1} \cdot$ft$^{-1} \cdot$ft$^{-1} \times 0.679$. By comparing wells drilled on and off lineaments and fracture traces for similar hydrogeologic settings (valley bottom, valley wall, upland), and from different categories of bedrock dip, the probability distributions for the maximum, minimum, and 50% productivity could be established respectively for "on lineament" wells (6.3, 0.001, and 0.2$\ell \cdot s^{-1} \cdot m^{-1} \cdot m^{-1}$) and the "off lineament" wells (1.5, 0.002, and 0.05 $\ell \cdot s^{-1} \cdot m^{-1} \cdot m^{-1}$) (69). It is apparent that some wells are highly productive regardless of whether they were on or off lineaments. Nearly all of the more productive wells were located on fracture concentrations. The slope of the line of wells considered to be on lineaments, approximately 1 km wide, regardless of whether they are on or off fracture traces, is less steep than the slope defined by wells remote from lineaments. This implies that wells will be more consistent in their yield, and display less variability,

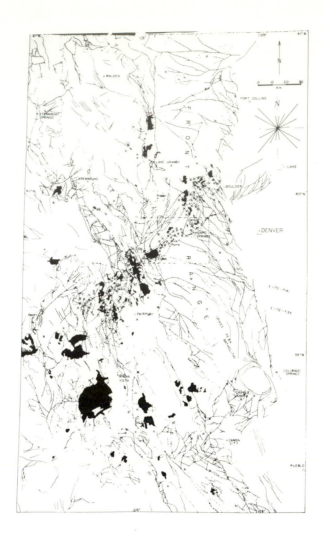

FIGURE 14.23. (a) Geological map of the Front Range in Colorado, showing Laramide intrusions (black) and dikes (dashed lines), faults (solid heavy lines), mineral occurrences (x), areas of Precambrian rocks (stippled), as well as the limits of the Landsat frames E-1172-17132-7 and E-1172-17135-7 (b) used to illustrate the effects of a Ronchi grid northeast enhancement (c), and west-northwest enhancement (c). The Ronchi grid is rotated slowly (about 1 cycle in 10 seconds) until an enhanced direction becomes apparent (Pohn 62, 1970), and its orientation is fixed by oscillating the grid to the sharpest line. This direction is noted on the image, and the terrain features showing the alignment are checked for geologic authenticity by removing the screen for a photogeologic interpretation. A study of the central Colorado mineral belt by Offield 61 (1975) showed a dominant northeast linear trend, and a pervasive west-northwest trend, among six subordinate or local trends. These trends are depicted vectorially, in an ordinal scale of decreasing prominence, on the modified rose diagram, located in the northeast corners of Figures 14.23a and b. There is a geometric association of the three mineral belts exposed in the Front Ranges with the well-developed northeast lineaments, which are thought to represent a preferential reactivation of older northeast trending fractures during the Laramide orogeny. In addition, the spatial relationship of mineral occurrences and Laramide intrusions within the structurally disturbed central mining belt, suggests a cross-strike structural control to epizonal and perhaps mesozonal depths. The northwest trending lineaments are not as conspicuous: many coincide with the Precambrian faults exposed west of Denver and Boulder, and others with the Front Range faults (Laramide age) north and west of Fort Collins. The intrusive relationships of the rocks, exposed in the rectangular area around Silver Plume, are shown in greater detail in Figure 14.11. The boundary of the Pikes Peak batholith (coarse stippled area in Fig. 14.23a) has been highlighted on the Landsat image (Fig. 14.23b). The epizonal nature of this 1 b.y. old pluton and its satellite stocks is apparent from the sharp contacts and relatively smooth boundary, and this characterization is compatible with the 1.6- to 6-km emplacement depth established by other criteria. The circular patterns within the batholith represent discrete late potassic intrusive phases.

FIGURE 14.23. (*b*) Landsat frames E-1172-17132-7 and E-1172-17135-7 of the Front Range in Colorado.

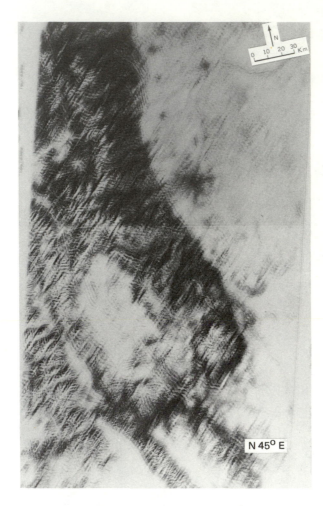

FIGURE 14.23. (c) The Front Range in Colorado illustrating the effects of a Ronchi grid northeast enhancement.

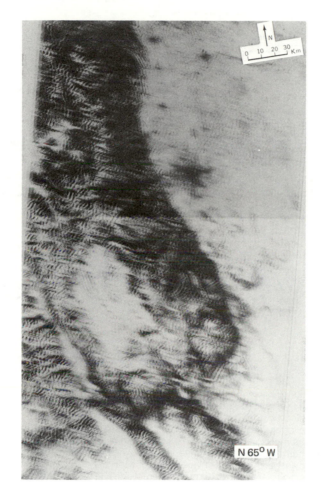

FIGURE 14.23 (*d*) The Front Range in Colorado illustrating the effects of a Ronchi grid west-northwest enhancement.

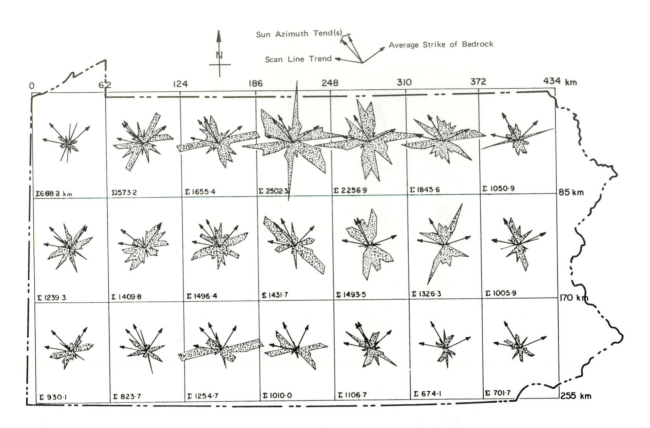

FIGURE 14.24. Short to intermediate length lineament orientations summarized in rose diagrams for cells (62 km by 85 km) on a grid across Pennsylvania (10). The scan line direction (west-northwest arrow), sun azimuth (2 arrows for cells on overlapping images with different sun angle), and general strike of bedrock (northeast trending arrow) are superimposed on each rose diagram. Note the general paucity of lineaments near these arrows, due to inherent biases (scan line effect and sun azimuth shadow effects) in the system, and the subjective filtering of lineaments coinciding with bedding strike by the mapper. The density of lineaments in each cell can be judged by the sum of their lengths, recorded in kilometers in the lower left corners. The lack of coincidence of the short lineaments with the dominant west-northwest and north-northwest directions for the longer lineaments (c.f. Figure 14.17a and b) suggest a complex mechanical relationship between fracture forms of different size[65]. [The diagram from Kowalik and Gold, 1976 (10), is reproduced with their permission.]

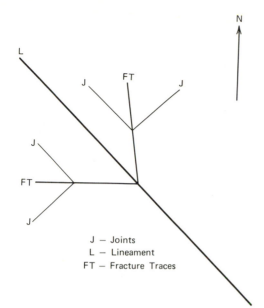

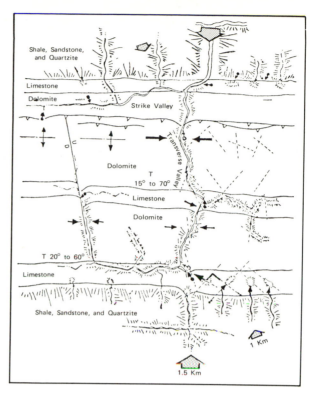

FIGURE 14.25. Geometric relationship between "fracture forms" mapped on a meso-, macro-, and megascopic scales in the vicinity of the Tyrone-Mount Union lineament in central Pennsylvania (67). The strike lineaments and fold axes trend northeast, and these have been omitted from this size-orientation summary diagram. [After Canich, 1976 (67); reproduced with permission.]

FIGURE 14.26. Summary of regional and local controls on permeability and porosity development in folded and faulted carbonate rocks of the central Appalachian type (69). This includes the interaction of rock type where textures favor intergranular permeability preservation and weathering, and the structural settings (tilt and sequence of beds, location on and type of folds, proximity to and type of fault or zone of fracture concentration) that favor regional patterns of anisotropic permeability distribution. Interaction of these factors influence recharge, flow, weathering, and solution, and contribute to the development of the surface morphology (69). The trace of the main lineaments are marked by the trajectory of the short, broad-headed arrows. The small solid arrows indicate the direction of ground-water movement. Dashed lines are fracture traces that are genetically linked to the ground-water flow. Although anomalous fracture conditions, as much as 1.5 km wide, are associated with some major lineaments, the distribution of fractures (joints) or zones of fracture concentration (fracture traces) within this broad zone is not even, and probably is analogous to joints in rocks on a mesoscopic scale. [The diagram is taken from Parizek, 1976b (69), with the author's permission.]

when located on fracture traces in a lineaments zone in favorable topographic and bedrock settings (69).

Internal drainage is common to most limestone terranes, and the coarse-textured drainage pattern, mottled soil tones, and sink holes are characteristic photographic-keys (see Chapter 6 and 12). A structural control of groundwater movement in some karst regions is apparent from the linearity of solution depression, as well as collapse and sink features (see Fig. 14.28). In the relatively flat-lying carbonate rocks of West Virginia, the aligned solution depressions (Fig. 14.28) that are discernable on Landsat and Skylab imagery, define logical sites for water wells.

Conduits for or concentrations of groundwater can be a hazard in underground mining operations at dam sites, and in excavations associated with highway and building operations. Construction of a part of the Route 200 bypass near Tyrone in central Pennsylvania revealed structurally disturbed and friable bedrock conditions coincident with some of the lineaments mapped on Landsat images by Krohn (59). Elsewhere along the steeply dipping

northwestern limb of the Sinking Valley anticline, the conventional road cut slopes were stable, despite the steep dips and the thin units encountered. However, at three lineament sites the artificial slopes become unstable; hillside creep was active and landslides developed on the graded slopes, especially during the spring thaw and runoff, or at

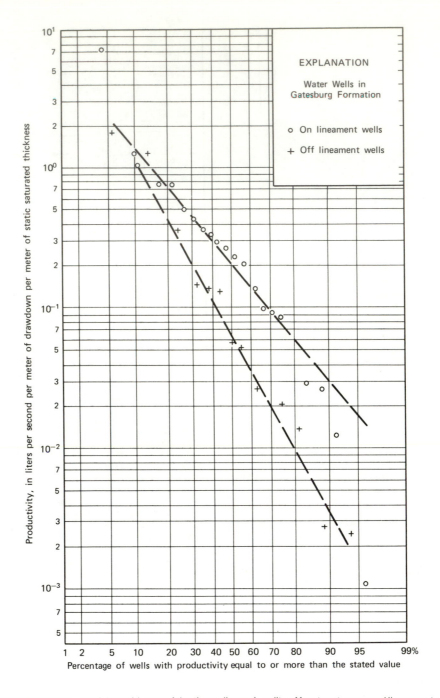

FIGURE 14.27. Much of the evidence of the three dimensionality of fracture traces and lineaments is gained indirectly, for example, from records on water well yields. The graphs above show a comparison of productivity frequency for wells located on short and intermediate length lineaments about 1 km wide with those from off-lineament sites. Only data from wells located in similar topographic settings in the Gatesburg Formation (Cambrian) are compared. The correlation of fracture trace and lineament sites to higher yields is rationalized in the greater fracture density shown to underlie some of these features (69). (After Parizek, 1976b, with the author's permission).

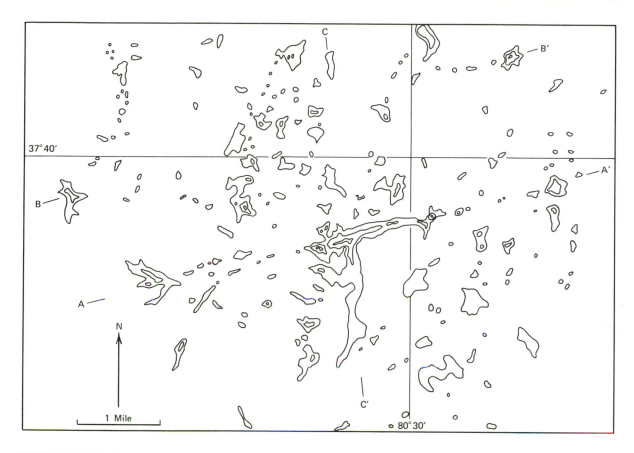

FIGURE 14.28. Distribution of depressions on the USGS 15-ft topographic quadrangle maps of the Anderson and Ronceverte areas, West Virginia. Depressions are shown by the closure on the 50-ft contour intervals; areas below the third contour are indicated in solid black. Werner has identified the alignments in *AA'*, *BB'*, and *CC'* as three prominent "photolineaments" in this region (71).

times of heavy and prolonged precipitation. More than a million cubic yards of material was excavated from one site. At another site, quick conditions encountered in fractured sandstone were stabilized with drains, but during the spring thaw in 1975, the concrete roadbed broke and heaved more than 30 cm. Test holes drilled through and adjacent to the roadbed showed a rise in piezometric surface of about 1 m, and artesian water flowed through the top of the pipes (72). It is apparent that these lineaments are underlain by fractured material and that they represent the trace of zones of high porosity and permeability for collecting and channeling groundwater. The periodic increase in pore pressure promotes landslide development on the artificial slopes.

14.4.4 Circular Features

Few geometric forms excite a geologist's curiosity as much as the linear and circular patterns or features in the topography. Perhaps it is because they are the forms most readily perceived by the human eye and brain, and because their origins are linked to special and unusual events. The search for circular features on Landsat imagery has the advantage of nearly global coverage and small scale. The type of geological features that exhibit quasicircular to circular outlines include volcanic craters and calderas, structural and/or topographic basins and domes, impact craters and other cryptoexplosion structures, and sinkholes. The volcanic craters generally occur in clusters (Fig. 14.13), rarely exhibit regular circular outlines, and are associated with other volcanic features such as cinder cones and lava flows.

The cryptoexplosion structures and meteorite impact sites are atypical circular features in their random distribution, variability in size and form, and apparent lack of any tectonic association. Apart from circularity, the only factor in common with structural basins (folds) is that they are "bottomed" by discontinuities. These structures are recognized by the state of the "target" rocks, and the shape and form of the "crater." Some of the metamorphic effects that reflect the ultrahigh pressure regime are the presence of shatter-cones, veins of pseudo-tachylite or clastic fragments in any impact melt, and fallback breccias in the crater. The other primary criteria are morphological in nature and can be measured (shape) or deduced (form) from remote sensing data if the analyst understands the model for crater formation on different scales. The morphological form of "craters" on different scales is illustrated in Fig. 14.29, by using empirical and conceptual data from shield terranes in Table. 14.3, as well as some ground-based data. The morpho-tectonic parameters of (a) degree of circularity, (b) structural contrast with the country rocks, (c) drainage (annual or radial), and (d) size, has been used for determining potential impact sites on the Canadian shield.

There is a strong correlation between crater form, size, and the depth to the nearest major seismic (shock wave) discontinuity. Apparently, the compressional shock wave is reflected by a discontinuity and returns to the surface as a tensional wave which combines with slump from the crater walls and a longer term isostatic rebound to produce a central uplift or mound in the crater floor. It should be noted from Table 14.3 and Fig. 14.29, that the rimmed pit morphology of the smaller craters is completely inverted in the large, annular trough type "craters," such as Manicouagan (see Fig. 14.29). There is a relationship in the shield areas between crater diameter and the size of the central uplift, but the dimensions of the uplift and the disturbed zone are likely to be much larger for an event of equivalent magnitude in layered rocks, where the basement unconformity is likely to act as a major reflecting horizon. The twin craters of the Clearwater lakes in central Quebec (see Fig. 14.30) are remarkably circular in outline, and have been described variously as calderas or impact sites. The western lake is about 30 km across

and contains a central uplift, exposed as a ring of island containing either volcanic breccias (?) or impact melt rocks (73). Drilling has revealed a small, submerged central uplift beneath the eastern lake.

14.4.5 Plate Tectonic Settings and Ore Deposits

In the quest for new ore deposits, the exploration geologist may search for anomalies by: (a) a blanket coverage of geology, geophysics, and geochemistry; (b) seeking only those areas where past geological environments were conductive to the processes of ore concentration; of (c) drilling on a grid basis (74). One approach is from a study of the genesis of mineral deposits (metallogeny), which emphasizes their relationship, in time and space, to regional petrologic features or events. The metallogenic map is an important exploration tool to the geologist for "area selection," that is, the most favorable environment for specific types of ore deposits. The scale of the structural domains and rock bodies (interior basins, sedimentary and volcanic clastic wedges, ophiolite slabs and alpine peridotite masses, thrust slices, folds, fault systems, plutons), is similar to the imaging scale of some remote sensing systems; many "tectonic bodies" are amenable to small scale characterization and mapping by remote sensing techniques. The objective of this section is to identify some of those terranes readily amenable to remote sensing interpretation, where tectonic conditions were favorable for the generation and location of ore deposits.

The metallogenic map attempts to combine geologic, tectonic, mine, and mineral location maps in simplified, yet critical and distinctive parameters so that ore deposits with similar components but different genesis, age, size, shape and tectonic setting can be distinguished. For example, iron may occur in sedimentary taconite deposits, hydrothermal replacement deposits, magmatic, lateritic, or black sand detrital deposits (75). The restriction of certain ores to specific rock types (for example, Pb-Zn to carbonates; Li to pegmatites, and so forth) and ages [for example, nickel sulfides, stratiform chromite, and titanium in anorthosites to Precambrian rocks, or the preponderance of porphyry copper and molybdenum, antimony, and mercury to Mesozoic or Tertiary orogenic belts (75)] has long been used in exploration strategy. The

TABLE 14.3   FORM OF PROBABLE IMPACT "CRATERS" AS A FUNCTION OF SIZE [a]

Types	Form	Example
Ia	Bowl or pit with a raised rim	New Quebec crater; Meteorite Crater, Arizona
Ib	Rimless bowl or pit	Brent crater; West Hawk Lake
IIa	Broad or shallow basin	Deep Bay; Lac La Moinerie
IIb	Island rimmed shallow basin	Lac Couture
IIIa	Basin with central mound (or island)	Lake Mistastin; Clearwater Lake (east)
IIIb	Basin with a central elevated smaller basin	Clearwater Lake (west); Sudbury (reconstructed)
IVa	Annular depression (syncline) around a central mound	Charlevoix
IVb	Annular trough	Manicouagan
Va	Annular syncline (locally overturned) around a structural dome (not elevated)	Vredefort Dome; Carswell Lake structure
Vb	Annular ridge	Flooded lunar crater
VI	Concentric "rings" of ridges and depressions	Mare Orientale
VII	Circular pattern of lineaments	Nastapoka arc in Hudson Bay; dome margin of Sudbury

[a] For size and morphotectonic expression, see Fig. 14.29. There is a general increase in size from Type I to VII.

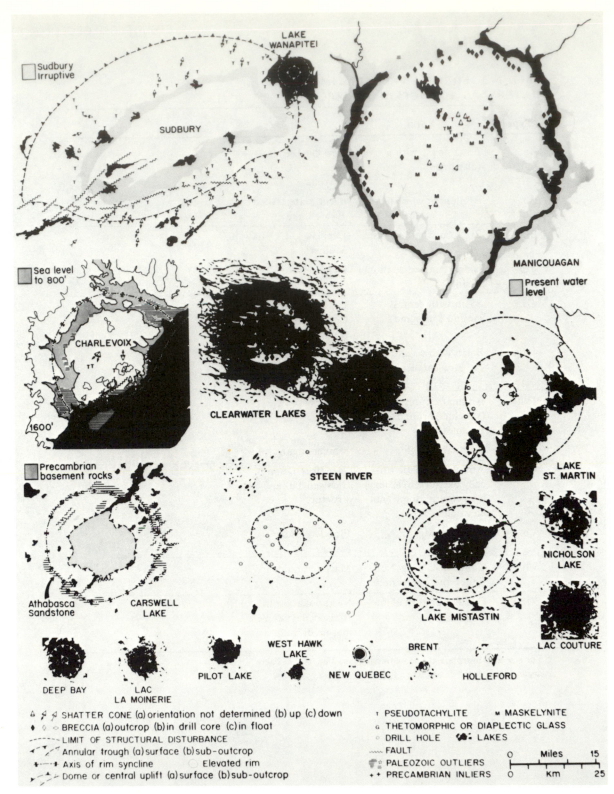

FIGURE 14.29. Diagram to illustrate the shape and form of some Canadian impact sites, using morphotectonic expression and ground-based data. Each "crater" is drawn to scale from topographic maps, and the form (see Table 14.3) is indicated by appropriate structural symbols. Other symbols are used to show the position of shock deformation features such as breccia bodies, maskelynite, diaplectic glass, pseudotachylite veins and dikes, as well as the direction of shatter cones. (This diagram was prepared by D. P. Gold in 1972, at the Earth Physics Branch of the Department of Energy, Mines and Resources, in Ottawa, Canada.)

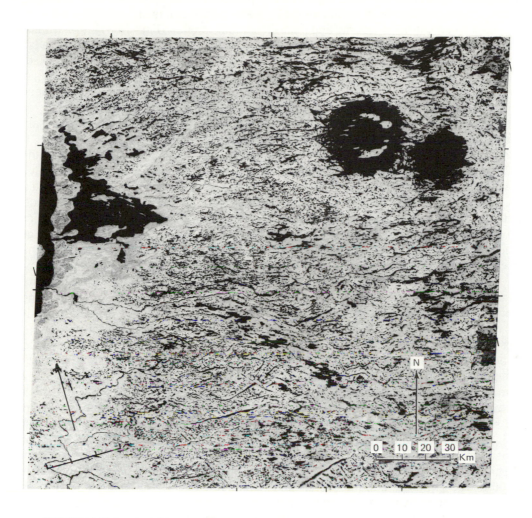

FIGURE 14.30. Landsat-1 image of Precambrian shield terrain in northern Quebec, showing East and West Clearwater Lakes (the two conjoined circular lakes in the upper right corner) and part of the Richmond Gulf (left center). Clearwater East Lake is essentially a bowl-shaped crater about 20 km across, in which the small central uplift is completely submerged. In contrast, the 33-km diameter Clearwater West Lake crater has a type IIIB morphology of a shallow basin with an inner "uplift" basin revealed by the annular group of islands. These "craters" are superimposed on the granitic gneiss terrane by a high-energy event, probably impacting meteorites—although others have interpreted the aphanitic "melt rocks" as lavas. Except for shallow dipping Proterozoic sedimentary rocks in the Richmond Gulf area (west-central margin of image), the rest of the terrain is underlain granitic and gneissic rocks of the Superior Province. The fracture patterns so characteristic of the shield areas are apparent in this summer scene. Landsat image E-1732-15285-7, July 25, 1974.

uneven distribution of ores in the crust has been attributed to unique conditions of sedimentation, weathering, deposition, intrusive activity and metamorphism, but only with the development of the "plate tectonics model" has a unifying cause become evident.

The implication of plate tectonics on the generation, distribution, and location of ore deposits has led to a renewed interest in metallogenic maps, and a classification of ore deposits in a geologic-tectonic framework (see Table 14.4). Various forms of divergent and convergent plate boundaries and associated "ore" settings are shown conceptually in Fig. 14.31. The distribution of the main active and recent plate margins (converging, diverging, and transform faults) are portrayed on the global tectonic map by Lowman (see Fig. 14.32) along with those volcanic centers that are considered to have formed from mantle plumes (hot spots). The refinements introduced by "plate tectonics" concepts to "continental drift" theory include the recognition of the Benioff zone as a subducting plate margin, correlating the age of oceanic magnetic anomalies to sea floor spreading from the midoceanic ridges, and the mechanisms of transform faults.

The plate tectonics model involves about a dozen plates and subplates, generally consisting of both continental and oceanic crust (see Fig. 14.31), that comprise the crust and are capable of moving about the Earth. These semirigid lithospheric plates, between 100 to 150 km thick, apparently maintain a constant mass balance by the generation of new crust from the mantle at the ocean ridges (diverging margins) and the consumption of the crust in subduction zones (converging margins) beneath island arcs and deep-sea trenches. The transgressive "lineaments" defined by offsets in the ocean ridges and the magnetic reversal patterns are another form of plate boundary, representing strike-slip motions within and between plates on socalled transform faults (76). Most of the transform faults are associated with the oceanic ridges (see Fig. 14.32), where they occur as minor transgressive irregularities on the divergent plate boundary. Those that link arc to arc (for example, Caribbean and Scotia arcs), or a ridge to an arc (for example, Cocas and Nazca plates) represent major boundaries to the plate. These faults are postulated to be a form of tear fault whose differential motions as a result of ocean floor generation and

spreading are of opposite sense to the ridge separations (see discussion in Section 14.4.6). In a conceptual model (Fig. 14.31) the transform faults would be analogous to tear faults in a large gravity slice, where tensional conditions would prevail in the trailing edge, or root zone (diverging margin) and compressional conditions in the leading edge, or nappe zone (converging margin).

Each plate moves relative to the adjacent plates, about an individual axis of rotation. In addition to the change in angular momentum with increasing distance from the rotation axis, plate geometry contributes to appreciable time lags along an orogen (75). With progressive divergence, incipient rifts in a continent (for example, East African rift valley) may evolve into incipient oceans (for example, Red Sea) and ocean rifts (Atlantic Ocean). In the same way, converging margins may change from continent/ocean floor collisions to continent/island arc, and arc/arc confrontations, arc accretion onto the continent, and even the start of a new cycle; at least five cycles are apparent in the Bay of Islands region of Newfoundland (77), and between the Red Sea and the Zagros Mountains (78).

Ore deposits formed at or near plate margins exhibit similarities that define metallogenic provinces oriented and zoned parallel to the plate margins. In contrast, ore deposits within plates generally are secondary in origin and depend on sedimentation conditions in the appropriate tectonic setting, such as in internal basins and trailing edge continental margins, or association with deep-seated igneous activity in regions of crustal tension (rift valleys) or on fracture zones (transgressive lineaments). The apparently randomly distributed layered intrusive complexes in the stable shield areas are unexplained by this model. The relationship of ore deposits to tectonic settings in and events involving lithosphere plates have been summarized by Guild (75) and are reproduced in Table 14.4.

In detail the plate boundaries are rather complex. The number of settings produced by plate collisions depend on the nature of the converging margins, that is, the distribution of continents, ocean floor, and volcanic island arcs within the plates, and whether both or only one of the plates is in motion. The main types of margins are: continent/ocean crust (Cordilleran type); ensialic island arc/ocean crust (Japanese or New Zealand arcs); ensimatic

TABLE 14.4 TECTONIC SETTING OF SOME ORE DEPOSITS IN LITHOSPHERIC PLATES (modified after Guild, 1971)[75]

Deposits formed	Types, possible examples
1. At or near plate margins	Orientation of deposits, districts, and provinces tends to parallel margin
a) Accreting (diverging)	Red Sea muds; *deep sea brines and fossil analogs? Certain cupriferous-pyrite (massive sulfide) ores, Cyprus? Newfoundland? Podiform Cr (may be carried across ocean and incorporated in island arc or continental margin) *Sulfur and petroleum, salt domes on continental shelf in rift sea setting
b) Transform	Podiform Cr, Guatemala? Cu and Mn, Boleo, Baja California
c) Consuming (converging)	Chiefly of continent/ocean or island arc/ocean type; deposits formed at varying distances on side opposite oceanic, descending plate Podiform Cr, Alaska Fe-S$_2$-Cu-Zn-Pb stratabound massive sulfide in *back-arc basins, e.g., New Brunswick, Japan (Kuroko ores), California, British Columbia Mn of volcanogen type associated with marine sediments, Cuba, California, Japan Magnetite-chalcopyrite skarn ores, Puerto Rico, Hispaniola, Cuba, Mexico, California, British Columbia, Alaska Cu(Mo) porphyries, Puerto Rico, Panama, SW United States, British Columbia, Philippines, Bougainville Ag-Pb-Zn, Mexico, western United States, Canada Au, Mother Lode, California; Juneau Belt, Alaska Bonanza Au-Ag, western United States W, Sn, Hg, Sb, western North and South America *Sn, Malaya *Asbestos in obducted ophiolite slabs *Petroleum in interarc basins
2. Within plates	Deposits tend to be equidimensional, distribution of districts and provinces less oriented (may be along transverse lineaments). Other settings include rift valleys, interior basins, leading and trailing edges of continental margins
a) In oceanic parts	Mn-Fe(Cu, Ni, Co) nodules Mn-Fe sediments in small ocean basins with abundant volcanic contributions? Evaporites in newly opened or small ocean basins *Petroleum in newly opened or small ocean basins
b) At continental margins of Atlantic (trailing) type	Black sands, Ti, Zr, magnetite, etc. Phosphorite on shelf *Peat (coal) in coastal plain swamps, e.g., Florida
c) In continental parts *(interior basin setting)	Au (U) conglomerates, Witwatersrand; *Blind River, Ontario Mesabi and Clinton type iron formation Evaporites, Michigan Basin, Permian Basin; salt, potash, gypsum, sulfur *Red-bed Cu (U); Catskill delta deposits in Pennsylvania; Kupferschiefer and Katangan Cu-Co
*(rift setting)	U, U-V deposits, Colorado Plateau Mississippi Valley type deposits, Pb-Zn-Ba-F-(Cu, Ni, Co) Carbonatite-associated deposits of Nb, V, P, RE, Cu, F; *Oka and St. Honoré in Quebec, Kimberlites, diamonds (inter rifts), Araxa in Brazil, Palabora in South Africa *Pb-Zn sulfide deposits in aulacogens, Sullivan, British Columbia *Oil and gas in aulacogens
*(other settings)	Fe-Ti-(V) in massif anorthosites, Canada, United States Stratiform Cr, Fe-Ti-V, Cu-Ni-Pt, Bushveld Igneous Complex, *Stillwater Complex (Montana) Kiruna-type Fe-(P), SE Missouri (?)

*Economic mineral deposits or types not listed by Guild. Reproduced with the authors permission.

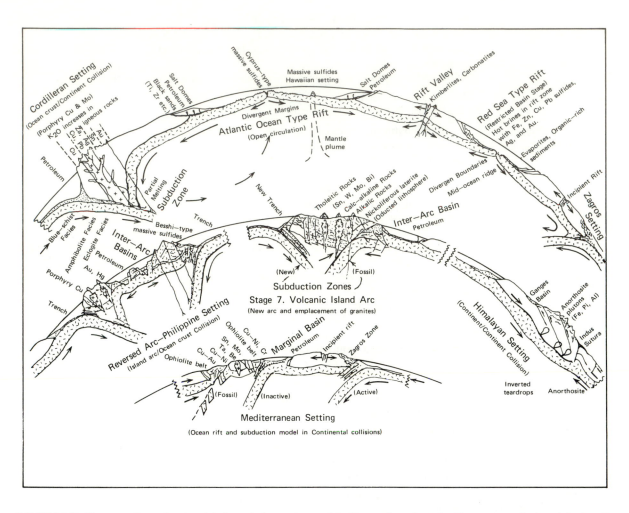

FIGURE 14.31. Cross-sections of some plate boundaries and associated types of ore deposits. These conceptual models show the tectonic setting of some of the types of ore deposits listed in Table 14.4. The rift ocean and subduction model for the Mediterranean type of continent/continent collision could also be drawn as an obducted ophiolite slab, such as the Indus suture, in a Himalayan-type setting. In this model the Hawaiian-type setting is generated by volcanic activity over a mantle plume or "hot spot." Volcanic island chains and seamounts are thought to represent periodic erruptions from a mantle plume into a moving lithospheric plate.

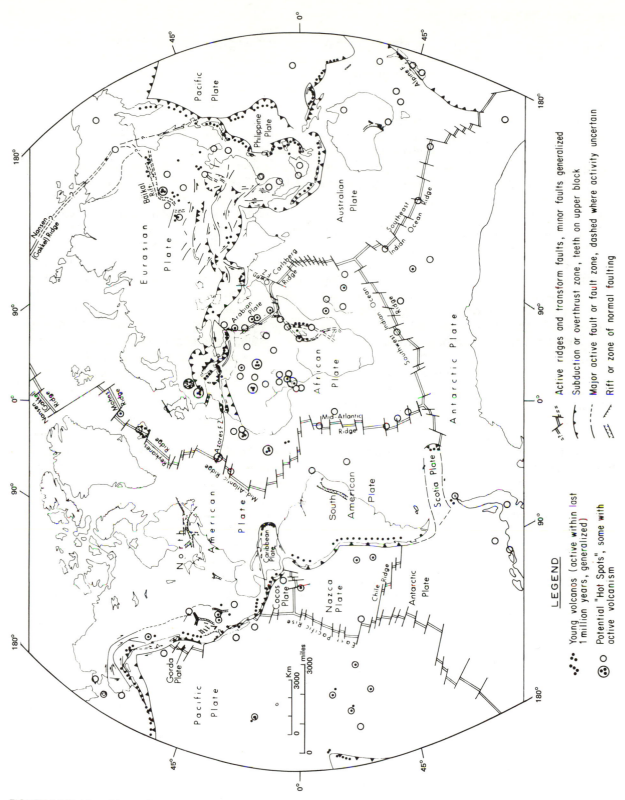

FIGURE 14.32. Map of the world, showing the distribution of modern plate boundaries, young volcanoes, and "hot spots" over inferred mantle plumes. This plate tectonics map, which was first compiled in 1978 by P. D. Lowman Jr., of the Goddard Space Flight Center on a National Geographic Society base map (The Physical World, 1975), was modified to incorporate additional tectonic features (95) and potential "hot spot" centers (89).

island arc/ocean crust (Yap, Bonin, Marianas arcs); continent/island arc (Mediterranean type); continent/continent (for example, the Himalayas); and arc/arc (for example New Caledonia, Phillipines, and Solomon Islands). Arc reversals that produce the latter types, and some of the other types may represent stages in the evolution of continent/ocean floor collisions (79) and/or combinations of different plate margins in active or stationary modes. Ore deposits are associated with each type as well as with specific stages, particularly with respect to the evolution of island arcs (79).

The "target selection" phase of locating the ore deposit requires a study, in greater detail, on any anomalies defined from the regional survey. Except in remote and relatively unexplored areas, the chances of finding an exposed "ore deposit" are very slim. However, Rowan (80) has used remotely sensed data to locate alteration halos around mineralized plutons in Arizona and Nevada. Generally, the effort needed to define the target is proportional to its depth below the surface, and it is from the character of the surface manifestations (structural, geophysical, chemical dispersion aureoles, and so forth) that the body is drilled to establish the grade, shape, size, and mineralogy. The identification of ore deposits by spectral signatures is discussed in Chapter 17.

Some of the indirect uses of remote sensing data for target selection have been discussed in groundwater exploitation in fracture traces and lineaments zones. Photogeologic analysis has been done successfully to define some of the structures that might act as traps for migrating fluids such as oil and gas (81, 82, 83). By analyzing stereo pairs of aerial photographs it is possible not only to identify anticlines and domes, but in favorable terranes, also to draw both structure contour and isopach maps (82). Even where rocks are not so well exposed, structural information can be deduced from outcrop trends, morphology, streams, soil tone, and vegetation patterns (84). Buried "structures" are more difficult to detect but regional fracture patterns may be deflected or refracted by subsurface inhomogeneities (85, 86) or faults (87) in a sufficiently predictable manner for fracture density and pattern to be used as an exploration tool.

The use of Landsat imagery for locating the optimum terranes for mineral exploration is gaining popularity because of the similarity in scale of coverage of the features sought, the worldwide extent of coverage available in four spectral bands, and the low user cost of about $8.00 an image. A number of tectonic settings that are particularly favorable for photogeologic analysis are discussed in the following sections.

14.4.6 Hot Spots, Mantle Plumes, and Aulacogens

An alignment of isolated volcanic centers within plates define "volcanic lineaments" on a mega- and even gigascopic scale. Most can be distinguished by their tholeiitic nature and progressive emplacement age from one end to the other, or from the alkaline affinity of rift valley volcanism. Because the distance/time value is similiar to the 5 cm/yr rate for plate motion deduced from the sea-floor magnetic anomalies, Morgan (88) proposed the concept of intermittent volcanic activity from an essentially stationary hot spot or plume from deep in the mantle, periodically intruding an over-riding plate. One of the best examples of this trail of "hot spots" is the alignment of volcanic centers from the Emperor seamounts (75 m.y.) through the Midway Islands (25 m.y.) and seamounts to the Hawaiian chain (4 m.y., in the northwest to active volcanoes in the southeast). The implications are important because if the "hot spots" do represent fixed points in the mantle, then they could be used as references to determine absolute motions on a single plate, rather than the relative motions between plates (89). The uneven distribution of the 122 hot spots so far identified (see Fig. 14.32) appear to be related inversely to the rate of plate movement (89). Their studies on the absolute motions of plates suggest that island arcs form only from under-riding interactions of a stationary continental plate margin by an advancing ocean floor plate margin, whereas the Cordillera-type of continent/ocean floor collision develops from an over-riding continental plate margin over an essentially stationary ocean floor plate margin.

While most of the plate junctions are planar, a spherical shell requires that there be points where three plates meet, namely, a triple junction. The geometry of triple junction interactions and the complex spreading patterns, as well as their similarity to the hexagonal outline of convection cells in a fluid have been studied by McKenzie and Morgan (90). The correspondence of the three-pronged rift fractures of spreading centers to some of the patterns in the East African rift system, for example, the Afar triangle, suggest that this configuration is fundamental to the continental fragmentation process. Their relationship to doming processes has been demonstrated experimentally (91), and their tectonic association to the broad domes in the continental crust of a temporarily static plate over a mantle plume has been suggested by Burke and Wilson (89). A reconstruction of Gondwanaland reveals many such three-armed rifts between the American and African plates in which two arms become active rifts and open to form an ocean, and the third arm remains an aborted rift valley as much as 800 km long. Sediments deposited in the aborted rift depression are favorable sites for petroleum accumulation. If the plate motions are reversed the active rifts close with the development of fold mountains along the old continental margin. This setting of a narrow trough of thick sedimentary deposits extending into the continental platforms almost perpendicular to the fold mountains, is termed an aulacogen, a word coined from the Greek words meaning "born furrows." The evolution of an aulacogen is illustrated conceptually in Fig. 14.33. Aulacogens are thought to represent the aborted arm of the three-armed rift system (89). Aulacogens of great antiquity are known; they have been recognized in 2-b.y. old rocks underlying the eastern arm of Great Slave Lake. The narrow belt of discordant structure and rock types, almost perpendicular to a fold mountain belt is the signature of the aulacogen. Generally these features can be discriminated on Landsat imagery (see Fig. 14.34 of the Ouachita aulacogen).

Successive outpourings of lava in isolated volcanic centers in North Africa, such as Tibesti and Air (see Fig. 14.12) suggest that the continent has been stationary with respect to a mantle plume for the past 30 m.y. The broad domes associated with the African centers, some of which occcur in three-armed fracture patterns in existing rifts, may be the precursor of a new continental fragmentation. Satellite imagery, in particular, is useful in characterizing the triple junction setting on a hot spot, as well as the "volcanic lineament" (the fossil signature of hot spots on a moving plate). Both the volcanic center and any associated aulacogens are favorable explorations sites, the former for stratified massive sulfide deposits of the Hawaiian-type that occur in the volcano-clastic sediments on the flanks of the volcanoes, and the latter for the

oil and gas trapped in Paleozoic and younger sediments.

14.4.7 The San Andreas Fracture System

The first color composite image produced from Landsat-1 data was of the Monterey Bay area (Fig. 15.1), and the structural interpretation (Fig. 15.2) was completed by Lowman (92) one week after the satellite was launched. This image covers part of the Coast Ranges, Great Valley, and Sierra Nevada physiographic provinces, and the San Andreas fault. The setting is of a fossil (Late Mesozoic) trench, in which the poorly sorted, immature graywacke assemblages of the Franciscan formation (eugeosynclinal basin) were dragged down the subducting east Pacific plate into an Abakuma-type metamorphic regime, which produced blue schists, and locally eclogite facies rocks. These rocks underlie the dark toned area of the Diablo Range (see Fig. 15.1) in which the east-west trending macroscopic folds are apparent. The serpentinite bodies that intrude these metasediments probably are remobilized slabs of ocean floor crust obducted into the melange deposits. The northwest trending linear feature on the image represents the trace of the San Andreas fault, an active fault immortalized in song and popular writing because of its association with destructive earthquakes. The terrain that includes the Santa Lucia, Gabilan, and Santa Cruz Mountains west of the San Andreas fault is underlain by high grade metamorphic rocks and granites. The quartzites and marbles in this "Salinian block" represent the miogeosynclinal deposits of an interarc basin that formerly existed several hundreds of kilometers southeast in the Sierra Nevada. The heavily cultivated area (light mottled rectangular pattern) of the San Joaquin valley, owes its flat texture and relief to the Tertiary sedimentary fill in a terrestrial intermontane basin.

A similar scene of the northern Sierra Nevada, around Sacramento, shows the Great Valley as a light toned, cultivated terrain in marked contrast to the darker toned and dissected pattern of the crystalline rocks of the Sierra Nevada mountains. Many of the "linear features" here have been related to faults and/or joint directions, and to fractures that were important for the localization of the hydrothermal mineral vein deposits of the Mother Lode belt. In southern California (see Fig. 14.35), the Salton Sea and the cultivated region of

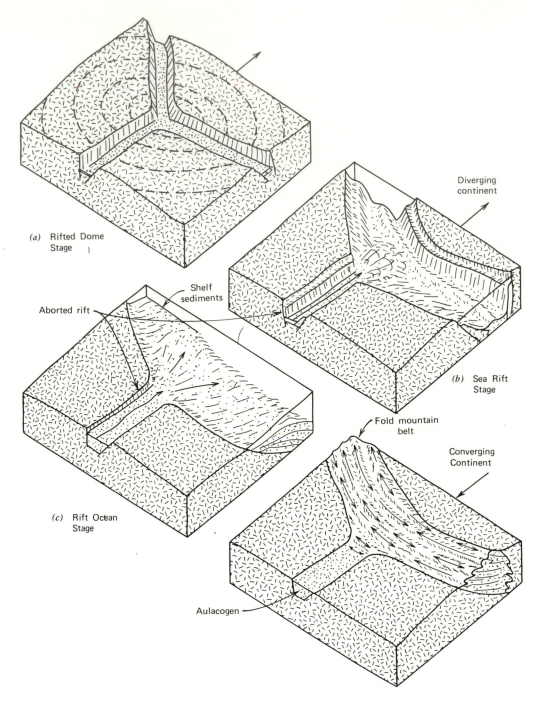

(a) Rifted Dome Stage

Shelf sediments

Aborted rift

(c) Rift Ocean Stage

Diverging continent

(b) Sea Rift Stage

Fold mountain belt

Converging Continent

Aulacogen

FIGURE 14.33. Conceptual diagrams showing the association of domes and rifts to a "hot spot" over a mantle plume, and the evolution of an aulacogen in an aborted rift valley. This model requires a divergent triple junction, in which only two arms have remained active, and a change in deformational style from tensional (spreading) to compressional conditions (convergence) with the development of a fold mountain belt transverse to the aulacogen. The breakup of continents may well take place along "hot spot" induced rift zones (89).

472

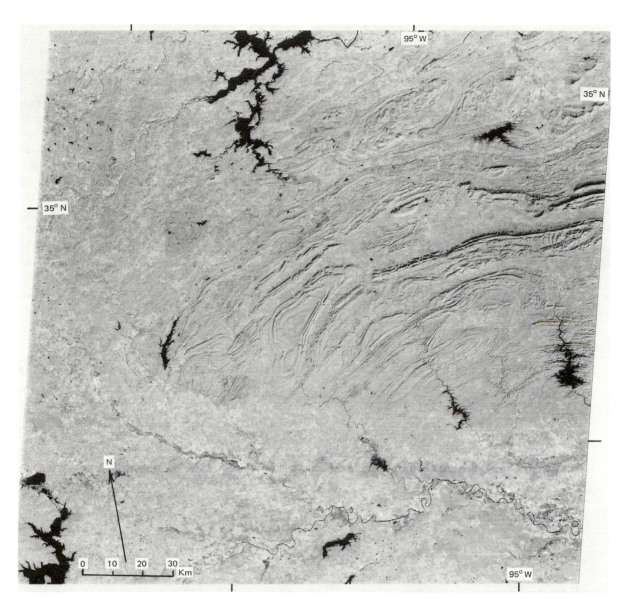

FIGURE 14.34. Landsat image of an ancient aulacogen in the Ouachita Mountains in southeastern Oklahoma (89). The convoluted pattern in the center and eastern past of the scene represents folded and faulted Paleozoic strata of the Ouachita fold belt, with a decrease in intensity of deformation northward into the gently folded terrane of the Ozark Plateau. Transverse to the fold trends in the southern part of the scene is the relatively flat terrain of the Ouachita aulacogen, which formed when a triple junction rift system opened to the south and east approximately 600 m.y. ago. Convergence with the African plate during the Upper Paleozoic closed the active rifts, folded the Paleozoic rocks, and set the stage for additional deposition into the aborted rift valley trough. This aulacogen extends for 400 km to the west, and with contributions of Cretaceous and Gulf Coastal Plain sediments it forms a narrow trough of sediments up to 15 km thick (89). [Landsat-1 image E-1128-16300-7, November 28, 1972.]

the Imperial Valley contrast with the darker tones and hackle-textured terrain of the Peninsular Ranges. The former is interpreted as a structural depression or graben over a modern spreading center where the North American plate has over-ridden the East Pacific ridge (93). This center is separated from the northeast extension of the spreading ridge (see Fig. 14.32) by breaks co-incident with the Banning-Mission Creek fault and the San Jacinto fault (see Fig. 14.35). These faults are essentially strike-slip in nature and their right lateral motion from left lateral ridge separations suggest they may be a continental expression of a transform fault (93). The crystalline rocks of the Peninsular Ranges, in a similar setting to the Salinian block farther north, are structurally and lithologically the southern extension of the Sierra Nevada. The recognition of the northeast trending valley "lineaments" in the Peninsular Ranges as possible fracture and fault zones that are not off-set by the Elsinore fault, is crucial evidence for the essentially dip-slip motion on this fault (92).

The San Andreas fault is part of a fracture system of first and lower order faults that trend northwesterly through southern California and the adjacent states for more than 1000 km. This system is well portrayed on the Landsat mosaic (Fig. 15.13) compiled by the United States Soil Conservation Service, and the major faults and folds are labeled in the companion map (Fig. 15.14). Most of the faults are considered to be nearly strike-slip in nature. The majority of faults were initiated during late Cenozoic times, and some currently are active. The most prominent of these faults, the San Andreas fault, is a nearly continuous zone of fracture(s) whose trace includes discrete fractures as well as zones, as much as 10 km wide, of subparallel fractures. Seismic activity indicates brittle deformation to depths of 7 to 12 km on essentially vertical fractures. Along the San Andreas system, the crust to the west is moving irregularly northwest-ward, and displacements of at least 300 to 450 km have been recorded for the creation of the Gulf of California during the past 4 to 6 m.y. (94). The rate of 2 to 5 cm/yr indicated from dated offsets is compatible with the angular rotation of 6×10^{-7} degrees/year of the Pacific plate from North America, derived from global considerations.

Fractured rocks, breccia, and gouge in the fault zone weather and erode more readily than the adjacent rocks, so that its trace is commonly marked by broad and shallow troughs, which locally are incorrectly termed the "San Andreas rift." Within the trough, fault landforms such as fault scarps, sluces, sag ponds and shutter ridges, deflected stream channels, and trellis drainage patterns, attest to recent movements. Many geomorphic features are direct expressions of differential lateral movements that cause differential vertical displacements (sags, grabens, horsts and medial ridges, scarplets, linear ridges, and land-slides), with zones of compression (en echelon fold ridges), and tension (en echelon lineaments, sags, and shutter ridges) along strike. Most of the recent displacement is along the San Jacinto fault; its trace southeastward from the Transverse Ranges is apparent almost to the Mexican border (see Fig. 15.13). Differential elevation on the San Andreas fault in the San Bernadino Mountains block during late Quaternary time locally produced a plateau dissected by deep, youthful canyons, as well well as an outwardly tilted erosion surface into the Mojave Desert. Although recent displacements are apparent at the western end, and in its southeastern extension in the Coachella Valley, Pleistocene alluvial fans are not displaced in the trench-like gash through the San Bernadino Mountains. Here, the northern branch of the San Andreas fault has an oblique slip, with the west block moving down and to the northwest.

FIGURE 14.35. (a) Landsat-1 image of the Salton Sea and Peninsula Ranges in southern California. The traces of the San Diego River fault (S.D.R.F.) and Sawtooth Range fault are seen, on the image, to extend across the Elsinore fault without offsets. [Landsat-1 image E-1106-17504-7, November 6, 1972.]

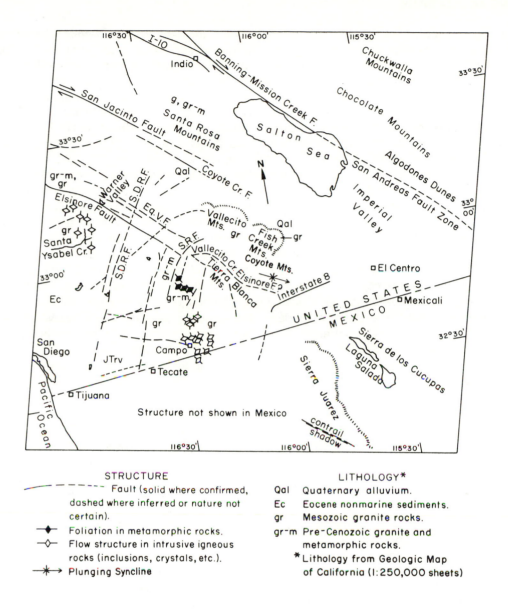

STRUCTURE

– – – – – Fault (solid where confirmed, dashed where inferred or nature not certain).

◆— Foliation in metamorphic rocks.

◇— Flow structure in intrusive igneous rocks (inclusions, crystals, etc.).

※→ Plunging Syncline

LITHOLOGY*

Qal   Quaternary alluvium.
Ec   Eocene nonmarine sediments.
gr   Mesozoic granite rocks.
gr-m Pre-Cenozoic granite and metamorphic rocks.

*Lithology from Geologic Map of California (1:250,000 sheets)

FIGURE 14.35. (*b*) A sketch map of the Salton Sea and Penninsula Ranges in southern California showing aspects of the geology and structure of a diverging zone. [Map after Lowman, 1976 (92).]

That the region encompassing the San Andreas fault system is an active orogenic zone is evidenced by the seismic activity, destructive earthquakes, landslides, subsidence, and the alignment of recent and modern volcanoes in belts of appropriate chemical polarity tholeiitic—calc alkaline—alkalic) inland. The San Andreas fracture system has been interpreted (93) as a transform fault emanating from a spreading center beneath the Salton Sea and the Gulf of California, and the seismic and volcanic belts, and the zones of anomalous crustal movements (uplifts and subsidence, landslides, and so forth) are thought to overlie active plate boundaries (92). Fossil subduction zones are apparent in the blue schist facies rocks and in the obducted ophiolite slabs. Not only are the tectonic basins sites of important oil and gas production, but the serpentinites in the ophiolite complexes also have yielded economic deposits of asbestos. Landsat imagery has been a valuable tool for regional analysis of this structurally complex terrane.

14.4.8 A Tectonic Analysis of Central Asia

One of the last regions to be explored geologically includes the Himalayan Mountains, the Tibetian Plateau, and associated mountain ranges of eastern Russia and China (see Fig. 14.36). Severe earthquakes are known to originate from this region, but the immense size of some structural features (many span political boundaries) and the inaccessible terrain ensure a paucity of geographical and geological knowledge matched only by the coverage of Antarctica and the Amazon Basin. The first accurate geographic map of this region was generated from satellite images, which also provided the base for mapping topographic and structural features. From a study of earthquake epicenters and the first motion data recorded on the worldwide seismic network, a structural and tectonic interpretation (Fig. 14.36) has been made by Molnar and Tapponnier (95) that is compatable with the plate tectonics model.

That the Himalayan Mountains and Hindu Kush chain are a manifestation of collision between a drifting Indian subcontinent and Eurasia was suggested by the early proponents of Continental Drift (96, 97). From a synthesis of geological maps and other structural, stratigraphic, and paleontological considerations such as: the late Cretaceous blocks (about 70 m. y.) in the ophiolite slabs of the Indus

suture; the late Eocene termination of an uninterrupted sequence from Cambrian to Eocene of uplifted shelf and continental slope deposits; the appearance of mammalian fossils about 45 m. y. ago; and the initiation of major mountain building during the Oligocene (35 m. y.), Molnar and Tapponnier (95) have placed the initial collision between India and Eurasia at 40 to 60 m. y. ago. Since then, if an average spreading rate of 5 cm/yr is maintained, India would have traveled northward some 2000 km into Asia, dissipating energy in seismic activity and cataclastic fracturing that extends some 3000 km north and east of the Himalayas (see Fig. 14.36), by mountain building, and "crustal thickening" by piling up successive thrust slices with the oldest to the north. Crustal shortening of some 300 km is apparent in the Tien Shan mountain ranges, which were elevated 30 to 40 m. y. ago; the same probably is true for the Nan Shan mountain ranges in China (95), but this is insufficient to account for 2000 km of continental crust "consumed" in the collision, even if another 700 km of crustal shortening has taken place in the Himalayas and Tibet.

A solution to the crustal shortening problem is suggested from an analysis of satellite imagery. By combining fault plane solutions with observed displacements following earthquakes, and the sense of separations from drag features and offsets mapped on Landsat imagery (see Fig. 14.37), Molnar and Tapponnier (95) were able to map the sense of motion on the main faults (see Fig. 14.36). Fault plane solutions on 75 earthquakes in this region show a predominance of strike-slip faults, followed by normal (extension) faults, and thrust (compression) faults in specific areas. In China and central Mongolia, the strike-slip faults are left lateral and exhibit a general east-west strike, whereas in Tien Shan and western Mongolia the faults strike north or northwest and exhibit right lateral motions. These patterns are consistent with a north to northwest compressional axis, and suggest that the northward movements of India are accommodated not only by piling up of crustal material, but also by a mainly eastward lateral shift, with displacement of China by left lateral strike-slip faults towards the Pacific. The Lake Baikal rift system and the Shansi graben system in China parallel the compression directions and are thus more consistent with tensional fracturing in collision tectonics than they are to rifting over a "hot spot" dome.

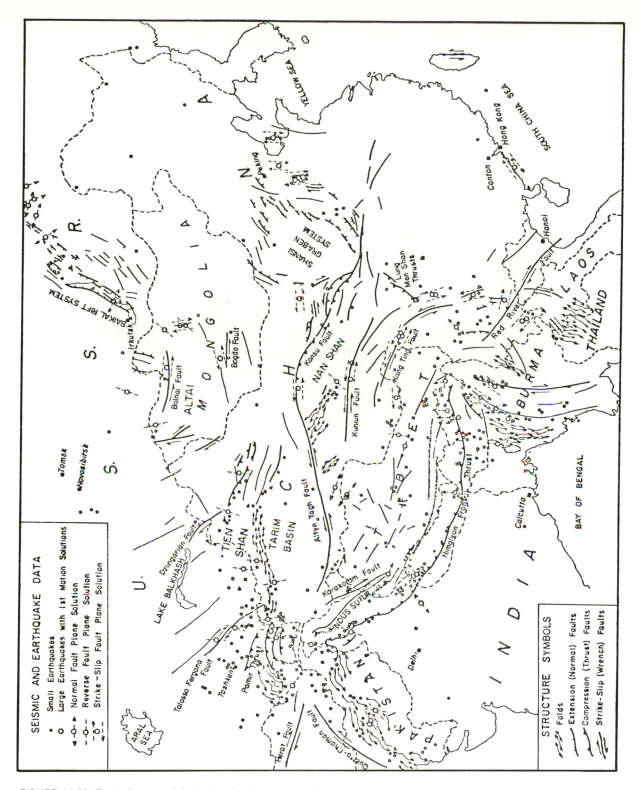

FIGURE 14.36. Tectonic map of Asia, showing the structural features interpreted to be associated with the northward drift of India into Asia during the past 40 to 60 million years (95). From photogeologic studies of folds and faults (see Figure 14.37) that were apparent on satellite images and seismological data (epicenter and magnitude of earthquakes, and sense of slip from first motions) Molnar and Tapponnier (95) were able to construct this tectonic map, and also to deduce an eastward shift of China, along major strike-slip faults such as the Altyn Tagh, to accommodate a suture zone model (no consumption of crustal material into the mantle), rather than a subduction of India beneath Asia. [Map modified after Molnar and Tapponnier, 1977 (95).]

477

FIGURE 14.37. Landsat-1 image of the Tien Shan Mountains in western China, showing an east-southeast trending fault scarp with an apparent left lateral displacement of approximately 50 km. A pediment surface of coalescing alluvial fans slopes southward from the fault for 20 to 30 km to the Tarim River (lower portion of the picture). Another east-west trending fault trace is apparent about 25 to 30 km to the north. Note the northeast trending dune ridges of the Takia Makan Desert in the lower left corner, and the sand dunes along the southern shore of lake Baghrash Kol (upper left). This scene is located north of the "C" in China in Figure 14.36. (Annotated photograph G-74-00196 of Landsat image E-1128-04250, November 28, 1972, released by NASA on October 11, 1973).

14.5 SUMMARY AND CONCLUSIONS

By varying the altitude of the platform carrying a modern sensing system, the Earth's surface can be imaged directly in a number of different wavelengths, on scales available before only by mosaicking or by compiling mesoscopic scale field data on a map. Even if spectral signatures cannot be developed to map the lithostructural units, some inferences can be made on the bedrock state and attitude, from an analysis of the landforms, particularly if large features are involved. These small scale orthoimages have added a new dimension to the analysis of structural and tectonic features, because each scale contains information best observed or sensed on that scale. They have given the geologist the ability to look at and to map directly on a macro-, mega-, and even gigascopic scale.

There has been a fortuitous blending in time between a unified theory of earth movements (plate tectonics) and the vehicle for mapping the results (remote sensing) on the appropriate scale. Plate tectonics theory provides a mechanism and an origin for structures and their tectonic setting in space and time, as well as an association of ore deposits of specific types in specific tectonic settings. Fortunately, many of these "settings" can be identified efficiently and quickly on remotely sensed data: Landsat images have been used as a mapping base to augment structural and tectonic mapping and analysis in active regions, and as an exploration tool for seeking the favorable "fossil tectonic settings" for ore deposition during bygone cycles. Some interplate lineaments, apparent on satellite imagery, may represent the continental expression of fossil transform faults, or megafractures periodically reactivated by reversals in plate motions, or the trail left by plate migration over a mantle plume.

Most dynamic analyses probably will be done by comparison of structural style and tectonic setting with a known terrane, or by analogy with predicted settings from conceptual plate tectonics models. These interpretations will be no better than the accuracy and validity of the model, or the significance of the analogy. Because of the complex mechanical relationships and boundary conditions that are difficult to specify, dynamic interpretations from two-dimensional remotely sensed data are as likely to be misapplied as they are to be properly used.

REFERENCES

1. Brock, B. B., 1960, A philosophy of mineral exploration: Optima, v. 10, no. 3, p. 143-158.
2. Carey, S. W., 1962, Scale of geotectonic phenomena: Geol. Soc. India, v. 3, p. 97-105.
3. Turner, F. J., and Weiss, L. E., 1963, Structural analysis of metamorphic tectonites: New York, McGraw-Hill, 545 p.
4. Melton, F. A., 1959, Aerial photographs and and structural geology: Jour. Geology, v. 67, no. 4, p. 351-370.
5. Hills, E. S., 1961, Morphotectonics and the geomorphological sciences with special reference to Australia: Quart. Jour. Geol. Soc., London, v. 117, no. 465, p. 77-89.
6. Hobbs, W. H., 1911, Repeating patterns in the relief and in the structure of the land: Bull. Geol. Soc. Amer., v. 22, p. 123-176.
7. Sonder, R. A., 1947, Discussion of "shear patterns of the Earth's crust": F. A. Vening Meinesz. Amer. Geophy. Union Trans., v. 28, p. 939-945.
8. Lattman, L. H., 1958, Technique of mapping geologic fracture traces and lineaments on aerial photographs: Photogrammetric Eng., v. 24, no. 4, p. 568-576.
9. Brock, B. B., 1957, World patterns and lineaments: Trans. Geol. Soc. South Africa, v. 60, p. 127-160.
10. Kowalik, W. S., and Gold, D. P., 1976, The use of LANDSAT -1 imagery in mapping lineaments in Pennsylvania: Proc. First Int. Conf. on the New Basement Tectonics, Utah Geol. Assoc. Pub. no. 5, p. 236-249.
11. Wise, D. U., 1968, Regional and sub-continental sized fracture systems detectable by topographic shadow techniques, in Conf. on Research in Tectonics (Kink Bands and Brittle Deformation), Baer, A. J., and Norris, D. K., eds.: Geol. Surv. Canada, Paper 68-52, p. 175-199.
12. Hills, E. S., 1956, A contribution to the morphotectonics of Australia: Jour. Geol. Soc. Australia, v. 3, p. 1-15.
13. Strahler, A. N., 1963, The earth sciences: New York, Harper, 681 p.
14. Gold, D. P., Alexander, S. S., and Parizek, R. R., 1974, Application of remote sensing to natural resource and environmental problems in Pennsylvania: Earth & Mineral Sciences Bull., The Pennsylvania State University, v. 43, no. 7, p. 49-53.

15. Hardy, R. H., Jr., 1971, Rock mechanics — A kit of tools for the structural geologist in Appalachian structures: origin, evolution, and possible potential for new exploration frontiers: West Virginia Geol. Surv., p. 257-312.

16. Griffiths, J. C., 1961, Measurement of the properties of sediments: Jour. Geol., v. 69, p. 487-498.

17. Twidale, C. R., 1971, Structural landforms: Cambridge, Mass., M.I.T. Press, 247 p.

18. Oberlander, T., 1965, The Zagros streams: A new interpretation of transverse drainage in an orogenic zone: Syracuse Geogr. Ser. no. 1, 168 p.

19. Podwysocki, M. H., 1974, An analysis of fracture trace patterns in areas of flat-lying sedimentary rocks for the detection of buried structure: NASA: Goddard Space Flight Center, Document X-923-74-200, 67 p.

20. Ramsay, J. G., 1964, The uses and limitations of beta-diagrams and pi-diagrams in the geometrical analysis of folds: Quart. Jour. Geol. Soc. London, v. 120, p. 435-454.

21. Ramsay, J. G., 1967, Folding and fracturing of Rocks: New York, McGraw-Hill, 568 p.

22. Short, N. M., Lowman, P. D., Jr., Freden, S. C., and Finch, W. A., Jr., 1976, Mission to Earth: Landsat views the world: NASA Pub. SP-360, 459 p.

23. Buddington, A. F., 1959, Granite emplacement with special reference to North America: Bull. Geol. Soc. Amer., v. 70, p. 671-747.

24. Lowman, P. D., Jr., 1976, A satellite view of diapiric archean granites in western Australia: Jour. Geol., v. 84, p. 237-238.

25. Lovering, T. S., and Goddard, E. N., 1950, Geology and ore deposits of the Front Range, Colorado: U. S. Geol. Survey, Prof. Paper 223, 319 p.

26. Carr, M. H., 1976, The volcanoes of Mars: Scientific American, v. 234, no. 1, p. 34-43.

27. Black, R., and Girod, M., 1970, Late Paleozoic to recent igneous activity in West Africa and its relationship to basement structure, in African magmatism and tectonics, Clifford, T. N., and Gass, I. G., eds.: Edinburgh, Oliver and Boyd, p. 185-210.

28. Hills, E. S., 1963, Elements of structural geology: New York, Wiley, 483 p.

29. Vening Meinesz, F. A., 1947, Shear patterns in the Earth's crust: Trans. Amer. Geophys. Union, v. 28, p. 1-61.

30. Sonder, R. A., 1956, Mechanik der Erde, Stuttgart.

31. Carey, S. W., 1962, Folding: Jour. Alberta Soc. Petrol. Geologists, v. 10, p. 95-144.

32. O'Driscoll, E. S., 1962, Experimental patterns in superimposed similar folding: Jour. Alberta Soc. Petrol. Geologists, v. 10, p. 145-167.

33. Whitten, E. H. T., 1966, Structural geology of folded rocks: Chicago, Rand McNally, 663 p.

34. Knowles, R., and Opdyke, N., 1968, Paleomagnetic results from the Mauch Chunk Formation: A test of the origin of the curvature in the folded Appalachians of Pennsylvania: Jour. Geophys. Res., v. 73, p. 6515-6526.

35. Gold, D. P., and Voight, B., 1966, The Mecatina Basin, eastern Quebec: astrobleme or interference fold depression: Absts. Geol. Soc. Amer., Northeast Section, p. 30-31.

36. Gold, D. P., Aitken, F. K., Voight, B., and Genencher, J. J., 1967, Report on the Mecatina structure, in Study of structural and mineralogical significance of meteorite impact craters, including mineral paragenesis, high pressure polymorphs, microfractures, and quartz lamellae: Fourth Annual Rept., NASA, p. 6-15.

37. Podwysocki, M. H., Gunther, F. J., and Blodget, W. H., 1977, Discrimination of rock and soil types by digital analysis of LANDSAT data: NASA-Goddard Space Flight Center Document X-923-77-17, 37 p.

38. O'Leary, D. W., Friedman, J. D., and Pohn, H. A., 1976, Lineaments, linear, lineation: some proposed new standards for old terms: Geol. Soc. Amer. Bull., v. 87, p. 1463-1469.

39. Hobbs, W. H., 1904, Lineaments of the Atlantic border region: Geol. Soc. Amer. Bull., v. 15, p. 483-506.

40. Hobbs, W. H., 1912, Earth features and their meaning: New York, Macmillan, 506 p.

41. Sonder, R. A., 1938, Die lineamenttektonik und ihre probleme: Eclogae Geol. Helvetiae, v. 31, p. 199-238.

42. Popelar, J., 1972, Gravity interpretation of the Sudbury area: Geol. Assoc. Canada, Spec. Pub. no. 10, p. 103-124.

43. Dennis, J. G., ed., 1967, International tectonic dictionary — English terminology: Amer. Assoc. Petroleum Geologists, Mem. 7, 196 p.

44. Barbier, E., and Fanelli, M., 1975, Attempts at correlating Italian long lineaments from Landsat-1 satellite images with some geological

phenomena: possible use in geothermal energy research: Proc. NASA Earth Resources Survey Symposium, v. I-B, p. 1079-1086.

45. Norman, J. W., Price, N. J., and Peters, E. R., 1977, Photogeological fracture trace study of controls of kimberlite intrusion in Lesotho basalts: Trans. Inst. Min. & Metal. Tec. B, v. 86 , p. B78-B90.

46. Gwinn, V. E., 1964, Thin-skinned tectonics in the Plateau and northwestern Valley and Ridge provinces of the central Appalachians: Geol. Soc. Amer. Bull., v. 75, p. 863-900.

47. Gwinn, V. E., 1967, Lateral shortening of layered rock sequences in the foothills regions of major mountain systems: Earth & Mineral Sciences Bull., The Pennsylvania State University, v. 36, no. 7, p. 1-7.

48. Rogers, J., 1970, The tectonics of the Appalachians; New York, Wiley-Interscience, 271 p.

49. Bloomfield, K., and Cepeda-Davila, L., 1973: Oligocene alkaline igneous activity in NE Mexico: Geol. Mag., v. 110, no. 6, p. 551-555.

50. Gold, D. P., 1967, Alkaline ultrabasic rocks in the Montreal area, Quebec, in Ultramafic and related rocks,: Wyllie, P.J., ed., New York, Wiley Interscience, p. 288-302.

51. Marsh, J. S., 1973, Relationship between transform directions and alkaline igneous rock lineaments in Africa and South America: Earth and Planet. Sci. Letters, v. 18, p. 317-323.

52. Heyl, A. V., 1972, The 38th parallel lineament and its relationship to ore deposits: Econ. Geol., v. 67, p. 879-894.

53. Werner, E., 1976, Photolineament mapping in the Appalachian plateau and continental interior geological provinces — A case study: Proc. Fifth Annual Remote Sensing Conf., Univ. of Tennessee Space Inst. Tullahoma, p. 403-417.

54. Wheeler, R. L., Mullennex, R. H., Henderson, C. D., and Wilson, T. H., 1974, Major, cross-strike structures of the central sedimentary Appalachians: Progress report: Proc. West Virginia Acad. Sci., v. 46, p. 196-203.

55. Finch, W. I., and Wright, J. C., 1970, Linear features and ground-water distribution in the Ogallala formation of the southern High Plains, in The Ogallala aquifer: a symposium,: Mattox, R. B., and Miller, W. D., eds.: Texas Tech. Univ. Int. Center for Arid and Semi-arid Land Studies, Spec. Rept. no. 39, p. 49-57.

56. Reeves, C. C., Jr., and Parry, W. T., 1969, Age and morphology of small lake basins, Southern High Plains, Texas and eastern New Mexico: Texas Jour. Sci., v. 209, no. 4, p. 349-354.

57. Reeves, C. C., Jr., 1970, Drainage pattern analysis, Southern High Plains, West Texas and eastern New Mexico, in The Ogallala aquifer: a symposium Mattox, R. B., and Miller, W. D., eds.: Texas Tech. Univ. Int. Center for Arid and Semi-arid Land Studies, Spec. Rept. no. 39, p. 58-71.

58. Parizek, R. R., 1975, On the nature and significance of fracture traces and lineaments in carbonate and other terranes, in Karst hydrology and water resources, v. I: Proc. of the U. S.-Yugoslavian Symposium, Dubrovnik, 1975, p. 3-1 to 3-62.

59. Krohn, M. D., 1976, Relation of lineaments to sulfide deposits and fractured zones along Bald Eagle Mountain, Centre, Blair, and Huntingdon Counties, Pennsylvania: Unpub. M. S. Thesis, Dept. of Geosciences, The Pennsylvania State University, 104 p.

60. Siegal, B. S., 1977, Significance of operator variation and the angle of illumination in lineament analysis on synoptic images: Modern Geology, v. 6, p. 75-88.

61. Offield, T. W., 1975, Line-grating diffraction in image analysis: Enhanced detection of linear structures in ERTS images, Colorado Front Range: Modern Geology, v. 5, p. 101-107.

62. Pohn, H. A., 1970, Remote sensor application studies progress report, July 1, 1968 to June 30, 1969: Analysis of images and photographs by a Ronchi grating: U. S. Geol. Surv. Rept., PB 197-101, 9 p.

63. Podwysocki, M. H., and Lowman, P. D., Jr., 1974, Fortran IV programs for summarization and analysis of fracture trace and lineament patterns: NASA Goddard Space Flight Center Document X-644-74-3, 39 p.

64. Gay, S. P., Jr., 1973, Pervasive orthogonal fracturing in Earth's continental crust: Tech. Pub. No. 2, American Stereo Map Co., Salt Lake City, Utah, 124 p.

65. Gold, D. P., Parizek, R. R., and Alexander, S. S., 1973, Analysis and application of ERTS—1 data for regional geological mapping, in Symposium of significant results obtained from the Earth Resources Technology Satellite-1: NASA Document SP-327, p. 231-245.

482   REFERENCES

66. Wheeler, R. L., Trumbo, D., Mullennex, R., Henderson, C. D., and Moore, R., 1976, Field study of the Parsons lineament, Tucker Co., West Virginia: Abst., Geol. Soc. Amer., v. 8, no. 2, p. 298.

67. Canich, M. R., 1976, A study of the Tyrone-Mount Union lineament by remote sensing techniques and field methods: Unpub. M. S. Thesis, The Pennsylvania State Univ.

68. Lattman, L. H., and Parizek, R. R., 1964, Relationship between fracture traces and the occurrence of ground-water in carbonate rocks: Jour. Hydrol., v. 2, p. 73-91.

69. Parizek, R. R., 1967, Application of fracture traces and lineaments to ground-water prospecting, in Field guide to lineaments and fractures in central Pennsylvania, Second international conference on the new basement tectonics, Gold, D. P., and Parizek, R. R., eds.: Department of Geosciences, The Pennsylvania State University, p. 38-59.

70. Parizek, R. R., 1976, Lineaments and ground-water, in Interdisciplinary application and interpretations of EREP data within the Susquehanna River basin, McMurtry, G. T., and Petersen, G. W., eds.: SKYLAB EREP Investigation #475, NASA Contract No. NAS 9-13406, ORSER-SSEL, The Pennsylvania State University, p. 4-59 to 4-86.

71. Werner, E., 1975, Long lineaments in southeastern West Virginia: Proc. West Virginia Acad. Sci., v. 47, no. 2, p. 113-118.

72. Gold, D. P., and Krohn, D. M., 1976, Stop 12—Bald Eagle Road cut, in Field guide to lineaments and fractures in central Pennsylvania, Second international conference on the new basement tectonics, Gold, D. P., and Parizek, R. R., eds.: The Pennsylvania State University, p. 70-72.

73. Dence, M. R., 1966, Shock zoning at Canadian craters: petrography and structural implications, in Shock metamorphism of natural materials: Baltimore, Mono Book Corp., p. 169-184.

74. Griffiths, J. C., 1966, Exploration for natural resources: Operations Research, v. 14, no. 2, p. 189-209.

75. Guild, P. W., 1971, Metallogeny: A key to exploration: Mining Engineering, v. 23, no. 1, p. 69-72.

76. Wilson, J. T., 1965, A new class of faults and their bearing on continental drift: Nature, v. 207, p. 343-347.

77. Strong, D. F., ed., 1976, Metallogeny and plate tectonics: Geol. Assoc. Canada, Spec. Paper no. 14, 660 p.

78. Garson, M. S., and Shalaby, I. M., 1976, Precambrian-Lower Paleozoic plate tectonics and metallogenesis in the Red Sea region: Geol. Assoc. Canada, Spec. Paper no. 14, p. 573-596.

79. Mitchell, A. H., and Bell, J. D., 1973, Island-arc evolution and related mineral deposits: Jour. Geol., v, 81, no. 4, p. 381-405.

80. Rowan, L. C., 1975, Application of satellites to geologic exploration: Amer. Scientist, v. 63, no. 4, p. 393-403.

81. Gol'braikh, I. G., Zabaluyev, V. V., Lastochkin, A. N., Mirkin, G. R., and Reinin, I. V., 1968, Morfostrukturnye metody izucheniya tektoniki zakrytykh platformennykh neftegazonosnykh oblastei (Morphostructural methods for the study of tectonics in covered platform oil and gas bearing regions), NEDRA, 151 p.

82. Landes, K. K., 1975, Petroleum geology: Huntington, New York, Robert E. Kreiger Pub. Co., 2nd ed. 445 p.

83. Saunders, D. F., 1969, Airborne sensing as an oil reconnaissance tool in Unconventional methods in exploration for petroleum and natural gas Heroy, E. B., ed.: Southern Methodist Univ. Press, p. 105-125.

84. Bench, B. M., 1948, Discovery of oil structures by aerial photography: Oil & Gas Jour. Aug. 26, 1948, p. 98-100, and 146, 150, 152.

85. Gol'braikh, I. G., Zabaluyev, V. V., and Mirkin, G. R., 1968, Tectonic analysis of megajointing: a promising method of investigating covered territories: Internat. Geol. Rev., v. 8, no. 9, p. 1009-1016.

86. Podwysocki, M. H., and Gold, D. P., 1974, The surface geometry of inherited joint and fracture trace patterns resulting from active and passive deformation: NASA-Goddard Space Flight Center Document X-923-74-222, 38 p.

87. Norman, J. W., 1976, Photogeological fracture trace analysis as a subsurface exploration technique: Trans. Inst. Mining & Metal. sec. B., v. 85, p. B52-B62. Discussion on above paper, sec. B, v. 86, p. B58- B60.

88. Morgan, W. J., 1972, Deep mantle convection plumes and plate motions: Amer. Assoc. Petrol. Geol. Bull., v. 56, no. 2, p. 203-213.

89. Burke, K. C., and Wilson, J. T., 1976, Hot spots on the Earth's surface: Scientific American, v. 235, no. 2, p. 46-57.

90. McKenzie, D. P., and Morgan, W. J., 1969, Evolution of triple junctions: Nature, v. 224, p. 125-133.

91. Cloos, H., 1939, Hebung-Spaltung-Valkanismus: Geol. Rundsch., v. 30, no. 4a, p. 405-527.

92. Lowman, P. D., Jr., 1976, Geologic structure in California: three studies with LANDSAT-1 imagery: California Geology, v. 29, no. 4, p. 75-81.

93. Elders, W. A., Rex, R. E., Meidav, T., Robinson, P. T., and Biehler, S., 1972, Crustal spreading in southern California: Science, v. 178, p. 15-24.

94. Dibblee, T. W., Jr., 1966, Evidence for cumulative offset on the San Andreas fault in central and northern California: Calif. Div. of Mines and Geology, Bull. 190, p. 375-384.

95. Molnar, P., and Tapponnier, P., 1977, The collision between India and Eurasia: Scientific American, v. 236, no. 4, p. 30-41.

96. DuToit, A. L., 1937, Our wandering continents: Edinburgh, Oliver and Boyd, 366 p.

97. Carey, S. W., 1958, Continental drift: a symposium: Hobart, Tasmania, University of Tasmania Press, 363 p.

CHAPTER 15

GEOMORPHOLOGY
PAUL D. LOWMAN, JR.
LAWRENCE LATTMAN

15.1 INTRODUCTION

The geologic value of orbital photography, or orbital remote sensing in general, was recognized early in the Space Age. However, despite the fact that some of the best early orbital studies were done by geographers (1), there has been relatively little application to date of orbital remote sensing to geomorphology. This chapter illustrates several specific fields of geomorphic education and areas of potential research, and highlights the unique advantages of orbital remote sensing for the study of geomorphic processes in physiography. Use of aerial photography in geomorphology studies is not addressed here; the reader is referred to numerous standard texts.

15.2 REGIONAL PHYSIOGRAPHY

Probably the most direct application of orbital imagery is the objective depiction of the physiography of large regions, a task traditionally carried out by a few highly skilled scientist-artists, of whom the late Erwin Raisz and A. K. Lobeck were perhaps the best known. Raisz sketched his striking physiographic diagrams from aerial photographs and photomosaics, producing maps of states, regions, and entire countries that are familiar features of many geology and geomorphology textbooks. Orbital imagery, either as mosaics or individual pictures from extremely high altitudes, has proved remarkably useful for showing regional physiography, as the following three examples demonstrate.

Figure 15.1(2) is a Landsat mosaic of 42 color-composite images covering the state of California and adjacent Nevada. An index map (Fig. 15.2) shows the major physiographic and structural features actually visible on it; a few well-known but concealed features, such as the Newport-Inglewood fault, are also included. Because of its great size and degree of Cenozoic tectonism, California has several unusually well-defined physiographic provinces, although on physiographic maps of the entire United States, such as those of Fenneman, the state is generally subdivided into only three of the standard provinces. The Raisz diagram of the entire United States also suggests such a generalized subdivision. However, the Landsat mosaic, because of its high resolution and the inherent objectivity of photographs (or electronically-produced images), shows clearly that the California Coast Ranges, for example, are strikingly different geomorphically from the Peninsular Ranges. The prominent role played by Cenozoic faulting in producing the various physiographic boundaries is obvious from the mosaic, although some faults (for example, the Garlock and San Andreas) have been inadvertently accentuated by the mosaic construction. One obvious advantage of the mosaics over the best hand-drawn maps is the clarity with which they show the influence of geomorphology on vegetation distribution, through control of precipitation. The rain-shadow effect of the Sierra Nevada is especially obvious; a comparable effect is visible to the south in the Salton trough, just east of the Peninsular Ranges.

Figure 15.3, also a Landsat mosaic, covers the opposite coast of the United States. The band 7 mosaic has been chosen because it has higher contrast between land and water than band 5. The area covered, the central Appalachians and adjacent regions, is unusually suitable for relief depiction with Landsat imagery because much of the topography is normal to the sun azimuth (from the southeast). The physiographic provinces covered include, first, the Coastal Plain (A), the inward boundary of which, the Fall Line, is not directly visible, but can be inferred with some certainty from the sudden narrowing of rivers, such as the Potomac and the Delaware, where they cross it. One of the reasons for the siting of so many major cities on the Fall Line, such as Baltimore, Philadelphia, and Trenton, was the difficulty or impossibility of bridg-

FIGURE 15.1. Mosaic of Landsat images constructed by General Electric Co., Beltsville Photographic Engineering Laboratory (used with permission). Original in color. See Short et al. (2), for discussion.

486

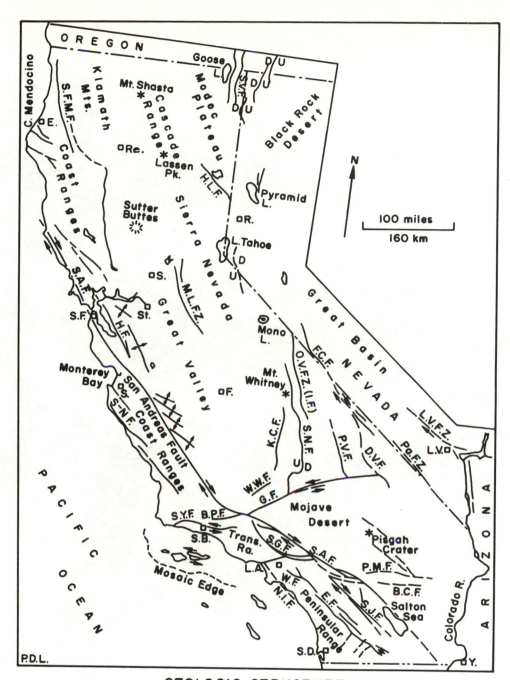

**GEOLOGIC STRUCTURE
CALIFORNIA AND WESTERN NEVADA
Based on General Electric Co. ERTS-I Mosaic
Paul D. Lowman, Jr., Goddard Space Flight Center
November, 1974**

FIGURE 15.2. Map of Fig. 15.1. mosaic. Major faults: S. A. F., San Andreas; E. F., Elsinore; G. F., Garlock; S. J. F., San Jacinto; N. I. F., Newport-Inglewood; W. F., Whittier; S. Y. F., Santa Ynez; S. G. F., San Gabriel; W. W. F., White Wolf; S. N. F., Sierra Nevada; O. V. F., Owens Valley; P. V. F., Panamint Valley; D. V. F., Death Valley; Pa. F. Z., Pahrump Valley; L. V. F., Las Vegas Shear Zone; S-N F., Sur-Nacimiento; K. C. F., Kern Canyon; M. L. F. Z., Mother Lode; S.F.M.F., South Fork Mountain; H. L. F., Honey Lake.

487

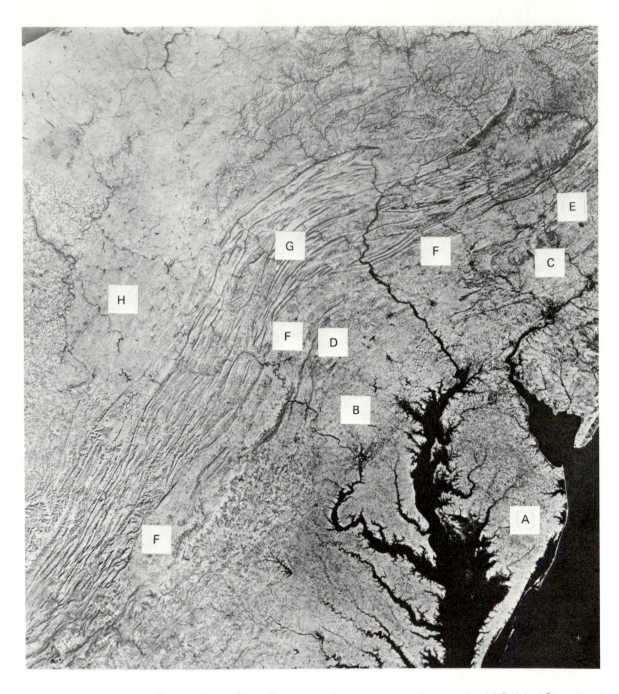

FIGURE 15.3. Soil Conservation Service (United States Department of Agriculture) Landsat mosaic of NE United States, band 7. See text for explanation.

ing such wide rivers in their lower portions, a geomorphic effect well illustrated here. The Piedmont province (B), shows no particular physiographic expression, although many of its individual features, such as the Triassic basalt flows (C), are visible even at this scale on a mosaic, unlike hand-drawn maps of comparable coverage.

Still farther inland, the Blue Ridge (D) in Pennsylvania and Maryland is conspicuous, and its unity with the Reading Prong (E) strongly suggested. The Great Valley (F) is easily distinguished from New York state to southern Virginia. The Ridge and Valley province (G) dominates the mosaic, the high resolution of which makes the folded structure of the Ridge and Valley province quite evident in contrast, for example, to Raisz's physiographic diagram of the United States. Furthermore, the general structural concordance of the Piedmont, Blue Ridge, and Ridge and Valley is also strongly suggested. Finally, the local section of the Appalachian Plateaus (H) is well displayed, although low-quality imagery mars part of the mosaic (left center). The largely dendritic drainage pattern of the plateau region of northern Pennsylvania and southern New York is conspicuous, and the upper reaches of the Susquehanna basin show the subtle continuation of low-amplitude Ridge and Valley folds.

The physiography of a somewhat smaller region, the Adirondack Mountains, is well-displayed by another Landsat mosaic (Fig. 15.4). The Adirondacks are a Precambrian outlier of the Canadian Shield, consisting of intensely folded and metamorphosed rocks of Grenville age. Many lineaments, expressing faults, joint concentrations, and dikes, are obvious. The possible role of the Adirondacks as a buttress against the deformation of the Taconic orogeny is suggested by their relation to the folds and overthrusts of the Green Mountains to the east. The fundamentally tectonic control of the Hudson-Champlain and St. Lawrence Valleys flanking the Adirondacks is also suggested by the mosaic.

The last example, a now well-known color mosaic of the conterminous United States, is reproduced here in black and white (Fig. 15.5). Figure 15.6 is a sketch map identifying the major visible features. The major physiographic provinces and features are generally well displayed, particularly west of the Rocky Mountains. However, the superiority of the Landsat mosaic to Raisz's 1957 map, "Landforms of the United States," is by no means

clear, especially in the low-relief central parts of the country. Features such as the Balcones escarpment of southern Texas, which Raisz depicted by artistic exaggeration, are nearly invisible on the mosaic; the Mescalero escarpment is similarly invisible. Furthermore, low quality imagery in portions of the mosaic introduce artificially hazy patches and, locally, apparent boundaries that have no physiographic meaning. On the other hand, the Landsat mosaic shows albedo differences, generally resulting from vegetation patterns, not shown on the Raisz map. As mentioned in relation to the California mosaic, these may be of some geologic value by calling attention to areas or features with distinctive vegetation patterns. For example, the Wallowa Mountains, prominent on the Landsat mosaic, would never be noticed in a casual examination of the Raisz map. Perhaps the greatest advantage of the mosaic is that it provides a more objective representation of the entire country than can any map. It is clearly a valuable complement to such maps.

15.3 INTERCONTINENTAL COMPARISONS

The orbit of an artificial satellite describes a great circle on the earth's surface, regardless of its inclination or altitude, and thus orbital remote sensing coverage is global (subject to latitude limits). This characteristic, coupled with the typically wide swath of orbital sensors, permits intercontinental comparisons of physiography and structure difficult or impossible with other techniques.

A good example of the use in geomorphology of this intercontinental coverage is provided by the deserts of North Africa and the southwestern United States, specifically the Basin and Range province. It was noticed on early Gemini photographs (Figs. 15.7, 15.8) of North Africa (3) that wind erosion appeared much more effective in sculpturing large areas than conventional geomorphologic opinion would suggest. Detailed examination of these pictures, with reference to some of the lesser-known but authoritative publications on desert morphology, in particular those of H. T. U. Smith, showed that wind erosion in the Sahara is an important, though not necessarily dominant, erosive (and depositional) agent. Re-examination of Gemini and Apollo pictures (Fig. 15.9) color plate 29 (4) of the southwestern United States, however, confirmed the

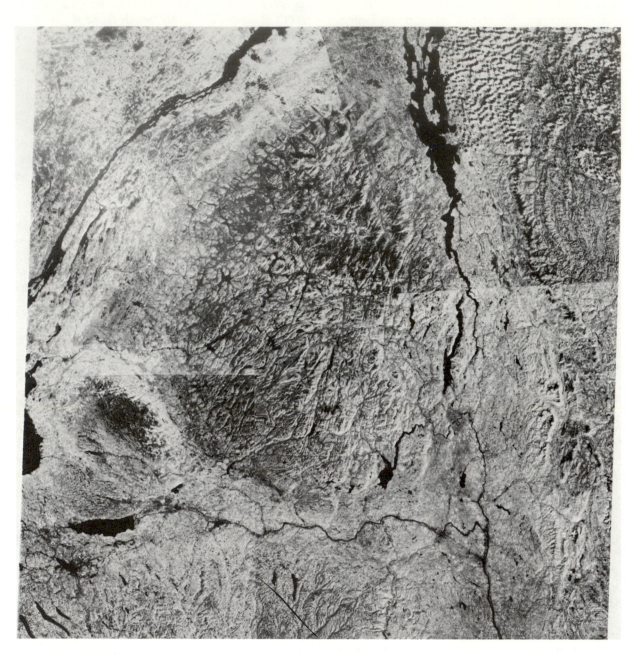

FIGURE 15.4. Landsat mosaic of Adirondack Mountains constructed by New York State Museum and Science Service. St. Lawrence River at upper left.

FIGURE 15.5. Landsat mosaic of United States, MSS band 5, constructed by Soil Conservation Service, exclusive of Alaska and Hawaii.

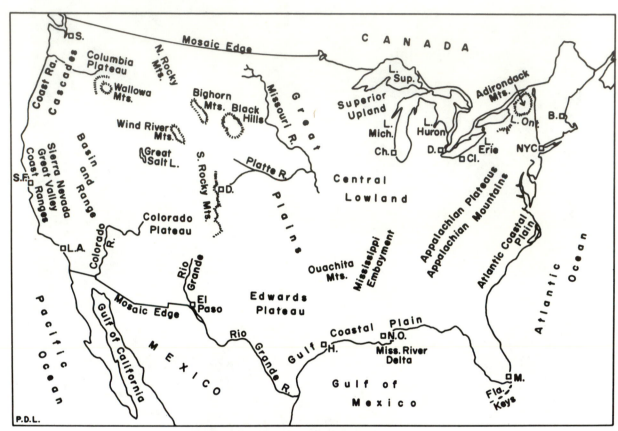

PHYSIOGRAPHIC SKETCH MAP
LANDSAT-1 UNITED STATES MOSAIC

Paul D. Lowman Jr.
Goddard Space Flight Center
1975

FIGURE 15.6. Map of Fig. 15.5. mosaic.

general opinion that fluvial erosion in that area is indeed the major land-sculpturing process. The reason for the great difference between the two continents was easily found with the aid of orbital photographs and other information. The North American deserts are topographically much rougher than those of North Africa, with the prevailing winds interrupted by the north-trending block-faulted mountains of the Basin and Range province (Fig. 15.5). However, in the Sahara there are only a few roughly equidimensional massifs such as the Tibestis, which have little effect on the prevailing winds. In addition, the Sahara is much drier than the North American deserts, a fact also illustrated by the orbital photographs, which show virtually no vegetation in the Sahara.

Another intercontinental comparison can be made between two folded mountain chains at about the same latitude but 150° apart in longitude: the Appalachians and the Zagros Mountains of Iran. The Appalachians have already been illustrated in Fig. 15.3; a closer view is provided by Fig. 15.10 color plate 30 (5). The Zagros Mountains are covered by many orbital pictures, one of the best of which was taken from Apollo 7 (Fig. 15.11) color plate 31. Both mountain belts are composed of more-or-less parallel linear ranges and appear to be the same type of tectonic feature, but the orbital photographs, even when studied with the aid of regional geologic maps, show that from a geomorphological viewpoint they are quite different. The Applachians of today are the result of differential erosion on rocks

FIGURE 15.7. Gemini 4 hand-held 70-mm photograph taken by E. H. White II, in 1965, showing Tibesti Mountains, Libya; view to northwest: Concentric features in foreground are wind-eroded grooves and sand dunes.

FIGURE 15.8. Gemini 7 hand-held 70-mm photograph taken in 1966, showing central Algeria; view to northeast, Ahaggar at upper right. Concentric features in foreground are deflation basins, some occupied by sebkhas (salt flats), formed on Paleozoic sedimentary rock.

of varying resistance, acting on structures dating chiefly from the Paleozoic (6). This results in the frequent occurrence, for example, of synclinal ridges, in which structural lows are topographic highs. The Zagros Mountains, however, are much younger (still, in fact, being actively folded); the topography reflects the structure, ridges being almost invariably anticlines. Part of this difference may be due to the different ages of these two mountain ranges (4), but other factors are doubtless important. In any event, study of the Appalachians and Zagros Mountains through orbital photography should be invaluable to students of geomorphology.

15.4 TOPOGRAPHY AND TECTONICS

The study of the relation between topography and geologic structure is a well established part of geomorphology, and aerial photographs have long been used for this purpose. But to study topography and tectonics (regional structure) with such photographs is difficult because of the small coverage of individual vertical photos. Aerial obliques, such as the classic pictures taken from a light plane by John Shelton (7), are more useful, at least for illustration. However, altitudes of 10,000 m or more provide especially useful coverage, as shown by the following examples.

Southern California, the boundary between two active plates, is an ideal place to study the relation between topography and tectonics by aerial and orbital photography. For the southern part of the San Andreas fault the view from about 5000 m (Fig. 15.12) is impressive, but from Landsat altitudes (Figs. 15.13, 15.14) almost as much detail is shown as in the air view and over several thousand square kilometers of land area. Furthermore, we see little of the distortion inherent in obliques. The regional perspective thus gives a much better idea of how the San Andreas fault compares physiographically with other strike-slip faults of this region (9, 10) and how it is related to the regional topography.

From higher altitudes, a Landsat mosaic (Figs. 15.15 and 15.16) of the entire Southwest (11) shows almost the entire exposed length of the San Andreas fault. Of course there is loss of resolution, but the wider field of view reveals physiographic relations not visible in individual scenes. For instance, the apparent topographic offset on the San Andreas in the San Gabriel-San Bernadino Mountains area is only about 15 or 20 miles, very different from the usually cited estimates of several hundred miles. Work by Baird et al. (12) shows that offset of some bedrock features, such as foliation and lithologic trends, is of the same order. The relative ages of the Sierra Nevada-Peninsular Range normal faulting and the San Andreas system are well illustrated in the mosaic, which suggests that the latter is a younger system diagonally cutting the normal faults making up the east faces of the Sierra Nevada and Peninsular Ranges. This relationship is also supported by independent data, such as the age of the Gulf of California as estimated from magnetic reversals.

15.5 ILLUSTRATING GEOMORPHIC FEATURES

Orbital photographs are especially useful for the great variety of landforms that can be shown in a fairly small number of pictures. A few examples illustrate this, but the reader is also referred to *Mission to Earth: Landsat Views the World* (2), a collection of over 300 Landsat pictures.

Figure 15.17 color plate 32 (2, 13) shows southern Iceland as imaged by Landsat-1. Located on the Mid-Atlantic Ridge, Iceland consists of new crust forming at the present time and spreading from the central rift (right side); thus, the morphology of the youngest part of the Earth's crust may be observed: fissures, lava flows, and faults. The picture also shows a variety of glacial features: snow-covered volcanic mountains, from which issue valley glaciers and numerous valleys carved by glaciers that have since retreated.

Figure 15.18 color plate 33 shows a quite different landscape. Covering part of the Sonoran Desert along the northeast Gulf of California, it illustrates a remarkable variety of landforms on a single picture. Dominating the scene is the Pinacate volcanic field, a remarkable collection of recent volcanoes (many of them maars) and associated flows and pyroclastics. East of the Pinacate volcanic field is a portion of the Basin and Range province, with several well developed sediments. At upper left in the picture are a variety of sand dunes of the Gran Desierto of Sonora, including barchans and transverse dunes, evidently formed by prevailing northerly winds from the Gulf of California. The coastline shows a well developed barrier island, and spits formed by north-flowing long-shore currents.

The Mississippi River valley has been studied by innumerable geomorphologists, and pictures of individual oxbow lakes and meanders are found in all text books, but an Apollo 9 picture (Fig. 15.19) color plate 34 shows dozens of these features at once, in all stages of development. The broad pink swath at left of center represents the Teche Mississippi, the course of the river before about 1000 A.D., a subtle feature of slight topographic relief not shown on the Raisz physiographic maps, and probably not visible on air photomosaics because of its relatively low contrast and great size. However, it is conspicuous in photographs on color infrared film, taken from orbit. [A more detailed description of this picture is given by Lowman (4)].

An equally well-known river valley is the Grand Canyon of the Colorado River, illustrated in Fig. 15.20 color plate 35 by a Landsat-1 image that covers some 10,000 square miles of the Colorado Plateau. Apart from its sheer visual impact, the picture shows numerous geomorphic features. The Grand Canyon is shown as a youthful valley, cut in a structural plain (6) formed on the more-or-less horizontal strata of the Colorado Plateau. Many well-known faults, folds, and monoclines are obvious; for example, the Kaibab Upwarp at center is conspicuous because of its greater elevation and con-

FIGURE 15.12. Air view of San Andreas fault from about 5000 m, looking northwest along fault toward Palmdale; Tehachapi Mountains on horizon at upper right, San Gabriel Mountains at lower left.

FIGURE 15.13. Landsat image, band 7, of Los Angeles area. See Fig. 15.14 for details. Image no. 1090-18012, MSS band 7, 21 Oct. 72.

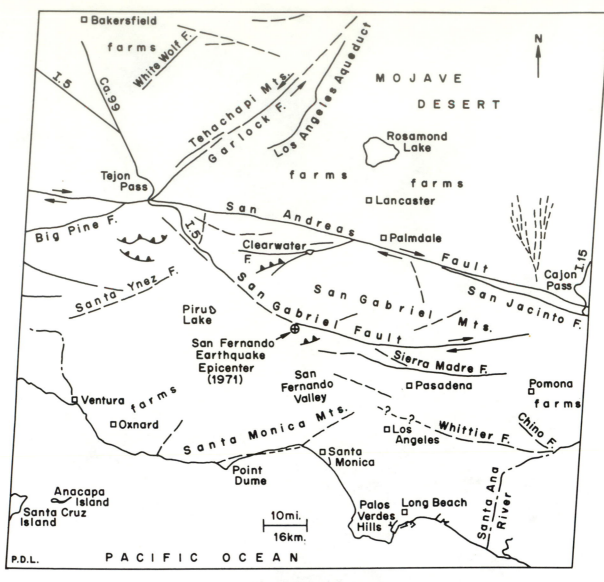

GEOLOGIC SKETCH MAP
TRANSVERSE RANGES, CALIFORNIA
FROM LANDSAT-1 IMAGE 1090-18012

LEGEND:

FAULT (STRIKE-SLIP IF SHOWN WITH ARROWS)

THRUST FAULT (BARBS ON UPPER PLATE)

Paul D. Lowman Jr.
1975

FIGURE 15.14. Map of Fig. 15.13. Nomenclature from Jennings (8).

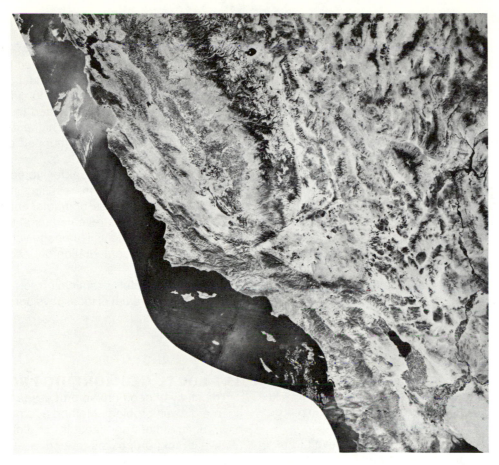

FIGURE 15.15. Section of Soil Conservation Service Landsat mosaic of United States, MSS band 5. See Fig. 15.16 for details.

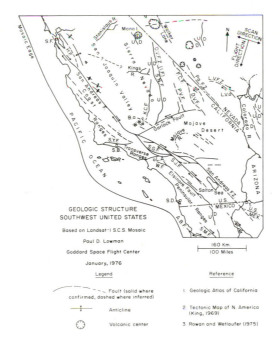

FIGURE 15.16. Map of Fig. 15.15.

sequent heavier vegetation. The evolution of the Colorado River drainage, a long-standing problem, has been studied with Landsat pictures. The report by Goetz et al. (14) includes excellent discussions of Colorado Plateau geomorphology.

Stream piracy is often referred to and described in geomorphologic literature, but there are few if any examples as easy to understand as that shown in Fig. 15.21 (15), a Gemini photograph of the remote Hadramaut Plateau on the southern Arabian Peninsula. The dendritic drainage pattern of the Wadi Hadramaut (lower right) dominates the picture; like the drainage of the Grand Canyon, it is carved in essentially flat-lying sedimentary rocks. However, at the center of the picture is a geomorphically radical disturbance of this pattern: A tributary of the Wadi Hadramaut has abruptly come under structural control (probably a fault) and begun cutting headward rapidly. In doing so, it has begun to behead other tributary valleys, aided in this by similar valleys being cut elsewhere along the supposed fault. Because of the remoteness and great size of the region, this example of stream piracy, first noticed by Karen M. Lowman, would almost certainly never have been noticed.

Various kinds of craters, most believed to be of impact origin, are the most common geomorphic features in the inner solar system (5). For obvious reasons, large terrestrial impact craters are not common, but there are a few comparable to lunar craters such as Copernicus. One, shown in Fig. 15.22, a Skylab 70-mm photograph, is the Manicouagan structure in Quebec, an immense 40-mile-wide ring (now expressed by the lake formed by the damming of the Manicouagan River), generally considered a volcanic feature. However, the weight of evidence, summarized by John Murtaugh (personal communication), favors an impact origin, probably in the late Paleozoic. The picture itself throws no definitive light on the problem, but it will undoubtedly stimulate geologists to wonder how many other circular landforms may be impact features (meteorite craters or astroblemes).

15.6 LOCAL PHYSIOGRAPHY

Although the advantages of the regional view of physiographic features have been stressed, it was not intended to slight local applications of orbital imagery. Many recent papers, too numerous to list, have used orbital imagery as a supplementary map.

For example, in arid regions alluvium of different ages or sources may be distinguished by different tone on one or more Landsat bands. Some local alluvial tonal differences are due to differences in microrelief, which, in turn, is related to age. In the eastern United States, third and higher order drainage lines can commonly be mapped on Landsat imagery.

Future plans for Landsat call for increased spatial resolution, perhaps to 50 m, resulting in increased local detail. In some cases, computer enhancement of Landsat images has revealed local tonal differences related to subtle soil drainage variations and varying degrees of cementation of unconsolidated deposits.

Generally, careful examination of a Landsat image reveals a wealth of local physiographic detail over a large area.

15.7 LOCAL GEOMORPHIC PROCESSES

The study of geomorphic processes has always been a difficult problem. Methods commonly used have been laboratory models, or field study of relatively rapid processes in local areas, such as solifluction, stream activity, windblown sand, and so forth. It has been difficult to study processes over a large region, because detailed maps made repetitively over several decades are extremely few. Some studies have been made of topographic changes, such as channel migration, by repetitive aerial photography. Systematic orbital repetitive imagery began with the launch of Landsat 1 in 1972 and will presumably continue for many years. Thus, for the first time geomorphologists will have the opportunity to locate areas of relatively rapid change easily and to monitor these changes. Of particular interest will be changes due to modern tectonism and volcanism, climatic variation, and the works of man.

Modern tectonic effects on topography may not be readily noted if they occur in remote areas. For example, a new fault scarp cutting alluvium in Nevada was first noted by an airline pilot who regularly flew the same route. If such a feature is sufficiently marked it will appear abruptly on repetitive Landsat imagery and its erosional history may be monitored. Detailed analyses of stream-channel shifting and avulsion possibly due to modern tectonism can also be made. Surging glaciers may be

FIGURE 15.21. Gemini 4 hand-held 70-mm photograph taken in 1965, showing Hadramaut Plateau and Hadramaut Wadi in southern Arabian Peninsula; view to southeast. Note example of stream piracy at right center. See Lowman (15) for discussion.

monitored on Landsat imagery; coastline changes are particularly obvious.

As imagery continues to be obtained for many years, geomorphic processes too slow to be studied easily now will become amenable to analysis. Two additional advances will enhance the usefulness of Landsat imagery for such uses. First, increased spatial resolution of the imagery will be available from later satellites; second, techniques for producing stereoscopic Landsat views exist and their use will enhance the geomorphic application of the imagery, particularly in regions of moderate to high relief.

15.8 REGIONAL GEOMORPHIC PROCESSES

The large area covered by a Landsat image produces a striking potential for the study of geomorphic processes on a regional scale. For example, the effect of man's activities on rivers could be monitored throughout an entire drainage system. The same is true for natural and artificial vegetation changes over large areas. Studies of the geomorphic effects of regional drought conditions can be undertaken with orbital imagery. Undoubtedly many new regional problems will be investigated that heretofore have been difficult or impossible to study.

FIGURE 15.22. Skylab 2 hand-held 35-mm photograph of Manicouagan structure (original in color), 65-km wide, 560 km NE of Quebec City. Structure generally considered largest verified impact structure on Earth, on basis of shock metamorphism and other evidence.

15.9 CONCLUDING REMARKS

The first and perhaps most important advantage of orbital remote sensing is the *global coverage* provided by satellites. Geomorphology is concerned with the processes that sculpture the face of the earth. These processes vary greatly with location and climate, and when combined with the bedrock complexity typical of many regions, result in a variety of landforms that can only be appreciated by seeing all of the earth's land area. Darwin's theory of evolution was stimulated initially by his observation of finches in the remote Galapagos, and it is quite possible that a comprehensive theory for evolution of the earth's surface will result from views of inaccessible areas, provided by satellites. A second advantage is the *wide field of view* characteristic of orbital remote sensing devices. It is now possible for the first time to see large physiographic features and entire provinces hitherto shown only by sketches that are still subjective representations, no matter how skillfully drawn, or by maps that are eclectic abstractions. Photographs or electronic images have their own limitations, but they are objective, and do not depend on skill, knowledge, or biases of the image producer. This property has already made orbital imagery extremely useful in tectonic studies; it may be similarly valuable in geomorphology. Finally, the *repetitive coverage* provided by earth satellites should eventually prove useful in geomorphology, particularly in the study of processes. Landforms themselves change noticeably over the years, even over periods of months. Aerial photography has been used to monitor such changes for years in some areas (coastlines, for example), but orbital photography should make it possible to do this on a regional or even global basis. Collectively, these advantages are likely to make orbital remote sensing as valuable in geomorphology as aerial photography is now.

REFERENCES

1. Morrison, A., and Chown, M. C., 1964, Photography of the western Sahara desert from the Mercury MA-4 spacecraft: NASA Contractor Rept. CR-126.
2. Short, N. M., Lowman, P. D., Jr., Freden, S. C., and Finch, W. A., 1976, Mission to Earth: Landsat views the world: NASA SP-360, 459 p.
3. Lowman, P. D., Jr., and Tiedemann, H. A., 1971, Terrain photogaphy from Gemini spacecraft, final geologic report: Goddard Space Flight Center, Document X-644-71-15, Greenbelt, Maryland, 75 p.
4. Lowman, P. D., Jr., The third planet: terrestrial geology in orbital photographs: Zurich, WELTFLUGBILD Reinhold A. Muller, 170 p.
5. Short, N. M., 1975, Planetary geology: New York, Prentice-Hall, 361 p.
6. Thornbury, W. D., 1969, Principles of geomorphology, 2nd ed.: New York, Wiley, 618 p.
7. Shelton, J. P., 1966, Geology illustrated: San Francisco, California, Freeman, 434 p.
8. Jennings, C. W., ed., 1969, Geologic atlas of California: California Division of Mines and Geology, Sacramento, California.
9. King, P. B., 1968, Tectonic map of North America, scale 1:5,000,000; USGS, Washington, D.C.
10. Rowan, L. C., and Wetlaufer, P. H., 1973, Structural geologic analysis of Nevada using ERTS-1 images: a preliminary report, *in* Proceedings of the symposium on significant results obtained from Earth Resources Technology Satellite-1: NASA SP 327, v. I., p. 413-424.
11. Lowman, P. D., Jr., 1976, Geologic structure in California: three studies with Landsat-1 imagery: California Geology, v. 29, no. 4, p. 75-81.
12. Baird, A. K., Morton, D. M., Baird, K. W., and Woodford, A. O., 1974, Transverse Ranges province: a unique structural-petrochemical belt across the San Andreas fault system: Geol. Soc. Amer. Bull., v. 85, no. 2, p. 163-174.
13. Williams, R. S., Jr., and Carter, W. P., 1976, ERTS-1, A new window on our planet: USGS Prof. Paper 929, 362 p.
14. Goetz, A. F. H., Billingsley, F. C., Gillespie, A. R., Abrams, M. J., Squires, R. L., Shoemaker, E. M., Lucchitta, I., and Elston, D. P., 1975, Application of ERTS images and image processing to regional geologic problems and geologic mapping in Northern Arizona: JPL Tech. Rept. 32-1597, California Institute of Tech., Pasadena, California, 188 p.
15. Lowman, P. D., Jr., 1968, Space panorama: Zurich, WELTFLUGBILD, Reinhold A. Muller, 142 p.

16.1 INTRODUCTION

Glaciers cover about 10% of the land surface, nearly 15 million km² (1). Of this area 85% is in the south polar region, and 1% is mountain glaciers in other latitudes. Former (Pleistocene) glaciers covered as much as 30% of the land, mainly in the high and midlatitudes of the northern hemisphere (Fig. 16.1). Thus about 20% of the land and parts of northern continental shelves show the effects of one or more episodes of glaciation.

Glaciology, the study of ice, is an interdisciplinary field for specialists with initial basic training in geology, geophysics, physics, geography, meteorology, or hydrology. The kinds of ice include snow, lake and sea ice, glaciers, and ground ice (mainly permafrost). Glaciers are important hydrologically, as recorders of climatic changes, and agents of erosion and deposition.

Glacial geology is the study of landforms and sediments resulting primarily from the action of glaciers. Its objectives are the interpretation of the climatic history of the earth during the Quaternary Era, mapping the distribution of glacial landforms and sediments, and determining their mode of origin and source. Most glaciated areas have been mapped at least on a reconnaissance scale. Vertical aerial photographs are used routinely; they often permit resolutions of objects a few centimeters in size, in contrast to most remote sensing devices where resolution is limited to several meters. In order to directly contact glacial materials, sensors commonly must penetrate vegetation as well as a zone of weathered material (soil). Initial experience seems to show that images made by several new techniques give little more information than aerial photo interpretation supported by field studies, hence glacial geologists have been slow to take advantage of recent advances in remote sensing. Glaciers and glaciated areas have genetic features ranging in size from kilometers to fractions of a millimeter. Images that resolve objects 0.5 m in diameter (such as boulders) are best as they provide adequate detail and reasonable areal coverage; smaller objects are likely to be concealed by vegetation. Much useful morphological information may be lost if the instantaneous field of view (IFOV) of the sensor is less than about 6 m as in radar; such imagery is useful if it presents physical properties rather than landforms or provides imagery not available photographically because of cloud cover or darkness. Aerial photographs of a scale 1:65,000 or larger are useful and scales of about 1:20,000 (± 5000) are optimal. Small-scale images such as Skylab photographs, with IFOVs ranging from 15 to 70 m and Landsat images, with an IFOV of approximately 80 m, give synoptic views not obtained by making mosaics of larger scale images, and are invaluable even though they lack detail. Satellite images are made in one instant or in a few moments and have uniform lighting, whereas photographs compiled into mosaics have a variety of lighting conditions that disrupts the continuity of the composite image.

The most used images for studies involving glacial geology are:

1. Conventional vertical aerial photographs with stereoscopic coverage. These give excellent details of landforms, drainage, and vegetation. Panchromatic minus blue photographs are commonly used. Infrared in black and white or in color enhance surface water and soil moisture differences. Color photographs are better than panchromatic black-and-white photographs, but do not generally yield enough extra information to justify the increased cost.

2. Multispectral scanning images (2) are available as Landsat pictures scanned in four bands in all seasons except as limited by cloud cover. Images made after spring snowmelt but before the growing season give the most information. The next best images are taken after the growing

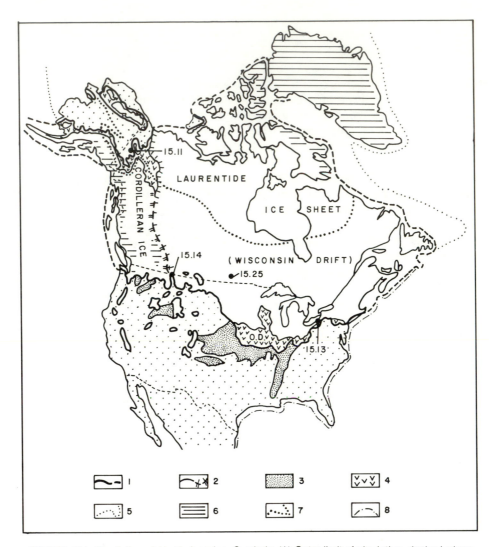

FIGURE 16.1. Glaciation of North America. Symbols: (1) Outer limit of glaciation, dashed where presumed. (2) Boundary between Laurentide ice (single ticks) and Cordilleran ice (double ticks). (3) Loess. (4) Older drifts of pre-Wisconsin age. (5) Outer limit of Pleistocene drift (sea) ice. (6) General areas of present glaciers. (7) Southern limit of continuous or extensive permafrost. (8) Limit of continental shelf exposed by lowering of sea level. The large numbered dots show the locations of Landsat imagery used as illustrations in this chapter. Map generalized from references 24, 34, and several other sources.

season and before snowfall. Images of unforested snow-covered terrain made when the sun elevation is lower than about 13° show features not otherwise visible. High-resolution multispectral scanning images made from aircraft should be useful when available. Multispectral (multiband) photography has been disappointing for geological studies, but multispectral scanning images provide data in a form readily manipulated by computer, facilitating enhancement and ratioing.

3. Thermal infrared images represent ground temperatures and emissivities which can be interpreted to represent soil and rock types, and moisture content (2). If two bands in the middle infrared are recorded at the same time and are suitably processed, maps showing ground emissivity and temperature can be derived. This technique is under development for the detection of massive ice in permafrost.

4. Examples of side-looking airborne radar (SLAR) images show glacial features of moderate and high relief and are useful for the study of glaciers, especially when cloud cover is a problem (2); resolution cell size ranges systematically from 6 to 18 m within a image (2, p. 446). Better resolution is obtained by synthetic-aperture radar (SAR). Few images of low-relief glaciated areas are yet available. Radar images show glacial geology best if made at depression angles of less than 15°, but most available imagery involves larger depression angles.

5. Underwater sidescanning SONAR images have geometry similar to SLAR images requiring interpretation of illumination and shadows, but have much shorter practical range (3). SONAR images of lake and sea floors up to 1500 m wide have revealed morphologic details such as boulders, depressions and mounds, and are used in conjunction with high-resolution seismic profiles that supply stratigraphic information.

6. Passive microwave images in several bands (0.5 to 3.0 cm) which are received from satellites (Tiros, Nimbus, and NOAA) have resolution cells of about 600 m and show some large features of glacial origin (2).

This chapter briefly reviews remote sensing applications in glaciology because glaciologists have advanced far in this field (4, 5), and some of their techniques could be applied to glacial deposits.

Glaciers are discussed and their significance in glacial geology is emphasized. The objectives of glacial geology are explained together with the aspects that might benefit from remote sensing data. These include the identification of glaciated areas, the interpretation of directional features, the mapping of nonsorted drift and stratified drift, and the characteristics of drifts of different ages. Illustrations are limited to aerial photographs and selected portions of Landsat images.

16.2 GLACIOLOGY

The quantity and condition of snow, a major source of soil moisture, and groundwater and stream discharge in high and midlatitudes can be evaluated by remote sensors. The importance of sea ice information for travel in polar regions is obvious, and the design of engineering structures for these regions requires data on ice movements, strengths, and pressures. Similarly, engineering design and environmental protection in permafrost regions requires knowledge of ice masses in the ground. Glaciers contain most of the world's fresh water and knowledge of changes in their mass is of increasing importance. Space limitations preclude full discussion of snow and floating ice here (see Chapter 18), but several techniques that might be applied to geological materials will be mentioned.

16.2.1 Snow

Snow distribution, thickness, mass (water content), and temperature are used to predict flood runoff and water supplies. Snow thickness of less than 10 cm may be interpreted from variations of tone in the visible spectrum. Snow shows well on bands 4 and 5 of Landsat imagery [Fig. 16.14(c)], but a closed forest canopy conceals it (Fig. 16.26). Meier (4) using Landsat images, determined the radiance ratios of red to infrared responses of snow and of forest, then used intervening ratios as a signal of snow in trees. The data were processed to generate a three-toned image showing snow, snow in trees, and forest.

Experiments with radar in snow research (2, p. 1031-4) show that in radar images old snow and firn can be distinguished from fresh snow, but the boundary between fresh snow and bare ground can not be reliably interpreted because radar penetrates fresh snow and more energy is returned from the base of the snowpack than from its surface. The dielectric characteristics of snow that influence the radar return depend on the complexities of its structure and degree of metamorphism.

Thermal IR images have been used to detect crevasses of glaciers through snow, but there is little literature on their use in general snow surveys. Photographic and satellite images seem to be more useful at present.

Passive microwave (1.55 cm) data can be used to determine snow distribution and temperature (6, p. 23-39). In an airborne study around South Cascade Glacier, Meier (in Ref. 4) obtained radiation brightness temperatures clustered around 200°K for dry snow, 255°K for wet snow, and 285°K for snow-free vegetated ground. Microwave resolution is poor (about 2.5° of arc) but average brightness values for each resolution cell give the proportion of the cell covered by snow. Digital summation is possible, making the delineation and measurement of areas unnecessary.

16.2.2 Floating Ice (Sea Ice)

Remote sensing of sea ice is an advanced technology (4). SLAR images are used to determine the character of sea ice and to track its movement for the immediate needs of ship navigation. Sea ice becomes rougher as it ages and so causes more radar backscatter. It is difficult, however, to distinguish smooth young ice from water. Landsat images show major structures of sea ice and reveal regional movements. Sea ice also grows colder as it ages so that thermal infrared images distinguish ice of different ages very well. Laser altimetry is used to measure roughness as small as a few centimeters, and the heights of pressure ridges. The latter may be as high as 3.0 m, and have keels extending as deep as 6.5 times the ridge height as shown by upward-looking underwater SONAR (4, p. 329, 349).

16.2.3 Permafrost

About a quarter of the Earth's land surface is perennially frozen beneath an active layer that freezes and thaws annually (7). The ground may be rocks or sand and gravel cemented by interstitial ice (dry permafrost), that undergoes no change in volume when it thaws; or it may be fine-grained soil rich in silt, that contains ice segregations in the form of lenses, wedges, and irregular masses resulting from the migration of water to a freezing front. The ice may comprise a greater proportion of the soil than the mineral matter, and thawing of this ground

(detrimentally frozen ground) results in local slumping, cavities, and mudflows. In polar regions and high altitudes frozen ground is continuous except under large rivers and lakes. It attains thicknesses of several hundred meters in the coldest regions. Toward the equator permafrost becomes thinner and then discontinuous until there are only sporadic patches a few meters thick (Fig. 16.1).

Surface features diagnostic of permafrost include forms produced during its development as well as forms produced when it melts (thermokarst). Permafrost persists in many places where it is protected by an insulating cover of vegetation; if the cover is destroyed or modified the ground may thaw. Permafrost forms (7) include polygonal ground of various types, stone and vegetation stripes, solifluction lobes (Fig. 16.7), scars of wet snow flows (Figs. 16.4 and 16.5), pingos, palsas, aufeis, rock glaciers, and altiplanation terraces. Melting permafrost is revealed by thermokarst features such as beaded drainage, thaw lakes (Fig. 16.24), collapsed pingos, alases, and residual positive features such as fields of low circular mounds, and isolated flat-topped (haystack) mounds and plateaus. Some thermokarst involves a recycling of topography and not necessarily the complete disappearance of permafrost. Lakes deeper than about 3.0 m do not freeze to the bottom so that the latent heat of fusion of the water blocks the egress of heat from the ground beneath the lake during the winter, and summer thaw predominates. However, sediments entering the lake cause it to shoal so that eventually it does freeze to the bottom and then the winter loss of heat exceeds the summer gain and permafrost develops. Alases are complex basins formed by this cycle (7). Many of these forms can be identified on aerial photographs at scales of 1:20,000, but most patterned ground (stone circles, nets, and stripes) can be seen adequately only at larger scales because lines less than 30 cm wide and circles of a similar diameter must be distinguishable. Stony features should be recognizable by their roughness characteristics (backscatter) on SLAR images (8).

Vegetation patterns represent permafrost distribution in some areas. In peripheral regions many north-facing slopes are frozen and have a pauperized spruce forest or park tundra cover, whereas nonfrozen south-facing slopes have normal boreal forest. Other combinations exist, and interpreters must be aware of the spurious effects of fire and the subsequent vegetation succession.

Some thermokarst areas in Alaska have been identified on enhanced (linear contrast stretched) low-sun Landsat images, and are suggested as analogous to certain features on Mars (9). Permafrost also exists under shallow Arctic seas, and mounds believed to be pingos have been identified in the Beaufort Sea at depths as great as 70 m from seismic reflection data and side-looking SONAR images (10).

In areas of permafrost massive gound ice occurs as lenses, vertical wedges, and horizontal sheets with thicknesses ranging from a few millimeters to several meters and linear dimensions from a few centimeters to many meters. Most are the result of segregation at freezing fronts in silty soil but occasionally aufeis, river, or lake ice is buried under alluvium, and residual glacier ice may persist under a drift cover. Pingos and palsas have cores of massive ice.

Photographic evidence of massive ground ice includes trees tilted by the collapse of supporting soil around the edges of thaw lakes or palsas; cusps formed by ice wedges which melt more slowly than frozen soil and project from low cliffs around lakes or from flat-topped residual mounds; polygonal cracks surrounding low-centered polygons; zig-zag beaded drainage, formed as melting ice wedges link the expansions at the junctions of ice wedges where melting is more rapid; abundant small circular embayments in lakes (Fig. 16.24); and fields of circular low silt mounds 1 or 2 m high and 7 to 10 m in diameter.

Thermal infrared images were used by R. J. E. Brown of the National Research Council of Canada to map palsas in a poorly drained area in northern Manitoba. The massive ice in palsas supports low plateaus standing 1 or 2 m above the surrounding swamp, and is thermally isulated by a thick layer of peat moss and the forest canopy. These islands of permafrost therefore paradoxically give a warm signature in contrast to the surrounding low-lying wet area.

A sophisticated approach to detecting massive ice is attempted by LeSchack and others (in Ref. 2). They use low altitude (230 m) simultaneous predawn thermal IR images in the 4.5- to 5.5-μm and 8-to 12-μm bands made early in the fall. At this time maximum temperatures occur at the ground surface and are strongly influenced by the geothermal gradient. This gradient is disrupted by massive ground ice so that the surface above the ice is cooler than the adjacent areas at this time of year. By scanning in two bands, data are obtained that enable calculation of both emissivity and temperature. Computer processed product imagery greatly enhances polygonal structures only partly visible in the 8- to 12-μm images, and not visible in panchromatic or infrared photographs.

16.2.4 Glaciers

Accurate maps are essential for most glacier studies (11). Thermal and hydrologic characteristics, thickness, volume, mass balance, nature and rate of movement, and response to climatic change are primary subjects of glacier research. Contour maps showing the boundaries of snow, firn, ice, and rock are needed. Parts of glacier margins are concealed under transient snow or under rock debris much or all of the year. Sequential contour maps with intervals 3, 5, or 10 m are used to measure changes in volume (Meier, in Ref. 11). The movements of glaciers, especially those that surge, are monitored on all available imagery including aerial photographs, Landsat, and SLAR. Glacial geologists are concerned with the sources, modes of acquisition, transport, comminution, sorting, and deposition of rock debris and the relationships between shifts of the terminus and climatic changes; for them glacier maps should show the distribution of debris.

The entire surface of a glacier receives mass as snow and loses mass by ablation. The upper part of a glacier, commonly more than 60% of its area, is the accumulation zone [Figs. 16.2(a) and (b)] and has a net mass gain each year (snow), whereas the lower area, the ablation zone, has a net loss of mass (snow and ice). In a steady-state glacier a mass equal to the net gains and losses is transferred by flow from the accumulation zone to the ablation zone and the total mass does not change. The line separating these zones is the equilibrium line. In temperate glaciers the equilibrium line is the average position of the lower boundary of the transient snow at the end of the summer melting season. However, on polar and subpolar glaciers snow is transferred down the glacier by slush avalanches and may freeze to become superimposed ice. This may lower the equilibrium line below the snowline, and so create patches of accumulation zone within the upper ablation zone. The accumulation zone is an area of continual burial of glacier structures, here mainly layering and crevasses. The ablation zone is an area of continual exposure and destruction of structures (foliation, folds, faults, and crevasses) by melting.

The velocity of a glacier is greatest at the equilibrium line and decreases to zero at both ends, but variations in this pattern are caused by differences in slope and cross section. Flow vectors have a downward component with respect to the surface of the glacier in the accumulation zone and an upward component in the ablation zone [Fig. 16.2(b)]. The overall effect is rotation about a hypothetical transverse axis above the equilibrium line. The surface of a glacier is continually in motion in three dimensions, and points on it are in their relative positions only at the instant of observation.

MAPPING

Glacier features commonly mapped (Blachut and Müller in Ref. 11) are shown schematically in Fig. 16.2 and are listed in Table 16.1. Contours are drawn from aerial photographs using high-order photogrammetric instruments. Because of the brilliant reflectance of snow and ice, photographs designed to show details of snow and ice rather than surrounding terrain must be used. Sometimes dye markers are placed on the glacier to provide control points for photogrammetric operations (Østrem in Ref. 11). Photographs using the blue end of the visible spectrum enhance structures and variations in snow, ice, and firn, and record details in areas of shadow which are illuminated by diffuse skylight.

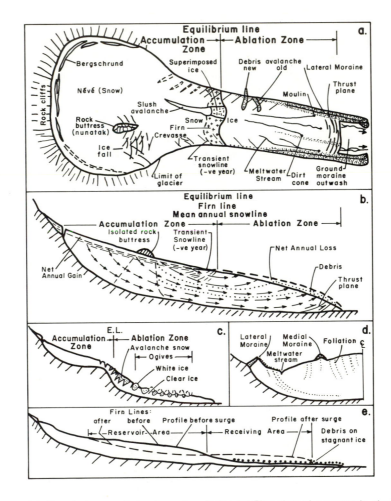

FIGURE 16.2. Terminology and structure of a glacier. (a) Plan view of a mountain glacier with a negative mass balance. (b) Longitudinal section near the axis of (a); arrows are flow vectors, light dashed lines are sedimentary layers, and dotted lines show the paths of debris derived from the isolated rock buttress. (c) Section showing the formation of ogives below an ice fall. (d) Transverse section in the ablation zone showing foliation and the nature of moraines. (e) Profiles of a surging glacier.

TABLE 16.1. GLACIER FEATURES OBSERVED IN BLACK-AND-WHITE PHOTOGRAPHS AND OTHER IMAGES.

Situation	Feature	Criteria	Special Techniques
General, mainly upper glacier	Transient snowline	Tone	Color IR photography
	Firn line (boundary between firn and glacier ice)	Form, tone	Color IR photography
	Ice divide	Topography	
	Snow avalanche cones	Form, tone	
	Crevasses	Tone, form, pattern	Thermal IR, SLAR, optical filtering
	Ice pinnacles, seracs	Form	
	Ice cliffs	Form	
Accumulation Zone	Transient snowline	Tone	Color IR photography
	Firn line	Tone	Color IR photography
	Equilibrium line (is approximated by the firn line)	Field data	Not imaged
	Slush avalanches	Tone, pattern	IR photographs
	Bergschrund	Form	Thermal IR
	Sedimentary layering	Tone, pattern	
Ablation zone	Active ice	Form, tone	SLAR; radio-echo fading
	Stagnant (dead) ice	Form, tone	SLAR
	Meltwater: lakes and ponds, streams	Tone	IR photography, thermal IR
	Moulins	Form	
	Foliation banding, blue bands	Tone, pattern	Photography involving blue light; optical filtering
	Ogives	Tone, pattern	Photography involving blue light
	Debris covered ice: medial moraines, dirt cones	Form, tone	Photography involving blue light; SLAR
	Thermokarst depressions: Ice exposed in walls	Form, tone	
	Walls covered by debris	Form	
Detached from main ice mass	Ice-cored moraines	Form	Thermal IR; SLAR
	Rock glaciers	Form	SLAR

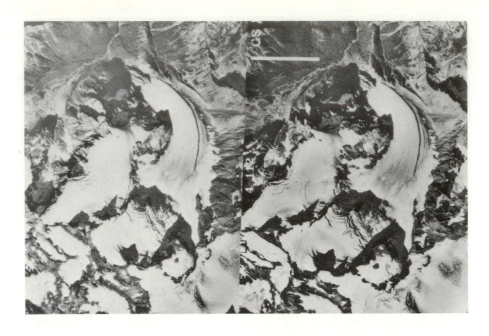

FIGURE 16.3. Stereogram of Blue Glacier, Olympic Mountains, Washington, showing accumulation and ablation zones, ice falls, crevasses, and medial and lateral moraines. White bar is about 1 km long. Photographs courtesy of United States Geological Survey.

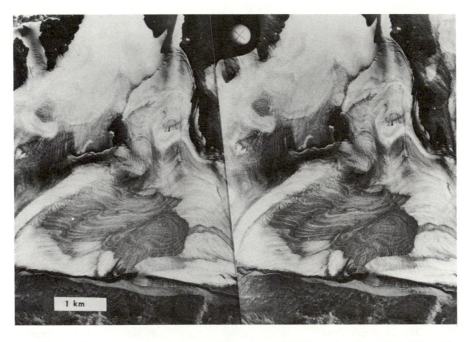

FIGURE 16.4. Stereogram of part of a glacier system SE of Penny Ice Cap, Baffin Island (lat. 66° 20'N, long. 63° 55'W) showing the snowline, slush avalanches (dark lines descending from a dark band in the upper right), and distorted sedimentary layers some of which may be enhanced by dust accumulations. Some layering in the lower area may be superimposed ice. The steep margin is typical of cold glaciers and has receded from a trim line visible in the bottom of the picture. Photographs courtesy of Surveys and Mapping Branch, Department of Energy, Mines and Resources, Canada.

512

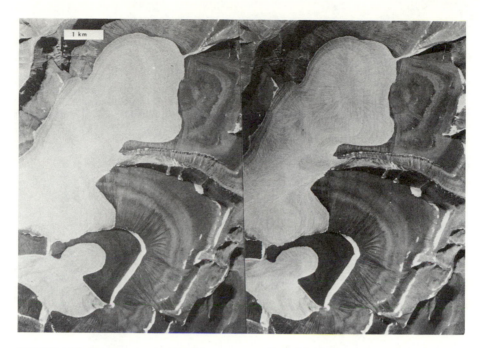

FIGURE 16.5. Stereogram of a small ice cap near Burnell Inlet, Devon Island (lat. 74° 35'N, long, 86° 30'W) showing undeformed sedimentary layering. The fine lines trending downslope across the glacier and adjacent land are the result of runoff during the period of spring snowmelt. A number of perennial snowbanks occupy nivation hollows. Solifluction lobes are visible on the right side, especially where they cross the bedding of bedrock. Photographs courtesy of Surveys and Mapping Branch, Department of Energy, Mines and Resources, Canada.

Infrared black-and-white photographs show water and wet snow well, but sacrifice detail in shadows. Color infrared photographs are excellent for all aspects of glacier mapping (12, 13).

Accumulation Zone

The limits of the accumulation zone are seen best at the end of the summer melting season. It has the lightest tone and the smoothest surface (Figs. 16.3 and 16.4). In mountains the highest permanent snow is frozen to the mountains and is separated from the actively moving part of the accumulation zone by a tension crack, the bergschrund. Snow metamorphoses to firn, a weakly cemented, granular substance resembling cube sugar with a density of 0.55, at a depth of a few tens of meters, depending on temperature and the amount of melting involved. Firn has sedimentary layering (Fig. 16.5). It has low permeability to water and some tensile strength. Crevasses form in firn oblique to the sides of actively flowing zones and across longitudinal convexities in the glacier bed [Fig. 16.2(a)]. They commonly are concealed by snow bridges. Many covered crevasses give distinctive warm signatures on thermal

infrared images (14), but some have cold signatures and others do not show at all.

In the lower part of the accumulation zone impermeable firn is close to the surface, and water in the form of small streams, ponds, and areas of saturated snow (slush) may be visible (Fig. 16.4). Water gives black tones on black-and-white infrared photographs, but also shows well (dark) on most panchromatic photographs because its albedo is low compared with snow and ice. Thermal infrared images show glacier ponds clearly because they are warmer than surrounding snow and ice (Poulin and Harwood in Ref. 11).

The density of firn increases with deeper burial until it becomes glacier ice with a density of about 0.83 at depths ranging from about 30 m in temperate glaciers to more than 100 m in cold glaciers. It initially has seasonal layering that is well preserved in cold glaciers. This is truncated obliquely by the ablation surface below the snowline (Figs. 16.2(b), 16.4, and 16.5). Layering shows because snow saturated during the summer becomes mostly ice and has a lower albedo than ice formed from underlying cold winter snow, which contains trapped air and

FIGURE 16.6. Part of the ablation zone of a glacier west of Tingin Fiord, Baffin Island (lat. 69° 00′N, long. 69° 30′W) showing complex sedimentary layering (left center), crevasse patterns (lower left and center), septa dividing ice streams, ogives (right), and ice-cored terminal and lateral moraines from which the terminus has receded forming a small glacier lake. Photograph courtesy of Surveys and Mapping Branch. Department of Energy, Mines, and Resources, Canada.

scatters light rather than absorbing it (Fig. 16.5). Windborne dust contributes to the darkening of some layers (Fig. 16.4). In slow-flowing cold glaciers (Figs. 16.4 and 16.6) the sedimentary layering may survive into the ablation zone; in temperate glaciers it is commonly destroyed before the firn is exposed at the snowline and the firn may have instead a longitudinal foliation. Sedimentary layering disappears in ice falls (Fig. 16.6).

The altitude of the snowline (the highest position of the lower limit of the accumulation zone) is an important parameter (see mass balance below). For most glaciers available contour maps are adequate for estimating the height of the snowline if it can be located horizontally. Meier and others (12) found that color infrared photographs most clearly discriminate snow with its smooth light tones from old firn which is darker, may retain the sedimentary banding, and often has incipient foliation. Ice is still darker and shows strong foliation. The summer melt line of the Greenland ice sheet, probably not far inland from the equilibrium line, and closely related to the snowline, has been mapped by using brightness temperatures and emissivity criteria determined from Nimbus passive microwave data (4, p. 385).

Ablation Zone

The ablation zone shows structures that originate deep within the glacier. The most pervasive of these is foliation consisting of thin (millimeter and centimeter) bands, each with a grain size distinct from adjacent bands. Overall, foliation theoretically forms a series of nested troughs parallel to the channel boundary, but where exposed it mostly dips vertically or steeply toward the centerline [Fig. 16.2(d)]. It gives a longitudinal texture to the ablation zone (Figs. 16.3 and 16.6) even though individual bands are not resolved except in large scale photographs. Septa consisting of distinctive bands (aggregates) of layers may show the boundary between ice streams flowing at differing velocities (Fig. 16.6).

Drag folds and small faults are common. Normal faults sometimes displace the glacier surface causing discontinuities in foliation. Reverse and thrust faults appear where the slope decreases and also at the terminus where compression dominates. These may displace the surface, but melting is generally more rapid than fault movement during the summer imaging season. Thrust faults near the terminus bring up basal rock debris that spreads out on the ice downslope from the trace of the fault. Bands of clear ice known as blue bands, 2 to 10 cm thick, are transverse to the foliation and dip steeply upglacier. Crevasses are common; where closely spaced they produce a chaotic surface of blocks and pinnacles. Meltwater streams are easily seen on most images but are enhanced on infrared images.

Linear features in the ice often form such a confused jumble of crossing lines that it is useful to study them with optical filtering procedures (Chapters 5 and 6). Bauer and others (15) used coherent light to obtain from a photograph a Fourier spectrum (diffraction pattern) of a crevassed area. The diffraction pattern was masked to eliminate an overwhelming linear trend caused by ridges formed by the ablation of alternating layers of clear and bubbly ice (possibly ogives). The image reconstituted from the filtered Fourier spectrum revealed crevasse patterns not noticed on the original image.

Undulations in the topography of broad, low-relief ablation zones of large, high-latitude glaciers such as Malaspina Glacier in Alaska and ice caps in Iceland, become visible as subtle tonal variations in Landsat images made under low sun illumination when snow cover conceals all other extraneous tonal patterns. Krimmel and Meier (in Ref. 4), enhanced the texture of such features on Malaspina Glacier by slightly offsetting a negative superimposed on a positive image. The wavy patterns and linear features that appear are thought to reflect subglacial bedrock topography.

Below ice falls, glaciers have a pattern of arcuate bands convex down-glacier known as ogives (Fig. 16.6). These bands are surface waves of nearly constant wavelength, with an amplitude that decreases away from the ice fall. The crests of the waves are light (high albedo) and the troughs are dark. The albedo differences are due to bubbly ice (light) and clear ice (dark) [Fig (16.2(c)]. Ogives are spaced one year's flow apart. Their origin is a subject of controversy, and the various theories center on differential ablation of metamorphosed avalanched snow and ice in the compressive zone below the ice fall.

THICKNESS

Glacier thickness traditionally has been measured by seismic and gravity methods. In the last decade radio echo-sounding systems that give continuous profiles of the bedrock floor have been developed (11). Within cold glaciers internal layering and other discontinuities interpreted as morainal debris and brine-soaked zones also are recognized.

Water and a roughness attributed to morainal boulders were detected at the base of the Antarctic ice sheet (Oswald in Ref. 11). In temperate glaciers, the effect of water on the dielectric properties of ice make radio sounding difficult, but low frequency (5 Mhz) systems are proving successful (Watts and others, in Ref. 11).

TEMPERATURE

Surface temperatures of the polar ice sheets have been mapped since 1961 by weather satellites using infrared radiometers in bands ranging from 0.8 to 12.5 μm. The smallest pixel is a circle 660 m in diameter (Gloerson and Salomonson, in Ref. 4). The temperature at a depth of 10 m, unaffected by seasonal variations, is sought as it represents the mean annual temperature of the glacier. Passive microwave emissions depend on emissivities and temperatures of a layer of snow with a thickness of from 10 to 100 times the wavelength (Chang and others, in Ref. 6), complicated by snow crystal size and other factors. The 1.55-cm band is in general use; wavelengths of 2.8 cm and longer eventually should provide information on the upper 10 m of glaciers.

MASS BALANCE

The mass balance of a glacier can be determined accurately only by field measurements of accumulation and ablation at points on representative traverses. Once the regime of the glacier is known for several substantially different years representing both positive and negative budgets, the position of the transient snowline at the end of the melting season becomes a useful indicator of the mass balance. Østrem (in Ref. 11) found a correlation between the altitude of the snowline and the annual specific net mass balance for several glaciers in Norway that supply hydroelectric reservoirs. He identified snowlines on aerial photographs and on Landsat images.

The long-term trend of mass balance of nonsurging glaciers can be judged qualitatively from the relation of ice margins to moraines and trimlines (Figs. 16.3, 16.4, and 16.7), and from the character of the lower ablation zone. If the glacier has sharp contact with the regional vegetation pattern and a moderate slope (Fig. 16.5) the net mass balance is probably zero, an equilibrium condition. Moraines may be accumulating at the margin. If the ice margin is steeply convex and there are tilted trees or pushed-up ridges the net mass balance is positive.

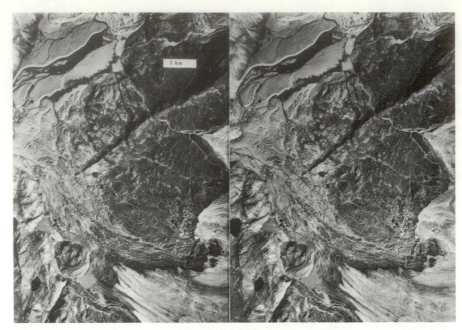

FIGURE 16.7. Stereogram of the stagnant remnants of an outlet glacier from a small ice cap on Baffin Island (lat. 69° 07′N, long. 69° 07′W). The retreating ice, lower right, has left behind ice-cored lateral and end moraines and other remnants which are collapsing as ponds form and accelerate melting. A kame terrace will remain in the lower left. Mounds and ridges (below right center) are probably ice-cored glacier stream deposits (kames). An outwash plain being dissected in the upper left was deposited partly as a delta when glaciers blocked the main valley. Lateral drainage channels now occupied by small streams are outside the ice-cored lateral moraines. Solifluction lobes are abundant in the right center, and solifluction stripes in the top center. Photographs courtesy of Surveys and Mapping Branch, Department of Energy, Mines, and Resources, Canada.

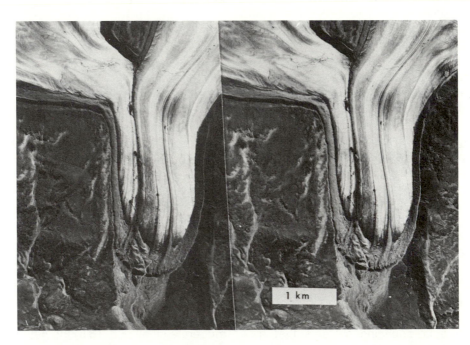

FIGURE 16.8. Stereogram of a glacier outlet tongue adjacent to Fig. 16.4, showing dirt cones, superglacial streams, ogives and a pond (top right), that can be used to monitor glacier flow rates. High ice-cored lateral and end moraines indicate a negative mass balance. Photographs courtesy of Surveys and Mapping Branch, Department of Energy, Mines, and Resources, Canada.

Negative mass balance is shown by a depression between the ice and its end moraine (Figs. 16.6 and 16.8); by high ice-cored lateral moraines (Fig. 16.3); by bare soil or an immature vegetation pattern between the ice and a former position marked by moraines or trimlines (Figs. 16.3 and 16.4); and by residual masses of drift-covered ice or ice-cored moraines (Fig. 16.7) beyond the margin. A concave cross profile in the ablation zone is diagnostic of a negative mass balance. Surging glaciers (Fig. 16.9) cannot be judged by these criteria. Glaciers that have a continuous cover of surficial debris near the terminus commonly have a negative mass balance; however if the cover extends far up the glacier from the terminus the glacier may be quiescent between surges.

All the features described above are easily seen on aerial photographs and many show on SLAR images (4, p. 176, 259). Shifts of the ice margin greater than 100 m (one Landsat pixel) and debris covered terminal zones can be seen on Landsat images.

Although the forgoing phenomena enable qualitative appreciation of the trends of glacier mass balance, field methods still are required for quantitative information.

GLACIER FLOW

Velocities of normal glaciers are greatest at the equilibrium line, and the center of a valley where the ice is thickest. Typical maximum velocities range from 40 m/a for glaciers in the Alps to 200 m/a for medium sized valley glaciers in Alaska. Small cirque glaciers move as little as 15 m/a and large outlet glaciers of the Greenland ice sheet as much as 2 km/a. These rates are for ice moving from accumulation zone to ablation zone and do not directly involve shifts of the terminus, which depend on the mass balance.

Velocity studies ordinarily require repeated surveys of lines of stakes along and across glaciers preferrably in relation to land-based reference points; the stakes are used for mass balance as well as flow studies. Recently, velocity measurements have been made from isolated stations on the surface of glaciers by means of a radio echo-fading technique (Doake in Ref. 4; 17). The method depends on the characteristics of the reflected radio

FIGURE 16.9. Stereogram of the lower debris-covered part of Steele Glacier, Yukon Territory, Canada, Sept. 6, 1955. This surging glacier is shown in the quiescent (stagnant) condition, but it surged in 1966. The tributary in the upper right has medial moraines, and the one in the left has ice-cored lateral moraines. The right side of the main valley (toward the top of the picture) has a kame terrace onto which the left tributary has advanced (surged?) and since retreated. The drift-covered stagnant ice in the main valley has chaotic topography; its surface has lowered at least 100 m since the previous surge. White rectangle in the left image is about 1 km long. Photographs courtesy of Surveys and Mapping Branch, Department of Energy, Mines, and Resources, Canada.

pulse which is governed by the roughness of the underlying bedrock. By moving an antenna along a series of lines 20 cm apart parallel to the long axis of an 8 × 2 m rectangle, a fading pattern unique to the underlying bedrock surface at that point is obtained. The observations are repeated and the shift of the pattern gives a measure of the movement of the ice surface during the interval between measurements. Movements ranging from about 40 cm/d to 258 cm/a have been observed over intervals of from one day to a year. So far the method has been used only on cold glaciers.

Sequential photographs and images that resolve stable features on glaciers, and preferrably include reference points on land, give data on flow rates. The spacing of ogives is equal to the annual movement. Reference features that last a year or more include large meltwater streams with terminal moulins (Fig. 16.8), supraglacial ponds, the bands in old firn (Figs. 16.4 and 16.6), and some, but not all, crevasses and blocks (seracs) between crevasses. Moving features likely to survive for several years include ice-cored dirt cones (Fig. 16.8), rock debris falls and avalanches that have moved part way across the glacier, and distinctive bends or junctions of medial moraines (Fig. 16.9). Most medial moraines show in SLAR images (2, p. 474, 475; 16), and the largest are resolved in Landsat images (4, p. 253, 393, 396, 400). Contorted moraines can be used to plot vectors of ice movement. Large boulders or blocks on the glaciers are distinctive but also may move toward the sun by toppling from the pinnacles formed under them as the uninsulated ice around them melts.

SURGING GLACIERS

Some glaciers are quiescent and have very low velocities for periods of 20-30 years, but ranging from 15-100 years for different glaciers; then they flow rapidly (surge) for periods of about 2-3 years, but ranging from 1-6 years. Velocities during surges are 10 to 100 times those during the quiescent phases. Typical surge velocities are 1 to 3 km/a, but rates of up to 6.6 km/a have been observed (Meier and Post in Ref. 18). The surge is the transfer of a large mass of ice from a reservoir area (Figs. 16.1(e) and 16.10), not necessarily within the accumulation zone, to a receiving area that stagnates during the quiescent phase (Figs. 16.2(e) and 16.9). Occasionally, a terminus advances in a spectacular manner. More than 200 of the several thousand glaciers in northwestern North America are known to surge;

many are known in other parts of the world. Landsat images have greatly facilitated their observation. Glacier surges have not been explained satisfactorily. They may be responsible for some dead ice deposits of Wisconsin age (18).

Glaciers that surge can be recognized during the quiescent phase by a combination of the following criteria: (a) medial moraines have loops and folds, some of which can be recognized as parts of tributary glaciers that were severed by a surge of the main glacier; (b) contorted foliation; (c) peculiar pits on the surface [Fig. 16.10(a)]; (d) the lower part is stagnant; (e) near the end of the quiescent phase, the long profile of the glacier is steeper than the profile of older lateral moraines and trimlines.

Glaciers actively surging have (a) a chaotically crevassed surface with new crevasses opening. The reservoir area separates into blocks [Fig. 16.10(b)] which break up into jagged pinnacles as they move into the receiving area. (b) The sides of the glacier are sheared off from tributary glaciers and ice-cored lateral moraines. (c) The surface of the reservoir area sinks as much as 100 m, and that of the receiving area rises by a corresponding amount. (d) There are large horizontal displacements. (e) The terminus has a bulging profile.

Landsat images are useful for monitoring surges (Krimmel and Meier in Ref. 4) because many contorted medial moraines are large enough to be resolved, and so are advances of the terminus. When crevasses develop, the glacier becomes darker in Landsat images because of shadows in the rough area; the breakup of the surface ice in the surging area can be observed expanding up and down glacier. SLAR images show the distortion and movement of medial moraines (16, Fig. 3).

DEBRIS-COVERED ICE

A cover of rock debris commonly conceals parts of the glacier margin in the ablation zone. It retards ablation and locally preserves thicker ice, reversing some slopes and sometimes even flow directions. Rock debris is derived from two main sources: Frost shattered rock falling from valley walls accumulates on the sides of mountain glaciers to form lateral moraines moving with the ice; lateral moraines of tributaries become medial moraines of trunk glaciers. Other debris is derived by erosion of projections and irregularities of the bedrock floor and some may appear in midglacier downstream from a submerged rock pinnacle or

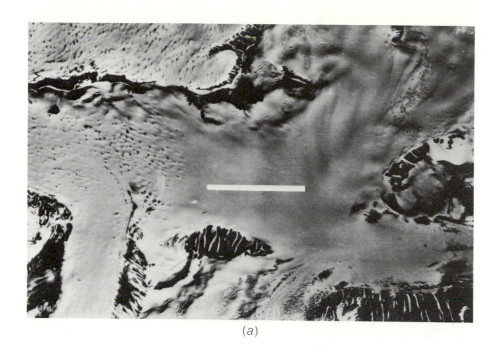

(a)

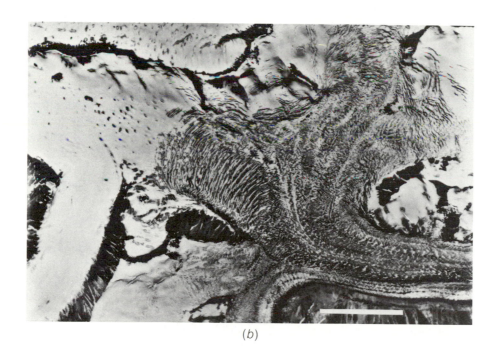

(b)

FIGURE 16.10. Sequential photographs of part of the accumulation zone of Susitna Glacier, Alaska, a surging glacier. (a) August 29, 1949, before surging. (b) July 3, 1954 during a surge, showing intense crevassing and lowering of the surface of the reservoir area. White scale bars are about 1 km long. Photographs courtesy of the United States Geological Survey.

isolated buttress [Figs. 16.2(a) and (b)]. At the terminus debris is brought to the surface by thrust faults. Basal debris is comminuted by grinding under the ice, and is fine-grained compared with rock fall debris which is subject only to physical weathering. Both kinds of debris accumulate on the stagnant part of a quiescent surging glacier and this accumulation is incorporated into the ice during the next surge together with new material from up-glacier.

Debris less than about 1.0 cm thick absorbs heat and accelerates melting of underlying ice; if it is thicker than about 2.0 cm it insulates the ice and retards melting. As a result, moving moraines are ice-cored ridges covered by a few centimeters or more of debris, usually bordered by shallow troughs where the drift (debris) cover is less than 1.0 cm thick. Ice-cored lateral moraines have slopes of 38° to 42° and a drift cover of 0.5 m or more on the distal side; debris continually sloughs off the proximal side. These moraines commonly stand as much as 30 m above the surface of glaciers with a negative mass balance. As the glacier retreats, they become isolated (Figs. 16.2(a) and 16.7) and may last for decades; some resemble moraines of earlier glaciations (Fig. 16.28). Newly deposited eskers and moraine ridges often contain ice cores, especially in the Arctic (Fig. 16.7). Melting of buried ice may be so slow and uniform that forests grow on the debris which also may support small lakes and streams containing a fauna and flora. If such an accumulation is disturbed by erosion, irregular melting, or glacier readvance, the ice is exposed again and rapid melting and collapse ensues. The chaotic topography (Fig. 16.9) results in redistribution of the drift and topographic inversions. The debris is eventually lowered to the ground retaining its chaotic topography but with subdued relief.

The interpretation of buried glacier ice generally is based on morphology seen in aerial photographs. However, thermal infrared images taken in late summer should show surface temperatures that are colder above the buried ice than above adjacent ice-free moraine or bedrock. The two-band technique used for massive ice detection in permafrost should be effective.

Areas of coarse-grained rockfall debris should be discriminated from areas of finer grained debris of basal origin in SLAR images of suitable wavelength (25 cm) because of differing backscatter characteristics of surfaces of different roughness (8).

16.3 REMOTE SENSING IN GLACIAL GEOLOGY

Immediate objectives of glacial geology include the identification and mapping of glacial materials and landforms (19), some of which are now submerged on continental shelves. Recognition and delineation of glaciated areas characterized mainly by erosion forms and drainage anomalies are also essential. The determination of directions of ice movement, and hence the sources of glacial drift, is useful in the search for mineral resources. The ultimate objective is to discover the relative and absolute ages of glacial events and interpret the history of climatic fluctuations during the Pleistocene Epoch.

16.3.1 The Role of Remote Sensing in Glacial Geology

The importance of resolution and the uses of various remote sensors in glacial geology were discussed in the introduction. Available aerial photographs were routinely used for mapping glacial geology in the 1940s (20), because landforms of younger (Wisconsin) glacial drift (Table 16.2) are very well shown by standard panchromatic stereoscopic coverage, enabling identification of most materials, flow directions, and the relative ages of some deposits. This preceeded the systematizing of aerial photography analysis by D. J. Belcher and his associates at Cornell University in 1951. Their method has become the substance of current textbooks (21), but no author has been a specialist in glacial geology, and the subject is presented with the bias of engineers or structural geologists and the brevity required in general references.

Early interest of some glacial geologists in remote sensing was curtailed by the inconclusive results of projects that apparently were conceived with more enthusiasm than background knowledge and forethought. The literature on nonphotographic remote sensing specific to glacial geology is sparse. Remote sensors that provide only morphological information compete with photographs which have much better resolution and cost less. Side-scan images have a perspective that requires a different mode of interpretation from stereoscopic images. The main value of nonphotographic form-imaging sensors is their ability to penetrate clouds (SLAR) and water (SONAR), and for this they are uniquely useful. Thermal infrared images are composite responses to ground temperatures and emissivities and their patterns combine not only these properties, but other elements due to illumination (terrain enhancement), vegetation responses, and cold air drainage (22). Their potential value will be realized only after much more experience has been gained. Multispectral scanning in middle infrared wavelengths (Section 16.2.3) and microwave bands enable computer-processed emissivity maps and ratioed images. These techniques aid identification of geological materials but are still in experimental stages.

In order to obtain a direct response from glacial materials of Pleistocene age, remote sensors must penetrate atmosphere, vegetation, and soil. The soil comprises an upper, usually permeable, organic-rich horizon as thick as 30 cm, and an underlying more dense horizon of reduced permeability, choked with clay minerals, carbonates, and hydrated oxides. These two horizons together range from about 0.3 to 2.0 m in thickness and usually have chemical and physical properties quite different from those of the underlying parent material. Soils depend on climate, vegetation, and time available for development as well as on parent material; hence they differ from region to region on the same type of deposit, and therefore no ubiquitous standard remote sensor signatures can be expected from the parent materials beneath soils. Some radar bands form images that show roughness (grain-size) of surface materials if unimpeded by vegetation (8). At present, interpretation of morphology, erosion forms, drainage patterns, and secondary terrain characteristics such as vegetation patterns and soil moisture, as shown by tone in aerial photographs, still gives the best results.

Multispectral scanning images of reflected light in the visible- and near-infrared ranges are now available for most areas and seasons in Landsat images. These provide overviews of poorly known regions and synoptic coverage of known areas, and occasionally show previously unrecognized relationships. Lineback (23) found that Landsat and Skylab images provided interesting coverage of Illinois, but yielded little new information. He detected many features only because of previous knowledge.

Good quality Landsat images will form stereo-images if the relief is greater than 300 or 400 m (the width of three or four pixels). In many pictures, a few scan lines are offset from adjacent lines by one or two pixels, creating spurious parallax that gives low relief surfaces an apparent wave-like relief greater than that of the terrain. True parallax results from the normal variation of 5 to 15 km in the positions of different overflights of the same flight path, and gives stereomodels in areas of moderate and high relief. Much better stereoimages are obtained from overlapping adjacent flights in high latitudes. Images are positioned for stereoviewing simply by aligning scan lines containing the same reference point. The use of images of different bands for stereoviewing has the advantage of superimposing the tonal patterns of both images on a three-dimensional model. It is sometimes useful to view different bands of the same exposure through a stereoscope; no relief model is obtained, but the best features of both image patterns are combined.

In summary, stereoscopic aerial photography is the main sensor used by glacial geologists, who depend on form, tone and spatial relationships for their interpretations. Most other sensors have not yet received fair trial, and their potential for differentiating different materials is inadequately tested. For instance, although the general synoptic value of Landsat images is recognized, few attempts have been made to use the seasonal changes to identify and delineate certain materials by the vegetation response to hydrologic changes. The ability of radar to provide images in spite of cloud cover is well known but its use, especially of longer wavelengths, for identifying materials is poorly known. SONAR provides information on underwater morphology that is otherwise inaccessible. Thermal IR imagery depends on a complex ground and atmospheric characteristics and needs much more research. The computer processing of images in two or more IR wavelengths to yield emissivities and temperatures should be extremely useful for identifying glacial deposits.

16.3.2 Extent and Type of Glaciation

Glaciated areas have distinctive landscape styles compounded of erosion forms, deposits, and drainage patterns that are influenced by bedrock lithology, preglacial relief, and postglacial landscape evolution. In areas central to former ice sheets the drift is thin and erosion forms tend to dominate.

TABLE 16.2. GLACIAL FEATURES VISIBLE ON LANDSAT IMAGES. FEATURES ARE RESOLVED IF THEY CAN BE IDENTIFIED WITHOUT PRIOR KNOWLEDGE, AND DETECTED IF THEY CAN BE LOCATED ONLY ON THE BASIS OF OTHER AVAILABLE INFORMATION.

Landforms, Deposits	Diagnostic Features	Optimum Conditions		Notes
		Season	Landsat Bands	
Glacial Erosion Forms				
Cirques	Semicircular basins bounded by sharp ridges, in mountainous areas	Summer	6, 5	Show clearly
Troughs	U-shaped valleys, moderately high relief, commonly contain lakes	Avoid low sun	6, 5	Show clearly
Crag and tail (large)	Rock knob with tapering ridge extending down-glacier; usually in groups	Low sun, snow	6 or any other band	Not detected under high sun or with forest cover
Giant grooves	Parallel lines extending down-glacier from an irregular bedrock ridge transverse to ice flow	Low sun, snow Spring	6, 5 5, 6	Well shown by shadows on snow with low sun; vegetation pattern in spring
Meltwater Erosion Features				
Marginal channels	Closely-spaced subparallel slightly arcuate, occasionally sinuous lines trending obliquely down hill slopes and valley sides; in groups of 2 to 15; visible in open country, may be concealed by forest	Any season, enhanced by low sun on snow	All bands	Commonly the side or floor is in shade
Spillways	Straight or slightly sinuous trenches with few, small tributaries; 1 to 3 km wide; may contain long narrow lakes dammed by alluvial fans; others are dry or contain underfit streams	Spring, fall	5, 6	Unmistakable features; contain underfit streams
Till, with Some Stratified Drift				
End moraine	Ridges and hummocky belts, straight or broadly arcuate; usually light tones; forested in other-wise cultivated regions; abundant ponds; obstruct drainage lines; outwash on one side, till plain on the other	Spring or fall	5, 7	Commonly not detected or resolved. In forested regions some show better on Landsat images than on conventional photographs
Lateral moraine (mountains)	Ridge, steep sides and sharp crest, extend from mountain front; may join an end moraine; usually in pairs	Any season	5, 6	Many too small to be seen; enhanced by low sun altitude

Landform	Characteristics	Season	Bands	Detectability
Interlobate moraine	Hummocky or smooth ridges, sinuous or straight, may have crack-like central depressions parallel to the ridge; some are joined obliquely by eskers	Any season	5, 6, 7	Resolution similar to end moraines
Dead-ice moraine (hummocky ground moraine)	Areas of abundant depressions (ponds) or knobs of irregular size and shape; some in arcuate or sinuous alignments but most are random; located on broad topographic highs	Spring Low sun, snow	7, 6, 5 First quality only	May be detected, rarely resolved; obscured by forest or lack of moisture contrast; low sun might enhance relief; easily confused with ground moraine
Till plains				
Ground moraine	Plains of low relief, may have irregular ponds or insequent drainage; tone often nonuniform; commonly natural pattern obscured by cultivation	Spring, fall	5, 6	May be confused with lake plain if no ponds present and with dead-ice moraine if ponds are present
Fluted and drumlinized ground moraine	Straight parallel bands (ridges or grooves) of variable widths, lengths many times the widths; small elongate lakes with parallel sides; lenticular islands in large lakes; elliptical hills (drumlins) in clusters or fields	Spring Low sun, snow	5, 6, 7	Large forms resolved; drumlins detected but not always resolved, especially in summer; enhanced by vegetation and moisture contrasts
Transverse forms (minor moraines)				
Ribbed moraine	Ridges of irregular width from 50 to 500 m, similar spacing; broadly arcuate pattern; many crossed by streamlined ridges, transitional to drumlins	Spring	5, 6, 7	Resolved easily in lake areas
Corrugated moraine	Rarely resolved; detected as lines of transverse elongate ponds; form broadly arcuate patterns	Spring	7, 6 First quality only	At or below resolution limit; commonly not detected; ridges are only one pixel wide
De Geer moraine (washboard moraines)	Not resolved; detected as regularly spaced opposed points on opposite sides of lakes	Spring, summer	7, 6 First quality only	Most not detected; ridges are narrower than one pixel in width
Ice-thrust moraines	Arcuate belts of uncultivated land with parallel lines of minute ponds (between ridges); located on small isolated uplands or on the sides of valleys and escarpments	Spring	5, 7 First quality	Some individual ridges may be detected

(Continued)

TABLE 16.2 CONTINUED

Ice-contact Stratified Drift

Feature	Description	Season	Bands	Remarks
Eskers	Gently sinuous ridges crossing lakes; local expansions; elongate depressions in the middle of some ridges; light tone or distinctive vegetation; double lines of small lakes and ponds; lines of gravel pits, locations of some roads	Spring Low sun, snow	5, 7 First quality	In unsettled regions most eskers are neither detected nor resolved; in settled areas many are detected only
Kames, Kame terraces	None observed or detected to date			

Proglacial Stratified Drift

Feature	Description	Season	Bands	Remarks
Outwash (gravel and sand)	Terraces in major mountain valleys; fans at the mouths of mountain valleys; fans extending away from end moraines on plains; may have a braided pattern; tones medium to light, or distinctive vegetation; tone lighter than till or lake clay plain; may have gravel pits	Spring, fall Low sun, snow	5, 6	Difficult to distinguish from other sand deposits; braided pattern shows during times of maximum soil moisture contrast
Lacustrine and marine clay plains	Plains with widely spaced sinuous or meandering main streams; some rivers in narrow valleys without flood plains, having straight reaches between smooth bends; tones uniform, usually darker than other glacial deposits; commonly intensely cultivated; where clay is silty modern lakes have prominent arcuate bays; De Geer moraines may be present	Spring, fall, summer	6, 7, 5, 4	Clay has dark tone; near-shore deposits (silt, sand) may have a tone similar to till, but more uniform; Summer is poorest season for discrimination
Deltas	Similar to outwash, usually bisected by a river; large deltas may give pattern of gravel at apex and sand, possibly with dunes, farther from apex, and silt with a pattern of gullies at the periphery; commonly bounded by wave-cut scarp or a steep foreset slope	Spring, fall	5, 6	
Strandlines (abandoned)	Wave-cut scarps form an abrupt discontinuity between tone and drainage patterns; strandline features are straight or arcuate and may have cusps pointing toward the former lake. Beach ridges are narrow lines of contrasting tone or vegetation, commonly light in tone; may occur in clusters and as spits and bars	Spring, fall, summer	5, 6, 7	Not all are detected or resolved; may be enhanced in early spring by a narrow snowbank

Eolian Deposits

Loess	Examples not seen; expected appearance—a dense, nonuniform dendritic drainage pattern along major valleys; a uniform medium tone	Probably spring	Probably 5 and 6	Probably detectable only where thick
Sand dunes	Stabilized dunes have light tones if grass covered, medium to dark tones if forested; contrast with other deposits; tones nonuniform with tongues of one tone invading another; rarely some small lakes and ponds; active dunes white, tongue- or U-shaped	Spring, fall Low sun, snow	5, 6	Small areas not detected; most individual dunes are too small to be resolved; many areas detected but not resolved

Recent Deposits

Alluvial fans	Arcuate fans with radial drainage debouching from the mount of a valley in an escarpment or tributary to another large valley. Tone light if grass or if soil is exposed; may be forested. Interrupts other patterns; in valleys may displace rivers to the opposite side of the valley floor	Spring, fall	5, 6	Many are resolved
Landslides	Subparallel lines of small ponds or elongate small lakes along the sides of valleys and scarps; lobate protrusion at the base of the slope	Spring, fall	7, 6 First quality only	Only large landslides detected; few are resolved

525

Deranged drainage that shows strong preglacial influence of bedrock is typical (Figs. 16.17, 16.19, and 16.24). In peripheral belts where ice margins were stable, the drift deposits are thick and modify or conceal the preglacial relief. In some areas, diversion of drainage across, or oblique to regional slopes (Fig. 16.13) is the only evidence of early glaciations, the deposits of which exist only as small remnants on divides. The relief of mountains may be sufficient to initiate a glacierization sequence starting with cirque glaciers that expand into valley glaciers as the climate cools and the snowline lowers. As snow and ice accumulate, valley glaciers thicken and overflow divides and coalesce to form transection glaciers and then ice sheets, through which the highest mountain summits may project as nunataks. During this phase, glacially submerged ridges and peaks are rounded by glacial erosion and projecting peaks and ridges are sharpened by physical weathering (frost action). Deglaciation of mountains is a reversal of the glaciation sequence. Landsat imaging often gives good regional views of glaciation styles; the examples given here are in order from high to low relief.

MOUNTAIN GLACIATION

The contrast between glaciated and unglaciated mountains is well shown in Yukon Territory, Canada, where the glaciated Ogilvie Mountains with peaks above 1500 m altitude are juxtaposed across the Tintina Trench with the Klondike area to the south, where ridge summits are between 900 and 1200 m altitude (Fig. 16.11). General relief is about 750 m in the Klondike area and about 1200 m in the Ogilvie Mountains.

The Ogilvie Mountains have sharp serrated ridge crests flanked by cirques, and straight valleys with broad floors. Several cols have been breached by glacial erosion to form through valleys, segments of which now contain only minor tributaries of the present river system. Lateral moraines project south from two major valleys into the Tintina Trench where outwash fans are superimposed on older (Tertiary) alluvial sediments. Details of the summits of similar glaciated mountains in British Columbia are shown in Fig. 16.12. The summits shown probably formed a nunatak during the glacial maximum.

In contrast, the unglaciated Klondike goldfields area has a well integrated dendritic drainage pattern. Valley sides slope uniformly and valley floors are narrow. Spurs on valley sides between tributaries typical of other regions are uncommon here, possibly because of strong periglacial activity on these slopes. Ridge crests and summits are rounded and lack the sharp serrated or cuspate crests of glaciated mountains.

Local "anomalies" in Landsat images invite misinterpretation. In Fig. 16.11, dark tones around the Tintina Trench are the result of recent forest fires. Thin cloud obscures some ground in band 5 but is mostly penetrated by band 7, although with some degradation of the image. Band 7 has a dark, straight linear feature crossing the eastern part of the Tintina Trench that is the shadow of the contrail of a high-flying aircraft. The contrail shows as a white line 20 km farther south in band 5. Offsets of the shadow could be used to calculate relief.

Mountain glaciation is shown also in the west half of Fig. 16.14 where relief is about 1500 m. Snow obscures the ridge crests in the band 5 image, but they show clearly in the simultaneous band 7 image (see Section 16.2.4 concerning blue and infrared light). When viewed stereoscopically, the two band 7 images show moraine systems associated with valleys containing trough lakes on both sides of the mountains. Very broad mountain valleys such as the Flathead valley in the west of Fig. 16.14 commonly contain streamlined forms (grooves, drumlins) that show well under low sun illumination and in SLAR images. Breckenridge (in Ref. 25) used Landsat images to compile a comprehensive preliminary map of mountain glaciation in northwestern Wyoming.

GLACIATION IN AREAS OF MODERATE RELIEF

Moderate relief is here defined as 100 to 400 m. Glaciers advanced into these areas rather than originating in them. South of Lake Erie (Fig. 16.13) successive continental glaciers moved up the north slope of the Appalachian Plateau (24). Relief enhanced by moderately low sun illumination (23°) in (b) is barely apparent with sun elevation of 49° (a). In band 5 (c) relief shading is subordinate to tones of vegetation and land use. The original images of (a) and (b) form a useful stereomodel in the southeast half but not in the northwest where relief is low and several scan lines are displaced. Preglacial trunk valleys can be traced from the southeast into the drift-covered area to the northwest. End moraines (d) are not obvious, unlike their continuations in Illinois (23) where relief is less; they may be evident on other bands at other seasons. Band 7 images

(a)

FIGURE 16.11 (a). Landsat image showing the boundary between glaciated and unglaciated mountains in Yukon Territory, Canada. The Tintina Trench separates the glaciated Ogilvie Mountains in the north from the unglaciated, dissected Klondike Plateau in the south. Band number is shown in white rectangle 10 km long. Image no. E1421-20145, Sept. 17, 1973, sun elevation 27°. Courtesy of the Canda Centre for Remote Sensing.

seldom show vegetation and land-use patterns as well as bands 4 or 5, but in (b) snow shows in open fields and is concealed by forest canopy. In the area of Wisconsinan drift the drainage pattern is immature and only the largest valleys show clearly, except for several major tributaries parallel and close to the southeastern drift border. Some major preglacial valleys such as the one containing Chautauqua Lake in the northeast are aligned in the direction of glacier flow and were widened and deepened by glacial erosion similar to the Finger Lakes in New York. Transverse valleys were nearly filled with drift. Close to Lake Erie, land-use patterns and route

locations show the position of a moraine and beaches of the high level predecessors of Lake Erie.

No glacial morphology remains here in Illinoian drift, which is present as remnants on divides. Several major streams that trend across or oblique to the regional slope are thought to be the result of glacial diversions in Illinoian time. The Allegheny River follows a similar course that may have resulted from an even earlier glaciation (Kansan). The drainage density in the unglaciated area seems to be slightly higher than in the area of Illinoian drift; however, this could be partly the result of differences in the underlying bedrock.

(b)

FIGURE 16.11 (b). Landsat image showing the boundary between glaciated and unglaciated mountains in Yukon Territory, Canada. The Tintina Trench separates the glaciated Ogilvie Mountains in the north from the unglaciated, dissected Klondike Plateau in the south. Band number is shown in white rectangle 10 km long. Image no E1421-20145, Sept. 17, 1973, sun elevation 27°. Courtesy of the Canada Centre for Remote Sensing.

GLACIATION IN AREAS OF LOW RELIEF

The eastern part of Fig. 16.14 has local relief of less than 100 m superimposed on very broad hills and valleys of greater magnitude. The Laurentide Ice Sheet entered this area from the east and northeast and the Cordilleran Ice Sheet advanced from the mountains in the west. Both ice sheets alternately covered some of the area east of the mountains north of the International Boundary at different times, but south of the boundary an irregular corridor was not glaciated.

Both ice sheets were of Wisconsin age, but some deposits of earlier Cordilleran glaciations occur on high-level pediment and terrace remnants near the mountain front [Fig. 16.14(d) cross-hatched]. Wisconsin age lobes that extended east from the mountains are outlined by moraines and by ice-marginal streams. Within the Two Medicine Creek lobe ground moraine is dissected by parallel drainage lines probably resulting from glacial fluting or grooves not visible at the scale of these images. The drainage pattern in the unglaciated area is dendritic, more dense, and better integrated than elsewhere. Glacial Lake Cutbank was ponded by Laurentide ice and its deposits have low relief and a nonuniform, immature drainage pattern. On the original Landsat images, morainal belts are shown by small lakes and ponds so abundant [Fig. 16.14(a)] that they are designated erroneously as "karst" on some aeronautical charts. The high level deposits of early glaciations form forested areas (dark) adjacent to the mountain front. The youngest Cordilleran ad-

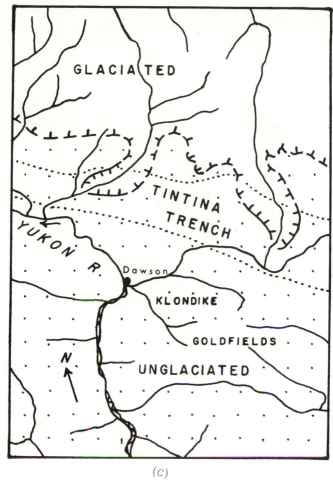

(c)

FIGURE 16.11 (c). Sketch map showing the boundary between glaciated and unglaciated mountains in the Yukon Territory, Canada.

vance (Pinedale) is represented by the moraines already mentioned.

The Laurentide drift in the north also has abundant ponds and lakes. Some are aligned parallel to the drift border as though between moraine ridges. In the east, the drift margin is more difficult to locate. Ponds are scarce there but the lack of integrated drainage is diagnostic. The eastern drift may be thinner and older than that in the north. The margin may be more obvious in images in other bands and at other seasons.

The dark, slightly curved west-east line just north of Cut Bank Creek is probably the track of a small isolated rainstorm that occurred a few hours before the image was made; it is not a flaw and does not appear on Figs. 16.14(a) or (c).

In terms of more recent glacial events, Landsat images show the former distribution of thin perennial snow and ice fields on the low relief uplands of northern Baffin Island (26). Light-toned areas in band 5 and 6 images indicate where moss and lichen were destroyed by perennial snowfields which existed from about 400 to 70 years ago.

16.3.3 Directions of Glacier Flow

Many regions lack large-scale indications of the direction of ice movement, but elsewhere directional features are visible in Landsat and SLAR images. The problem commonly is to determine which of two directions the ice moved. Crag and tail (Fig. 16.15) are rock knobs with ridges of till extending down glacier. Many streamlined ridges (drumlins),

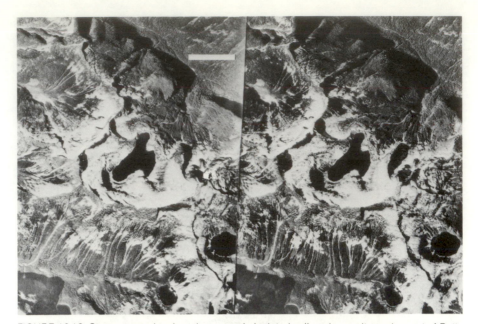

FIGURE 16.12. Stereogram showing cirques and glaciated valleys in granite rocks east of Butte Inlet, British Columbia, Canada. Several horns and arêtes rise from slopes above cirques instead of forming their headwalls; these probably projected as nunataks above the Cordilleran Ice Sheet. White line is about 1 km long. Photographs courtesy of Surveys and Mapping Branch, Department of Energy, Mines, and Resources, Canada.

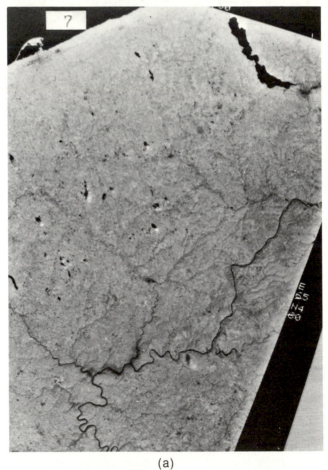

(a)

FIGURE 16.13 (a). Landsat image showing the limit of glaciation south of Lake Erie in western Pennsylvania and New York. Image (E1407-15350) taken Sept. 3, 1973, sun elevation 49°. Band number is shown in white rectangle 10 km long. Image courtesy of the Canada Centre for Remote Sensing.

(b)

FIGURE 16.13 (b). Landsat image showing the limit of glaciation south of Lake Erie in western Pennsylvania and New York. Image (E1497-15332) taken Dec., 1973, sun elevation 23° Band number is shown in white rectangle 10 km long. Image courtesy of the Canada Centre for Remote Sensing.

fluting, grooves, and crag and tail are visible in low-sun snow-cover Landsat images (Fig. 16.16) and in synthetic-aperture radar images (26). Illumination must be transverse or oblique to show these forms. The crag and tail is as clear in Fig. 16.16 as in small-scale aerial photographs.

Drumlins (Fig. 16.17) show in many Landsat images especially under snow-covered low-sun illumination and where they form islands. They seldom show where illumination is parallel or where there is a complex land-use pattern (Fig. 16.13). The broad, steeper end of a drumlin is usually, though not always, the up-glacier end. In SLAR images of drumlins near Syracuse, New York (2, p. 1235) a like-polarized (HH) image gives better definition of upstream and downstream ends than a cross-polarized (HV) image, but the reason for this is not obvious; perhaps the smooth topography and extensive cultivation are factors. Synthetic-aperture radar gives excellent detail of drumlins north of Anchorage, Alaska (27, Image G-140).

The directions and concentrations of linear features such as fluted moraine and striations on bedrock can be analyzed by means of the Fourier transform (diffraction pattern) of an image created by coherent light (28, Figs. 6 and 9). The intensity of light transmitted through the transform and a slit rotated around its center of symmetry is plotted against its angular position to give a graph called a wiener spectrum. This curve has peaks at right angles to the directions of corresponding linear features in the image. The positions of the peaks give the mean orientations of linear forms, and their widths indicate the deviations about the means.

(c)

FIGURE 16.13(c). Landsat image showing the limit of glaciation south of Lake Erie in western Pennsylvania and New York. Image (E1407-15350) taken Sept. 3, 1973, sun elevation 49°. Band number is shown in the white rectangle 10 km long. Image courtesy of the Canada Centre for Remote Sensing.

16.3.4 Glacial Deposits and Landforms

Routine aerial photo interpretation of glacial geology is explained in other books (21, 29) and is not repeated here. Evidence for glacial landforms and deposits visible in Landsat images is summarized in Table 16.2, and additional information on optimum bands and seasons is in Table 16.3. Most of this information is from study of selected Landsat coverage of Canada and adjacent United States. The accessible literature is limited to several papers on satellite imagery, a couple on thermal infrared, and passing references to radar. There is additional information in in-house and "near" publications with limited circulation.

This discussion is organized genetically: (a) unstratified drift (till) which is deposited directly from glacier ice; (b) stratified drift which is sorted, water-laid sediment classified as (i) ice-contact stratified drift if it is deposited on or against ice, or (ii) as proglacial drift if it is carried away from the glacier and deposited subaerially as outwash or subaqueously in lakes or the sea; (c) wind transported sediment derived from proglacial sediments and now forming dunes and loess; (d) meltwater channels, which are part of the system of proglacial transport.

Normally photo interpretation is used to identify materials on the basis of form, tone, and spatial relationships. It has been suggested that tone could be quantified and the spectral signatures of surface materials might be diagnostic. This approach was tested quantitatively by Welch (19) who studied a

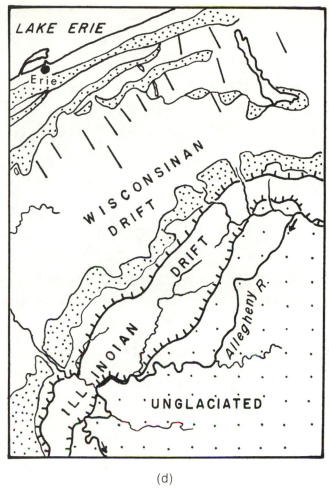

(d)

FIGURE 16.13 (d). Geology sketched from Ref. 24; stippled areas are moraines and straight lines trending N to NW are drumlins or groups of drumlins.

glacial landscape in Iceland. He measured luminance (reflectance) of various surfaces in the field, and also the densities of negatives of the same materials using various film and filter combinations, including multispectral exposures in bands 20 to 50 μm wide in the visible and near-infrared ranges. He found that the extra effort of this approach produced little information additional to that obtained from ordinary photo interpretation. His work showed that color and color infrared photographs are more useful and economic than a series of narrow-band multispectral black-and-white photographs.

NONSTRATIFIED DRIFT (TILL)

Till accumulates as ridges of end moraine at stable ice margins or as sheets of ground moraine left behind retreating ice. Ground moraine has low relief (generally less than 5 m) and may have either random depressions and knobs or a variety of patterns ranging from systems of ridges and grooves parallel to the direction of ice flow to systems of transverse ridges, or combinations of these. The patterns are commonly too fine to be resolved in Landsat images but may affect plant distributions or the illumination of snow cover so that areas with different homogeneous, although unknown, slope characteristics can be outlined; in general, the rougher the ground, the darker is the tone.

End moraines have greater relief and withstand erosion longer than the smaller forms on ground moraine. End moraines represent the limits of glacial advance and are especially important in the

(a)

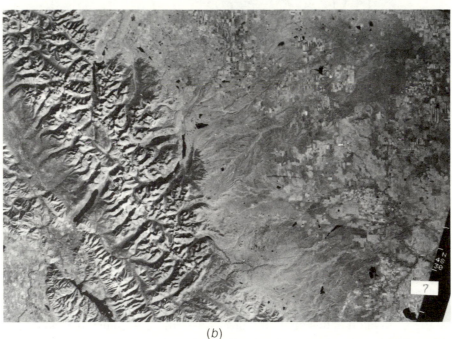

(b)

FIGURE 16.14. Landsat images of a region straddling the International Boundary at the eastern front of the Rocky Mountains. The Rocky Mountains were glaciated by the Cordilleran Ice Sheet, and the Great Plains to the east by the Laurentide Ice Sheet; and intervening area is unglaciated. (a) Image E1702-17495 taken June 24, 1974, sun elevation 57°. Snow on the mountains masks the cirque topography in band 5 but is barely discernible on band 7 where cirques show clearly. (b) Image E1432-17555 taken Sept. 28, 1973, sun elevation 36°, gives a stereo model with (a). In (a and b) rectangles containing band numbers are 10 km long; images courtesy of the Canada Centre for Remote Sensing.

(c)

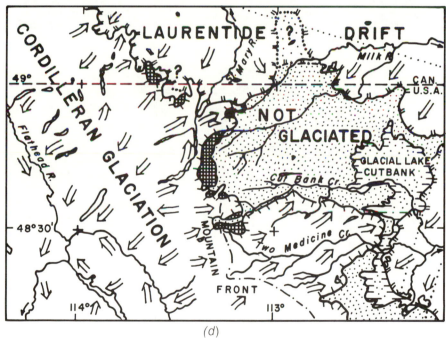

(d)

FIGURE 16.14 (continued). (*c*) Image E1702-17495 taken June 24, 1974, sun elevation 57°. Snow on the mountains masks the cirque topography in band 5 but is barely discernible in band 7 where cirques show clearly. The rectangle containing the band number is 10 km long; image courtesy of the Canada Centre for Remote Sensing. (*d*) Sketch map of glacial geology showing ice margins as ticked lines; broad arrows are flow directions; cross-hatched areas are old (pre-Wisconsin) Cordilleran drift; dotted lines are borders of images. Sketched from Ref. 24 and published works of Alden, Horberg, Richmond, and Stalker.

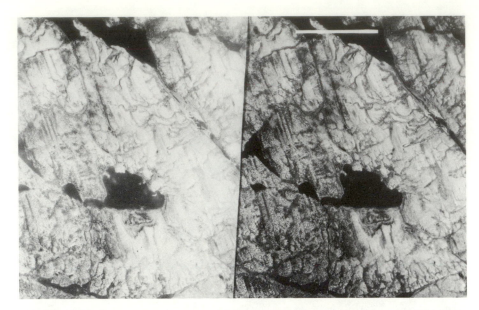

FIGURE 16.15. Stereogram of crag and tail forms in an area of gneiss near Kidd River, Quebec, Canada (lat. 54° 30'N, long. 67° 25'W) showing ice movement from SSE to NNW. The lobate hollow south of the lake is a nivation hollow formed by a perennial snowbank during a postglacial climatic episode colder than the present. Sinuous and arcuate ice-margin channels occur on the left side of the valley in the upper right and in the upper left corner. The white line is 1 km long. Photographs courtesy of the Surveys and Mapping Branch, Department of Energy, Mines, and Resources, Canada.

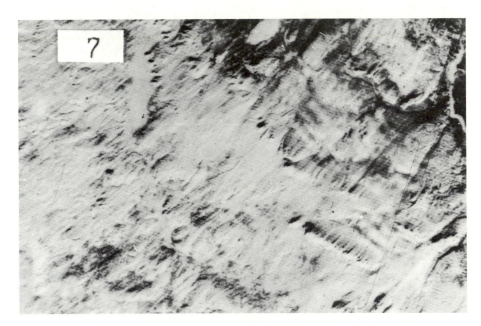

FIGURE 16.16. Part of Landsat image E1135-15201 taken Dec. 5, 1972, near Kaniapiskau River, Quebec, Canada (lat. 57° 20'N, long. 69° 30'W) showing snow-covered ground with a sun elevation of 9°. The white rectangle containing the band number is 10 km long. Crag and tail forms appear in the right part of the picture, and hummocky moraine is adjacent to the river in the extreme right. An esker trends NE across the middle of the picture; streamlined ridges show in the left side. Small valleys similar in size to ice-marginal channels appear in the left as sinuous dark lines. Image courtesy of the Canada Centre for Remote Sensing.

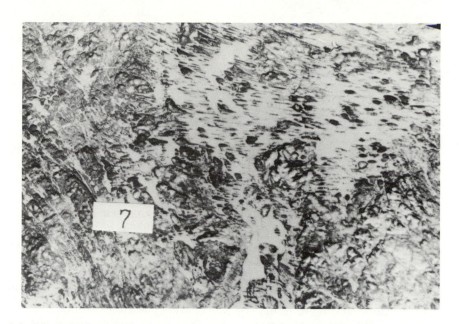

FIGURE 16.17. Part of Landsat image E1490-14493 of Michikamau Lake in the Quebec-Labrador region, Canada (lat. 55°N, long. 64°W). Drumlins forming islands show glacier movement from W to E. Image taken Nov. 25, 1973, sun elevation 13°. Typical deranged drainage patterns of the Precambrian shield show in the W and SE. Rectangle containing the band number is 10 km long. Image courtesy of the Canada Centre for Remote Sensing.

TABLE 16.3. DISCRIMINATION OF GLACIAL LANDFORMS IN SELECTED LANDSAT IMAGES. BANDS ARE LISTED IN ORDER OF USEFULNESS (MOST USEFUL FIRST). MISSING NUMBERS INDICATE THAT THOSE BANDS DO NOT SHOW GLACIAL FEATURES WELL; ASTERISK MEANS THAT ONLY THE LISTED NUMBERS WERE EXAMINED. UNDERLINED BANDS GIVE EXCELLENT DISCRIMINATION. BRACKETED NUMBERS EQUALLY USEFUL.

Area	Major Features	Vegetation, Land Use	Season	Response
St. Lawrence lowland, Quebec	Marine clay plain, sand areas, abandoned channels	Broadleaf and mixed forest; agriculture	Winter, low sun, snow	4, 7*
			Midsummer	6, 7, 5, 4
			Late summer	7, 5, 6, 4
Manitoba lowland (Red River valley)	Lacustrine clay plain, delta, sand dunes, beach ridges, till plain, alluvial fans	Agriculture (grain); aspen parkland; spruce woodland	Early spring	<u>6</u>, 5, 4, 7,
			Midsummer	<u>5</u>, 6,*
			Early fall	5, 4
Drift plains; southwestern Manitoba, North Dakota	End moraines, minor moraines, till plains, outwash	Agriculture (grain); aspen parkland, prairie	Early spring	6, 5
			Early summer	<u>5</u>, 6
			Midsummer	all poor
			Winter low sun, snow	<u>6</u>,*
Cretaceous escarpment, northeastern Saskatchewan	Dead-ice moraine, glacial lake beaches, spillways, till plain, Lacustrine clay plain, streamlined ridges	Broadleaf forest, spruce forest; aspen parkland; agriculture (grain)	Late winter, snow, high sun	6,*
			Early spring	5, 4, 6
			Midsummer	5, 4
			Fall	7, 6, (5, 4)
Rocky Mountains Alberta, British Columbia, Montana	Cirques, glaciated valleys, glaciers, streamlined ridges, outwash	Forest; alpine tundra	Midsummer	6 (poor)*
			Fall (snow on peaks)	7, 6, 4
			Early winter, low sun	unsatis-factory

537

midwest United States, a classical region to which Quaternary events elsewhere in North America are correlated. In the first studies of Landsat images (25) the detection of end moraines received first priority, but success was limited.

Many diagnostic characteristics of end moraines (Fig. 16.18) disappear in small scale, low-resolution images (Figs. 16.13 and 16.19). The detection of end moraines in Landsat images depends on inference from land-use and drainage patterns and relationships with other features. Young (Wisconsinan) end moraines in Illinois are easily traced on Landsat and Skylab pictures (23). Similarly, end moraines on Long Island are detected, but not the Valley Heads moraine in central New York State (Isachsen and others in Ref. 25). Morrison and Hallberg (in Ref. 25) interpreted positions of older moraines in Nebraska from stream dissection and stream divide patterns. End moraines in the Precambrian Shield form discontinuities in the background pattern of linear features (Fig. 16.19) and support vegetation with a tone pattern different from adjacent deposits (Fig. 16.24).

Radar images have responses depending on scale, depression angles, and view direction. The end moraine in Fig. 16.18, in an area of moderate bedrock relief, is hardly detectable in X band (3 cm) SLAR images with the scan lines parallel to the moraine (30), but nearby where the same moraine is a simple narrow ridge transverse to the scan lines it is very distinct. End moraines at the foot of the Grand Teton Mountains in Wyoming (2, p. 1234) show on SLAR images as arcuate ridges with a roughness due either to forest cover or to topography. Synthetic aperture radar (27, Image G-140) shows small end moraines in Alaska as clearly as do photographs of similar scale.

Nonstratified drift is most abundant as ground moraine. It shows more variations in morphology and tones than do lake basin deposits, though local relief may be almost as small. In peripheral areas of continental glaciation, differences in drainage density, integration, and degree of dissection show relative ages of ground moraine areas. Drumlins and similar streamlined forms (Section 16.3.3) are a type

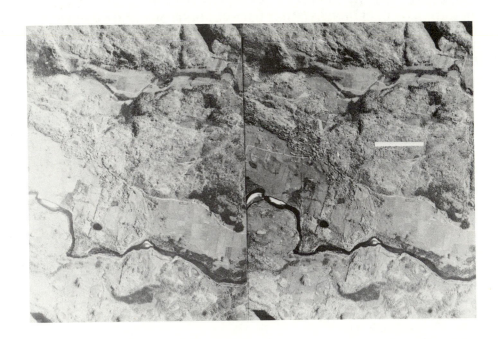

FIGURE 16.18. Stereogram showing an end moraine that crosses the valley of Rivière Rouge south of St. Jovite, Quebec, Canada. The proximal side (left) has a gentle slope with several low ridges; the distal slope (right) is steep and is the upper boundary of an outwash train. Though not large this moraine is part of the St. Narcisse moraine system and can be traced for more than 300 km. White line is 1 km long. Photographs courtesy of Surveys and Mapping Branch, Department of Energy, Mines, and Resources.

FIGURE 16.19. Part of Landsat image E1265-15463 north of Sudbury, Ontario, Canada (Lat. 47° 30'N, long. 81° 30'W), showing the interruption of drainage and lineament patterns by an end moraine. The rectangle containing the band number is 10 km long. Image taken April 14, 1973, sun elevation 47°. Courtesy of the Canada Centre for Remote Sensing.

of ground moraine. Transverse ridges referred to as minor moraines but of uncertain origins also are mostly varieties of ground moraine. The largest type of minor moraines, called ribbed moraines, are gradational into drumlins [Fig. 16.20(a)] and are recognizable on some Landsat images [Fig. 16.20(b)]. Smaller minor moraines (Table 16.2; Fig. 16.21) are at or below the resolution limit of Landsat imagery, but occasionally the patterns of intervening small ponds reveal their presence. On low-sun snow-enchanced Landsat pictures minor moraine ridges produce a slightly darker tone than ordinary ground moraine because of the reduced illumination on northern slopes. Minor moraines form prominent broadly arcuate or lobate patterns in aerial photographs (Fig. 16.21) and should be obvious in radar and thermal infrared images.

Dead-ice moraine (hummocky ground moraine) (Fig. 16.22) is deposited from stagnant ice (Fig. 16.9). Its chaotic topography of random knobs and depressions, with relief ranging from about 5 to 25 m, results from topographic inversions during the final stages of ice wastage (see "Debris-covered Ice," above). Thick superglacial drift accumulations occur now in Alaska, and similar conditions existed in parts of the Great Plains about 13,000 years ago. In the midcontinent, dead-ice moraine is on uplands

against which flowing ice was compressed so that thrust planes formed and carried subglacial debris up to the surface. In other places the dead-ice moraine is attributed to surges of discrete lobes of ice sheets (18, p. 899). Landsat images of the prairies show dead-ice moraine as mainly uncultivated areas (bands 4 and 5) randomly speckled with abundant small lakes and ponds (bands 6 and 7). Low-sun, snow-enhanced images show the chaotic knob and kettle topography (Fig. 16.16, east side).

Representative midsummer thermal infrared (8-14 μm) imagery of an area of mainly hummocky moraine under prairie vegetation in southern Alberta was obtained by Holmes and Thompson (31). The images show the diurnal warming and cooling of the ground. Qualitative evaluations of the thermal inertia of various features, mostly related to agricultural practice, were made. Differential heating and cooling of north and south facing slopes results in a thermal image resembling a low sun angle photograph. Moist areas are warm at night and cool during the day relative to dry areas, causing reversals of tone. More information about hydrology and agriculture than about geology was obtained in this experiment; for geology, the authors recommend additional springtime thermal IR imagery and the use of more than one band (see Section 16.2.3).

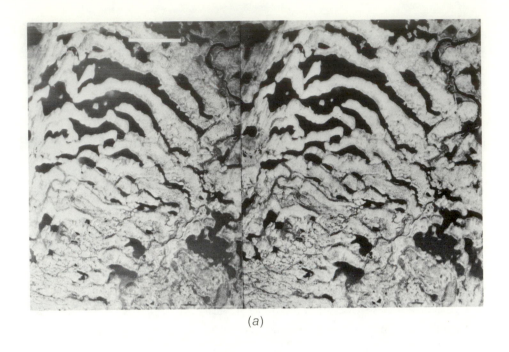

(a)

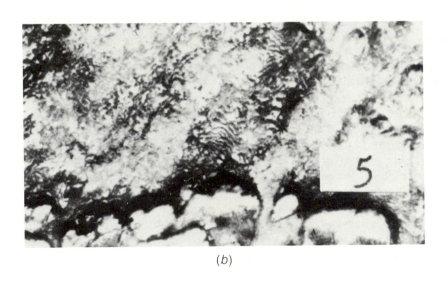

(b)

FIGURE 16.20. Ribbed moraines west of Cow Lake, Quebec, Canada (lat. 52° 30'N, long. 70° 30'W). (a) Stereogram showing ribbed moraines merging into streamlined forms on the left; a small esker winds S and SW along the right side. White line is 1 km long. Photographs courtesy of Surveys and Mapping Branch, Department of Energy, Mines and Resources. (b) Part of Landsat image E1133-15100 with the area of (a) just below the center; taken on Dec. 3, 1972, sun elevation of 13°. The ribbed moraines, and hummocky moraine to the west and north are visible but the esker is not. The rectangle with the band number is 10 km long. Courtesy of the Canada Centre for Remote Sensing.

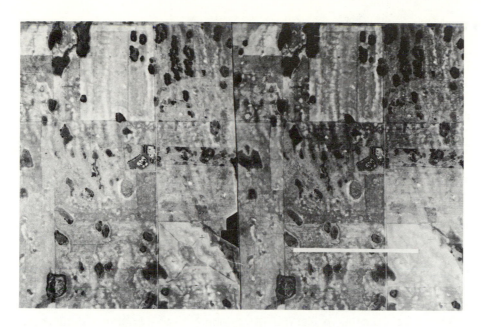

FIGURE 16.21. Stereogram of corrugated moraine east of Stoughton, Saskatchewan, Canada, deposited by ice retreating from left to right. White line is 1 km long. Several areas of these moraines are in Fig. 16.26 but are not detectable. Photographs courtesy of Surveys and Mapping Branch, Department of Energy, Mines, and Resources, Canada.

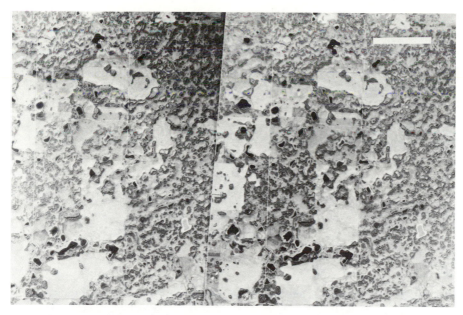

FIGURE 16.22. Stereogram showing hummocky dead ice moraine with moraine plateaus deposited in ice-basin lakes, near Kinsella, Alberta, Canada. White bar is 1 km long. Photographs courtesy of Surveys and Mapping Branch, Department of Energy, Mines, and Resources, Canada.

STRATIFIED DRIFT

Ice-Contact Stratified Drift

Eskers are the most common form of ice-contact stratified drift. These sinuous gravel and sand ridges typically are 3 to 50 m high, and from 0.5 to 150 km long. They may be joined by tributaries (Fig. 16.23), and some have regularly or irregularly spaced expansions pitted by kettles. A few eskers terminate in gravel mounds (kames) as much as 2 km in diameter. In populated areas a line of gravel pits may be all that remains of an esker. Eskers should give good SLAR and thermal IR signatures. In Landsat images, they are surprisingly difficult to detect, in view of their prominence in aerial photographs (Fig. 16.23) and widespread distribution (34). They are visible as chains of islands in some lakes, and are shown by sinuous shadows in low sun angle (13° or less) snow enhanced images where they are transverse or oblique to the direction of illumination (Fig. 16.16). They are not visible where both sides are equally illuminated. Eskers generally are only one to five Landsat pixels wide. The regional vegetation pattern commonly masks the narrow distinctive tone of most eskers.

Kame terraces are the deposits of rivers flowing along the sides of ice tongues in valleys (Fig. 16.9, north side of main valley) and occur where ice sheets retreated across area of moderate relief (500 m) as well as in mountains. They form uniform terraces bounded by steep irregular scarps, and have numerous kettles adjacent to the former ice margin. Most are too narrow to be detected on Landsat images but should show well in SLAR and thermal IR images (see following section).

Proglacial Stratified Drift

Outwash sand and gravel occurs in valley trains, terraces (Figs. 16.7, 16.18), and fans (Fig. 16.28) that have uniform surfaces and distinctive vegetation or tones. In populated areas with dry climates the drought-prone soils of outwash are seldom cultivated. The smooth surface may have a relict braided channel pattern. Extensive outwash shows on Landsat images (Figs. 16.11 and 16.26, extreme left). The smooth terrace surfaces with steep scarps facing the present rivers are distinct in SLAR images (2, p. 1234). Details of relict channels are visible in synethetic-aperture radar images (27, Images G-

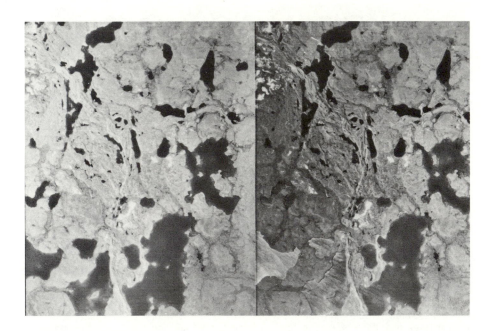

FIGURE 16.23. Stereogram of an esker near Darrell Lake, Northwest Territories, Canada (lat. 63° 45′N, long. 105° 45′W). A tributary from the south joins the main esker which has been traced for at least 400 km E-W. The picture is about 4.5 km long from top to bottom. Photographs courtesy of Surveys and Mapping Branch, Department of Energy, Mines, and Resources, Canada.

140, G-142). Thermal infrared images (8-14 μm) enabled Waldrop (22) to distinguish well drained outwash from bedrock and till by its darker tone on evening images in Yellowstone National Park. The thinner forest on gravel permits more rapid cooling, giving the darker tone. The effects of terrain enhancement (aspect and slope), forest cover, cold air drainage, and relative temperature reversals complicates thermal IR interpretation to the extent that most geological features there show more clearly on vertical aerial photographs and radar images (22).

Proglacial drift is deposited in lakes and shallow marine basins as clay, varved clay, and silt; marginal deposits include near-shore sand and silt, deltas, and beach deposits. The fine-grained basin deposits retain moisture and where not masked by vegetation give dark tones on photographs and Landsat images (Fig. 16.25). Most glacial lake basins of appreciable size are smooth low areas visible in Landsat images (23), though their limits are often indistinct (Fig. 16.14). In the Precambrian Shield, major rivers crossing glacial lake basins have a sinuous or irregularly curved course, are usually incised, and commonly lack wide flood plains. They have straight reaches that result from superposition onto bedrock structures. In some northern (boreal) areas, modern lakes in areas of thick clay and silt have semicircular embayments (Fig. 16.24), and superficially resemble thaw lakes, but most are not now within the permafrost zone.

Many, though not all, relict shoreline scarps and beach ridges appear in suitable Landsat (Fig. 16.25) and radar images. Commonly, shoreline scarps separate areas with different tones or topographic textures. Beaches are narrow bands, often in a parallel series. Like some eskers, some relict beaches are now represented only by a line of gravel pits. Relict deltas are triangular or lobate sand and gravel benches best seen on moderate to steep slopes on the sides of basins; some extend far out into the basins. They typically give light tones in the visible spectrum and have distinctive vegetation. Usually remnants of one or more abandoned river channels are visible. Large deltas (Fig. 16.25) have a sediment size gradation ranging from gravel at the apex through sand to silt in the distal part. Where not concealed by vegetation, sand and gravel have light tones, and silt has medium tones on photographs and Landsat images. Seasonal changes in vegetation (leafing, spring and fall coloration) cause some reversals in tone which help delineate areas

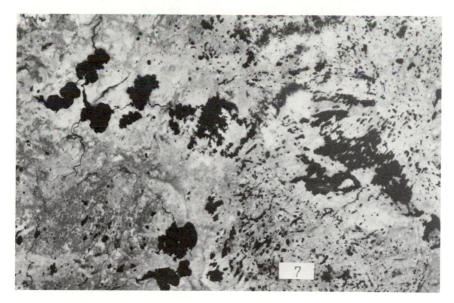

FIGURE 16.24. Part of Landsat image E1400-16345 of the Windigo-North Caribou Lake area, Ontario, Canada (lat. 52° 25'N, long. 91°W). The lakes with round bays in the upper left are in an area of glacial lake clay; a larger esker shows as a dark line trending NE between the lakes. An end moraine arcs NW from the large lakes in the lower central part of the picture; drumlins trend SW in the east centrl part. In the lower right the drift is thin and bedrock structure controls the pattern of lakes and streams. The rectangle containing the band number is 10 km long. Image taken Aug. 7, 1973, sun elevation 43°. Courtesy of the Canada Centre for Remote Sensing.

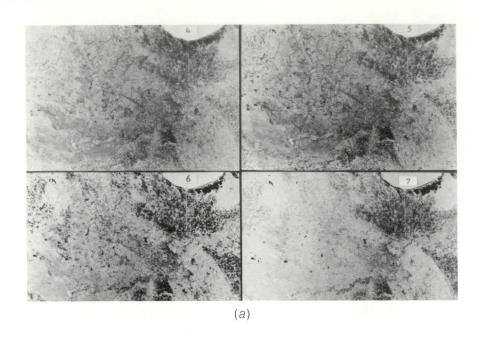

(a)

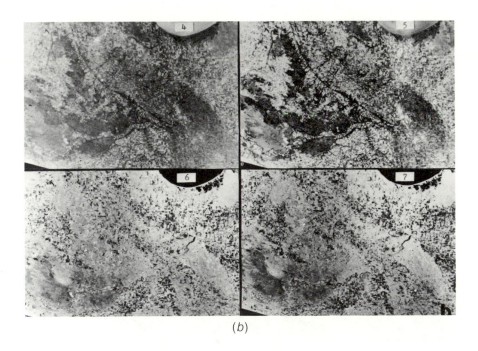

(b)

FIGURE 16.25. Parts of a series of Landsat images of an area in southern Manitoba, Canada (lat. 50°N, long. 99°W) showing multispectral responses of glacial lake sediments before and after the growing season. (a) April 27, 1973 (E1278-17001). (b) September 18, 1973 (E1422-16581). The white rectangles containing the band numbers are 10 km long. Images courtesy of the Canada Centre for Remote Sensing.

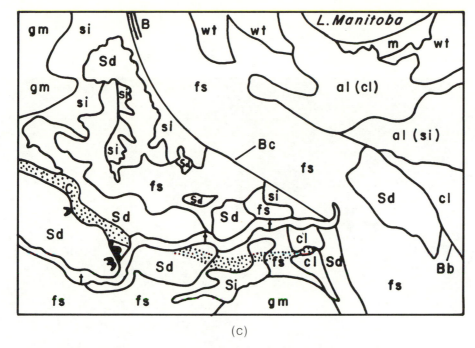

(c)

FIGURE 16.25 (c). Sketch map showing deposits of this area in southern Manitoba, Canada. Symbols as follows: al, alluvium; B, beaches of Glacial Lake Agassiz (Bb, Burnside strandline; Bc, Campbell strandline); C, relict channel (stippled); cl, clay; fs, fine sand; gm, ground moraine; m, marsh; Sd, sand dunes (active dunes shown black); si, silt; t, river terraces; wt, wave-washed till.

of sediments with diagnostic vegetation. Landsat images in bands 4 and 5 show a sand dune area in the southeast of Fig. 16.25(a) with light tones in April before leafing of the aspen woodland, and with dark tones in the same bands in September [Fig. 16.25(b)] under full foliage. Corresponding infrared bands (6 and 7) show light tones at both seasons. Sand dunes in the central and western parts of this area have a mixed coniferous (spruce) and deciduous (poplar) woodland vegetation and give medium tones in bands 4 and 5 in the spring and darker tones in the fall. The gravelly apex of this delta has a smooth surface well shown by a low-sun snow-cover Landsat image (Fig. 16.26). Beach ridges and shoreline scarps also are distinct with low-sun snow enhancement. The basin of Glacial Lake Souris is the smooth, light toned area in the southwest of Fig. 16.26; areas of sand dunes within it have dark tones formed by coniferous woodland.

Some lake basins and areas of marine submergence have complex patterns of intersecting straight and curved grooves with low bordering ridges. They are several kilometers long and have a relief of 1-2 m. They are not obvious in aerial photographs through much of the year, but are prominent during wet seasons because of soil moisture contrasts and even standing water. These grooves are thought to be furrows made by floating ice (icebergs and the keels of pressure ridges in drift ice) driven by the wind. Some are resolved in Landsat images (Fig. 16.26, northeast). They should be discernible on thermal infrared imagery. Similar grooves have been found on northern continental shelves by side-scanning SONAR.

Glacio-marine clays are subject to a nearly unique type of earth-flow landslide that usually leaves a pear-shaped scar with a narrow exit. These occur along river banks and other scarps more than about 15 m high and many superficially resemble meander scars. A few of these can be detected but rarely resolved in Landsat images. Most are smaller than two or three pixels. These slide scars are prominent in aerial photographs and SLAR images.

Eolian sediments are derived from outwash, deltaic, and littoral deposits. Loess masks pre-existing topography, especially on the downwind side of large rivers that carried glacial outwash. It gives the landscape a very uniform tone. Erosion of loess

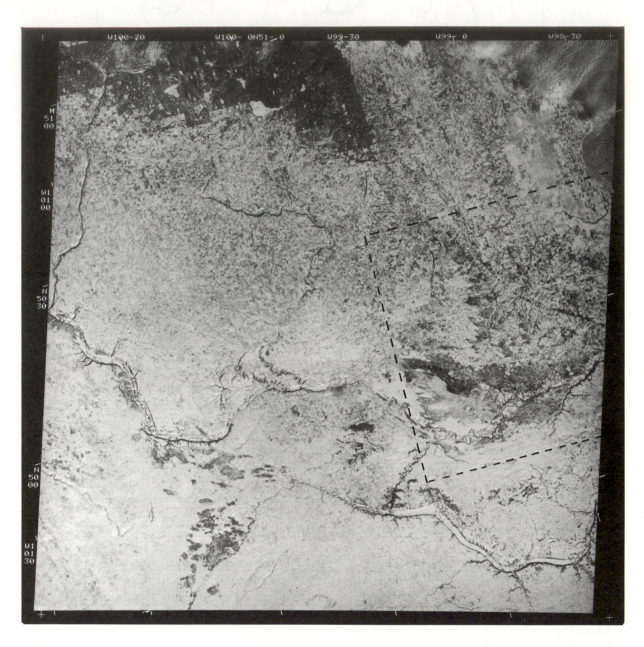

FIGURE 16.26. Landsat image of the region around Brandon, Manitoba, Canada, sun elevation of 13°. The west part of the area shown in Fig. 16.25 is outlined by dashed lines. Glacial meltwater channels, sand dunes, and hummocky moraine (below the forested area in the upper left) are enhanced. The dark areas are mainly spruce forest. Minor (corrugated) moraines are extensive in this region, but are not detected. Part of the basin of Glacial Lake Agassiz, with several visible beaches and ice furrows is in the northeast. Part of the basin of Glacial Lake Souris occupies the central-southwest part of the image. Image courtesy of the Canada Centre for Remote Sensing.

produces diagnostic vertical-sided, flat-floored gullies that form a dense dendritic drainage pattern that is nonuniform on young loess. Most of these are less than a pixel in width. Where loess is abundant in Illinois, Lineback (23) apparently did not recognize any of it in Landsat and Skylab pictures. Small gullies in loess might cause a darker average tone but additional criteria would be needed to attribute this to loess. The dissected topography should be shown well by synthetic-aperture radar [possibly (27), Image N-1086, area unidentified].

Most sand dunes in glaciated areas are stabilized by vegetation except for locally active dunes along lake shores and where vegetation has been damaged by fire, overgrazing, or drought. The dunes are the parabolic type and may be as high as 25 m. Many occur in groups with long limbs extending several kilometers upwind (Fig. 16.26). Though many dunes are in grassland, others in subhumid climates support woodland or forest because the sand functions as a reservoir of groundwater accessible to many tree species, especially in interdune depressions (Figs. 16.15 and 16.26). The infrared bands of Landsat images show hydrologic conditions in the grass-covered sand dunes of Nebraska (25, p. 356). Landsat images with low-sun snow enhancement show much detail of dune systems not otherwise visible (Fig. 16.26). SLAR images and synthetic-aperture radar should be useful also, especially for monitoring the movements of active dunes, which commonly give such a bright return in the visible spectrum that most detail is obscured in photographs.

MELTWATER CHANNELS

Steep-sided, flat-floored channels 1 to 2 km wide and 10 to 150 m deep containing small lakes and underfit streams are prominent features in many glaciated areas, especially where waning ice sheets retreated down regional slopes (24, 34). These spillways drained ice margins and some were the outlets of glacial lakes. They show well in any type of imagery and their function during deglaciation is best appreciated on the extensive aerial coverage of Landsat pictures. They are strongly enhanced in low-sun snow-cover images (Fig. 16.26). Much smaller lateral and submarginal channels eroded into both bedrock and drift occur in subparallel series almost parallel to the contours of slopes in areas of moderate relief (Fig. 16.15). They show the shift of receding ice margins and it

has been speculated that many may represent annual ice recession. Some are resolved in Landsat images (Fig. 16.27). Meltwater channels are obvious in SLAR images of central New York State (2, p. 1235) and are well displayed in thermal infrared images in southern Alberta (31).

UNDERWATER GLACIAL GEOLOGY

Glacial geology of the eastern continental shelf of North America (32) and of the Great Lakes (3) has been studied by seismic profiling, acoustical sounding, and side-scanning SONAR, which produces an image formed by back-scatter and shadows, similar to side-scanning radar. On the continental shelves morphology is less important than stratigraphy derived from seismic profiles because strong wave action erodes hills and fills in depressions. The character of a bottom echo gives information on the nature of the sea floor (soft, stony, and so forth) and side-scan images show forms produced by the processes now active. Glacial stratigraphy interpreted from continuous seismic reflection profiles enabled King and others (32) to map an end moraine complex 30-40 km off the coast of Nova Scotia for a distance of about 850 km. Elsewhere on the Labrador shelf and in the Beaufort Sea, side-scanning SONAR has revealed the depths to which icebergs and the keels of pressure ridges scour the bottom, information vital for petroleum engineering installations in these places.

In Lake Michigan, side-scan SONAR shows depressions, sand ridges, rocky and bouldery areas, and parallel ridges of till or sand aligned in the direction of glacier movement. Ice-scour grooves formed during earlier glacial lake phases have been observed on the floor of Lake Superior.

AGE OF GLACIAL DEPOSITS

Deposits of mountain glaciers belonging to seven worldwide but not necessarily comtemporary episodes of glaciation are classified according to extent, form, degree of weathering, and erosion of deposits (Table 16.4) by Gage (33). The shape and size of mountain glaciers changes as mountain valleys are enlarged by erosion. The criteria of weathering and erosion apply equally to continental glacial deposits. Weathering generally increases the clay content of the soil and, with other products of chemical alteration, reduces infiltration capacity and increases runoff and erosion. The oldest drift deposits are now represented by isolated patches

FIGURE 16.27. Part of Landsat image E1438-15012 in northern Quebec, Canada (lat. 55° 30'N, long. 67° 20'W), showing ice-marginal channels similar to those in Fig. 16.15. White rectangle bearing the band number is 10 km long. Image courtesy of the Canada Centre for Remote Sensing.

of deeply weathered material on stream divides (Fig. 16.13). Intermediate age drifts are dissected by streams, but may still have some recognizable end moraines. Young drifts have well preserved depositional landforms and show different degrees of drainage integration that depend on relative ages. Hence, sensors showing form and drainage patterns, or detecting soil characteristics, can aid the delineation of boundaries between drifts of different ages. For example, the soil on Illinoian drift in Illinois has a light gray clayey horizon (gley) a meter or two thick. In dissected areas, the upper valley slopes of expanding stream systems bevel this horizon which is then exposed by cultivation. Thus, the eroded Illinoian drift shows a light-toned area in Skylab photographs (23). In tectonically active localities, differences in the ages of moraines are shown by the relative amounts of fault displacement (Fig. 16.28) as well as by the degree of erosion. Thermal IR images should aid delineation of drift boundaries characterized by differences in soil development.

16.4 SUMMARY

Glacial geologists used vertical aerial photographs to obtain information about the ground long before modern operational remote sensing systems became generally available. For glacial geology most remote sensing studies still are experimental and have not provided very much information that is not obtained more economically from aerial photo-

graphs already available. In contrast, glaciologists use several sophisticated operational remote sensing systems that provide crucial data available by no other means. They use remote sensing to acquire information on snow that facilitates water resource management and flood control, and on the distribution, character, and movements of floating ice vital to polar navigation. Techniques for determining the distribution of ice in permafrost now under development, eventually will be applicable to geology. Glacier thickness is measured rapidly by radio echo-sounding, and ice temperatures are interpreted from passive microwave data. Landsat images enable estimates of glacier mass balance from positions of snowlines relative to mean positions established by field studies, and facilitate the monitoring of glacier movements, especially of surging glaciers.

The operational sensors used by glacial geologists respond mainly to morphology and to secondary phenomena such as vegetation patterns, most of which are shown with better resolution in aerial photographs. Radar provides morphological images through cloud and darkness in areas where optical systems seldom can function because of atmospheric conditions. Landsat images provide synoptic coverage with a perspective not achieved by assembling photographs. This perspective is invaluable for understanding regional relationships and planning research. Seasonal changes shown by the two visible and two near-infrared bands of

TABLE 16.4. DEPOSITS OF PLEISTOCENE MOUNTAIN GLACIERS CLASSIFIED BY FORM AND AGE (MODIFIED FROM GAGE, 1965)

Age	Ice Extent	Situation of Deposits	Surface Form and Weathering	Faulting (if applicable)
A (oldest)	Uncertain: probably short broad glaciers	On interfleuves, fragmentary, may be buried	Never preserved; intensively weathered; often not exposed	Appreciable or severe (maximum)
B	Extensive, long and broad; in some places the most extensive	On ridges, summits, and high terraces of the present valley	Little of original surface preserved. Very deeply weathered; large boulders only present	Appreciable displacement
C	Usually the most extensive; longer and narrower than B	Intermediate terraces, piedmont slopes, high plains	Gross features present but dissected; drainage integrated; ridges broadly rounded; a few boulders present; outwash channel patterns absent; deeply weathered.	Appreciable displacement
D	Narrow, long glaciers, less extensive than C or B	Lower terraces, piedmont slope, lower valleys	Features well preserved, little dissection; drainage insequent; ridges rounded; large boulders common; moderately weathered; outwash has visible channel patterns	Some fault displacement
E	Similar to D	Lower terraces, lower valleys	Features very well preserved without dissection; drainage unintegrated; ridges have sharp crests; boulders abundant; outwash has distinct channel pattern, weathering superficial	Not faulted
F	Less than E	Within the valley, nearer to present glacier than E	Sharp crested ridges, bouldery; vegetated; not weathered; little or no outwash	Not faulted
G (youngest)	Small extent	Near existing glaciers and higher cirques	Sharp crested bouldery moraines; not vegetated; not weathered; outwash negligible	Not faulted

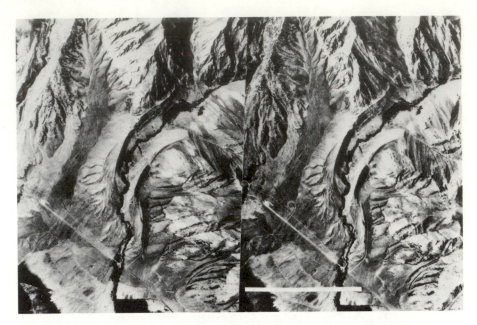

FIGURE 16.28. Stereogram of moraines deposited by Pleistocene McGee Creek glaciers in the east side of Sierra Nevada, California. The longest, oldest pair of lateral moraines is dissected by gullies. A pair of inner, shorter moraines lacks gullies. Both systems are faulted downward along the Hilton Creek fault at the mountain front. The end moraine is represented by the bulbous terminus of the oldest left lateral moraine. An outwash fan lies beyond the moraines. The white scale bar is about 1 km long. Photographs courtesy of the United States Geological Survey.

Landsat delimit several glacial deposits, and as experience accumulates, will be useful for identifying materials. Low-sun snow-cover Landsat pictures of nonforested regions have potential for showing landform details that have not yet been fully appreciated. Thermal infrared image interpretation is still experimental and gives more hydrological than geological information; computer manipulation of data from two or more bands should give results of greater geological significance. Side-scanning radar and synthetic-aperture radar provide images with synoptic value similar to Landsat images. Side-scanning SONAR has obtained information on underwater glacial geology not otherwise accessible.

REFERENCES

1. Flint, R. F., 1971, Glacial and Quaternary geology: New York, Wiley, 892 p.
2. Reeves, R. G., ed., 1975, Manual of remote sensing: Falls Church, Virginia, American Society of Photogrammetry, 2 vols., 2144 p.
3. Berkson, J. M., Lineback, J. A., and Gross, D. L., 1975, A Side-scan SONAR investigation of small-scale features on the floor of southern Lake Michigan: Environmental Geology Notes, v. 74, Illinois State Geol. Survey.
4. Glen, J. W., Adie, R. J., and Johnson, D. M., eds., 1975, Symposium on remote sensing in glaciology, Cambridge, 16-20 September, 1974: Jour. Glaciology, v. 15, no. 73, p. 1-482.
5. Bock, P., ed., 1974, Approaches to Earth survey problems through the use of space techniques: Berlin, Akademie Verlag.
6. Glen, J. W., Adie, R. J., and Johnson, D. M., eds., 1976, International symposium on the thermal regime of glaciers and ice sheets, Burnaby, 8-11 April, 1975: Jour. Glaciology, v. 16, no. 74, p. 1-316.
7. Washburn, A. L., 1973, Periglacial processes and environments: London, Edward Arnold, p. 320.
8. Schaber, G. C., Berlin, G. L., and Brown, W. E., 1976, Variations in surface roughness within Death Valley, California: Geologic evaluation of 25-cm-wavelength radar images: Geol. Soc. America Bull., v. 87, no. 1, p. 29-41.
9. Gatto, L. W., and Anderson, D. M., 1975, Alaskan thermokarst terrain and possible Martian analog: Science, v. 188, p. 257-275.

10. Shearer, J. M., McNab, R. F., Pelletier, B. R., and Smith, T. B., 1971, Submarine pingos in the Beaufort Sea: Science, v. 174, p. 816-818.

11. Gunning, H. C., ed., 1966, Symposium on glacier mapping, Ottawa 1965: Can. Jour. Earth Sci., v. 3, no. 6, p. 737-915.

12. Meier, M. F., and others, 1966, Multispectral sensing tests at South Cascade Glacier, Wash., in Proceedings fourth symposium on remote sensing of environment 1966: Willow Run Laboratories, Inst. Sci. and Technology, Univ. Michigan, Ann Arbor, Mich., p. 145-159.

13. Welch, R., 1968, Color aerial photography applied to the study of a glacial area, in Smith, J. T., ed., Manual of color aerial photography: Falls Church, Virigina, American Society of Photogrammetry, p. 400-401.

14. McLerran, J. W., 1965, Airborne crevasse detection, in Proceedings third symposium on remote sensing of environment 1964: Willow Run Laboratories, Inst. Sci. and Technology, Univ. Michigan, Ann Arbor, Mich., p. 801-802.

15. Bauer, A., Fontanel, A., and Grau, G., 1967, The application of optical filtering in coherent light to the study of aerial photographs of Greenland glaciers: Jour. Glaciology, v. 6, no. 48, p. 781-893.

16. Elachi, Charles, and Brown, W. E., Jr., 1975, Imagery and sounding of ice fields with airborne coherent radars: Jour. Geophys. Research, v. 80, no. 8, p. 113-119.

17. Doake, C. S. M., Gorman, M., and Paterson, W. S. B., 1976, A further comparison of glacier velocities measured by radio-echo and survey methods: Jour. Glaciology, v. 17, no. 75, p. 35-38.

18. Ambrose, J. W., ed., 1969, Papers presented at the seminar on the causes and mechanics of glacier surges, St. Hilaire, Quebec, Canada, September 10-11, 1968 and the symposium on surging glaciers, Banff, Alberta, Canada, June 6-8, 1968: Can. Jour. Earth Sci., v. 6, no. 4, pt. 2, p. 807-1018.

19. Welch, R., 1969, Reflectance characteristics of a glacial landscape and their relation to aerial photography, in Anon, ed., Seminar proceedings, new horizons in color aerial photography: Am. Soc. Photogrammetry and Soc. Photog. Sci. and Eng., June 9-11, 1969, p. 17-35.

20. Thwaites, F. T., 1947, Use of aerial photographs in glacial geology, Photogrammetric Eng., v. 13, no. 4, p. 584-586.

21. Way, D. S., 1973, Terrain analysis: Stroudsburg, Pa., Dowden, Hutchinson and Ross, p. 197-283.

22. Waldrop, H. A., 1971, Thermal infrared detection of glacial gravel, Yellowstone National Park, Wyoming, geological survey research 1971: U.S. Geol. Survey Prof. Paper 750-B, B-202-B206.

23. Lineback, J. A., 1975, Illinois geology from space: Environmental Geology Notes, v. 73, Illinois State Geol. Survey, Urbana, Illinois.

24. Flint, R. F., and others, 1959, Glacial map of the United States east of the Rocky Mountains: Geol. Soc. America.

25. Freden, S. C., Mercanti, E. P., and Becker, M. A., eds., 1973, Symposium on significant results obtained from the Earth Resources Technology Satellite-1; NASA SP-327, Washington, D.C.

26. Barry, R. G., Andrews, J. T., and Mahaffy, M. A., 1975, Continental ice sheets: conditions for growth: Science, v. 190, p. 979-981.

27. Goodyear Aerospace, Developing Earth resources with synthetic aperture radar: Goodyear Aerospace, Litchfield Park, Ariz., GIB-929OE, (no date).

28. Nyberg, Sten, Orhaug, Torleiv, and Svensson, Harold, 1971, Optical processing for pattern properties: Photogrammetric Eng., v. 37, no. 6, p. 547-554.

29. Colwell, R. N., ed., 1960, Manual of photographic interpretation; Falls Church, Virginia, American Society of Photogrammetry, p. 185-192; 373-382.

30. Parry, J. T., 1974, X-Band radar in terrain analysis under summer and winter conditions, in Proceedings second Canadian symposium on remote sensing, Thompson, G. E., ed.: Canadian Remote Sensing Soc., Ottawa, p. 471-485.

31. Holmes, R. M., and Thompson, D., 1973, Infrared remote sensing in Quaternary research: Geoexploration, v. 11, p. 249-267.

32. King, L. H., MacLean, Brian, and Drapeau, Georges, 1972, The Scotian shelf submarine end-moraine complex, in Gill, J. E., ed.: 24th Int. Geol. Congress, Sec. 8, Marine Geology and Geophysics, p. 237-249.

33. Gage, Maxwell, 1965, Accordant and discordant glacial sequences, in International studies on the Quaternary INQUA 1965, Wright, H. E., Jr., and Frey, D. G., eds.: Geol. Soc. Am., SP-84, p. 393-414.

34. Prest, V. K., Grant, D. R., and Rampton, V. N., 1968, Glacial map of Canada: Geol. Surv. Canada, Map 1253A.

CHAPTER 17

MINERAL EXPLORATION
LAWRENCE C. ROWAN
ERNEST H. LATHRAM

17.1 INTRODUCTION

In the last several decades, there has been an increasing awareness of a constantly expanding demand for minerals to support an industrial economy, and a constantly decreasing supply of easily locatable and exploitable sources of those minerals. New methods to search for concealed deposits, and new ideas on where they may occur are being actively sought. Remote sensing, as a direct adjunct to field lithologic and structural mapping, has played an important role in the study of mineralized areas since aerial photography became readily available in the early 1950s. As the techniques and data of remote sensing have become more varied and sophisticated in the intervening years, and remote sensing from satellites has been developed, two avenues of study have shown most promise of benefit to mineral exploration—the regional study of linear features and analysis of areas of alteration through the use of multispectral data.

Because of the complexities of both approaches and rapidly evolving technology, neither approach is treated fully here. Instead, emphasis is placed on the methodology currently being used to detect and analyze these features. Although linear features and altered rocks are treated separately, integrated study of these features is more promising than either approach taken separately.

17.2 STUDY OF LINEAR FEATURES

The study of linear features in images and photographs is the study of structural geology, as most of these features represent the surface expression of faults, joints, folds, lithologic contacts, or other geologic discontinuities, all of which are clues to the internal geometry of the lithosphere. Park and McDiarmid have observed that structural studies "have unquestionably led to the discovery of more ore than has any other directed effort; this is because the movements of fluids underground is controlled by permeability, a function of the original character of the rock, plus the elements of superimposed structure" (1). Although they were referring to detailed structural studies, by extension their statement may be applied more generally to metallogeny (the study of the genesis of ore deposits), as most deposits are related to some type of deformation of the lithosphere, and most theories of ore formation and concentration embody tectonic, or deformational concepts. As linear features shown on remote sensing imagery of increasingly smaller scale [greater extent] reflect increasingly more fundamental structures, their study will provide insights not only to the location of ore bodies and mineralized districts, but also to metallogenic theories as well.

17.2.1 Nature of Linear Features

Although numerous efforts have been made to classify linear features in terms of their origin, mode of expression, or size, a more useful concept for mineral exploration is one based on the scope of deformation these features reflect. Hence, some linear features indicate the character of individual structures or local groups of structures, others delineate the general features of a regional structural regime (foldbelt, platform, basin), and still others signify fundamental tectonic elements in the crust that have governed the response of the lithosphere to earth stresses.

LINEAR FEATURES INDICATING LOCAL STRUCTURES

These features are clearly discernible on aerial photographs, but are poorly seen in satellite imagery. They are generally 10 km or less in length, and their study is best accomplished through aerial photo interpretation as a direct adjunct to field mapping. These linear features indicate the form and position of individual folds, faults, joints, veins, lithologic contacts, and other detailed geologic features that may lead to the location of individual ore bodies. In general, they reflect only immediate surface and near-surface conditions, and are poor guides to concealed deposits. Their geologic significance is readily determined from surface and near-surface geologic and geophysical data. The only discernible pattern in these linear features is the pattern of the structure or structures with which they are associated.

In most studies of the nature and distribution of linear features in readily available satellite imagery, these features are excluded.

LINEAR FEATURES INDICATING REGIONAL STRUCTURAL REGIMES

These are linear features generally mapped in studies of Landsat imagery. They are mostly from 10 to 200 km in length, although in some areas they are longer, and some may be parts of much longer linear features of more fundamental significance. For example, the Midas Trench system in Nevada (2) is a part of a Landsat linear feature extending from Montana to western California, a feature that coincides with a basement structure that Eaton and others (3) believe controlled the emplacement of the Snake River basalts (I, Fig. 17.6).

Linear features of regional structural regimes indicate the general geometry of folds, faults, and other structures that characterize a foldbelt, or the joints and other structural patterns of a platform area, and the contacts of the gross lithologic units involved. They are less abundant than the shorter linear features by probably an order of magnitude, and are more abundant than the much longer ones to about the same degree. In Alaska, the trends of these linear features are also much more varied than the trends of the longer ones. Comparative plots of the linear features observed in a Landsat mosaic of the Yukon-Tanana Upland of eastern Alaska illustrates these characteristics (Figs. 17.1 and 17.2).

A regional pattern is commonly observed in linear features indicating regional structural regimes. This pattern, however, is determined by the internal character of the tectonic element, and is commonly different in adjacent or nearby tectonic elements of differing structural style. Carter (4) has prepared a preliminary Landsat lineament map of the conterminous United States (Fig. 17.3); a relative change in density and pattern of linear features from area to area is readily apparent. A count of 1585 linear features recognizable on Landsat images of the

United States was also made by Haman (5), as well as a series of contour maps of linear feature density and trend. By comparing these contour maps to maps showing the general distribution of gross tectonic elements, he demonstrated that both linear feature density and trend distribution vary from element to element. Haman also found the average length of the linear features he measured to be 156 km.

Linear features of this group can generally be related to surface geologic data ("ground truth") or to near-surface magnetic data.

These linear features are useful in defining target areas—local settings in which ore bodies may be concentrated, and which merit more detailed study in the field. A strong positive spatial correlation of mineral districts with these linear features has been noted in many areas, among them eastern Alaska (6), Nevada (2), Montana, Colorado, New Mexico, and central Canada (7). In all these areas, a high percentage of the known mining districts occurs within 2-7 km of linear features. Albert (6) also noted that in the Nabesna area of eastern Alaska, a large number of anomalous areas not previously considered to be mineralized, which may represent alteration zones, also lie along or near the linears. In South Africa, the distribution of foldbelts surrounding Precambrian nuclei has been mapped by study of these linear features, and in one area the extension of a zone of potential mineralization resulted (8).

LINEAR FEATURES INDICATING CRUSTAL TECTONIC ELEMENTS

Linears that indicate crustal tectonic elements are more than 200 km long, and generally are 1000 km or more in length. They commonly mark all or parts of the boundaries of regional structural elements, or zones separating areas within these elements in which local or regional patterns of type or trend of structures are divergent. On the other hand, they may traverse the entire extent of a regional structural element, or transgress the boundaries between elements having structural regimes of differing types, with little major change in the surface geology.

These linear features occur as alignments that are combinations of surface geologic structures, linear valleys or ridges, and linear changes in tonal contrast marking differences in soil type, moisture, or vegetation. Most are broad and diffuse, some

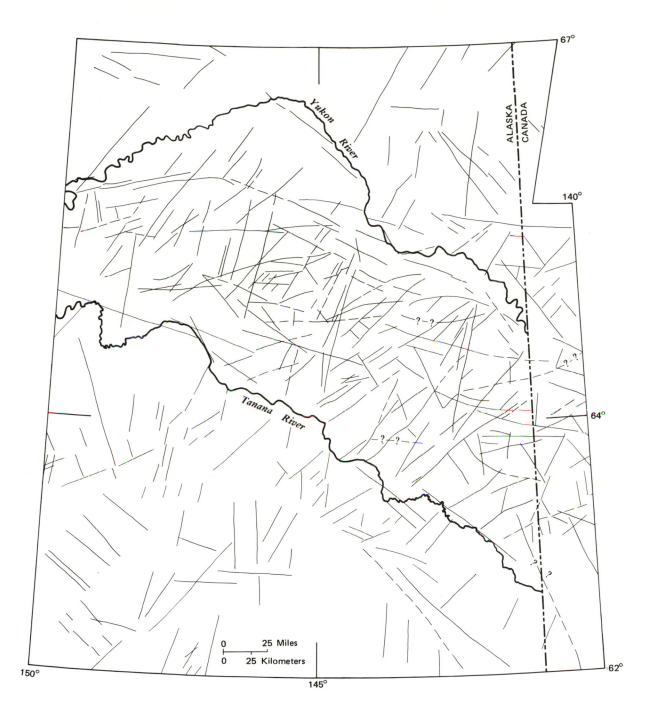

FIGURE 17.1. Map showing linear features less than 160 km in length in the Yukon-Tanana Upland area.

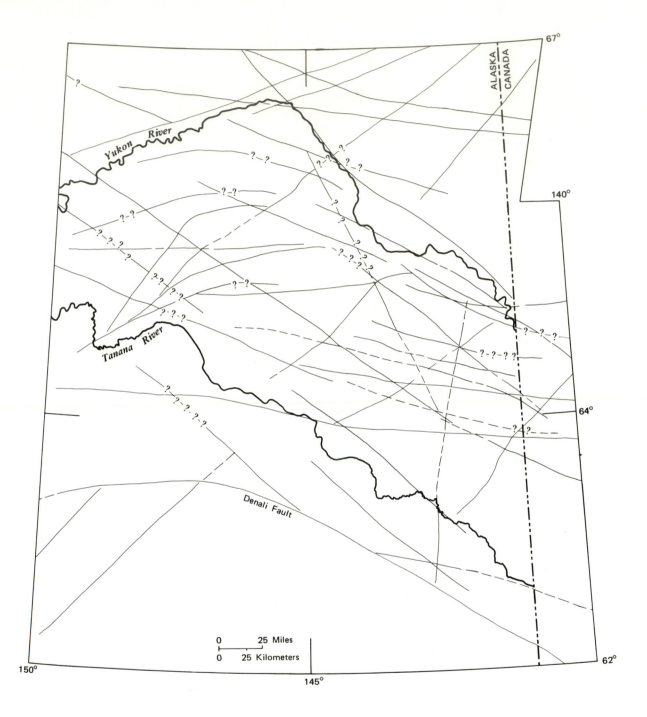

FIGURE 17.2. Map showing linear features greater than 160 km in length in the Yukon-Tanana Upland area.

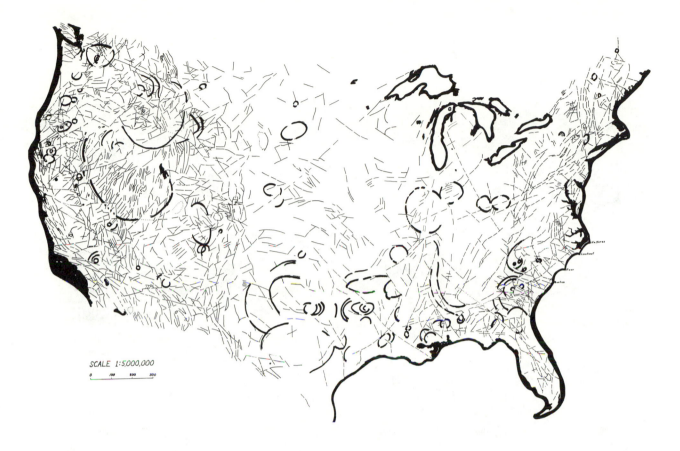

SCALE 1:5,000,000

0   100   200   300

FIGURE 17.3. Preliminary map of linear curvi-linear features on Landsat imagery of the conterminous United States.

being poorly identifiable for short stretches along their length. Parts of the trace of many of these linear features coincide with the trace of major known faults. A portion of the Landsat mosaic of the United States prepared by the Soil Conservation Service, Department of Agriculture, on which some of these features can be seen (Fig. 17.4), illustrates their subtle nature.

Because of the great length of these linear features, they are most effectively studied on mosaics of Landsat imagery, or on smaller-scale images from other satellites. The number of these features identifiable increases from Nimbus to NOAA to Landsat views, that is, from smaller-scale, low resolution images to larger-scale, higher resolution images. This is readily apparent from the linear features that can be seen on views from Alaska (Fig. 17.5). Most of the features visible on Nimbus and NOAA images can be identified on Landsat images,

but on the smaller-scale images they tend to be more diffuse and can be less accurately located. The more synoptic view of the smaller-scale images, however, permits recognition of the very longest linears, which may not be apparent on Landsat images, and also shows that some of the shorter features are in reality probably segments of a much longer one that is only intermittently perceptible. On the other hand, the sharpness of features in the higher resolution images may lead to selection of unrelated shorter linear features as representing the trace of the crustal feature, and cause its mislocation. Some linear features visible on the low resolution images are less easily recognized on the Landsat images. Examples are the curvilinear and circular features shown on the NOAA image of Alaska [(Fig. 17.5(b)]. They can be identified on the Landsat mosaic, but were recognized only after they were seen on the NOAA image.

FIGURE 17.4. Landsat mosaic (Band 5) of parts of California, Oregon, Idaho, and Nevada, showing appearance of selected linear features indicating crustal elements (compare with Fig. 17.6), and examples of circular features of different sizes and settings. Annotations (A-A', etc.) indicate trends of individual features but are not discussed in detail in text. Mosaic excerpted from Landsat mosaic of United States by Department of Agriculture, Soil Conservation Service.

558

A systematic pattern is discernible in the distribution of these linear features. In Alaska, the pattern is a nearly orthogonal one [Fig. 17.5(d)] and not only characterizes the whole state, but is recognizable throughout the northern and central parts of the North American Cordillera (Fig. 17.6). Parts of similar patterns have been recognized in local areas in other parts of North America, and on other continents as well. Hence, these linear features seem to be globally ubiquitous, and everywhere display a nearly orthogonal pattern. It is tempting to assign firm compass measurements to the trends of these linear features, and this is commonly done in local studies. However, the trends of most in the North American Cordillera are not sufficiently consistent, and the orientation of the overall pattern seems to vary slightly from area to area. This may partly be due to map projection. Although comparisons of these linear trends to those reported in other areas suggest a variation of as much as ±10°, still there seems to be a fairly consistent pattern of compass directions comparable to those of the North American Cordillera. These are roughly N40E, N60E, N50W, N80W, and N.

17.2.2 Origin

A concensus has been reached among those studying linear features that these very long elements reflect discontinuities within the crust. Terms such as "crustal flaws," "deep tectonic zones," "zones of crustal weakness," and "tectonic block" or "crustal block" boundaries have been employed. Numerous examples exist of the coincidence of these features with alignments of anomalies or discontinuities in contour maps of the gravity and magnetic response of deep materials, and with major tectonic discontinuities such as the Grenville Front, the western boundary of the Idaho batholith, and the postulated north-trending line along it separating Sr [86-87] ratios, indicative of differing crustal nature on the two sides (J, Fig. 17.6), the Lewis and Clark (K, Fig. 17.6) and Texas Lineaments and the MacDonald fault, and associated tectonic discontinuity between the Churchill and Slave provinces in the Precambrian shield of Canada (L, Fig. 17.6). Russian geologists have also shown that some of these features are related to differences in the mantle, reflected in variations in the Mohorovicic discontinuity (Sergei Strelnikov, oral communication, 1975). In all of these examples, and along other linear features as well, there is evidence of recurrent movement since Precambrian time and presumably since the origin of the crust. An examination of many of the examples of lineaments believed to be crustal in origin, which were described from known surface and subsurface geology prior to the flights of the satellites (Fig. 17.7) (9-24), shows a very strong coincidence between these and the pattern of space image linear features indicating crustal structure in the Cordillera.

A common misconception is that, since many of these long linear features (both those on satellite images and those observed by study of various maps) can be demonstrated to coincide with crustal discontinuities, *all* such linear features have similar characteristics. In fact, many of the linear features cannot be clearly demonstrated to be related to deep structures. A single prime example will suffice to illustrate this point—the Rocky Mountain Trench in southern Canada. This well-known feature has been studied by many geologists for many years, but as yet all evidence indicates that the Precambrian basement and overlying early Paleozoic sediments dip smoothly beneath the trench without evidence of disruption. Strong evidence of deep discontinuities or fundamental tectonic changes are necessary to identify significant individuals of this group of linear features. The viability of tectonic or metallogenic concepts developed through interpretation of these features accordingly will depend on the careful selection of those that are really significant and will require extensive geologic, ore deposit, and tectonic knowledge and skill in interpreting their history.

Most interpretations of the character of linear features reflecting crustal structures refer to, or modify, the thesis of the Global Regmatic Shear pattern developed by Sonder, Vening Meinesz, and others (25). According to this concept, lineaments in the surface of the Earth indicate a regular fracture pattern that pervades the globe, generated by the forces of the rotation of the body on its axis. Flattening of the earth by north-south compression along its axis would produce a set of north-trending tensional and east-trending compressional weakness zones, and northwest- and northeast-trending shears. These would be propagated throughout the life of the planet. Katterfeld and Charushin (26) later recognized such a pattern on satellite imagery of the moon and other planets, and reaffirmed its existence on Earth through study of 1:1,000,000 scale top-

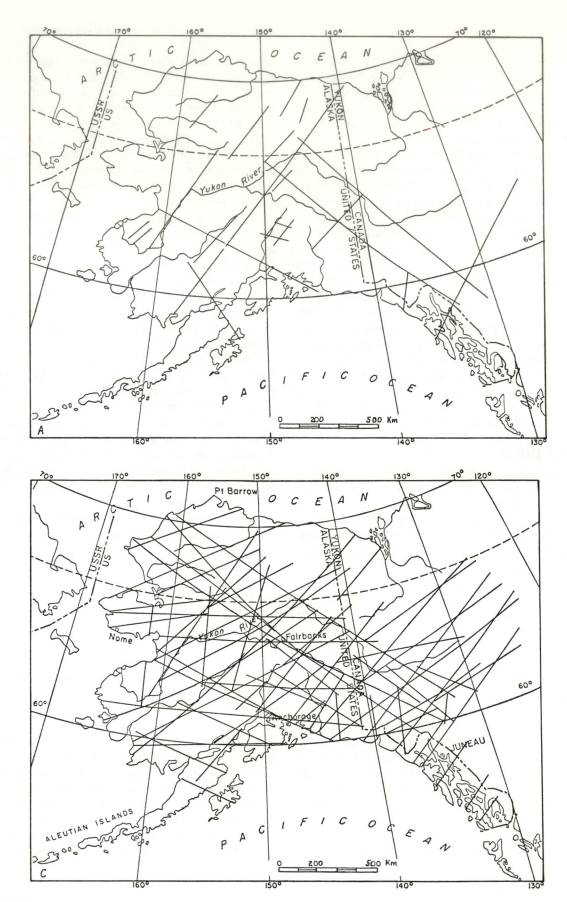

FIGURE 17.5. Linear features indicating crustal tectonic elements in Alaska on (*b*) NOAA, (*a*) Nimbus, and (*c, d*) Landsat imagery.

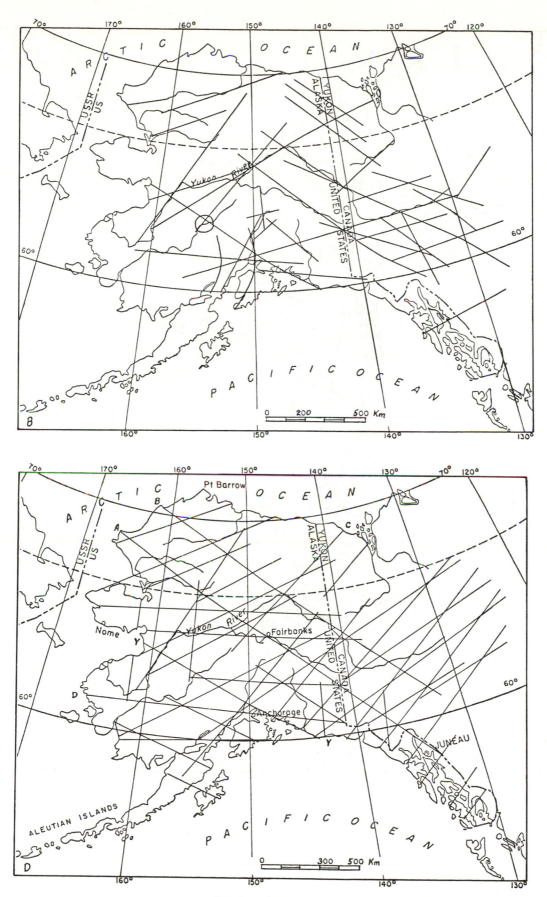

FIGURE 17.5 Continued

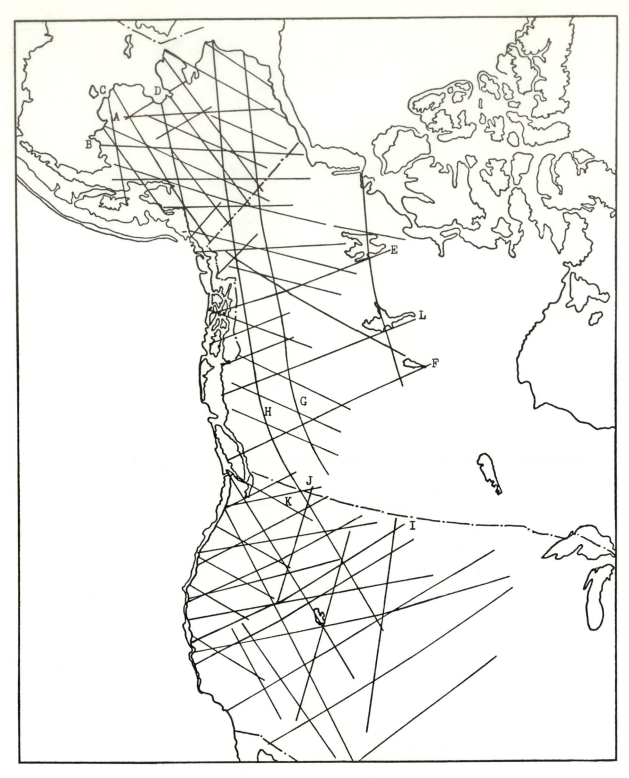

FIGURE 17.6. Pattern shown by selected linear features indicating crustal tectonic elements in the North American Cordillera. See text discussion for an explanation of symbols.

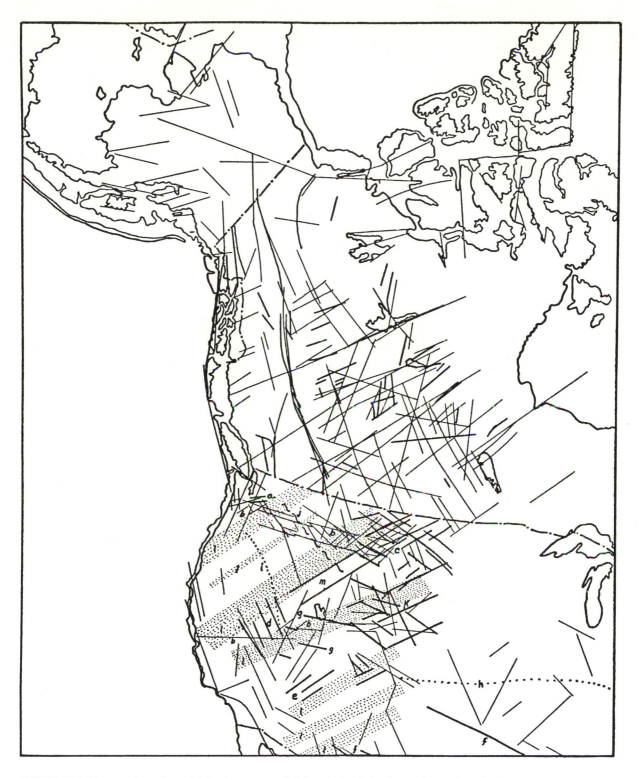

FIGURE 17.7. Lineaments and crustal fractures suggested in selected inductive analyses. Compiled from: Moody and Ozoray, (most long lineaments, fine line); and (*a*) Grant, (*b*) Zietz and others, (*c*) Thomas, (*d*) Roberts, (*e*) Shoemaker and others, (*f*) Muehlberger, (*g*) Cook, (*h*) Heyl, (*i*) Stewart and others, (*j*) Yates, (*k*) Maughan, (*l*) Landwehr, (*m*) Eaton and others, (in western Canada) Haites.

ographic maps. However, not one system, but many, of compressive tensional and shear structures has been recognized on Earth, whose trends are not easily reconcilable with the basic regmatic shear pattern.

Moody and Hill (27), analyzing the known great fault systems of the world and the structures associated with them, developed the concept of Wrench Fault Tectonics. They postulated that lineaments on the Earth's surface marked basement zones of weakness developed as conjugate shear fractures during Precambrian orogenies by the imposition of pure-shear north-south compressive stresses. Continued imposition of nearly the same direction of stress during later orogenies would develop up to eight orders of associated folds and fractures at differing angles to the primary shear, depending on the order of the system observed locally. While this concept rationalized numerous deviations from the pattern expectable in the regmatic shear concept, it has not been widely accepted.

Saunders and others (7) examined these concepts, and postulated a system of simple-shear movements parallel to the lineaments which they believe reflect the movement of crustal blocks laterally past each other. Lineaments they consider as boundaries of fundamental crustal blocks reactivated by differing directions of tectonic stress at different times. In simple-shear block coupling, the differential stress along the bounding primary shears would produce drag folds, shears, and tension fractures at angles to the primary shears, angles that would decrease as the intensity of stress increased. In analysis of Landsat linear features in northern Alaska, Montana, Colorado, New Mexico-west Texas, and in the Superior province of Canada, they were able to demonstrate a positive correlation between known structures and those expectable by interpreting Landsat linears indicative of crustal structures as primary shears—the boundaries of crustal blocks involved in block-coupling. This concept may be a viable approach to analyzing the linear features of a structural regime disrupted by a crustal linear feature.

Garson and Krs (28), using primarily a wealth of geophysical data, and building on the concept of "deep tectonic zones" marking the boundaries of crustal blocks developed by Slavic geologists and expressed by Stovickova (29), identified a pattern of northwest- and northeast-trending linear features in the Red Sea area, many of which they also recognized on Landsat images. From their geophysical data, they concluded that these linear features expressed the boundaries of blocks within the crust. In considering the rift origin of the Red Sea, they pointed out that the plate divergence occurred along one of the northwest-trending linear features, a pre-existing fracture in the continental crust. Further, they showed that the northeast-trending linear features cross the Red Sea and its central rift, and many, where they intersect the rift, are the sites of mineral-rich brine pools or sediments. At the same time, E. H. Lathram and R. G. H. Raynolds (30) suggested that the pattern of linear features noted on the Landsat images of the North American Cordillera (Fig. 17.6) indicated a mosaic of crustal blocks whose vertical as well as horizontal adjustments through time had guided tectonic development in the Cordillera. They presented evidence of disruption in elements of the Aleutian Arc complex in Alaska, suggesting that the forces of plate convergence in that area have been relieved along northwest-trending pre-existing crustal fractures marked by Landsat linear features that coincide with crustal block boundaries.

The new concept of global tectonics, embodying drifting and rotating continents, raises some challenging and as yet unsolved problems with respect to satellite linear features and crustal block boundaries. Continental movement would help to explain some of the angular deviations from a regmatic shear pattern that have been noted. At the same time, the evidence of coherent patterns of space image linear features covering large areas, parts of which have been considered to be extremely mobile, and of the general congruence of the trends of such linear patterns in various parts of the Earth, imply a less mobile lithosphere.

17.2.3 Linear Features and Metallogenic Models

Earlier metallogenic models have been based on the classical geosynclinal theory, and generally have related mineral deposits to crustal downwarping, plutonism, and associated hydrothermal activity, primarily in mobile zones. Most recent models are based on contemporary plate tectonic theory and associate mineral deposits with subduction or mantle upwelling along plate margins. Repeated tectonism and plutonism, changes in the rate of subduction, the development of new plate margins, or the migration of plates over mantle plumes are believed to generate differing epochs of mineralization.

Lineament studies of continental masses have generally presumed mineral deposits to be related to crustal features, or zones of weakness, which have provided pathways for mineralizing agents to rise from a general source in the lower crust or upper mantle and form deposits wherever the host conditions were propitious.

Although the application of the study of satellite linear features to the interpretation of tectonic history and processes and to metallogeny is in its infancy, the results to date are encouraging. This is particularly true of the study of linear features that indicate crustal structure, and of the concept that these features represent the boundaries of crustal blocks whose differential vertical and horizontal movements have significantly affected response to tectonic stress within the lithosphere.

17.2.4 Linear Features as "Plumbing Systems"

In an early study, Lathram and Gryc (31) related some of the Nimbus and Landsat image linear features to the distribution of mineralized areas in Alaska, and compared this with the conclusions of Sutherland-Brown and others (32) that areas of mineral concentration in British Columbia were related to the intersection of major through-going orthogonal structures (Fig. 17.8). Comparison of the space image linear feature pattern of the entire Cordillera to the empirical "shear-stress network" developed by Kutina (33) in the western United States, shows a striking correlation between the two patterns in orientation and spacing of individual elements. In addition, a large number of the linear features of the space image linear pattern cross a number of highly mineralized areas, and a large number of the intersections of the space image linear features coincide with metal-rich areas. The trends and spacing of linear features in the space image pattern correspond closely to the trends and spacing of the "deep-seated tectonic zones" recognized by Garson and Krs (28) in the Red Sea area. Garson and Krs showed these zones to be related to mineralized areas on land as well as to brine pools in the Red Sea.

In Nevada, Rowan (2) pointed out that the preponderance of mineralized districts occurs along two Landsat linear features, the northeast-trending

Midas Trench system and the northwest-trending Walker Lane.

Study of Landsat linear features in South America employing computer enhancement to accentuate poorly visible linear features, resulted in the discovery by T. Offield (34) of a significant east-trending ore-controlling linear feature. This feature extends from the continent across the Atlantic continental shelf, and is coextensive with a transform fault crossing the southern Atlantic Ocean. In Africa, coextensive with the transform fault, is a linear zone of mineralization whose pattern of distribution of minerals is comparable to that along the linear feature in South America.

17.2.5 Linear Features as Guides to Crustal Block Tectonics

Consideration of space image linear features only as indicators of crustal discontinuities that have provided pathways for the movement of mineralizing agents seems too simplistic. Also, it fails to take into account the known affinity of particular ore suites with certain lithologic associations and tectonic settings. However, the concept of linear features as boundaries of crustal blocks whose varied movements may have resulted in varied tectonic histories has great promise for metallogenic analysis.

Many metallogenic experts, especially Clark and others (35) in Alaska, Sutherland-Brown (36) in British Columbia, Sillitoe (37) in the Andes, and Guild (38) have pointed to the existence of broad, linear zones of like mineralization, parallel to major tectonic belts, in many areas. Sutherland-Brown studied the regional distribution of metals in British Columbia, as shown by geochemical sampling, and concluded that the similarity between the pattern of the mineral deposits and that of the metal background means that the deposits owe their origin to the level of background in the terrain they inhabit— "the rocks have the ore deposits they deserve." Many others have considered mineral deposits as formed by "lateral secretion"—the deposits have been "sweated out" of surrounding rocks and concentrated (39).

Sillitoe examined the known linear zonation of mineral deposits in the Andes, and recognizing geologic discontinuities transverse to the length of the zones, subdivided the Andes into tectonic segments, each with a metallogenic character different from that of its neighbors. In Alaska, the lower

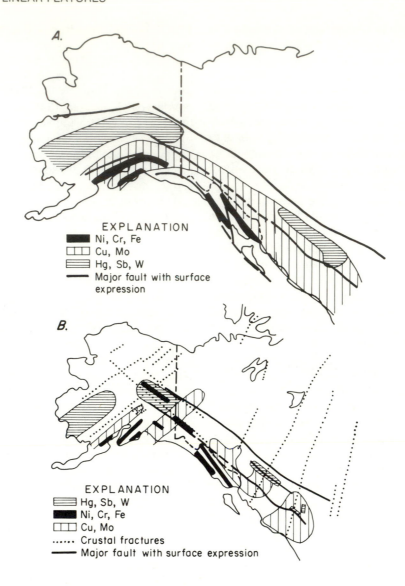

FIGURE 17.8. Favorable areas in Alaska and western Canada for location of selected metals.

Kuskokwim River area is exceedingly rich in mercury as compared to the rest of the state. The area of mineralization is not related to a single linear feature, but is largely confined between two northeast-trending and two northwest-trending long linear features (Fig. 17.6 A, B, C, D). In British Columbia, the rich mercury area is similarly confined between two northeast-trending features and between the northwest-trending Rocky Mountain Trench and the Denali-Yalakom-Fraser fault system (Fig. 17.6 E, F, G, H).

Although only suggestive, these observations indicate that a differing history experienced by individual lithospheric blocks as a result of the vertical or horizontal relief of internal or external tectonic forces along the boundaries of crustal blocks may also determine the metal fecundity of an area.

17.2.6 Circular Features

Circular features may be considered as a special case in the study of linear features. They are identified by the same kind of features as others—morphology and tonal contrast. They do not seem to be arranged in regional patterns. Some occur on or near major linear features, and some seem disrupted by such features. Many others, however, have little or no relation to long linear features.

The geologic identity of circular features tens of kilometers or less can generally be determined by the surface geology, or inferred by the tectonic setting in which they occur. Those in volcanic settings, such as in Oregon and Washington [some previously unknown near Portland] (W. D. Carter, written communication, 1972) are probably concealed calderas, as are those in Nevada. Others possibly mark the outline of subjacent igneous intrusives, as in Arizona, and still others describe the outline of alteration zones around igneous bodies, as in eastern California (40). A number of authors have considered many of these features to be impact craters, and some undoubtedly are.

The origin and nature of circular features hundreds of kilometers in diameter, or features of such size suggested by large-radius curvilinear features is more questionable. Many of these have been indicated by Carter (4) (Fig. 17.3). Lowman (41) has suggested a theory of evolution of the earth's crust based on observations of the Moon, Venus, and Mars that would be commensurate with an interpretation that these features represent Archean craters similar to those visible today on other planets. Smith (42) on the other hand, interprets curvilinear and circular features in the Basin and Range province of the western United States as logical structural resultants of the segmentation of the continental crust into smaller plates. Except in central Nevada, little evidence of congruity between magnetic and gravity data and the large circular features has been adduced as yet, but a correlation between many of these features and the distribution of basement tectonic elements is evident on the Basement Map of the United States (43).

The known affinity of mineral deposits for volcanic necks, pipes, ring dike complexes, caldera, diatremes, impact craters, alteration zones around intrusive bodies, and near-circular porphyritic intrusives makes the study of these important for mineral exploration. At present, however, no general systematic classification, or theory of occurrence and relation to mineralization has been presented. Each must be studied individually in the context of its local setting.

17.2.7 Some Observations on Linear Features and Mineral Exploration

Multistage remote sensing techniques provided by satellite plus aircraft sensors permit a deductive approach to mineral exploration—a successive "zeroing-in" on potential targets—and a more effective utilization of costly surface and subsurface exploration methods, replacing outmoded prospecting and inductive techniques.

Imaginative interpretation of space-image linear features by individuals well versed in traditional as well as contemporary tectonic thought, and experienced in the associations and habitats of mineral concentrations can produce effective alternative hypotheses for mineral exploration.

Consideration of linear features indicative of crustal structure as not only pathways for mineral movement and loci for their concentration, but also as indicators of fundamental boundaries indicating blocks that may have experienced differing tectonic histories and may in consequence possess differing metallogenic character will provide a powerful tool for metallogenic analysis.

17.3 DETECTING AND MAPPING ALTERED ROCKS

Many factors must be considered in formulating a remote sensing strategy for mapping altered rocks, but most important is selection of a diagnostic physical property that can be measured by airborne and satellite-borne instruments. Although the properties traditionally used in the field for studying altered rocks—mineralogical composition and small-scale texture—cannot be measured directly from remote platforms, several related properties provide bases for remotely *distinguishing* most altered and unaltered rocks. In many places, studies of these physical properties can also be useful for placing bounds on the surface and near-surface mineralogical compositions. *Identification* without some *a priori* knowledge of the rocks present is rarely possible, however.

Results of aircraft and satellite experiments and evaluation of laboratory and field measurements indicate that the most useful physical properties are:

1. Visible and near-infrared spectral reflectivity.
2. Thermal inertia.
3. Thermal-infrared spectral emissivity.
4. Gamma-ray spectral emission.

Techniques based on analysis of visible and near-infrared spectral reflectivity are most important because their usefulness has been more clearly established for mapping certain types of alteration. Evaluation of the other techniques is still in the initial stages, but these techniques appear to have considerable potential for detecting and mapping some alteration products that do not have distinctive visible and near-infrared spectral reflectivities. Thus, this discussion progresses from examining the clearly established techniques to considering the promising but less fully evaluated methods.

These techniques are based on measurement and analysis of phenomena that are related to the composition of the surface and near-surface materials. Because each technique provides a different type of compositional information, they are complementary rather than redundant. Ideally, all four techniques would be used in surveying an area, but because in most places this approach would be prohibitively expensive, some selection of methods is desirable. In areas of reasonably good exposures, the subject of this discussion, selection of the most effective technique, should be based on the expected mineralogical differences between the altered and unaltered rocks and soils and the types and magnitudes of measurable related physical properties.

Systematic studies of the relationship between these physical properties and mineralogical composition have not been conducted for all, or even most, types of alteration. The most extensive studies have dealt with hydrothermal alteration, but these studies are far from complete. Recently, alteration associated with sedimentary uranium deposits has received considerable attention because of the national uranium/thorium resource assessment program. The knowledge gained thus far from the hydrothermal alteration studies is especially important because of its broad economic importance, and when supplemented with laboratory measurements of the physical properties listed above, it provides a basis for evaluating the potential of these remote sensing techniques for detecting and mapping other types of altered rocks.

Thus, this discussion will concentrate on hydrothermally altered rocks, with emphasis on the relationship between their mineralogical composition and the related physical properties listed above. Discussion of sedimentary uranium deposits will be limited to their pertinent visible and near-infrared spectral-reflectivity characteristics because the ap-

plication of gamma-ray radioactivity surveys has been described in many papers and texts. Although the vigor, type, and areal distribution of vegetation are potential indicators of alteration and metallization, this topic is not discussed here because it is treated in detail in Chapter 15. Vegetation is mentioned only when its presence obscures measurement of the physical properties of rocks and soils and influences interpretation of the resulting data.

Discussion of interpretive methods will be keyed to the analysis of the visible and near-infrared spectral-reflectance characteristics of hydrothermally altered rocks because many of these analytical techniques are also applicable to other wavelength regions. Emphasis will be placed on the limitations and capabilities of each technique for dealing with various types of hydrothermally altered rocks.

17.3.1 Hydrothermal Alteration

MINERALOGY

Studies of hydrothermal ore deposits often concentrate on the compositions and textures of the ore minerals and are carried out because of the genetic as well as obvious economic importance of the ore minerals. However, many ore minerals, especially metals, are not exposed at the Earth's surface because they are unstable or only metastable. Furthermore, concentrations of most ore-forming minerals are too low to be resolved by remote sensing systems such as the Landsat MSS. Therefore, mapping hydrothermally altered rocks from remote platforms is an indirect approach to exploration for ore bodies that is based on studies of the gangue minerals, that is, the nonmetalliferous or nonvaluable metalliferous minerals associated with ore deposits. Although rarely of economic value, these minerals are commonly widespread in alteration zones and, as will be described, some have physical properties that contrast with those of many unaltered rocks.

The mineralogy of hydrothermally altered rocks mainly depends on the prevailing physical conditions at the time of alteration, especially temperature, chemistry of the fluid phase (pH, salinity, fugacities of oxygen and sulphur) and composition of the original host rocks (44). Variations in these factors within large hydrothermal systems, such as porphyry copper deposits, commonly give rise to zones characterized by a particular mineral or mineral assemblage. Ideally, the degree of alteration grades progressively both vertically and laterally from intense alteration near the source to minor changes in the outermost zone, but complete zonation is not usually clearly exposed in a single deposit because of incomplete topographic transection and variations in conditions within the system.

Additional mineralogical complexity results from processes operating in the zone of weathering, including oxidation, leaching, and hydration. Oxidation is especially important from the remote sensing standpoint because interaction with oxygenated meteoric water causes conversion of iron-bearing sulfide minerals (pyrite, marcasite, pyrrhotite, arsenopyrite, and calcopyrite) and subsequent deposition of iron-oxide minerals (hematite, geothite, and lepidocrocite) on fracture and joint surfaces. These minerals, along with other secondary minerals, such as jarosite, are responsible for the distinctive red, orange, brown, and black colors that have been used for many years as a guide for locating hydrothermal alteration zones with gossans. However, intense leaching in the zone of weathering commonly results in removal of these distinctive, but not unique, surface stains. Furthermore, many important alteration zones initially lacked sufficient iron sulfide minerals to produce these ocherous colors.

Because no single hydrothermally altered area encompasses all the important mineralogical variations, our present understanding of the pertinent remote sensing properties must be obtained by assembling data from many individual deposits. The most detailed remote sensing studies of hydrothermal alteration deal with epithermal deposits which have weak to very intense alteration in south-central Nevada and in the Virginia Range southeast of Reno (45). Many visible and near-infrared spectral reflectance measurements have been recorded in the field for the various types or grades of altered rocks and for the most common unaltered rocks. The categories of altered rocks described below are generalized from a detailed classification formulated by Roger P. Ashley, United States Geological Survey, Menlo Park, California, using field observations and laboratory analyses.

Propylitization is commonly present in the outer fringes of alteration zones where more advanced alteration is present (44). The mineralogy of propylitic rocks depends on the composition of the host rock, but the most common changes are the addition of H_2O in hydroxide-bearing minerals, such as chlorite and montmorillonite, and of CO_2 to form

carbonates. Other minerals included in the Nevada rocks are quartz, epidote, albite, and locally, pyrite (Table 17.1) (46). In most places, propylitic rocks are dark green, and therefore lack the conspicuous color, mineralogical, and textural features that typify the higher grade rocks. However, Ashley points out (1977, written communication) that propylitic rocks are commonly bleached where pyrite is abundant.

Major mineralogical and chemical changes that are less dependent on host rock composition appear in *argillic* (clay rich) rocks, as Ca, Na, and Mg, are removed in general proportion to the amount of H introduced (44). In the Nevada argillic rocks, quartz, montmorillonite, kaolinite, K-mica, hematite, goethite, and jarosite are the dominant minerals (Table 17.1). Important variations within the argillic category are the relative proportions of montmorillonite, kaolinite, and K-mica.

Phyllic rocks, which include quartz-sericite assemblages, are commonly gradational with argillic rocks. The most distinguishable mineralogical characteristic of phyllic rocks is the dominance of K-mica relative to clay minerals (Table 17.1).

In *advanced argillic* rocks, alunite, kaolinite, pyrophyllite, K-mica, and diaspore are present in significant amounts (Table 17.1); montmorillonite is absent, and hematite is the dominant iron-oxide mineral. In contrast with the higher grade opalite and silicified rocks, the quartz content of advanced argillic rocks is generally less than 50%.

Opalites, such as those in the Cuprite mining district south of Goldfield, Nevada, are characterized by large amounts of opaline quartz. They may also have alunite, kaolinite, pyrophyllite, and calcite as major components, but the calcite is nearly everywhere leached from the typically vuggy exposed surfaces. Iron-oxide minerals are generally deficient, resulting in very high albedos. Opalites are considered to be intensely altered because of intense cation leaching.

The most intensely altered rocks, *silicified rocks*, are dominated by quartz, the amount commonly being greater than 50% (Table 17.1). Other important components include alunite, kaolinite, pyrophyllite, and diaspore. The iron-oxide mineral content is highly variable, but in some places, such as the silicified ledges in the Goldfield district, the quartzose substrate is commonly obscured by a coating of these minerals. Because of the high quartz content, silicified rocks are generally resistant to erosion.

TABLE 17.1 MAJOR AND MINOR (+) MINERAL COMPONENTS IN CATEGORIES OF ALTERED ROCKS STUDIED IN SOUTH-CENTRAL AND WESTERN NEVADA (AFTER ASHLEY, 1977, WRITTEN COMMUNICATION).

Propylitic Rocks	Epidote, chlorite, albite, carbonate, montmorillonite, goethite + K-mica, pyrite, zeolites
Argillic Rocks	Montmorillonite, kaolinite, quartz, K-mica, goethite, hematite, jarosite + chlorite carbonate
Phyllic Rocks	Quartz, K-mica, kaolinite, mixed-layer clays, hematite, jarosite + K-feldspar, albite
Advanced Argillic Rocks	Quartz, pyrophillite, alunite, kaolinite, opal, K-feldspar, K-mica, hematite + anatase, relict zircon
Opalite	Quartz (opaline), alunite, kaolinite, pyrophyllite, calcite + anatase
Silicified Rocks	Quartz, alunite, kaolinite, diaspore, pyrophyllite, hematite, goethite, jarosite + anatase, rutile, relict zircon, opal, K-mica

(Left margin, with downward arrow: INCREASING INTENSITY)

Quartz or opal is ubiquitous in all these rocks, but generally increases with intensity of alteration. The amount of iron-oxide minerals varies from outcrop to outcrop, although it is often more abundant in the most intensely silicified rocks. In general, hematite is more common in the zones of more intense alteration, whereas goethite and jarosite are dominant in the lower grade rocks (Table 17.1). Note that iron-oxide minerals are not abundant in the opalite deposits, a property which is also common to some argillic and phyllic rocks.

Several important types of alteration are not represented in this discussion because they have not been studied with regard to their pertinent remote sensing properties. Most important are potassium-silicate rocks, which mainly consist of K-feldspar and quartz and only minor amounts of iron-oxide and hydroxyl-bearing minerals. Although biotite or chlorite may be present in the more hydrous varieties of potassium silicate rocks, the amounts are typically not abnormally high relative to many unaltered rocks.

VISIBLE AND NEAR-INFRARED SPECTRAL REFLECTANCE

Our understanding of the relationship between mineralogical composition and visible and near-infrared spectral reflectance is mainly derived from extensive measurements of selected laboratory specimens (see Chapter 2). However, laboratory studies of the spectral reflectance of altered rocks have been limited to a few of the constituent minerals, the altered rocks *per se* being essentially unstudied. A more direct approach is analysis of *in situ* spectral reflectance measurements coordinated with mineralogic and petrologic analyses. This approach is especially productive when reflectance measurements of undisturbed soils, as well as outcrops, can be made of areas viewed by airborne and satellite-borne instruments.

Analysis of representative *in situ* spectral reflectance measurements made from 0.45-2.5 μm using a field-portable spectrometer developed at the Jet Propulsion Laboratory, Pasadena, California (47) shows that most of the altered rocks described above are characterized by decreasing reflectance toward wavelengths shorter and longer than 1.6 μm (Fig. 17.9). The spectral features responsible for this shape are an intense absorption band in the ultraviolet region (not shown in Fig. 17.9), several minor bands in the visible, a broad band centered near 0.90 μm, and a sharper band centered at approximately 2.2 μm. All the absorption bands shown short of 1.6 μm in these spectra are due to electronic processes in the constituent iron ions of such minerals as hematite, goethite, jarosite, and montmorillonite. The 2.2-μm feature is a combination band related to OH bond stretching and Al-O-H bond bending vibrations; clay minerals, alunite, and pyrophyllite contain these bonds. These electronic and vibrational processes are discussed in more detail in Chapter 2. The 2.2-μm feature is more weakly expressed in the limonitic silicified rocks [Fig. 17.9(h)] because the rocks are mainly composed of quartz and limonite, neither of which has a 2.2-μm absorption band.

As previously mentioned, some hydrothermal alteration zones lack significant iron-oxide surface staining. The effect on the visible and near-infrared spectrum is illustrated in Figs. 17.9(a), (d), and (f). The rocks represented in Fig. 17.9(f) are highly siliceous, commonly vuggy opaline rocks of the Cuprite district. Because of the low iron content, their albedo is very high, and the iron absorption bands are weakly expressed. The 2.2-μm feature is evident, however, because of the presence of alunite, kaolinite, and pyrophyllite. A prominent 2.2-μm band is also present in the spectra for iron-deficient argillic and phyllic rocks [Figs. 17.9(a) and (d), respectively].

All the spectral features described above are absent or weakly expressed in most unaltered rock spectra. In the Nevada areas, for example, the most widespread unaltered rocks are andesite, rhyolitic flows and tuffs, basalt, rhyodacitic and latitic flows and associated intrusive rocks, and granodioritic and quartz monzonite plutonic rocks. Some iron-oxide stain present on the surfaces of these rocks causes iron-absorption bands and generally decreasing reflection intensity to wavelengths short of 1.6 μm (Fig. 17.10). However, the rate of decrease is generally less than that observed in the altered rock spectra. In addition, the 2.2-μm OH-absorption band is essentially absent.

Propylitized rocks are commonly spectrally similar to their unaltered equivalents, except that a very subtle band may be present near 2.2-2.3 μm, which is apparently due to their chlorite, epidote, and montmorillonite contents; consequently, spectra for these rocks are not shown in Fig. 17.9.

The reflectance of all the altered and unaltered rocks in the Nevada areas has a large variance; ranges overlap at all wavelengths (Fig. 17.11). The largest separation occurs near 1.6 μm, but in the most commonly used spectral region between 0.45

and 1.1 μm, the overlap is very large. Therefore, we should not expect albedo (the percent of incident solar illumination reflected from the surface) to be very useful for distinguishing these broad categories of rocks. The shapes of the spectra, as determined by the iron- and OH-absorption bands, are more diagnostic.

Because of mineralogical similarities, some unaltered rocks are spectrally similar to these alteration products. In south-central Nevada, large areas are underlain by bright pink tuffs, which have conspicuous iron-absorption bands at wavelengths shorter than 1.3 μm, but the 2.2-μm band characteristic of the altered rocks is not present in the tuff spectra (Fig. 17.12). The pink color of these rocks is attributed to the presence of numerous microscopic hematite particles in the groundmass. Absence of the 2.2-μm feature in the tuff spectrum is due to the general lack of hydroxyl-bearing phases in the tuffs, which are devitrified to cristobalite and alkali feldspar, and their residual soil, which is mainly a product of mechanical rather than chemical weathering. Examination of laboratory spectra (48) also shows the presence of intense iron-absorption bands in the blue part of the spectrum and near 0.90 μm in some spectra for glassy volcanic rock spectra, as well as in the spectra of all obviously ferruginous rocks.

The 2.2-μm band is present in many laboratory spectra of unaltered rocks containing hydroxyl-bearing phases, but the intensity of the absorption is usually lower than that seen in the altered rocks described above. Exceptions can usually be related to the presence of muscovite or clay minerals as a primary constituent, such as in shales, or to the products of intense late-stage alteration, exemplified by development of clays in rocks such as volcanic glasses. When, as in ferruginous shales, both iron-oxide and clay minerals are principal constituents, the visible and near-infrared reflectance spectra are indistinguishable from the spectra of limonitic altered rocks. This situation is illustrated by comparing the field spectra for red and tan limonitic shales [Figs. 17.12(b) and (c)] with those of the limonitic altered rocks (Fig. 17.9) in Nevada.

The data recorded by most visible and near-infrared remote sensing instruments are ultimately used to form images. However, nonimage forming airborne spectrometers are being tested on an experimental basis. In order to record sufficient energy to form images in the 2-μm region, it is

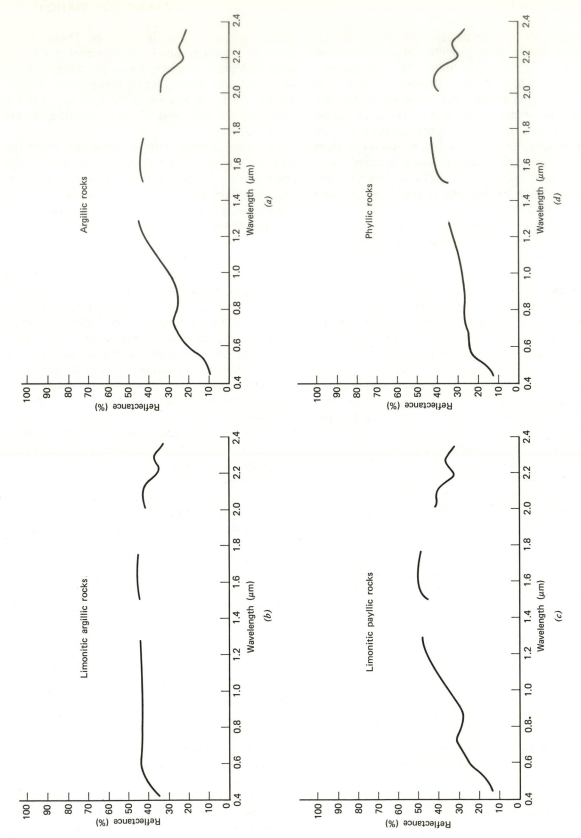

FIGURE 17.9. *In situ* reflectance spectra for altered rocks studied in Nevada. Gaps in spectra are regions of atmospheric absorption.

572

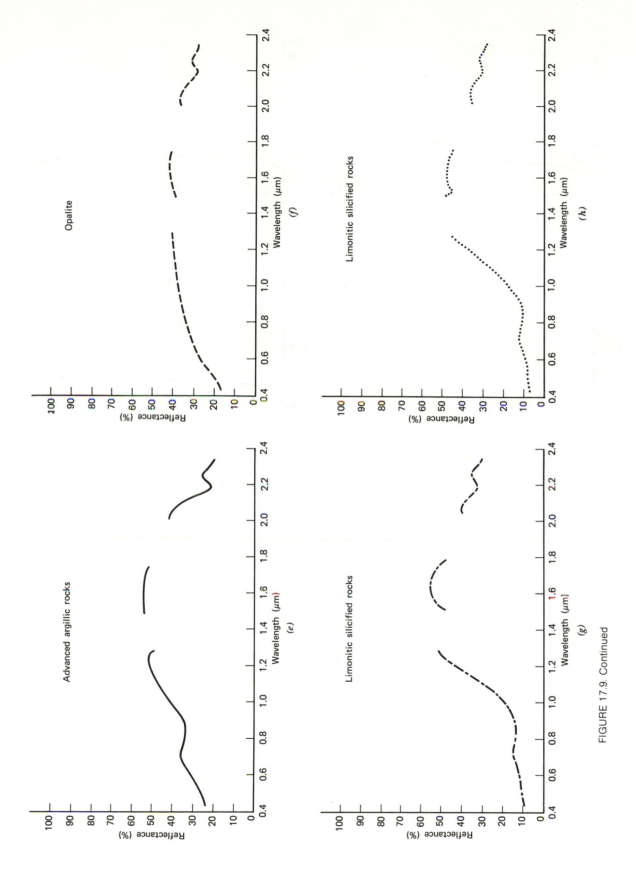

FIGURE 17.9. Continued

573

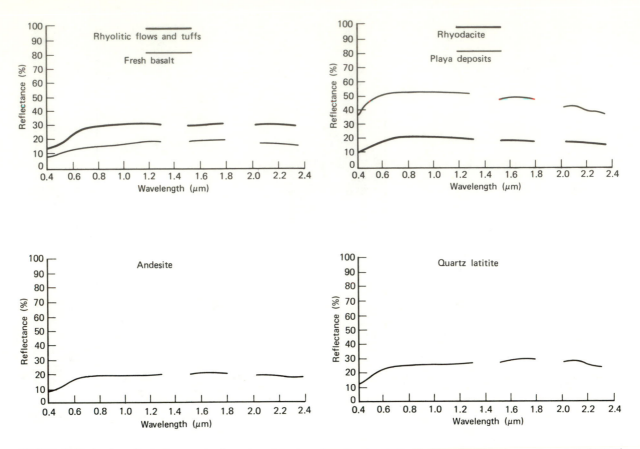

FIGURE 17.10. *In situ* reflectance spectra for categories of unaltered rocks studied in Nevada. Gaps in spectra are regions of atmospheric absorption.

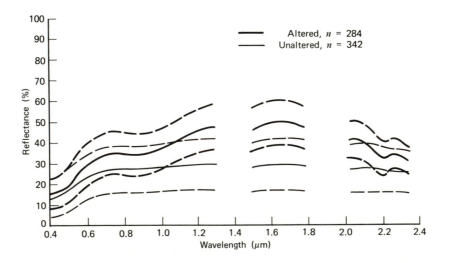

FIGURE 17.11. Average *in situ* reflectance spectra for all altered and unaltered rocks studied in Nevada. Solid and dashed lines represent the mean and 1 standard deviation values, respectively, with heavy lines for altered rocks and light lines for unaltered rocks. Gaps in spectra are regions of atmospheric absorption. *n* = number of measurements.

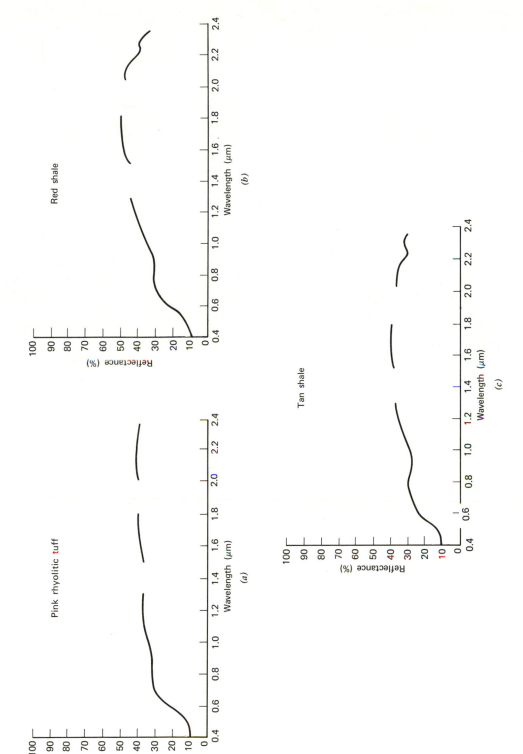

FIGURE 17.12. *In situ* reflectance spectra for problematical unaltered rocks studied in Nevada. Gaps in spectra are regions of atmospheric absorption.

currently necessary to use broadband filters. Consequently, if the 2.2- μm OH feature is to be useful for identifying altered rocks, then other features in this wavelength region become of interest. Some minerals such as most amphiboles, show a strong absorption band near 2.3 μm instead of 2.2 μm because of trioctohedral rather than dioctohedral coordination of the metal cations (see Chapter 2). Carbonate rock spectra are commonly characterized by sharp intense CO_3^{-2} absorptions near 2.35 μm (see Fig. 2.13, Chapter 2). These absorption bands further complicate utilization of the 2.2- μm region for discriminating altered and unaltered rocks because they cause a general decrease in the radiance recorded in a broadband in this region.

In summary, hydrothermally altered rocks are characterized by hydroxyl mineral phases such as clay minerals, sericite, alunite, pyropyllite, and by iron-oxide minerals where pyrite was present. Vibrational processes cause a sharp absorption band near 2.2 μm, whereas electronic processes in iron ions result in broad absorption bands in the ultraviolet, visible, and short wavelength near-infrared spectral regions. These features are absent or weak in many unaltered rocks, but they are not unique to altered rocks. Another factor that limits the usefulness of the 2.2- μm feature for detecting alteration when using wide bandpasses is the presence of absorption bands near 2.3 μm and 2.35 μm in some unaltered hydroxyl rock and carbonate rock spectra, respectively. Although all these features are distinct in individual spectra, they might not be separable in images. In addition, the 2.2- μm feature is absent from the spectra of the most intensely altered, nearly pure silica rocks, and neither the Fe- nor the OH-absorption bands are well displayed in the Nevada propylitic rocks. This feature would not be expected in some other important altered rocks, including the quartz-K-feldspar assemblage typical of some copper-porphyry alteration zones.

Thus, the 2.2- μm absorption band is indicative of the presence of hydroxyl phases, but it is not uniquely diagnostic of hydrothermal alteration. Similarly, iron-absorption bands are related to the presence of iron-oxide minerals on the surfaces of some unaltered, as well as pyritic, hydrothermally altered rocks. Nevertheless, these spectral features provide a basis for discriminating most hydrothermally altered from most unaltered rocks. Furthermore, they allow delineation of limits on the surface mineralogy.

17.3.2 Alteration Associated with Uranium Deposits

Uranium is concentrated to form potential ore deposits by processes operating in two contrasting geological settings: (1) primary vein deposits associated with fractionation of dominantly felsic magmatic intrusive bodies, and (2) secondary enriched zones in sedimentary rocks. Most exploration is being conducted in sedimentary basins, especially in the western United States because of their high-volume potential and favorable extraction conditions. Consequently, remote sensing surveys focused attention on sedimentary deposits, although the techniques used also appear to be applicable to detecting limonitic primary magmatic deposits.

In Wyoming, the current focal point of much uranium exploration, the ore deposits appear to have been leached either from sediments derived from felsic intrusive rocks or from felsic volcanic rocks and transported as soluble uranyl ions in oxygenated, alkaline groundwater to favorable sites of deposition. Conditions leading to deposition include the presence of some reducing agent and relatively coarse-grained, usually poorly sorted sediments where the groundwater becomes concentrated. Commonly the reductant appears to have been H_2S, produced either by anaerobic bacteria converting sulfate to sulfide, or by pyrite (49, 50). Under some conditions, the H_2S may be related to hydrocarbon microseepage (51, 52, 53).

From the geometrical standpoint, sedimentary uranium deposits are commonly roll-type deposits, tabular, or ellliptically shaped bodies; some are linear, especially where the ore is concentrated along fault zones. Roll deposits are one of the most important types and have been described by several workers (49, 50, 54, 55, 56, 57, 58). Ideally, they are crescentic in cross section [Fig. 17.13(a)] and curvilinear to sinous in plan view [Fig. 17.13(b)]. However, the exact shape depends on many factors, including the initial configuration of the fluvial channel, relative permeability of the sediments, distribution of organic material, and direction and gradient of groundwater flow. Salmon and Pillars (58) stress the importance of the erosional level for distinguishing roll-type deposits using remotely sensed data.

The mineralogy of most roll deposits is that of arkosic to subarkosic sandstones. Mineralization in the preferred environment results in deposition of pyrite, uranium minerals—commonly uraninite and coffinite—and, in some deposits, chamosite and nontronite (50). Also, calcite is commonly found in the upper part of the altered zone and, to a lesser extent, in the lower parts. In the concave part of the roll, which commonly extends well back from the roll front, the altered rocks are typically oxidized to various shales of red, yellow, and yellowish brown [Fig. 17.13(a)]. In general, colors in the ore-bearing zone are darker and dominantly red. On the convex side, the host rocks are unaltered and therefore lack these ocherous colors. Similar mineralogical and color relationships are found around tabular ore bodies, although their shapes are less distinctive. Elliptical bodies are commonly entirely enclosed by the oxidized bleached zone.

Although few spectral-reflectance measurements are available for the altered rocks and soils associated with sedimentary uranium deposits present, emphasis is being placed on the distinctive reflectance spectra of limonite and hematite. The spectral-reflectance characteristics of these deposits would be essentially identical to those of the limonitic rocks in hydrothermal alteration zones. Spectral reflectance differences arising from the alteration of plagioclase feldspar to hydroxyl phases have not been assessed, but for these country rocks, they are likely to be less than those seen in the volcanic rocks in Nevada owing to the prevalence of clay minerals in these sediments.

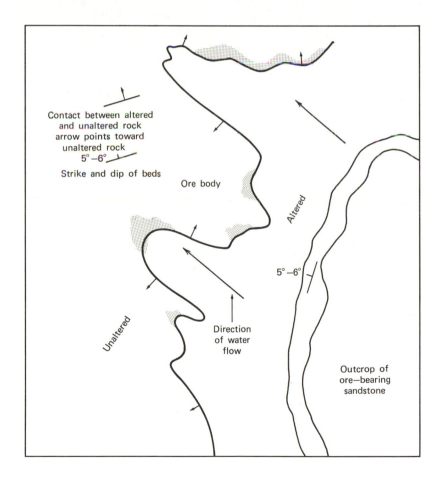

FIGURE 17.13. Schematic diagrams showing (a) cross section through a roll-type uranium deposit (11), and (b) plan view of sandstone type uranium deposit (12).

17.3.3 Methods for Analyzing and Interpreting Visible and Near-Infrared Images

Methods for analyzing and interpreting visible and near-infrared images may vary widely with respect to complexity, effectiveness, and cost. Some methods are essentially extensions of conventional photo interpretation techniques, whereas others are more akin to geophysical methods requiring digital computer modeling of the phenomena being measured. In general, the effectiveness and cost increase with the degree of sophistication of the methods used.

Most of the analytical methods in current use can be placed in two broad categories: (1) image interpretation and (2) computer classification or automatic recognition maps. Both approaches are outgrowths of research conducted using aircraft data, but the success achieved using Landsat 1 and 2 Multispectral Scanner (MSS) data has resulted in much progress in the application of these techniques, especially in mineral exploration. Therefore, this discussion will concentrate on the analysis and interpretation of MSS data for detecting and mapping altered rocks and soils. These same techniques are applicable to analysis of airborne-scanner visible and near-infrared reflectance data and to the data collected in other spectral regions, including the 1.5-2.5 μm region.

INTERPRETATION OF IMAGES

The simplest and least expensive form of image analysis for mapping altered rocks is based on the fact that many altered areas have either high albedos or distinctive ocherous colors. Argillization and silicification, either in the absence of iron sulfide minerals or where intense leaching has taken place, result in a general increase in the albedo of the altered rocks. Thus "bleaching" of the host rocks is commonly used in conventional photo interpretation as a guide to alteration.

In south-central Nevada, many of the known altered areas have very high albedo in all the MSS bands; several of the most conspicuous areas, including the Cuprite mining district (A, Fig. 17.14) and two other areas to the north and northeast (B, Fig. 17.14), are identified in Fig. 17.14, a standard MSS band 7 image. However, the largest and most profitable mining district, Goldfield (59), appears as a uniform light-gray region, and the image gives little indication of the extensive alteration zones there (C,

Fig. 17.14). Other altered areas are also indistinct (D, Fig. 17.14). In addition, several large areas of unaltered outcrop as well as alluvium have radiance levels that are equivalent to the bleached altered areas (E, Fig. 17.14). Thus, analysis of albedo differences in standard MSS images offers an inexpensive straightforward approach to searching for altered areas, but its effectiveness is limited by the widely observed facts that many important altered areas are not bleached, and many unaltered rocks have high albedos (60).

The most widely used single parameter in conventional photo interpretation of color aerial photographs is the ocherous colors of limonitic rocks. Although these colors are commonly vividly displayed in aerial photographs, they become less apparent from higher altitudes because of increasingly severe atmospheric scattering. At satellite altitudes, where the energy detected has penetrated the entire Earth's atmosphere twice, only the most intense color differences are apparent in photographs and multispectral images. Hundreds of excellent color photographs were recorded by the 6-in. (15-cm) and 18-in. (45.7-cm) focal-length cameras of the Skylab S190A and S190B experiments, respectively. Although these photographs are of considerable use for many geological studies, their value for mapping altered rocks appears to be quite limited.

An S190A color photograph color plate 36 was analyzed by a photo interpreter who is experienced in analyzing satellite images but unfamiliar with the region. Only about half the known altered areas were delineated, and most of these were distinguished on the basis of textural differences seen in the stereoscopic photographs. A few of the altered areas, such as the one northeast of Goldfield (A, Fig. 17.15 color plate 36), appear reddish in the Skylab photograph, but most are not distinctive. Note in particular that, although the Goldfield district has a few rust-colored areas (B, Fig. 17.15, color plate 36), similarly colored areas present elsewhere are unaltered (C, Fig. 17.15, color plate 36). Most of the Goldfield alteration zone is an indistinct tan to brown in the photograph. In fact, the most obvious colors are seen in the areas of the pink hematitic tuff mentioned above (D, Fig. 17.15, color plate 36).

Several other disadvantages of satellite color photographs for mapping altered rocks deserve mention: (1) the response range of color film emulsions is limited to approximately 0.4-0.7 μm,

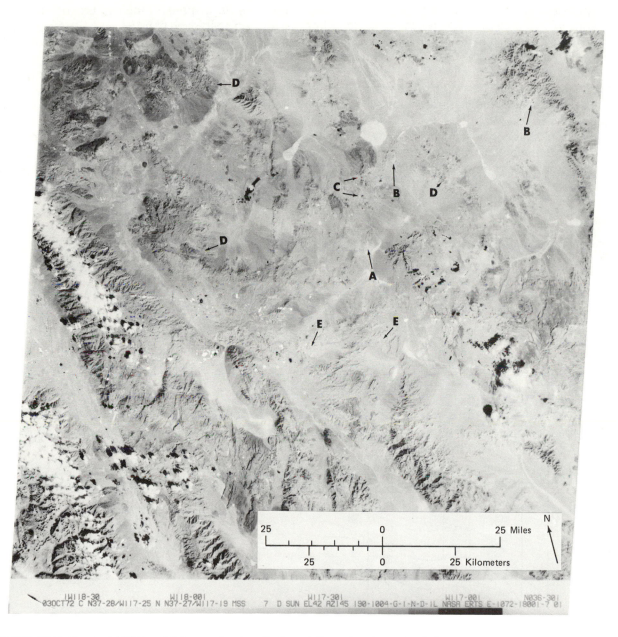

FIGURE 17.14. Standard MSS band 7 image showing albedo variations of rocks and soils in south-central Nevada. *A*, sinter deposits of Cuprite mining district; *B*, hematitic altered area; *C*, alteration zone in the Goldfield mining district; *D*, other moderate to low albedo altered areas; *E*, high albedo unaltered areas. Image number E1072-18001, acquired Oct. 3, 1972.

which precludes recording the 0.9- μm absorption band typical of iron oxide minerals; (2) the film must be digitized to enhance the limited multispectral information contained in the photographs; (3) because of atmospheric attenuation and film's response range, albedo differences are suppressed; and (4) in the case of Skylab, the total cloud-free coverage is limited. Color-infrared photographs, which are sensitive to 0.5-0.9 μm energy, partially alleviate problems number 1 and 3, but still lack adequate spectral reflectance contrast for distinguishing altered and unaltered rocks. It should be emphasized, however, that the high spatial resolution and stereoscopic coverage provided by the Skylab photographic experiments are critical for many geologic purposes, including geomorphologic studies that pertain to mineral exploration.

The spectral reflectance differences between the altered and unaltered rocks described above are pronounced in the field spectra, except in the few places where their mineralogical compositions are similar. These differences are unavoidably subdued, however, in multispectral images because of the necessity for broad spectral bands. Of the four MSS spectral bands, band 7 (0.8-1.1 μm) is most problematical for mapping limonitic rocks because its response range is broader than the 0.90- μm absorption band (Fig. 17.16). Most important, of course, is the lack of radiance data in the 1.6- μm and 2.2- μm regions.

These spectral limitations of the MSS mean that only large spectral radiance differences can be detected between altered and unaltered rocks and these differences are greatly subdued as recorded on the CCTs. Therefore detection of alteration in this wavelength region is largely restricted to limonitic altered rocks. Evidence of the small magnitude of the spectral radiance differences among rocks is given by the fact that no tonal differences are apparent in band-to-band comparisons of the standard MSS images of south-central Nevada (Fig. 17.17).

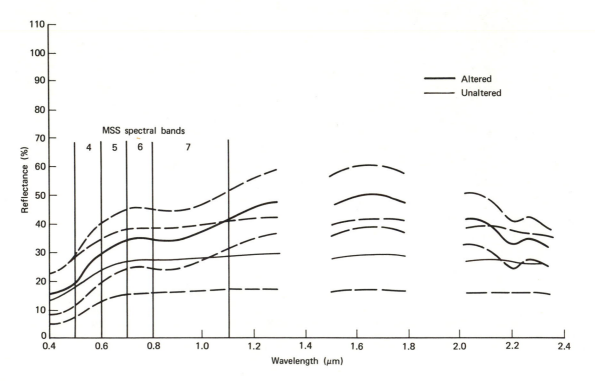

FIGURE 17.16. Average *in situ* reflectance spectra for all altered (heavy lines) and unaltered (light lines) rocks shown in Fig. 17.3, showing locations of Landsat MSS bands.

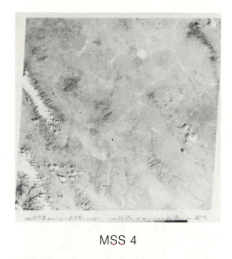

MSS 4

MSS 5

MSS 6

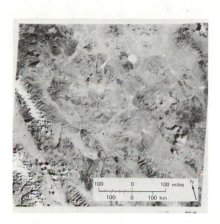

MSS 7

FIGURE 17.17. Comparison of standard MSS band images of south-central Nevada showing lack of contrast among rock types, including altered and unaltered rocks. Image E1072-18001, acquired October 3, 1972.

Contrast stretching of the computer compatible tape (CCT) data enhances the overall level of detail and some of the bleached zones noted in Fig. 17.14, but most of the same problems regarding distinctions based on albedo are still evident in the contrast-enhanced image (Fig. 17.18).

Color-infrared composite images of the MSS bands have been widely used for reconnaissance geologic mapping and therefore are useful for establishing a geologic framework for mineral exploration. Where intense limonitic stains are accompanied by high albedo (A, Fig. 17.19 color plate 37), a faint yellow to yellow-orange is seen in enhanced color-infrared composite images. However,

because of the subtleness of the spectral radiance differences between most altered and unaltered rocks in the MSS bands and the effects of topographic slope and albedo variations, these composite images appear to have few advantages over standard color and color-infrared photographs.

Ratioing of the spatially registered picture elements (pixels) that make up the images for each MSS band is a highly effective means of minimizing the first-order effects of brightness variations due to topographic slope (see Chapter 5 for a more detailed discussion). Other advantages of ratioing are: (1) the difference between slopes of reflectance spectra of the bands can be displayed in a

FIGURE 17.18. Linearly stretched MSS band 7 image of south-central Nevada showing areas annotated in Fig. 17.6. Image E1072-18001, acquired October 3, 1972.

single image; (2) the responses of the members of like classes of materials having similar reflectance curve shapes but varying albedos are normalized; and (3) the width of the frequency distribution of ratio values is reduced so that a greater contrast stretch can be applied to a selected part. Selection requires knowledge of the ratios for a few pixels representing the materials of interest.

Because of the low ratio values of limonitic materials, they appear dark in positive MSS 4/5 ratio prints and transparencies [A, Fig. 17.20(a)]; they are light in the inverse ratio image product, MSS 5/4. Only some alluvial deposits are as dark as the limonitic materials [B, Fig. 17.20(a)]. All other materials appear lighter and vegetation is very light [C, Fig. 17.20(a)].

It is this distinctively low ratio image tone that led Vincent (53) and several other workers to extensive use of MSS 5/4 images for mapping limonitic zones in uranium exploration in sedimentary terrain. Density slicing can be used to further enhance the subtle tonal differences related to the type and amount of iron-oxide minerals present in different parts of the area.

Some mineralogical differences are not detectable in any single ratio image, however. In the south-central Nevada area, for example, the northern part of the Goldfield alteration zone is represented by the same level of density in the MSS 4/5 image [A, Fig. 17.20(a)] as many of the unaltered rhyolitic rocks [D, Fig. 17.20(a)] and is lighter than most of the alluvial areas [B, Fig. 17.20(a)]. Also, note the Cactus Spring district, one of the largest and most intensely altered areas, is not distinctive in the MSS 4/5 image [E, Fig. 17.20(a)]. The opalite deposits and rhyolitic rocks commonly have indistinguishable MSS 4/5 image densities [F and D, respectively, Fig. 17.20(a)].

Some of these problems can be overcome by analyzing color composites of ratio images, referred to as color-ratio composite images. Evaluation of the nine possible combinations of the three MSS ratio images of south-central Nevada shows that the most effective combination for mapping altered rocks consists of MSS 4/5, MSS 5/6, and MSS 6/7 using blue, yellow, and magenta diazo films, respectively (Fig. 17.21, color plate 38) (45, 60). In this composite, linear contrast stretches have been used, although logarithmic and CDF stretches (see Chapter 6) have proven to be beneficial for some areas. MSS 4/6, instead of MSS 5/6, provides

improved enhancement for dinstinguishing some altered rocks, especially in more densely vegetated areas, but is commonly less useful for discriminating among the unaltered rocks.

Corrections for atmospheric scattering and absorption have not been made because they tend to decrease the signal-to-noise ratio and therefore reduce the spatial information on the ratio images. However, such corrections are useful in areas of rugged topography and are necessary for quantitative analysis of the spectral radiance (see Chapter 6 for descriptions of methods).

Color composites of ratio images provide subtle distinctions that are not apparent in individual black-and-white ratio images, and diazo films are more flexible than other color-compositing techniques tested by the author. In the diazo color-ratio composite prepared for south-central Nevada, most of the altered areas appear green (A and B, Fig. 17.21, color plate 38), the spectral region to which the eye is most sensitive. These areas are predominantly argillic, phyllic, advanced argillic, and silicified rocks, in which the limonite is mainly light to dark brown or black. Altered rocks represented by brown in the composite (C, Fig. 17.21, color plate 38) are commonly bright red hematitic silicifed rocks. Red-brown areas, such as the Cuprite district (D, Fig. 17.21, color plate 38) are consistently very bright opalized or unaltered silicic volcanic rocks. Argillic rocks that are deficient in iron-oxide are not widespread in the study area; they appear as blue or colorless in the composite (E, Fig. 17.21, color plate 38). Where limonite is present, they appear green.

It must be emphasized that colors seen in the color-ratio composite image express spectral radiance, brightness variations due to albedo and topographic slope having been minimized through ratioing. The relationship between these colors and spectral reflectance can be generally understood by approximating the ratios for the representative spectra (Fig. 17.9), and by remembering that low ratio values resulting in dark gray tones on the original ratio images are portrayed in the color of the diazo film used; conversely, light or white areas yield no color. Progressively dark areas in the ratio image are removed from the color film, allowing considerable flexibility for tailoring the film to show specific differences.

The green color of the pink tuff and shales (K and L, respectively, Fig. 17.21) and of the limonitic

FIGURE 17.20 (a). Linearly stretched MSS ratio of south-central Nevada, MSS 4/5. In this image A, northern part of Goldfield district alteration zone; B, alluvial areas; C, vegetated areas; D, unaltered rhyolitic rocks; E, Cactus Spring-district; E, silicarich opaline (sinter) deposits of the Ralston district. Image E1072-18001, acquired October 3, 1972.

584

FIGURE 17.20 (*b*). Linearly stretched MSS ratio image of south-central Nevada, MSS 4/6. Image E-1072-18001, acquired October 3, 1972.

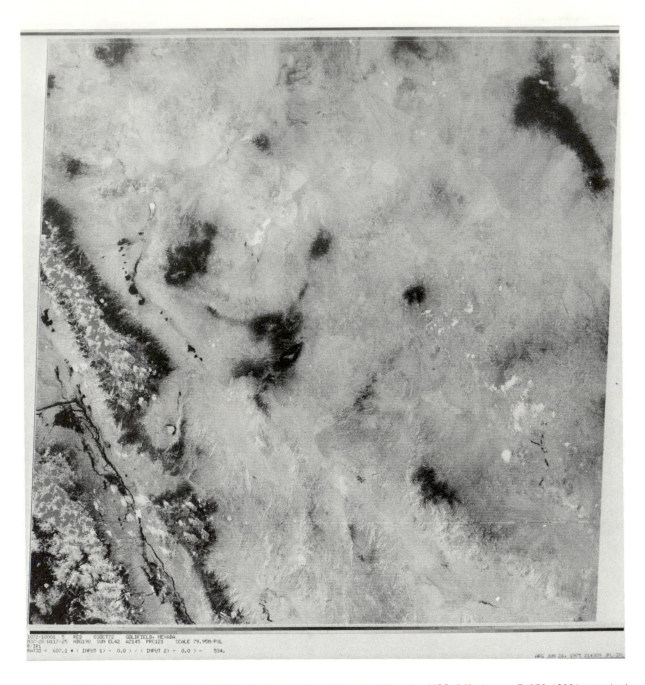

FIGURE 17.20 (c). Linearly stretched MSS ratio image of south-central Nevada, MSS 5/6. Image E1072-18001, acquired October 3, 1972.

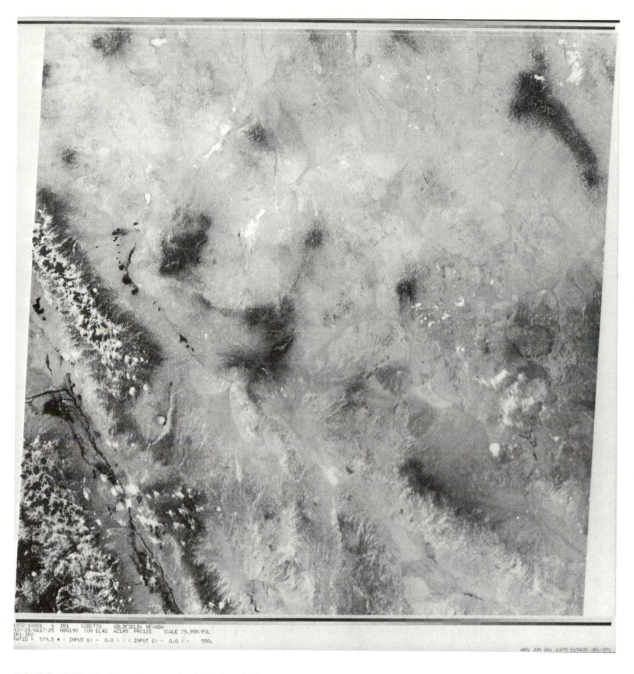

FIGURE 17.20 (*d*). Linearly stretched MSS ratio image of south-central Nevada, MSS 6/7. Image E1072-18001, acquired October 3, 1972.

altered rocks in the color-ratio composite (A and B, Fig. 17.21) stems from their comparatively low MSS 4/5 and MSS 5/6 ratio values and high MSS 6/7 values. Consequently, intense blue and yellow in the respective diazo films composite to yield green in these areas. Magenta is added where the MSS 6/7 ratio is relatively low. However, low MSS 6/7 ratios may represent markedly different rocks. For example, some highly limonitic rocks have low MSS 6/7 ratios because of a steep rise in band 7, even though a well-defined 0.90- μm band is present [Figs. 17.9(g) and (h)]; these rocks are commonly brown in the color-ratio composite image because of the combination of blue, yellow, and magenta diazo colors (C, Fig. 17.21). In contrast, the red-brown of the opalite deposits of the Cuprite district indicates dominance of magenta due to a low MSS 6/7 ratio, but includes some blue and yellow contribution from MSS 4/5 and MSS 5/6, respectively (D, Fig. 17.21). These general proportions are consistent with the moderately steep slope of the representative spectrum throughout the MSS range. The blue image color for playa deposits (G, Fig. 17.21) is due to the steep falloff toward the ultraviolet, for an otherwise generally flat spectrum (Fig. 17.10).

Similar, but less clear relationships can be deduced from the colors of the other altered and unaltered rocks in this scene. The value of this exercise lies in the construction of empirical spectral radiance curve shapes for particular areas of interest. For example, the presence of the green color in a similarly processed color-ratio composite of an unknown area indicates low MSS 4/5 and MSS 5/6 ratios and relatively high MSS 6/7 values; these values describe a spectrum diagnostic of limonite. Red-brown suggests a steep spectral curve slope in the MSS 6 and MSS 7 region relative to the two visible wavelength bands, and probably Fe-poor rocks. Therefore, general spectral shapes can be derived from the colors of color-ratio composites, and in some places these shapes can be related to some aspects of the mineralogical compositions. Of course, more precise shapes can be determined by examining the actual ratios for specific areas.

Detailed examination of the color-ratio composite image (Fig. 17.21) shows that, although a particular color dominates an area, pixels with other colors are present. For example, green pixels are most numerous in the Goldfield alteration zone, but brown, blue, yellow, and magenta pixels are scattered throughout the area. These spectral radiance variations stem from several sources, including changes in vegetation density and type, atmospheric scattering, grain size variations, second-order albedo and topographic slope effects, and MSS system noise, which becomes very apparent in highly enhanced products. The most important factor, however, is variations in surface mineralogy, especially iron-oxide concentrations. These sources of spectral radiance variations combine to make quantitative treatment of the MSS data difficult.

Vincent (61) has made some progress in relating the colors in color-ratio composite images by using laboratory measurements of spectral reflectance and modeling of atmospheric and vegetation variations. Although this approach appears to provide approximate calibration, it is limited by the inadequacy of laboratory measurements for representing field conditions, especially soils having varying compositions and grain sizes.

Field evaluation of the color-ratio composite of south-central Nevada (Fig. 17.22) shows that, except for propylitic rocks, all known altered areas were detected, being represented by green, brown, and red-brown in the composite. However, as pointed out in the preceding pages, several unaltered rocks are similarly portrayed. Large areas in the western part of the scene are underlain by bright pink tuff (K, Fig. 17.21) and tan limonitic shale (L, Fig. 17.21), which appear green in the composite. It is important to note that *slightly* pink, tan, and gray silicic tuffs and flows are common in the study area, but are consistently pink to light orange in the composite (F, Fig. 17.21). Another limitation of this technique arises from the tendency for iron-oxide encrustations to form on fragments derived from some unaltered rocks (N, Fig. 17.21). Delineation of alluvial deposits is more readily achieved using Skylab photographs rather than MSS images because of their higher resolution and stereoscopic capability. Yet another problematical rock type is an andesite north of Mud Lake (Fig. 17.21); spectral measurements are not yet available for this area.

Thus, color-ratio composite images offer a powerful technique for compiling maps of hydrothermal alteration, such as the one shown in Fig. 17.22 for the eastern part of the south-central Nevada scene, if the limitations described above are kept in mind. In summary, limonitic altered and unaltered rocks are not consistently discriminable on the basis of MSS

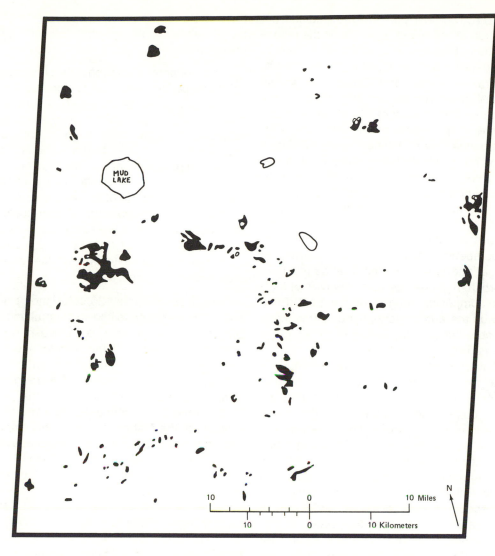

FIGURE 17.22. Alteration map of part of south-central Nevada scene.

data alone. Moreover, Fe-poor argillic rocks, propylitic rocks, and opalite lack distinctive Fe-absorption bands. The 2.2-μm region has considerable potential for distinguishing hydroxyl-bearing altered rocks and for identifying limonitic unaltered rocks without hydroxyl-bearing minerals. Included in the latter category are hematitic tuffs and alluvial deposits derived from unaltered rocks that have surficial iron-oxide encrustation.

Analysis of NASA aircraft scanner images of the Cuprite district that include 1.6-μm and 2.2-μm bandpasses, as well as shorter wavelength bands, shows that the opalite is readily distinguishable from all of the unaltered rocks in color-ratio composite images (62). Furthermore, in the East Tintic Mountains, Utah, use of color-ratio composite images prepared from these data has resulted in discrimination of limonitic altered and limonitic unaltered rocks (63). However, alteration products lacking both anomalously high ferric iron and hydroxyl-bearing mineral contents such as potassium-silicate rocks require other techniques for detection.

Vegetation cover poses an obvious limitation, but the magnitude of the effect depends on the reflectance contrast between the rocks and specific types of vegetation and on wavelength. Analysis of field spectra by Siegal and Goetz (64) shows that only 10% green vegetation cover obscures the spectral features of dark rocks such as andesite (see Chapter 13). However, the iron- and OH-absorption bands typical of limonitic argillized rock spectra are recognizable when there is as much as 30% green vegetation, although the iron-absorption bands are more muted. The effect of desert plants, such as manzanita, is less pronounced than that of green vegetation, approximately 50% cover being the upper limit for recognition of the spectral features of limonitic argillized rocks. Siegal and Goetz suggest that as much as 60% dry brush might be tolerated. Analysis of MSS color-ratio composite images of the East Tintic Mountains by the author generally confirms these values. A more detailed discussion of the effects of vegetation is given in Chapter 12.

Several other types of image enhancement have been tested on an experimental basis in the Goldfield area, but none of them are as effective as the color-ratio compositing approach. One of these types deserves some discussion, however, because of the apparent success with which it has been applied to mineral resource analysis in Alaska. This type of enhancement, referred to as simulated natural color, attempts to predict the missing blue part of the spectrum by placing each MSS pixel in the general categories of vegetation, water, or rock and soil according to its MSS 5/6 ratio value after haze removal (65). A different algorithm is then used to predict the theoretical radiance for the blue part of the spectrum. These values are used, along with MSS 4 and MSS 5, to form a simulated natural-color composite.

Analysis of a simulated natural-color composite image of the Nabesna quadrangle in south-central Alaska (Fig. 17.23, color plate 39) by Albert and Chavez (66) resulted in identification of 69 color anomalies. Prior to study of a standard color-infrared composite yielded 72 anomalously colored areas; only 21 areas were common to the two composite images. Interestingly, Albert and Chavez reported that no color anomalies were identified in a color-ratio composite produced by filtering MSS 5/4 with red light, 6/4 with green, and 4/5 with blue. However, Albert (6) pointed out that the con-

trast stretches used enhanced the vegetation at the expense of the spectral radiance of the exposed areas.

The color anomalies seen in the simulated natural-color composite image range from reddish-orange (A, Fig. 17.23) to bluish-green, yellow being the dominant color; gray and brown are present but subordinate. In the color-infrared composite, the anomalous areas are chiefly various shades of yellow and green. In general, anomalies in the same rock unit tend to be of similar color and commonly are seen in one or the other image, but not both (6).

The yellow, orange, and reddish-orange in the simulated color composite suggest iron-oxide stains, but the significance of the green image colors is not clear; they may be related to low-albedo rocks, both altered and unaltered.

Evaluation of these 120 color anomalies shows that "39 correspond to known mineral occurrences. Of the 81 that do not, 17 correspond to geochemical anomalies. Of the 64 color anomalies that correspond to neither known mineral occurrences nor geochemical anomalies, 27 were in areas that were not geochemically sampled. Thus, at least 56 (47%) of the color anomalies observed on Landsat imagery of the Nabesna quadrangle appear to be related to mineralization." (6) The number of altered areas that were not detected is unknown.

These results are encouraging but somewhat perplexing, because they are essentially diametrically opposed to the author's findings in Nevada and Utah, to those of Spirakis and Condit in Arizona (67), and to results of other workers in Mexico, Iran, and Bolivia. Analysis of a simulated natural-color image mosaic of Nevada (Fig. 17.24, color plate 40) shows that only the most vividly colored limonitic areas appear reddish-brown and that most of the altered areas are indistinct. In fact, the color distinctions are similar to those seen in the previously discussed color-infrared composite image (Fig. 17.19) and Skylab color photograph (Fig. 17.15). Assuming that the Nevada simulated natural-color mosaic was properly processed, one must conclude that either the altered rocks in the Nabesna quadrangle are more vividly colored than those in Nevada or that many altered areas went undetected in the Alaska area. Judging from descriptions of the Alaska altered rocks (6) and our knowledge of the south-central Nevada rocks, the latter explanation appears to be more plausible.

Review of the literature and evaluation of ongoing research by the United States Geological Survey by Offield (68) shows that most of the image analysis findings concerning limonitic hydrothermally altered rocks also pertain to iron-stained sedimentary uranium deposits: (1) color photographs and color-infrared MSS composite images are generally inadequate; (2) MSS 4/5 ratio images generally distinguish red hematitic areas but are less useful for showing yellow and yellow-brown dominantly limonitic areas; and (3) color-ratio composite images are most effective for discriminating between the altered and unaltered sedimentary rocks and, in some places, for distinguishing subtle differences among the altered and unaltered rocks.

Offield describes varying degrees of success using ratio images, the most important limiting factors being vegetation cover, presence of ferruginous unaltered sedimentary rocks and soil, cultivation patterns, and the wavelength position and width of the MSS bands. In the Crooks Gap, Wyoming district, an MSS 4/5 ratio image faithfully differentiates the most productive and promising dominantly hematitic altered areas, as well as some altered but less promising areas of uranium mineralization (Fig. 17.25). Some ambiguities exist in the MSS 4/5 ratio image, however. Highly reflective areas, such as sand dunes, tailing piles, and cleared mining areas are also dark in the image and therefore cannot be distinguished from the reddish altered rocks. Color-ratio composite images were found effective for alleviating these problems.

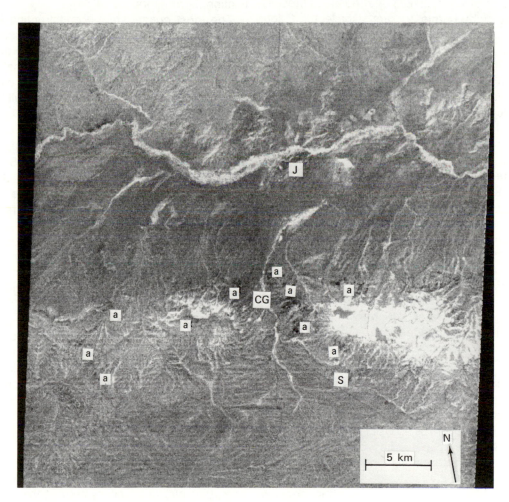

FIGURE 17.25. Linearly stretched MSS 4/5 ratio image of the Crooks Gap, Wyoming district.

Image processing of an MSS scene showing the Monument Hill-Highland Flat-Box Creek, Wyoming area in the southern Powder River basin of Wyoming was less successful. Although density slicing of an MSS 4/5 ratio image shows the mapped red altered ground with impressive accuracy, cultivated areas east of the district consisting of yellowish, apparently unaltered soil appear equally dark (Fig. 17.26). A color-ratio composite consisting of MSS 4/5 in cyan and MSS 6/7 in magenta removes some of the ambiguous areas, but many others remain, requiring extensive field checking for verification. Offield (68) stressed the need for spectral bands positioned to distinguish between yellowish and reddish or reddish-brown soils.

An interesting comparison of color-ratio composite and simulated natural-color composite images has been provided by studies in the Cameron, Arizona uranium district (67). Altered areas present within the light purple to gray Chinle formation sandstones are light brownish-yellow. The chief alteration products are limonite, jarosite, alunite, and gypsum. In the simulated natural-color composite image [Fig. 17.27(a), color plate 41], the altered areas are light colored and indistinct from the Chinle rocks. Marked improvement resulted when MSS 5/4, MSS 6/4, and MSS 7/4 ratio images were composited, using blue, green, and red filtered light, respectively. In this image [Fig. 17.27(b), color plate 41], the altered areas appear bluish and correspond closely in exposed areas with uranium mines. The altered areas are not discriminated from the Chinle mudstones, however. Nevertheless, these results confirm the author's findings regarding the relative merits of these two types of enhanced color images for portraying limonitic altered areas.

R. K. Vincent and his associates have been responsible for several important innovations in using multispectral images for uranium exploration. Some of these will be described in the next section.

COMPUTER CLASSIFICATION MAPS

Classification of surface materials by analog and digital computer analysis, sometimes referred to as automatic recognition, using visible and near-infrared spectral reflectance data, has been successfully used in crop inventory surveying for several years, but until recently only moderate success has been achieved in applying these techniques to mapping rocks and soils. Several factors combine to make accurate classification of rocks and soils more difficult than that of crops. Spectral contrast is generally higher for cultivated crops than for rocks and soils. Also, crops are generally on nearly horizontal or gently undulating surfaces and therefore lack the large variance factor introduced by topographic slope. Another factor is that the temporal changes attending the growth of crops can commonly be used as a distinguishing parameter.

As described in Chapters 6 and 13, two general approaches have been followed in compilation of computer classification maps, supervised and unsupervised classifications. Supervised classification, in which known areas are used to establish classification parameters for the materials of interest, is more widely used in geological investigations because the small spectral reflectance difference among rocks necessitate carefully defined classification parameters.

The most critical step in constructing a supervised classification scheme for automatically mapping altered rocks and soils is selection of training sites for establishing diagnostic parameters. Because of surface mineralogy variations over small distances, training site selection is difficult in hydrothermally altered areas, especially for epithermal deposit such as those in the Goldfield district.

Figure 6.25, color plate 10 shows a digital classification map based on analysis of MSS radiance data for all four bands using the Purdue University Laboratory for Application of Remote Sensing (LARS) program and an additive color-ratio composite prepared by the Jet Propulsion Laboratory (47). Compared with the color-ratio composite, which is a reasonably accurate representation of altered areas, the classification map is inferior in several respects. Although limonitic argillic to silicified rocks are accurately classified in the southern part of the mining district (red, Figure 6.25, color plate 10), only small parts of the northern zone are shown. More importantly, many large areas consisting of unaltered rocks are incorrectly classified as limonitic altered rocks. Discrimination of the opalites is roughly equal in both Figure 6.25, color plate 10 presentations, but poorer than in the diazo color-ratio composite (see Fig. 17.21, color plate 32).

Processing of the DN (data number) radiance values on the General Electric Image 100 (GE-100) computer classification system at the Jet Propulsion Laboratory, using iterative selection of training

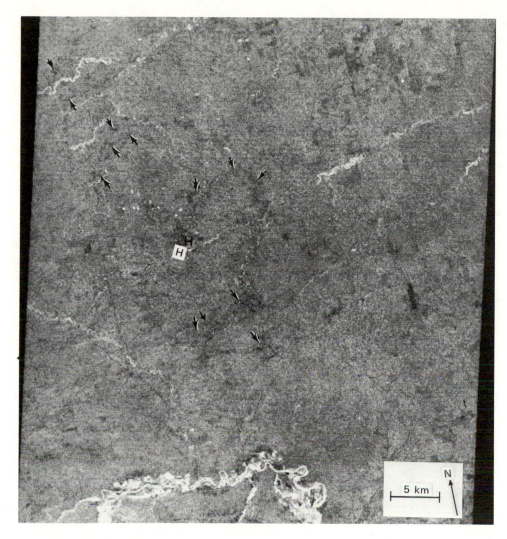

FIGURE 17.26. Linearly stretched MSS 4/5 ratio image of the Monument Hill-Highland Flat-Box Creek, Wyoming area.

areas, substantially improved the overall accuracy, especially through correct classification of the unaltered materials. However, little improvement was achieved in the northern part of the Goldfield alteration zone or for distinguishing the limonitic and hematitic altered rocks. The GE-100 does offer interactive processing, which is especially important for progressive refinement of classification of reasonably well-known areas. In addition, statistical information is readily available for analysis. A disadvantage at this and most other GE-100 installations is the difficulty of obtaining suitable hard copy for use in laboratory and field study.

The results of an experiment conducted by Schmidt and his colleagues (69, 70) in western Pakistan are more encouraging than the classification studies of the Goldfield region. The training area, Saindak, has many characteristics of the ideal porphyry-copper model proposed by Lowell and Guilbert (71). A group of small copper-bearing porphyritic quartz diorite stocks is surrounded by zones of contact-metamorphosed and hydrothermally altered lower Tertiary sedimentary rocks. According to Schmidt (69, p. 12), "The stocks are enclosed in a sulfide-rich envelope that contains as much as 15 volume percent pyrite; the envelope in turn is surrounded by a zone of propylitic alteration in which pyroclastic rocks in particular are altered to a hard dark epidote-rich hornfels. The most highly altered central part of the deposit is not fully defined, but there is a pervasive quartz-sericitic alteration, and recent studies by Shahid Noor Kahn are outlining the potassic alteration containing much hydrothermal biotite." Erosion of the Saindak area has resulted in formation of a valley which is encircled by a symmetrical rim of hills consisting of the more resistant hornfels and propylitic rocks. The valley is light toned and red to orange, but eolian and alluvial deposits obscure much of the colored soil. Outcrops within the highly altered central part are intensely leached to bright surfaces nearly free of iron-oxide minerals.

Construction of the classification table was based on the DN radiance values for pixels within the following classes of materials at Saindak: (1) mineralized rocks, including quartz-sericitic, quartz diorite and pyritic rock; (2) eolian and alluvial deposits, consisting of drywash materials, boulder-fan deposits, and bright windblown sand; and (3) a broad category of dark materials that included two types of hornfelsic rocks and desert varnished lag gravel deposits. Radiance limits for these classes were determined by iteratively examining computer classification maps of the known area and adjusting the bounding values until the falsely classified pixels reached an apparent minimum. A more restricted range was also constructed for some of the classes to provide a means of testing the confidence of the classifications made in the broader ranges. Because of the expense of these modifications, only a few revisions were made to produce the initial table (Table 17.2).

TABLE 17.2   DIGITAL CLASSIFICATION TABLE USED IN 1974 MINERAL EVALUATION IN THE CHAGAI DISTRICT, PAKISTAN (70).

Class Number	Rock Type			Symbol	Multispectral Scanner Bands			
					4	5	6	7
1	Mineralized rock	Quartz diorite	High reliability	O	46-50	52-60	50-60	18-22
2			Low reliability	θ	44-45	52-60	45-49	18-19
3		Pyritic rock	High reliability	X	41-45	47-54	39-44	16-17
4			Low reliability	X	41-45	47-54	39-44	16-19
5	Dry wash alluvium		High reliability	=	39-46	39-46	35-44	14-17
6			Low reliability	-	41-45	46-51	42-49	18-19
7	Boulder fan			+	33-40	39-46	30-35	9-16
8	Eolian sand		High reliability	.	38-44	46-54	42-51	18-22
9			Low reliability	,	45	46-127	42-127	18-63
10	Various dark surfaces including hornfelsic rock outcrops, desert-varnished lag gravels, and detrital black sand			|	33-36	28-38	29-35	11-15
11				||	24-35	19-27	20-32	8-12
12				H	29-36	28-38	20-28	9-14

Application of this classification scheme to the 2100-km² Chagai Hills, a promising porphyry copper district, resulted in about 50 areas with concentrations of pixels classified as mineralized quartz diorite and pyritic rock. Of these areas, 30 were judged by Schmidt to have the highest probability of correct classification and the most favorable geologic setting for porphyry copper deposits. The number of sites to be examined in the field was further reduced by study of 1:40,000-scale black-and-white aerial photographs, as seven areas were found to consist dominantly of windblown sand.

Ultimately, 19 areas were examined in the field and five were found to be dominantly hydrothermally altered sulfide-rich rock (Fig. 17.28); two additional areas contain altered rock and some sulfide, but appear to be less promising prospects. Prior to

field evaluation, Schmidt constructed a general order or priority that was based mainly on the morphologic expression of the 19 areas and his knowledge of the geology (69). He stated that "of the five sites found to have outcrops of hydrothermally altered sulfide-rich rock, two had been rated as medium priority prospects and three as low priority" (69, p. 20). The five promising altered areas have eroded to a soft-appearing, indistinct morphology and therefore do not have the prominent outcrop patterns characteristic of intrusive rocks, a feature used in assigning the high priority.

The mineralogical basis for discrimination of the altered and unaltered rocks is not clear, as neither detailed mineralogical or spectral reflectance studies have been conducted. Although some iron-oxide staining is present on the altered surfaces, the

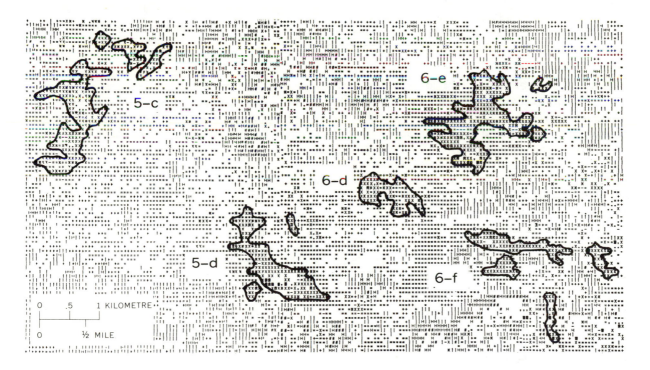

FIGURE 17.28. Digital classification map of the Chagai Hills area, Pakistan.

dominant reflectance characteristic appears to be high albedo due to intense leaching. This conclusion is supported by the spectral reflectance similarities between the altered rock and windblown sand deposits, as well as observations on the ground and in aerial photographs.

Another noteworthy result of Schmidt's study is that the digital classification experiment was preceded by visual analysis of a very high quality MSS color-infrared composite of this arid, extraordinarily well-exposed area. Although the colors and densities were adjusted to enhance the contrast between eolian deposits and other bright materials, this distinction was still particularly difficult. Some light-toned areas shown as felsic intrusive bodies on published maps seemed to be promising targets, but field evaluation showed these areas to be resistant unaltered bodies. Since that initial evaluation and the digital classification work described above, Schmidt (1976, personal communication) has reassessed the value of this composite and concluded that most of the five potential porphyry copper deposits would be selected by an experienced interpreter if few restrictions were placed on the number of areas to be delineated. However, a large number of false anomalies would result, thereby substantially reducing the overall practicality of this approach.

Since this initial study, Schmidt and Bernstein (72) have formulated a new classification table designed to resolve some of the problems encountered in the scheme shown in Table 17.2. The main problems were that many incorrect classifications resulted because of topographically controlled radiance variations, and windblown sand was commonly classified as mineralized rock. Most of these problems have been overcome by narrowing the class limits and selecting more training areas, especially on opposing topographic slopes. Ratioing to minimize slope effects has not yet been attempted.

These results demonstrate the effectiveness of digital classification for mineral exploration in perhaps the most convincing manner, that is, location of five promising altered *and* mineralized areas, although their potential as ore bodies must still be determined through detailed mapping and drilling. Of course, there is no reason to believe that the mineralization itself was detected; the alteration serves as a guide to *potentially* mineralized areas.

Vincent (53) has made extensive use of supervised classification maps, in addition to color-ratio composite images, for mapping limonitic areas associated with sedimentary uranium deposits. The classification maps (automatic recognition maps) appear to be especially useful for detecting and mapping small areas that might be missed in interpreting ratio images. Many of these maps are generated at a scale of 1:18,000. Because there are only five pixels per inch in this scale, nearly all of the spatial relationships are greatly obscured or lost to the interpreter. He has developed a recognition contour map to partially overcome this problem. In this type of map, contours enclose areas containing equal percentages of the target of interest, thereby portraying to the interpreter the general spatial relationships of the target materials.

The digital computer classification approach is regarded by many workers to be competitive with image enhancement and interpretation methods, but in practice they are complementary. Each approach has inherent advantages and disadvantages which vary in importance according to the problem of interest and the resources available. For broad area reconnaissance, to which the Landsat MSS data are most applicable, interpretation of enhanced images offers the advantages of essentially unsupervised classification while retaining important spatial information usually lost in classification maps. As more detailed analysis of small areas is required, the advantages of classification maps become apparent. Considerable knowledge about the materials present is necessary, however, for selection of adequate training areas. Otherwise, costly repetitive computer processing is commonly required to establish the optimum spectral radiance parameters. Such costs probably are similar to those expended for generating enhanced images, although no actual figures have been published.

Precisely registered, calibrated color-ratio composite images now being used by the author are nearly optimum MSS products. The principal advantages of both approaches are combined into a single product, while overcoming most of the disadvantages. The color-ratio composite is used first for reconnaissance mapping of limonitic areas. Then, by enlargement of selected areas, precisely registered pixels are classified individually according to their colors; in calibrated color-ratio composite images, the colors represent spectral reflectance.

17.3.4 Thermal-Infrared Emission

Two thermal-infrared techniques, thermal inertia and emissivity ratioing, have been developed for distinguishing various rock types (see Chapters 8 and 9). Neither of these techniques has been evaluated for their ability to detect altered rocks, but results of experiments dealing with unaltered rocks and consideration of laboratory measurements suggest that these techniques should greatly complement the other techniques in some situations.

THERMAL INERTIA

Measurements of thermal conductivity, specific heat, and density necessary for calculating thermal inertia have not been made for altered rocks or for the most important constituent minerals. Therefore, quantitative thermal inertia determinations are not yet possible for specific alteration products. General trends can be predicted, however, for different grades of hydrothermal alteration by considering the changes expected in thermal conductivity, specific heat, and density. Of these three factors, density is most important for reasonably dry materials because it is generally linearly correlated with thermal inertia (73) (Fig. 17.29), and a few density measurements are available for altered rocks. The density of surface and near-surface deposits is dependent on mineralogical composition and porosity, the latter property being especially important in altered rocks.

In general, thermal inertia appears to increase with grade of hydrothermal alteration, although many secondary processes affect this relationship. Mineralogical and textural changes associated with propylization and argillization should result in an overall reduction of thermal inertia. In a detailed study of compositional and related specific gravity variations across a subsurface vein at Butte, Montana, Sales and Meyer (74) showed that quartz monzonite was progressively altered by hydrothermal fluids to green montmorillonite as plagioclase and ferromagnesium minerals were hydrolized in the distal zone, kaolinite nearer the vein where Fe and Mg were leached from the montmorillonite, sericitic rocks still closer to the vein because of the breakdown of orthoclase, and silicified rocks adjacent to the vein (Fig. 17.30).

Specific gravity variations from fresh unaltered quartz monzonite across the increasingly intense zones of alteration are directly related to the percentages of the alteration minerals present (Fig. 17.30). Specific gravity decreases roughly pro-

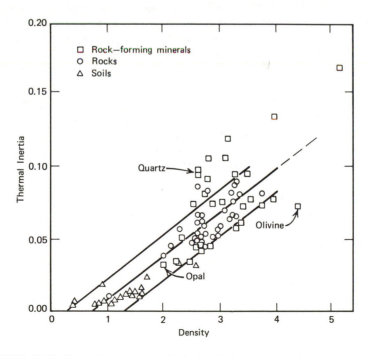

FIGURE 17.29 Graph showing general relationship between density and thermal inertia.

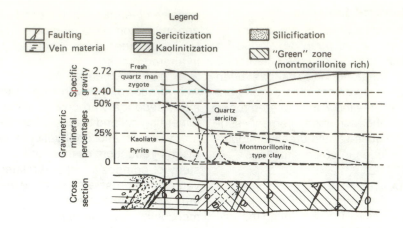

FIGURE 17.30. Variations of specific gravity and percentages of alteration minerals across zones of alteration in Butte, Montana.

portionally to increasing amounts of montmorillonite from 2.72 for fresh quartz monzonite to 2.40 where montmorillonite constitutes 25% of the wall rock. As the montmorillonite content declines to less than 5% and kaolinite increases to near 25%, specific gravity remains at 2.40. Rapid decline of the kaolinite content and increasing amounts of sericite and quartz result in a specific gravity of approximately 2.72 adjacent to the vein. Thus, argillic alteration of the quartz monzonite resulted in about a 12% reduction in specific gravity. On the other hand, the sericitic and silicified rocks and quartz monzonite specific gravities are similar. It is important to realize that these data relate only to mineralogical changes and do not take porosity variations into account. The porosities of the altered rocks, especially the leached kaolinite rocks, might be significantly higher than those of the fresh unaltered rocks, thereby further reducing their bulk density.

This trend toward lower density materials in argillic and perhaps some propylitic zones should be reinforced at and near the surface because of the prevalence of expansive, highly porous clay-rich soils. Soils such as these should contrast sharply with the mechanically disintegrated soils typical of unaltered rocks in Nevada. Leaching at the surface is yet another factor that drives hydrothermally altered rock toward lower bulk density; it creates numerous voids at the previous sites of unstable minerals, such as iron sulfide, the iron-oxides, and, in the most severe cases, clay minerals.

Therefore, four factors cause argillic rocks to trend toward lower density and consequently lower thermal inertia: (1) replacement of plagioclase and ferromagnesium and opaque minerals by phases of lower specific gravity; (2) generally increased porosity due to net loss of material during metasomatism and fracturing; (3) increased porosity due to leaching in the zone of weathering; and (4) presence of low density clay-rich soils. The resulting contrast between these altered rocks and the country rocks depends on the mineralogies and textures of the unaltered rocks. Contrast will be highest where the country rocks are low-porosity mafic or intermediate rocks and lowest for highly porous country rocks such as tuffs, marls, and lakebed deposits.

As the grade increases to silicified rocks, thermal inertia should increase substantially, mainly because of the dominance of quartz. Although quartz has a density of only 2.65, its high thermal conductivity results in a moderately high thermal inertia [$0.095 \, cal/cm^2 \cdot s^{1/2}$)] (Fig. 17.29). Another factor that contributes to higher thermal inertia of these rocks relative to argillic rocks is the lower content of low-density clay-rich soils in most silicified areas.

Where intense alteration yields dense quartz veins and ledges that are only locally vuggy, as in the Goldfield district, the thermal inertia should be high, except perhaps in very vuggy or highly fractured zones. Presence of hematitic coating on the quartz substrate, which is common at Goldfield, would further increase the thermal inertia because of its very high thermal inertia (Fig. 17.29). It is important to realize, however, that, although the thermal energy radiates from the exposed surface, the net flux is governed by the thermal properties of the near-surface layer. Hence, even a thick hematite coating would not totally dominate measurements of these rocks. Nonetheless, the moderate to high thermal inertia of these rocks should make them readily detectable in thermal inertia maps or images, particularly if they have been identified by other methods as being part of an altered area. It is interesting to note that goethite has a lower thermal inertia then both hematite and quartz. Although the presence of goethite would subdue rather than enhance the thermal inertia of quartz ledges and veins, the techniques may provide a method for distinguishing between the two types of iron-oxide minerals in some situations, such as gossans, assuming similar substrate mineralogies and textures.

In contrast to the high thermal inertia expected for silicified rocks, hot springs deposits should have low thermal inertias. Such deposits commonly have very high porosities and contain opaline quartz as a major constituent. In the Steamboat Springs, Nevada area, for example, Schoen and White (75) noted that calculated porosities range from 7.9% in fresh "Steamboat Basaltic Andesite" to 58.7% in completely altered samples. That opaline, rather than alpha quartz, is the main silica phase is important because the thermal inertia of alpha quartz [0.095 cal/(cm^2•s$^{1/2}$)] is substantially larger than that of opal [0.034 cal/(cm^2•s$^{1/2}$)] (Fig. 17.29). Combined porosity and mineralogical composition might reduce the thermal inertia of these altered rocks to a level on the order of 0.01 cal/(cm^2•s$^{1/2}$), similar to most soils (Fig. 17.29). Deposits such as these should be anomalous in thermal inertia maps or images and distinguishable from most exposed rocks, except perhaps high-porosity tuffs and marls.

Where the albedo range of the altered rocks and the topographic slope variations are small, predawn thermal-infrared images or images showing day-night temperature differences (see Chapters 8 and 9) might be useful for detecting the altered areas, thereby eliminating the need for producing a thermal-inertia image. For example, high-albedo porous argillized rocks should be cool in predawn images because most of the solar radiation would be reflected and they would cool rapidly during the night. However, modeling will probably be needed in most areas.

Thermal-infrared images have been applied to uranium exploration on a limited basis by Offield (68), the main objectives being discrimination of geologic units, detection of wet or silicified fault zones, and lithologic changes associated with roll-front alteration. Thus far, good results have been achieved in the Freer-Three Rivers, Texas area, where conventional geologic mapping is hampered by lack of outcrops, nondistinctive lithologies, a thin partial cover of lag gravel, and difficulty of land access. Analysis of predawn images shows that conglomeratic channel-fill units can be defined as relatively warm areas because of their high thermal inertia relative to the surrounding clayey units. The importance of this result is that the channel-fill is a favorable site for deposition of uranium-rich materials, but is difficult to map by conventional methods. Although other anticipated lithologic distinctions were not achieved and few lineaments were noted in this initial aircraft survey, better results are expected under more favorable weather conditions and when the analysis of the available data is completed, by using mathematical models to formulate thermal-inertia images from the predawn and midday images.

THERMAL EMISSIVITY

Another approach that promises to augment the methods already discussed is thermal-emissivity ratioing. This technique is based on the observation that the emissivity of silicate minerals and rocks departs significantly from black-body emission in the 8-12 μm region, resulting in a minimum that shifts to shorter wavelengths as the silica content increases (see Chapters 2 and 8). This spectral feature is absent or weakly expressed in nonsilicate rocks. Aircraft images that monitor radiative energy variations in spectral bands in this region have been ratioed to display the distributions of silicate and nonsilicate rocks (76) and of the various silicate rocks (77).

The most straightforward application of thermal-emissivity ratioing to mineral exploration appears to be for distinction between hydrothermally altered rocks, which are mainly composed of silicate minerals, and nonsilicates that have similar visible and near-infrared spectral reflectances and thermal inertias. Examples are bleached nonlimonitic argillic rocks versus reasonably pure high-albedo limestone, and nonlimonitic higher-grade altered rocks versus dolomite. In both pairs of rocks, the visible and near-infrared reflectance spectra of the altered and unaltered rocks would probably appear similar in multispectral images, even though the OH- and CO_3-absorption bands are, as previously noted.

separated by 0.15 μm, and the thermal inertias would not likely be distinctive. On the other hand, the emittance spectra for the altered and unaltered rocks in these pairs should be significantly different because of the high silica content of the altered rocks. Although carbonate rocks have an emissivity minimum near 11.5 μm, it is relatively weak. Thus, these pairs of rocks should be discriminable in thermal-emissivity ratio images, assuming that the carbonate rocks do not have significant silica or clay fractions, which would reduce the spectral contrast. In addition to these hypothetical cases, thermal-emissivity ratio images should also be useful for identifying highly siliceous deposits, such as the quartz veins in many districts, silicified zones in the Goldfield district, and the opalite deposits at Cuprite.

Viewed in a broader way, this technique is most applicable for mapping rock types that will contribute to establishing a framework for mineral exploration in poorly mapped areas. Furthermore, compositional information can be obtained that is not available from other remote-sensing sources. Another spectral region that should be investigated lies between 2.5 μm and 8 μm.

17.3.5 Gamma-Ray Spectral Emission

Some hydrothermal alteration zones lack all the potentially distinctive physical properties discussed above; in others, the contrast with adjacent country rocks may be inadequate for consistent discrimination. Zones enriched in K but lacking significant pyrite and hydroxyl-bearing mineral content are especially important because this type of alteration is typical of the central parts of some porphyry copper deposits. Sillitoe (78) pointed out that at Chuquicamata, Chile, one of the world's largest porphyry copper deposits, an early phase of alteration is represented by a feldspar-quartz assemblage in the core of the deposit. Although much of the feldspar was replaced by sericite during a later phase of alteration, the amount of sericite is small in large parts of this zone. A similar assemblage is present in the buried core of the Kalamazoo body in southeastern Arizona, except that secondary biotite is also an important constituent (71).

Laboratory measurements show that the visible and near-infrared spectral reflectance of quartz, K-feldspar and biotite generally lack prominent Fe^{+3} and OH-absorption bands. Of the K-silicate assemblages described above, only sericite has any

important spectral features (2.2-μm OH band), and in places its content is low. None of these minerals appear to be distinctive in the thermal-infrared region, although adequate laboratory measurements are admittedly sparse.

Radiometric surveys and chemical analysis of four porphyry copper deposits in Arizona and New Mexico show that the K content is 1.2 to 3 times higher in the altered rocks than in nearby temporally and petrogenetically associated unmineralized stocks (79). The K distribution is zoned so that the highest values are near the center of the intrusive body. In general, uranium does not appear to be abnormally higher than it is in other regions. It is important, therefore, that high K be coupled with low Th values in the alteration zones because the K/Th ratio should be the most useful means for distinguishing K-bearing barren intrusive and K-silicate altered rocks. This approach should also be valuable for mapping the zones within other altered areas.

Moxham and others (80) have indicated that a three-channel airborne instrument recording Tl^{208}, Bi^{214} and K^{40} photopeaks should be adequate for detecting the differences between the altered and unaltered rocks studied in Arizona. This capability is now commercially available. The airborne approach has several disadvantages, however. Flight altitudes must be limited to approximately 150 m because of severe attenuation of the signal at higher levels. Even at these low altitudes, the spatial resolution is low, roughly four times the altitude. Survey costs are high when compared on an equal-area basis with the other techniques discussed here. However, costs should be significantly less than the geochemical sampling and analyses required to gain the equivalent information. Although aerial gamma-ray spectrometry surveys for mapping K-silicate alteration have not been reported, the national program being sponsored by the Energy Research and Development Agency (ERDA) should provide the data needed for evaluating this promising technique.

17.4 SUMMARY

The main visible and near-infrared spectral reflectance characteristics of limonitic hydrothermally altered rocks are broad Fe^{+3} absorption bands short of 1.3 μm, a sharper OH-absorption band near 2.2 μm, and a window between these bands near 1.6

μm where the reflectance is highest. In general, these features are absent or weakly expressed in the unaltered volcanic rocks of south-central and western Nevada volcanic rocks and in their mechanically disintegrated soils.

In spite of the breadth and limited response range of the Landsat MSS bands, limonite can be detected and mapped in large areas with a high level of confidence by using color-ratio composite images. Smaller better known areas can be mapped in detail by using digital classification schemes. These capabilities have been clearly established and are being used in commercial mineral exploration programs. The main limitation in dealing with limonitic alteration is that vegetation cover greater than 40-50% obscures the spectral radiances of the underlying rocks. Higher spatial resolution, obtained either by means of improved satellite-borne systems or aircraft scanners, would partially overcome this problem, although data processing requirements would increase substantially. The capability of distinguishing small variations in limonitic material is also pertinent to uranium and petroleum exploration.

Mineralogical considerations place additional constraints on the applicability of the MSS data to mineral exploration. First, iron-oxide coatings are common on the surfaces of some unaltered rocks. In some places, however, these unaltered rocks lack the OH band near 2.2μm, thereby suggesting a means of distinguishing some limonitic but unaltered rocks from limonitic altered rocks. Second, many altered rocks are not characterized by anomalously high concentrations of limonite, but many of these do have a prominent 2.2-μm absorption due to the presence of hydroxyl-bearing phases. The importance of the 2.2-μm region for resolving these problems has been demonstrated in the Cuprite, Nevada mining district and in the East Tintic Mountains, Utah. Unaltered rocks that are mineralogically similar to altered rocks are probably not separable in the visible and near-infrared region because of their similar spectral reflectances.

Results of aircraft experiments dealing with unaltered rocks and consideration of laboratory measurements suggest that thermal-inertia mapping should augment the visible and near-infrared spectral reflectance approach. Abnormally low thermal inertias are anticipated in argillic and propylitic rocks because of increased porosity of the soils as well as the rocks, and because of conversion of ferromagnesian and opaque minerals to less dense

phases. With increasing grade of hydrothermal alteration and higher proportions of quartz, moderately high thermal inertia is expected, but increased porosity due to leaching would reduce the level. Nonporous quartzose ledges and veins should have the highest thermal inertia. An important exception to this trend are intensely altered opaline sinter deposits in which both texture and mineralogy should contribute to low values. Another promising thermal-infrared technique monitors the presence and position of the emissivity minimum between 8.4 μm and 11μm by ratioing multiple spectral bands. Although this technique presently appears limited to detection of silica-rich altered rocks, it has considerable potential for distinguishing commonly occurring rock types and for placing additional limits on their mineralogical composition.

Some hydrothermally altered rocks do not appear to have distinctive visible and near-infrared or thermal-infrared properties and might not be detectable using state-of-the-art remote sensing techniques. However, judging from field measurements, some of these, especially potassium silicate assemblages, may be detectable from analysis of airborne gamma-ray spectrometric data. This technique is limited to low-altitude aircraft platforms, however.

17.5 CONCLUDING REMARKS

Lineament analysis and limonitic alteration mapping will yield the largest benefits to mineral exploration programs where they are used together and the results are integrated with those of more conventional geologic mapping, geophysics, and geochemistry methods. The benefits that can be expected depend largely on the geologic and environmental setting and on the level of knowledge about ore deposits in a given area. Because of the synoptic view of the MSS and the possibility of discovery, highest benefits are expected in reconnaissance of large, poorly known, well exposed areas. As the size of the area decreases, vegetation becomes more abundant, and geologic understanding increases, the benefit will decline. However, this approach has substantial potential even in reasonably well-known and exposed areas such as the western United States of America because, in spite of detailed mapping in many small areas and general understanding of the regional framework, adequate metallogenic models still have not been formulated for most areas. Such models are crucial to exploration of the areas that are most promising for new discoveries. Determination of the regional spatial arrangement of major lineaments and limonitic alteration zones are important to development of these models.

An excellent example of combining these remote sensing capabilities for developing an exploration model is provided by studies being carried out in northern Sonora, Mexico (81). Four northeast-trending shear zones, spaced at 30 to 50 km intervals, have been defined using MSS images. These lineaments, which appear to be of Precambrian ancestry, intersect major north-northwest-trending lineaments believed to be Basin and Range faults. Areas of limonitic alteration are concentrated along the flanks of the northeast-trending shear zones, mainly near intersections with north-northwest-trending lineaments. Raines (81) points out the similarity of this tectonic pattern with that of the Colorado Plateau, the porphyry copper province of Arizona, and central Mexico. Geophysical and geochemical surveys conducted subsequent to the remote sensing studies strongly support the general validity of Raines' model. The remote sensing results were obtained in a much shorter time and at less expense than could have been achieved using conventional procedures.

Several areas of needed research are apparent. In lineament analysis, the most critical need is development of a better understanding of the origin of these features and their place in regional tectonic frameworks. The chief research tools will undoubtedly be geologic mapping and geophysical surveys, but other remote sensing techniques, such as thermal-infrared, passive and active microwave, and stereoscopic images will also prove useful in certain situations. Another critical need is a systematic program to determine the effects of various human and technical factors in the compilation procedure, thereby reducing the presently considerable subjectivity.

Improvement of alteration mapping methods depends on additional physical property measurements, especially in the near-infrared to midinfrared region, and field evaluation of aircraft data in carefully selected settings. Contrasting textures are commonly very important in alteration studies, but higher spatial resolution is required to allow the necessary measurements. Ultimately, models must be developed, probably using pattern recognition

techniques, to analyze data for a large number of parameters measured using conventional as well as remote sensing techniques.

REFERENCES

1. Park, C. F., and MacDiarmid, R. A., 1970, Ore deposits, 2nd ed.: San Francisco, Freeman, 522 p.
2. Rowan, L. C., 1975, Application of satellites to geologic exploration: American Scientist, v. 63, p. 393-403.
3. Eaton, G. P., Christiansen, R. L., Iyer, H. M., Pitt, A. M., Mabey, D. R., Blank, H. R., Jr., Zietz, I., and Gettings, M. E., 1975, Magma beneath Yellowstone National Park: Science, v. 188, p. 787-796.
4. Carter, W. D., 1974, Tectolinear overlay of the United States: U.S. Geol. Survey Open File map, 1:5,000,000.
5. Haman, P. J., 1975, A lineament analysis of the United States: West Can. Res. Publ., ser. 4, no. 1, 27 p.
6. Albert, N. R. D., 1975, Interpretation of Earth Resources Technology Satellite imagery of the Nabesna quadrangle, Alaska: U.S. Geol. Survey Misc. Field Studies map MF-655J, 2 sheets, scale 1:250,000.
7. Saunders, D. F., Thomas, G. E., Kinsman, F. E., and Beatty, D. F., 1973, ERTS-1 imagery use in reconnaissance prospecting: Dept. Commerce, Natl. Tech. Inf. Service, N74-20948, 129 p.
8. Viljoen, M. J., and Viljoen, R. P., 1973, ERTS-1 imagery as an aid to the definition of the geotectonic domains of the Southern African crystalline shield, in Symposium on significant results obtained from the Earth Resources Technology Satellite-1, volume I, Freden, S.C., Mercanti, E. P., and Becker, M. A., eds.: Technical Presentations: Natl. Aero. Space Admin. SP-327, p. 483-491.
9. Moody, J. D., 1966. Crustal shear patterns and orogenesis: Tectonophysics, v. 3, p. 479-522.
10. Ozoray, G. F., 1972, Tectonic control of morphology on the Canadian interior plains, in International geography, 1972 Adams, W. P., and Helleiner, F. M., eds.: Res. Council Alberta, Canada Contr. 547, p. 51-54.
11. Grant, A. R., 1969, Chemical and physical controls for base metal deposition in the Cascade Range of Washington: Washington Div. Mines and Geol. Bull. 58, p.

12. Zietz, I., Bateman, P. C., Case, J. E., Crittenden, M. D., Jr., Griscom, Andrew, King, E. R., Roberts, R. J., and Lorentzen, G. R., 1969, Aeromagnetic investigation of crustal structure for a strip across the western United States: Geol. Soc. America Bull., v. 80, p. 1703-1714.

13. Zietz, I., Hearn, B. C., Jr., Higgins, M. W., Robinson, G. D., and Swanson, D. A., 1971, Interpretation of an aeromagnetic strip across the northwestern United States: Geol. Soc. America Bull., v. 82, p. 3347-3372.

14. Thomas, G. E., 1974, Lineament-block tectonics: Williston-Blood Creek Basin: Am. Assoc. Petroleum Geologists Bull., v. 58, p. 1305-1322.

15. Roberts, R. J., 1964, Economic geology, in Mineral and water resources of Nevada: Nevada Bu. Mines Bull. 65, p. 39-45.

16. Shoemaker, E. M., Squires, R. L., and Abrams, M. J., 1974, The Bright Angel and Mesa Butte fault systems of northern Arizona, in Geology of northern Arizona with notes on archeology and Paleoclimate, I, Karlstrom, T. N. V., Swann, G. A., and Eastwood, R. L., eds.: Geol. Soc. America Rocky Mt. Section Meeting, Flagstaff, Arizona, p. 355-391.

17. Muehlberger, W. R., 1965, Late Paleozoic movements along the Texas lineament: New York Acad. Sci. Trans., v. 27, p. 385-392.

18. Cook, D. R., ed., 1957, Geology of the East Tintic Mountains and ore deposits of the Tintic mining district: Utah Geol. Soc. Guidebook to the geology of Utah, no. 12, 183 p.

19. Heyl, A. V., 1972, The 38th parallel lineament and its relationship to ore deposits: Econ. Geol., v. 67, p. 879-894.

20. Stewart, J. H., Walker, G. W., and Kleinhampl, F. J., 1975, Oregon-Nevada lineament: Geology, v. 3, p. 265-268.

21. Yates, R. G., 1968, The trans-Idaho discontinuity, in Internat. Geol. Cong., 23rd, Prague, 1968, Proc., sec. 1, Upper mantle (geol. processes): Prague, Academia, p. 117-123.

22. Maughan, E. K., 1966, Environment of deposition of Permian salt in the Williston and Alliance Basins, in Second symposium on salt: Cleveland, Ohio, Northern Ohio Geol. Soc., p. 35-47.

23. Landwehr, W. R., 1967, Belts of major mineralization in western United States: Econ. Geol., v. 62, p. 494-501.

24. Haites, T. B., 1960, Transcurrent faults in western Canada: Alberta Soc. Petroleum Geologists Jour., v. 8, p. 33-79.

25. Sonder, R. A., 1947, Discussion of "Shear Patterns of the Earth's Crust" by F. A. Vening. Neinesz: Am. Geophy. Union Trans., v. 28, p. 939-945.

26. Katterfeld, G. N., and Charushin, G. V., 1970, Global fracturing on the earth and other planets: Geotectonics, no. 6, 1970, p. 333.

27. Moody, J. D., and Hill, M. J., 1956, Wrench-fault tectonics: Geol. Soc. America Bull., v. 67, p. 1227.

28. Garson, M. S., and Krs, Miroslav, 1976, Geophysical and geological evidence of the relationship of Red Sea transverse tectonics to ancient fractures: Geol. Soc. America Bull., v. 87, p. 169-181.

29. Stovickova, N., 1973, Hlubinna zlomova tektonika a jeji vztah k endogennimgeologickym procesum (Deep-fault tectonics and its relation to endogenous geological processes): Praha, Academia, p. 1-198.

30. Lathram, E. H., and Raynolds, R. G. H., 1979, Relationship between selected space image lineaments and primary crustal segments in Alaska: in Podwysocki and Earle, eds., Proceedings of the Second International Conference on Basement Tectonics, Newark, Del., July 13-17, 1976, p. 191-197.

31. Lathram, E. H., and Gryc, George, 1973, Metallogenic significance of Alaskan geostructures seen from space: Proceedings, 8th Internat. Symposium on Remote Sensing of Environment, Ann Arbor, Mich., p. 1209-1211.

32. Sutherland-Brown, A., Cathro, R. J., Panteleyev, A., and Ney, C. S., 1971, Metallogeny of the Canadian Cordillera: Canadian Inst. Mining and Metallurgy Trans., v. 74, p. 121-145.

33. Kutina, Jan, 1969, Hydrothermal ore deposits in the western United States: a new concept of structural control of distribution: Science, v. 165, p. 1113-1119.

34. Offield, T. W., Abbott, E. A., Gillespie, A. R., and Loguercio, S. O., 1977, Structure mapping on enhanced Landsat images of southern Brazil: Tectonic control of mineralization and speculations on metallogeny: Geophysics, v. 42, no. 3, p. 482-500.

35. Clark, A. L., Berg, H. C., Cobb, E. H., Eberlein, G. D., MacKevett, E. M., Jr., and Miller, T. P., 1974, Metal provinces of Alaska: U.S. Geol. Survey Misc. Inv. map I-384, 1:5,000,000.

36. Sutherland-Brown, A., 1974, Aspects of metal abundances and mineral deposits in the Canadian Cordillera: Canadian Inst. Min. Metal. Trans., v. 77, p. 14-21.

37. Sillitoe, R. H., 1974, Tectonic segmentation of the Andes: implications for magmatism and metallogeny: Nature, v. 250, p. 542-545.

38. Guild, P. W., 1972, Metallogeny and the new global tectonics, in Mineral deposits, section 4: Proc. Internat. Geol. Cong. 24th, p. 17-24.

39. Stanton, R. L., 1972, Ore petrology: New York, McGraw-Hill, 713 p.

40. Bechtold, I. C., Reynolds, J. T., and Wagner, C. G., 1975, Application of Skylab imagery to resource exploration in the Death Valley region, in, Proceedings of NASA Earth resources symposium, vol. 1B, Geology information systems, Smistad, Olav (coord.): Natl. Aero. Space Admin., TM X-58168, p. 665-672.

41. Lowman, P. D., Jr., 1976, Crustal evolution in silicate planets: implications for the origin of continents: Jour. Geology, v. 84, no. 1, p. 1-26.

42. Smith, M. R., 1976, Arcuate structural trends and Basin and Range structures (based on a study of ERTS-1 imagery), in Proceedings of the First International Conference on the New Basement Tectonics, Hodgson, R. A., Gay, S. P., and Benjamins, J. Y., eds.: Utah Geol. Assoc. Pub. 5, p. 626-634.

43. Bayley, R. W., and Muehlberger, W. R., 1968, Map of basement rocks of the United States: U.S. Geol. Survey, scale 1:2,500,000.

44. Hemley, J. J., and Jones, W. R., 1964, Chemical aspects of hydrothermal alteration with emphasis on hydrogen metasomatism: Economic Geol., v. 59, p. 538-569.

45. Rowan, Lawrence, C., Goetz, Alexander, F. H., and Ashley, Roger P., 1977, Discrimination of hydrothermally altered and unaltered rocks in visible and near-infrared multispectral images: Geophysics, v. 42, no. 3, p. 522-535.

46. Ashley, Roger P., 1974, Goldfield mining district: in Guidebook to the geology of four Tertiary volcanic centers in central Nevada: Nevada Bur. Mines and Geol., no. 19, p. 49-66.

47. Goetz, A. F. H., Billingsley, F. C., Gillespie, A. R., Abrams, M. J., and Squires, R. L., 1975, Application of ERTS images and image processing to regional geologic problems and geologic mapping in northern Arizona: Jet Propulsion Laboratory Tech. Rept. 32-1597, p. 188, Pasadena, California.

48. Hunt, Graham R., Salisbury, John W., and Lenhoff, Charles J., 1973, Visible and near-infrared spectra of minerals and rocks: VII acidic igneous rocks: Modern Geol., v. 4, p. 217-224.

49. Rackley, R. I., 1972, Environment of Wyoming Tertiary uranium deposits: The Mountain Geol., v. 9, p. 143-157.

50. Langen, Raymond E., and Kidwell, A. L., 1974, Geology and geochemistry of the Highland uranium deposit, Converse County, Wyo.: Mountain Geol., v. 11, no. 2, p. 85-93.

51. Klohn, M. L., and Pickens, W. R., 1970, Geology of the Felder uranium deposit, Live Oak County, Texas: Amer. Inst. of Min. Eng. (AIME) Annual Meeting, preprint no. 70-I-38, Feb 15-19, 1970, Denver, Colorado.

52. Donovan, Terrence J., 1974, Petroleum-microscopage at Cement, Oklahoma: evidence and mechanism: Amer. Assoc. Petroleum Geol., v. 58, p. 429-446.

53. Vincent, Robert K., 1977, Uranium exploration with computer-processed Landsat data: Geophysics, v. 42, no. 2, p. 536-541.

54. Armstrong, Frank C., 1970, Geologic factors controlling uranium resources in the Gas Hills District, Wyoming, in Wyoming Geol. Assoc. Guidebook, 1970, Wyoming Sandstone Symposium, p. 31-44.

55. Germanov, A. N., 1960, Main genetic features of some infiltration-type hydrothermal uranium deposits (English translation): Akad. Nauk. SSSR Izv. Ser. Geol., no. 8, p. 60-71.

56. Anderson, D. C., 1969, Uranium deposits of the Gas Hills, in Wyoming Contributions to Geol.: Wyoming Uranium issue, v. 8, no. 2, p. 93-103.

57. Houston, R. S., 1969, Aspects of the geologic history of Wyoming related to the formation of uranium deposits, in Wyoming Contributions to Geology: Wyoming Uranium issue, v. 8, no. 2, p. 67-79.

58. Salmon, Bette C., and Pillars, William W., 1975, Multispectral Processing of ERTS-A (Landsat) data for uranium exploration of the Wind River Basin, Wyoming: A visible region ratio to enhance subtle alteration associated with roll-

type uranium deposits: Environ. Res. Inst. of Mich., Rept. 110400-2-F, Ann Arbor, Mich.

59. Ashley, R. P., 1975, Preliminary geologic map of the Goldfield mining district, Nevada: U.S. Geol. Survey Misc. Field Studies Map, MF-681.

60. Rowan, L. C., Wetlaufer, P. H., Goetz, A. F. H., Billingsley, F. C., and Stewart, J. H., 1974, Discrimination of rock types and detection of hydrothermally altered areas in south-central Nevada by use of computer-enhanced ERTS images: U.S. Geol. Survey Prof. Paper 883, 35 p.

61. Vincent, Robert K., 1975, Commercial applications of geological remote sensing: Inst. of Elec. and Elec. Eng. (IEEE) Conference on Decision and Control, preprint TA 1-5, p. 258-263.

62. Abrams, M. J., Ashley, R. P., Rowan, L. C., Goetz, A. F. H., and Kahle, A. B., 1977, Mapping of hydrothermal alteration in the Cuprite mining district, Nevada, using aircraft scanner images for the spectral region 0.46 to 2.36μm: Geology, v. 5, no. 12, p. 736-738.

63. Rowan, L. C., and Abrams, M. J., 1978, Mapping of hydrothermally altered rocks in the East Tintic Mountains, Utah using 0.4 to 2.3 μm multispectral scanner aircraft images; abst., International Association on the Genesis of Ore Deposits, Snowbird, Alta, Utah, Aug. 1978, Programs and Abstracts, p. 156.

64. Siegal, B. S., and Goetz, A. F. H., 1977, Effect of vegetation on rock and soil type discrimination: Photogrammetric Engr. and Remote Sensing, v. 43, no. 2, p. 191-196.

65. Eliason, Eric M., Soderbloom, Laurence A., and Chavez, Pat S., Jr., 1975, Simulating true color images of Earth from ERTS data: in Proceedings of the American Society of Photogrammetry, 41st Annual Meeting, March 9-14, 1975, Washington, D.C., p. 785-788.

66. Albert, Nairn, R. D., and Chavez, Pat S., Jr., 1976, Computer-enhanced Landsat imagery as a tool for mineral exploration in Alaska: U.S. Geol. Survey Open File Rept. no. 76-65.

67. Spirakis, Charles S., and Condit, Christopher D., 1975, Preliminary report on the use of Landsat-1 (ERTS-1) reflectance data in locating alteration zones associated with uranium mineralization near Cameron, Arizona: U.S. Geol. Survey Open File Rept. no. 75-416, 20 p.

68. Offield, Terry W., Remote sensing in uranium exploration: Amer. Inst. Min. Eng. (AIME) proceedings of meeting, March 30-April 2, 1976, Vienna, Austria, (in press).

69. Schmidt, Robert G., 1975, Exploration for porphyry copper deposits in Pakistan using digital processing of ERTS-1 data: U.S. Geol. Survey Open File Rept. no. 75, 29 p.

70. Schmidt, R. G., Clark, B. B., and Bernstein, R., 1975, A search for sulfide-bearing areas using Landsat-1 data and digital processing techniques: in Proceedings of the NASA Earth Resources Survey Symposium, NASA, TM X-58168 (JSC-09930), v. I-B, p. 1013-1038, June 1975, Houston, Texas.

71. Lowell, J. D., and Guilbert, 1970, Lateral and vertical alteration-mineralization zoning in porphyry ore deposits: Econ. Geology, v. 65, no. 4, p. 373-408.

72. Schmidt, R. G., and Bernstein, Ralph, 1975, Evaluation of improved digital processing techniques of Landsat data for sulfide mineral prospecting: U.S. Geol. Survey Open File Rept. no. 75-632, 29 p.

73. Watson, K., 1975, Geologic application of thermal-infrared images: Jour. Inst. Elec. and Elec. Eng. (IEEE), v. 63, no. 1, p. 128-137.

74. Sales Reno H., and Meyer, Charles, 1948, Wall rock alteration at Butte, Montana: Amer. Inst. of Min. and Metal. Eng., v. 178, p. 9-35.

75. Schoen, Robert, and White, D. E., 1965, Hydrothermal alteration in GS-3 and GS-4 drill holes, main terrace, Steamboat Springs, Nevada: Economic Geol., v. 60, p. 1411-1421.

76. Vincent, R. K., and Thomson, F. J., 1972, Rock-type discrimination from ratioed infrared scanner images of Pisgah Crater, California: Science, v. 175, p. 986-988.

77. Vincent, R. K., Thomson, F. J., and Watson, K., 1972, Recognition of exposed quartz, sand and sandstone by two-channel infrared imagery: Jour. Geophys. Res., v. 77, p. 2473-2477.

78. Sillitoe, Richard H., 1973, The tops and bottoms of porphyry copper deposits: Economic Geol., v. 68, no. 6, p. 799-815.

79. Davis, Jerry D., and Guilbert, John M., 1973, Distribution of the radioelements potassium, uranium and thorium in selected porphyry copper deposits: Economic Geol., v. 68, no. 2, p. 145-160.

80. Moxham, R. M., Foote, R. S., and Bunker, C. M., 1965, Gamma-ray spectrometer studies of hydrothermally altered rocks: Economic Geol., v. 60, no. 4, p. 653-671.

81. Raines, Gary, 1978, A porphyry copper model for northern Sonora, Mexico: U.S. Geol. Survey Jour. of Research, v. 6, no. 1, p. 51-58.

CHAPTER 18

WATER RESOURCES

VINCENT V. SALOMONSON
ALBERT RANGO

18.1 INTRODUCTION

The concept of the hydrologic cycle, as presented schematically in Fig. 18.1(1), depicts a very dynamic and fascinating process involving one of our basic resources. The various major environments and components of the hydrologic cycle can be broken down into the atmospheric, surface, and subsurface environments where water resides or is stored and fluxes of precipitation, evapotranspiration, runoff, and groundwater outflow or inflow occur. Another way to view the water deposition and redistribution process is to recognize that the watershed or catchment area responds to the deposition of rainfall by continually, although slowly, adjusting its stream network and surface cover to transport the water out of the watershed via runoff, evapotranspiration, and groundwater fluxes. As such, the watershed geology, soils, and vegetation modulate or control the rate at which rainfall is manifested as runoff or the other fluxes.

A growing population and desires for improved quality of life are placing increasing demands on our water resources. This places greater demands on scientists and engineers to provide means to better observe and inventory the amounts of water in the various environments mentioned above. Considerable effort and progress has been made in improving the conventional means of observing water resources and associated watershed characteristics. In most cases, however, these measurements are made at one point only, or at widely separated points. Thus, it is difficult to adequately monitor the spatial variability in phenomena such as soil moisture or rainfall. In view of these difficulties there is a need to utilize and understand the advantages offered by remote sensing approaches and sensors. Remote sensing approaches, as indicated in previous chapters, basically offer a capability for repetitively monitoring large areas with a high observational density that can accurately depict spatial and temporal variability.

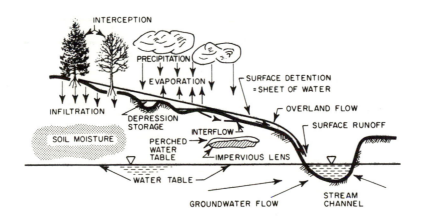

FIGURE 18.1. Components of the hydrologic cycle (1).

This chapter will discuss monitoring the various environments and fluxes in the hydrologic cycle, the most advantageous portions of the electromagnetic spectrum for observation, available sensors and systems (and their associated spatial and temporal resolution capabilities, accuracies and precisions), and the general data processing and interpretation techniques that have proven useful.

Before proceeding further, a short description of the major data sources is provided. The emphasis in this section will be on high altitude aircraft and spacecraft. Table 18.1 summarizes the characteristics of the most applicable sensors on these platforms.

As can be noted in Table 18.1, these data sources have been divided into Operational, Quasi-operational (some applications are using these data now, but the vehicles are not operational; that is, coverage is not guaranteed) and strictly Research and Development. Most of the applications to be described in forthcoming sections will come from the sensors in the first two categories, but some macroscale geophysical studies and insights can be gained from examining information contained in the other data sources. As these applications and studies are discussed reference will be made to these sensor systems. A brief description of each system as viewed in a hydrological context is provided in the next few paragraphs.

Various types of remote sensing instruments and platforms are available (some especially tailored for water problems) to water resources investigators, but presently the most widely applicable sensors are the Multispectral Scanner Subsystem (MSS) on the Landsat satellites, the Very High Resolution Radiometers (VHRR) on the NOAA meterological satellites, and basic multispectral camera arrays flown on high altitude aircraft such as the NASA U-2 and RB-57. These platforms and their attendant sensors tend to complement each other over a wide range of basin size and watershed management activities.

Remote sensing flights from low and medium altitude aircraft can generally be tailored to suit

TABLE 18.1   SELECTED HIGH-ALTITUDE AIRCRAFT AND SPACECRAFT SYSTEMS THAT HAVE PRODUCED OR ARE PRODUCING DATA APPLICABLE FOR WATER RESOURCES MONITORING AND HYDROLOGIC SYSTEMS STUDIES.

VEHICLE/SENSOR	SPECTRAL BANDS	NOMINAL SPATIAL RESOLUTION	APPROPRIATE IMAGE/SCENE AREAL COVERAGE	FREQUENCY OF COVERAGE	PERIOD OF DATA AVAILABILITY	DATA CENTER
OPERATIONAL						
NOAA/VHRR	0.6 - 0.75 μm 10.5 - 12.5 μm	1 KM	SUB-CONTINENT	1/DAY - VISIBLE 2/DAY - I.R.	1972 TO 1978	SUITLAND, MD.
ESSA-NOAA/ AVCS-SR	VISIBLE (AVCS) 0.5 - 0.75 μm (SR) 10.5 - 12.5 μm	4 KM	SUB-CONTINENT	1/DAY - VISIBLE 2/DAY - SR	1966 - PRESENT	SUITLAND, MD.
SMS-GOES/ VISSR	0.6 - 0.7 μm 10.5 - 12.5 μm	1 KM	1/3RD OF GLOBE (WESTERN) HEMISPHERE)	SEVERAL TIMES PER DAY	1974 - PRESENT	SUITLAND, MD.
TIROS-N/NOAA AVHRR	4 -5 BANDS VISIBLE NEAR INFRARED THERMAL INFRARED	1.1 KM	SUB-CONTINENT	1/DAY VISIBLE/ NEAR INFRARED 2/DAY THERMAL INFRARED	1978 - PRESENT	SUITLAND, MD.
QUASI-OPERATIONAL						
LANDSAT 1 - 3/MSS	0.5 - 0.6 μm 0.6 - 0.7 μm 0.7 - 0.8 μm 0.8 - 1.1 μm 10.5 - 12.5 μm (L-3)	80 M 240 M	34,000 KM	ONCE EVERY 18 DAYS (9 DAYS WITH TWO SATELLITES)	1972 - PRESENT	SIOUX FALLS, S.D.
HIGH ALTITUDE NASA AIRCRAFT	VISIBLE AND INFRARED PHOTOGRAPHY	10 METERS (APPROX)	400 - 900 KM^2	VARIABLE	OCCASIONAL COVERAGE IN SELECTED AREAS FOR 10 YEARS OR MORE	SIOUX FALLS, S.D.
RESEARCH AND DEVELOPMENT						
SKYLAB-EREP/ MULTISPECTRAL CAMERAS SPECTROMETERS	VISIBLE AND NEAR-INFRARED THERMAL INFRARED	10 - 70 M	10,000 - 30,000 KM^2	VARIABLE	1973 - 1974 (THREE FLIGHTS)	SIOUX FALLS, S.D. SUITLAND, MD. SALT LAKE CITY, UT
SKYLAB-EREP/ MICROWAVE SCATTEROMETER- RADIOMETER	13.9 GHz	11 KM	0 - 18° INCIDENCE ANGLES	VARIABLE	1973, 1974	SIOUX FALLS, S.D. SUITLAND, MD. SALT LAKE CITY, UT
SKYLAB-EREP L-BAND RADIOMETER	1.4 GHz	124 KM	–	VARIABLE	1973, 1974	SIOUX FALLS, S.D.
NIMBUS 1 - 7	MULTISPECTRAL RADIOMETERS (VISIBLE/I.R.)	4 - 55 KM	SUB-CONTINENT	DAILY IN SELECTED PERIODS	DISCONTINUOUS COVERAGE SINCE 1964	GREENBELT, MD.
NIMBUS 5 - 7 MICROWAVE RADIOMETERS	0.81 - 4.54 GHz	20 - 150 KM	SUB-CONTINENT	DAILY	DISCONTINUOUS COVERAGE SINCE 1972	GREENBELT, MD.

the user and can be contracted for use from a variety of private concerns and governmental agencies. High altitude missions (15-20 km altitude) are flown by NASA's Earth Resources Aircraft Program (ERAP) in support of various satellite missions and other earth science related research projects. The high altitude remote sensing platforms combine the advantages of satellite sensors and low altitude aircraft by being able to obtain high resolution data over medium size watersheds (<1000 km²). Observations over watersheds of specific events such as floods are usually provided at infrequent intervals and on a request basis. Highly repetitive observations (more often than once per month, for example) over long periods of time and for several locations are rather difficult to obtain, but when observations are taken they usually provide good detail about information on the ground because of the high spatial resolution that can be provided.

The Landsat MSS observations can be used successfully on watersheds covering areas from 10-20 km² in size up to watersheds covering approximately 30,000 km². Although this system is described in considerable detail in earlier chapters it should be noted that the relatively high spatial resolution (80 m) and the cartographic quality of the imagery from Landsat have enhanced the use of satellite data for water resources monitoring. The major limitation of these Landsat 1 and 2 systems for hydrological applications has been the frequency of coverage (nominal frequency is once every 18 days) and the speed of data delivery. It is expected that these deficiences will be reduced with the implementation of more advanced Earth Resources Satellite Systems.

Since 15 October 1972 the National Oceanic and Atmospheric Administration has launched a series of environmental satellites providing daily, 1-km spatial resolution observations. These satellites are in a near-circular, sun-synchronous orbits at nominal altitudes of 1500 km. NOAA-satellites have provided two views of North America daily, one at approximately 1000 and one at about 2200 (local time). The payload of the NOAA environmental satellites include a number of sensors, but the one of major interest is the Very High Resolution Radiometer (VHRR). The VHRR is a dual-channel scanning radiometer sensitive to energy in the visible spectrum (0.6-07 μm) and in the infrared (10.5-12.5 μm). The instantaneous

field of view is designed to be 0.6 mrad for both channels. This sensor provides observations of greatest utility on basins larger than approximately 1000 km². Beginning with TIROS-N and succeeding NOAA (6,7,etc.)polar orbiting satellites, and advanced VHRR (AVHRR) will be operating and obtaining observations in several spectral bands. The TIROS-N AVHRR, for example, has bands covering the 0.55-0.68, 0.725-1.10, 3.53-3.93 and 10.5-11.5 spectral regions. All observations will be available at a full resolution of 1.1 km.

As one considers these various systems it is helpful to keep in mind the time and space scales in which various phenomena occur and the degree to which available remote sensing systems are compatible with the requirements dictated by these phenomenological scales. Figure 18.2 shows a diagram illustrating these considerations. Of course, aircraft systems can also observe these phenomena over short periods of time, but for real understanding of the physics of these phenomena, long period, repetitive observations as provided by satellites are very helpful.

Once a basic knowledge of remote sensing capabilities is acquired, probably the best way to specifically become acquainted with advantages of remote sensing for a particular watershed problem is to obtain some data over the area of interest. These data can then be processed and compared to previous knowledge and conventionally available data to develop a familiarity with the potential uses. The purpose of the following sections in this chapter is to convey insight as to the applications of aircraft and, in particular, satellite data that have been made and the techniques and principles involved in interpreting and utilizing these data.

18.2 WATERSHED GEOMETRY AND LAND USE

18.2.1 Applications of Watershed Physiography and Land Use

The drainage basin is the basic areal unit in which physiographic and land use features can be measured and studied. The measurement of these watershed characteristics is important because they are intimately related to basin water and sediment yield. Physiographic observations such as basin area and shape, stream network organization,

Figure 18.2

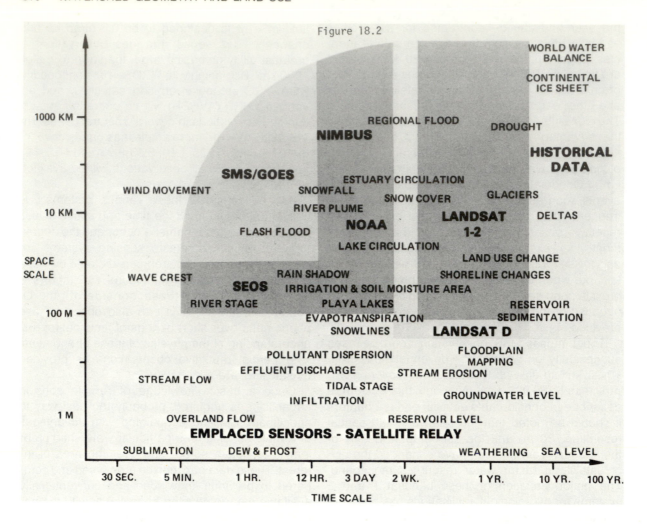

FIGURE 18.2. Time and space scales for representative hydrologic parameters with relevant observational capabilities of satellite systems represented by shaded areas.

drainage density and pattern, and specific channel characteristics can enable an investigator to estimate the mean annual discharge and mean annual flood flows from a watershed, as well as the rapidy of watershed response to a particular rainfall event. In the northeastern United States it has been demonstrated that streamflow and its distribution are primarily functions of climate, physiography, and land use (2).

In addition to being useful for estimating mean streamflow characteristics, physiography and land use data are required as calibration inputs to numerical watershed models that are commonly used for simulating or estimating daily, weekly, monthly, and annual flows. Physiographic parameters necessary for the operation of most of these models include stream length, drainage density, overland flow length and roughness, and watershed area and shape. Area estimates of land use categories including surface water, forest, and watershed imperviousness are additionally necessary. Some models also require that the riparian nature of the various land use categories be identified.

In order to meet requirements for water quality planning under Section 208 of the Federal Water Pollution Control Act Amendments of 1972, regional planning agencies must make predictions of future water quality based on projections of land use patterns. Land use surveys from current remote sensing data provide up-to-date baseline data from which to make the required projections. Various models for predicting current and future non-point source pollution in rivers and lakes based on current and projected land use have been developed. As an example, sixteen land use categories were provided using Landsat data for the Ohio-Kentucky-Indiana Regional Council of Governments that were applied to 225 drainage areas in the region (3). The machine processing techniques employed permitted the nine-county region to be mapped to one-acre detail in 90 days for $20,000. The remote sensing techniques used were concluded to be more timely, economical, and rapid than conventional methods.

In addition to water quality, the United States Soil Conservation Service (SCS) empirical methods for the prediction of water quantity require the input of land use and cover type, as well as hydrologic soil type (4). Landsat permits the necessary land cover estimates and may even be useful for the necessary soil characterization in vegetation-sparse areas. (See Section 18.2.3)

18.2.2 Physiography Measurements

In the past, low altitude photography has been shown to be an effective way to measure various physiographic parameters such as watershed area and shape (for small watersheds), drainage pattern, stream length and order, drainage density, and sinuosity and meander wavelength in combination with or in lieu of topographic maps. Depending on stream order, channel width measurements can often be obtained. In some cases, aircraft imagery is the most effective way to detect subtle feature such as paleochannels (5). Measurements of these same parameters from high altitudes and space are theoretically feasible, but resolution is usually significantly lower so that the detection of only selected features is possible.

Landsat imagery has been evaluated as a means of providing useful watershed physiography information (6). In this analysis physiographic parameters such as drainage basin area and shape, stream length, drainage density, and sinuosity were obtained in three study areas: (1) southwestern Wisconsin; (2) eastern Colorado; and (3) portions of the middle Atlantic States.

The 0.6-0.7 μm and 0.8-1.1 μm Landsat bands were selected for the physiographic mapping of drainage basins. The near-infrared band (0.8-1.1 μm) reveals the most contrast between land and water features and is best suited for mapping large streams, whereas the visible red band (0.6-0.7 μm) shows good contrast between different land features and vegetated and nonvegetated areas and is most useful for network delineation where streams are not wide enough to be detected in the near-infrared band. Color composites using various spectral bands may also be used to enhance drainage network features. In order to facilitate comparisons with USGS topographic maps, Landsat images were enlarged photographically or with the aid of a zoom transfer scope (which superimposes an image onto a map with the aid of a system of lenses, mirrors, and scale adjustments with an additional capability for removing image distortion) to the scale of the map. Generally, study enlargements were made to 1:250,000 scale for comparison, but areas of particular interest were also enlarged to a 1:100,000 scale.

Drainage basin area and shape, stream length, drainage density, and sinuosity were calculated from maps and images with the assistance of a compensating planimeter and a map measurer. The drainage basin boundary for a particular stream was outlined by identifying divides between small headwater streams on Landsat imagery or by using topography in the case of maps. The basin area was then measured by running a compensating planimeter along the drainage basin boundary. The drainage shape was subsequently quantified by using the following formula:

$$\text{basin shape} = \text{basin area}/C$$

where C = area of a circle having the same perimeter length as the basin. A basin shape of 1.0 indicates a perfectly circular watershed. The perimeter of the basin and the total length of streams within a watershed were calculated by running the map measurer along the line in question. Drainage density was then easily obtained by dividing the drainage area into the total stream length. The stream sinuosity was determined by measuring the length of the main stream in the basin and dividing by the axial length of the drainage basin (a sinuosity of 1.0 would indicate a perfectly straight river with no meanders).

Using Landsat imagery enlarged to 1:250,000 and 1:100,000 scales it has been found that drainage basin area and shape and stream sinuosity were comparable (within 10%) in all study areas to physiographic measurements derived from conventional topographic maps at the same scales (6). Improved drainage network and density information was obtained from Landsat imagery in dissected areas such as southwestern Wisconsin, but in heavily vegetated areas (middle Atlantic States) or areas with little physical relief (eastern Colorado) low order streams were difficult to detect and the derived drainage densities were significantly smaller than those obtainable from standard maps. The analysis of several seasonal Landsat scenes or, even better, the use of high altitude U-2 photography (at 20 km), however, permitted the extraction of greater drainage network detail. Considering the limitations of the data, Landsat and high altitude photography can be used as valuable ancillary data in the United States and most effectively as the primary data source in unsurveyed or data sparse regions.

18.2.3 Measurements of Soils, Cover Type, and Land Use in Relation to Water Yield Prediction

Hydrologic characteristics of soils are particularly important for predicting the amount of runoff to be expected from a particular storm event. The SCS specifies four groups of soil necessary for hydrologic analysis—groups A (low runoff potential), B (moderate rate of water transmission), (C) (slow rate of water transmission), and D (high runoff potential) (4). Various soil associations delineated from Landsat data may be useful for estimating the hydrologic soil type, but the widespread availability of soil surveys (in the United States) provides a more pertinent data source (7).

The SCS method for prediction of peak flows is used for the planning of small retention dams in the United States. Rather than use Landsat data to actually measure the land use, cover type, and soils necessary for generating the peak flows, attempts have been made to use the Landsat average watershed reflectance ratios (bands 5 and 7) and correlate them directly with the SCS curve numbers used for transforming rainfall to rainoff (8). Some results indicate that a high correlation exists, but in general the approach applies to only a specific area and may not be widely applicable.

Curve numbers necessary for the generation of flow volumes in the SCS methods can be more consistently and reliably obtained by actually delineating various cover types and working through the various tables provided by the SCS to obtain a curve number. Prediction of flood volumes to be expected in urbanizing areas has become more important in recent years. The SCS has recently published a manual on the techniques for estimating flows from urban areas (4). The percent of the basin that is impervious is a critical factor to be estimated. Following SCS guidelines imperviousness can be estimated from Landsat residential, urban, and commercial-industrial categories. Impervious area on the Anacostia River watershed in Maryland was calculated to be 24% from conventional techniques and 19% from Landsat data (9). Approximately 94 man days were required for the conventional estimate and less than four man-days for Landsat. In the calculation, residential area was assumed to be 35.6% impervious, urban 90%, and commercial-industrial 90%. Agreement between the conventional method and the Landsat approach was excellent for subwatershed areas as small as 1.48 km^2. Using Landsat data for studies of this kind seems to be a proven and effective approach warranting strong consideration in applications of this type.

In addition to the study referred to above, a sensitivity analysis has been performed which has identified the input parameters in the Kentucky Watershed Model that are amenable to current remote sensing systems (10). This model is representative of more complex, parametric, continuous simulation models. The input parameters that can presently be obtained with remote sensing at an acceptable accuracy include watershed area, fraction of impervious area, water surface fraction of the basin, vegetation maximum interception rate, mean overland flow surface length, overland flow roughness coefficient, and fraction of the watershed in forest. Other parameters have been identified as potentially extractable as improvements in image interpretation and analysis techniques become available and new remote sensing methods are developed. Using up-to-date remote sensing-derived estimates of impervious areas, as opposed to conventional estimates from topographic maps, it has been shown that remote sensing-based model calibrations provided more accurate streamflow simulation on the Patuxent Western Branch watershed in Maryland, for example, than did calibrations based on conventional data (10). It appears, overall, that remote sensing data can be a valuable tool in water yield prediction, especially in areas experiencing rapid land use changes.

18.2.4 Special Land Use Categories of Interest to Water Resources

FLOODPLAIN DELINEATION

Because of the National Flood Insurance Act, there currently is a need for surveys of floodprone areas in both urban and rural areas. These surveys are very detailed and as a result only a limited number can be completed each year. As a result various techniques for rapid preliminary planning surveys have been investigated. Remote sensing delineations of floodplains using aircraft data have proven successful because of a capability of the detection of various natural and artificial indicators (11). The alternative method of using remote sensing to produce a floodprone area map is to acquire color infrared imagery of a significant flood in progress or soon after the peak has passed. Such an approach directly provides a pictorial extent of the area inundated by a flood event with a specific return period. Although both of these approaches have been attempted with aircraft, some of the best examples can be shown by using Landsat data which possess relatively high resolution, cartographic fidelity, and near-infrared sensors.

The direct approach, that is, mapping floods as they occur, has been reported several times (12, 13, 14, 15). Areas inundated are detected in the near-infrared Landsat bands as areas of reduced reflectivity due to standing water, excessive soil moisture, and vegetation moisture stress. Most important is the fact that Landsat near-infrared observations as late as two weeks after the flood crest still show the characteristic reduced reflectivity of the previously inundated areas, thus reducing the need for obtaining satellite observations at the time of peak flooding. Working at 1:250,000 scale, it has been shown that Landsat observations are directly comparable with conventional low altitude flights over the flooded East and West Nishnabotna Rivers in Iowa (13). Figure 18.3 shows a comparison between Landsat and the low altitude flight. Only the East Nishnabotna River was flown because of monetary constraints. Over

80% of the area mapped as flooded with both the Landsat and the aircraft data agreed. In the remaining area minor discrepancies resulted from (1) incomplete low altitude coverage and resulting faulty area extrapolations, and (2) the fact that the aircraft imagery was acquired before the peak had reached the lower portion of the basin. The advantages of the satellite include that it is a low cost way to obtain large area coverage of a flooded region. Landsat does have a problem equaling the detail available from aircraft data, however. In the case of the Nishnabotna Rivers, a floodprone area map for the 100 year flood event was produced, but at a scale of 1:250,000. Digital processing of Landsat data would be required to approach a more workable map overlay at a data scale of 1:24,000.

Other investigators (16, 17) have shown that areas likely to be flooded, known as floodprone areas, tend to have Landsat multispectral signatures which are at times different than the signatures of surrounding nonfloodprone areas. These floodprone areas have distinctive natural vegetation and soil characteristics as well as different cultural features acquired over a long period of time in response to increased flooding frequency that enable them to be distinguished from the nonfloodprone area maps in areas that may not receive above have also shown that the Landsat floodprone area signatures have, as yet, a relatively unexplained, but fortuitous, correlation with the 100 year flood engineering and legal boundaries. Hence, it is believed that Landsat can provide preliminary floodprone area maps in areas that may not receive a detailed survey for years, or they may be used as another source of data to check already existing surveys and perhaps accentuate areas of discrepancy that may merit further detailed ground surveys. In addition, Landsat data have the potential for being used to monitor adherence to developmental or land use restrictions in already delineated floodprone areas and for being used in hydrologic analyses to derive coefficients necessary for generating the 100 year discharge as indicated in the previous section.

Flood and floodprone area observations from Landsat are indeed promising, but only on a regional basis with the current 80-m resolution. Digital Landsat flood and floodprone area maps have been produced at 1:24,000 and 1:62,500 scales, but they do not meet national map accuracy standards. For many legal requirements it is necessary to generate

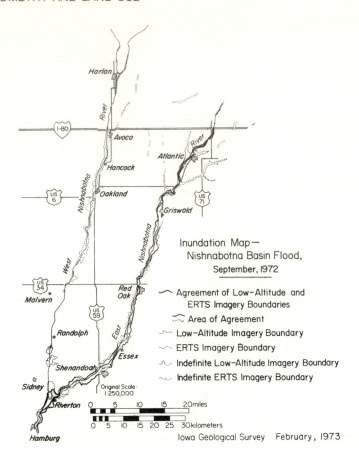

FIGURE 18.3. Comparison of Landsat (ERTS-1) flood map boundaries and low-altitude aircraft-determined boundaries.

products at even larger scales. As a result, flood assessment on small watersheds must generally be done using high resolution, color infrared photography such as available from the U-2. Such imagery provides the needed resolution for mapping inundated areas. The detection of floodprone areas at required legal scales has also been performed using aircraft data (18). It appears, overall, that for most large-scale flood and floodplain applications, high resolution aerial photography is the basic and necessary tool.

IRRIGATED AGRICULTURE

Most irrigated agricultural activities take place in the arid and semiarid western United States. This area of the country, because of a marked lack of natural vegetation to obscure the soil surface, is very amenable to remote sensing techniques. Irrigated lands, for example, tend to stand out in sharp contrast with natural range and dryland farming. Several of these irrigated areas are evident in Fig. 18.4 color plate 42 as bright red regular patterns to the north and east of Denver, Colorado. Under close inspection most of the irrigated fields are rectangular whereas a few, using center-pivot water application, are strikingly circular. Not only can the extent of irrigated lands be monitored on a regular basis, but with suitable interpretive techniques, the type of irrigation in use may be estimated. Irrigation canals can be easily located, and, by checking for anomalous vegetative growth along the distribution system, the presence of significant leaks may be located. Whenever large area surveys of total acreage devoted to irrigation are needed on a yearly basis, such as for the states of California, Idaho or other Western states, Landsat data appear to provide adequate and valuable sources of information.

18.3 SURFACE WATER

18.3.1 Methodology

Standing water on the surface of the earth, in general, has a rather low reflectance unless substantial amounts of sediment or organic materials are present. This reflectance is consistently low in the near-infrared portion of the electromagnetic spectrum (unless algal material is present) while the reflectance of other portions of the spectrum can be relatively high. If one must choose a spectral band from an array of sensors such as that given in Section 18.1, the bands in the region from 0.7 to 3.0 μm region are best. The Landsat 0.8-1.1 μm band is the best choice among those available on the Landsat Multispectral Scanner Subsystem (MSS), for example, if one wishes to locate or measure the area of water bodies. For observing variations in turbidity or delineating various classes of turbidity or water color in water bodies, bands in the 0.4-0.7 μm region are recommended based on theoretical light transmission and reflectance considerations as well as empirical evidence (19). If one wishes to monitor variations in depth in water bodies the shorter wavelengths in the visible region are preferable. The compromise that must be considered, especially when dealing with spacecraft sensors, concerns the effect of the atmosphere. Here the shorter wavelengths are most affected due to absorption and scattering processes caused by the atmosphere. Ways of approaching this problem will be discussed in later paragraphs.

When using remote sensing approaches to monitor water bodies there are some features that can be confused with water. Among these are cloud shadows, shadows produced in rough terrain, plowed fields involving dark-colored soils, lava fields, and, occasionally, dense urban areas will appear so dark as to be identified with water. These problems are most evident if one is attempting to use simple density slicing techniques applied to near-infrared observations. Errors can be reduced considerably if simple photo interpretation can be used to check the results, or if completely clear scenes are obtained to eliminate cloud shadows, or if a multispectral classification/identification approach is used. One approach that does provide some advantage, yet is computationally simple, uses the ratio of a visible and near-infrared bands and a display of the results that separate water

from other features. The low reflectance of water in both bands will help to separate urban areas, fields, and, sometimes, cloud shadows, and ambiguities produced by variations in atmospheric transmissions (20). In general, one will simply have to choose methods that are most commensurate with their data-processing capabilities and desires for accuracy. The principal item to keep in mind is that although open water is a relatively easy parameter to detect using remote sensing, significant errors or misclassifications can occur if care is not taken.

The ability to detect water bodies with remote sensors is largely a function of the attendant sensor spatial resolution. Because picture elements (pixels) in digital images often represent regions on the ground which are partially filled with land and water surface around the boundary of a water body, a water body must be approximately four times larger in surface area, or twice as long and twice as wide in dimension as a pixel, (the "instantaneous field of view") before it can be reliably detected. Given the resolutions provided in Table 18.1, it follows that 10-m spatial resolution should enable one to measure water bodies of 400 m² in size (0.04 ha), 100-m spatial resolution should allow the identification of water bodies 4 ha in size, and 1-km spatial resolution is commensurate with detecting water bodies, or changes in water bodies that are 4 km² in size.

Measurement of the size of a water body may, of course, be accomplished by either photo interpretation or computer classification. Because of the border pixel problem, the use of computer classification to measure the area of water bodies is not as straightforward as it might at first appear. Photo interpretation can easily adjust for this difficulty and, as a result, can map water bodies rather well and with an accuracy that is competitive with advanced computer processing methods if photo processing allows an interpreter to retain all information in an image. If a computer classification approach is used to measure the area of water bodies and only "pure" water pixels are counted in the process, the results will be a consistent underestimate.

The mixed pixel or partial pixel approach can and has been successfully applied in several instances to measure surface water area (21, 22, 23). A description of the fundamental concepts behind the partial pixel approach is provided by Malila and Nalepka (22). The approach will work particularly well where water bodies are surrounded by dense vegetation. The vegetation has a high reflectance in the near infrared and the water has a low reflectance. Therefore, where a pixel reflectance falls between these two extremes the proportion of water and vegetation creating this reflectance can be estimated and converted to appropriate area estimates. For a specific accuracy figure, it is estimated (21) that when using a mixed pixel approach one may measure water bodies from the Landsat spacecraft to ± 1% for water areas 500 ha or more in size and ±8% for water areas 5 ha in size.

In nearly every application of remotely sensed data for water resources monitoring there is a need to prepare graphics, overlays, or maps showing the spatial distribution of a parameter. This is particularly true in the case of surface water. Using photo processing, overlays that are of near cartographic quality can be obtained from the Landsat spacecraft at scales up to 1:250,000. Using photo processing, and more successfully, digital processing, overlays that add valuable information for updating maps or regional planning can be provided up to scales of 1:24,000. To be successful at the larger scales digital processing approaches must carefully account for geometric distortions and spacecraft or aircraft altitude variations and/or use ground control points. Prepackaged programs designed to map surface water are available such as the Detection and Mapping (DAM) package available from the NASA Johnson Space Center in Houston, Texas (24) and from the Computer Software Management Information Center (COSMIC), University of Georgia, in Athens, Georgia.

18.3.2 Surface Water Inventories and River System Studies

Using the techniques and approaches already suggested high altitude aircraft or spacecraft sensors can be used to make repetitive inventories over large areas or regions. For instance, the change in surface water area over several thousand square kilometers can be observed to change in response to climate variations or to water usage. It has already been noted in Section 18.2 that flooded areas along rivers can be monitored and measurements of flood inundated area obtained rapidly.

The principal advantage of high altitude/spacecraft sensors is the ability to synoptically monitor and inventory water bodies in areas or in situations that are difficult to attack using conventional methods. For instance, it has been desirable for many years to inventory the many playas and lakes in the Southern High Plains of Texas. Using Landsat data, an inventory of all lakes larger than approximately 4 ha can be rapidly accomplished using either photo interpretation or computer processing approaches. A repetitive inventory of this type offers an index of the water balance and relative locations of water bodies in this large region and a valuable, inexpensive ancillary data source for conventional surveys. If more detailed surveys are needed high-altitude and low-altitude aircraft can be deployed using the spacecraft data as a guide. These data surveys can be accomplished in a timely fashion and provide water resources planning data.

Remote sensing capabilities have been used by the United States Army Corps of Engineers to satisfy statuatory requirements to map water impoundments (Public Law 92-367) (25). Mapping surface water area in reservoirs can be used to estimate the volume of water in storage. Either an area/volume relationship may be developed or topographic features can be used to estimate the water stage in the reservoir and then relate the stage to volume. Agencies can use this approach not only to supplement data they obtain for the reservoirs they manage, but also to maintain cognizance of reservoirs they do not control, but which may affect their own reservoir management strategies under extreme conditions such as the flooding that occurs along the Mississippi, Missouri, and Ohio River systems each year.

In the case of states with extensive shorelines, it is often important to keep track of the shoreline gain or loss from year to year. Remote sensing provides a means of doing this rapidly and inexpensively, particularly in regions when the shoreline is very complex as it is in Louisiana, for example.

Another exemplary situation where surface water surveys can be greatly facilitated by remote sensing is in wetland or swamp areas such as exist in southern Florida. Surface water surveys are difficult to obtain using conventional approaches, not only because of the vast area covered by surface water, but also because of the general inaccessibility of the region. Landsat data have been used to measure the surface water area and combined with depth/stage measures acquired by satellite data collection and telemetered measurements to make estimates of water volume on a much more repetitive basis (monthly) than previously possible (26). Simply having the synoptic view of the surface water distribution to combine with conventional point measurements makes it possible to more confidently process, interpret, and present the conventional data.

The distribution of vegetation and wildlife is very much dependent and related to surface water distribution. The use of remote sensing on high altitude and spacecraft platforms have made it possible to study and inventory large wetlands, marsh, and swamp systems much more effectively. Not only can the surface water area dynamics be monitored, but the change in the associated vegetation species can be monitored successfully (27). Spacecraft observations seem to be most applicable to studies of coastal wetlands and large swamps such as the Great Dismal Swamp on the Virginia/North Carolina border and the Okefenokee Swamp on the Georgia/Florida border. For the observation of inland wetlands where the vegetation communities and water bodies are much smaller and spatially complex and the statuatory requirements are commensurately more exacting, remote sensing capabilities associated with high and low altitude aircraft may be necessary. In the case of wetlands it is also important to note the extent of dredging, lagooning, and other cultural practices in that they so often impact the wetlands environment significantly. This again can be done expeditiously from remote sensing because of the ability to provide the information in a readily appreciated, visual form that can be combined with conventional data sources.

18.3.3 Surface Water Quality Surveys

The principal advantage of remote sensing in the case of water quality surveys is, again, the synoptic, large region overview and high observational density capability. As has already been noted the reflectance of tonal changes observed in remote sensing data are a result of scattered/ reflected light interacting with particulate matter in the water. The most practical advantage of remote sensing is to serve as a means of extrapolating point measurements of water quality parameters over large areas. For example, variations in turbidity due to sediment variations in the near-surface layers (<1 m) of water bodies can be easily seen in Landsat data, particularly in the 0.5-0.6 and 0.6-0.7 μm bands (28). When combined with point measurements, or used to guide the placement of point measurements, remote sensing is a valuable tool. Persistent, anomalous regions of turbidity such as are associated with industrial plant effluent can often be detected also.

Some progress has been made in developing empirical relationships between radiometric reflectance measurements and sediment load in the near-surface layers. These relationships must be developed for each local, environmental condition encountered as they change because (a) the type and concentration of the constituents of the material in the water changes, (b) the reflectance geometry changes, that is, the relationship of the sensor to the sun and the water body changes, and (c) the atmospheric composition varies significantly. Using remote sensing (aircraft and spacecraft) this approach has been successfully applied (29, 30) for monitoring and estimating the extent of surface sediment load in small streams and large water bodies, including inlets and rivers.

18.3.4 Water Depth Estimation

In applying remote sensing measurements for estimating water depth and variations in bottom topography, the blue and green wavelengths in the visible tend to get the greatest penetration in clear and turbid waters. In clear water the maximum penetration occurs near 0.45 μm. This maximum penetration tends to shift toward longer wavelengths in turbid waters. The penetration of light in clear waters such as those surrounding the Little Bahama Bank is approximately 4-15 m and no more than 20 m. In moderately turbid waters average penetration depths are reduced to 1-2 m.

The ability of Landsat data for this purpose is dependent on selecting scenes that involve skies that are as clear and as uniform in atmospheric transmission as possible. Furthermore, success will be greatest in combining reliable, conventional depth information with the Landsat data, wherein the Landsat information is employed for interpolating between the conventional information. The principal application would come in updating navigation charts in coastal areas and near atoll-like islands in many of the more remote ocean areas of the world.

18.4 SNOW AND ICE

18.4.1 The Contribution of Remote Sensing in Snow and Ice Monitoring

The extraction of meaningful information about our snow and ice reserves is possible because various important properties or characteristics of snow and ice are amenable to remote sensing. Not the least of these is the fact that both snow and ice, although relatively common, tend to occur in great quantities in relatively remote and (at least in winter) inaccessible regions. Conventional snowpack and ice measurements are generally difficult to obtain and provide only point estimates of a few snow or ice characteristics which are not easily extendable to entire watersheds or water bodies. With remote sensing, the needed large-area monitoring capability is available; the major question is whether enough meaningful information can be extracted from the remote sensing data for operational applications. From initial experiments, it is apparent that the extraction of snow and ice information is possible using several spectral regions, namely, the microwave, thermal infrared, visible and near-infrared, and the gamma-ray spectral ranges.

Because snow (and ice to a degree) accumulates over a long period, from late autumn to spring, and then melts away in a matter of weeks, the use of remote sensing is logical because both repetitive and timely coverage is available. Not possessing the same rapid transistory characteristics of rainfall, snow can be effectively monitored by using most of the repeat cycles of existing satellites. If a particularly critical melt or breakup situation occurs, aircraft flights can be employed to collect the appropriate information.

18.4.2 Identification of Snow and Ice

The remote sensing of various earth surface materials are sometimes fraught with problems of location, recognition, and measurement. This is not generally true when snow and ice are the targets, however. Snow and some forms of ice generally have a high albedo in the visible portion of

the spectrum, thus making them easily distinguished against the darker backgrounds associated with other natural earth materials. In the near infrared (0.8-1.1 μm) snow and ice albedo is reduced, especially when liquid water is present, but reflectivities are still much higher than other substances. Further out in the near infrared (1.55-1.75 and 2.10-2.35 μm), the reflectivity of snow drops drastically so that it is less than nearly all other substances and provides an opportunity for automatically separating snow from highly reflective clouds (31).

During the melt season when snow observations tend to be most important, 273°K is the maximum temperature attained by snow. With the use of a sensor in the thermal infrared, snow reaching that temperature can be identified as potentially ready to melt. In the microwave region of the spectrum the presence of water in the snowpack causes a significant increase in the dielectric constant when compared to dry snow. Wavelengths longer than 3 cm are sufficient to penetrate a dry snowpack and may provide an indication of snow depth. Terrestrial gamma-ray radiation on the other hand is strongly absorbed by the snow and, when compared to snow-free gamma-ray measurements, the absorption provides a way to estimate the snow water equivalent (32).

Considering each of the ways to identify or measure snow and ice in the various portions of the spectrum, it appears that snow and ice monitoring is one of the most applicable for remote sensing.

18.4.3 Snow-covered Area Measurements and Application

TECHNIQUES

Several techniques are available for the mapping of snow-covered area using satellite data. The most straightforward approach is the use of simple photo-interpretation techniques. The most sophisticated involves the use of automatic data processing for digital analysis. Electronic, interactive analysis combines both photo-interpretive and digital analysis methods.

PHOTO INTERPRETATION

Even if more sophisticated methods are eventually used photo interpretation is extremely valuable for gaining familiarity and experience with snow mapping and the study areas. In situations where technical manpower is available and computer facilities are limited, photo interpretation has many advantages because the actual procedures are straightforward and generally do not require any sophisticated equipment. Several photo-interpretation options are available and are discussed in the following paragraphs.

SNOW LINE DISCRIMINATION AND PLANIMETERING OF SNOW COVERED AREA

This particular technique in its simplest form involves delineation of the watershed boundary on original 1:1,000,000 scale Landsat images and tracing the snow line on an overlay. The area snow covered is then planimetered manually and expressed as a percentage of total watershed area. Once the interpreter has gained experience with the study area, this particular technique can be used rapidly and requires very little training time, usually less than a half-day. When the number and area of watersheds is small this direct approach is to be recommended. Variations of this include enlargement of the Landsat scenes to scales as large as 1:100,000 and subsequent mapping and area measurement. In cases where shadows or trees partially obscure the snowpack, the snowline can be enhanced by a diazo overexposure technique (33). Subsequent enlargement and planimetering would still take place. In order to more precisely co-locate the imagery and watershed map boundaries a zoom transfer scope can be employed. The snow line is subsequently located at the desired scale and the snow-covered area planimetered. Further improvement can be obtained by using an automatically recording planimeter that immediately prints out snow-covered area. The complexity of these simple photo-interpretive procedures depends on the specific method and instruments used. The primary disadvantages are encountered when numerous basins are to be surveyed. Manpower becomes a problem and the transition to automated methods should be considered. Additionally, subjective interpretations are involved with these methods and consistent measurements between different operators and different basins are difficult to obtain, especially in areas where heavy forest cover seriously obscures the snowpack. The potential for human error is at a maximum when using these methods.

MEAN SNOW LINE ALTITUDE CONVERSION TO SNOW-COVERED AREA

This method indirectly obtains snow-covered area by producing an estimate of mean snow line altitude in a watershed. The mean snow line altitude is derived by interpolation of the elevation of the snow line at specific locations off of a transparent overlay with topographic contours (34). Twenty or more prints are then averaged to produce the mean snow-line altitude and converted to equivalent snow-covered area by using an area-altitude curve for the watershed. This method is fast, avoids problems of interpreting the actual location of the snow line in forest or shadow areas, and is relatively precise.

INTERPRETATION OF GRIDDED IMAGERY

This technique of using artificial grid elements or boxes as the base for interpretation of snow-covered area was first used by applying 1.98 × 1.98 km grid-overlay boxes (35). These same techniques have been used on the Feather River in California (36) and on basins in the Northwest using 2.5 × 2.5 km boxes (34). The 2.5 × 2.5 km box size corresponds to the Universal Transverse Mercator (UTM) metric grid printed on topographic maps. Basically the amount of snow cover in each grid element (or a sample of elements) is recorded and stored. The amounts for each unit are summed over the entire watershed for the total snowcovered area. In some situations winter snow scenes are compared with similar gridded summer scenes to reduce confusion between snow, trees, rocks, and shadows. The method is much more time consuming than the simple photo-interpretation methods but is precise, provides a good way to store and retrieve data, and seems to be a good transition step to the use of digital processing. Comparison between operators is easily facilitated by use of the grid system.

OPTICAL ENHANCEMENT DEVICES—DENSITY SLICING

In this method Landsat images are mounted on a device that attempts to separate snow from other features on the basis of an arbitrarily selected reflectance level. These devices can accept watershed masks and provide an indication of the percentage of the watershed above or below a particular reflectance level. Snow-covered area can be obtained automatically using this procedure if a particular reflectance level corresponding to the snow/no snow discontinuity can be found. It is difficult to arrive at the proper slicing level across an entire scene or watershed and hence there is room for considerable error. Repetitive attempts to slice the same scene at the same level very often gives different answers due to machine error. Such estimates are useful, however, for quick snow-covered area determinations that may be acceptable on certain watersheds. The cost of acquisition of the density slicing equipment is high and the lack of measurement precision may not merit such a purchase.

DIGITAL ANALYSIS

The use of Landsat digital tapes for snow and ice delineation has the immediate advantage of analysis of individual resolution elements rather than the average of several pixels which is the case of photo interpretation. More information is therefore available and maximum resolution is achieved. In digital processing, the data are generally classified into various types of ground features. In classification a choice must be made between two considerably different approaches. If some prior (ground truth) or concurrent knowledge of the imaged features on the ground exists, the computer can be trained to recognize the known features over the entire study area or watershed. This "supervised" classification approach is very effective for identifying snow and snow and ground cover mixtures, but depends very much on the accuracy of the ground truth used in the training process. The Purdue University LARSYS or Pennsylvania State University ORSER systems are good examples of maximum likelihood classifiers that produce very detailed snow classifications, but consume large amounts of computer time doing so. Since snow is relatively easy to identify this sophisticated classification approach may not be necessary.

Alternative supervised classification systems, such as the G.E. Image 100, employ relatively coarse parallelepiped classifiers that seem generally adequate considering the ease with which snow can be identified. Although classification results from the Image 100, when compared with LARSYS, initially seem to be slightly less accurate, the results are very promising when one considers that the Image 100 computer processing time is about

an order of magnitude less than the maximum likelihood approaches and only limited ground truth is required (37).

Digital processing of snow data, although automatic in concept, still requires the presence and interpretation of a skilled analyst for selection of training areas and distinctions between snow, snow in trees, shadows, and water. Neither type of computer processing is feasible for isolated watersheds and to be effective should be used over large areas. The selection of the kind of data analysis method to use depends on users' needs, manpower and equipment available, number and size of watersheds, and data analysis turn-around time requirements.

COMPARISON OF LANDSAT ANALYSIS WITH CONVENTIONAL TECHNIQUES

Up until the launch of Landsat, the most efficient and accurate way to obtain watershed snow-covered area was to fly low altitude aircraft for photographic or visual surveys. Such flights are expensive, time consuming, cover limited area, and possess an element of danger. Comparisons of Landsat derived snow-covered area and similar areas from low altitude visual techniques and high altitude photography were made in order to assess the applicability of the Landsat data.

Landsat snow extent area compares favorably with area estimates derived from aerial snow survey charts in California and Arizona (38). In California Landsat produces snow extents that are consistently greater than the Corps of Engineers aerial surveys by about 6%, most likely because the aerial observers do not map light snowfalls as snowpack whereas Landsat includes these areas in the estimate. In Arizona, personnel of the Salt River Project map all areas of trace and greater snow accumulation and Landsat produced about a 7% less snow area assessment in most cases. The comparative maps indicate that more detail in the snow line can be mapped from Landsat data than by the aerial observer because the aerial observer often views the terrain at an oblique angle and most often draws a "smoothed" snow line (38).

An example of Landsat and high altitude U-2 (flown at about 20 km) photography comparisons are presented in Fig. 18.5 for the Dinwoody and Bull Lake Creek watersheds in the Wind River Mountains of Wyoming. Both the Landsat and U-2 photography were acquired on 29 May 1975. Similar

Figure 18.5

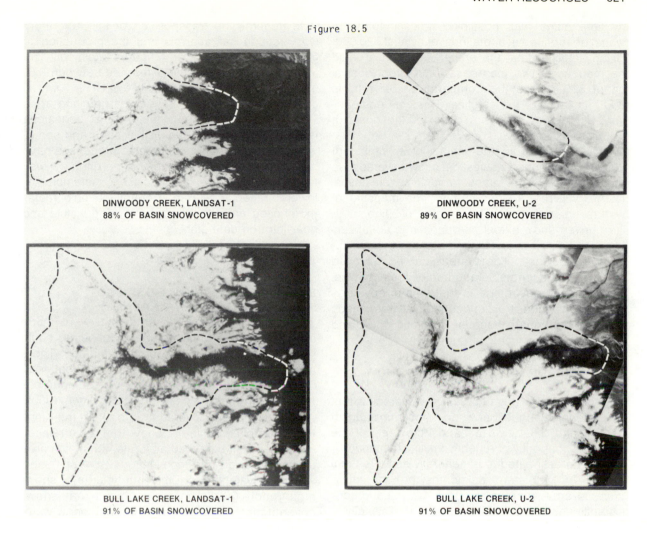

FIGURE 18.5. Comparative U-2 aircraft and Landsat views of Dinwoody and Bull Lake Creek watersheds in the Wind River Mountains, Wyoming.

photo-interpretation techniques using a zoom transfer scope were used on both images. On the Dinwoody Creek watershed, Landsat snow extent in percent of watershed area was 88% whereas the U-2 value was 89%. For Bull Lake Creek, both values were 91%. In all Landsat and U-2 comparisons more detail in the snow line is available from the U-2 than Landsat; however, even though more detailed patterns can be identified in the U-2 aircraft data, the information content of the Landsat image with regard to mapping snow cover is equal to that of the higher resolution photog-raphy (38). It was also apparent that snow-line determinations from Landsat were considerably easier to obtain than from the U-2. In this regard it has been found that Landsat snow mapping is six times faster than U-2 snow mapping (39).

NOAA-2 VHRR snow mapping is almost as fast as Landsat, more timely, and applicable on large watersheds. A cost-comparison figure has been derived for NOAA-2 satellite measurements of Sierra Nevada snow cover versus conventional air-craft measurements (39). Assuming that 20 basins were of interest and that a simple altimeter survey by light plane was possible, at least 40 hours would be required at a total cost of at least $20,000. Using satellite data, the entire Sierra block could be mapped in two man-days for a direct cost of about $100. Thus the comparative cost ratio is 200:1 in favor of the satellite. This cost comparison, of course, does not consider the costs of the plane or satellite.

It is evident from the results of a number of the abovementioned studies that snow-covered area measurements from Landsat and, in certain cases, NOAA VHRR are adequate for the mapping of watershed snow cover. Advantages of using satellite derived snow extent as opposed to aerial surveys include that satellite mapping is less expensive and faster, large areas can be surveyed at once, and acceptable accuracies are achieved. The major disadvantage is that time of coverage and repeat periods from space on small watersheds may be inadequate because of cloud cover and therefore require timely aircraft flights.

DETECTION OF METAMORPHOSED SNOW COVER

In addition to detection of snow areal extent it has been observed that when comparing Landsat 0.6-0.7 and 0.8-1.1 μm bands the area of the snowpack currently or recently melting can approximately be delineated. It is generally accepted that the 0.6-0.7 μm band delineates the total snow-covered area whereas the 0.8-1.1 μm band consistently indicates less snow cover. This difference, as shown for seven Wind River Mountain, Wyoming watersheds in Table 18.2, was attributed to the reduced near-infrared reflectance associated with melting or refrozen previously melting snow (metamorphosed snow). This same effect of the presence of water in the near infrared has been observed for melting ice (40).

In laboratory experiments in New Hampshire it has been found that snow which is even slightly melting, when compared with nearly fresh cold snow, has a distinct lower reflectance in the red visible region and an even greater reflectance decrease in the near-infrared region (41). It is noted that there may be an exception to this, however, in that reflectance may stay nearly the same as that of fresh snow (or even increase slightly) in the 1.9-2.0 μm wavelength region. Refrozen snow that has been previously exposed to melting temperatures has a reflectance curve generally resembling that of melting snow.

In general, the transposition of this laboratory work to Landsat analysis and the production of metamorphosed snow area figures is difficult. The major problem is that the difference between 0.6-0.7 μm snow-covered area and 0.8-1.1 μm snow-covered area is not entirely due to melting area. The near-infrared band under normal conditions is less reflective than the visible band and somewhat more so when metamorphic processes have taken place. At present there is no effective way to distinguish between the normal and metamorphic-lowered reflectance. Hence, quantification of metamorphosed area is not possible using visible and near-infrared data alone.

RELATION OF SATELLITE-DERIVED SNOW-COVERED AREA TO RUNOFF

If timely snow-covered area data were made available to water management agencies, could they make effective use of these data? There are a number of investigative results that indicate that snow-covered area is a meaningful parameter for the forecasting of snowmelt runoff. Analyses of snow-covered area data on the Kings River Basin in California show that a correlation exists between observed melt rate, remaining area of snow cover, the conventional 1 April snow water equivalent, and remaining volume of snowmelt runoff (42). The indication is that snow-covered area could be combined with snow water equivalent to improve runoff forecasts and, where necessary, such as in wilderness areas, be substituted for conventional snow water equivalent data. It is felt that the snow-covered area data can be used effectively late in the snowmelt season to update forecasts and reduce procedural forecast error.

In Arizona areal snow-cover estimates from Landsat for 1973 were used to derive a statistically significant relationship between snow cover and subsequent runoff during the snowpack depletion period (43). They conclude that measurement of areal snow cover from Landsat may become a valuable tool in forecasting snowmelt runoff in Arizona. In Wyoming it was found that the snow-covered area on a particular date was related better to the ratio of accumulated runoff/total seasonal runoff than to just the seasonal runoff in a statistically significant expression (44). Such relationships are not only useful for total flow forecasts but also

TABLE 18.2   A COMPARISON OF OBSERVATIONS IN THE LANDSAT 0.5-0.6 μm BAND (BAND 5) AND THE 0.8-1.1 μm BAND (BAND 7) AND THE APPARENT PERCENT OF DESIGNATED BASINS COVERED BY SNOW

1973

	May 21		June 8		June 25-26	
	Band 5	Band 7	Band 5	Band 7	Band 5	Band 7
Bull Lake Creek	86%	85%	58%	56%	53%	46%
Dinwoody Creek	80%	78%	61%	56%	49%	43%
Pine Creek	93%	88%	80%	76%	70%	62%
Big Sandy River	71%	66%	23%	21%	19%	15%
East Fork River	91%	90%	40%	39%	24%	21%
Wind River	41%	39%	16%	11%	Cloudy	
Green River	55%	49%	27%	23%	Cloudy	

1974

	May 16		June 2-3		June 21	
	Band 5	Band 7	Band 5	Band 7	Band 5	Band 7
Bull Lake Creek	89%	89%	79%	78%	57%	51%
Dinwoody Creek	85%	84%	76%	76%	56%	46%
Pine Creek	96%	95%	93%	90%	78%	71%
Big Sandy River	76%	73%	45%	41%	21%	18%
East Fork River	87%	83%	74%	66%	29%	25%
Wind River	67%	63%	Cloudy		23%	a
Green River	86%	85%	Cloudy		34%	31%

a Contrast in this band is not suitable for snow mapping.

for short term forecasts of time distribution of seasonal runoff. Further experiments in Wyoming's Wind River Mountains have shown that there is a statistically significant correlation between the extent of snow cover on 15 May and the seasonal runoff from 15 May to 31 July when several adjacent watersheds are grouped on the basis of elevation (45). Figure 18.6 shows the derived relationship for the four low elevation watersheds. Although two to three years of data are insufficient to derive a meaningful relationship for an individual watershed, the results indicate that snow-covered area could be incorporated into empirical runoff prediction methods as an additional index parameter after enough years of data have been accumulated.

In large data sparse areas the snow-covered area data may be the only available hydrologic information. Five to seven years of low resolution meteorological satellite data were used Indus (162,100 km²) and Kabul (88,600 km²) river basins in Pakistan to derive regression relationships early April snow covered area to April-July streamflow (45). The relationships are significant and provide perhaps the only way of estimating future runoff for these remote inaccessible areas. Predictions of 1974 seasonal streamflow using the regression equations were within 7% of the actual 1974 flow. Such meteorological satellite data is available from 1966 to the present, but the data are only applicable to large watersheds commensurate with the 4-km resolution capability of such satellites.

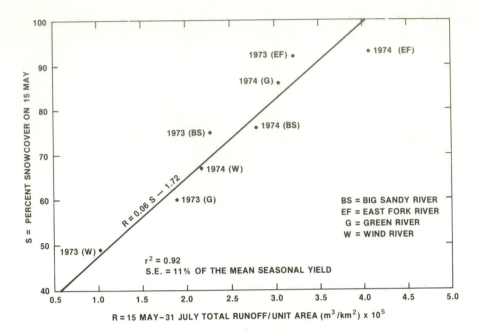

FIGURE 18.6. Percent snow cover from Landsat versus seasonal runoff for watersheds in the Wind River Mountains of Wyoming.

18.4.4 Snow-cover Applications of Thermal Infrared Data

The NOAA satellites possess a thermal infrared band (10.5-12.5 μm) with the same resolution as the visible channel. Because radiative temperatures of ground surfaces, such as snow, are measured rather than reflectance, NOAA VHRR thermal data can be acquired during both the daytime and nighttime passes. In most instances it appears that snow cover can be delineated in the VHRR thermal data because of its lower temperature, although the thermal gradients associated with snow boundaries are greater during the spring than during the winter (46). Fortunately, the data are most pertinent in spring during the melt period. The thermal gradients are reported to be on the order of $7° \pm 2°K$ in the spring. Difficulties in interpretations of thermal infrared data in mountainous terrain have been encountered where temperature differences due to variations in elevation may obscure the temperature differences associated with snowcover (38). Careful comparisons of infrared mesurements made with and without snow cover or in combination with topographic maps and atmospheric lapse rates can resolve most problems.

With VHRR thermal data external calibration must be made before absolute temperatures can be determined.

Currently, although the VHRR resolution of 1 km is a marked improvement over earlier scanning radiometer data (4 km), only general thermal snow-pack information can be obtained. In the near future, however, better resolution thermal infrared sensors will be flown on satellites. The improvement to be expected is one of greater thermal detail in extensive snow-cover areas that may be converted to the quantitative determination of melting areas which could not be accomplished with visible and near-infrared data alone. The capability of the improved thermal data can be used to show temperature differences across the snowpack with 0°C area being fairly easy to locate in spring when the edge of the snowpack can be assumed to be at this temperature. Once the area of the snowpack of 0°C is defined, the area that can potentially be melting is also defined. Using the near-infrared and visible in these areas to locate abnormally low reflectance compared to nearby areas of the snowpack at temperatures less than 0°C, areas currently or recently melting can more easily be located and quantified in a manner not presently possible.

18.4.5 Lake Ice Monitoring

Just as with the successes in the monitoring of snow cover from space, lake ice surveillance has been proven to be very successful. Landsat and NOAA satellites have been used for observing various ice features, detection of melting ice, monitoring of lake ice dynamics, and measurement of lake ice temperature (47). Many ice features were identifiable on Lake Erie using the high resolution Landsat data and include shuga, light and dark nilas, fast ice, icefoot, ice breccia, brash ice, fracturing, ridging, rafting, sastrugi, thaw holes, rotten ice, ice floes, dried ice puddles, hummocked ice, and leads. It was felt by the authors that Landsat coverage of once every 18 days can present problems for monitoring ice dynamics. Shallow lakes are subject to rapid freezeup, melting, and breakup and as a result significant changes occur in a very short time span, so rapid as to go unrecognized by the Landsat sensors. When day-to-day sidelap imagery is available, however, ice movement analysis can be performed. The thermal sensor on the NOAA VHRR was used to record temperatures of the ice surface with an absolute accuracy of 2°C and a relative accuracy of 1°C. Future study of these data, which had previously been unavailable, will assist in lake studies on local winter weather effects, energy balance, ice formation and thickness, evaporation, and effluent dilution. Many of these satellite ice studies have great importance for lake navigation and dependent industries during winter. NOAA VHRR thermal data appears adequate for thermal ice monitoring now, but Landsat resolution on a more timely basis is required for total shipping route planning and forecasts.

In many of the northern lake areas weather conditions are often not optimum for the use of visible and thermal data and an all weather system capable of penetrating clouds would be desirable. The aircraft prototype of such a system employing X-band SLAR imagery was tested in 1974 and 1975 by NASA, the United States Coast Guard, NOAA/National Weather Service, and the United States Army. Data flown over the Great Lakes was relayed rapidly to the Ice Information Center in Cleveland where it was interpreted and complied with conventional data into an ice chart. As shown

in Fig. 18.7, this ice chart together with the radar image was transmitted in near-real time to ship and shore installations possessing facsimile equipment. Initial results indicate a truly comprehensive lake ice picture that could not previously be duplicated. Data from an aircraft mounted, short-pulse radar ice thickness profiler were also obtained and incorporated into the ice chart, thus providing both area coverage and thickness data. Such techniques appear to be feasible from space platforms in the future and when combined with visible and thermal data should provide a truly comprehensive lake ice monitoring system. Until that time aircraft obtained radar data are an excellent aid to winter navigation planning.

18.5 SUBSURFACE WATER

The parameters discussed in the first sections of this chapter in many instances can be considered as being demonstrated capabilities that are simply evolving into operational application as the necessary process of familiarization and use by the water resources/hydrologic communities occurs and as the commensurate organizations are ble to acquire the training and trained personnel needed to employ this technology in accomplishing their mandated duties or in gathering information needed to explore the fundamental properties of hydrologic systems.

The observation of subsurface waters such as soil moisture monitoring and groundwater exploration involve efforts that are yet to be demonstrated conclusively enough to recommend remote sensing as a tool to utilize routinely. The presence of soil moisture does clearly effect the characteristics of electromagnetic radiation reflected or emitted from the soil because of the basic effect that moisture has on the reflectance, heat capacity, and dielectric properties of the soil, air, and moisture complex. The challenge that still faces the remote sensing community is that of providing data interpretation methods for quantitatively specifying the soil moisture at usable depths in the soil profile. In nearly all cases empirical relationships can be provided for specific environments or situations, but unambiguous, readily interpreted remote sensing results will remain goals to be reached.

There are four major regions of the electromagnetic spectrum that have been shown to be

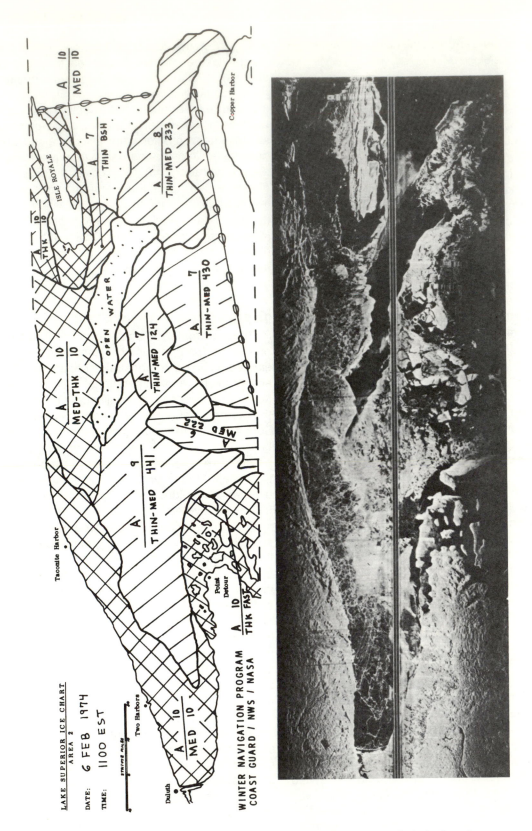

FIGURE 18.7. Side-looking radar image and interpreted ice chart for Lake Superior.

626

sensitive to variations of moisture in the soil. These are the gamma-ray region (< 0.1 Å), the visible and near-infrared region (0.4-3.0 μm), the atmospheric window regions in the emitted, terrestrial infrared region (for example, 3.4-4.3 μm, 8-12 μm), and the microwave regions of the spectrum between 1 and 30 cm wavelength.

The presence of water in the soil affects the amount of natural gamma radiation emitted from the soil. A good quantitative relationship does appear to exist between the levels of background gamma radiation and observed from appropriate instrumentation on low flying aircraft (≈ 100 m altitude) and soil moisture in the near surface (5 cm) layers of the soil (32). Instrument platforms must be flown near the surface because of the obscuring effect of the atmosphere. However, the technique represents a substantial advance over ground-based methods involving collection of soil samples and subsequent drying and weighing approaches or neutron soil moisture measurements in terms of its being a technique to sample much larger areas rapidly.

It is widely recognized that soils change their visual appearance when they become wet. This relationship has been quantified in terms of albedo versus soil moisture content (48). This relationship has been further explored and found to be a smoothly varying function of the surface soil moisture content. Furthermore, the character of the albedo variability can be explained in terms of three major stages of drying where the drying is a function of the atmosphere's drying capability, the hydraulic conductivity of the soil, and the attraction of the moisture to the soil particles themselves (49).

Most recently the fact that soil surface temperatures have a diurnal variability that varies as a function of moisture content has been exploited to obtain quantitative measures of soil moisture content in the layers of the soil between 0 and 4 cm depth (49).

There are several difficulties encountered in the practical applications of solar reflectance and terrestrial emitted infrared approaches to measuring soil moisture that should be kept in mind. First of all, these radiances observed by remote sensing systems originate from the first few microns of the soil profile and are, fundamentally, only related directly to soil moisture content in that region.

Estimates of soil moisture at greater depths must be established on an empirical basis and these relationships vary considerably as a function of soil composition, type, and roughness. It has been shown that one can reduce the soil characteristics dependence by carefully choosing how the soil moisture is expressed (for example, pressure potential or percent of field capacity) (40). Furthermore, the utility of the technique is greatly reduced or altered by the presence of vegetation on the soil surface. The reflectance or temperatures of plant communities themselves can serve as indicators of soil moisture because of stress induced changes. However, great care in the interpretation must be exercised in order to isolate other factors that may provide the same symptoms, that is, plant disease and soil differences such as salinity content. Finally, the presence of clouds will completely obscure the view of the sensors and render useless solar reflectance and terrestrially emitted infrared measurements from space or high altitude aircraft observations.

The microwave region of the electromagnetic spectrum is much less sensitive to clouds and less sensitive to vegetation cover than other spectral intervals. For applications purposes the wavelength region between 1 and 30 cm is being examined in great detail. Results have been produced indicating what can be done with "passive" microwave instrumentation in monitoring the naturally emitted microwave radiation (50). The fundamental principle that makes this portion of the spectrum attractive is the effect that water has on the dielectric properties of the soil, subsequently causing the emitted energy to decrease as the water content increases. Most success using passive microwaves has been achieved using the longer "L-band" measurements in the neighborhood of 20 cm. At this wavelength the response to soil moisture is approximately 2°K per one percent change in soil moisture and the observation is only modestly dependent on soil type. Furthermore, sensitivity to soil moisture is retained through light and moderate vegetation canopy densities although the sensitivity is reduced compared to that achieved over bare soil. Another strength of this approach is that the microwave radiation is emitted from depths of a few centimeters (for example, 1-5cm) and hence provides a better soil profile sample than the solar reflectance and thermal infrared wavelengths.

Much more work needs to be done on utilizing the microwave portion of the spectrum to ascertain what degree of specificity will be obtained over the range of conditions to be observed from space. Passive microwave measurements are somewhat limited by the amount of spatial resolution (mostly a function of antenna size) obtainable from space. Spaceborne sensors that appear to be feasible for implementation will provide spatial resolution on the order of tens of kilometers. Active microwave, synthetic-aperture approaches will provide quite useful spatial resolution (for example, 100 m) but the information content is relatively undetermined at this point in time.

Considerable amounts of water are stored in the form of groundwater. This reservoir is being utilized very extensively in many portions of the world. As populations expand and the demand for higher agriculture output grows a greater dependency is placed on this resource. As a result it is increasingly imperative that new sources of groundwater be located and utilized wisely.

Searching for deep groundwater is essentially a hydrogeological inference task. Remote sensing and associated image interpretation and image enhancement activities have helped the geologist to get an overview of large regions that is unattainable from ground survey and to locate potential groundwater drilling areas with a much higher probability of success.

Two approaches to finding groundwater seem to exist. One approach applies to limestone/karst terrain where fractures and the intersection of linear features or lineaments seem to correlate with the occurrence of groundwater. In this approach image interpretation and enhancement methods are particularly useful. The second approach amounts to finding outcrops or sedimentary rock regions which are known to overlay groundwater locations. These identifications have been most successful when using the ratioing of observations in discrete wavelength intervals and multispectral classifiction approaches (51).

Remote sensing has also proven useful for locating shallow aquifers. In this realm of endeavor the near-surface presence of the water may be manifested in the reflectance and distribution of overlying vegetation or in the temperature characteristics of the surface. Multispectral classification methods applied to high altitude aircraft data and Landsat data will help in separating out anomalous regions of vegetative cover that can be related to the presence of buried river valleys or near-surface water in aquifers or aquicludes. Flying a thermal scanner or imager just prior to sunrise has helped to identify near-surface aquifers because the presence of the water tends to keep the surface warmer (owing to the higher heat capacity) than surrounding dry surfaces.

One final approach to locating groundwater deposits is to note anomalous areas of temperature in streams or coastal regions that may be due to groundwater outflow. In the winter open ice areas along generally frozen streams may also offer clues to the presences of groundwater and these open areas are easily detected by visible and thermal infrared wavelength observations. High altitude aircraft and spacecraft again offer the ability for readily surveying long reaches of streams. What spacecraft systems presently lack is adequate spatial resolutions. Therefore, this remote sensing task must be done using aircraft at the present time.

PRECIPITATION MONITORING
One climatic variable which profoundly affects the magnitude and timing of runoff is the amount and distribution in time and space of precipitation. Furthermore, precipitation serves as the basic input to the drainage basin and is modulated and distributed by the watershed soil and surface cover, geology and drainage character until it is manifested as the basic output parameter—the runoff. Although it is of fundamental importance, precipitation is quite variable in space and time with attendant sharp gradients; therefore, it is difficult to observe with conventional rain gauges. Considerable effort has been devoted to utilizing the basic advantage of satellite coverage; namely, synoptic coverage, to better observe precipitation (52).

Again all portions of the spectrum can be used to attempt observations of precipitation. The visible observations can be enhanced to delineate the heaviest and most dense clouds such as those associated with cumulonimbus clouds. This kind of work has been accomplished using synchronous meteorological satellite observations and the results have been shown to very closely resemble nearly concurrent ground-based radar observations (53).

Precipitation must occur with substantial amounts of vertical motion in the atmosphere.

This condition is clearly met in cumulonimbus clouds and towering cumulus such as that associated with severe storms. The magnitude of the vertical motion is many times correlated with the height of the convective cloud systems and the height of the cloud top is correlated with the temperature. As a result, cloud top temperatures and the rate of change of the cloud top temperature may be related to the severity of the storm and the precipitation character. Cloud turrets protruding above the cumulonimbus anvil-cirrus layer can be associated with severe storms. Synchronous meteorological satellite observations, because of the ability of these spacecraft to monitor such features every few minutes have the most potential for observing these features, but much more work needs to be done to corroborate this possibility.

Polar-orbiting and geosynchronous satellite data are being used to supplement rain gage measurements over mountainous areas and areas in the Great Plains (54, 55). Meteorologists skilled in satellite imagery interpretation classify the storm type and/or cloud type in images and put these into various classes of precipitation amount (56). Over large areas and several days of time the inferred amounts of precipitation appear to correlate with independent estimates of rainfall obtained from conventional rain gauge networks, but, on the other hand considerable discrepancies in comparisons with rain gauge measurements are observed and more research is necessary.

One of the more exciting technological advances as it relates to global hydrological and meteorological studies is the recent demonstrated capability for the monitoring of rainfall over the oceans. The oceans represent a case where the rain gauge network is extremely sparse. Examination of the Nimbus 5 and 6 Electrically Scanning Microwave Raiometer (ESMR) observations over the oceans have shown that rainfall amounts can be observed and estimates of weekly, monthly, and seasonal rainfall amounts acquired via satellite over all the oceans of the world (57). Inference of rain over the oceans is possible with microwave radiometry because the observed brightness temperature of the ocean is so low due to the low emissivity of water. Rain occurring over the ocean raises the brightness temperature according to a relationship such as shown in Fig. 18.8. Similar observations are much more difficult over land because of the higher and more variable emissivities of land surface. These same ESMR observations over oceans have also made it possible to usefully and readily observe total precipitable water in the asmosphere. This is a significant contribution in terms of providing a better estimate of the total water in the atmosphere and reducing uncertainties in assessing the world water balance.

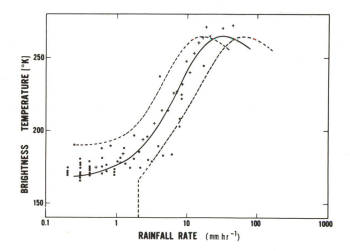

FIGURE 18.8. Rainfall rate versus microwave brightness temperature (Nimbus 5) relationship used to estimate rainfall over oceans.

18.6 CONCLUDING REMARKS

There are other parameters in the hydrologic cycle which have not been touched upon in the previous sections. Among these parameters are evapotranspiration, river stage and flow rates, and groundwater depths. In the case of evapotranspiration remote sensing can delineate various land uses and densities of vegetation which are related to relative amounts of evapotranspiration. However, this relationship has not been well quantified. For those parameters that are not amenable to remote sensing, the data collection and relay technology on satellites can facilitate rapid collection of uniform quality data over large, remote regions involving, on occasion, severe environments. This technology has had a generally positive impact for measuring river stage, certain water quality parameters, and obtaining meteorological information in mountainous environments. As a result this capability can be considered nearly operational. This is particularly true, in view of the fact that the data relay capability is incorporated on the Geostationary Operational Environmental Satellite (GOES) operated by the National Oceanic and Atmospheric Administration.

In summary, remote sensing capabilities obtainable from spacecraft and high altitude aircraft do demonstrably offer valuable data for improved management of water resources and more comprehensive regional, inter-regional, continental, and global hydrologic studies. In the majority of cases remote sensing does not replace conventional data, but instead complements these data by permitting better interpolation between conventional data station observations and extension or interpretation of these data over larger regions. This capability can increase confidence in the input to physical watershed models and watershed system management models or operational procedures. At this point in time a substantial percentage of the work with remote sensing may be considered research or quasi-operational, but it seems safe to predict that as a result of more pressure being placed on our finite water resources, the technology will be used more and more with the passage of time and eventually become an integral part of operational procedures.

REFERENCES

1. Gray, D. M., 1970, Principles of hydrology: Port Washington, N. Y., Water Information Center, Inc., 560 p.
2. Sopper, W. F., and Lull, H. W., 1970, Streamflow characteristics of the northeastern United States: Pennsylvania State University, Agricultural Experiment Station, Bull. 766, 129 p.
3. Rogers, R. H., Reed, L. E., Schmidt, N. F., and Mara, T. G., 1975, Landsat-1: automated land-use mapping in lake and river watersheds: Proceedings of American Society Photogrammetry; Fall Convention, Phoenix, Arizona, p. 660-672.
4. Soil Conservation Service, 1975, Urban hydrology for small watersheds: Technical Release no. 55, Engineering Division, U. S. Department of Agriculture, 48 p.
5. Schumm, S. A., 1968, River adjustment to altered hydrologic regimen — Murrumbidgee River and Paleochannels, Australia: U. S. Geological Survey Prof. Paper 588, Washington, D. C., 65 p.
6. Rango, A., Foster, J., and Salomonson, V. V., 1975, Extraction and utilization of space acquired physiographic data for water resources development; Water Resources Bull., v. 11, no. 6, p. 1245-1255.
7. Westin, F. C., and Frazee, C. J., 1975, Landsat-1 data, its use in a soil survey program, in Proceedings of Earth Resources Survey symposium, June 8-13, 1975: NASA TM X 58168, v. I-A, Houston, Texas, p. 67-95.
8. Blanchard, B. J., 1974, Measuring watershed runoff capability with ERTS data, in Third Earth Resources Technology Satellite-1 symposium: NASA SP-351, v. I-B, p. 1089-1098.
9. Ragan, R. M., and Jackson, T. J., 1975, Use of satellite data in urban hydrologic models: Journal of the Hydraulics Division, ASCE, v. 101, no. HY 12, p. 1469-1475.
10. Salomonson, V. V., Ambaruch, R., Rango, A., and Ormsby, J. P., 1975, Remote sensing requirements as suggested by watershed model sensitivity analyses, in Proceedings of the tenth international symposium on remote sensing of environment: Ann Arbor, Michigan, Environ-

mental Research Institute of Michigan, v. II, p. 1273-1284.

11. Burgess, L. C. N., 1967, Airphoto interpretation as an aid in flood susceptibility determination: International Conference on Water for Peace, Washington, D. C., v. 4, p. 867-881.

12. Deutsch, M., and Ruggles, F., 1974, Optical data processing and projected applications of ERTS-1 imagery covering the 1973 Mississippi River Valley floods; Water Resources Bull., v. 10, no. 5, p. 1023-1039.

13. Hallberg, G. R., Hoyer, B. E., and Rango, A., 1973, Application of ERTS-1 imagery to flood inundation mapping, *in* Proceedings of the symposium on significant results obtained from the Earth Resources Technology Satellite-1: NASA SP-327, v. I, p. 745-753.

14. Rango, A., and Salomonson, V. V., 1974, Regional flood mapping from space: Water Resources Res., v. 10, no. 3, p. 473-484.

15. Williamson, A. N., 1974, Mississippi River flood maps from ERTS-1 digital data: Water Resources Bull., v. 10, no. 5, p. 1050-1059.

16. Clark, R. B., and Altenstadter, J., 1974, Application of remote sensing techniques in land use planning: floodplain delineation: Final Rept. NAS5-21807, NASA, Goddard Space Flight Center, Greenbelt, Maryland, 23 p.

17. Rango, A., and Anderson, A. T., 1974, Flood hazard studies in the Mississippi River basin using remote sensing: Water Resources Bull., v. 10, no. 5, p. 1060-1081.

18. Harker, G. R., 1974, The delineation of flood plains using automatically processed multispectral data: Technical Rept. RSC-60, Texas A & M University, Remote Sensing Center, College Station, Texas, 227 p.

19. Maul, G. A., and Gordon, H. R., 1975, On the use of the Earth Resources Technology Satellite (Landsat-1) in optical oceanography: Remote Sensing of Environment, v. 4, p. 95-128.

20. Kritikos, G. B., Sahair, B., and Triendlo, E., 1974, Mapping of water bodies in Northern Germany from ERTS tapes, in B. T. Battrock and others, eds., European Earth Resources of Satellite Experiments, DFVLR Institute for Satellite Electronics, 469 p.

21. Barker, J. L., 1975, Monitoring water quality from Landsat: NASA document X-923-75-190, Goddard Space Flight Center, Greenbelt, Maryland, 33 p.

22. Malila, W. A., and Nalepka, R. F., 1974, Advanced processing and information extraction techniques applied to ERTS-1, *in* Third Earth Resources Technology Satellite-1 symposium: NASA SP-351, v. 1, p. 1743-1722.

23. Work, E. A., Gilmer, D. S., and Klett, A. T., 1974, Utility of ERTS for monitoring the breeding habitat of migrating waterfowl, *in* Third Earth Resources Technology Satellite-1 symposium: NASA SP-351, v. 1, p. 1671-1685.

24. Earth Observations Division, 1975, Water mapping from satellite data: an automated procedure: Johnson Space Center Publication, Johnson Space Center, NASA, Houston, Texas, 16 p.

25. McKim, H. L., Marlar, T. L., and Anderson, D. M., 1972, The use of ERTS-1 imagery in the national program for inspection of dams: Special Rept. 183, U. S. Army Corps of Engineers, Cold Regions Research and Engineering Laboratory, Hanover, New Hampshire, 20 p.

26. Higer, A. L., Coker, A. E., and Cordes, E. H., 1974, Water management models in Florida using ERTS-1 data, *in* Third Earth Resources Technology Satellite-1 symposium: NASA SP-351, v. 1, p. 1071-1088.

27. Anderson, R., Alsid, L., and Carter, V., 1975, Comparative utility of Landsat 1 and Skylab data for coastal wetland mapping and ecological studies, *in* Proceedings of Earth resources survey symposium, June 8-13, 1975, Houston, Texas: NASA TM-X-58168, v. I-A, p. 469-477.

28. Scherz, J. P., Sydor, M., and Van Domelen, J. F., 1974, Aircraft and satellite monitoring of water quality in Lake Superior near Duluth, *in* Third Earth Resources Technology Satellite-1 symposium, NASA SP-351, v. 1, p. 1619-1636.

29. Rosgen, D. L., 1975, The use of color infrared photography for the determination of suspended sediment concentrations and source areas: Unpub. manuscript, U. S. Forest Service, Fort Collins, Colorado, 13 p.

30. Yarger, H. L., and McCauley, J. R., 1975, Quantitative water quality with Landsat and Skylab, *in* Proceedings of Earth resources survey symposium, June 8-13, 1975, Houston, Texas: NASA TM-X-58168, v. I-A, p. 347-376.

31. Barnes, J. C., and Smallwood, M. D., 1975, Synopsis of current satellite snow mapping techniques, with emphasis on the application of near infrared data, *in* Proceedings of a workshop on operational applications of satellite snowcover observations: NASA SP-391, p. 199-213.

32. Peck, E. L., Bissell, V. C., Jones, E. B., and Burge, D. L., 1971, Evaluation of snow-water equivalent by airborne measurement of passive, terrestrial gamma radiation: Water Resources Res., v. 7, p. 1151-1159.

33. Foster, J. L., and Rango, A., 1975, A method for improving the location of the snowline in forested areas using satellite imagery: NASA Document X-910-75-41, Goddard Space Flight Center, Greenbelt, Maryland, 8 p.

34. Meier, M. J., and Evans, W. E., 1975, Comparison of different methods for estimating snowcover in forested, mountainous basins using Landsat (ERTS) images, *in* Proceedings of a workshop on operational applications of satellite snowcover observations: NASA SP-391, p. 215-234.

35. Lauer, D. T., and Draeger, W. C., 1974, Techniques for determining areal extent of snow in the Sierra Nevada Mountains using high altitude aircraft and spacecraft imagery. *in* Proceedings of the symposium on advanced concepts and techniques in the study of snow and ice resources: National Academy of Sciences, Washington, D. C., p. 532-540.

36. Katibah, E. F., 1975, Operational use of Landsat imagery for the estimation of snow areal extent, *in* Proceedings of a workshop on operational applications of satellite snowcover observations: NASA SP-391, p. 129-142.

37. Itten, K. I., 1975, Approaches to digital snow mapping with Landsat-1 data, *in* Proceedings of a workshop on operational applications of satellite snowcover observations: NASA SP-391, p. 235-247.

38. Barnes, J. C., and Bowley, C. J., 1974, Handbook of techniques for satellite snow mapping: Final Rept. NAS5-21803, NASA, Goddard Space Flight Center, Greenbelt, Maryland, 95 p.

39. Wiesnet, D. R., and McGinnis, D. F., 1973, Snow-extent mapping and lake ice studies using ERTS-1 MSS together with NOAA-2 VHRR, *in* Proceedings of the third Earth Resources Technology Satellite-1 Symposium: NASA SP-351, v. I-B, p. 995-1009.

40. Rango, A., Greaves, J. R., DeRycke, R. J., 1973, Observations of arctic sea ice dynamics using the Earth Resources Technology Satellite (ERTS-1): Arctic, v. 26, no. 4, p. 337-339.

41. O'Brien, H. W., and Munis, R. H., 1975, Red and near-infrared spectral reflectance of snow, *in* Proceedings of a workshop on operational applications of satellite snowcover observations: NASA SP-391, p. 345-360.

42. Brown, A. J., and Hannaford, J. R., 1975, Interpretation of snowcover from satellite imagery for use in water supply forecasts in the Sierra Nevada, *in* Proceedings of a workshop on operational applications of satellite snowcover observations: NASA SP-391, p. 39-51.

43. Aul, J. S., and Ffolliott, P. F., 1975, Use of areal snow cover measurements from ERTS-1 imagery in snowmelt-runoff relationships in Arizona, *in* Proceedings of a workshop on operational applications of satellite snowcover observations: NASA SP-391, p. 103-112.

44. Thompson, A. G., 1975, Utilization of Landsat monitoring capabilities for snowcover depletion analysis, *in* Proceedings of a workshop on operational applications of satellite snowcover observations: NASA SP-391, p. 113-127.

45. Rango, A., Salomonson, V. V., and Foster J. F., 1977, Seasonal streamflow estimation in the Himalaya region employed meteorological satellite snow cover observations: Water Resources Research, v. 13, pp. 109-112.

46. Barnes, J. C., Bowley, C. J., and Cogan, J. L., 1974, Snow mapping applications of thermal infrared data from the NOAA satellite Very High Resolution Radiometer (VHRR): Final Rept. Contract No. 3-35385, NOAA, National Environmental Satellite Service, Suitland, Maryland, 72,p.

47. Wiesnet, D. R., McGinnis, D. F., and Forsyth, D. G., 1974, Selected satellite data on snow and ice in the Great Lakes Basin 1972-73 (IFYGL): Proceedings 17th Conference Great Lakes Research, Internat. Assoc. Great Lakes Res., p. 334-347.

48. Bowers, S. A., and Hanks, R. J., 1965, Reflection of radiant energy from soils: Soil Science, v. 100, p. 130-138.

49. Idso, S. B., Jackson, R. D., and Reginato, R. J., 1975, Detection of soil moisture by remote surveillance: American Scientist, v. 63, p. 549-557.

50. Schmugge, T., Gloersen, P., Wilheit, T., and Geiger, F., 1974, Remote sensing of soil moisture with microwave radiometers: Jour. of Geophysical Res., v. 79, p. 317-323.

51. Goetz, A. F. H., Billingsley, F. C., Gillepsie, A. R., Abrams, M. J., Squires, R. L., Shoemaker, E. M., Lucchitta, I., and Elston, D. P., 1975, Application of ERTS images and image processing to regional geologic problems and geologic mapping in Northern Arizona: Technical Rept. 32-1597, Jet Propulsion Laboratory, Pasadena, California, 188 p.

52. Martin, D. W., and Scherer, W. D., 1973, Review of satellite rainfall estimation methods: Bull. American Meteorological Soc., v. 54, p. 661-674.

53. Sikdar, D. N., 1972, ATS-3 observed cloud brightness field related to meso-to-synoptic scale rainfall pattern: Tellus, v. 24, p. 400-413.

54. Davis, P. A., and Serebreny, S., 1974, Application of satellite imagery to estimates of precipitation over northwestern Montana: Final Rept. U. S. Geological Survey Contract, 14-08-0001-13271, Stanford Research Institute, Menlo Park, California, 90 p.

55. Merritt, E., 1976, Earthsat spring wheat yield system test 1975: Final Rept., NASA Contract, NAS9-14655, Earth Satellite Corporation, Washington, D. C., 300 p.

56. Follansbee, W. A., 1973, Estimation of average daily rainfall from satellite photographs: NOAA Technical Memorandum, NESS 44, U. S. Department of Commerce, Washington, D. C., 39 p.

57. Wilheit, T. T., Rao, M. S. V., Chang, T. C., Rodgers, E. B., and Theon, J. S., 1975, A satellite technique for quantitatively mapping rainfall rates over the oceans: NASA Document X-911-75-72, Goddard Space Flight Center, Greenbelt, Maryland, 28 p.

CHAPTER 19

ENVIRONMENTAL GEOSCIENCE
WAYNE A. PETTYJOHN

19.1 INTRODUCTION

Remote sensing techniques permit an evaluation of physical, chemical, and biological environments far beyond the imagination and capabilities available only a few years ago. In particular, data obtained from sophisticated electronic sensors aboard spacecraft now permit study of large areas viewed during a single instant in time. Furthermore, the same area can be reviewed repetitively, the time span depending only on the frequency of the spacecraft overflight.

In certain types of investigations or situations, particularly those requiring high resolution or immediate examination, aircraft-mounted sensors or cameras utilizing various types of films and filters are superior to spacecraft platforms. At other times ground base sensors are required.

From an environmental viewpoint, remote sensing technology allows us to examine various parts of the ecosystem—close at hand, at a moderate distance, or from great altitude. This permits detailed studies combined with regional examinations that, in turn, provide a broad understanding of our environment. For example, a few years ago air pollution monitoring was carried out largely by visual examination of smoke plumes at their source. The spread and dilution of plumes were also monitored by aircraft, but again observation was limited by the ability to actually see the smoke. Because of these serious limitations, it was generally assumed that the area of smoke plume influence was local, at least on a relative scale. Following the examination of early Landsat data in 1972, it became abundantly clear that smoke plumes could be mapped from spacecraft and that, under certain meteorological conditions, some extended with little dilution far beyond their source. Moreover, particulate matter from smokestacks and water vapor from cooling towers have caused inadvertent weather modification exemplified by above normal snowfalls, smog, and fog miles from the source of pollution. Most of these important discoveries are directly related to the space program and the development of remote sensing technology.

Environmental geoscience includes a wide spectrum of study that overlaps many disciplines, all of which are interdependent. In this chapter we will deal with several broad subjects including water pollution, effects of atmospheric contamination, strip mining, and natural disasters. Many other applications are possible. Future environmental studies need be limited only by the scope of imagination or technical expertise of the investigator.

19.2 WATER POLLUTION

19.2.1 Introduction

Despite the wealth of information available in the literature, it is still not possible either to accurately delineate all areas of water pollution or to quantify the impacts of many environmental stresses.

The body of quantitative water-quality data generally available consists almost entirely of point source analyses of physical, chemical, or biological parameters. The study of hydrology is concerned largely with changes in water quantity and quality characteristics. To measure the continuing impact of environmental stress on water quality, however, it is necessary to observe the distribution of quality characteristics spatially and also to observe changes in these characteristics with time.

Although traditional water-quality monitoring techniques commonly include repetitive, or sometimes continuous measurement of selected water-quality parameters, the problem in observing quality distribution spatially has not been solved except in certain special instances involving physical characteristics such as oil spills, color changes, or floating debris. Remote sensing technology combined with modern, computerized data processing, appears to offer a potential for monitoring environmental conditions on a spatial, continuous, and real-time basis.

The most notable application of the new technology has been in the detection and measurement of thermal anomalies involving the infrared portion of the electromagnetic spectrum. It has been possible, for example, to analyze the imagery collected over the New York sector of the Lake Ontario shoreline to detect and delineate thermal plumes from power plants and discharging streams. Point data of the type traditionally measured serve as so-called "ground truth" in interpreting the remotely sensed data.

At present, it is possible to relate remotely sensed data spatially only to the point measurements of selected physical parameters such as temperature, color, and turbidity. One can also infer the distribution of certain chemical quality characteristics from the physical data, for example, where the discharging fresh river water and the brackish or saline water of an estuary are of different temperatures. In this case, the distribution of radiant energy is used to determine the surficial distribution of chloride. Although it would be more desirable to remotely measure chloride or any other dissolved chemical constituent directly, such a capability does not presently exist. Ground data are still necessary in this case because it is impossible by use only of remotely sensed data, showing reflected or emitted radiation, to distinguish differences in the chloride content of fresh river water, brackish estuary water, seawater, or brine discharging from a desalinization plant.

It is not always necessary to measure the traditional quality parameters directly in order to assess the effects of a specific stress on the environment, because remote sensing technology in many instances, provides the capability of observing such effects. For example, polluted water commonly has detrimental effects on the ecosystem. In certain instances these effects can be measured by changes in reflectance with time, or by noting the difference between stressed and unstressed species.

Rogers and others (1) have attempted to determine relative values of dissolved substances in water by means of computer processing of Landsat computer compatible tapes (CCTs). Their study includes Saginaw Bay, a shallow extension of Lake Huron in southeastern Michigan. The Saginaw River,

which enters the bay at its extreme southwestern end, contributes approximately 90% of the pollutants found in the bay. These pollutants have been increasing for many years and the consequent levels of turbidity and algal production are consistently high. The resulting eutrophication has led to major declines in commercial fish yields, waterfowl population, and aesthetic values.

The investigators found that Landsat data and ground truth measurements can be correlated by stepwise linear regression techniques and the resulting equations used to estimate some dissolved water-quality constitutents in nonsampled areas. Ground truth measurements included temperature, Secchi disk depths, conductivity, chloride, chlorophyl a, sodium, potassium, magnesium, calcium, total dissolved phosphorus, total phosphorus, and total nitrogen. The authors noted that these parameters need not be directly detectable by imagery provided the distribution can be correlated with some water characteristic that is detectable on the imagery. Although not mentioned in their report, considerable care must be exercised in use of techniques of this type because there may be no relationship whatsoever between measurable paramaters, such as turbidity and the concentration of specific ions.

Blackwell and Boland (2), Boland (3), and Boland and Blackwell (4) have classified a series of lakes in Wisconsin, Minnesota, and New York in a study similar to that of Rodgers. Their approach consisted of developing relative trophic indices from the ground-truth measurements for each lake. The trophic indices were established by principal component analysis (as discussed in Chapter 6) of up to seven water-truth measurements. The trophic indices were then utilized as parameters for spectral classifications using Landsat CCTs. These authors in turn produced color coded lake classification photographs. This imagery depicts the spatial distribution of the various classes of water in each lake (Fig. 19.1, color plate 43).

19.2.2 Detection of Oil-Field Brine Damage

Generally, crude oil in its natural state is intimately associated with highly mineralized brine. When pumped from the ground, the mixture is mechanically separated; the crude oil is pumped to a storage tank, while the brine, being of no economic value, must be disposed of. Where governmental controls did not prevent it, brine has been dumped directly into streams, spread over roads and fields, or placed in evaporation ponds. Some brine is pumped deep underground for disposal or to repressurize the oil-producing zone.

Brine stored in holding ponds or pits may slowly infiltrate into the ground, contaminating groundwater supplies. Brine poured on the ground sterilizes the soil and stresses or kills the vegetation. Dead or stressed trees and other vegetation commonly mark areas where brine-contaminated springs occur and places where groundwater discharges into streams. Because of this effect on vegetation, the presence of brine can sometimes be detected by remote sensing of secondary indicators. For example, trees affected by oil-field brine, although outwardly appearing to be unaffected may indicate ecologic stress on color infrared film. In conjunction with known oil-field production, holding ponds, other dead or stressed vegetation, and some water-quality information, infrared photography would be sufficient to prove contamination.

A study along these general lines was conducted in the vicinity of Alum Creek, a central Ohio stream by Pettyjohn (5). The investigation showed that the major source of chloride contamination was the disposal of oil-field brines in the densely drilled upper part of the basin. Alarmingly high concentrations of chloride, recorded periodically for several years at the water-treatment plant at Westerville, produced short-term but acute water-quality problems. These episodes resulted from either the deliberate dumping of brine directly into the stream or one of its tributaries, or failure of a retaining wall surrounding a brine-holding pond. The water quality at Westerville also deteriorated in periods of low stream flow due to infiltration of brine into the ground and its ultimate discharge into a tributary.

On June 1, 1973, water samples were collected at 25 sites between Westerville and the headwaters of the basin. During this interval the stream was photographed from an aircraft with cameras using color and color infrared film and an 11-channel multispectral scanner. The imagery and photography were correlated with the ground-survey data. Dead and stressed vegetation were abundantly clear on the color infrared photographs, and these areas correlated closely with the high concentrations of chloride found in the streams or the adjacent brine-contaminated aquifers. The affected vegetation appeared in two general locations—in the near vicinity of brine holding ponds that had overflowed,

and along the floodplain of the river, particularly adjacent to the stream channel (Fig. 19.2 color plate 44). In the latter case, the vegetation was being adversely affected by highly contaminated groundwater that was discharging directly into the stream in the vicinity of the dead trees. By means of this survey it was possible to detect (1) the exact sites where wells had been drilled more than 10 years previously, (2) some areas of groundwater contamination, (3) that part of the basin most adversely affected by brines during the longest interval of time, and (4) the major sites of contamination that have the greatest long-term effect on the quality of the stream water.

It is possible that color infrared studies could also determine when the groundwater in a region has been flushed clear of brine. Because vegetation will not recover as long as the brines are present in harmful concentrations, repetitive color infrared photography by revealing vegetation recovery, could indicate areas where quality of groundwater has improved.

19.2.3 Evaluating Trace Element Pollution and Turbidity

Trace elements are those substances that normally occur in water at concentrations of less than 1 milligram per liter (mg/l). Little is known of the potential adverse or beneficial physiological effects on living organisms, particularly man, of several trace elements that are present in commonly used or consumed products. However, in some cases there is only a small range between requirement and toxicity.

Trace element studies in recent years have shown that most trace elements occur in barely detectable concentrations in water, in concentrations two to three orders of magnitude greater in streamside or lake deposits and that plants receiving their nutrients from sediment may contain several hundred or even several thousand parts per million more than the sediment. Apparently, most trace elements that occur in water are barely detectable because, in addition to dilution and dispersion, they become attached to fine-grained sediments, which are subsequently deposited. These elements may then be removed from the sediment and concentrated by aquatic plants or bottom-feeding organisms.

The discovery of mercury in Lake St. Clair, the Detroit River, and Lake Erie in the late 1960s led to the local banning of sport and commercial fishing. The ban was related to the experience at Minamata Bay, Japan, where scores of people were poisoned after consuming mercury-rich seafood caught in the bay. The bay had been contaminated by the disposal of mercury compounds originating from a plastics factory.

The major mercury contributors to Lake Erie were identified as chemical plants in Wyandotte, Michigan, and Sarnia, Ontario. In these plants, mercury cells are used in the manufacture of chlorine gas and caustic soda and the rate of accidental loss of mercury was estimated to be about 23 kg per day from 1950 to 1970, but greatly reduced thereafter. The normal concentration of mercury in Lake Erie water is less than 0.0002 mg/l, while in bottom sediments it averages less than 0.07 mg/l.

The distribution of mercury in the upper 2 cm of Lake Erie bottom sediments is shown in Fig. 19.3.

The Detroit River is the major source of contamination, carrying the mercury, probably attached to suspended sediment, into the shallow western basin of Lake Erie. Notice that the concentration of mercury decreases significantly with increasing distances from the mouth of the river. Examination of a satellite scene permits a determination of the distribution of suspended sediment during a major spring runoff event (Fig. 19.4). Notice the close similarity between turbidity patterns and mercury concentrations in Fig. 19.3 and Fig. 19.4, indicating that satellite data can be used to predict the paths that sediment will take in a large lake or reservoir as well as the major areas of deposition. Lake-sampling programs, if based on satellite imagery, would allow a greater refinement of ground-truth collection than would otherwise be possible.

Scherz and others (6) mapped changes in turbidity (suspended solids) in the southwestern part of Lake Superior using different shades of brightness

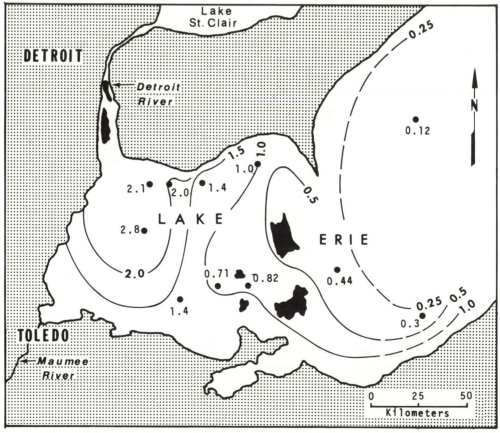

FIGURE 19.3. Distribution of mercury in the upper 2 cm of sediment on the bottom of the western part of Lake Erie. Dots are sampling sites, numbers indicate concentration, in parts per million.

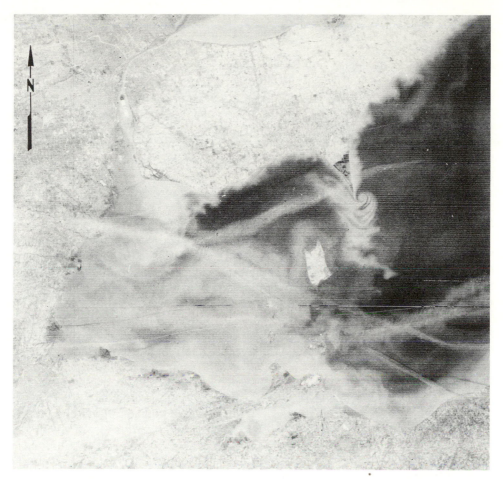

FIGURE 19.4. Sediment patterns in the western end of Lake Erie during a major period of runoff on 14 April, 1973 (Landsat scene E 1265-15480, band 5).

on satellite imagery. It was pointed out that had engineers and planners examined low-altitude aerial photographs (available as far back as 1939) or satellite imagery, an $8 million water-supply intake for the City of Duluth would have been constructed in a different area. Because they did not, the intake was installed in an area of excessive turbidity and over 50% of the time yielded a water too turbid for drinking.

19.2.4 Detection of Paper and Sawmill Wastes

Paper manufacturing along the New York shoreline of Lake Champlain has an adverse effect on water quality due to effluent discharge directly into the lake. Lynn and others (7) used Landsat imagery (bands 4 and 5) to distinguish and monitor the pollution plume, which is darker than the characteristically turbid water in the receiving part of the lake.

The paper mill effluent, originating from aeration lagoons, was originally discharged into the lake at a rate of about 80,000 m^3 per day through underwater discharge lines. The humic, dark-reddish brown effluent, contains suspended solids, high concentrations of sodium, dissolved solids, and phosphates, and has a high biochemical oxygen demand. The nutrients stimulate algal blooms, while the other wastes deteriorate the lake's quality and have an adverse effect on drinking-water quality, aquatic life, and recreation.

The State of Vermont initiated a legal action in the United States Supreme Court against the State of New York and the paper company, alleging that the paper waste deteriorates Lake Champlain's quality to such an extent that it exceeds Vermont water-quality standards and that the effluent crosses the New York-Vermont state boundary. This action was a milestone in legal and space technology because satellite imagery was in this case for the first time accepted as evidence in court.

Sawdust and wood chips have been settling along the banks and bottoms of tidal rivers in logging areas for more than a century. Biological samplings of a few of these rivers have demonstrated the general unsuitability of sawdust-covered substrates for the reproduction of bottom-dwelling organisms, such as clams and flounder, and the migration of species such as salmon. Constant vertical and horizontal movement of bottom materials, induced by tidal action, coupled with a high biochemical oxygen demand brought about by the decaying sawdust, contribute to the near sterility of sawdust-covered river bottoms.

The ecological stress leading to decreased clam reproduction causes a substantial loss of revenue, amounting to millions of dollars each year. In order to reduce this loss, sawdust contaminated areas first need to be found and mapped and then the deposits removed by dredging. Determination of the areal extend of these accumulations, parts of which are commonly exposed during low tide, have been quickly and adequately accomplished by remote sensing methods.

Chase, Conrod, and Imhoff (8) described the techniques used to map sawdust accumulations along the Penobscot River in Maine. In their study an aircraft was mounted with a 70-mm framing camera and an eight-channel multispectral scanner (MSS). A morning flight along the Penobscot River was made at low tide when the banks and mudflats were exposed. Color photographs from the framing camera were used as guides in selecting parts of the flight line that were thought to be targets of interest—that is, sawdust, mudflats, and muddy water. These parts were then used to train a computer to combine the data, recorded on magnetic tape, from each MSS channel. By training the computer with spectral signatures of known sites of sawdust accumulations and statistical analyses of the data, it was possible to quickly and accurately locate sawdust deposits and distinguish them from muddy water and mudflats.

The imagery formed a map showing the areal extend of the sawdust exposed during low tide. In order to determine the volume of the deposits, thickness measurements were made with a boat-mounted coring device. Once the volumes were determined, cost-benefit ratios provided planners with sufficient information to examine the feasibility of sawdust removal.

19.2.5 Detection of Oil Spills and Ocean Dumping

Several types of sensors have been used to examine oil spills in coastal waters. Their purposes are to record the rate of oil-slick drift, determine its areal extent, and predict when and where the contaminant will reach the shore so that plans for containment and cleanup can be initiated. In many instances the differences in temperature between petroleum discharging from a leaking well and the receiving water are sufficient to be measured by thermal scanners. Ultraviolet detectors have shown even greater promise for monitoring oil spills.

A synchronous aircraft and satellite overflight off the New York-New Jersey coastline permitted Wezernak and Roller (9) to detect areas of municipal and industrial ocean dumping of waste. This convenient method of waste disposal has not received adequate investigation, and no doubt has a measurable adverse effect on marine and near-shore ecosystems. By means of thermal properties and other spectral characteristics, these workers were able to identify acid-iron wastes dumped approximately 20 km south of Long Island. The wastes contained approximately 8.5% H_2SO_4, 10% $FeSO_4$, and small quantities of various metallic elements. After dumping, oxidation of the iron-rich effluent produced a suspension that tended to remain in distinct patterns for considerable periods of time. Also evident on the imagery were New York harbor dredging spoil and construction debris ocean dumping sites.

Maps constructed from the satellite data have been used to determine the surface movement of the wastes, the location of the major water mass boundaries, and dilution effects. Similar methods can also be used to detect unauthorized dumping or accidental discharges, thus providing some safeguards on ocean disposal. Such data can be used not only for enforcement proceedings, but also as location maps for underwater biological surveys. The effect of a great variety of wastes on aquatic organisms in the vicinity of the large ocean-floor dumps is unknown.

19.2.6 Uses of Temperature Variation Detection

Variations in water temperature can be due to both natural and artificial causes. These variations can be detected by thermal scanners, in some cases by a black-and-white aerial photograph and, indirectly, by multispectral scanners. Determination of the causes and areal extent of temperature anomalies in water are important from a variety of environmental viewpoints, only a few of which will be discussed here.

In past years it was common practice to dispose of liquid wastes by installing effluent outfalls on or near the bottoms of rivers or lakes. Presumably, this method of disposal resulted in greater mixing and more rapid dilution between the effluent and the receiving water. The inaccessibility of these outfalls also made it difficult for enforcement agencies to detect violations. In most cases, however, there are sufficient differences in temperature between the receiving water and the effluent to be detected by thermal scanners. Furthermore, a scanner mounted in an aircraft can be used to rapidly investigate an entire river, either by day or night. Thus, thermal scanner flights can be used to detect underwater effluent discharge points so that ground survey teams can minimize field time collecting water samples.

Thermal scanners have also been used to map the areal extent and movement of thermal plumes originating from the discharge of heated cooling waters at electric generating plants. Repetitive measurements, reflecting a variety of meteorological, particularly wind, conditions have proven to be of significant aid in the design and location of cooling-water outfalls. Moreover, maps showing the temperature anomalies can be of considerable value to aquatic biologists in their studies of the effect of temperature changes on aquatic life.

Although commonly not realized, most perennial streams flow throughout the year because groundwater discharges into them (Fig. 19.5). The rate of groundwater discharge, however, ranges widely from one part of the stream to the next. Generally, the greatest increase in stream flow occurs in zones of high permeability in the stream channel. These zones may consist of deposits of sand and gravel, solution openings in limestone, an abundance of fractures in the rocks, or merely by local facies changes that cause increases in permeability.

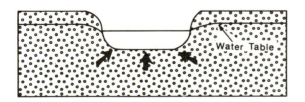

FIGURE 19.5. Perennial streams flow throughout the year because groundwater continually discharges into the channels.

The temperature of groundwater is nearly uniform, reflecting the mean annual temperature of the region. Consequently, during the wintertime those stream stretches receiving a significant amount of groundwater discharge should be warmer than adjacent areas, which may be near freezing. Conversely, in the summertime the relatively cold groundwater, as it mixes with the warm surface water, tends to produce colder water (Fig. 19.6).

Detection of areas where groundwater discharges into a stream is important for three major reasons; (1) the water in that stretch is of a more nearly constant temperature than in adjacent regions, (2) the chemical quality of the water may be significantly different in that stretch than elsewhere, and (3) wells constructed in the vicinity might produce greater yields.

Those stretches where the stream gains in discharge because of groundwater inflow can be detected in a variety of ways, not all of which are related to remote sensing. Field examination or winter aerial photography may show places that are either free of ice or places where the ice is very thin. The reason for this phenomenon is the warmer stream water resulting from the discharge of groundwater, which may be several tens of degrees warmer than the stream. In the summertime, it is possible to float by boat down a river, periodically measuring the temperature. Groundwater discharge sites are detected by a temperature decrease. A third method of detection is by means of an aircraft-mounted thermal scanner. Being a much more sophisticated instrument, this technique, because of its ability to detect slight differences in temperature, might be much more accurate than thermometry or low-altitude aerial photography.

In most cases, groundwater is more highly mineralized than surface water and it contains practically no dissolved oxygen. Although apparently never investigated, it is strongly suspected that in some streams these chemical deviations may have a subtle effect on the aquatic ecosystem. Investigations might show that the more active fish, such as trout, would be more abundant in groundwater discharge zones because of the relatively constant temperature, which during summertime, could be significantly colder than adjacent stream segments. Other plant and animal organisms that are highly temperature sensitive might also be present.

In addition to the effects brought about by stream-water quality, areas that provide groundwater to streams are also capable of providing water to wells. More than likely, wells that would produce larger

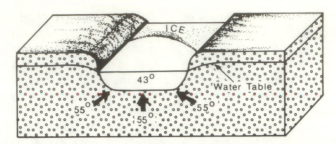

FIGURE 19.6. During winter groundwater is warmer than surface water and where it flows into a stream in sufficient quantity, the surface will remain unfrozen. These same reaches are colder during summer.

yields could be installed in the stream valleys in the vicinity of groundwater discharge zones. On the other hand, stream-side wells commonly derive part of their discharge by inducing water to flow from the stream into the ground; the quality of the well water will eventually approach the quality in the stream. Bearing this in mind, it is evident that if wastewater outfalls were installed just upstream from the well field, the deterioration of the stream's quality would soon be evident in the well water.

Many islands suffer from a shortage of fresh water. Part of the rain that falls on the island infiltrates to become groundwater and eventually discharges at near-shore underwater springs. It is possible to detect these underwater springs due to thermal differences between the spring water and the ocean water. Well supplies could be established in their near vicinity and capture the fresh water that normally flows to waste.

One of the earliest studies along these lines took place in the Hawaiian Islands. Thermal scanner data pinpointed underwater springs and seepage zones. Care must be exercised in studies of this nature, however, particularly in volcanic islands, because the fresh groundwater may range widely in temperature, being warmer than seawater in one area, and colder in another.

19.3 EFFECTS OF ATMOSPHERIC POLLUTION

19.3.1 Introduction

Air pollution not only decreases visibility but it may also have a profound effect on the chemical quality of rainfall, thus increasing ecologic stress. This is exemplified by the decreasing pH of rain that is related to the increasing concentration of atmospheric gaseous contaminants, such as sulfur and nitrogen oxides, released by the burning of high sulfur fuels. Although acid rainfall is not of immediate danger to humans it does considerable damage to human-made structures and has serious implications for ecologic systems.

From the ground or even from aircraft it is difficult to visualize the effects of atmospheric contamination, map the areal extent of smoke plumes, or detect areas of possible inadvertent weather modification. Without an accurate mapping capability it frequently proves difficult to determine the average or maximum area of influence of smoke or vapor plumes, thus adding to the problem of adequately locating the ground stations that are needed to evaluate the effect of atmospheric pollutants on the ecosystem.

For years the atmosphere has been examined by sensors mounted on balloons, kites, aircraft, and satellites. Increasing sensor resolution is convincing many workers of the value of these instruments, particularly in satellites, for detecting the causes and effects of air pollution, its regional spread, and subtle, real or potential, long and short term effects.

Data obtained from satellites can be useful for the regional study of smoke plume dispersion and inadvertent weather modification detection, and can provide several other potentially important bits of information as well. They can also be used to determine areas where ground surveys should be conducted in order to evaluate the effects of air pollution on vegetation.

19.3.2 Detection of Smoke and Its Effects

Along the Chicago-Gary industrial complex that fringes the south shore of Lake Michigan are 16 major sources of particulate matter (10). One of the principal point sources is a cement plant that discharges into the atmosphere more than 130,000 tonnes/y of particulate matter, far surpassing the total suspended particulate emission of many heavily industrialized areas. Smoke plumes drifting from the industrial complex across Lake Michigan are readily evident on Landsat imagery; they have been traced for distances of more than 60 km. Smoke plumes are more easily observed over water bodies than over land. Lyons (10) reported that ice nuclei emanating from the Chicago-Gary steel mill complex were traced to the vicinity of Battle Creek, Michigan, some 180 km downwind. Layers of red iron oxide smoke observed drifting past Milwaukee originated at steel mills at the southern end of Lake Michigan almost 165 km away.

A plume from a large elevated point source, such as a power plant, may travel over water for long distances with relatively little dilution, arriving above a downwind shoreline in a highly concentrated condition. If such a plume arrives during midday and solar radiation is sufficiently intense, the plume is rapidly mixed and brought to the land surface by the turbulence forming over the heated ground. This may cause high air pollution levels several kilometers inland. Conditions such as these generally occur from spring to late fall when lake water is cooler than the overlying air mass. It is due to this atmospheric mixing process that Benton Harbor, Michigan, is periodically affected by air pollution originating in the Chicago-Gary complex.

High pressure systems commonly stagnate over wide regions for several days bringing about reduced mixing of pollutants in the atmosphere and the accumulation of a wide variety of airborne particulate and gaseous matter. Inversion conditions such as these and the resultant accumulation of SO_2 and particulates in the atmosphere caused more than 400 deaths at Thanksgiving time in 1966 in the New York City area (10).

A cursory examination of several Landsat scenes was carried out by Pettyjohn and McKeon (11, 12) in an attempt to evaluate their usefulness for monitoring smoke plumes. Photographic prints, made from several 70-mm negatives of Landsat scenes, were enlarged to a convenient scale (approximately 1:500,000). The areal extent of visible smoke plumes, as best seen on the green and red bands (4 and 5, respectively), was recorded on a base map.

The major areas of interest included two sites on the Muskingum River in southeastern Ohio, and an industrial complex along the Mahoning River between Warren and Youngstown in northeastern Ohio. The areas along the Muskingum River lie in the vicinity of Philo and Beverly. In the Philo area (Site A, Fig. 19.7) major smoke-producing industries include a ferro-alloy plant (Fig. 19.8) and, downstream, a coal-burning electric generating plant.

At Philo and near Beverly, smoke was evident on seven Landsat scenes, not evident on four scenes, and obscured by clouds on the other four scenes examined, which were imaged during the period 21 August 1972 to 23 October 1973. As indicated in Fig. 19.9, the plumes range from short, high-density to long, low-density features and are controlled by local atmospheric conditions. The visible plume originating from the high stack at the Beverly plant is most often the largest and may exceed 32 km in length. On 30 July 1973, the visible part of the Philo plume, originating from a low level at the ferro-alloy plant extended northward about 48 km and reached a width of 6 km.

A computer processed, density-sliced image of the smoke plume originating at Philo is in Fig. 19.10. The different shades of gray indicate differences in particulate loading. Although it is a relatively simple process to map smoke plumes directly from Landsat imagery, it is not presently possible to determine what concentrations of particulate or gaseous matter the visible plumes represent. In order to be a truly viable monitoring technique, it would be necessary to determine, directly from the imagery, the total suspended particulate mass loadings and size concentrations in the plumes. Very likely this could be accomplished by aircraft, instrumented with sensitive air pollution and meteorological sensors flying through the plumes during several Landsat overpasses. The measured concentrations could then be compared with differences in the spectral response evident on the imagery.

Smoke plumes, per se, were not evident on any of the Warren-Youngstown scenes. On two scenes (12 January and 21 September, 1973), however, long narrow clouds appear to originate at several sites just north of the Mahoning River but downwind of the industrial complex (Fig. 19.11). As these clouds are traced downwind, they become significantly larger while the clouds on either side of them remain about the same size. It is strongly suspected that the change in cloud density and size is caused by particulate matter originating upwind at 10 or 12 sources, probably steel mills.

FIGURE 19.7. Partial Landsat-1 scene showing smoke plumes, trending southwest, originating at Philo (site A) and Beverly (site B). Bottom edge of figure is approximately 50 km. Landsat scene 1408-15410, band 5.

644

FIGURE 19.8. Ferro-alloy plant near Philo, Ohio (site A)

Murtha (13) visually examined several Landsat scenes, enabling him to map different intensities of forest damage in an area north of Wawa, Ontario, where vegetation was damaged by high concentrations of SO_2 from stack emissions. The area of total vegetation kill extended about 21 km, heavy damage about 31 km, and moderate and light damage extended about 35 and 48 km, respectively, in the prevailing downwind direction from the point source.

The red band (band 5) was the most useful for assessing the extent of damage, at least during the growing season. Zones of damage were delineated by image brightness, areas of total kill being brightest and decreasingly stressed areas characterized by successively darker tones. Discrimination between the SO_2 area and areas affected by fires or logging was based on the fact that the SO_2 damage has a strong point source, expands downwind, and the boundaries are indistinct as compared to those associated with logging areas and forest fires.

A prospector looking for gold in southeastern Tennessee in 1843 panned the first metallic minerals from what was to become known as the Burra

Burra lode in the Ducktown copper district. Production began in 1847 and has continued with little interruption to the present day. By 1878 about 130 km^2 of the Copper Basin had been stripped of timber, which was used both in the mines and for roasting the ore. It was the nineteenth Century open-roast process that caused much of the denudation of a vast region surrounding the mine areas. The vegetation that was not cut, died as SO_2 fumes from smelters settled in the basin leaving the bare soil vulnerable to erosion. The smelting process was modified in 1904 in such a manner that SO_2 was no longer a severe problem.

For the past 25 years reforestation has been practiced and annually thousands of pine seedlings are planted along the outer margin of the affected area; grasses are planted in the once barren central zone.

The Ducktown, Tennessee mining area can be clearly delineated on Landsat imagery because of its distinct reflectance, but the area has not been examined in detail by means of remotely sensed data. It offers a classic site of environmental deg-

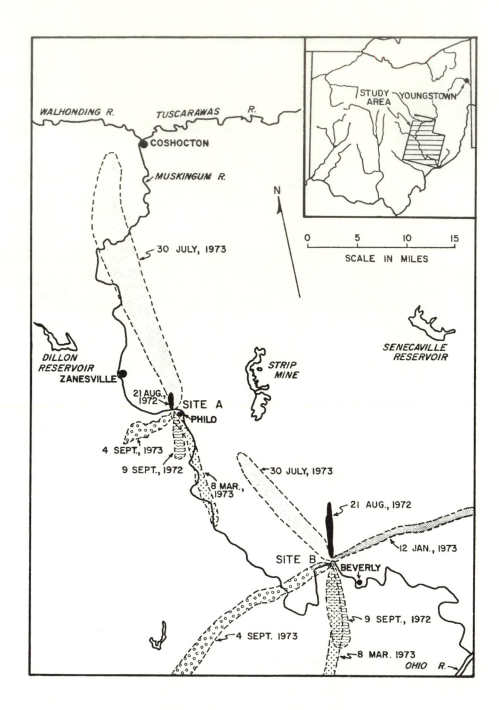

FIGURE 19.9. Map based on Landsat-1 data showing areal extent of smoke plumes originating at Philo and Beverly. From Pettyjohn and McKeon (12).

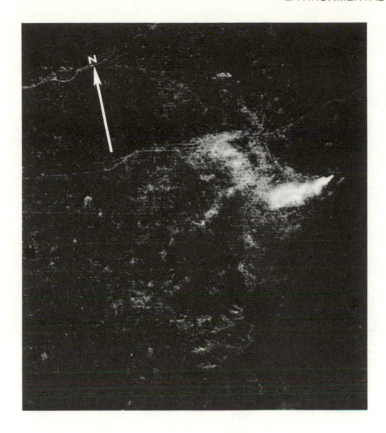

FIGURE 19.10. Density sliced smoke plume at Philo (site A) produced from satellite imagery. The lighter tones indicate the densest smoke. Bottom edge is approximately 48 km. From Pettyjohn and McKeon (12).

radation and subsequent concern for reclamation. Certainly reclamation techniques and success could be evaluated by means of satellite imagery.

19.3.3 Inadvertent Weather Modification

Inadvertent weather modification is caused by man's activity and although not well studied, it has been widely documented. One type of potential modification includes jet aircraft contrails, which may have an effect on both local and global climates. These high cirrus clouds, even though very thin, are highly reflective and return to space a significant amount of the sun's radiation. They may diffuse over a large region at altitudes in excess of 6000 meters, particularly in areas of heavy jet traffic. Other examples of inadvertent weather modification include localized fogs, clouds, light rain, and snow produced by cooling ponds and towers, and fog-filled valleys caused by particulate matter from

industrial sources that serve as a nucleus around which moisture accumulates. It has long been recognized that cities, as compared to rural areas, are hotter, have more rain, more cloudy days, and more contaminated air than surrounding rural regions.

Several interesting phenomena are shown on a partial Landsat scene acquired on 12 January 1973 over eastern Ohio and adjacent regions in West Virginia (Fig. 19.12). The Ohio River, along which lie many industrial sites and power plants, nearly bisects the scene. An abundance of clouds, with their long axes trending northeast-southwest, appear in the eastern part of the image. The light area in the northwest is a thin cover of snow on the ground, while the light bands extending diagonally represent trails of snow that probably fell only a few hours before the image was recorded. The snow clouds were moving southeast; the wind at the time

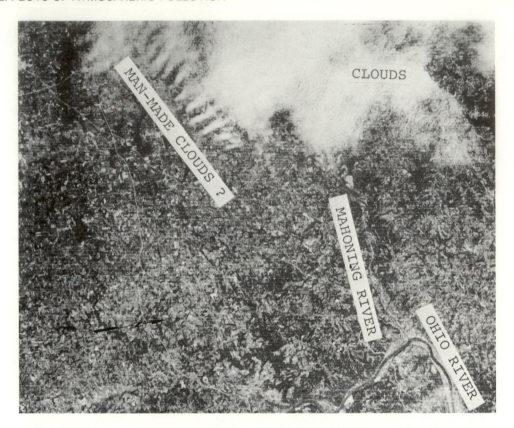

FIGURE 19.11. Partial Landsat-1 scene of the Warren-Youngstown, Ohio area showing man-made clouds (?) originating in the vicinity of major industrial sites (steel mills) along the Mahoning River. Bottom edge of figure represents approximately 80 km. Landsat scene 1425-15350, band 5. From Pettyjohn and McKeon (12).

of satellite overpass was blowing to the east as indicated by the plumes at sites A, C, and D.

Site A is an electric generating plant and the large white cloud originating there is caused by condensing water vapor from the adjacent cooling tower (Fig. 19.13). About 20 km to the east is a large snow-covered area (site B). This area of snowfall appears to reflect inadvertent weather modification caused by the large supply of hot water vapor generated at the cooling tower.

The clouds in the vicinity of site C appear to be less dense and larger than those in surrounding areas. Here also is a cooling tower that lies on the floodplain of the Monongahela River; it is evident because of the condensing water vapor. It is suspected that the large quantity of water vapor and heat released at the cooling tower is directly related to the differences in the apparent density of the clouds.

19.4 STRIP MINING

19.4.1 Introduction

Surface extraction of coal can be classified as either contour or area strip mining. In the contour method a bench is cut into a coal seam that crops out along a hillside (Fig. 19.14). Earth removal continues into and around the hill until the overburden becomes too thick for economical mining. The spoil stripped from above the seam is dumped 180° from the cut filling the adjacent valley. At the cessation of mining the area is characterized by a steep face (the highwall), a relatively flat bench that follows the contour of the hill, and an adjacent spoil bank consisting of the overburden previously removed. Lakes commonly form on the bench between the spoil and the highwall.

Area strip mining is practiced where the overburden is relatively thin. The area to be stripped is

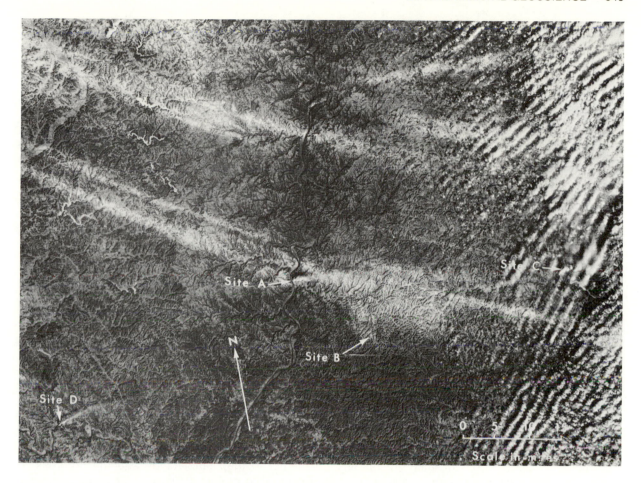

FIGURE 19.12. Partial Landsat-1 scene of eastern Ohio and western West Virginia showing snow belts, smoke plume (site D), cooling towers (sites A and C), and large area of snow caused by cooling tower vapor (site B). Landsat scene 1173-15362, band 5. From Pettyjohn and McKeon (12).

cleared, drilled, and blasted prior to overburden removal, generally by huge earth movers (Fig. 19.14). The overburden is deposited in long, narrow, steep banks that lie opposite the highwall. Once the overburden is removed, smaller equipment is used to remove the coal from the stripping bench.

Strip mining has caused severe environmental problems. These are related to (1) the erosion of bare or thinly vegetated spoil banks and (2) the discharge of highly mineralized waters.

Earth materials forming unreclaimed spoil areas are easily and quickly eroded from the steep, bare banks to accumulate in lakes and impoundments, along stream channels and floodplains, and on agricultural lands. This, in turn, has caused flooding, road and ditch maintenance problems, smothering of vegetation, and the inundation and eventual

abandonment of farm lands.

Iron sulfide minerals, which occur both in the coal and adjacent strata, are abundant in spoil banks and break down to form water-soluble compounds. As water infiltrates through the spoil, these by-products are leached out forming highly mineralized acidic water. The unpotable, corrosive waters contain large concentrations of dissolved solids, sulfate, iron and hardness, as well as heavy metals. The acid-mine drainage may eventually flow into streams by means of groundwater or spring discharge and overland flow.

Acid-mine drainage from strip mines is only a part of most regional acid-water problems, which are primarily caused by abandoned underground coal mines discharging highly concentrated fluids throughout very small areas. The overall effect on

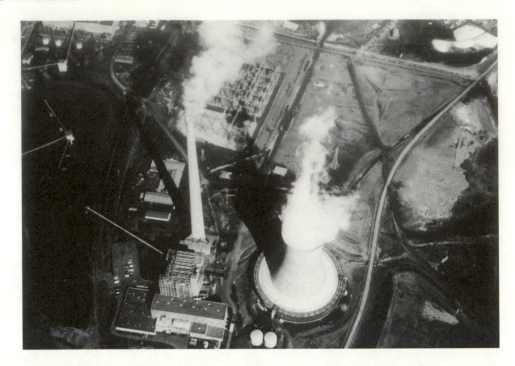

FIGURE 19.13. Cooling tower and a tall smoke stack at a coal burning electric generating plant in Ohio.

water resources due to drainage from underground mines may be far greater than that derived from a strip mine of much larger size.

In strip-mined areas contaminated streams almost invariably have headwaters either adjacent to or in stripped lands. Obviously detection of every contaminated stream in a stripped area, which may encompass several square kilometers is an ominous task, requiring a great number of man-hours and expense. Moreover, the data are only of limited value because (1) in active mines the area of stripping expands daily thus increasing the acid drainage and sediment load and (2) reclamation in many areas closely follows land disruption, which, in turn, tends to minimize or reverse the environmental degradation.

Because of the difficulty and time consuming nature of detailed field investigations and the general lack of up-to-date aerial photographs, satellite data can be utilized to survey strip-mined lands on a repetitive basis at minimal cost. Automatic processing of Landsat data provides a readily available and accurate method, not only for locating strip mines but also for monitoring stripping progress, reclamation, and determining potential sources of water pollution.

19.4.2 Strip-Mine Monitoring

Nearly 7500 km² were investigated in east-central Ohio by Pettyjohn and others (14), in an attempt to map strip mines and the progress of mining and reclamation by means of computer processing of satellite computer compatible tapes (CCTs). The area has been disrupted by coal mining since the early 1800s and by strip mining since the early 1920s.

The data processing procedure these workers applied uses computer target spectral recognition techniques as a basis for classification. Products developed included printout tables showing the area covered by each target category and map overlays. The target classifications included (1) stripped earth and major areas of erosion, (2) partially reclaimed earth and minor areas of erosion, (3) vegetation, (4) deep or clear water, (5) shallow or turbid water, and (6) unclassified areas. The stripped earth and partially reclaimed category areas are also the major sites of origin of acid-mine drainage.

To produce data that would directly relate to a map, a method for correcting the CTT for earth rotation and other geometric errors was developed. The modified tape, when played back on a computer in a format suitable for driving a plotter, produced a

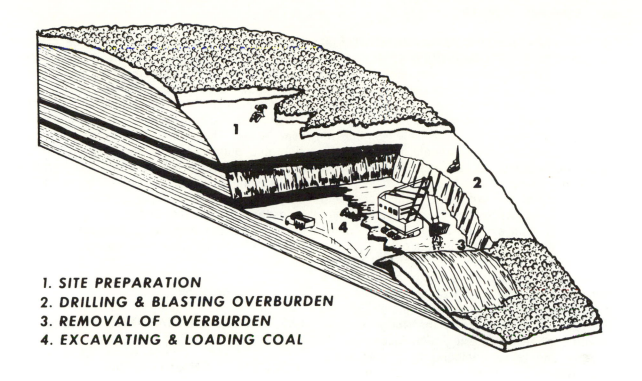

1. **SITE PREPARATION**
2. **DRILLING & BLASTING OVERBURDEN**
3. **REMOVAL OF OVERBURDEN**
4. **EXCAVATING & LOADING COAL**

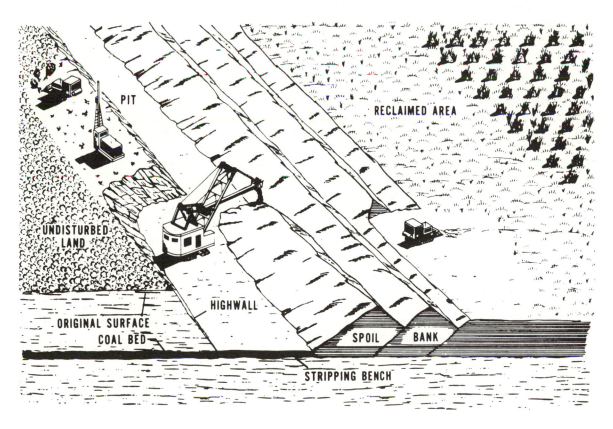

FIGURE 19.14. Surface mining methods, USEPA.

geometrically corrected map of each target category to be drawn on film at any scale specified by the computer operator. The pen drawings were used to produce color-colded overlays at a scale of 1:250,000. A combined series of color-coded overlays for a part of east-central Ohio is shown in Fig. 19.15 color plate 45. Forests and rangeland, shown as separate overlays on the figure are merged as natural vegetation in the area printout table (Table 19.1).

Accuracy of the automatic data processing techniques was ascertained by comparing computer maps with existing maps, ground-survey data, and aerial photographs. A direct comparison between an aerial photograph taken in 1973 and the computer generated stripped earth map of a part of a mine based on a 21 August 1972 CCT, is shown in Fig. 19.16, color plate 46. This illustration shows not only that there is close agreement between the two techniques (satellite imagery and aerial photography), but also that there were some significant changes in mining areas between August 1972 and September 1973.

It is readily evident that satellite data can be used to develop strip-mine maps. Unfortunately, water-quality information cannot be obtained directly from the imagery, although certain chemical characteristics can be inferred. For example, streams affected by acid-mine drainage from strip mines must have headwaters that lie either in or near a mine. If a mine or part of a mine has been successfully reclaimed, water discharging from it should be less mineralized than that derived from unreclaimed land.

In order to assess the area, field work is essential. This time-consuming task can best be handled by generating computer category maps (stripped earth and partially reclaimed categories) at a convenient scale. The transparent overlays can then be placed directly on any map or photograph of the same scale showing the drainage system. Streams, too small to be evident on the imagery, can then be viewed directly and their proximity to potential sources of acid-mine drainage is readily evident. The field investigator then is merely required to examine those streams whose headwaters lie in or adjacent to stripped areas.

Automatic data processing of computer compatible tapes has the advantage of rapid map preparation and calculation of the area of each target category. Moreover, it requires no manual interpretations or measurements. Other investigators have used various printout schemes, direct interpretation of images, and color enhancement-density slicing techniques for evaluating mine areas. These procedures require a considerable outlay of time and rely on the skill of individual photo interpreters. Furthermore, with several of these systems one set of data must be physically compared with another set in order to detect changes. It is commonly not possible by manual techniques to detect subtle reflectance difference caused by mining or reclamation.

19.5 NATURAL DISASTERS

19.5.1 Fire and Heat Detection

Remote sensing techniques can be used by planners and scientists alike to protect the population from natural disasters or to evaluate disasters after they have occurred. Aircraft or satellite-mounted scanners, particularly thermal scanners, can be used to monitor volcanoes that are beginning to show signs of increased activity. In addition to possible eruptions and the pouring forth of lava, volcanic activity can lead to other types of natural disasters. In snow-covered terrain the increased heat may produce landslides caused by pyroclastic materials becoming saturated, causing the deposit to catastrophically flow down the sides of the volcano.

Remote sensing tools can be used to locate geologic "hot spots." These areas, which are commonly characterized by heated rocks near land surface, can lead to severe construction problems. Asphalt roads, for example, constructed over an abnormally hot area might tend to deteriorate. Noxious gases and odors could render the site aesthetically unpleasing. Building structures may tend to become weakened, requiring expensive maintenance.

Landsat imagery has been used to detect and map grass fires and to monitor their rates and direction of burning in southern Africa. Most of the fires are set by man to clear areas for subsistence

TABLE 19.1. AREA OF EACH OF FIVE CATEGORIES IN THE STUDY AREA BASED ON THE AUGUST 21, 1972, ERTS-1 CCT

ERTS Processor

ERTS Scene 10- 1029-15361

Date of Scene - August 21, 1972

Center of Scene - N39-25/W081-00

Sun coordinates - EL53°
              AZ130°

Spacecraft Heading - 191°

Tape Number - 1

Starting Scan Line = 450

Ending Scan Line = 950

October 15, 1973   9:01:32

Category	Percent of Total	Acres	KM²
Unclassified	4.47	20293.15	82.12
Stripped Earth	0.74	3343.62	13.53
Dirty Water	0.80	3610.80	14.61
Reclaimed Earth	3.95	17928.80	72.55
Vegetation	89.66	406771.91	1646.13
Clean Water	0.38	1704.79	6.90
Totals	100.00	453653.07	1835.84

Program Run Time = 00:23:23

crops, provide early grazing, or force wild animals from their grassy habitat.

Although generally intended to be small, the fires may rapidly burn over vast areas, greatly upsetting the delicate ecologic balance. Wightman (15), using satellite data, showed that a fire in Botswana burned approximately 181 km² of grassland during a single 24-hour period. Elsewhere the rate of burning ranged between 52,611 and 80,940 m²/min.

Although aircraft are used for fire control and monitoring, it is simply impossible to cover extremely large, sparsely populated areas on a sufficiently repetitive basis for this method to be effective.

In a few areas in the lignite fields of the northern Great Plains, underground seams have been burning uncontrolled for scores of years. Little concern has been expressed for the loss of this natural resource, and practically no attempts have been made to extinguish the underground fires. One problem in extinguishing such fires is detecting the burning front. Once located, ditches separating the burning coal from the unaffected parts can be excavated.

Following light snowfalls, heat from the underground fires rapidly melts the overlying snow, distinctly marking the area of greatest heat generation. As a result, aerial photographs, if obtained at the optimum time, could be used to map the damaged area. However, it is unlikely that the melted snow would precisely mark the burning front; vehicle or aircraft-mounted thermal scanners could be used during any season and would probably more accurately locate the fire boundaries.

Literally thousands of gob or waste piles lie near the entrance of underground coal mines in Appalachia. Oxidation of iron sulfide minerals in the wastes generate heat and, in many instances, has been sufficient to cause spontaneous combustion. Because of the heat generated in these materials, it would be advantageous to locate and monitor them in order to reduce fires and perhaps reclaim them to reduce acid-mine drainage. Rapid surveys and monitoring could be accomplished with aircraft-mounted thermal scanners.

In logging regions, large sawdust piles are common in the vicinity of existing or long-abandoned sawmills. These wastes may also burst into flame because of spontaneous combustion or often by careless individuals who attempt to extinguish their smoking materials by shoving them into the soft sawdust. Once ignited, the sawdust may smolder and slowly burn for months. Even when detected

such fires are exceedingly difficult to extinguish and may cause forest fires. Thermal scanners could be used to detect and monitor temperature fluctuations of these waste piles.

One of the major problems encountered in attempting to control forest fires is the difficulty of accurately locating the boundary of the fire, which may be totally obscured by dense smoke. In this regard thermal scanners have also proven their value because they have the capability to "see" through the smoke.

19.5.2 Flood Inundation Mapping

Flood losses in the United States exceed $1.5 billion annually and cost scores of lives. It has been estimated that about 12% of the population of the United States live in areas subject to periodic flooding. This, coupled with the unpreparedness of the population, the lack of local flood protective works, inadequate flood-plain planning and regulations, and the inability to provide quick warnings account for the excessive flood losses.

Conventional flood mapping procedures are well established but time consuming and expensive. Partly for this reason, remote sensing methods have been used in several flood studies. The disastrous effects of the June 9, 1972 flood that hit Rapid City, South Dakota, were almost immediately evaluated by Myers, Waltz, and Smith (16) with color and color infrared aerial photography and thermal imagery. Morrison and Cooley (17) portrayed the limits of Gila River flooding from Landsat and found good agreement from maps prepared from ground surveys and aerial photography. Hallberg, Hoyer, and Rango (18) mapped a September, 1972 flood on the Nishnabotna River in western Iowa. They were able to delineate inundated areas from Landsat data, the results of which compared favorably with aerial black-and-white, color, and color infrared photography.

Particularly well documented was the flooding that occurred in the Mississippi River Valley during March and April, 1973. In much of the Mississippi River Valley region, March, 1973, was characterized by severe storms, rain and flooding. On April 2, the entire river from Iowa to Louisiana was above flood level and by late April more than 4.5 million ha of land were inundated. At least 25,000 persons were forced from their homes. It was soon evident that the spring floods of 1973 were the most disastrous in the recorded history of the river valley. Lowland areas were inundated along the mainstem and its tributaries in an area stretching from St. Louis to New Orleans. On April 28 the river reached its crest of 13.2 m at St. Louis, which was 4.1 m above flood level. The previous record flood crest of 13.1 m occurred in 1785. The river remained above flood stage for 76 days.

Deutsch and Ruggles (19) used satellite-derived temporal composites to differentiate between dry soil, saturated soil, and standing water in the Mississippi valley. The temporal composites consisted of color enhanced satellite imagery created from preflood and flooding or postflood data. In this technique preflood water distribution is shown by one distinct color while flooding or postflooding conditions are demarked by an overlay that depicts the inundated area in red. Preparation of temporal two-color composites are described by Deutsch and others (20).

The value of NOAA-2 satellites for regional flood mapping was described by Wiesnet and others (21). This operational satellite contains a Very High Resolution Radiometer (VHRR), a two-channel scanning radiometer that records in the visible (0.6-0.7 μm) and thermal infrared (10.5-12.5μm) parts of the spectrum. Although ground resolution is only about 1 km, the 12-hour overpass frequency presents a powerful monitoring tool that can furnish rapid synoptic information. Furthermore, only a few existing nonmilitary satellites contain thermal scanners, the only method available for detecting conditions during hours of darkness. Of course, laser and radar imaging devices are as effective at night as thermal scanners, but are not commonly used at present.

Although not a replacement for typical engineering flood evaluations, aircraft and spacecraft sensors readily provide comparatively simple means of flood inundation mapping. Furthermore, these data can be used to generate flood-prone area maps, show locations where new or additional flood works are needed, indicate time of travel of floodwaves, and permit determination of length of time of inundation.

As pointed out by Deutsch and Ruggles (19), regional flood mapping can provide disaster relief agencies with a near real-time overview from which can be determined early estimates of damage thus permitting the expeditious dispersement of assistance. In addition, regional planners can use the information to ascertain the best use of the floodplain, housing authorities, to determine areas sub-

ject to flooding, private lending agencies, to consider the flooding potential of proposed urban developments, engineers, to deduce effectiveness of flood control works, and flood insurance agencies, to assess flooded areas and assist their customers more quickly.

19.5.3 Violent Wind Systems

Hurricanes generally create havoc in coastal areas and sometimes inland regions as well. Damage estimates amount to millions of dollars, and coupled with loss of life, it is understandable why they are so greatly feared. Little can be done to control these violent wind systems; therefore only adequate planning, constant weather monitoring, and rapid disaster relief operations can provide comfort to those affected. Hurricanes hit the Texas coast on an average of once every two years. Since 1900 they have caused more than 7000 deaths and property damage in excess of $1.3 billion dollars. Hurricane damage is related to three types of phenomena (1) excessive winds, (2) heavy rains and freshwater flooding, and (3) saltwater surges.

Celia (1970) was characterized by strong wind gusts that flattened nearly everything within the area of influence. Hurricane Beulah (1967) produced intense rainfall that, coupled with winds in excess of 225 kph led to widespread, freshwater flooding in Texas. Carla (1961) with her 251 kph winds created surge tides as high as 6.7 m, while Camille (1969) developed 7.6-m surges that inundated lowlying areas with saltwater. All of these phenomena can cause rapid shoreline erosion or accretion.

Although weather satellites provide adequate and generally accurate early warnings, only suitable planning can reduce hurricane damage. Remote sensing can be used to evaluate vulnerable areas, provide rapid analyses for disaster relief needs, and monitor approaching weather systems.

19.5.4 Erosion

About two million people live along the narrow 322 km long Ohio shoreline of Lake Erie. Deposits forming the shoreline consist of lake clay, sand, silt, and glacial till. Consequently, the zone is easily erodible and constantly changing. The topography is near lake level along the south-western shoreline and flooding is common during spring storms. Eastward, the shoreline gradually changes from low relief to bluffs, 3 to 20 m high. Although the bluffs are not subject to flooding, the steep faces retreat as much as 6 m/yr.

Between 1796 and 1838 the shoreline eroded its way inland an average of 40 m in the reach stretching from the Pennsylvania line to Marblehead. The erosion was due entirely to natural causes. Locally, rates of erosion have been reduced and shoreline deposition increased due to the construction of groins, jettys, and breakwaters, but erosion still creates havoc.

Aerial photographs and satellite imagery provide methods for determining areas of shoreline erosion and buildup, and the paths taken by the eroded sediment. These data can be used in planning for the design, construction, and maintenance of human-made structures for the control of erosion.

19.6 SUMMARY

It is evident that remote sensing technology offers a practical and relatively inexpensive method that can be used in a variety of environmental studies. Satellite data are particularly valuable for analysis of regional stresses brought about by natural phenomena (storms, floods, earthquake damage, and so forth) or man's activities that have effected wide areas (smoke plumes, inadvertent weather modifications, sediment runoff, and so forth.) On the other hand, aircraft data are generally more useful in local studies that require high resolution, such as detecting underwater outfalls, evaluating small areas of vegetative stress, and precisely delimiting flooded areas for insurance or relief purpose.

During any investigation using remotely sensed data, it is essential that ground truth information be collected and used. Furthermore, care must be exercised in order to reduce the number of assumptions and analyses that are based on ignorance. For exampe, one investigator noticed that the number of lakes in a large active strip mine detected on Landsat imagery, was greatly reduced over a 12-month period. He assumed that these acid lakes disappeared because the water infiltrated and, therefore, each lake was a major source of groundwater pollution. This assumption, however, was far from the truth. The number of strip-mine lakes certainly decreased from one year to the next, but the reason was simply that the area was being reclaimed and in the process, the lakes were filled!

Remotely sensed data are not being used to their maximum potential, perhaps because their capabilities are not well-known. They could be used by a homeowner to detect stressed vegetation (infrared film) in his lawn, long before it is evident to the unaided eye, and thus initiate application of pesticides or fertilizers. Remotely sensed data could be used by state and federal regulatory agencies to detect illicit liquor stills, monitor the spread of atmospheric contaminants, or track ocean dumping. Other investigators might examine the linear features so evident on satellite imagery, in order to relate them to continental drift, earthquakes, or the location of mineral deposits or water supplies. The only limit of usefulness of remotely sensed data is the imagination of the investigator.

REFERENCES

1. Rogers, E. H., Shah, N. J., McKeon, J. B., Wilson, C., Reed, L., Smith, V. E., and Thomas, N. A., 1975, Application of Landsat to the surveillance and control of eutrophication in Saginaw Bay: 10th International Symposium on Remote Sensing of Environ., Oct. 6-10, 1975, Ann Arbor, Michigan, v. 1, pp. 437-446.

2. Blackwell, R. J., and Boland, D. H. P., 1976, The trophic classification of lakes using ERTS multispectral scanner data: Proceedings American Society of Photogrammetry 41st Annual Meeting, March 9-14, 1976, Washington, D.C., p. 393-414.

3. Boland, D. H. P., 1975, An evaluation of the Earth Resources Technology Satellite (ERTS-1) multispectral scanner as a tool for the determination of lacustrine trophic state, unpub. Ph.D. Dissertation, Oregon State University, Corvallis, 311 p.

4. Boland, D. H. P., and Blackwell, R. J., 1975, The Landsat-1 multispectral scanner as a tool in the classification of inland lakes, in Proceedings of Earth resources survey symposium, Houston, Texas; NASA TM X 58168, v. I-D, p. 419-442.

5. Pettyjohn, W. A., 1973, Sources of chloride contamination in Alum Creek, central Ohio—a remote sensing survey: Ohio Dept. Nat. Res., 53 p.

6. Scherz, J. P., Sydor, M., and VanDomelen, J. F., 1974, Aircraft and satellite monitoring of water

quality in Lake Superior near Duluth, *in* Third Earth Resources Technology Satellite-1 symposium: NASA SP-351, v. 1, sec. B., p. 1619-1636.

7. Lind, A. O., Henson, E. B., and Pelton, J., 1973, Environmental study of ERTS-1 imagery: Lake Champlain and Vermont, *in* Proceedings symposium on significant results obtained from Earth Resources Technology Satellite; NASA SP-327, v. 1, sec. A, p. 643-650.

8. Chase, P., Conrad, A., and Imhoff, E., 1971, Location of sawdust by a multispectral scanner: Ann Arbor, Michigan, Bendix Aerospace Systems Div., 14 p.

9. Wezernak, C. T., and Roller, N., 1973, Monitoring ocean dumping with ERTS-1 data, *in* Proceedings symposium on significant results obtained from Earth Resources Technology Satellite: NASA SP-327, v. 1, sec. B, p. 635-642.

10. Lyons, W. A., 1974, Satellite detection of air pollutants: Symposium Proceedings of Remote Sensing Applied to Energy-Related Problems, Dec. 1974, Univ. Miami, p. WAL 1-WAL 30.

11. Pettyjohn, W. A., and McKeon, J. B., 1974, Smoke plume detection by means of satellite data: Ohio Environ. Prot. Agency Newsleaf, v. 2, no. 16, p. 4-5.

12. Pettyjohn, W. A., and McKeon, J. B., 1976, Satellite detection of smoke plumes and inadvertent weather modification, *in* Proceedings symposium on acid precipitation and the forest ecosystem: USDA Forest Service General Tech. Rept. NE-23, p. 337-347.

13. Murtha, P. A., 1974, SO_2 damage to forests recorded by ERTS-1, *in* Third Earth Resources Technology Satellite-1 symposium: NASA SP-351, v. 1, sec. B., p. 137-144.

14. Pettyjohn, W. A., Rodgers, R. H., and Reed, L. E., 1974, Automated strip mine and reclamation mapping, *in* Third Earth Resources Technology Satellite-1 symposium: NASA SP-356, v. 2, p. 87-101.

15. Wightman, J. M., 1973, Detection, mapping, and estimation of rate of spread of grass fires from southern African ERTS-1 imagery, *in* Proceedings symposium on significant results obtained from Earth Resources Technology Satellite: NASA SP-327, v. 1, sec. A, p. 593-601.

16. Myers, V. I., Waltz, F. A., and Smith, J. R., 1972, Remote sensing for evaluation flood damage conditions—the Rapid City, South Dakota flood, June 9, 1972: South Dakota State Univ., Remote Sensing Inst. Rept. 72-11, 29 p.

17. Morrison, R. B., and Cooley, M. E., 1973, Assessment of flood damage in Arizona by means of ERTS-1 imagery, *in* Proceedings symposium on significant results obtained from Earth Resources Technology Satellite: NASA SP-327, v. 1, sec. A, p. 755-760.

18. Hallberg, G. R., Hoyer, B. E., and Rango, A., 1973, Applications of ERTS-1 imagery to flood inundation mapping, *in* Proceedings symposium on significant results obtained from Earth Resources Technology Satellite: NASA SP-327, v. 1, sec. A, p. 745-753.

19. Deutsch, Morris and Ruggles, F. H., 1974, Optical data processing and projected applications of the ERTS-1 imagery covering the 1973 Mississippi River Valley floods: Water Resources Bull., Amer. Water Res. Assoc., v. 10, no. 5, 1023-1039.

20. Deutsch, Morris, Ruggles, F. H., Guss, P., and Yost, E., 1973, Mapping of the 1973 Mississippi River floods from the Earth Resources Technology Satellite: Proc. No. 17, Am. Water Resources Assn. Symp. on Remote Sensing and Water Resources Management, p. 39-55.

21. Wiesnet, D. R., McGinnis, D. F., and Pritchard, J. A., 1974, Mapping of the 1973 Mississippi River floods by the NOAA-2 satellite: Water Resources Bull., Amer. Water Res. Assn., v. 10, no. 5, p. 1040-1049.

CHAPTER 20

EXTRATERRESTRIAL GEOLOGY

R. STEPHEN SAUNDERS
THOMAS A. MUTCH

20.1 INTRODUCTION

Remote sensing of the Earth is a matter of choice; for other bodies in the solar system it is a necessity. For the forseeable future, we will be restricted to deciphering the geologic properties of other planets from a distance. This chapter describes the techniques and some of the progress in understanding the geology of planets, satellites, and asteroids from Earth-bound observatories as well as from spacecraft.

The past two decades have been singularly exciting for planetary geoscience. For the first 5000 years of recorded history man charted the stars and the courses of planets with his unaided eye. Following the discovery of the telescope were 350 years of development of that instrument to its practical limit. Now, in the recent decades, man has transported himself and the technological equivalent of his senses far beyond the Earth. The reader is invited to formulate his own extrapolations for the future.

Extraterrestrial geology is defined as geology applied to the other planets and solid bodies. The science is at least as broad in scope as the parent science including all the traditional branches such as geophysics, structural and historical geology, stratigraphy, volcanology, sedimentology, mineralogy, and perhaps even paleontology. Planetary geoscientists, however, often speak in somewhat broader terms than do other geoscientists about goals of determining the origin and evolution of the solar system. This discussion will be restricted to description of the techniques of remote sensing of the "terrestrial" objects.

Remote sensing in this context is the data-collecting, instrument oriented aspect of the science. It is the experimental rather than the theoretical approach. More than that, it is experimental not in the laboratory sense but in the old fashioned field sense. The approach, as in any science, is dictated by goals. In this case, we design experiments that allow us to identify, characterize, and

interpret the processes involved in planetary evolution. Basically we ask: (1) What is the present nature of the planetary object? (2) What processes contribute to the evolution of the object? These are purely descriptive questions, the answers to which provide information for building and testing all the other hypotheses in the other aspects of planetary geology. This approach is a direct extension of the geologic approach to the understanding of the Earth—go and see. Usually geologists test hypotheses continuously in the course of their observations, altering both hypotheses and direction of observation as required. Sometimes we just collect data, although seldom is this the sole justification for a mission or experiment. In our discussion the hypotheses will be given at bit less importance and we shall emphasize the techniques of remote sensing for extraterrestrial geology.

The general procedure will be to describe each data type and the information revealed by the various techniques. Examples for various planets are included. Following the description of techniques is a planet by planet summary of the status of exploration. Finally, we will discuss briefly the future including expected missions, new instruments, and techniques for managing the ever increasing quantities of data that are accumulating.

Several remote sensing methods have been omitted from the discussion. They are all remote sensing techniques in a real sense. These techniques are laser altimetry, far ultraviolet spectrometry, magnetometry, seismic studies, heat flow determination, and electrical conductivity as determined from solar wind interaction with the Moon.

Throughout the following descriptions the term resolution will be used. The reader should be aware that in planetary remote sensing resolution is defined in various ways. We will follow the convention of using the term resolution for line pair resolution. When describing television imaging systems and other nonimaging methods the terms pixel (picture element) or resolution element are used when giving the dimensions of a single line or resolution cell. In the case of pixel dimensions, we give the dimensions of the picture element projected onto the ground surface. The line pair resolution would be approximately twice this dimension. Finally, the term "identification resolution" refers to the effective resolution or the dimension of the smallest topographic feature that can be identified as a depression or hill. For high contrast targets the value of identification resolution approaches the line pair resolution.

20.2 EARTH-BASED OPTICAL TELESCOPES

The earliest instrument used for planetary study has no doubt long ago crumbled into the floodplain soil between the Euphrates and the Tigris along with the ancient Sumerian observatories. The observations that resulted in the five millenia old clay tables of the ancients and all subsequent records up to and including those of the last and the greatest observer to use nonoptical instruments, Tycho Brahe, were limited by the angular resolution of the human eye. Galileo in 1609 became, as far as is known, the first to use the telescope, a glass curiosity discovered the previous year, to observe the planets. The eye has a spatial resolution of about one minute of arc. Galileo probably obtained about a factor of 10 improvement with his simple telescope. This instrument in its numerous variations remained the principal tool for observing the planets until recently when it has tended to be replaced by advanced imaging devices used on spacecraft and has been joined by the giant radiotelescopes that have the capability of peering through the Venerian clouds. It almost seems that many planetary scientists (mostly nonastronomers) tend to assume that the usefulness of Earth-based telescopes for planetary observations has about ended. However, over the same brief span that planetary exploration by spacecraft has captured and dominated attention, important new applications for optical observations have also been developed. One such development is the technique of acquiring high resolution spectra, in the visible and near infrared (IR), of the Moon, planets and other objects in the solar system. Previously, telescopes were used in planetology largely for images of surfaces. The fundamental nature of the Moon's surface was deduced from such observations. The significance of the recent "narrow band" techniques is described at the end of this section.

Predictably, an Earth-based optical telescope will provide more geologic information for those bodies that are closest to the Earth. For the Moon, approximately 356,000 to 407,000 km away depending on its exact position in the eccentric orbit, features smaller than 1 km can be resolved. For Mars, approximately 60×10^6 km distant when Mars and Earth are in opposition, only surface markings 100 km or larger can be delineated. Although Venus is slightly closer to Earth than Mars, a dense unbroken cloud cover completely obscures the surface. For the remaining inner planet, Mercury, and for satellites of outer planets, virtually no surface markings can be resolved.

A distinction between "brightness resolution" and "spatial resolution" should be made. The spatial resolution of Earth-based telescopes generally is limited by air movement within the Earth's atmosphere, the same phenomenon that causes scintillation, or twinkling, of stars. Beyond a certain critical diameter, increasing the size of telescopic mirrors and lenses carries no reward of increased spatial resolution. However, the light-gathering capacity of the telescope *is* increased. Very dim objects, undetectable with small telescopes, can be identified and more important, details of their spectra can be resolved, with large telescopes. For this reason, very bright markings on planetary surfaces can be seen, even though their dimensions are less than the limit of spatial resolution. In effect, their true shapes have been distorted and enlarged.

Disregarding problems of surface mapping, telescopic observations of distant planets are still important for determining overall brightness, visible and IR spectra, and size of the bodies.

The character of telescopic photographs of the Moon varies greatly according to phase angle. This is defined as the angle, measured at the target point (Moon), between illumination source (Sun) and detector (telescope) (Fig. 20.1). When the phase angle is close to 0°, a condition corresponding to a full Moon, variable brightness of surface materials is emphasized, reflectivity measured at 0° phase angle are termed geometric albedos. In general, the cratered lunar highlands are brighter (higher albedos) than the mare plains. Youthful craters are surrounded by bright halos and rays. The rays around Tycho are especially prominent.

Since the major types of surfaces on the Moon, maria and highlands, are distinguished by albedo, it might be suspected that albedo may be an important indicator of rock type. Albedo is controlled by the opacity of the surface and on the mean distances between boundaries that scatter light. Composition strongly affects these properties. The Apollo 15 and 16 x-ray fluorescence experiment determined the Al/Si and Mg/Si ratios along the ground tracks by measuring the value of production of scattered x rays having energies characteristic of these elements. These quantities are a fairly good indicator of rock type. Not unexpectedly, it is found that albedo and Al/Si ratio correlate extremely well— perhaps well enough to allow extrapolation of the Al/Si ratio over the rest of the Moon (Fig. 20.2).

Lunar photographs taken at very low Sun angle contain prominent shadows which emphasize topographic relief. Maximum relief is displayed directly along the terminator which divides the sunlit shadowed hemispheres of the Moon. Subtle features such as degraded craters with slopes of a few degrees or mare ridges with elevations of several meters can be identified by the shadows they cast. At high Sun angles these landforms are indistinguishable (Fig. 20.3).

The crispness in detail of low-Sun lunar pictures results from the absence of light-scattering atmosphere. However, the lunar guideline for pictures generally applies to remotely acquired pictures of any planetary surface. High Sun emphasizes variable brightness of surface materials. Low Sun emphasizes topographic detail.

Most telescopic photographs are black and white. However, there has been considerable experimentation with color pictures. These are acquired either by exposing the image on conventional color film or by taking three separate pictures in red, green, and blue light, and then combining them to reconstruct a color image. Useful color pictures of Mars have been acquired but the colorimetric distinctions on the Moon are so subtle that an

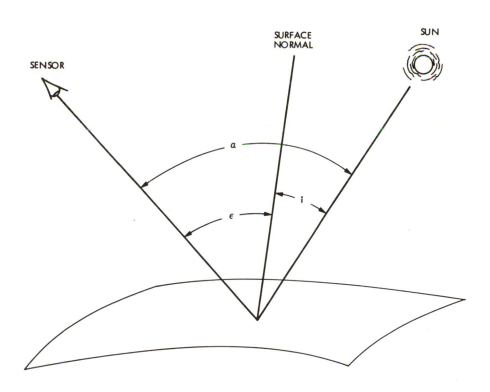

FIGURE 20.1. Illumination and viewing geometry, showing relationship between phase angle (a), emission angle (ϵ), and incidence angle (i).

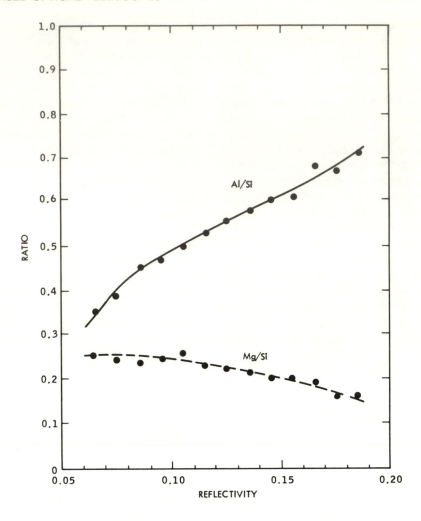

FIGURE 20.2. Relationship of reflectivity and lunar surface geochemistry. Notice the apparent correlation between reflectivity and Al/Si ratio. Figure adapted from Shevchenko (30).

accurate rendering of the surface is difficult. However, extremely informative "false color" images have been acquired by differing brightnesses as recorded in blue and red light. This technique reveals several colorimetrically distinct areas within mare regions that appear to be homogeneous in albedo and crater density.

More recently, attempts have been made to take advantage of the property of most substances that their reflectance varies with wavelength. Where we speak of albedo, we normally are considering the reflectance of an object integrated over the entire spectrum of light energy received by the detector

(usually 0.3 to 1.1 μm). It is possible also to look at the energy reflected by the object within very narrow wavelength bounds. Ordinary color pictures provide an approximate representation of color by taking simultaneous images through rather broadband filters centered at red, green, and blue wavelengths. Much important early telescopic work used such broadband filters to classify astronomical objects by color. However, such filters do not provide detailed spectral resolution, they only allow a relative assessment of which objects in a scene are redder or bluer, and so forth. We can more accurately describe the spectral character of a surface if we exam-

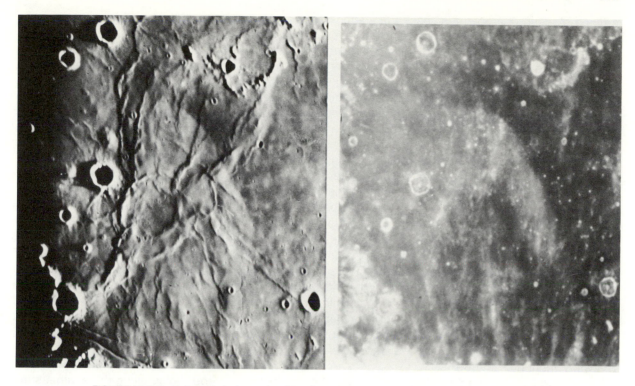

FIGURE 20.3. Earth-based photographs of southern Mare Tranquillitatis of the Moon. On the left, subtle topographic features are emphasized by the low early morning Sun. On the right, the same area is seen near zero phase angle or full Moon, emphasizing variations in surface reflectivity. Photos from Kuiper et al. (31).

ine the reflectance over a large number of points using special filters that transmit light in a band only a few hundredths of a micrometer wide.

A few years ago it was hoped that unique identification of minerals could be obtained by comparing the spectral reflectance of planetary surfaces with that of substances measured in the laboratory. A number of dips in the spectral curves, absorption bands, occur in these laboratory spectra. Unfortunately, these bands are not well defined in many cases and because natural surfaces usually contain complex mixtures the initial hope was not borne out. Most early work suggested that remote compositional determinations would remain ambiguous at best. A few groups, however, refused to give up. (For a recent review and additional references see Ref. 1.) Convincing correlations between certain compositional parameters obtained from Apollo samples and details of both laboratory and remotely acquired spectra have been developed into remote mapping tools that will allow wide-scale mapping of the lunar surface.

One of the best examples that has been developed allows mapping of the TiO_2 content of mature lunar mare soil based on the slope of the reflection spectrum between 0.40 and 0.56 μm. Since basalt units may be classified on the basis of TiO_2 content, this allows direct geologic mapping of the maria. One active group led by Thomas McCord has obtained a large number of spectra for representative lunar regions. Comparison of these curves shows that the important characteristics can be obtained with only three or four selected points on the spectrum. This characteristic allows the important parameters to be contained in a few images of large regions rather than require detailed spectral curves which for equivalent spatial resolution would require an enormous amount of data. Thus, we are able to extend the compositional information obtained at a few Apollo landing sites to wide regions of the Moon. The data are in the form of images that have resolution approaching 1 km.

It should be recognized that there are numerous complications that have been considered and examined. A serious problem is the Earth's atmosphere, which is not perfectly transparent and varies with time. These effects are removed by taking frequent observations on standard stars and comparing the observations to a standard region of the lunar surface. Another consideration is that anytime we look at a large part of the Moon we include regions with different conditions of illumination and viewing geometry (Fig. 20.1). The phase angle is nearly constant over a given observing time ranging from near 0° at full Moon to nearly 180° at new Moon; of course, nothing would be visible at either extreme since either the Moon or the Earth would be in eclipse. At the phase angles ordinarily used for observation, the effects at different wavelengths have been found to be small. The other angles are incident angle (i) and emission angle (ϵ). These angles, the wavelength, and the surface brightness are related by a photometric function. The analytic form of this function is not well-known but it appears that both phase angle and wavelength variations have little effect especially if the phase angle is held between about 30° and 60°

Lunar surfaces can be classified on the basis of their spectral reflectance into the basic compositional groups, maria and uplands. This distinction is useful where albedo or topography are inconclusive. The maria may be subdivided into a number of subgroups. The major differences, especially the difference between maria and uplands, are due to compositional differences. The other variations are controlled largely by the degree of aging and maturation of the surface. Exposure to micrometeoroid bombardment results in breakdown of the crystal structure and production of glass and agglutinates. This causes destruction of the major mafic mineral clinopyroxene which has an absorption band at approximately 0.95 μm, the wavelength of the Fe^{2+} absorption. Thus the depth of the 0.95-μm band indicates pyroxene abundance or, for a given rock type, the degree of maturity, hence the age, of crater ejecta. It has been suggested that departure from symmetry of the 0.95-μm band can be related to the presence of olivine. The maria subgroups are identified by subtle spectral differences at the blue end which correlate with amount of Ti^{3+} in the glass phase, hence the requirement that the technique be applied to mature soils having at least 70% glass and agglutinates. The development of glass with high Ti^{3+} content in the mare soils also causes the

albedo to decrease with time. Therefore, on an extensive mare surface with numerous craters in various stages of degradation, the brightness at 0.4 μm indicates the relative freshness of crater materials. This is a quantitative assessment that lunar geologists have previously made more subjectively in assigning craters relative ages based on the brightness of their rays. On a more regional scale, the changes in brightness at 0.4 μm in mare regions between craters gives a quantitative assessment of the Ti^{3+} content and thus allows delineation of individual units.

Details of mare spectral curves have also allowed correlation between Apollo sites and other specific mare areas. For example, the basalts that probably underlie the dark mantle at the Apollo 17 site are similar to those at the Apollo 11 site which are in turn identical to the material that forms the dark ring around Mare Serenitatis. The material of central Mare Serenitatis, on the other hand, is most like the Apollo 12 soil sample. The surface material at the Apollo 17 site, at least in part, consists of dark spheres that compositely make up a dark "mantle" that covers a large region along the southeastern boundary of Mare Serenitatis. Similar material can be identified spectrally in other mare areas largely on the basis of its much stronger relative reflectance at the blue and UV ends of the spectrum as well as in the near IR.

In addition to the general darkening of crater rays with time as an indication of age, it has been established that the spectra of fresh crater rims on maria differ from those of the uplands. This provides a potentially useful tool for determining substrate characteristics since craters excavate material from a depth proportional to their diameter. Small mare craters may display spectrally inhomogeneous but mature material adjacent to their rims which may help in the unraveling of local stratigraphic successions. Larger craters also exhibit this. The craters Copernicus and Kepler lie on mare material. Copernicus has upland material in its ejecta while Kepler has not.

An aspect of the lunar spectral mapping that has been extensively studied in recent years is polarization. The nature of asteroid surfaces, whether dusty or rocky, has been inferred. The albedo of small distant objects that are esentially point targets can be derived from polarization measurements (2). For a discussion of polarization measurements and their interpretation for planetary remote sensing see Dollfus and Geake (3) and other papers referred to in that work.

In the infrared, there are two "windows" in the Earth's atmosphere that admit radiation. These lie at about 8-14 μm and 17-24 μm. The lunar surface has been observed from Earth in the 8-14 μm region, the most complete observations were made at 11 μm during lunar eclipse. In general the infrared Moon is quite uniform. The hot spots generally correspond to blocky craters or to regions of high albedo. Most of the hotter craters occur in the maria. It is fairly certain that the presence of blocks at the surface produces most of the thermal anomalies. This topic is discussed further in Section 20.7.

20.3 PHOTOGRAPHY FROM SPACECRAFT

Pictures of planetary surfaces obtained from spacecraft permit improvements in spatial resolution of many orders of magnitude over Earth-based telescopes. For example, compare telescopic pictures of Mars (100-km resolution) with Viking Orbiter or Mariner 9 B-frame pictures (40-100-m resolution). The improvement is a factor of 1000. Similar improvements are represented by Apollo orbital film pictures (1-m resolution) over telescopic pictures (1-km resolution). Figure 20.4 compares the best photographic coverage available for each planet.

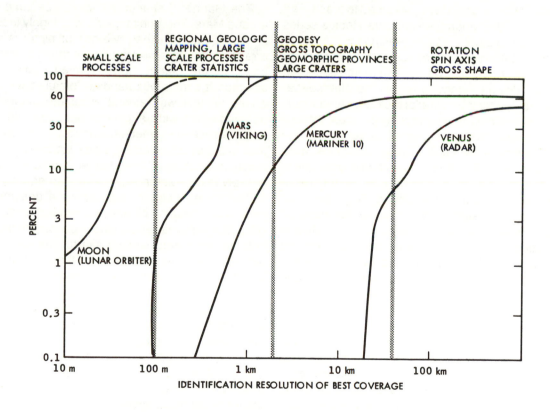

FIGURE 20.4. Percent of imagery coverage for the terrestrial planets at various resolutions. The identification resolution is approximately twice the pixel dimensions of the television and radar data types.

The principal components of most spacecraft camera systems are a telescope, photosensor, and movable platform. Constraints of weight and structural rigidity generally limit the size of the telescope, and hence spatial resolution. Photosensors are usually vidicons, as in conventional television applications. A motordriven platform, directed by command from Earth, points the camera in the desired direction.

The first spacecraft pictures of the Moon were acquired in 1959 by the Soviet Luna 3. The spacecraft circled the Moon and obtained, for the first time, images of the far side. The images were recorded on film inside the spacecraft and the information was transformed into a digital signal for transmission to Earth by scanning the negatives with a light beam and recording the opacity of the film along adjacent scan lines.

The first successful United States mission to the Moon was Ranger 7, a spacecraft that was aimed directly at the lunar surface, designed to crash. As it approached the Moon the spacecraft took a series of television pictures with increasing resolution, radioing the information back to Earth before crashing. Ranger 7 was followed by two more spacecraft of similar design, Ranger 8 and 9. The latter spacecraft, targeted for the center of the crater Alphonsus, provided particularly spectacular pictures, the last of which have resolution slightly better than 1 m.

Five unmanned Lunar Orbiter missions were flown during 1966 and 1967. Their chief goal was to obtain detailed pictures of the surface for use in selecting and certifying Apollo landing sites. The photographic system was similar to that used on Luna 3. Images were recorded on film which was developed onboard the spacecraft. The exposed film was "read" by a scanning light beam. A detector, located on the opposite side of the film, recorded the varying intensity in light related to opacity of the film. The resulting succession of signals was transmitted back to Earth.

Resolution limits of Orbiter missions varied with the height of orbit. Periapsis for Orbiters I, II, and III missions was about 50 km, resulting in ground resolution of about 1 m. Orbiter IV photographed the entire near side from periapsis of 2700 km with ground resolution of 100 m or better. Periapsis for Orbiter V was about 100 km.

Hand-held film cameras were operated by astronauts on all the Apollo missions. Starting with Apollo 15 and continuing through Apollo 17, two automated film camera systems were flown on the orbital modules. A mapping (metric) camera system was operated in association with a stellar camera and laser altimeter providing cartographic control necessary for establishing a lunar geodetic network. From an altitude of about 110 km, the resolution of the mapping camera was approximately 20 m. A panoramic camera provided even better ground resolutions, approximately 1-2 m.

Spacecraft pictures of Mars have been acquired by six United States vehicles: Mariners 4, 6, 7, and 9, and Viking 1 and 2, and several Soviet spacecraft: notably Mars 5. Mariner 4 was a flyby mission which acquired 22 television pictures with best resolution of 3 km. Mariners 6 and 7, also flyby missions, acquired approximately 60 pictures in the equatorial and southern latitudes. Both spacecraft had wide-angle (low resolution) and narrow-angle (high resolution) cameras. Best ground resolution with the narrow-angle camera was about 2 km.

The Mariner 9 spacecraft was placed into orbit around Mars. The oribital plane was highly inclined to the equator so that swathes of pictures were arranged in longitudinal strips. As the subperiapsis point on the planet slowly rotated beneath the spacecraft continuous bands of pictures were acquired. In this way almost all of the planet was mapped with a wide-angle camera at ground resolution of several kilometers. A narrow-angle camera took pictures of selected areas with best ground resolution of 100 m. The Mariner 9 spacecraft was actively taking pictures for almost an entire year.

The Viking Orbiter cameras were designed to take high resolution pictures of the Martian surface (Fig. 20.5). This capability was necessary in order to help locate a safe landing site for the two Viking landers. The best resolution of these cameras is about 40 m/pixel dimensions. The imaging system on Viking consists of two identical cameras with 475-mm lenses and a 38-mm selenium sensor. Color filters on the cameras provide the capability for broadband color imaging.

There has been one spectacularly successful space journey to the vicinity of Mercury (Fig. 20.6). In March of 1974 Mariner 10 flew by the innermost planet of our solar system. It then circled the Sun, returned for another flyby in September 1974, and a third flyby in March 1975. During these three passes by Mercury more than 3000 useful imaging frames were acquired. Approximately one hemisphere was photographed. Ten percent of the entire globe was

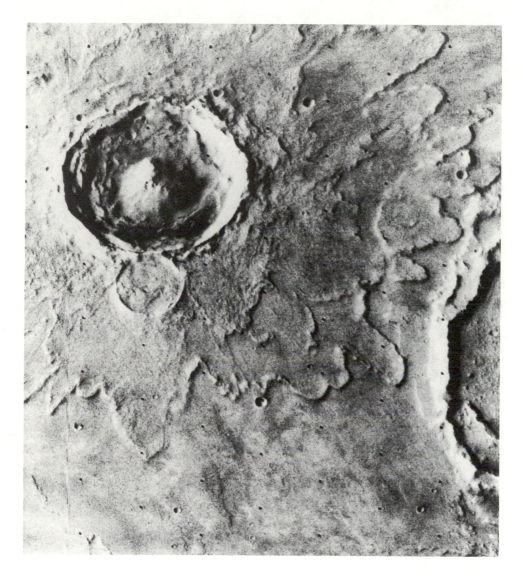

FIGURE 20.5. Viking 1 Orbiter photograph, taken on the third revolution about Mars, of a crater 18 km in diameter. While the lobate form of the ejecta blanket is typical of many Martian craters it is not observed on the Moon or Mercury.

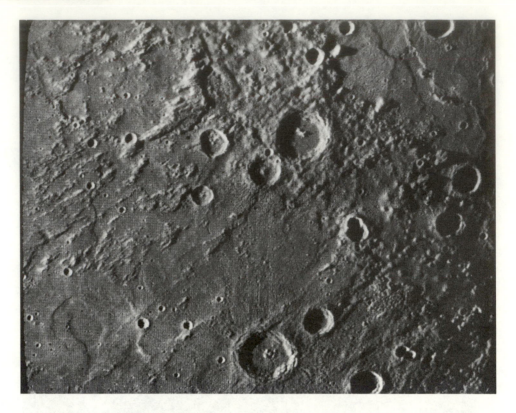

FIGURE 20.6. Mariner 10 image of a region near the Caloris basin of Mercury. The craters are similar in morphology to lunar craters.

photographed with ground resolution better than 1 km. The best resolution of near-encounter pictures approached 100 m.

Most of the cameras flown on United States spacecraft have been equipped with a variety of filters which can be interchangeably placed in the optical train. In theory, combinations of images taken in red, green, and blue light should permit construction of color pictures. In practice, very few useful color images have been constructed. One problem has to do with spatial matching. The different members of a given set are usually taken with slightly different perspectives, sometimes on different orbits. Rectification of images to remove distortion and scale effects is difficult. The second problem involves color balancing. The exact radiometric response of the vidicon is seldom known well enough to achieve appropriate blending of three images to reproduce the subtle colors which characterize the Moon and Mars.

Overlapping pairs of pictures can be used to perform stereoscopic analysis and construct topographic maps. Best results have been achieved with the Apollo mapping and panoramic cameras which were specifically designed to facilitate construction of accurate topographic maps.

20.4 SPECTRAL ANALYSIS

As discussed in Chapter 2, information can be acquired remotely throughout the electromagnetic spectrum. In studying the geology of planets other than Earth, conventional pictures taken with visible light are the primary source of data. However, other spectral measurements provide important supplementary information. Gamma-ray and x-ray surveys have been conducted from spacecraft orbiting the Moon. A gamma-ray spectrometer was onboard the Soviet spacecraft, Venera 8, 9, and 10 which descended to the Venerian surface. Ultraviolet (UV) spectrometers were carried on Mariners 6 and 7, which passed by Mars; Mariner 9, which orbited Mars; and Mariner 10, which made three passes by Mercury. Infrared surveys of the Moon, Mars, and Mercury have been accomplished with Earth-based instruments. Infrared radiometers and spectrometers were carried on Mariners 6, 7, 9, and 10 and on Viking. Radar study of the Moon, Mercury, Venus, and Mars has been carried out using Earth-based transmitters and receivers.

20.4.1 x-ray Spectrometers

x-ray spectrometers were carried on the orbital vehicles associated with the Apollo 15 and 16 missions. The instruments measured fluorescent x rays produced at the lunar surface by interaction of surface materials with incident solar x rays (4). The incoming energy causes transfer of electrons from one atomic orbital shell to another. This, in turn, results in emission of x rays which have characteristic energies for particular elements. In theory, the lighter elements, generally those with atomic number 20 or less, can be detected in this manner. Useful lunar results have been generally restricted to relative and absolute amounts of Al, Mg, and Si, although the current data are most accurately expressed as the ratios Al/Si and Mg/Si. The spatial resolution of the experiment is about 100×100 km on the lunar surface. As might be anticipated, the Al/Si ratio is lowest over the maria and highest over the uplands. There is a strong correlation between albedo and Al/Si as previously noted. This is probably caused by the higher content of plagioclase in the upland rocks. There is some suggestion that circular maria are lower in Al/Si than irregular maria. There is an inverse correlation between Al/Si and Mg/Si.

Obviously, an x-ray experiment of the type just described is inappropriate for planets with energy-absorbing atmospheres.

20.4.2 Gamma-ray Spectrometers

Gamma-ray spectrometers were included on the orbital scientific payloads of Apollo 15 and 16. The instruments measured radiation from the naturally radioactive elements Th, U, and K (5). There is a good possibility of developing the capability of detection of other elements, notably O, Si, Fe, and Mg, due to activation by cosmic rays.

Gamma-ray spectrometers have been proposed for lunar- and Mars-orbiting spectrometers. Preliminary calculations indicate the surface emission will not be severely degraded by atmospheric interactions (6).

20.4.3 Infrared Radiometers

Infrared radiometers measure the integrated thermal emission, or temperature, in selected spectral bands, characteristically chosen between wavelengths of 1 to 100 μm. The usual experimental strategy is to measure temperature at a given locality repeatedly. In this way a diurnal cycle is determined. Temperature variations are controlled by the thermal inertia of surface materials, a parameter that depends upon thermal conductivity, density, and specific heat. For most materials of geologic interest the controlling variable is conductivity, which, under high vacuum, varies by as much as several orders of magnitude between dispersed powder and rock (7).

The discussion above assumes that the temperature of surface materials is due exclusively to absorption and emission of solar radiation. In addition, if large internal heat sources exist, they could be identified by anomalously high IR emission.

An Earth-based IR map of the eclipsed Moon has been constructed by Saari and Shorthill (8). Craters that are topographically distinct in conventional pictures appear as bright spots on the IR image. As previously outlined, this measure of relatively high thermal inertia may mean that the youthful craters are surrounded by rocky ejecta whereas other parts of the lunar surface are covered by finer dust.

The orbital data have been discussed by Mendell (9). The Earth-based IR eclipse maps show that Oceanus Procellarum is slightly enhanced relative to the other maria. The near-Moon Apollo data show many small anomalies. There are cooler halos around some craters, contrary to the idea that these areas should be blocky and hence show thermal anomalies. Mendell suggests that this observation is consistent with the hypothesis that Procellarum is a relatively young mare with a thin regolith. An explanation of the thermal data is that small (100 m) craters have many meter size blocks within them, but larger craters blanket their surroundings with fine-grain rubble.

Infrared brightness temperatures were measured over a third of the Martian surface on the Mariner 9 mission (10). Temperature profiles were obtained for selected regions. No obvious correlations were discovered between albedo and thermal inertia. Derived surface particle sizes range from 0.06 mm to 0.5 cm. Internal heat sources were not detected.

On the Viking orbiter, an infrared detector designed to observe a wavelength around 1.4 μm was used to measure the amount of water in the Martian atmosphere. This instrument measures the intensity of a water vapor absorption band and allows detection of as little as 1 μm of precipitable water as vapor in the atmosphere. Data from this experiment were used to help in the selection of landing sites, and provided additional clues to the nature of the surface and its possible roughness.

The infrared Thermal Mapper (IRTM) on the Viking orbiters measured the infrared temperature of the ground along the orbital track.

20.4.4 Infrared Spectrometers

Infrared spectrometers can serve a variety of scientific purposes, depending on the spectral region investigated and the spectral resolution. As discussed in an earlier section, useful Earth-based measurements of the Moon, Mars, and Mercury have been made by reflectance spectrometers operating in the range from visible light to wavelengths of 2 or 3 μm. Outside the Earth's atmosphere experiments are not limited to the atmospheric "windows." Depressions in the reflectance curves are caused by preferential absorption related to specific mineralogy and chemistry. Electronic transitions from one orbital shell to another, electronic transfer between orbital sites of adjacent ions, and molecular vibrations are all important energy-absorbing mechanisms. The absorption related to the presence of Fe^{2+} has proven to be an especially powerful geochemical discriminator.

There was an IR emission spectrometer, operating between 5 and 50 μm, onboard the Mariner 9 spacecraft that orbited Mars (11). In the spectral range under investigation, characteristic absorption curves are displayed by several atmospheric species, notably water vapor and CO_2. The instrument successfully mapped total pressure, as well as regional and seasonal distributions of water vapor. Temperature profiles were constructed for the atmosphere. During the early months of the Mariner 9 mission, when the planet was enveloped by a dust storm, IR measurements served to characterize the dust. Brightness temperatures suggested grain sizes between 1 and 10 μm. Some absorption structure may be related to the specific silica content of the dust, but the exact mineralogic implications are obscure.

20.4.5 Gravity

The detection and interpretation of variations of the gravitational potential must be included in any treatment of remote sensing techniques. Gravity methods have been extremely useful in determining the crustal structure of the Earth and so it is not surprising that much effort has been devoted to modeling the gravitational fields of the planets.

The first complete global gravity map of a planet other than Earth was constructed for Mars using tracking data of the orbit of Mariner 9 (12). Variations in the mass distribution of the planet reveal themselves through perturbations of the spacecraft orbit from a perfect ellipse. These perturbations are detected in the Doppler shift of the radio signal from the spacecraft as it speeds up and slows down in the line of sight direction from Earth. Conventional geophysical modeling allows construction of crustal models consistent with the data (13). These techniques earlier led to the discovery and explanation of lunar mascons (14, 15). The complete lunar gravity map was constructed later by a complex analysis of many orbits from many different spacecraft (16, 17). The horizontal resolution of gravity determined from spacecraft orbits is approximately equal to the height of the spacecraft above the surface.

20.4.6 Radar

All the remote sensing techniques just described can be classified as passive. The recorded energy is reflected or emitted naturally from the target materials. In contrast, radar instruments are active in the sense that a microwave signal is transmitted, and the characteristics of the returned echo are then observed. Radar instruments operate in the general wavelength range between 1 and 100 cm. Almost all measurements to date have been made with Earth-based instruments. The desirability of placing a radar instrument on a Venus-orbiting spacecraft in order to "see" beneath the dense cloud cover has long been recognized, but the weight and power requirements are difficult to overcome. A "lunar sounder" was operated from orbit during the Apollo 17 mission. This instrument, operating at frequencies of 5 MHz, 15 MHz, and 150 MHz, has the ability to make elevation profiles, probe several hundred meters below the surface, and image the scene with two-dimensional displays (18, 19). The concept of an orbiting imaging radar for planetary exploration has recently been given considerable impetus by the success of the Seasat imaging radar experiment in imaging land features.

Several types of radar measurements have been employed with Earth-based instruments. The simplest technique is to measure the elapsed time between transmission of a signal and reception of an echo from the sub-Earth point on the target planet. As the planet rotates an elevation profile can be constructed along a latitudinal belt. This has been a particularly useful technique for delineation of Martian topography (20).

It is also possible to look at the strength of the back-scattered echo which is controlled by many variables, including surface roughness at the scale of the radar wavelength and dielectric constant (a measure of capacity to store electric charge). Low dielectric constant usually indicates fine-grained, porous materials. Using this approach Evans and Hagfors (21) determined that the Moon was covered with a regolith of sand to boulder-sized material. This prediction essentially was confirmed by Apollo investigations. Analogous study of back-scattered echoes from Mars has been puzzling. Many regions have extremely low dielectric constant and, by inference, might be dispersed sediment with insufficient strength to support a landed vehicle.

By examining progressively delayed echoes along concentric annuli moving out incrementally from the sub-Earth point on the target it is possible to create a radar brightness map (Fig. 20.7). Such images superficially resemble conventional photographs (22). Radar images of Venus, even though having low spatial resolution, are particularly valuable since the Venerian surface is obscured from visible observation (23). Circular bright spots on the radar images suggest the presence of large impact craters. There is a variety of landforms visible in the recent Venus radar images (Fig. 20.8).

20.5 STATUS OF PLANETARY REMOTE SENSING

Much is now known about the planets, asteroids, and satellites, but many important questions remain. The following is a brief summary of what has been learned and some of the questions that can be identified at present as those with the highest priority.

20.5.1 Mercury

Mercury is the last terrestrial planet to be explored by unmanned spacecraft from Earth. It is nearest the Sun and has the highest density among the terrestrial objects, an observation consistent with current theories of compositional gradients in the early solar nebula. Mercury has a magnetic dipole field. The planet is seen to be strikingly Moon-like in the pictures returned by Mariner 10. However, detailed analysis of the Mariner 10 pictures by the investigators shows that the Mercurian crater morphology differs in significant ways from that of lunar craters as a consequence of Mercury's greater gravitational attraction. The surface gravity of Mercury is nearly the same as that of Mars. Although the mean density of Mercury is in close agreement with theories of planetary formation, the composition is, in fact, unknown. It is probable that Mercury is strongly differentiated as are the other planets, with a refractory-rich crust and a massive iron core. But the partitioning of elements that provide a continuing source of energy through radioactive decay is not known. This factor bears on whether Mercury can have the same types of volcanic activity either early or late in its history as do other terrestrial planets. Little is known of the time history of the events that shaped the surface of Mercury. There are parallels with the Moon in crater density and morphology that suggest that Mercury may have experienced a history of early bombardment followed by volcanism much as did the Moon, but the time history of events on Mercury is not known.

20.5.2 Venus

Venus is similar to the Earth in mass, radius, and position in the solar system. Because of it massive cloudy CO_2 atmosphere little else is known. Earth-based radar images have provided a glimpse of about ten percent of the surface revealing a variety of landforms. In the near future the resolution capability of Earth-based radar is expected to improve enough to allow a much better assessment of whether some of these features are impact craters, volcanoes, or tectonic in origin. The temperature at the surface of Venus is 750°K (nearly 900°F), precluding the possibility that there can be liquid water. Past conditions, however, may have permitted water to exist as a liquid. Some of the features seen in

MOSAIC OF POLARIZED 70CM RADAR MAPS OF THE MOON (CONTINOUS)

FIGURE 20.7. Mosaic of polarized 70-cm radar images of the Moon. Brightness variations are due, in part, to topography and, in part, to surface roughness on the scale of the radar wavelength. Images were obtained at the Arecibo observatory by Thomas W. Thompson.

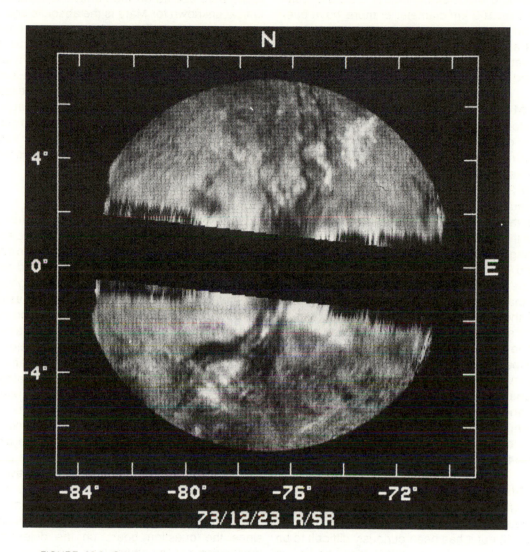

FIGURE 20.8. Small portion of the Venerian surface imaged at 12.6 cm at the Goldstone facility (23). The black band is a region near the radar equator where the data cannot be used. Notice the broad channel-like feature running from top to bottom of the image.

the radar images are morphologically similar to impact craters, especially the highly eroded craters on the Moon and Mars. Since impact craters of comparable size, 100 km diameter or more, have been erased on the Earth, positive identification of these features as impact craters would indicate a lower erosion rate and less pervasive tectonism on Venus compared to Earth.

20.5.3 Moon

The Moon has been the subject of far more study than any other planetary object and consequently we have many more unanswered questions. Several books have been written about the Moon (for example, 24, 25) so we shall not attempt a full review of the results of lunar remote sensing, but will outline some of the outstanding problems that remain. The origin of the Moon remains unknown and few of the competing hypotheses have been severely compromised. For a discussion of hypotheses of formation of the Moon see the book by Taylor (25). The shape, that is, global topography, and the gravity field of the Moon are not known well enough to definitely solve many interesting geophysical problems. The chemistry and mineralogy of the lunar surface materials as determined by the Apollo lunar sample investigators need to be extended over the entire lunar surface. This will help resolve the when, where, and how of magmatic differentiation and volcanism on the Moon. The precise characterization and determination of the origin of the lunar magnetic field remains to be accomplished.

20.5.4 Mars

Mars after Viking remains the most enigmatic of the planets. Again, books have been written on the subject (26) but it has been curiously difficult to be definitive and consolidate what is felt to be known about the geology of Mars. With each successive mission to Mars it has almost appeared that we were not returning to the same planet. The Viking mission was no exception. An almost incredible array of features not previously seen in enough detail to remark on have been photographed by the Viking orbiters. For many of the features, there are no known analogs on the Earth or the Moon, and each new photograph seems to generate new questions. The lander images are not so puzzling. In fact, the scene appears vaguely familiar to anyone who knows the desert. It is a wind-swept barren landscape strewn with small dune and angular rocks of all

sizes. The rocks have a thin reddish coating, which indicates that at some time, perhaps at present, the rocks were altered on the surface to limonite. The critical unknown for Mars is the absolute age of the geologic features. There are enormous channels that may once have held water. Olympus is the largest volcano that has been seen on any planet. Hellas appears to be the product of a massive impact by an asteroid. The age of none of these features is known. It is conceivable, indeed probable, that none are much less than one billion years old. This problem can only be solved by returning samples from the surface of Mars to laboratories on the Earth.

20.5.5 Asteroids

The Asteroids have only been studied from Earth by telescopic examination of sunlight reflected from their surfaces. Sizes and shapes have been inferred from the total amount of sunlight reflected and the way it varies as the asteroid tumbles through space. The reflectance spectra of about 100 asteroids have been reported. For detailed discussion see the work of McCord and Chapman (27). Many asteroids have spectra similar to the laboratory determined spectra of meteorites. Since particular families of asteroids can be determined to be mineralogically similar to a meteorite specimen, it seems reasonable to assume that many meteorites are in fact samples of the asteroids. Many interesting questions have evolved concerning the asteroids as more has been learned. Some meteorites and the asteroids that resemble them appear to have been melted and differentiated, while others are still in a primitive chemical and petrological state. The source of energy for this melting is of great interest since the question bears on the differentiation mechanisms of the other planets. An additional impetus for correlating meteorites with asteroids is that primitive meteorites may be similar to the initial material from which the planets were formed. The initial composition of the planets is a vital piece of information for models of their subsequent development.

20.5.6 Satellites

The Satellites of the giant planets Jupiter, Saturn, and Uranus have been studied from Earth by the same techniques that were used to analyze the asteroids. Images of the Jovian satellites recently obtained by Voyagers 1 and 2 reveal the four major satellites of Jupiter to be strikingly different from any of the other terrestrial planets. Io with its continually erupting volcanoes is unique, while Calisto appears to be the most highly cratered object in the Solar System. There are a number of curiosities among the satellites of the outer planets that make them worthy of study. Much of the interest stems from the fact that they offer the only solid surfaces between the asteroids and Pluto. Some of the satellites are bright because they are covered with ices of various gases, not necessarily water. Io is also bright but it has recently been suggested that it may also contain various salts, and abundant sulfur compounds. Titan has an atmosphere that rivals Earth's, although it is methane rich and highly reducing, unlike the relatively inert atmospheres of the Earth, Venus and Mars. Conditions on Titan are highly favorable for the formation and preservation of organic compounds. Although we now have images of the surfaces of these bodies, for a long time to come an important data source will be their reflectance spectra as seen from Earth.

20.6 POSSIBLE FUTURE EXPERIMENTS

20.6.1 Venus Imaging Radar

A future mission to orbit Venus may include a radar system. Such systems could operate from a controlled platform on a three-axis stabilized vehicle as a side-looking system or perhaps a system will be designed that can operate by sweeping out from a spin stabilized vehicle. Either way, radar is the only means of obtaining high resolution images of Venus, the planet most similar to Earth in mass and size. The expected ground resolution element could be as small as 20 m.

20.6.2 Radio Altimeter

Included on the Pioneer Venus Orbiter is an altimeter that provides high resolution altimetry along the ground track of the orbit. An instrument similar in concept may be flown on a future lunar orbiter (LPO or Lunar Polar Orbiter). Although a radar imaging device can be operated in an altimetry mode it is difficult to optimize the instrument for both altimetry and imaging. Incidentally, the LPO instrument package as presently envisioned would be adaptable to any of the terrestrial planets and would include many of the instruments mentioned in earlier sections.

20.6.3 Mars Surface Sample Return

A future experiment that rates very high among many planetary scientists is the return of a sample of soil and rocks from the surface of Mars. The amount of effort already expended on just thinking about this one is probably the equivalent of one skilled scientist thinking about all aspects of the problem for about 10 years. Of course, this work was spread among dozens of scientists and engineers over several years. The importance of this sample to remote sensing is seen by the great impetus given to lunar remote sensing by the acquisition of ground truth during the Apollo missions. And, of course, the whole concept of ground truth was developed during remote sensing of the Earth whereby it is possible to extend the detailed analysis of a small ground region over a much larger area using remotely sensed data. The number of laboratory experiments that would be performed is enormous, now that scientists have gained experience from the lunar program of working with very small irreplaceable samples. The most serious objection to collecting a sample from Mars is that no one can guarantee that we will not bring back and accidently release what at the same time gives the whole idea much of its appeal—Martian organisms

20.7 FUTURE TRENDS

Even a cursory reading of the above reveals some problems that may be holding up progress in extraterrestrial geology. Most obvious is that the data should be correlated and the parts related to the whole. Before that can be accomplished, the data need to be made available in some uniform format. Most of the data are available in some form from the National Space Science Data Center in Greenbelt, Maryland. The National Aeronautics and Space Administration has been pursuing a vigorous program in data analysis and synthesis. Some true synthesis involving several data sets has resulted (28, 29). However, many important data sets have yet to be fully reduced and made available in a form useful to other scientists who are not expert in the particular field. Examples are x-ray and gamma-ray spectrometry and lunar sounder data. But virtually every spacecraft and Apollo experiment suffers from a greater or lesser lack of meaningful expression and availability to other experimentors.

Eventually, of course, all the data will be incorporated into the geologic models of the planets. Much of the problem may be circumvented in the future. Those who design and plan future missions are speaking of "end-to-end" data systems, for example. While such concepts mean different things to different people, the goal is to have future experiments use common data formats and, as far as possible, design the experiment to be compatible with data reduction by ground computers using common input and output formats. The result would be to efficiently and quickly get the data into archives that are available to users. Not all experiments can be easily handled by a common system, but reduction methods will be adopted so that the final product is in a format consistent with other data types.

The exact form of the final products is not yet decided, but some of them can be anticipated because they are more useful. Radar images, photographic images, silicon imaging photometry data, topography, infrared eclipse cooling, and albedo have been put into the LAC (Lunar Aeronautical Chart) format. A subset of these data types are available on Mercury and Mars charts. These provide a useful means for intercomparison of the data sets. Unfortunately, maps are not suitable for computer analysis, so another approach that is being experimented with is to create digital arrays of data for each set for the various planets. Lunar data are put in arrays of 30 data points per degree. This is equivalent to one data point per kilometer. Lower resolution data are sampled at the appropriate interval. Higher resolution data can be smoothed and resampled for comparison with the low resolution data sets. It may be several years before this becomes a reality, but eventually all suitable "planetary geodata" can be put in a computer compatible image format. The advantage of this treatment is that many data sets can be efficiently stored and recalled. Programs exist to analyze data in this format and reveal correlations between two or many sets.

Remote sensing for studies in extraterrestrial geology has been advanced more in the last two decades than ever before in the long history of man's curiosity about the universe beyond the Earth. Many myths, such as the canals as evidence of intelligent life on Mars, have been disproved. The uncovering of myths that long passed as scientific truth serves as an objective warning that we may even now accept some hypotheses that are little more substantial than the canals of Mars. A more tangible benefit of planetary exploration is the better understanding of the Earth that results from study of bodies that have experienced other courses of development. Each planet is unique, but the ultimate hypothesis for the origin, evolution, and fate of the Earth must also take into account the other planets. Although we have made much progress, we are today barely beginning the most important and exciting phase of exploration that humankind has ever attempted.

REFERENCES

1. McCord, T. B., Pieters, C., and Feierberg, M., 1976, Multispectral mapping of the Lunar space using groundbased telescopes: Icarus, v. 29, no. 1, p. 1-34.
2. Veverka, J., and Noland, M., 1973, Asteroid reflectivities from polarization curves: calibration of the slope-albedo relationship: Icarus, v. 19, p. 230-239.
3. Dollfus, A., and Geake, J., 1975, Polarimetric properties of the lunar surface and its interpretation: part 7 — other solar system objects, in Proceedings lunar science conference 6th: New York, Pergamon Press, p. 2749-2768.
4. Adler, I., Trombka, J., Gerard, J., Lowman, P., Schmadebeck, R., Blodget, H., Eller, E., Yin, L., Lamothe, R., Osswald, G., Gorenstein, P., Bjorkholm, P., Gursky, H., Harris, B., Golub, L., and Harnden, F. R., Jr., 1972, Apollo 15 geochemical x-ray fluorescence experiment: preliminary report: Science, v. 175, p. 436-440.
5. Metzger, A., Trombka, J., Reedy, R., and Arnold, J., 1974, in Proceedings Lunar Science Conference 5th: New York, Pergamon Press, p. 1067-1076.
6. Metzger, A., Parker, R., Arnold, J., Reedy, R., Trombka, J., 1975, Preliminary design and performance of an advanced gamma-ray spectrometer for future orbiter missions, in Proceedings Lunar Science Conference 6th: New York, Pergamon Press, p. 2769-2784.
7. Wechsler, A., Glaser, P., and Fountain, J., 1972, Properties of granulated materials, in Thermal characteristics of the Moon, Lucas, J. W., ed.: Cambridge, Mass, MIT Press, p. 215-242.
8. Shorthill, R., 1973, Infrared atlas charts of the eclipsed Moon: The Moon, v. 7, p. 22-45.

9. Mendell, W., 1975, Infrared orbital mapping of Lunar features, *in* Proceedings Lunar Science Conference 6th: New York, Pergamon Press, p. 2711-2719.

10. Kieffer, H., Chase, S., Jr., Miner, E., Münch, G., Neugebauer, G., 1973, Preliminary report on infrared radiometric measurements from the Mariner 9 spacecraft: Jour. Geophys. Res., v. 78, p. 4291-4312.

11. Conrath, B., Curran, R., Hanel, R., Kunde, V., Maguire, W., Pearl, J., Pirraglia, J., Welker, J., and Burke, T., 1973, Atmospheric and surface properties of Mars obtained by infrared spectroscopy on Mariner 9: Jour. Geophys. Res., v. 78, p. 4267-4278.

12. Sjogren, W., Lorell, J., Wong, L., and Downs, W., 1975, Mars gravity field based on a short-arc technique: Jour. Geophys. Res., v. 80, p. 2899-2908.

13. Phillips, R., and Saunders, R., 1975, The isostatic state of Martian topography: Jour. Geophys. Res., v. 80, p. 2893-2898.

14. Muller, P., and Sjogren, W., 1968, Mascons: Lunar mass concentrations: Science, v. 161, p. 680-684.

15. Phillips, R., Conel, J., Abbott, E., Sjogren, W., and Morton, J., 1972, Mascons: progress toward a unique solution for mass distribution: Jour. Geophys. Res., v. 77, p. 7106-7114.

16. Ferrari, A., 1975, Lunar gravity: the first farside map: Science, v. 188, no. 4195, p. 1297-1300.

17. Ananda, M., 1975, Farside lunar gravity from a mass point model, *in* Proceedings Lunar Science Conference 6th: New York, Pergamon Press, p. 2785-2796.

18. Phillips, R., Adams, G., Brown, W., Jr., Eggleton, R., Jackson, P., Jordan, R., Peeples, W., Porcello, L., Ryu, J., Schaber, G., Sill, W., Thompson, T., Ward, S., and Zelenka, J., 1973, The Apollo 17 Lunar sounder, *in* Proceedings Lunar Science Conference 4th: New York, Pergamon Press, p. 2821-2831.

19. Peeples, W., Sill, W., May, T., Ward, S., Phillips, R., Jordan, R., and Abbott, E., 1978, Orbital radar evidence for lunar subsurface layering in Maria Serenitatis and Crisium: Jour. Geophys. Res., v. 83, p. 3459-3468.

20. Pettengill, G., Counselman, C., Rainville, L., and Shapiro, I., 1969, Radar measurements of Martian topography: Astron. Jour., v. 74, p. 461-482.

21. Evans, J., and Hagfors, T., 1968, eds., Radar astronomy: New York, McGraw-Hill, 620 p.

22. Rumsey, H., Morris, G., Green, R., and Goldstein, R., 1974, A radar brightness and altitude image of a portion of Venus: Icarus, v. 23, no. 1, p. 1-7.

23. Goldstein, R., Green, R., and Rumsey, H., 1976, Venus radar images: Jour. Geophys. Res., v. 81, p. 4807-4817.

24. Mutch, T., 1972, Geology of the Moon, a stratigraphic view, 2nd ed.: Princeton, N.J., Princeton University Press, 324 p.

25. Taylor, S., 1975, Lunar science: a post-Apollo view: New York, Pergamon Press, 372 p.

26. Mutch, T., Arvidson, R., Head, J., Jones, K., and Saunders, R., 1976, The geology of Mars: Princeton, N.J., Princeton University Press, 400 p.

27. McCord, T., and Chapman, C., 1975, Asteroids: spectral reflectance and color characteristics: Astrophys. Jour., part I, v. 195, p. 553-562; part II, v. 197, p. 781-790.

28. Thompson, T., Shorthill, R., Whitaker, E., and Zisk, S., 1974, Mare Serenitatis: a preliminary definition of surface units by remote observations: The Moon, v. 9, p. 89-96.

29. Moore, H., Tyler, G., Boyce, J., Shorthill, R., Thompson, T., Wilhelms, D., Wu, S., and Zisk, S., 1976, Correlation of photogeology and remote sensing data along the Apollo 14, 15, and 16 bistatic-radar ground tracks, part II—a working compendium, U.S. geological survey, interagency report: Astrogeology, 80 (Open File Report 76-298).

30. Shevchenko, V., 1974, The surface prevalence and stratigraphy of lunar rock from its albedo, *in* Proceedings Soviet-American conference on the cosmochemistry of the Moon and planets: NASA TT F-16, 033.

31. Kuiper, G., Whitaker, E., Strom, R., Fountain, J., and Larson, S., 1967, Consolidated lunar atlas: Air Force Cambridge Research Laboratories, Office of Aerospace Research, Published by the Lunar and Planetary Laboratory, Univ. of Arizona.

CHAPTER 21

GEOLOGICAL REMOTE SENSING IN THE 1980S

ALEXANDER F. H. GOETZ

21.1 INTRODUCTION

Authors who make predictions about the future of technology are rarely right. The only recourse an author has is to set forth the ground rules for making a prediction and convince his readers that at the time he made the prediction he did it for the right reasons. The readers are asked to sympathize with his plight because they have the advantage of 20/20 hindsight. This author is no exception to the rule, and he fully expects some of the predictions to be incorrect by the time this book is published. Nonetheless, certain trends have evolved in the science of remote sensing for geologic applications which will continue into the foreseeable future, and the identification of these trends is the basis for this chapter.

The aim of remote sensing in this discipline should be to make better geologic maps. The job in the future will be to continue to make better geologic maps.

Until now the greatest contributions of remote sensing to geologic mapping, both from aircraft and satellite, have been in structural interpretations, both local and regional. The synoptic view from satellite photography has been particularly important in regional interpretation. Much less success has been obtained in directly mapping lithologies, the other key component of a geologic map. Some notable exceptions have been the successes in identifying areas of hydrothermal alteration from computer enhanced Landsat photography (1). However, in general, direct lithologic and mineralogic identification has been hampered by the lack of uniqueness in the reflection spectra of rocks in the regions covered by color photography, color infrared photography, and the Landsat MSS.

The remote sensing geologists now find themselves in a position similar to that of the astronomers in the early 1960s. The astonomers were on the verge of an explosion in their understanding of the universe because of the new vistas that opened up through extension of the wavelength spectrum for measurement. Radio astronomy was just coming into its own and the discovery of quasars had a major impact on astronomical thinking. At the same time, astronomers were using newly available infrared detectors to extend their vision to the center of the galaxy and it led to the discovery of newly forming stars. Likewise, the extension of the spectrum to the ultraviolet, and eventually x rays, by use of satellite telescope systems, had a profound effect on astronomical theories. While the effects of extending the useful spectrum may not be as dramatic for the geologists as it was for the astronomers, certainly new insights will be gained, and new data sets with concomitant interpretation problems will become available.

In general, it can be expected that new data gathering techniques in new wavelength regions will facilitate better mapping of mineralogy and lithologies as well as expose new anomalies. New understanding of such anomalies, as secondary or tertiary surface manifestations of subsurface phenomena, is one of the major goals for a researcher in this discipline.

On the more mundane side, the next decade will see remote sensing take its logical place in the roster of tools and techniques that the geologist will use in his everyday work. The maturation and acceptance of remote sensing in geology will play as important a role in the advancement of the discipline as will the results expected from the new technologies.

The potential evolution of remote sensing over the next decade can best be described by separating the elements into two major categories—data collection and data interpretation.

21.2 Data Collection

The last decade has seen enormous advances in the art of data collection both in quality and in quantity. The most important development has been the move away from film systems for data acquisition to multispectral scanner (MSS) electronic systems. The advantages of the electronic approach were made clear to everyone in the remote sensing field with the advent of ERTS-1. The sheer quantity of data, returned at a rate of 15 megabits per second (Mbps), made it obvious that computer image processing techniques were mandatory and that the MSS data, having spectral band information in registered, digital form, were ideal for this analysis. The extraordinary usefulness of Landsat band 7, covering the range 0.8-1.1 μm, brought home the need for images at wavelengths outside the sensitivity range of photographic film and whetted the appetite of those desiring images at other wavelengths.

21.2.1 The Spectrum

Figure 21.1 shows the portion of the electromagnetic spectrum of most interest to remote sensing, plotted on a logarithmic scale. On this figure are plotted the regions of high atmospheric transmission through which remote sensing measurements can be made from both aircraft and spacecraft. In addition, the spectral region for which we already have a large amount of data is heavily hatched while those areas for which there is interest and some data are lightly hatched, and those

regions for which we have little or no reflectance or emittance data have been left blank. Figure 21.1 demonstrates how restricted in wavelength thorough remote sensing studies have been. The past concentration on this small wavelength region has not been because of a lack of technology but rather because of a lack of understanding of the value of these wavelength regions and the lack of resources to make significant studies in regions other than the visible and near infrared. Much of the resources have been concentrated on the obtaining and analyzing of the Landsat data and it is for this reason that so little has been done at other wavelengths.

What is it that we can expect to find at other wavelengths when the resources become available to apply to the problems? A number of different types of data can be obtained depending on both spatial and spectral resolution. Given sufficiently high spectral resolutions (bandwidths on the order of one hundredth of the wavelength) some direct mineralogical identification can be made. In the 1 to 2 μm region some of the minerals in mafic rocks, such as pyroxenes, can be identified. In the 2 to 2.5 μm region overtone bending-stretching vibrations (compare Chapter 2) strongly affect the shape of the reflection spectrum of many minerals. In particular, hydrous minerals, such as clays, exhibit unique spectra. The region from 3 to 5 μm is a

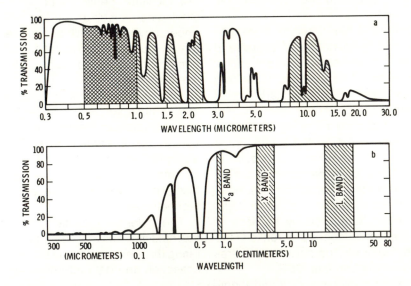

FIGURE 21.1. Nominal atmospheric transmission as a function of wavelength, extending from the ultraviolet to the microwave. The heavily hatched area is the spectral region already well studied by spacecraft, aircraft, and on the ground. Lightly hatched regions denote areas for which there is interest at present and some data. Those regions for which there is little or no reflectance or emittance data have been left blank. Fig. 1a adapted from Ref. 8 and Fig. 1b from Ref. 9 and 10.

particularly difficult portion of the spectrum from which to obtain data because of the very small amount of energy emitted by the Earth's surface. For this reason no field spectral measurements have been made. However, this region should be very rewarding because of the plethora of bending-stretching spectral features to be found. In particular, the sulfates and carbonates exhibit spectral features in this region. Unfortunately, atmospheric water and CO_2 absorptions strongly affect this region, adding to the difficulty in making measurements here.

The 8 to 14 μm atmospheric window has received considerable attention in the past, particularly since the advent of cooled semiconductor detectors. In this region the basic stretching vibrations in silicates play an important role in the emission spectrum of the Earth's surface. Unfortunately, the physical state of materials, such as particle size, plays an important part in affecting the spectral contrast observed in this region. Future work will concentrate on obtaining realistic emittance spectra in the field, and the development of multispectral imaging systems in this wavelength region. An increased use of thermal inertia images (Chapter 8) that exploit this region of the spectrum can be expected.

The region 16 to 25 μm has not been studied in any detail for remote sensing potential. Laboratory spectra demonstrate interesting silicate mineralogy identification potential, but this region suffers from the lack of energy emitted, as well as the difficulties with varying atmospheric absorption. Beyond 25 μm no significant work has been done on the use of passively emitted energy for lithologic determination.

In the microwave region beyond 0.8 cm emission measurements have been made but no significant uses for lithologic identification have been reported. The region beyond 0.8 cm has been explored to some extent with active systems. Three radar systems, K_a-band at 0.85 cm, X-band at 3 cm and L-band at 23 cm, have been used mainly as a synthetic light source for imaging the Earth's surface. Recently attempts have been made to use the scattering characteristics or the spectral information for lithologic discrimination purposes (2). Radar in this region has potential for mapping microscale geomorphologic differences that may be indicative of lithology. In the 1980s it is expected that major emphasis will be placed on this region and such properties as polarization and spectral scattering will be exploited. Passive measurements in this region will probably be limited to the determination of roughness, thermal inertia, and soil moisture.

21.2.2 Technology

Advances in technology over the next decade are the key to expanding the capabilities in remote sensing data collection. The requirements for higher spatial resolution, higher spectral resolution, and more spectral bands in imaging systems of the future are dependent on new technologies in three areas:

1. Data transmission
2. Detector arrays
3. Wavelength discriminators

The demand for higher quality data collection systems requires all three of the above technologies since any improvement in one will affect the demands on the other technologies in the data collection system. An increase in spatial resolution, a requirement that can be expected to escalate during the 1980s, requires higher data recording rates on the ground and in aircraft. Fig. 21.2 shows what can be expected from satellite transmission systems in the 1980s and beyond. This figure was taken from a NASA forecast document (3) which discusses, among other things, data transmission rates from synchronous orbits to Earth by microwave channels. Throughout the report, forecasts are split into two groups: "What will be" and "What is possible."

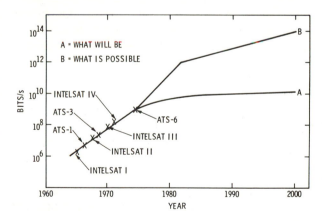

FIGURE 21.2. Data transmission rates from synchronous orbits to Earth by microwave channels. This curve, from Ref. 2, depicts "What is possible" in the way of data rates if significant advances in data compression are achieved and new high frequency allocations are made. "What will be" anticipates developments in the use of millimeter bands and specialized antennas.

Curve *A* in Fig. 21.2 shows "What will be" given developments in the use of millimeter bands and specialized antennas. The "What is possible" curve requires significant advances in data compression, requires new high frequency allocations, and is limited by the carrier frequency. "What will be" in the next decade is a data rate of approximately 10^{10} bps. At that data rate it will be possible to transmit the entire Bible, both Old and New Testaments, 300 times per second. In terms of the present Landsat MSS, which has a data rate of 15 Mbps, the resolution could be increased by approximately a factor of 25 to produce an image with an instantaneous field of view of 3 m.

In order to be able to utilize the transmission channel to its fullest capacity to realize the potential 3-m resolution from a Landsat orbit requires an advance in the present day detector technology. The Landsat MSS contains an array of six detectors per channel. These detectors are scanned across the scene perpendicular to the spacecraft track. An increase in resolution means that less light will fall upon the detector for a shorter length of time. Increasing the size of the telescope is required to obtain more light but another technique is required to offset the decreasing residence time of a particular point on the Earth sensed by the detector. The decreased residence or integration time for the detector means that the signal-to-noise ratio is decreased. By increasing the number of detectors simultaneously viewing the Earth, the signal-to-noise ratio can be increased by the square root of the number of detectors. Imaging systems employing linear detector arrays are called "pushbroom scanner" systems. Silicon detector arrays up to 2000 elements in length are presently available and in the near future undoubtedly larger arrays will be built.

The next step in increasing the potential signal-to-noise ratio is to use area arrays. A charge-coupled device (CCD) area array of 800 × 800 elements now exists and larger arrays can be expected in the future. The ultimate limitation lies in the speed at which the data can be read off from the detectors and in the requirement for image motion compensation, which becomes more significant as the resolution is increased.

As imaging systems are designed for wavelengths beyond 1 μm, materials other than silicon must be used. Intensive research is being carried out on the construction of detector arrays of mono-lithic indium-antimonide as well as hybrid CCD systems. In the next decade the detector systems will revolutionize imaging at longer wavelengths. Presently only broadband imaging systems are possible because of signal-to-noise constraints. Identification of minerals by their bending and stretching mode vibrations will require narrow wavelength band systems. Linear and area array detectors are required to produce data with high enough signal-to-noise ratios for proper interpretation.

Wavelength discrimination systems presently used for imaging include interference filters, prisms, and gratings. The latter two systems have the disadvantage that slits are required to establish wavelength resolution. Narrow band interference filters require temperature control in high resolution applications. They also have the disadvantage that they cannot be tuned to any other wavelengths than for which they were designed. A recently developed wavelength discrimination element called a Tunable Acoustical Optical Filter (TAOF) (4) holds promise to provide continuously tunable high resolution wavelength filtering for line or area array detectors. In the 1980s we will see new applications developing for such high resolution, versatile devices.

21.2.3 Techniques and Instrumentation

When higher data rates, better detectors, and better wavelength discrimination devices become available, then a number of present day laboratory techniques can be adapted to aircraft and satellite use. As discussed above, narrow spectral interval imaging will become practical and, for instance, discrimination among such minerals as montmorillonite and kaolinite, which have absorption features near 2.2 μm, will be possible. For this application, bandwidths of 0.02 μm or smaller will be required.

Active imaging systems at short wavelengths, using tunable dye lasers are now under development. If tunable laser systems for longer wavelengths become a reality, systems might be developed which have the capability of detecting specific mineral and rock types. Active systems have the advantage in that they are not dependent upon solar energy in the reflected portion of the spectrum and high signal-to-noise ratios are potentially achievable. Another active system presently under development uses an ultraviolet laser system to induce fluorescence at the surface. While the present development is aimed at detecting crude oil types, fluorescence spectra and decay times might well be used for lithologic discrimination. Much laboratory work is still required to verify the usefulness of the fluorescence method. Measurement of luminescence in vegetation and rocks induced by sunlight has been carried out on a point-by-point basis, and recently an imaging Fraunhofer Line Detector has been constructed. Passive systems such as this hold promise for adaptation to orbiting vehicles.

Higher resolution images from satellites will become available in the next decade. The ultimate resolution will not be limited by technology but most likely by national security considerations. Worldwide stereo coverage is feasible in the 1980s. It could be provided by Shuttleborne film systems or by a free-flying satellite, possibly using a pushbroom scanner system. Stereo data are important in photo interpretation and to this date coverage from orbit has been spotty. Skylab provided most of the stereo data presently available.

21.2.4 Satellite Systems

With the advent of Shuttle, a great opportunity exists in the next decade to develop and test new orbital remote sensing systems for geologic applications. When these experimental systems reach maturity they can be flown aboard free-flyer systems launched from Shuttle into polar orbit to obtain the worldwide coverage not available with one seven- or thirty-day Shuttle mission. Table 21.1 is a listing of satellite systems to be launched through the early 1980s. Only those satellite systems are listed that have applications to geological remote sensing.

OFT stands for Orbital Flight Test of Shuttle. Five test flights will carry aloft some instrumentation. In particular an imaging radar system borrowed from Seasat and a multispectral radiometer (SMIRR) for testing spectral bands for future imaging systems will be flown. In the 1980s Spacelab will be borne aloft by Shuttle several times a year. Cameras and other types of imaging systems are expected to be aboard Spacelab flights and Spacelab will act as a test bed for new systems to be flown later as free-flyers.

Landsat-D and in particular the thematic mapper, will provide 30-m IFOV data with the same aerial coverage as Landsats 1, 2, and 3. Of particular interest to geologists is the increased resolution capability as well as the fact that the thematic mapper will contain a channel in the 1.55 to 1.75 and 2.08 to 2.36 μm region. These bands potentially will provide a large amount of information on lithological differences since in this region the rock reflectances vary from 8 to 80%, twice the range found in the region 0.4 to 1.0 μm. The 2 μm band will be valuable for the detection of hydrous mineral phases.

Stereosat is envisioned as a free flying satellite in sun-synchronous orbit utilizing a three-camera electronic pushbroom scanning system. Two cameras are positioned 26° off nadir fore and aft and the third camera is pointed at the nadir. This configuration yields two different base/height ratios of 1.0 and 0.49. Horizontal and vertical resolution at 1.0 base/height is approximately 15 m. The data will be compatible with Landsat to such an extent that multispectral stereo pairs can be made using data from both systems.

21.3 INTERPRETATION

The flood of new data predicted in the previous sections will go unused if an equally serious effort is not mounted to increase our capabilities to interpret the data. The types of analysis and interpretation fall into three categories: (1) sample studies to understand the relationship betwen reflectance and emittance signatures and the *in situ* properties of materials; (2) image processing and image display; and (3) modeling studies in both the sense of understanding the effect of the environment on measurables, as well as determining the surface indicators of buried resources.

21.3.1 Sample Studies

The success of geologic remote sensing in the 1980s will be in its ability to detect and identify materials of interest. Some work has been carried out in identifying spectral reflectance characteristics of *in situ* rock and soil samples (5). However,

TABLE   21.1

Mission	Probable Launch Date
OFT (Radar and SMIRR)	1981
Spacelab	1982+
Landsat-D	1982
Stereosat	1985

much remains to be done to quantify the spectral signatures. This will require a detailed mineralogical and geochemical analysis as well as the proper application of crystal field theory to the data.

A similar set of studies is necessary to understand the effect of vegetation species and vigor on remote sensing measurements. The problem is difficult and therefore has not been studied in as great a detail as the spectral reflectance of rocks and soils. The reflectances and emittances of both vegetation and other Earth surface materials need to be measured at wavelengths of $8 \mu m$ and beyond. In particular the region 16 to 25 μm has not been studied at all, nor do we understand very thoroughly the response of materials to incident radiation at 1 cm wavelength and beyond. Quantitative data in these regions will steer the application of new technology and instrumentation for remote geologic mapping.

Evaluation of these spectral signatures will be made by statistical techniques to determine which spectral bands contain unique information for separating a material of interest from those in juxtaposition.

21.3.2 Image Processing

There should be no question now that geological remote sensing is moving out of the era of direct photo interpretation of unenhanced photography and into the era of computer manipulation of image data. The present systems simply generate too much data for the photo interpreter to use directly. Image enhancement, as described in Chapter 5, can display the most useful subset or subsets of image data in a form more directly usable by the geologist. The types of data and the quantities of data to be produced by new instruments in the next decade will most certainly require computer image processing. Some of the problems and limitations that presently exist with computer systems must be overcome if the anticipated data quantities are to be handled by more than the largest institutions with vast computing capabilities.

At present, the cost of computer manipulation and enhancement of one Landsat image is approximately 0.0003 cents per bit or about $500 per image. At this rate the processing of Landsat D images will cost approximately $3500 and at the ultimate data rate of 10^{10} bps, the manipulation of Landsat type images with 3-m IFOV could cost 100 times that figure. For wide-scale use of these data, computing prices must be reduced by orders of magnitude. Most likely, specialized hard-wired systems will be built to accommodate the high data rates and the specialized processing requirements. Still, substantial capital outlays will be required to support the development of such equipment.

The techniques of computer image processing are now quite highly developed. However, further development will be required to intelligently display varied and disparate data sets in a meaningful way. Data such as passive and active microwave, thermal inertia, gravity, magnetics and spectral as well as emittance reflectance data are examples of the kinds of data sets that must be displayed together in order for the best interpretation to take place.

Statistical techniques for data handling and data selection will play a greater role in the production of enhanced as well as classified multispectral images, as the number of bands or data sets increases. The processing of images will be directed by the quantitative data derived from sample studies. These data, along with greater understanding of optimum display criteria, will direct future computer image analysis.

Ultimately, data processing will be done at the sensor. The technology is presently available, but only in very special circumstances is enough known about the signatures of the targets in question to be able to design a foolproof analysis technique. The advantage of on-board processing is primarily the speed with which the end result is obtained. In addition, far less data needs to be stored and transmitted.

21.3.3 Modeling

In any type of analytical work the development of models is a necessary prerequisite to the interpretation of data as well as to develop new ideas and directions for obtaining data. A model for the radiative and conductive heat transfer at the surface of the Earth was developed for the construction and interpretation of thermal inertia images (6). This is an example of the type of modeling that will be necessary in the future to develop an understanding of the interaction of centimeter wavelength radar with the Earth's surface. Other models will be developed for the correlation of disparate data types so that they may be combined in the form of images for interpretation.

Another type of model, the resource exploration model, is now being developed and will be more thoroughly investigated over the next decade. Resource exploration modeling attempts to derive the surface observables for buried mineral and petroleum deposits. One such model (7) predicts geomorphological features and alteration patterns associated with leaking of the volatile fractions of hydrocarbon reservoirs. Although the reservoir is at 3000 m or more in depth, surface manifestations, observable by remote sensing, can be anticipated. We can expect the development of models for major mineral commodities, such as volcanogenic, massive sulfide deposits, that will aid in the detection of these deposits by remote means.

The importance of modeling to the advancement of the discipline of remote sensing cannot be overstressed. To some extent, remote sensing has been a method looking for a problem for a number of years. With increasing emphasis being placed upon resource exploration, remote sensing techniques are receiving more serious scrutiny than in the past. The development of models for the use of remote sensing in exploration will be the key to the advancement of the state-of-the-art in this discipline.

21.4 SUMMARY

Remote sensing in the 1980s will be a decisive decade for the discipline. New technologies and new space platforms will open up the full range of the spectrum and provide data at sufficiently high resolution to meet the objections of the most ardent, present-day critics. However, the analysis of that data will require significant advances in the art of computer image processing and in the ability of the geologist to interpret the data. To the latter end, the development of models to predict surface observables will become a key factor. The degree to which remote sensing becomes standard practice with geologists will determine to what extent government and industry will support the development and application of new technologies.

REFERENCES

1. Rowan, L. C., Wetlaufer, P. H., Goetz, A. F. H., Billingsley, F. C., and Stewart, J. H., 1974, Discrimination of rock types and detection of hydrothermally altered areas in south-central Nevada by the use of computer-enhanced ERTS images: U.S. Geological Survey Prof. Paper 883, 35 p.

2. Daily, M., Elachi, C., Farr, T., and Schaber, G., 1978, Discrimination of geologic units in Death Valley using dual frequency and polarization imaging radar data: Geophysical Research Letters, v. 5, p. 889-892.

3. National Aeronautics and Space Administration, 1976, A forecast of space technology 1980-2000: NASA SP-387.

4. Harris, S. E., Nick, S. T. K., and Winslow, D. K., 1969, Electronically tunable acousto-optic filter: Applied Phys. Letters, v. 15, p. 325-327.

5. Rowan, L. C., Goetz, A. F. H., and Ashley, R. P., 1977, Discrimination of hydrothermally altered and unaltered rocks in visible and near infrared multispectral images: Geophysics, v. 42, p. 522-538.

6. Kahle, A. B., 1977, A simple thermal model of the Earth's surface for geologic mapping by remote sensing: Jour. of Geophysical Res., v. 82, p. 1673-1680.

7. Donovan, T. J., 1974, Petroleum microseepage at Cement, Oklahoma: evidence and mechanism: Amer. Assoc. of Petrol. Geolog. Bull., v. 58, p. 429-446.

8. Goody, R. M., 1964, Atmospheric radiation, I. Theoretical basis, p. 4: London, Oxford University Press, 436 p.

9. Traub, W. A., and Stier, M. J., 1976, Theoretical atmospheric transmission in the mid- and far-infrared at four altitudes: Applied Optics, v. 15, p. 364-377.

10. Waters, J. W., 1976, Absorption and emission by atmospheric gases, in Methods of experimental physics, v. 12B Radio telescopes: New York, Academic Press, 309 p.

$\overset{\circ}{A}$ *(Angstrom)*: A measurement of length (10^{-10}m).

Absorption: A process by which radiation is converted to other types of energy (especially heat) by a material.

Absorptivity: The capacity of a material to absorb incident radiant energy. Absorptivity is a special case of absorptance, a fundamental property of material which has a specular (optically smooth) surface and which is sufficiently thick to be opaque.

Active Microwave Imager: A device for imaging reflected microwave energy; an imaging RADAR.

Albedo: Energy reflected from a surface over a broad region of the spectrum after effects of lighting intensity and geometry are removed.

Analog Mode: Continuous measurement in contrast to a digital mode where a signal is sampled and represented by integers.

Angle of Depression: (1, general) Any angle measured from the horizontal to an object below the observer. (2, radar) The angle formed by the horizontal plane and the line of the radar beam to a ground feature.

Angle of Incidence: The angle at which EMR strikes a surface as measured from the normal to the surface at the point of incidence (limits: 0° to 90°).

Antenna: The device that radiates EMR from a transmitter and receives EMR from other antennae or other sources.

Antenna Beam Pattern: Plot of power of transmitted signal versus the angle off boresight.

Antenna, Synthetic-Aperture (Radar): The effective antenna produced by storing and comparing the doppler signals received while the aircraft travels along its flight path. This synthetic antenna (or array) is many times longer than the physical antenna, thus sharpening the effective beamwidth and improving azimuth resolution.

Aperture: The opening of a lens or mirror system through which radiation passes.

A-scope: A mode of presenting data on a CRT in which the horizontal deflection represents time and vertical deflection the signal (observable) magnitude.

Atmospheric Windows: Wavelength intervals at which the atmosphere transmits most electromagnetic radiation.

Azimuth: (1, general) The direction of a line given as an angle measured clockwise from a reference direction, usually north. (2, radar) Direction at right angles to the antenna beam. In side-looking radar, the direction is parallel to ground track.

Backscattering: In radar usage, backscatter generally refers to that radiation reflected back toward the source.

Backscattering Coefficient: The radar cross-section (σ) divided by the minimum resolved area illuminated on the target.

Beamwidth: A measure of the concentration of power of a directional antenna. It is the angle in degrees subtended at the antenna by arbitrary power-level points across the axis of the beam. This power level is usually the point where the power density is one-half that which is present in the axis of the beam at the same distance from the antenna (half-power points). Also called beam angle.

Beat Frequency: One of the frequencies obtained when two sinusoidal oscillations of different frequencies f_1 and f_2 are superimposed in a nonlinear device. The beat frequency equals $f_1 - f_2$.

Bit: A digit (0 or 1) in binary notation.

Black body: An ideal emitter that radiates energy at the maximum rate per unit area at each wavelength for a given temperature.

Black-body Radiation: The electromagnetic radiation emitted by an ideal black body; it is the theoretical maximum amount of radiant energy of all wavelengths which can be emitted by a body at a given temperature. The spectral distribution of black-body radiation is described by Planck's law and related radiation laws.

Brightness Temperature: The temperature of a black body radiating the same amount of energy per unit area at the wavelengths under consideration as the observed body. Also called effective temperature.

Brute Force: A side-looking radar system that transmits and receives from a long physical antenna to narrow the beamwidth and increase azimuth resolution; the received returned EMR is used directly to produce an image. Also called real aperture. Compare synthetic aperture.

Christiansen Frequency: The frequency at which the refractive index of the material is equal to that of the medium in which it exists. It is located in geological materials to just longer frequencies than the maximum in the absorption coefficient and for particulate samples occurs as maximum in transmission and emission.

CIE: Commission Internationale de l'Eclairage or International Commission on Illumination. Also a system adopted by that commission which quantitatively describes the relationship of visible wavelengths in color space, and the position of various mixtures relative to the color primaries as based on experiments in additive color mixing.

Classification: A technique used in digital image processing in which the computer is programmed to assign each point in an image to a category or class of data usually identified by the analyst beforehand. A thematic map is thus created.

Coherent: For EMR, it means "in phase," so that waves at various points in space all act in unison.

Coherent Radar: A radar system in which the phase relation between transmitted and received signals is measured and utilized (normally to obtain Doppler information).

Combination Tone: Occurs when a vibrational transition occurs between the ground level ($\nu_i = 0$) to a level whose energy is determined by the sum of two or more fundamental or overtone levels.

Contrast: The difference between highlights and shadows in a photographic image. The larger the difference in density the greater the contrast.

Contrast Stretch: Modification of contrast in an image; usually used in reference to digital processing and usually implies a boost to the image contrast.

CRT: Cathode Ray Tube. A vacuum tube that generates a focused and movable electron beam directed onto a phosphorescent screen. A television image tube is a CRT.

Crystal Field Theory: A model of chemical bonding applicable to transition elements that describes the origins and results of the interaction of the surroundings on the orbital energy levels of the metal ions.

Depression Angle: See angle of depression.

Detector: The component of a remote sensing system that converts electromagnetic radiation into a signal that is recorded.

Diazo: Refers to a series of UV sensitive salts which when processed in ammonia vapors yields a specific dye in inverse proportion to the amount of UV exposure. A positive reproduction of a black-and-white transparency is produced when exposed in contact with a diazo film where the diazo film is exposed in direct proportion to the transparency of the black-and-white film.

Dichroic: A beam-splitting mirror that efficiently reflects certain wavelengths while efficiently transmitting others.

Dielectric: Substance containing few or no free charges and is consequently a poor electrical conductor.

Dielectric Constant, Complex: The combined effects of dielectric and conducting or other loss properties of a material may be stated together by use of the complex dielectric constant.

Dielectric Constant, Relative: The ratio of electric flux density to that in a vacuum for the same electric field. Also quoted as the ratio of the capacitance of a capacitor with the specified dielectric material to that for a geometrically identical capacitor with vacuum dielectric. It is therefore a measure of the amount of polarization that takes place in the molecules of a dielectric material, because the excess flux density is defined as equal to the polarization, and the excess capacitance is caused by the polarization. The dielectric constant is thus seen to be primarily a function of polarization; however, it is also dependent on temperature and frequency, as well as pressure and composition.

Diffuse Reflector: Any surface which reflects incident rays in a multiplicity of directions because of irregularities in the surface.

Digital Data: Data displayed, recorded, or stored in binary notation.

Dispersion: Separation of EMR into its spectral components by passage through or reflection from a diffraction grating, or by refraction by a prism, or by Fourier transforming an interferogram.

DN: *Data Number*: An integer used to represent a variable (usually brightness) in a digital image.

Drift Meter: A meter to measure the motion of an aircraft lateral to its heading caused by the force of cross-track wind.

Electrical Video: Electrical signal that contains imaging information, for example, the electrical signal transmitted to produce TV images.

Electromagnetic Radiation (EMR): Energy propagated through a vacuum or material medium in the form of an advancing interaction between electric and magnetic fields.

Electromagnetic Spectrum: The ordered array of known EM radiation extending from the shortest (cosmic rays) to the longest (radio waves).

Emission: With respect to EMR, the process by which a body emits EMR as a consequence of its temperature only.

Emission Angle: The angle, measured at the point of observation, between the observer and the normal to the surface.

Emissivity: The ratio of radiant flux from a body to that from a black body at the same kinetic temperature.

False Color Image: A color image in which the dye color is not the same as scene color. Infrared Ektachrome film produces false color images since the infrared exposure is represented as red, the red exposure as green, and the green exposure as blue.

Filter: Any physical device or mathematical function that is used to modify a spectrum.

Fractures: Planes of low cohesion along which a rock has ruptured or failed. A *fault* is a fracture that displays a relative displacement of the walls parallel to the fracture surface. A fracture in which the displacement is essentially oblique or perpendicular to the fracture plane is called a *joint* (displacements of up to 1 cm) or *fissure* (greater displacements).

Fracture Trace: A photogeologic term referring to a natural linear feature, discernable on conventional aerial photographs as an alignment of shallow depressions, stream segments, sinkholes, or springs. They may be expressed as vegetational or tonal alignments. Many have been shown to overlie a zone of increased fracture density.

Framing Camera: A camera that observes and records the scene in sections (or frames) in contrast, for example, to strip or panoramic cameras.

Fraunhofer Lines: Absorption lines in the spectrum of the Sun due to gases in the Sun's atmosphere.

Front-Surface Mirror: Mirror with the reflective coating on the top of the glass. Used in thermal IR scanners.

Fundamental Vibration: Normal mode of vibration occurring as a result of a transition from the ground state (quantum number $v_i = 0$) to a state where $v_i = 1$.

Gamma Rays: (γ-rays) High energy electromagnetic radiation emitted by radioactive decay. γ-rays have wavelengths shorter than 1 Å.

Geobotany: The study of vegetation to define geologic differences of the landscape.

Gray Body: A material that does not display strong absorption and emission features, but has an overall reduced emissivity (in comparison to a black body) which is practically constant over all wavelengths.

Ground Range: The distance from the ground track (nadir) to a given object.

Gyrostabilized Platform: A platform whose attitude is maintained stable by the use of signals from gyroscopes.

Hue: One parameter used to describe color. Blue, green, yellow, and red are different hues.

Image Rectification: The correction of an image for defects introduced by the camera. Also referred to as image restoration. The term may also be used to imply projection of images to some standard mapping format.

Incident Angle: (or incidence angle) The angle measured at the point of observation, between the illumination source and the normal to the surface.

Infrared: Pertaining to or designating that portion of the electromagnetic spectrum lying between the red end of the visible spectrum and microwave radiation.

Intensity: A measure of energy reflected from a surface.

Irradiance: The radiant power density incident on a surface (w/cm^2).

Irradiation: The impinging of EMR on an object or surface.

Joint Trace: A two-dimensional term used in a photogeologic sense for the expression of the intersection of joint planes with the topographic surface.

Kinetic Temperature: The internal temperature of an object which is determined by the molecular motion. Kinetic temperature is measured with a contact thermometer. Kinetic temperature differs from radiant temperature which is a function of emissivity and internal temperature.

Law of Additivity: Refers to the mixing of colors or wavelengths of light. If two colors are mixed in equal proportion the result will be a perceived color that is equivalent to a wavelength between the two principal wavelengths, for example, red and yellow light yields orange. Varying the proportions will produce colors on a continuous scale between the two principal wavelengths.

Layover: Displacement of the top of an elevated feature with respect to its base on an image.

Ligand: Anions and dipolar groups (treated as point negative charges in crystal field theory) that are situated on a lattice surrounding a transition metal ion.

Line Scanner: A device that produces a continuous strip image by means of scanning. When mounted in a moving platform, scanning is achieved by the forward motion of the platform and lateral movement of the imaging field by rotating or oscillating mirrors.

Lineament: A two-dimensional geomorphological term referring to a mappable, simple or composite linear feature of a surface, whose parts are aligned in a rectilinear or slightly curvilinear relationship and which differs distinctly from the patterns of adjacent features and presumably reflects a subsurface phenomenon. They are essentially megascopic scale features (with overlap onto macro- and gigascopic scales), and because of their varied nature (some appear only on derivative maps) appropriate qualifying adjectives should be used to indicate their mode of expression, for example, photo lineaments, topographic lineaments, magnetic lineaments, gravity lineaments, volcanic lineaments, geological lineaments, strike lineaments, cross or transverse lineaments, and so forth.

Linear Feature: A general two-dimensional term for any straight or slightly curved feature, or alignment of discontinuous features that are apparent on a map, image, or photograph of a region.

Lobe: An element of a beam of focused radio energy. Lobes define surfaces of equal power density at varying distances and directions from the radiating antenna.

Loss Tangent: Ratio of the imaginary to real part of dielectric constant. A measure of the relative loss per wavelength.

LTF: *Light Transfer Function.* The curve relating the actual intensity of light from a scene to the *DN* recorded by an imaging device.

Luminance: A measure of light related to the intensity emitted by a source in a given direction. Similar to radiance except that it applies only to visible wavelengths.

Microwave: A very short EM wave; any wave between 1 m and 1 mm in wavelength or 300 GHz to 0.3 GHz in frequency. The portion of the electromagnetic spectrum in the millimeter and centimeter wavelengths that is bounded on the short wavelength sides by the far infrared (at least 1 mm) and on the long wavelength side by very high-frequency radio waves. Passive systems operating at these wavelengths sometimes are called microwave systems. Active systems are called radar, although the literal definition of radar requires a distance-measuring capability not always included in active systems. The exact limits of the microwave region are not defined.

Microwave Imager: A device for imaging a scene by the microwave radiation reflected or emitted by it. Imaging is performed by means of electrical or mechanical scanning.

Mie Theory: Mathematical-physical theory of scattering of EMR by spherical particles that embraces all ratios of diameters of particles to wavelength. For very small particles it coincides with Rayleigh scattering and for very large particles with geometric optics.

Monochromatic: Pertaining to a single wavelength or a narrow band of wavelengths.

Mosaic: A synoptic image or picture formed by integrating overlapping images or photographs to give a greater regional coverage.

Mossbauer Effect: Phenomenon of recoilless resonance fluorescence of gamma rays from nuclei bound in solids.

MSS: An acronym for *Multi-Spectral Scanner* and now commonly, if not exclusively, used to denote the multispectral scanner of Landsat.

MTF: Modulation Transfer Function. A measure of the sensitivity of a sensing system to data as a function of spatial frequency.

Multispectral Scanner: A line scanner that simultaneously scans a scene in multiple wavelengths.

Normal Vibrational Modes: Are the limited number of ways in which a system of N atoms can vibrate, given by $3N-6$, or $3N-5$ for a linear arrangement.

Nyquist Frequency: The highest frequency sine wave that a discrete signal can represent faithfully: 0.5 cycles per sample.

Optical Constants: Fundamental constants given for each EMR frequency for a material, they are n, the refractive index and k, the absorption coefficient.

Orthographic View: An observation normal (perpendicular) to a plane. Hence all rays to the plane are parallel.

Overtone Band: Spectral feature that occurs as the result of a transition from the ground state to one where the energy corresponds to double or triple excitation of the fundamental frequency.

Parallax: The apparent displacement of position of a body with respect to a reference point or a system of coordinates, caused by a shift in the point of observation. Absolute stereoscopic parallax of a point in a pair of truly vertical aerial photographs is the algebraic difference, parallel to the air base, of the distances of the two images from their respective principal points.

Passive Microwave Imager: A microwave imager that does not use a cooperative illuminating source, that is, the scene is not illuminated specifically for the purpose of producing the image.

PDF: Probability Density Function. The frequency of occurrence of different *DN* in an image.

Periapsis: The nearest point of approach in the orbit of one object about another. For a spacecraft orbiting a planet, at periapsis the spacecraft is at its nearest point to the planet.

Phase Angle: The angle, measured at the point under observation, between the illumination source and the observer.

Photoconductive: The property wherein a material changes its electrical resistance when radiation is absorbed.

Photoconductive Effect: The increase in electrical conductivity of a semiconductor caused by an increase in free carriers produced by incident radiation.

Photoelectromagnetic Effects: A diffusion phenomenon that occurs when radiation falls on a semiconductor that is within a magnetic field.

Photoemissive: The property wherein a material emits electrons when radiation is absorbed.

Photogeology: The interpretation of the geology of an area from an analysis of landforms, drainage, tones, and vegetation distribution on aerial photographs.

Photovoltaic Effect: Takes place at a *pn* junction in a semiconductor as a consequence of photons causing hole-electron pairs at the junction that diffuse in opposite directions, producing charge separation and therefore producing a voltage.

Pixel: A single sample of data or "picture element" in a digital image.

Polarization: The direction of the electric vector in an EM wave (light or radio). A wave is said to be unpolarized if the direction of the electric vector is randomly distributed (has random orientation), so that the direction at any instant cannot be predicted.

Polygonal Ground: A general term here meaning patterned ground on a scale large enough to be resolved in aerial photographs; mostly caused by ice wedges in permafrost, but some are stone nets. Dessication cracks also form polygonal ground in deserts that are not permanently frozen.

Projection: Rearrangement of image geometry, usually to conform to a cartographic format.

PSF: *Point Spread Function.* The image of an impulse.

Pulse Duration: The time interval between the first and last instants at which the instantaneous amplitude reaches a stated fraction of the peak pulse amplitude.

Quantum Detectors: Semiconductors that essentially count the number of photons that strike the sensitive elements as a consequence of the photons interacting with the crystal lattice, freeing electrons or holes.

Quantum Theory: Theory stating that all EMR is emitted or absorbed in quanta, each of magnitude $h\nu$ (where h = Planck's constant and ν = frequency of radiation).

Radar: Radio Detection And Ranging. A method, system, or technique, including equipment components, for using beamed, reflected, and timed EMR to detect, locate, and (or) track objects, to measure altitude and to acquire a terrain image. In remote sensing of the Earth's or a planetary surface, it is used for measuring, and often, mapping the scattering properties of the surface.

Radar Beam: The vertical fan-shaped beam of EMR produced by the radar transmitter.

Radar, Brute Force: A radar imaging system employing a long physical antenna to achieve a narrow beamwidth for improved resolution.

Radar, Coherent: A radar system in which the relative phase between the transmitted and received signals is compared and used.

Radar Shadow: A no-return area extending in range from an object that is elevated above its surroundings. The object cuts off the radar beam, casting a shadow and preventing illumination of the shadowed area behind it.

Radiance: A measure of electromagnetic radiation related to the intensity emitted by a source in a given direction.

Radiant Energy: Energy carried by the electromagnetic radiation in units of joules or ergs.

Radiant Flux: The amount of radiant energy that is emitted, transmitted, or received in unit time.

Radiant Flux Density: The radiant flux arriving at, crossing, or leaving unit area.

Radiant Temperature: Concentration of the radiant flux from a material. Radiant temperature is the product of the kinetic temperature multiplied by the emissivity to the one-fourth power.

Radio Astronomy: The study of celestial objects through observation of EMR emitted or reflected by these objects.

Radiometer: An instrument for quantitively measuring the intensity of EMR in some band of wavelengths in any part of the EM spectrum. Usually used with a modifier, such as IR radiometer or microwave radiometer.

Rayleigh Scattering: Scattering by particles small in size compared with the wavelength being scattered, for example, scattering of blue light by atmospheric molecules.

Range Direction (Radar): The direction from radar to target. The perpendicular to this direction and parallel to the ground is sometimes called the azimuth direction.

RBV: *Return Beam Vidicon.* A type of television image tube.

Real Aperture: See brute force.

Reflectivity: A fundamental property of a material that has a reflecting surface and is sufficiently thick to be opaque.

Relief Displacement: A shift in position of the optical image of an object caused by height of the object above or depth below a datum plane.

Resolution: The ability of an entire remote sensor system, including lens, antennae, display, exposure, processing, and other factors, to render a sharply defined image. In radar, resolution usually applies to the effective beamwidth and range measurement width, often defined as the half-power points.

Resolution Cell (Radar): The element on the ground distinguishable on the image, usually consisting of the half-power beamwidth distance by the half-power pulse duration. As some systems use other discrimination techniques, however, different definitions may apply.

Reststrahlen (or residual ray): An almost metallic reflection that occurs in otherwise almost transparent materials where either the refractive index or the absorption coefficient is large.

Return (Radar): EMR reflected by an object back to the antenna. Strong (bright): Strong or bright EMR returns that appear as light toned areas on imagery. Weak: Weak EMR returns that appear as a gray-toned area on imagery.

Roll-Front Deposits: Sedimentary uranium deposits that are crescentric in cross section, curvilinear to sinuous in plan view, and arkosic to subarkosic in composition; commonly limonitic on the concave side.

Ronchi Grid: A diffraction grating that is used as an optical aid to preferentially enhance or suppress linear features on an image. A common spacing is 200 lines to an inch (78 lines to 1 cm).

Rose Diagram: A type of circular histogram showing the frequency of directional data (usually strike). It depicts two-dimensional data only, and generally they are plotted on some modification of an azimuthal or polar projection.

SAR: Synthetic Aperture Radar. A high resolution side-looking radar that achieves its resolution by synthetically generating the equivalence of a large antenna.

Saturation: (1) The relative purity of color measured by the absence of white light. White light decreases saturation or lightens the color. (2) The condition in which a signal exceeds the response range of an instrument.

Scale or Representative Fraction: The ratio of the size of an object in an image or map to its actual size in the same units, for example, a map to a scale of 1:63360 is one in which 1 in. represents 63390 in., or 1 in. equals 1 mile. The *scale number*, 63360, is the reciprocal of scale. A *small scale* map (for example, 1:1,000,000) represents a large area, and a *large scale* map (for example, 1:24,000) represents a small area, generally in greater detail.

Scanner: An optical-mechanical imaging system in which a rotating or oscillating mirror sweeps the instantaneous field of view of the detector across the terrain.

Scatterer, Isotropic: A theoretical device that, after intercepting a given amount of energy, reradiates it equally in all directions.

Scattering: The process by which a rough surface reradiates EMR incident upon it.

Scatterometer: An active microwave radiometer that measures the microwave reflection (scattering) as a function of angle.

Scatterometry: A method of using radar to measure the variation of radar scattering coefficient. These variations may be used by geoscientists to discriminate between surfaces with different roughness and materials. The scatterometer is distinguished from other radars by its ability to measure amplitude.

Side-looking Radar: An all weather, day/night remote sensor that is particularly effective in imaging large areas of terrain. It is an active sensor, as it generates its own energy, which is transmitted and received to produce a photo-like picture of the ground.

Slant Range Image: An image in which objects are located at distances corresponding to their slant range distances from the flight path. Compare ground range image.

SLAR: Side-Looking Airborne Radar. An active microwave imaging device.

S/N (SNR): Signal-to-Noise Ratio.

Specific Heat: The ratio between thermal capacity of a substance and thermal capacity of water.

Spectral Radiance: The radiance per unit wavelength interval.

Spectral Radiant Emittance: The radiant power density per unit wavelength interval emitted by a surface.

Spectral Signature: The spectral characteristics by which a material may be recognized.

Specular Surface: A surface that is smooth with respect to the wavelength of EMR incident upon it; reflects like a mirror.

Structure: The alignment of structural or fabric elements in a definite form or configuration.

Synthetic Antenna (Aperture): The effective antenna produced by storing and comparing the Doppler signals received while the aircraft travels along its flight path. This synthetic antenna is many times longer than the physical antenna, thus sharpening the effective beamwidth and improving azimuth resolution.

Texture: Refers mainly to the spatial relationships of drainage and those structural or fabric elements with morphological expression. Textural parameters include density and pattern.

Thermal Capacity: The ability of a material to store heat, expressed in cal/(g • °C).

Thermal Conductivity: The measure of the rate at which heat will pass through a material, expressed in $cal \cdot cm^{-1} \cdot s^{-1} \cdot °C^{-1}$.

Thermal Cross-Over: On a plot of radiant temperature versus time, this refers to the point at which the temperature curves for two different materials intersect.

Thermal Inertia: A measure of the resistance of a material to a change of temperature.

Thermistor: Infrared detector whose resistance changes with temperature.

Thermocouple: Temperature sensing element (constructed of two dissimilar metals) that converts thermal energy directly to electrical energy.

Tone: The magnitude of brightness or shade (gray-level) in an image.

Transition: The change from one discrete energy state or level to another as a consequence of absorption or emission of EMR, that is, $E_1 - E_2 = h\nu$.

Transmissivity: Transmittance for a unit thickness sample.

Tristimulus Values: The amount of each of the three primary colors (red, green, and blue) necessary to describe the color of an object.

UV: The abbreviation for ultraviolet, that portion of the electromagnetic spectrum lying on the short wave length of violet but longer than x rays.

Vidicon: One type of television image tube.

Vignetting: A gradual reduction in density in a negative from the center to the edge caused by uneven exposure due to a difference in intensity of light reflected at nadir and at large angles away from nadir.

AASHO soil ratings, 245, 250
Absorption, 23, 259, 344
Acid-mine drainage, 650
Active microwave, 297
Active sensor, 47
Adirondack Mountains, 489, 490
Aerial photography, 229, 421, 424, 437, 452
  distortion, 235, 237, 238
  multiband, 230, 234
  oblique, 232
  specifications, 235
  U2, 230, 231
  vertical, 230, 235
Agfa contour film, 127
Airborne geobotany, 370
Air Plateau, Niger, 432, 436, 437
Albedo, 120, 121, 660
Allegheny Front, 423, 424
Allegheny Plateau, 423, 424
Altered rocks, 567
Anamorphic system, 237
AN/APQ-97, 298, 317
AN/APQ-102, 299
Angle of illumination, 310, 311, 317, 334
  of incidence, 300-302, 306, 311, 312, 317, 318, 334
  of reflectance, 302
Antenna, 346
Apparent brightness temperature, 259
Apollo photography, 102-105
Apollo 17 Lunar Sounder Experiment (ALSE), 321
Apollo-Soyuz Test Project (ASTP) Photography, 105
Applications Explorer Mission (AEM), 113
APS/94D, 298
Arbuckle Mountains, 279
Arid Climate, 427

Artesian water, 461
ASCS, 234
Aspect angle, 301, 306
  ratio, 159
Asteriods, 674
Atmospheric attenuation, 234
  pollution, 642, 643
  radiation, 261
  scatter, 120, 129
  window, 681
Attenuation, 302
Aulacogen, 452, 470-473
Automatic classification, 409
Azimuth direction, 299, 333
  resolution, 298, 305-308

Backscatter, 299, 300, 302
Balcones escarpment, 489
Basin and Range province, 489, 491-493
Basin (structural), 417, 441, 461, 462
Bayesian classification, 193, 411
Beta Diagrams, 430
Bilinear interpolation, 162
Black body, 10, 11, 258, 340
  radiation, 339
Blue Ridge, 488, 489
Body tides, 435
Box filters, 177, 217
Breached anticlines, 427
Brent Crater, Ontario, 463, 464
Brewster's angle, 343
Brightness temperature, 337
Brute force, 298
Bushveld Igneous Complex, 429

Caldera, 461, 462
Cata zone intrusions, 432, 434

California, major faults of, 486, 487, 499
Cameras, 229
  aerial, 229
  comparison, 230
  framing, 52
  mapping, 229
  multiband, 230, 234
  reconnaissance, 229
  strip, 229
Canonical analysis, 408, 409
Carbonate signature, 40
Carrizo Plains, 278, 279, 281
Carswell dome, Alberta, 463, 464
Cartographic projection, 159, 170
Cathode-ray tube, 299, 307, 308
Characteristic curve, 121, 124, 126, 127, 132
Charge-coupled device, 682
Charge transfer, 31
Charlevoix crater, Quebec, 463, 464
Chlorosis, 374
CIE: color prediction model, 132-136
  color coordinates, 133
Circular features, 461, 557, 566, 567
Clearwater Lakes, Quebec, 462-465
Cleavage, 428
Climatic effects, 382
Clovis area, New Mexico, Texas, 450
Clutter, 299
Coal mines, 253
Coastal Plain (eastern U.S.), 485, 488, 491, 492
Color: additive, 123
  balance, 131, 237
  brightness, 123
  centers, 32
  complementary, 124

coordinate system, 203-205
enhancement, 119, 131
false color composite, 129
law of additivity, 127
relationships, Grassman's laws, 122
prediction model, 132-136
saturation, 122, 124, 132
separation, 129
simulated infrared composite, 129, 131
subtractive, 124
transformation, 202
tristimulus values, 132-133
Colorado Plateau, 491, 492, 495, 500
Color ratio composite images, 214, 583-589
Commission Internationale de l'Eclairage, 216
Comparative analysis, 239
Complex dielectric coefficient, 303, 321
constant, 302, 333
Compton effect, 7
Computer classification maps, 592-596
Conduction band, 31
Continental Drift, 466, 476
Contour maps, 209
Contrast: enhancement, 132, 401
image, 129, 131
index ratio, 121, 125, 127-130
scene, 120, 121, 124, 125, 129
stretching, 124, 184, 209-219
Converging margins, 466
Convolution, 141, 161, 177
Couture, Lac, Quebec, 463, 464
Crab, 235
Crater, 420, 424, 435, 461
Crater Lake, 294
Craters of the Moon Volcanic Field, 411
Cratons, 422, 432
Critical sampling, 149
Cross-correlation, 172
Cross-polarized, 303, 304, 321, 322
Cross strike, 422, 437, 450, 452
CRT 70, 299

Crustal blocks, 422
Crypto-explosion structures, 435, 461, 462
Crystal field effect, 30
Culminations, 435, 441
Cumulative distribution function, 209
Curve numbers, 612

Dasht-i-Lot, Iran, 99
Data transmission, 681
Debye formulae, 341
Deep Bay crater, 463, 464
Deformation, 425
Dendritic drainage, 488, 500, 501
Density, 263
color, 124, 129
definition, photographic, 120
film, 126, 127
slicing, 127, 214, 644
Depolarization, 303, 304
Depression angle, 300-302, 306, 310, 311, 313-315
Depressions, 435
Derivative images, 187
Detection, 149
Detector arrays, 681, 682
D log E curve, 121, 124, 126, 127, 132
Diapirs, 424
Diazo film, 128-130, 132, 133
Dicke radiometer, 347
Dielectric constant, 300, 302, 340-342, 348
Difference picture, 178, 183, 200
Diffuse scatter, 302
Digital elevation image, 185
Digital filtering, 175, 177
Digital image, 139, 140
Dikes (dykes), 384, 424, 428, 435
Discontinuities, 427, 435
Discontinuity Concept, 426
Distortion, 311
Divergent margins, 466
DN, 139
Domain boundaries, 430
Domains, 435
Domes, 427, 435, 440, 441, 461

Doppler frequency, 307
Drag folds, 430
Drainage analysis, 385
patterns, 389
Dynamic analysis, 425, 426, 432
Dynamic range, 209

Earth Resources Experiment Package (EREP), 101, 105, 106
Earth Resources Observation System (EROS), 101, 102, 234
Earth Terrain Camera, 105
Echo, 299, 300
Edge enhancement, 124
Eigenvalues, 195
El Nino, 353
Electric field vector, 303
Electromagnetic radiation, 299, 303
Electromagnetic spectrum, 9, 47, 680
Electronic energy, 13, 17
processes, 29
Emergent angle, 158
Emission, 23, 337
Emissivity, 259, 277, 339, 340, 349
En echelon folds, 435, 437, 441
Energy sources, 10
Enveloping surface, 424, 430
Environmental factors, 27
Eolian sediments, 545-548
loess, 545, 547
sand dunes, 546-548
Epizonal ore deposits, 449
Epizone intrusions, 432, 434, 435
Equivalent black body temperature, 259
Erosion, 656
Eskers, 430
Everett-Bedford lineament, 423
Exposure control, 68

Fabric (elements), 425, 428, 431

Fall Line, 485, 488, 489
Far range, 301, 310, 315, 317, 322
Fault blocks, 427
Faults, 420, 428, 435, 447, 485-487, 490, 491, 493, 495-499
Film, 232
  black-and-white, 232
  color negative, 232
  color positive, 232
  comparison of types, 232, 247, 251
  false color infrared, 232
  infrared, 232
  orthochromatic, 232
  panchromatic, 232
Filters, 234
  directional artifacts, 216-219
  HF-3 haze, 234
  minus blue, 234
  ringing, 216-218
  Wratten-12, 234
Filtering, of color pictures, 220
Fisher criterion, 201
Fissure eruptions, 435
Fissures, 447
Flight base, 241
  plan, 234, 236
  path, Gemini, 97
Flinders Range, South Australia, 441, 446, 447
Floating dot principle, 241
Flood mapping, 654
Floodplains, 613
Floodprone areas, 613
Floods, 613
Fluting scours, 430, 431
Fluvial erosion, 493, 500
Focal length, 229, 230, 240
Fold, 420, 430
Folded mountains, 488-494
Foliation, 428, 432
Foreshortening, 310-316, 326, 330
Foreslope, 314
Forest damage, 645
Fourier transform, 141, 177, 217
Fractures, 428, 435, 447
Fracture trace, 420, 422, 428, 429, 447, 452, 453-457

Fraunhofer line, 369
Fraunhoffer line discriminator, 88
Frequency response, 178
Fresnel relations, 343
Front Range, Colorado, 432, 434

Gamma, 121, 124, 125, 127, 132
Gamma-ray spectral emission, 600
Gamma-ray spectrometer, 68, 669
Garlock fault, 485-499
Gemini photography, 93-101
Geobotany, 365
Geoflex (orocline), 437, 441
Geometric analysis, 426, 441
  displacement, 311, 313
  distortion, 313
  rectification, 159
Geomorphology, 419, 425
Geothermal, 293
Gigascopic, 420, 424, 479
Glaciers, 495, 505, 509-520
  ablation zone, 509, 510, 514, 515
  accumulation zone, 509, 510, 513, 514
  crevasses, 508, 513, 515, 518
  debris-covered ice, 518, 520
  equilibrium line, 509
  flow, 517, 518
  mapping, 505, 510-513
  mass balance, 515-517, 548
  moraines, 515, 517, 518, 520
  ogives, 515, 518
  Pleistocene, types and extent, 505, 520, 521, 526, 549
    flow directions, 529, 531, 536-538
  snowline, 514
  surging, 518
  temperature, 515
  thickness, 515
Glacial deposits, 522-549
  age, 547-549
  glacial lake, 543-545
  glacio-marine clay, 545, 546
  ice-contact stratified drift, 520, 532, 536, 542, 543
  Illinoian age, 527, 548

spectral signatures, 532, 533
  till, 533-541
  Wisconsinan age, 520, 527, 528
Glacial geology, 505, 521-550
  underwater, 547
Glacial lakes, Pleistocene, 543-546
  beaches, relict, 527, 543, 546
  clay, varved, 543
  deltas, 543
  shoreline scarps, relict, 543
Glacial landforms, 522-525, 532-547
  craig and tail, 529, 531, 536
  dead-ice moraine, 539
  drumlins, streamlined ridges, 529, 531, 536-538
  erosion, mountain, 526, 530, 534, 535
  eskers, 520, 536, 542, 543
  kame terraces, 517, 542
  meltwater channels, 536, 546-548
  moraine, end, 526, 528, 533, 538, 539, 548
  fluted, 531
  ground, 526, 533, 538, 539
  hummocky, 536, 539, 541
  moraine ridges, 529, 548
  moraines, minor, 539-541
Glaciation, mountain, 526, 530, 534, 535
  Pleistocene, 505, 520, 521, 526
Glacier flow direction indicators, 529, 531, 536-538
Glaciology, 505, 507-520
  underwater, 547
Goldfield Nevada, 408
Grand Canyon, 495
Gran Desierto, 495
Graphical transfer instruments, 237
Gravity, 670
Gray body, 259
Gray level distribution, 188
Gray scale or shades of gray, 120, 124-126, 132
Grazing, 300, 310, 312, 326
Great Dyke, Rhodesia-Zimbabwe, 432, 438, 439
Great Valley, Appalachians, 488, 489

Great Valley, California, 486, 487
Great Valley, Pennsylvania, 423, 424
Green Mountains, 489, 490
Greenstone belts, 432, 438, 439
Grid unit overlays, 619
Griffiths formula, 426
Ground range resolution, 306, 309, 310, 317, 328
Ground studies, 235
Groundwater, 453, 460, 628
Gwinn-type lineaments, 433

Hadramaut Plateau, 500, 501
Heat capacity, 263
Heat Capacity Mapping Mission (AEM-1), 113
Heat transport, 261
HH, 303, 334
High-pass filters, 179, 216
Himalayan Mountains, 476, 477
Hogsback, 427
Horizontal electric field vector, 303
Hotspots, 469-471, 476
Hue, 203
Hurter-Driffield curve, 152
HV, 303
Hydrologic cycle, 607
    soil type, 612
Hydrothermal alteration, 568
    mineralogy, 568
    visible and near-infrared spectral reflectance, 570-576
Hydroxide signature, 41
Hyperaltitute photography, 91

Ice: floating, 546, 548
    lake, 505, 509
    pressure ridges, 508
    sea, 505, 507, 508
Iceflow features, 430, 431
Iceland, 495
IFOV, 77, 305
Image classification, 188
Image degradation due to:
    atmosphere, 119, 121, 129
    resolution, 120, 129

shadow, 120
Image enhancement, 139
Image magnification, 52
Image sharpness, 124
Imagery tone, 302, 323
Imler Road anticline, 284, 285
Impact structures, 500, 502
Imperviousness, 612
Incidence angle, 158, 300-302, 306, 311, 312, 317, 318, 326, 334
Indio Hills, 281
Infrared, 669
    color composite, 129
Intensity, 203
Interpolation, 161
Irradiance, 257
IR radiometers, 275, 276, 283
    scanners, 275, 276
Irrigated land, 614
Island arc, 420
Isostasy, 424
Isotatic rebound, 462
Isotropic reradiation, 300
    scatter, 302

Joints, 420, 447, 452

K-band, 297
Ka-band, 298, 303, 317, 320, 321, 323, 324, 333
Kinematic analysis, 425, 532
Kirchhoff's law, 259
Koksoak River, Quebec, 431

Labrador trough, 431
Lake-bed terrain, 247, 251-252
Lake ice monitoring, 625
La Moinerie Lake, 463, 464
Landform analysis, 382
    description, 243
Laser altimetry, 508
    profiler, 87
Latent heat, 262
    image, 52, 54
Layover, 310, 312-315, 326, 330
L-band, 297, 299, 321, 322, 334

Leaf area index, 368
Leaf spectral reflectance, 367
Light sources, 236-237
Light transfer function, 152
Like-polarization, 303, 304, 321
Lineaments, 94, 96, 99, 100, 109-111, 113, 208
Linear features, 553
    in Alaska, 555, 556, 558-561, 566
    in conterminous U.S., 557-559, 562-564
    as crustal block boundaries, 559, 565, 566
    crustal tectonic elements and, 554
    on Landsat imagery, 554-557, 560, 561
    local structures and, 553
    and metallogeny, 564-566
    on Nimbus imagery, 557, 560
    on NOAA imagery, 557, 561
    and plate tectonics, 564
    regional structural regimes and, 554
    systematic pattern in, 559, 561, 562, 564, 565
Lineation, 428, 441
Line scanners, 54, 55
Lithology, 239, 252, 253
Lithosphere, 466
Lobe, 346
Look direction, 317-319, 326, 327, 333
Low-pass filter, 187
Luminance, 216

McAlevys Fort-Port Matilda lineament, 423
Macroscopic scale structure, 420, 430, 432, 453, 479
Mars, 435, 439, 674
Micro-macro-, mega-structural elements, 428
Manicougan Crater, 462-464, 500, 502
Mantle plumes, 466, 468, 470, 471
Mantled gneiss domes, 432

Mapping symbols, 245, 248, 249, 250
Mare Orientale, 463
Mariner, 9, 435, 439
Marker horizons (beds), 427, 428
Maxwell equations, 7, 8
Mean image, 187
Mecatina crater, Quebec, 441, 448, 449
Megascopic scale structure, 420, 422, 429, 430, 432, 453, 479
Mercury, 671
Mercury photography, 93-96
Mescalero escarpment, 489
Mesoscopic scale structure, 420, 430, 453, 479
Mesozone intrusions, 432, 434
Metallogenic map, 462
Metamorphic aureole, 432
Meteorite Crater, Arizona, 463
Microscopic scale structures, 420
Mid-infrared spectra, 34
Migmatite, 432
Military radar, 297
Mineral exploration, 470
Minnaert photometric law, 158, 185
Minor folds, 430
Mission control factors, 234
Mississippi River, 495
Mistastin Lake, 463, 464
Mixed pixel approach, 615, 616
Modeling, 684
Modulation transfer function, 72, 141, 178
Monteregian Hill petrographic province, 462, 464
Moon, 674
Moraines, 430, 431
Morphotectonics, 422, 425, 430, 432, 463, 464
Mosaics, photographic, 239-240
Mud volcanoes, 447
Multiband photography, 54
Multispectral cameras, 54
  image enhancement, 129, 131
Munsell Renotation System, 216

Nadir, 313
Nastapoka arc, Quebec, 463, 464
National Flood Insurance Act, 613
Natural selection, 371, 373-375
Near range, 301, 313, 315-317, 322
Nearest neighbor algorithm, 160
Nighttime navigation, 280
Nimbus, 82, 92, 93, 337, 353, 356, 360
NOAA, 655
Noise equivalent reflectance difference, 72
Npose equivalent temperature difference, 72
Non-penetrative elements, 428
Nuees ardentes, 424
Nutrient availability, 366
Nyguist frequency, 149, 163

Oceanic plates, 420
Oil-field brine, 636, 637
Oil slick, 357-358
Oil spills, 640
Olympus Mons, 435, 439
Operator variability, 451
Optical filtering: glaciology, 515
  wiener spectra, 531
Optical imaging systems, 50
Optical transfer function, 141
Orbital photography, 91
Ore deposit, 402, 466, 467, 470
Original horizontality (Steno's Law), 428
Orocline, Appalachian, 423, 424, 437
Orthographic projection, 172
Ortho-images, 421, 428, 479
Outwash, proglacial, 516, 538, 540, 543, 550
Overprinting, 209, 437
Oxide signatures, 41

Parallax, 184, 185, 240, 330, 332
  bar, 242
  distortion, 74
Parallelepiped classifier, 192

Parallel polarization, 322
Particulates, 25
Passive microwave, 297
  use in glaciology, 507, 508, 514, 515, 548
Pattern analysis, 241-242, 323, 324, 330
PDF, 188, 200, 404
Penetrative structural elements, 428, 430
Peninsula Ranges, California, 474, 495, 499
Pennsylvania orocline, 443
Periodic noise removal, 178
Permafrost, 505, 507, 508, 509
  ground ice, 509
  palsa, 509
  patterned ground, 508
  pingo, 508, 509
  solifluction lobes, 508
  thaw lakes, 508, 543
  thermokarst, 508
  underwater pingo, 509
Perspective, 231
  one point, 231
  orthographic, 231
Phase angle, 660
  history, 307
Phosphate signatures, 40
Photo analysis, 229, 251
Photodetectors, 12
Photogeology, 419, 470
Photographs, aerial, 505, 510, 513, 520, 521, 548
  comparison of types, 232, 240, 247, 251
  infrared, 513
  Skylab, 505, 521, 547, 548
Photographic film/developer combinations, 52, 124, 126, 130, 131, 240
Photographic mission, 229, 234
Photo interpretation, 229, 396
Photolineament, 461
Photometric corrections for contrast and exposure, 129-131
Photomosaic, 171
Phreatophytes, 375, 376
Physiography, 609-612

Pi diagrams, 430, 435
Piedmont province, 488, 489
Pikes Peak, Colorado, 454, 455
Pinacate volcanic field, 495
Pisgah Crater, 288
Pitch, 58
Pixel, 139, 305, 659
Planck's law, 10, 65, 258, 339
Plant luminescene, 368
  radar reflectance properties, 369,
    370
  thermal properties, 369
Plateaux, 419
Plate boundaries, 466
  margins, 466
  movements, 470
  tectonics, 466, 476, 479
Playback devices, 219
Point spread function, 141
Polarization, 300, 303, 321, 322,
    334, 343
Polyphase deformations, 437
Precipitation measurement, 628,
    629
Primary discontinuities, 427
  structures, 428, 429, 447
Principal component analysis, 409,
    411
  transformation, 195-201, 220
Principal point, 240, 241
Proglacial stratified drift, 542, 547
Projection viewers, 238
Propagational vector, 300
Pseudotachylite, 462
Pulse duration, 306
  length, 305, 306
  rectangle, 305
Pumpelli's rule, 424, 429, 435

Qom Play, Iran, 405
Quantum theory, 8, 14

RADAM (Radar of the Amazon),
    330
Radars, 61, 65, 80, 81, 670, 681
  cross section, 300
  hologram, 308

SAR, 507, 538, 542, 547, 548,
    550
  shadow, 310
  SLAR: glacial geology, 521, 526,
    538, 542, 545-548, 550
    glaciers, 517, 518, 520
Radial displacement, 231
Radiant emission, 257
  energy, 257
  flux, 257
  flux density, 257
  intensity, 257
  temperature, 277, 279, 291
Radio: astronomy, 337
  echo-fading, 517
  echo sounding, 515, 548
Radiometer, 61, 337, 346-347
Radiometric correction, 152
Rain: rate, 353
  shadow, 485, 486, 491
Random noise, 178, 217, 219
Range resolution, 65
Ratio pictures, 202, 203
Ratioing, 406, 407
Rayleigh criterion, 64
Rayleigh-Jeans Approximation, 340
RBV, 82
Reading Prong, 488, 489
Real antenna, 308
  aperture, 298, 299, 305-308
Reconnaissance mapping, 243-247,
    251
Rectification, 139, 152
Reflectance angle, 302
Reflection, 23
Reflectivity, 340, 343, 357
Registration of images, 172
Regmatic shear, 422, 451
Resampling, 159, 160
Reseau marks, 143, 165, 182
Residual images, 156
Resolution, 71, 72, 141, 280, 304,
    305, 322, 425, 660
  cell, 306
Restoration filtering, 175, 177
Rhodesian shield, 432
Richat Dome, 435, 550
Ridge and Valley Province, 488,
    489

Rift Valley, 420, 422, 466-468,
    470
  zone, 449, 452, 468
Rock mechanics, 426
Rocky Mountain trench, 105
Roll, 58
Ronchi grid, 451, 452
Rotational energy, 13, 15
RT, 299

SO65 experiment, 102-105
S190 experiment, 105
Salt domes, 99, 427
Salton Sea, California, 471, 474,
    475
Sampling theory, 144, 149
San Andreas fault, 107, 111, 281,
    471, 474, 476, 486, 487,
    495-499
Sand dunes, 493-495
San Francisco volcanic field, 384
SAR, 299, 306, 330, 333, 334
Satellite Platforms, 419, 420, 425
Saturation, 203
Sayan fault, 92-94
Scale, 420, 422, 424, 426, 428,
    430, 453, 462, 479
Scattering, 350
  coefficient, 302, 303
Scatterometers, 88
Schroedinger equation, 7, 14
SCS methods, 612
Sea ice, 350-351
Seafloor spreading, 466
Seasat, 82, 113, 333, 356, 357
Seasonal changes, 543
Secchi disk, 636
Second order shear, 453
Secondary discontinuities, 427, 428
  structures, 429, 447
Section 208, Federal Water Pollu-
    tion Control Act Amend-
    ments of 1972, 610
Sensible heat, 262
Sensitivity, 72
Shadowing, 310, 312, 313, 317,
    319, 323, 324, 326, 330, 332
Shape, 323, 324, 330

Shattercones, 462
Sierra Nevada, 471, 474, 486, 487, 491, 492
Signal amplitude, 302
Signal to noise ratio, 178, 220
Silicate signatures, 34
Sinc function, 163, 164, 218, 219
Size, 323, 324
Skewing of images, 166, 170
Skylab, 82, 230, 327, 398, 401, 420
  photography, 105-109
Slant range, 307-309, 311, 312-314, 317
  distance, 306, 307
  resolution, 306, 328
SLAR, 297, 301, 303, 305, 306, 308, 310, 311, 313, 317, 321, 324, 325, 327, 331-333
Slope analysis, 245, 250, 253
SMMR, 356, 357, 360
S/N, 72
Snell's Law, 302
Snow, 505, 507, 508, 622
Snow and ice, identification, 618
Snow-covered area mapping, 618-622
Snowmelt runoff forecasts, 622, 623
Soil analysis, 239, 240
  moisture, 348-350, 625-628
  ratings, 245, 250
  texture, 245
Solar illumination, 302
  radiation, 261
SONAR, side-scanning, 507, 508, 509, 521, 545-547, 550
  underwater glacial geology, 547
  underwater permafrost, 509
Sounding rocket photography, 91, 92
Spaceborne Imaging Radar (SIR-A), 334
Spatial filtering, 216-219
  relationship, 324
  resolution, 425
Specific heat, 263
Spectral signature, 426, 441, 470

Spectrometers, 88
Speculor reflector, 302
Stefan-Boltzman law, 11, 258
Stereometer, 241
Stereostat, 683
Stereoscopes, 237
Stereoscopic analysis, 184, 240, 452
  model, 232, 240, 241
Strain, 425
  discontinuities, 430, 433
Stream priacy, 500, 501
Stress-strain relationships, 426
Strip mining, 649, 650
Structural analysis, 419, 426
Structure, 428, 441
Study guide, regional, 236, 238, 239
Sulphate signatures, 41
Sun angle, 234
Sun-synchronous orbit, 109, 113
Superimposed structures, 437
Superposition (Steno's Laws), 428
Superstition Hills fault, 284, 285
Surface water mapping, 615-617
Symmetry of structures, 425
Synthetic antenna, 307
Synthetic aperture radar, 65, 66, 298, 299, 305, 306, 308, 321, 330, 333, 334

Teche Mississippi, 495
Tectonic setting of ore deposits, 467, 469, 479
Telemetry, 70
Television cameras, 54
Temporal change, 401
Terrain backslopes, 302, 310-312, 314, 326
  foreslopes, 302, 311-313, 326
  illumination, 302, 326
  slope angle, 300-302, 310, 311, 312, 315
Terrestrial heat flow, 261
Texas lineament, 99-101
Texture, 182, 323, 324, 330, 428, 430
Thermal anomalies, 636

conductivity, 263
contrasts, 278, 279, 283, 289, 641
detectors, 12, 22
diffusivity, 263
inertia, 205-208, 263
radiation, 261
Thermal infrared images: glacial geology, 521, 539, 543, 546-548
  glaciers, 509, 513, 520
  permafrost, 507, 509
Thematic mapper, 87
Thin skinned tectonics, 433, 435
38th Parallel lineament, 449
Tibesti Mountains (Libya, Chad), 489, 493
Tibetan Plateau, 94-96
Tilt, 235
Tone, 323
Trace elements, 637
Teansform faults, 466, 467
Transmission, 21, 340
Trellis drainage, 422
Trend surface, 429
Triassic basalt flows, 488, 489
Trichromatic theory, 202
Triple junction, 471
Tristimulus chromaticity coordinates, 215
Trubidity, 638
Tunable acoustical optical filter, 682
Tyrone-Mount Union lineament, 423, 453

U-2, 401
Ultra-violet (UV) radiation, 234
Unconolidated materials, 247, 248, 252
Universal Transverse Mercator projection, 159

Valley, preglacial, 526, 527
  segments, 427
Valley and Ridge Province, 423
Variable threshold zonal filter, 217

Variance image, 187
Vegetation cover, 412
Velocity to height ratio, 73
Vents, 435
Venus, 671
Vertical electric field vector, 303
    exaggeration, 240
    incidence, 300, 302, 312
Vibrational energy, 13, 16, 32
Vignetting, 143, 156
Viking rocket photography, 91, 92
VHRR, 655
Volcanoes, 291, 420, 435
Volcanic breccia, 462
Vredefort dome, 463, 464

Wallowa Mountains, 489, 491, 492

Water depth, 617
    gaps, 422, 427, 447
    pollution, 635
    quality, 617
Watershed models, 610
Wavefront, 309
Wave incident, 303
Weather modification, 647
Weather Satellite photography, 92, 93
Wells (water), 453
West Hawk Lake crater, 463, 464
Westinghouse, 298
Wetlands, 617
Wiener filter, 178
Wien's displacement law, 258

Wind erosion, 489, 493
    gaps, 453
Windows, 48

X-band, 297, 299, 303, 320-324
X-ray spectrometer, 669

Yardangs, 99
Yaw, 58

Zargros Mountains, Iran, 383, 427, 437, 466
Zenith, 302
Zig-zag folds, 435, 437